BIODIVERSITY IN ECOSYSTEMS - LINKING STRUCTURE AND FUNCTION

Edited by **Yueh-Hsin Lo, Juan A. Blanco**
and **Shovonlal Roy**

Biodiversity in Ecosystems - Linking Structure and Function
http://dx.doi.org/10.5772/58494
Edited by Yueh-Hsin Lo, Juan A. Blanco and Shovonlal Roy

Contributors

Andre Pelser, Martin Reiss, Peter Chifflard, Franc Bavec, Martina Bavec, Victor Lionel Finot Saldías, Gloria Rojas, Juan A. Barrera, Alicia Marticorena, Clodomiro Marticorena, Julio Portela, Pedro Marín, Maria Barba, Raul Vilela, Teodoro Ibarrola, Virginia Polonio, Araceli Muñoz, Elena Tel, Jose Luis Del Rio, Santiago Parra, Juan Acosta, Pilar Rios, Javier Cristobo, Yoshitaka Oishi, Keizo Tabata, Ricardo Edvan, Carlo Aldrovandi Torreão Marques, Leilson Rocha Bezerra, Ana Ribeiro De Barros, Nalukui M. Matakala, David Nangoma, Stephan Syampungani, Natasha Sofia Ribeiro, Joana Falcão Salles Stephanie Jurburg, Juan A. Blanco, Mike Wang, Clive Welham, Yueh-Hsin Lo, Suzane Fank-De-Carvalho, Sônia Bao, Maria Salete Marchioretto, Nádia Silvia Somavilla, Graham Yapp, Richard Thackway,Carlos Oyarzun, Pedro Herve-Fernandez, Siu Mui Tsai, Acácio Aparecido Navarrete, Lucas William Mendes, Dennis Goss Souza, Antonio J Castro, Marina García-Llorente, Kelsey N. Bowman, Jason P. Julian, Caryn C. Vaughn, Shovonlal Roy, Jose Schoereder, Jeff Harvey, Miriama Malcicka, Bert Kohlmann, Alejandra Arroyo, Monika Springer, Danny Vásquez, Shinichi Watanabe, Enrica Roccotiello, Grazia Cecchi, Mirca Zotti, Mauro Mariotti, Simone Di Piazza, Pietro Marescotti, Márcio Viera, Suzane Bevilacqua Marcuzzo, Eduardo González Izquierdo, Alfredo Jimenez González, José Sánchez Fonseca, Martín González González, Barbarita Mitjans Moreno, Rogelio Sotolongo Sospedra, Gretel Geada López, Cesar Benetti, Josefina Garrido, Amaia Pérez-Bilbao

CBS, Edition **2017**

Published by InTech

Janeza Trdine 9, 51000 Rijeka, Croatia

Notice

Publishing Process Manager Iva Lipovic
Typesetting InTech DTP Team
Cover InTech Design Team

Additional hard copies can be obtained from orders@intechopen.com

Biodiversity in Ecosystems - Linking Structure and Function, Edited by Yueh-Hsin Lo, Juan A. Blanco and Shovonlal Roy
p. cm.
ISBN 978-953-51-2028-5

Contents

Preface

During the 20th century urban development was extended to all the regions of the world. With a booming human population and the intensification of economic development (first in Europe and North America and lately in the rest of the world) practically all the ecosystems in the world were impacted in one way or another. Therefore, it was just a matter of time that some of the iconic wildlife species of the world started to suffer from fast reductions in their populations, or even facing extinction. The danger of losing species such as whales, lions, tigers, elephants, panda bears, gorillas, brown bears, buffalos, sequoias, etc., was very real [1]. This danger was highlighted by scientists and environmental managers around the world, and the society responded with the creation of environmentalists groups, whose social pressure helped to create lists of endangered animal and plant species needing specific actions for conservation. This was the base to develop programs and activities focused on the protection of individual high-profile species. Many of these campaigns were supported by the public due to the easy sympathy or spiritual connection with some of these majestic species, and as a consequence, natural conservation was seen by the main public as "avoiding things getting worse". Some of these activities have achieved important successes, such as the halt in commercial hunting of whales, the breeding programs of panda bears or the increase in numbers of American buffalos [2]. However, in other cases the protection of the target species was not enough to prevent its decline or extinction (e.g. the Yangtze River dolphin, or the Pyrenean wild goat), or just the species were not interesting enough for the public opinion and therefore not the main focus of protection efforts, such in the case of "ugly" species such as amphibians, reptiles, insects, cacti, etc. Such species-specific conservation programs are becoming less important as the world realizes that biodiversity loss is not a matter of losing a few iconic species, but a large number of all kinds of species of plants, animals, fungi and even microbial organisms.

In the latest decades, there has also been an increase in the understanding on how the presence or absence of species could affect not only ecosystem structure, but ecosystem function as well. The discussion for formal maintenance and conservation of biological diversity (biodiversity) was first organized in a cohesive fashion by the United Nations Environment Programme in 1992 at the Rio Earth Summit. The following year, 168 countries signed the Convention of Biological Diversity (CBD) to protect and ensure conservation and sustainable use of biodiversity [3]. More recently the United Nations Secretary General initiated and completed the Millennium Ecosystem Assessment to assess the consequences of ecosystem change for human wellbeing and the scientific basis for action needed to enhance the conservation and sustainable use of ecosystems [4]. The assessment provided a reaffirmation that sustainable societies are dependent on the goods and services provided by ecosystems, including clean air and water, productive soils, and the production of food and fiber, and, more importantly, it propagated the ecosystem services paradigm upon which to assess and value biotic resources throughout the world [5]. The latest event at the policy-making level has been the creation of the Intergovernmental Platform on Biodiversity and Ecosystem Services (IPBES), a body created by the United Nations following the success of the Intergovernmental Panel on Climate Change, as it is becoming clear that the risk of massive biodiversity loss is at the same level as the risk of massive climate change.

During the first years of the 21st century, human-caused climatic and land use changes have caused that almost all ecosystems on Earth are under different varieties of stresses such as habitat loss and degradation, shortage of water and food supplies, toxic contaminants, and invasive species. These stresses have affected ecosystem structure or functions, altered the viability of species and communities, and therefore changed the quality and forms of ecosystem services [6]. The issue on how biodiversity support ecosystem functions is becoming increasingly important in ecological and biological studies. Up to date, to predict the ecological consequences of biodiversity loss, researchers have spent much time and effort quantifying how biological variation affects the magnitude and stability of ecological processes that underlie the functioning of ecosystems. Some of the most important ecosystem functions (biomass production, nutrient, water, and energy recycling, among others) could be affected by the presence or absence of plant, animal, fungi, and microbial species.

The current state-of-the-art confirms that biodiversity does indeed simultaneously enhance both the production and stability of biomass in experimental systems, and this is broadly true for terrestrial [7] and aquatic primary producers [8]. During the last few years an important number of experiments and field research has generated enough data to allow for extensive synthesis [9-17, among others]. From these synthesis studies, it has become clear that the strength of diversity effects on ecosystem functions such as biomass production or organic matter recycling is independent of diversity effects on temporal stability [17]. From such review, it has become evident that the highest levels of productivity in a diverse community are not associated with the highest levels of stability. Thus, on average, diversity does not maximize the various aspects of ecosystem functioning we might wish to achieve in conservation and management. In addition, the previous reviews have highlighted the issues related to researching the connections between ecosystem biodiversity and function, as knowing how biodiversity affects productivity gives no information about how diversity affects stability (or vice versa). Therefore, to predict the ecological changes that occur in ecosystems after extinction, we will need to develop separate mechanistic models for each independent aspect of ecosystem functioning [17]. Within this book, readers can find some of such models for aquatic (see chapter by Roy) and terrestrial (see chapter by Lo et al.) ecosystems.

Most of the previous studies, however, have been done in terrestrial ecosystems. For example, it is well established the role of non-tree species in the functioning, productivity, and stability of the ecosystem [7]. On the other hand, the aquatic habitats are of global importance due to their large spatial coverage, significant functional role in carbon fixation, oxygen generation, and high biodiversity. However, unlike the terrestrial world, our knowledge of biodiversity patterns across aquatic habitats is limited. Within the aquatic ecosystems, the relationships between ecosystem function and biodiversity are not always similar to that in the terrestrial or benthic ecosystems, and the relationships are often debated (e.g. [8]).

Theoretical as well as empirical studies addressed this question from different perspectives. The aquatic habitats can be divided into three major categories, primarily based on the salinity level of the water medium: freshwater systems, transitional and brackish waters, and marine systems. Although this categorization is rather crude, it helps to understand aquatic biodiversity on regional scales. The biodiversity in marine ecosystems is generally higher than that in freshwater systems, and the transitional waters are generally less diverse than freshwater and marine systems. The extreme diversity in aquatic or marine ecosystems is often puzzling in view of the established ecological theories of species coexistence. Theoretically, several mechanisms have been proposed to explain the biodiversity within aquatic ecosystems, which include temporal effects (e.g., environmental fluctuations, periodic forcing), spatio-temporal effects (spatial heterogeneity), self-sustaining cycles, deterministic chaos, spatio-temporal chaos, self-organized segregation, grazing and chemical signaling (e.g. [18]).

On the other hand, empirical and experimental studies have not only established some of the mechanisms proposed theoretically, but also put forward new scenarios. For example, experiments have shown how food supply may affect the function and dynamics aquatic ecosystems, and identified factors such as the effects of temperature fluctuation, inducible defense, autotoxin and density-dependent effects for the sustainability of simple aquatic ecosystems (e.g. [19]).

The first section of this book provides an overview of different concepts and theories to be taken into account when dealing with links between ecosystem biodiversity and function. Roy theoretically describes a mechanism termed as pseudo-mixotrophy, by which allelopathy among marine phytoplankton mediates nutrient competition and promotes the diversity of the primary producers. Much of his theoretical framework could be translated into terrestrial ecosystems. Jurburg and Falcão-Salles discuss the role of functional redundancy on ecosystem function. Sobrinho et al. explore the implications for ecosystem functioning of biodiversity in tropical forests. Harvey and Malcicka examine the interactions between climate change and modifications of species distribution shifts and their implications in multitrophic networks. Yapp introduces the use of criteria and indicators for managing the change and restoration of biodiversity in vegetated landscapes. Lo et al. review the different theoretical models available to examine the interactions between plant diversity and stand growth in ecological restoration projects. De Souza et al. analyze the implications at ecosystem level of microbial diversity and assembly. To finish this section, Bavec and Bavec review the management options available to increase biodiversity in agroecosystems.

The second section of this book provides a wide range of studies showcasing evidence and facts that support the need and importance of research on links between biodiversity and ecosystem function. A range of studies on different aquatic environments can be found, where the authors described new mechanisms of biodiversity, and patterns of biodiversity distribution across different habitat. Reiss and Chifflard address the issue of the fauna-microhabitat relationship of springheads, and have presented results on quantitative and qualitative assessment on substrate preferences of invertebrates. Pérez-Bilbao et al. study the invertebrate fauna in freshwater ponds, and investigate if those invertebrates could be used as good indicator of the environmental quality of temporary ponds. Acosta et al. study the effects of bottom trawling on benthic vulnerable marine ecosystems on the high seas of the Southwest Atlantic using several research cruises, and present preliminary results on the impacts of fishing activities on sensitive benthic organisms. Kohlmann et al. examine the possible impacts of human activities on the macro-invertebrate biodiversity along the length of a tropical river, and their influence on the structure and function of agrorural environments. Oyarzun and Hervé-Fernández discuss the implications that forest management has on the eco-hydrology of Chilean rivers.

Moving into terrestrial ecosystems, the next block of studies provides an extended view of the links between ecosystem biodiversity and function at different ecological scales. Watanabe explores the role of niche and prey diversity on the presence of wild cats in Pacific islands. Finot et al. discuss the diversity of grass species on Easter Island, and the implications for ecosystem function of the presence of invasive species. Fank-de-Carvalho et al. also explore the diversity of plant species but in the Brazilian neotropical savannah. The former studies provide facts and evidence for mostly natural ecosystems. The next block of chapters, however, provides evidence on the connections between management, biodiversity and ecosystem functions for a variety of terrestrial ecosystems. Castro et al. explore the connections between social perception of ecosystem values and biodiversity. On a similar line on the links between ecology and society, Oishi and Tabataanalyze discuss the importance of conserving large trees for spiritual purposes with the conservation of epiphytic bryophyte diversity. Ribeiro et al. review the state-of-the-art on research aimed to support the sustainable use of miombo woodlands in southern Africa.

The next three chapters of this block deal with the possibility of carrying ecological restoration to gain some of the biodiversity lost by human actions. In a methodological study, Marcuzzo and Viera introduce the approach to studying and carrying ecosystem restoration projects in ecological units with homogeneous features. González-Izquierdo et al. describe the steps taken and the results obtained in three different restoration projects on tropical Caribbean forests, highlighting the importance of the social component of such projects and the increased success rate when incorporating locals into the process. Roccotiello et al. provide an example of the complexities on restoring biodiversity in highly altered ecosystems such as abandoned mine sites.

The final block of chapters in this book deals with the links between ecosystem diversity, functions and services in agroecosystems, a highly important issue as they are the source of most of food for human societies. Bavec and Bavec review the effects of traditional and organic farming on agroecosystem biodiversity and its links with agricultural productivity. Pelser et al. discuss the possibilities to include local use of agricultural resources inside protected areas to achieve both local development and biodiversity conservation. Edvan et al. examine the consequences on grassland and herbivore production of shortages in grasses biodiversity, and the influence of exotic invasive grasses.

All things considered, these 26 chapters provide a good overview of the links between ecosystem biodiversity and ecosystem functions, which are then translated into ecosystem services useful for all humans. These chapters show the importance of biodiversity on the structure and function of terrestrial and aquatic ecosystems, and in managed or natural conditions, which that can be applied to all the regions of the world. They are an introduction to the research being done around the globe in connection to this topic. We hope the readers from academia, management, conservation, and any other stakeholders will enjoy reading this book and regard it as an initial source of information and study cases on what is the role that biodiversity plays in ecosystems.

The Editors want to finish this preface acknowledging the collaboration and hard work of all the authors. We are also thankful to the Publishing Team of InTech for their continuous support and assistance during the creation of this book. Special thanks are due to Ms. Iva Lipović for inviting us to lead this exciting project and for coordinating the different editorial tasks. Last but not least, we want to acknowledge InTech's generosity and social commitment by supporting scientists from developing countries and making their research freely available.

Dr. Yueh-Hsin Lo

Dr. Juan A. Blanco
Dep. Ciencias del Medio Natural,
Universidad Pública de Navarra,
Spain

Dr. Shovonlal Roy
Dep. Geography and Environmental Science,
University of Reading,
UK

References

[1] Laliberte AS, Ripple WJ. Range contractions of North American carnivores and ungulates. BioScience 2004;54 123–138.

[2] Lo YH, Lin YC, Blanco JA, Yu CH, Guan BT. Moving from ecological conservation to restoration: an example from central Taiwan, Asia. In: Blanco JA, Lo YH. (eds.) Forest Ecosystems: more than just trees. Rijeka: InTech; 2012. p339-354.

[3] Secretariat of the Convention on Biological Diversity, 2005. Handbook of the Convention on Biological Diversity, Including its Cartagena Protocol on Safety (3rd ed). Montreal: Secretariat of the Convention on Biological Diversity; 2005.

[4] Millennium Ecosystem Assessment. Ecosystems and Human Wellbeing: Our Human Planet. Washington; Island Press, 2005.

[5] Boykin KG et al. A national approach for mapping and quantifying habitat-based biodiversity metrics across multiple spatial scales. Ecological Indicators 2013;33 139–147.

[6] Grimm NB et al. Climate-change impacts on ecological systems: introduction to a U.S. assessment. Front Ecology Environment 2013; 11(9) 456–464.

[7] Blanco JA, Lo YH, editors. Forest Ecosystems: more than just trees. Rijeka; InTech; 2012.

[8] Duarte P, Macedo MF, da Fonseca LC. The relationship between phytoplankton diversity and community function in a coastal lagoon. Hydrobiologia 2006;555 3-18.

[9] Balvanera P, Pfisterer AB, Buchmann N, He JS, Nakashizuka T, Raffaelli R, Schmid B. Quantifying the evidence for biodiversity effects on ecosystem functioning and services. Ecology Letters 2006; 9 1146–1156

[10] Stachowicz, J, Bruno JF, Duffy JF. Understanding the effects of marine biodiversity on communities and ecosystems. Annual Review of Ecology, Evolution, and Systematics 2007;38 739–766.

[11] Cadotte MW, Cardinale BJ, Oakley TH. Evolutionary history and the effect of biodiversity on plant productivity. Proceedings of the National Academy of Sciences USA. 2008;105 17012–17017.

[12] Worm B et al. Impacts of biodiversity loss on ocean ecosystem services. Science 2006;314 787–790.

[13]   Schmid B et al. Consequences of species loss for ecosystem functioning: a meta-analysis of data from biodiversity experiments. In: Naeem S, Bunker D, Loreau M, Hector A, Perring C (eds). Biodiversity and human impacts. Oxford; Oxford University Press; 2009. p14–29.

[14]   Srivastava DS, Cardinale BJ, Downing AL, Duffy JE, Jouseau C, Sankaran M, Wright JP. Diversity has stronger top-down than bottom-up effects on decomposition. Ecology 200;90 1073–1083.

[15]   Quijas S, Schmid B, Balvanera P. Plant diversity enhances provision of ecosystem services: A new synthesis. Basic and Applied Ecology 2009; 11: 582–593.

[16]   Flynn DFB, Mirotchnick N, Jain M, Palmer MI, Naeem S. Functional and phylogenetic diversity as predictors of biodiversity–ecosystem-function relationships. Ecology 2011;92 1573–1581.

[17]   Cardinale BJ et al. Biodiversity simultaneously enhances the production and stability of community biomass, but the effects are independent. Ecology 2013;94(8) 1697–1707.

[18]   Roy S, Chattopadhyay J. Towards A Resolution of 'The Paradox of The Plankton': A Brief Overview of The Proposed Mechanisms. Ecological Complexity 2007;4(1) 26-33.

[19]   Roy S, Chattopadhyay J. The stability of ecosystems: A brief overview of the paradox of enrichment. Journal of Biosciences 2007;32(2) 421-428.

Concepts and Theories

Importance of Allelopathy as Peudo-Mixotrophy for the Dynamics and Diversity of Phytoplankton

Shovonlal Roy

Additional information is available at the end of the chapter

http://dx.doi.org/10.5772/59055

1. Introduction

Phytoplankton are responsible for oceanic primary production and oxygen generation; and essential for regulating the global carbon cycle [1]. The dynamics and diversity of phytoplankton are constrained by several top-down and bottom-up effects. Complexities further arise from inter-species interactions within phytoplankton communities. Resources available for the growth of phytoplankton (e.g., light and dissolved nutrients) are often limited. But, despite the presence of limited variety of resources, phytoplankton are capable of maintaining an extreme level of species diversity [1–3]. This diversity is paradoxical to the theory of competitive exclusion [3], which suggests that in the steady state the number coexisting species cannot exceed the number of limiting resources [4, 5]. The mechanisms proposed to explain phytoplankton diversity include environmental fluctuations, periodic fluctuations, spatial heterogeneity, deterministic chaos, life cycles, grazing, and chemical interactions (detailed in [6]). But, when the top-down effects and external factors are negligible, it is difficult to explain the 'building block' of the extreme diversity of phytoplankton, i.e., the stable coexistence of two phytoplankton on a single limiting resource.

There is a growing body of evidence, both theoretical and experimental, suggesting that allelopathic interactions among phytoplankton species have a major role in shaping phytoplankton-zooplankton dynamics and regulating phytoplankton diversity [6–23]. Some of these studies [20, 21] suggested that 'toxin-allelopathy' can prevent competitive exclusion in Lotkta-Volterra interactions. Further, the allelopathic effect can potentially mediate resource competition in a chemostat. Focusing on simple resource-competition models, Roy [22] proposed that two phytoplankton can stably coexist on a single resource in a homogeneous media without any external factors when allelopathy acts as 'pseudo-mixotrophy'. This chapter elucidates how this mechanism ('if you cannot beat them or eat them, just kill them by chemical weapons' [22]) determines the outcome of resource

competition between two phytoplankton, and how it potentially contributes to maintaining phytoplankton diversity in natural waters.

2. Mixotrophy and allelopathy

Mixotrophy is known to influence species interactions within a food web [24]. Mixotrophic algae that can combine phototrophy and phagotrophy are an important component of phytoplankton communities (e.g., [25]). Mixotrophy can be an effective strategy for securing essential carbon required for the survival of algae in adverse conditions, such as, low radiation, unfavourable temperature, salinity or pH [26, 27]. Studies further suggested that certain algae (e.g., Prymnesium) can simultaneously be toxin producer and mixotrophic to 'kill and eat' [28]. However, not many species are known to follow this dual strategy that combines allelopathy and mixotrophy. But, several species are known to be allelopathic as they produce toxic or allelopathic chemicals (e.g., [13]). Studies suggested that the dynamics of phytoplankton with competitors and grazers are modulated by the presence of toxic species (e.g., [21, 29–31]). Allelopathy of toxin producers affects the growth and competitive ability of sensitive species. Allelopathy alone can potentially overturn the outcome of interspecific competition by providing 'additional' competitive and growth advantages to the allelopathic species [20, 22]. Roy [22] proposed that theoretically the effect of allelopathy can be viewed as pseudo-mixotrophy for the survival or coexistence of phytoplankton in nutrient competition. In the rest of the chapter, this mechanism will be discussed.

3. Allelopathy mediating competition for a single nutrient

To demonstrate the allelopathic effect on nutrient competition, a standard resource-competition model (presented in Table 1) is considered, which is a generalised version of the model analysed by [22].

Eq. (1): Nutrient

$$\frac{dN}{dt} = \overbrace{d\,(N_0 - N)}^{\text{net nutrient input}} \overbrace{- \frac{1}{\eta_1}\,f_1\,(N)\,P_1}^{\text{uptake by } P_1} \overbrace{- \frac{1}{\eta_2}\,f_2\,(N)\,P_2}^{\text{uptake by } P_2} + \overbrace{\eta\,\alpha_1\,(m_1\,P_1 + \phi(P_1,P_2)\,P_1)}^{\text{recycling from } P_1} + \overbrace{\eta\,\alpha_2\,(m_2\,P_2)}^{\text{recycling from } P_2}$$

Eq. (2): Non-allelopathic species

$$\frac{dP_1}{dt} = \overbrace{f_1\,(N)\,P_1}^{\text{growth}} - \overbrace{m_1\,P_1}^{\text{loss}} - \overbrace{\phi(P_1,P_2)\,P_1}^{\text{loss by allelopathy}}$$

Eq. (3): Allelopathic species

$$\frac{dP_2}{dt} = \overbrace{f_2\,(N)\,P_2}^{\text{growth}} - \overbrace{m_2\,P_2}^{\text{loss}}$$

Table 1. Representation of allelopathy in a nutrient-competition model of two phytoplankton

Parameter	Meaning	Unit	Functions/values from [22]
$f_1(N)$	Nutrient uptake function for species 1	day^{-1}	$\frac{\mu_1 N}{K_1+N}$
$f_1(N)$	Nutrient uptake function for species 2	day^{-1}	$\frac{\mu_2 N}{K_2+N}$
$\phi(P_1, P_2)$	Loss rate of species 1 due to allelopathy	day^{-1}	$\gamma\, P_1 P_2^2$
μ_1	Maximum growth rate of species 1 (P_1)	day^{-1}	1.0
μ_2	Maximum growth rate of species 2 (P_2)	day^{-1}	1.1
K_1	Half-saturation constant for species 1	gL^{-1}	0.6
K_2	Half-saturation constant for species 2	gL^{-1}	1.5
m_1	Per capita loss rate for species 1	day^{-1}	0.012
m_2	Per capita loss rate for species 2	day^{-1}	0.01
d	Dilution rate	day^{-1}	0.25
N_0	Input nutrient concentration	gL^{-1}	0.11
γ	Allelopathy parameter	$cell^{-3}\,day^{-1}$	0.02
α_1	Nutrient content per cell of species 1	$gcell^{-1}$	5×10^{-5}
α_2	Nutrient content per cell of species 2	$gcell^{-1}$	1×10^{-5}
η	Recycling efficiency	dimensionless	0.5
η_1	Yield coefficient of species 1	dimensionless	1.0
η_2	Yield coefficient of species 2	dimensionless	1.0

Table 2. Functions and parameters with their meanings used in the nutrient competition model with allelopathic effect. The quantities N, P_1 and P_2 are the concentrations of the nutrient, non-allelopathic species and allelopathic species, respectively.

The nutrient (with concentration N) uptakes by the non-allelopathic species (with concentration P_1) and allelopathic species (with concentration P_2) are described by the functions $f_1(N)$ and $f_2(N)$. In particular, these functions can take the standard Michaelis-Menten forms (Table 3). The parameters of the model are described in Table (2). Allelopathy of species 2 imposes a higher mortality to the non-allelopathic species, which can be described by an 'additional' mortality term in the form of a phenomenological function $\phi(P_1, P_2)$. This function may be a high-order interspecific product of P_1 and P_2 (see, Table 3) - a particular case of which was considered in [22]. In the absence of allelopathy, the model takes the form of a standard resource-competition model, which predicts the persistence of one of the two species depending on the lowest minimum nutrient requirements (i.e., depending on minimum R^* [5]). So, if $\phi(P_1, P_2) = 0$, and if the non-allelopathic species has a lower minimum nutrient requirement, it will win over the allelopathic species in nutrient competition. However, if $\phi(P_1, P_2) \neq 0$, allelopathy provides advantage to species 2 by imposing a higher mortality to species 1.

3.1. Coexistence of two phytoplankton on single nutrient

As mentioned in the previous section, the loss rate of species 1 due to allelopathy of species 2 can be described by a high-order product of P_1 and P_2: $\phi(P_1, P_2) = \gamma P_1^{\beta_1} P_2^{\beta_2}$. A particular case was analysed in [22], where the exponents were taken as $\beta_1 = 1$ and $\beta_2 = 2$. For $\phi(P_1, P_2) = \gamma P_1^{\beta_1} P_2^{\beta_2}$, it can be derived (following the analysis of [22]) that there exist a critical value γ^c for the allelopathy parameter γ, such that, if $\gamma < \gamma^c$, no coexisting steady state is possible. However, if $\gamma > \gamma^c$, two alternative steady states are possible, and depending on the initial conditions the system will settle to one of the two steady states (see, Fig. 1). Therefore, for $\gamma > \gamma^c$ one can find suitable initial concentrations of P_1 and P_2 for which stable coexistence two phytoplankton on a single nutrient is possible: Fig. 1-(b) shows that in this case the ratio of P_2 to P_1 is stabilised to a non-zero value.

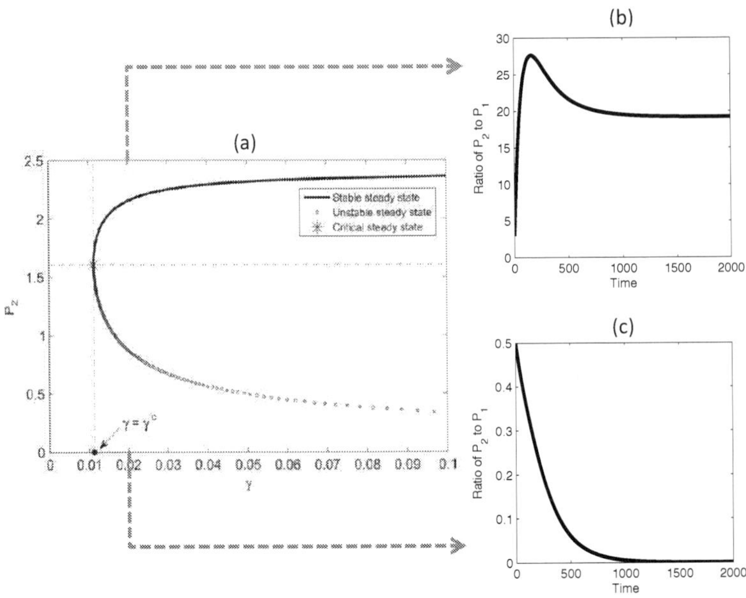

Figure 1. Dynamics two phytoplankton when allelopathy exceeds critical level. (a) Saddle-node bifurcation for the model system with γ as the bifurcation parameter. The figure was reproduced from [22] with permission from the publisher. (b) & (c) Ratio of P_2 to P_1 corresponding to the stable and unstable dynamics presented by (a), respectively. Condition for no recycling was used with other parameters and functions fixed at their default values/forms as in Table (2).

3.2. Critical conditions for coexistence

The critical level of allelopathy γ^c is a crucial quantity, which can be computed from the parameters of the model. Corresponding to γ^c, there exists an unique coexisting steady state $\left(N^*, P_1^c, P_2^c\right)$, where the magnitudes of P_1^c and P_1^c depend on the model parameters. Extending the analysis of [22] , the magnitudes of these quantities can be derived explicitly for all possible forms of the function $\phi(P_1, P_2)$ (Table 3). When $\gamma > 0$, the critical conditions for the existence of the unique steady state can alternatively be derived with respect to N_0 - the input nutrient concentration. The allelopathy parameter γ would depend on the inherent biological properties of the allelopathic species, and hence its magnitude cannot normally be altered using experimental conditions. However, the parameter N_0 associated with the experimental conditions can very well be controlled. Rearranging the expressions of γ^c (Table 3), one can derive the corresponding threshold magnitudes of the input nutrient concentration, say, N_0^c, so that, for $N_0 > N_0^c$ alternative steady states are possible leading to the stable coexistence of two phytoplankton. The explicit expressions of N_0^c for different forms of $\phi(P_1, P_2)$ are

$\phi(P_1, P_2)$	$\left(P_1^c, P_2^c\right)$	γ^c for a given N_0	N_0^c for a given γ
$\gamma\, P_2$	Does not exist	-	-
$\gamma\, P_1\, P_2$	$\left(\frac{c_3}{2c_1}, \frac{c_3}{2c_2}\right)$	$\dfrac{4\,A\,c_1\,c_2}{c_3^2}$	$N^* + \left(\dfrac{4\,A\,c_1\,c_2}{\gamma\,d^2}\right)^{\frac{1}{2}}$
$\gamma\, P_1\, P_2^2$	$\left(\frac{c_3}{3c_1}, \frac{2c_3}{3c_2}\right)$	$\dfrac{27\,A\,c_1\,c_2^2}{4\,c_3^3}$	$N^* + \left(\dfrac{27\,A\,c_1\,c_2^2}{4\,\gamma\,d^3}\right)^{\frac{1}{3}}$
$\gamma\, P_1^{\beta_1}\, P_2^{\beta_2}$	$\left(\frac{c_3\,\beta_1}{c_1\,(\beta_1+\beta_2)}, \frac{c_3\,\beta_2}{c_2\,(\beta_1+\beta_2)}\right)$	$\dfrac{A\,c_1^{\beta_1}\,c_2^{\beta_2}\,(\beta_1+\beta_2)^{(\beta_1+\beta_2)}}{c_3^{(\beta_1+\beta_2)}\,\beta_1^{\beta_1}\,\beta_2^{\beta_2}}$	$N^* + \left(\dfrac{A\,c_1^{\beta_1}\,c_2^{\beta_2}\,(\beta_1+\beta_2)^{(\beta_1+\beta_2)}}{\gamma\,\beta_1^{\beta_1}\,\beta_2^{\beta_2}\,d^{(\beta_1+\beta_2)}}\right)^{\frac{1}{(\beta_1+\beta_2)}}$

Table 3. Parametric conditions for stable coexistence when allelopathy acts as pseudo-mixotrophy in nutrient competition models. The allelopathic effect is denoted by $\phi(P_1, P_2)$, the critical steady state by (N^*, P_1^c, P_2^c), the critical level of allelopathy by γ^c , and the threshold level of N_0 by N_0^c. The quantities c_1, c_2, c_3 and A used in the table are defined as: $c_1 = \frac{1}{\eta_1}\, f_1(N^*) - \eta\,\alpha_1\,(m_1 + A)$, $c_2 = \frac{1}{\eta_2}\, f_2(N^*) - \eta\,\alpha_2\,m_2$, $c_3 = d\,(N_0 - N^*)$, with $A = f_1(N^*) - m_1$, and N^* is given by $f_2(N^*) = m_2$. In particular, $f_1(N) = \frac{\mu_1\,N}{K_1+N}$ and $f_2(N) = \frac{\mu_2\,N}{K_2+N}$, $N^* = \frac{m_2\,K_2}{\mu_2 - m_2}$.

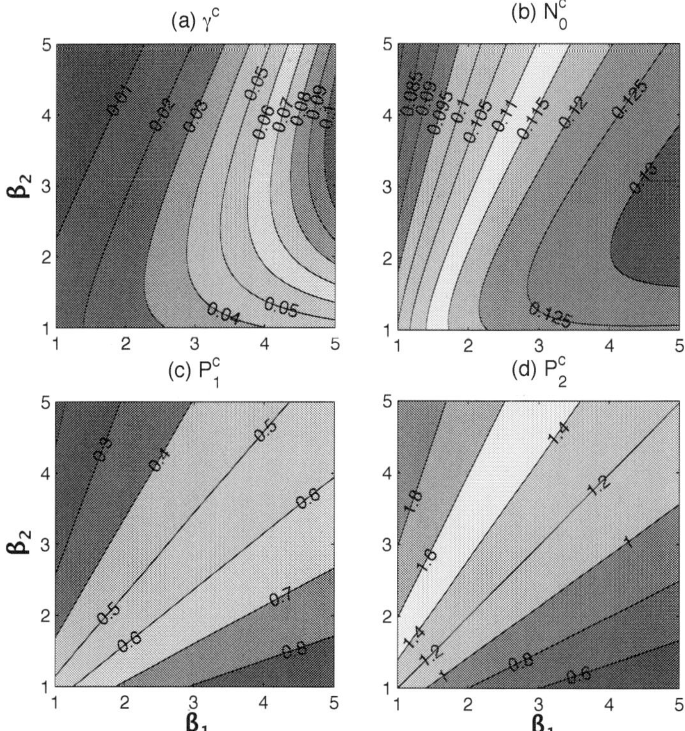

Figure 2. The magnitudes of γ^c, N_0^c, P_1^c and P_2^c are computed for a range of values of the exponents β_1 and β_2 corresponding to the function $\phi(P_1, P_2) = \gamma\, P_1^{\beta_1}\, P_2^{\beta_2}$. The parameters are fixed at their default values as in Table (2).

presented in Table (3). The results in Table (3) can be used to address how the critical values γ^c, N_0^c, P_1^c and P_2^c may change due to uncertainties in describing the allelopathic effect by a phenomenological function. Considering the general form $\phi(P_1, P_2) = \gamma\, P_1^{\beta_1}\, P_2^{\beta_2}$, the magnitudes of γ^c, N_0^c, P_1^c and P_2^c are computed for a range of values of the exponents β_1 and β_2 (Fig. 2). If the model parameters are fixed, γ_c or N_0^c would be minimum when β_1 is the lowest and β_2 is the highest (Fig. 2-a, b). The unique steady states of P_1 and P_2 depend on both β_1 and β_1: for a given β_1, P_1^c decreases but P_2^c increases with β_2 (Fig. 2-c, d).

3.3. Allelopathy as pseudo-mixotrophy

The function of allelopathy in mediating the coexistence can be understood from the Figs. (3) & (4). Under nutrient-limiting conditions, allelopathy of the weaker competitor helps increase the availability of nutrient the by killing the stronger competitors: an illustration of this process based on the model of [22] is given in Fig. (3-a). In the simplest scenario,

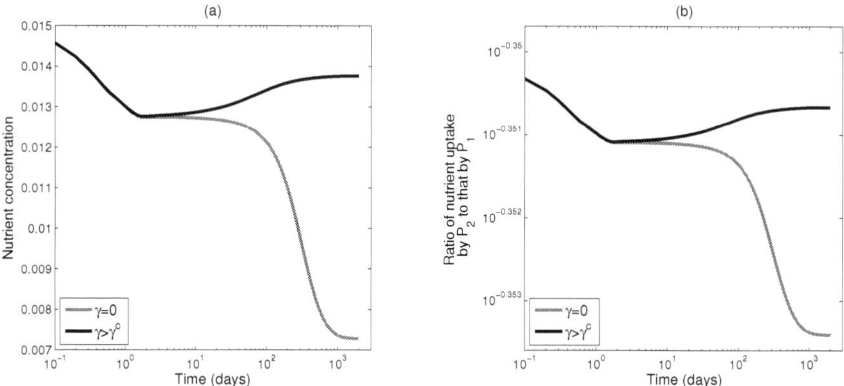

Figure 3. Allelopathy as pseudo-mixotrophy: (a) Enlargement of the available nutrient pool due to killing of competitors by allelopathy; (b) Increased ratio of per capita nutrient uptake by allelopathic species to that by non-allelopathic species. Red and black lines indicate conditions of no allelopathy and allelopathy beyond the critical level, respectively. Condition for no recycling was used with other parameters and functions fixed at their default values/forms as in Table (2).

when recycling of nutrient is 'turned off' in the model, and no killing by allelopathy takes place, the level of available nutrient decreases and stabilises to a low value where the non-allelopathic species alone survives eventually (Fig. 3-a, b). However, the extra (higher) mortality of species 1 (P_1) due to killing by allelopathy of species 2 (P_2) leads to elevation of the nutrient concentration (and further prevents it from decreasing gradually) (Fig. (3-a); the nutrient concentration eventually stabilises to a level where both species stably coexist (Fig. 3-a, b). The ratio of per-capita nutrient uptake by P_2 to that of P_1 decreases to a low value when killing by allelopathy does not take place (Fig. 3-b, Fig. 4-a); however, this ratio stabilises to a considerably higher value when allelopathy kills stronger competitors (Fig. 3-b, Fig. 4-b). When nutrient recycling is incorporated, killing by allelopathy increases the dead cells (Fig. 4-c,d), and the recycling process releases a portion of the nutrient quota of the dead competitors available for uptake (Fig. 4-d). The recycling process coupled with killing by allelopathy thus generates an extra amount of nutrient (Fig. 4-d) available for uptake by the species. Therefore, by imposing higher mortality to stronger competitor, allelopathy provides clear advantage to the weaker competitor. This mode of action of allelopathy

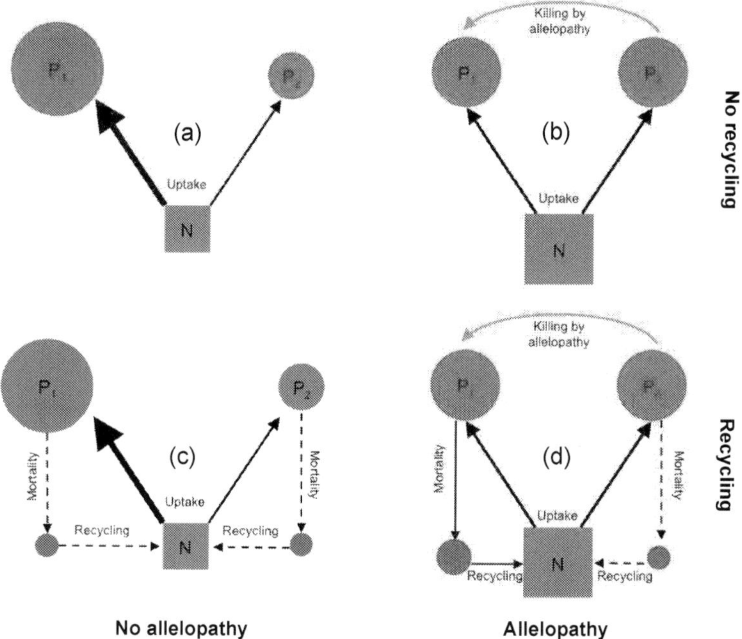

Figure 4. Schematic diagram showing the function of allelopathy as pseudo-mixotrophy. Competition between two phytoplankton (P_1, P_2) under a single nutrient (N) with or without the effect of allelopathy: (a) no nutrient recycling and no allelopathy; (b) no nutrient recycling, but killing by allelopathy; (c) nutrient recycling, but no allelopathy; and (d) nutrient recycling and killing by allelopathy. The thin and thick arrows indicate low and high values for nutrient uptake, respectively; dashed and continuous lines represent low and high level of recycling respectively; and the sizes of the circles and squares represent concentrations of the variables. The orange curved-arrows indicate when killing by allelopathy is incorporated. In (a) and (c), the weak competitor P_2 is excluded, whereas, P_1 survives. In (b) and (d), P_1 and P_2 stabilises with concentrations depending on the model parameters (which are not represented by the relative size of the circles).

that provides growth advantage to the allelopathic species, not through direct predation but through killing of the competitors (e.g., Fig. 4-b, Fig. 4-d), was termed as pseudo-mixotrophy [22]. In this process killing by allelopathy provides a positive feed-back by increasing of the growth limiting resource that reduces the competition pressure (e.g., Fig. 4). Clearly, this feedback loop provides crucial benefit to the growth rate of the allelopathic algae, and modulates the dynamics of the resource competition within a common trophic level.

4. Relevance to empirical and experimental studies

It is clear from the previous sections that allelopathy acting as pseudo-mixotrophy can theoretically stabilise nutrient competition of two phytoplankton on a single limiting nutrient. However, the applicability of this mechanism across natural phytoplankton is largely unexplored. An empirical or experimental evidence for pseudo-mixotrophy is still in demand. But, recent studies have shown promise that the role of allelopathy in maintaining biodiversity of natural phytopalnkton may be explored further. For example, chemical warfare has been shown to increases bio-diversity in microbial realm [32]; and [20] showed that allelopathy may be responsible for co-existence of the competing phytoplankton in the Bay of Bengal. The question of how much diversity of phytoplankton can be supported though allelopathy alone was addressed by [23], who derived a deterministic relationship between the abundance of the potential allelopathic species and the diversity of non-allelopathic phytoplankton (see, Fig. 5). The abundance-diversity relationship in Fig.

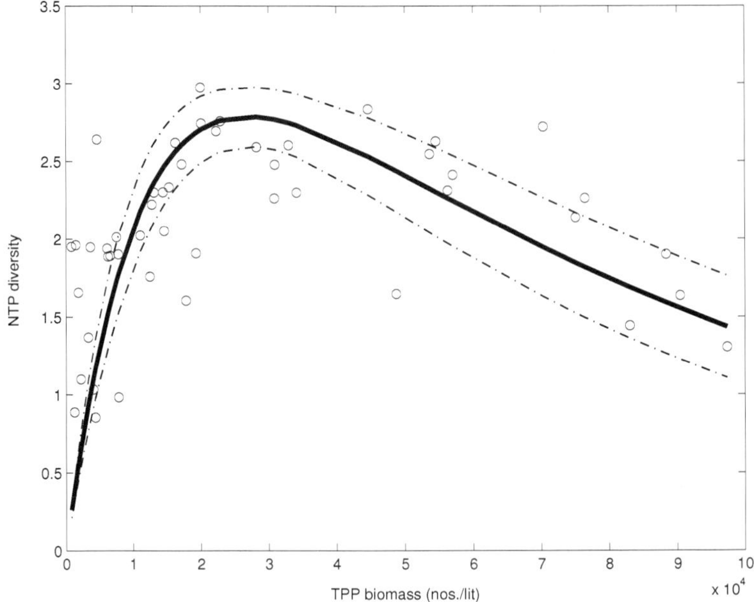

Figure 5. Deterministic relationship between the abundance of toxin-producing phytoplankton (TPP) and the diversity of non-toxic phytoplankton (NTP) in the Bay of Bengal. The figure is reproduced from [23] with permission from the publisher. The Shannon diversity of non-toxic species is plotted as a function of the abundance (nos./l) of toxin-producing phytoplankton defined by [23]. The solid line represents the fitted model of with the data presented in open circles. The dashed lines are the predicted model at 95% confidence level.

(5) shows a unimodal pathway through which the abundance of allelopathic phytoplankton regulates the diversity of non-allelopathic phytoplankton [23].

5. Concluding remarks

This chapter elucidates how phytoplankton allelopathy may function as pseudo-mixotrophy in determining the dynamics of nutrient-phytoplankton models, and how phytoplankton diversity is maintained in those systems. Firstly, the ecological conditions under which allelopathy functioning as pseudo-mixotrophy overturns the outcome of nutrient competition between two phytoplankton (e.g., [22]) is presented explicitly in terms of the model parameters. Secondly, the difficulties in mechanistically describing the allelopathic effect of a phytoplankton on its competitors is addressed by considering a phenomenological function, and the ecological conditions for the coexistence of phytoplankton species and stability of competition dynamics are derived. Thirdly, the competition dynamics is explored under the assumptions of 'no nutrient recycling' and 'continuous nutrient recycling'; and the effects of changing initial nutrient pool in culture media is explored. Therefore, a comprehensive set of constraints is derived under which allelopathy acts as pseudo-mixotrophy in nutrient-phytoplankton models. Finally, the evidences of allelopathic effects in determining the diversity of phytoplankton in natural systems are presented. In particular, how the increasing abundance of allelopathic species may regulate the diversity of phytoplankton (e.g., [23]), is discussed. The mechanism presented here would be useful for better understanding of the biodiversity and function of marine ecosystems. Allelopathy functions as 'pseudo-mixotrophy' in nutrient-phytoplankton models, which are often the basis of marine biogeochemical and ecosystem models. This mechanism has not been explored in relation to ocean biogeochemical models, which are generally used to predict phytoplankton species composition, and estimate the scale of oceanic carbon sink. Given the complexities in representing phytoplankton functional types in global biogeochemical models (e.g., [33]), it would be useful to understand how allelopathy or pseudo-mixotrophy of a phytoplankton type may affect the dynamics of the other types. The ecological conditions derived will be useful for investigating the role of 'pseudo-mixotrophy' in marine ecosystem models. The current challenges in monitoring, controlling and managing harmful algal blooms (HAB) (e.g., [34]), and predicting their consequences in aquatic ecosystems require better understanding of the dynamics of toxic or allelopathic species. Recent studies have also reported other roles of phytoplankton allelochemicals, e.g., defence against predators [35], and 'casual parasitism' that helps supplying organic nutrient to the mixotrophic donors by lysis of prey [36, 37]. It will be worthwhile to further explore the mechanism presented here in relation to the succession of phytoplankton taxa that are known to form HABs. It is noteworthy that currently the mechanism has been explored using simple resource-competition models that can be tested in an experimental (chemostat) set up. Such an experiment will be helpful in formulating and parameterising resource-competition models including allelopathy, and for better understanding of the constraints of phytoplankton diversity.

Author details

Shovonlal Roy

Address all correspondence to: shovonlal.roy@reading.ac.uk

Department of Geography and Environmental Science, University of Reading, Whiteknights, U.K.

References

[1] Falkowski P. The power of plankton. Nature. 2012;483:S17–S20.

[2] Scheffer M, Rinaldi S, Huisman J, Weissing FJ. Why phytoplankton communities have no equilibrium: solutions to the paradox. Hydrobiologia. 2003;491:9–18.

[3] Hutchinson GE. The paradox of the plankton. American Naturalist. 1961;95:137–145.

[4] Hardin G. The competitive exclusion principle. Science. 1960;131:1292–1298.

[5] Tilman D. Resource Competition and Community Structure. Princeton, NJ: Princeton University Press; 1982.

[6] Roy S, Chattopadhyay J. Towards a resolution of 'The Paradox of the Plankton' : A brief overview of the existing mechanisms. Ecological Complexity. 2007;4(1-2):26–33.

[7] de Freitas M, Fredrickson A. Inhibition as a factor in the maintenance of the diversity of microbial ecosystems. Journal of General Microbiology. 1978;106:307–320.

[8] Maestrini SY, Bonin DJ. Allelopathic relationships between phytoplankton species. Canadian Bulletin of Fisheries and Aquatic Sciences. 1981;210:323–338.

[9] Mason CP, Edwards KR, Carlson RE, Pignatello J, Gleason FK, Wood JM. Isolation of chlorine-containing antibiotic from the freshwater cyanobacterium Scytonema hofmanni. Science. 1982;215:400–402.

[10] Arzul G, Seguel M, Guzman L, Denn EEL. Comparison of allelopathic properties in three toxic alexandrium species. Journal of Experimental Marine Biology and Ecology. 1999;232(C11):285–295.

[11] Rengefors K, Legrand C. Toxicity in peridenium aciculisurum - an adaptive strategy to outcompete other winter phytoplankton? Limnology and Oceanography. 2001;46:1990–1997.

[12] Schagerl M, Unterrieder I, Angeler DG. Allelopathy among cyanoprokaryota and other algaeoriginating from Lake Neusiedlersee (Austria). International Review of Hydrobiology. 2002;87:365–374.

[13] Cembella AD. Chemical ecology of eukaryotic microalgae in marine ecosystems. Phycologia. 2003;42(4):420–447.

[14] Granéli E, Johansson N. Increase in the production of allelopathic *Prymnesium parvum* cells grown under N- or P-deficient conditions. Harmful Algae. 2003;2:135–145.

[15] Hulot FD, Huisman J. Allelopathic interactions between phytoplankton species: The role of heterotrophic bacteria and mixing intensity. Limnology and Oceanography. 2004;49:1424–1434.

[16] Fistarol G, Legrand C, Selander E, Hummert C, Stolte W, Granéli E. Allelopathy in alexandrium spp.: effect on a natural plankton community and on algal monocultures. Aquatic Microbial Ecology. 2004;35:45–56.

[17] Solé J, García-Ladona E, Ruardij P, Estrada M. Modelling allelopathy among marine algae. Ecological Modelling. 2005;183:373–384.

[18] Schatz D, Keren Y, Hadas O, Carmeli S, Sukenik A, Kaplan A. Ecological implications of the emergence of non-toxic subcultures from toxic *microcystis* strains. Environmental Microbiology. 2005;7:798–805.

[19] Roy S, Alam S, Chattopadhyay J. Competitive effects of toxin-producing phytoplankton on overall plankton populations in the Bay of Bengal. Bulletin of Mathematical Biology. 2006;68(8):2303–2320.

[20] Roy S, Chattopadhyay J. Toxin-allelopathy among phytoplankton species prevents competitive exclusion. Journal of Biological Systems. 2007;15(01):73–93.

[21] Roy S, Bhattacharya S, Das P, Chattopadhyay J. Interaction among nontoxic phytoplankton, toxic phytoplankton and zooplankton: inferences from field observations. Journal of Biological Physics. 2007;33(1):1–17.

[22] Roy S. The coevolution of two phytoplankton species on a single resource: Allelopathy as a pseudo-mixotrophy. Theoretical Population Biology. 2009;75(1):68–75.

[23] Roy S. Do phytoplankton communities evolve through a self-regulatory abundance-diversity relationship? BioSystems. 2009;95(2):160–165.

[24] Davidson K. Modelling microbial food webs. Marine Ecology Progress Series. 1996;145:279–296.

[25] Wiedner C, Nixdorf B. Success of crysophytes, cryptophytes and dinoflagellates over blue-greens (cynobacteria) during an extreme winter (1995/96) in euphotic shallow lakes. Hydrobiologia. 1998;370:229–235.

[26] Bird DF, Kalff J. Phagotrophic sustenance of a metalimnetic phytoplankton peak. Limnology and Oceanography. 1989;34:155–162.

[27] Hammer AC, Pitchford JW. Mixotrophy, allelopathy and the population dynamics of phagotrophic algae (cryptophytes) in the Darss Zingist Bodden estuary, southern Baltic. Marine Ecology Progress Series. 2006;328:105–115.

[28] Tilmann U. Kill and eat your predator: a winning strategy of planktonic flagellate *Prymnesium parvum*. Aquatic Microbial Ecology. 2003;32:73–84.

[29] Ives JD. Possible mechanism underlying copepod grazing responses to levels of toxicity in red tide dinoflagellates. Journal of Experimental Marine Biology and Ecology. 1961;112:131–145.

[30] Nielsen TG, Kiorboe T, Bjornsen PK. Effect of Chrysochromulina polylepis sub surface bloom on the plankton community. Marine Ecology Progress Series. 1990;62:21–35.

[31] Kozlowsky-Suzuki B, Karjalainen M, Lehtiniemi M, EngstrÃûm-Ost J, Koski M, Carlsson P. Feeding, reproduction and toxin accumulation by the copepods Acartia bifilosa and Eurytenora affinis in the presence of the toxic cynobacterium Nodularia Spumigena. Marine Ecology Progress Series. 2003;249:237–249.

[32] Lenski RE, Riley MA. Chemical warfare from an ecological perspective. Proceedings of the National Academy of Sciences USA. 2002;99:556–558.

[33] Anderson TR. Plankton functional type modelling: running before we can walk? Journal of Plankton Research. 2005;27:1073–1081.

[34] Anderson DM. Approaches to monitoring, control and management of harmful algal blooms (HABs). Ocean and Coastal Management. 2009;52:342–347.

[35] Wolfe GV. The chemical defense ecology of marine unicellular plankton: constraints, mechanisms, and impacts. The Biological Bulletin. 2000;198(2):225–244.

[36] Stoecker D, Tillmann U, Granéli E. Phagotrophy in harmful algae. In: Ecology of harmful algae Springer berlin Heidelberg: Springer; 2006.

[37] Jonsson PR, Pavia H, Toth G. Formation of harmful algal blooms cannot be explained by allelopathic interactions. Proceedings of the National Academy of Sciences USA. 2009;106(27):11177–11182.

Chapter 2

Functional Redundancy and Ecosystem Function — The Soil Microbiota as a Case Study

Stephanie D. Jurburg and Joana Falcão Salles

Additional information is available at the end of the chapter

http://dx.doi.org/10.5772/58981

1. Introduction

Understanding ecosystem functioning has been a main focus of ecological studies due to its importance for the maintenance of ecosystem integrity and human livelihood. While identifying and measuring relevant ecosystem functions may be a seemingly straightforward task, isolating the biota responsible for the provision of a particular function is far more complicated. In this context, understanding how biota influence ecosystem functioning remains a very active area of research in ecology, known as Biodiversity-Ecosystem Function (BEF) [1]. Given the accelerating rates of biodiversity loss [2] and predicted increases in the intensity and duration of extreme climate events [3], understanding how species interact to provide ecosystem functions is crucial for anticipating change as well as for establishing appropriate biodiversity buffers in order to minimize the risk of functional loss and maintain ecosystem integrity.

Functioning can be evaluated in the short-term, in which case the magnitude of the process is of interest, or in the long-term, measured as the probability that this is maintained in the face of environmental change. In both cases, functioning is an emergent property of ecosystems: interactions between the system's members and coevolution result in functioning which deviates from that expected from a system in which functioning was simply additive. In the case of environmental change, redundancy—the phenomenon in which a function is carried out by multiple species in an ecosystem—buffers functioning, as for any given environmental state there will be multiple organisms within a functional group which can perform optimally at a range of environmental conditions.

It has been suggested that concerns for the maintenance of biodiversity cannot be extended to microbes [4]. The implicit assumption is that microbial community composition is not relevant for determining function because microbes are endlessly diverse, so that the only filter

determining their function is the environment. Specifically, in microbial systems, where diversity and abundance are extreme and growth rates are rapid, it was formerly assumed that redundancy is so high that diversity and community composition are decoupled from functioning due to the following observations: 1) most microbial species are ubiquitous and present in very low densities, awaiting an opportunity to "bloom"; 2) the rapid adaptability of microbes means that such a system will never be so impoverished as to cease functioning; and 3) the microbial system is so tightly linked to its physical environment that it cannot be studied within the context of cause-effect that is generally necessary for BEF studies. However, recent studies have shown that community composition matters to function [5,6]: in soil, microbial communities exhibit a home-field advantage in decomposing endemic vs. foreign litter [7,8] and different communities do not become more similar when exposed to the same environment [9]. This ongoing discussion has been particularly important in the realm of ecosystem models, where stable physical parameters or very coarse microbial parameters (such as total biomass) are assumed to accurately represent microbial contributions to ecosystem function [10].

Despite the current gaps in knowledge of microbial communities, this is an extremely attractive system for the study of BEF: the ease of manipulation, wide range of metabolic diversities, and availability of direct links between genetic diversity and function (i.e. functional gene analyses) allow for a range of experiments which would not be possible in other ecosystems. Particularly, the high turnover rate and diversity allow for studies which target the effect of redundancy on long-term function. A wide range of studies regarding this relationship are now available (for in-depth reviews, see [11,12]), but the results of microbial BEF studies have often been contradictory. The purpose of this chapter is to provide a comprehensive analysis of redundancy in microbial communities, paying special attention to the intricacies of these systems, in order to understand why these contradictions arise, and shed light on how redundancy might bolster ecosystem function in these extremely diverse ecosystems.

2. Microbial diversity and its contribution to ecosystem function

Microbial systems are responsible for the provision of a wide range of crucial ecosystem services, but little is known about the role of diversity in maintaining this function. This is mostly due to the overwhelming complexity found in them: the study of microbial communities has been likened to the study of solar systems [13]. This diversity is still not properly constrained: the lack of an ecological species definition for prokaryotes [14] has led to the usage of the operational taxonomic unit (OTU), defined as 97% sequence similarity in the 16S rRNA gene, is used as a threshold for prokaryotic species, however this threshold may not be comparable to the eukaryotic definition of species [14]. This means that most prokaryotes can be identified based on their sequences alone, which makes distinguishing rare species from sequencing errors nearly impossible [15], and obscures the definite measurement of prokaryotic diversity. Nevertheless, it is agreed that microbial diversity is extremely high: one gram of soil may contain 10^3-10^6 unique taxa [16,17]. Furthermore, the link between phylogeny and function is truncated for prokaryotes, where horizontal gene transfer allows for the acquisition of functions—particularly those associated with adaptability to new environments—further complicates analyses of function through genes [18].

Despite these obstacles, microbial BEF—particularly for soil microbial communities — demands much attention. In addition to serving as repositories of genetic information [19], they provide ecosystem services which are fundamental for human persistence, including the maintenance of agricultural systems and waste recycling [20]. In an assessment of the economic benefits of biodiversity, the soil microbiota was partly or fully responsible for waste recycling, soil formation, nitrogen fixation, bioremediation of chemicals, biotechnology, and biocontrol of pests. These services amounted to an estimated $1.16 trillion dollars per year globally, which was over a third of the estimated annual contribution of terrestrial ecosystem services to the worldwide economy [21]. This study contrasts sharply with another estimate which, while considering both terrestrial and marine ecosystem services, differed in its estimate of the total annual value of these services by more than an order of magnitude[22]. This discrepancy illustrates the prevailing lack of consensus regarding the economic weight of ecosystem services, which is particularly problematic the face of biodiversity loss [20] because it obscures the value of preserving biodiversity for the sake of the services it provides. It also illustrates how functional classifications may be considered arbitrary: depending on the functions selected, how they are measured, and how they are valued, very different views of the same system can be obtained.

Novel technologies are beginning to open the door for the pursuit of deeper ecological understanding of microbial systems, but these advances are not accompanied by an increase in ecological theory. High-throughput sequencing has greatly accelerated the rate at which new microbial species can be detected, but their ecological properties remain a mystery [19]. Thus, although we know increasingly more about *"who is there?"*, this information is not accompanied by characterization of the new species' niche spaces (*"what are they doing?"*), which precludes the understanding of how additional species affect function at an ecosystem level. Instead, the large majority of BEF studies in microbial ecology tend to focus on a single or few ecologically relevant functions, and often measure the abundance and diversity of functional groups or genes associated with those functions. For example, the soil microbiota play a crucial role in the nitrogen cycle and studies trying to understand the link between N associated functions and soil microbiota use functional genes associated with different steps of the cycle, such as those associated with nitrification and denitrification, as a way to cut through the overwhelming diversity found in soils, and focus on functionally relevant microbial community dynamics, which may scale up and affect functioning at the ecosystem level [23].

3. Microbial BEF: A world of contradictions

Due to their rapid generation times and the large diversity found in small volumes, microbial systems are ideal settings to probe BEF relationships, particularly in controlled laboratory microcosm experiments [19]. Indeed, while much remains unknown about the world's microbiota [24], microbial BEF research has seemingly kept pace with macroecological research [25]. The former, however, has been riddled with contradictory results, and evidence for a positive BEF relationship has not been as strong as for the latter. Some of these discrepancies may arise from the heterogeneity which is unique and inherent to the microbial system. From an environmental perspective, the extremely heterogeneous soil matrix may unevenly buffer the effect of environmental change, reducing the homogeneity of the community's response. It is also important to note that the phenomena occurring in microenvironments within which the soil microbiota exist are of necessity averaged out for measurement, as current methodologies require soil to be homogenized before studying [26]. Furthermore, while positive BEF relationships are expected [1], a negative relationship resulting from antagonistic interactions has been documented [27,28].

Many contradictions have been attributed to differences in experimental setup. A recent meta-analysis indicates that most microbial BEF research has relied on comparative approaches, which test the BEF relationship across environmental gradients or treatments, rather than explicitly manipulating biodiversity [25] (Figure 1). The more common, comparative approaches are potentially riddled with hidden variables, and thus do not allow for the drawing of a direct link between diversity and function. For this reason, here we focus mainly on experiments which involve direct manipulation of diversity, which tend to find a strong, positive BEF relationship [13].

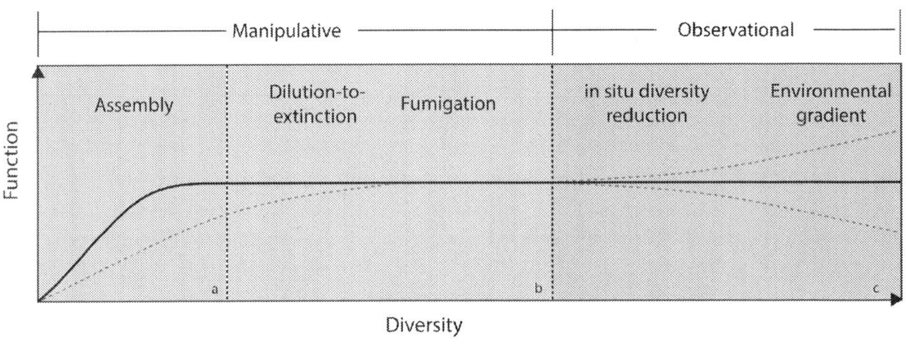

Figure 1. The relationship between diversity and function is asymptotic; different experimental approaches target different levels of species richness [13]. By greatly reducing diversity and environmental variability, assembly experiments seek mechanistic insight into the direct effect of diversity on process rates under minimized redundancy, that is, short-term function (a). Dilution-to-extinction and fumigation experiments retain greater species richness, and tend to emphasize the relationship between diversity and stability (i.e. long-term function) under otherwise stable environmental conditions (b). These experiments focus on systems in which the functioning asymptote is approached, although some dilution experiments cover broader ranges of diversity, as in [29] (b, dotted line). In observational studies, diversity is not manipulated, and the focus is rather on the effect of environmental change on the community's ability to maintain process rates (c). In this case, the level of redundancy is high enough to ensure no effect of diversity on functioning, although both positive and negative effects (c, dotted lines) have been observed for this type of experiments [28,30]

The manipulative experiments fit within two categories. In **assembly experiments**, a community is experimentally assembled to test the effect of each additional species or community structure on the community [31]. By studying overly-simplified communities, these studies tend to target the ecological functioning that arises from minimally redundant systems—that is, right before functioning begins to 'saturate' (Figure 1a). This approach has been criticized because it can only include culturable bacteria, which may represent less than 1% of soil microbes [32], and because the diversity levels achieved are always unrealistically low, and effects observed at such low diversity levels may not be relevant or applicable to more realistic scenarios and thus is not representative. Furthermore, this setup generally ignores the effect of historical selection patterns on community composition, which seems to be related to functioning as well [7]. Nevertheless, studying only culturable microbes allows for a full functional characterization of each population introduced into the system, and in this way over-yielding of the community as an emergent property of biodiversity can be studied mechanistically. For example, by characterizing 16 species of denitrifying bacteria in terms of their use of 6 carbon resources found in soil, Salles and colleagues created a model to predict CO_2 production and denitrification based on the added functioning of each individual in the system. In this way, they were able to detect over-yielding and potential antagonism within their assembled communities [33]. This body of work has found a strong, positive BEF relationship, but has also stressed that it is the diversity of the functional traits in the community—not the number of taxa present—which affect functioning: for example, a recent 12-strain assembly experiment found that the best predictor of function was the phylogenetic diversity of each microcosm [34], which agrees with previous findings [35]. The ability to manipulate genotypic and functional diversity as well as the distribution of species in assembled communities has been crucial for this [36,37]. Unfortunately, assembly experiments represent less than 1% of microbial BEF studies, and long-term studies using assembly experiments are non-existent: the lack of further mechanistic insight is one of the greater gaps in microbial BEF research [13,25].

A second approach is to erode a large part of the microbial population selectively (e.g. using heat or chloroform) or randomly (reinocculating sterile soil with serial dilutions of the original community), in the so called **removal experiments**. These systems seem to maintain redundancy and a large part of their complexity, and much of the extant long-term BEF research has depended on removal microcosms (Figure 2b). The first studies on microbial BEF used these approaches [38], and together with subsequent works have found that broad microbial functions, such as organic matter decomposition, are not affected by large decreases in diversity, but that soils with lowered diversity seem to be less resistant to invasion and less resilient to disturbance [38,39]. Nevertheless, these studies have also yielded contradictory results. For example, in one case, microbial diversity was reduced by inoculating sterile soil with serial dilutions of its original community, but the rate of carbon mineralization, nitrification and denitrification enzyme activity were not related to the diversity treatments, even after diversity reductions of more than 99% of the soil biota, suggesting no BEF relationship [40]. Using the same serial dilution approach, another experiment found that while a 10^{-5} dilution led to a 75% decrease in estimated richness, the potential denitrification rates of these soils was reduced by about 75% as well, pointing at a strong, positive BEF relationship [29].

Soil microbes are intricately tied to their environment and to each other. The complexity of the system requires that it be simplified for study, but in doing so in ways which maintain an ecosystem which is representative of the natural one has been incredibly challenging [13]. The three approaches discussed here—comparative gradient analysis, assembly, and removal experiments—target the study of the effect of the environment, diversity, and redundancy on functioning, respectively.

4. Functional redundancy and diversity

Redundancy is a characteristic of ecological systems which arises when "different species perform the same functional role in ecosystems so that changes in species diversity do not affect ecosystem functioning", and must be defined relative to the system being studied [41]. The term was first developed in an attempt to optimize conservation efforts and direct them towards the most ecologically relevant species, highlighting the importance of diversity in maintaining functional stability and the integrity of the ecosystem in the face of environmental fluctuation [42], and was later taken up as a way to calculate how much biodiversity could be lost before it affected function [43].

Functional redundancy emerges from the functional classification of its individuals. In contrast to taxonomic classifications, functional ones group organisms based on their contribution to ecosystem functioning rather than phylogeny. This classification paradigm has several advantages: functional diversity is generally a better indicator of ecosystem functioning than the direct measurement of species richness [34,44–46], and functional classifications implicitly account for environmental and biotic interactions by measuring only the outcome of community composition, thereby overcoming the oversimplification which stems from studying individual species in a laboratory setting. While this classification scheme is not universally applicable in the sense that functions must be defined relative to the system, it allows for the comparison between ecosystems that contain different species [47].

A major obstacle in applying functional classifications is the different interpretations of what constitutes a functional group, functional guild, or functional type. While functional classifications are not new to ecology, they became popular fairly recently, with the definition of the functional guild as a conceptual tool: "…a group of species that exploit the same class of environmental resources in a similar way. This term groups together species, without regard to taxonomic position, that overlap significantly in their niche requirements. (…) A species can be a member of more than one guild" [48]. Since then, new terms (e.g. functional group, strategy, trait, etc.) emerged and were used to define slightly different, yet overlapping concepts (for an in-depth discussion, see [49]). While the concept was rapidly adopted by ecology, it was not applied rigorously during the development of classification schemes, rendering them incomparable in many cases. Perhaps the biggest problem has been differentiating between *functional response groups* (groups of organisms which respond similarly to changes in environmental factors) and *functional effect groups* (groups of organisms species which contribute in a similar way to ecosystem function) [50]. In order to understand the link between ecosystem functioning and biodiversity, both of these classifications are necessary: under a given environmental condition, knowing which organisms are in their optima and which are out of their functioning range precludes the understanding of how biodiversity affects function, as much of this diversity may be apparent in terms of functioning if the organisms are diverting resources from growth to persistence. Classifying organisms into functional response groups becomes even more important if the functions in question are long-term, and environmental variability is a factor (see section 5.2).

Nowhere is the need for functional effect classifications more important than in the soil microbiome, where it is estimated that 85% of microbial cells and over 50% of microbial OTU's are inactive at any given time [51]. This means that a majority of the soil microbial diversity is only apparent with regards to short term functioning (the long term implications of these 'microbial seed banks' are discussed in a later section). Despite the need, to our knowledge only one experiment has classified a set of microbes based on their response to environmental change [52]. In this study, respiration—which is related to growth—was used both as an indicator of function (functional effect trait) and fitness (functional response trait) for 23 individual strains of bacteria and 22 strains of fungi across a range soil moisture contents. While for some organisms the wettest soil coincided with the highest respiration, many strains exhibited optimal respiration rates at intermediate moisture contents. Different niche breadths —tolerance to a wide range of environmental change—were observed. There was a strong phylogenetic signal associated with moisture tolerance: closely related strains performed more similarly that would be expected if the relationship between phylogeny and functional response were random. Finally, it was observed that biofilm-producing organisms performed better at low moisture content and had a wider tolerance range, but grew more slowly, highlighting the fact that environmental adaptation requires trade-offs [52].

The above study created the first microbe-focused functional response classification, but did not further study whether these strains, when combined, behave similarly, or whether the behavior changes with increasing community diversity. To our knowledge, no such studies exist. The novel practice of seeking the 'core microbiome' of an environment—that is, to distinguish between microbial species which change in response to the environment [53]— alludes to the need to group organisms based on their response traits, but it is generally measured in natural environments, and as such is riddled with confounding factors. One factor which distinguishes prokaryotes from other organisms is the ability to acquire mobile genetic elements (i.e. plasmids), which often contain genes that facilitate survival in a wider range of environmental states [18]. The potential change in response trait classification resulting from the acquisition of mobile genetic elements also remains unexplored.

5. The additive effect of biodiversity

The primary concern of BEF research is not the individual capacity of an organism to function, but rather the emergent properties that arise from biodiverse communities. This improvement in functioning may be an increase in functional output—known as the short term effects of biodiversity—or an increase in the probability that this level of functioning will be maintained given environmental change, known as the long term effects (Figure 2). These emergent properties are particularly hard to measure in complex systems due to the difficulty of partitioning and attributing changes in community function amongst a plethora of individuals.

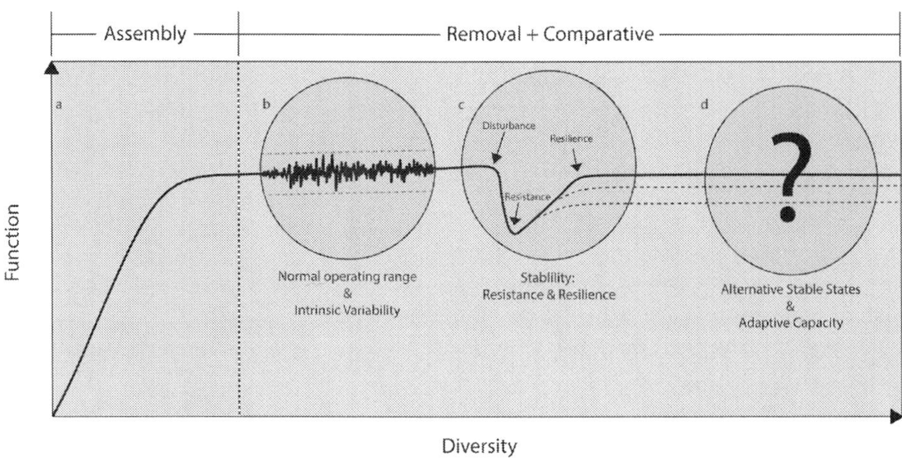

Figure 2. The short and long-term effects of biodiversity are studied in systems where diversity is simplified to different levels: for the former, the assembly approach discussed in section 3 is generally optimal (a), as simple systems are more tractable and it is easier to link an individual to increases in function. For the study of the long-term effects of biodiversity on ecosystem function—namely stability and adaptive capacity—more diversity is preserved. The emphasis is on monitoring the variability of functional parameters over time, if the goal is to determine intrinsic stability (b); or to measure resistance and resilience of the system to disturbance, if the focus is on functional stability *sensu* Pimm 1984 [54]. The study of alternative stable states and adaptive capacity is in its infancy, and even less is known regarding the redundancy on these two ecological properties in microbial systems.

5.1. Short term effects: Productivity

The idea that biodiversity increases ecosystem function was engraved in Darwin's original work "...if a plot of ground be sown with one species of grass, and a similar plot be sown with several distinct genera of grasses, a greater number of plants and dry herbage can be raised in the latter than in the former case (...) the truth of the principle that the greatest amount of life can be supported by the great diversification of life, is seen under many natural circumstances" [55]. At the most basic level, BEF research seeks to understand which characteristics arise from the presence of additional species in an ecosystem before ecosystem function begins to saturate (Figure 2a). These emergent properties—also known as biodiversity effects—are broadly categorized as selection or complementarity [25], and are considered to be the mechanistic processes by which more diverse ecosystems exhibit higher process rates.

Selection refers to the phenomenon in which a more diverse community will have a higher probability of containing more productive organisms. The better-performing organism tends to outcompete the rest for resources, returning the system to a monoculture in which its productivity dominates the entire system's productivity; interactions between competing species are not considered to be significant contributors to changes in function. Here, the maximum functioning for the community is determined by the rate of functioning of the most productive species [25,56]. In cases where the most competitive species is the less productive one, selection can lead to a negative BEF relationship.

Complementarity on the other hand, results from the competition for resources within a community, which may result in specialization and niche differentiation: as two species compete for a resource, they become specialized in exploiting the resource in different ways or times in order to minimize competitive pressure. In time, a greater efficiency is expected from the system as resources are used more thoroughly. Facilitation is a special case of complementarity, where mutualisms arise among organisms in a community, and result in higher ecosystem productivity [25]. While complementarity also predicts an asymptotic relationship between diversity and function, in this case the maximum productivity of the system may be higher than the productivity of any single member species—a phenomenon termed overyielding. In this scenario, the productivity of the system should be superior from the added productivities of the component species [57,58].

Evidence for resource-use complementarity in the soil microbial system is scant: in one case, microcosms containing up to 8 strains of cellulolytic bacteria were assembled and monitored over 25 days. Greater species richness supported more individuals and faster decomposition rates than any monoculture. Furthermore, the initial frequency distribution of inoculated organisms was maintained in the richest microcosms, suggesting coexistence, but it was not possible to distinguish whether this coexistence was due to niche differentiation or facilitation, although the authors suggest both mechanisms were present [59]. Similarly, in the assemblage experiment with denitrifying bacteria mentioned earlier, the expected function of an assembled community ('community niche') was calculated by summing the functioning of each of its members, and this was compared to the realized function. The most productive species in terms of CO_2 did not coincide with the most productive denitrifiers, illustrating the danger of underestimating relevant species when a single function is used to study the community. In addition, community niche had a much greater explanatory power for the observed functions than species richness alone. The positive relationship between community niche and function suggested that the pattern of resource utilization of the species in a community are a major driver of the increased functioning resulting from higher diversity (i.e. complementarity). The authors also found a minor selection effect, where certain species had a greater effect on community functioning than others, but they argue that in such dynamic communities, teasing out the influence of selection from complementarity is irrelevant, as these are tightly intertwined [33]. In contrast, a study using a similar experimental approach found that respiration in assembled bacterial microcosms was lower in pairwise cultures than expected from the monocultures, and even lower in multispecies cultures, suggesting a predominance of negative interactions in this system [27].

5.2. Long-term effects: Stability and resilience

Ecosystems are dynamic, and communities must maintain ecological processes in the face of environmental change (stability), recover from radical environmental change (resilience), and adapt to constantly changing environments (self-organization) in order to persist. These three properties of diverse systems arise from the interplay between functionally redundant organisms in the community: species within a functional effect group might belong to different functional response groups. When environmental change occurs, it is the presence of organisms with different response patterns that allows for the maintenance of function, as species with more favorable responses to environmental change can compensate for the loss of function by the more sensitive species. In a similar way, the presence of functionally redundant organisms allows for other, tolerant individuals to maintain function when sensitive ones die or go dormant in response to disturbance.

Redundancy may be particularly important for the highly dynamic soil microbial system where, while diversity may be extreme, it may be necessary to buffer environmental change and guarantee the maintenance of function. The most well studied long-term BEF effect is functional stability. The notion that redundancy results in stability is not new, however interest in the development of mathematical models which mechanistically explain *why* this occurs did not become popular until the late 1990's. The importance of redundancy to ecosystem performance was initially modeled by applying concepts of reliability engineering to the stability of function [59]. In this model, ecosystem functioning was defined as "the biogeochemical activities of an ecosystem or the flow of materials and processing of energy", complexity as the number of functional groups in the system, and reliability as the probability that the system will provide enough services to perpetuate the cycle. Here, diversity increases the stability within a functional group through **compensatory growth**, by which one species within a functional group increases when another is reduced. This refers to the difference in environmental tolerances between organisms, which suggests that in redundant systems, there is a higher probability that some organisms will be unaffected by the environmental change, and these will be able to use the resources left behind by the more sensitive species. Interestingly, this model looks at each functional group in the system as a compartment that feeds into the others, and so collapse of the system may come about if a single functional group becomes unstable.

The insurance hypothesis, developed a year later [60], builds on the previous model, and attributes the increase in functioning and decreased variability to the **positive selection** of the more productive species and the **temporal asynchronicity** of species responses to environmental fluctuation, respectively. Here, stability arises because the dynamics of the diverse systems are less dependent on individual species. This is particularly important in soils, which exhibit a very high species turnover rate: in one case, the bacterial and archaeal ammonia oxidizing communities in a range of Dutch agricultural soils showed above 50% change in community structure between seasons [61,62]. In another, it was shown that when colonizing a novel environment, the microbial community undergoes drastic rearrangement, and draws heavily from members of the 'rare biosphere' [9,63], a strategy which may be crucial for stress-response [51].

While the intrinsic variability of soils and the mechanisms that support it may be of interest to understanding how redundancy contributes to microbial ecosystem function (figure 2b), soil research rarely focuses on this aspect of stability. Instead soil stability is measured by applying a disturbance to soils with naturally or artificially differing levels of diversity and testing whether the microbiota are able to maintain function in the face of disturbance (resistance), and the time it takes the function to be restored to its pre-disturbance levels (engineering resilience, figure 2c) [54]. Redundancy can be measured as the diversity within a functional group, which is often assessed through functional gene markers that allow for the inclusion of unculturable organisms. As a whole, the results emerging from this area of research are hard to interpret: the usage of disturbances of different identity, duration, and intensity as well as the different time intervals between the measurements of resistance and resilience render these studies incomparable [64].

Nevertheless, this body of work has yielded important insights into the relationship between diversity and stability. For example, one study found that the diversity of both nitrite oxidizing and denitrifying bacteria in soil was not significant in determining the rate of functional recovery from experimental heating; rather, the main factor affecting this phenomenon was the abundance of the genes responsible for the functions tested [65]. In this case, it was not diversity, but sheer abundance which was responsible for stability. In another case, the recovery rate of two soils with naturally differing levels of diversity was compared: while mineralization of a labile carbon source (^{14}C-labeled wheat shoot) remained unaffected, mineralization of a recalcitrant substrate (^{14}C-labeled 2,4 dichlorophenol) was impaired. The more diverse soil was able to recover within the 9 weeks of the experiment, while the less diverse soil did not [30], suggesting here diversity mattered not only for stability, but also for the decrease in function.

Generally, narrower or less redundant functions have been found to be less stable following disturbance than broad functions[66], supporting the notion that biodiversity acts to buffer the system against fluctuations. In one case, respiration in serially diluted soil microbial microcosms exhibited no change in basal respiration or decomposition despite the large reductions in diversity, but nitrification was progressively retarded with each dilution [38]. Changes in community composition may affect function when, following disturbance, an abundant and efficient species is replaced by a redundant, but less efficient yet tolerant one. For example, monitoring potential nitrite oxidation (PNO) on soils that were treated with the cessation of tillage on tilled land or the establishment of tillage on untilled land, it was possible to detect a switch from *Nitrobacter*-like nitrite oxidizers to *Nitrospira*-like nitrite oxidizers with tillage, which explained the decrease in PNO [67]. The cessation of tillage did not result in a restoration of the *Nitrobacter* community within the 17 months of study, suggesting that long-term function might have been permanently affected by treatment. In an assembly experiment comparing the recovery from heating or metals in microbial communities of 1-12 bacterial species, biodiversity increased stability, measured as community biomass, but this stability was closely associated to the number of tolerant species in the community, a phenomenon analogous to the selection effect [34]. In a separate experiment, altering the pH in mixed culture fermentation reactors was shown to bring about the dominance of different species of *Clostridium* and elicit slight changes in the reactor's chemical output in accordance with the dominant species' preferences [68].

The distribution of species abundances within a community also affect stability: more evenly distributed communities are generally more stable than communities characterized by one or two dominant species [37,69]. In one case, the effect of selective stress on the stability of assembled denitrifying communities of up to 18 species was highly dependent on initial community evenness [37]. Even excluding the effect of the presence of tolerant species on the community's response, evenness played a significant role in maintaining stability.

Perhaps the clearest results have been obtained from studies looking at invasion resistance as an indicator of functional stability [11]. In general, diversity decreases invasibility in microbial systems [36,39,45]. By using both assembly and dilution in bacterial microcosms, a strong, negative correlation was shown between diversity and invasibility of an invader *E. coli* strain [39]. In a more recent experiment, the authors were able to attribute this decrease in invasibility to a reduction in easily available resources and reduced competitive advantage in the more diverse treatments. This result was confirmed by applying a resource pulse to the community following invasion, which led to an increase in the abundance of the invading species [70]. An analogous result was found in assembled communities consisting of different strains of *Pseudomonas fluorescens*, where genetic dissimilarity within a community increased productivity and decreased the success of the invader *Serratia liquefaciens* by decreasing the amount of resources available to the invader [71].

While it seems that theoretical predictions of a positive relationship between diversity and stability are somewhat in agreement, a large gap in the literature arises from differences between the definition of stability employed in these two fields: while experimental microbial ecology uses a functional definition of stability, which depends on resistance and resilience, theory often relies on intrinsic functional stability, which is a stand-alone parameter that measures the reduction of variability when there is no change in environmental parameters (Figure 2b). It is expected that more diverse communities will be more functionally stable and less compositionally stable, yet this has received little attention. The measurement of intrinsic stability requires the repeated measurement of an unperturbed community over time. Instead, a measurement is made immediately before disturbance to determine "normal" levels of functioning, immediately after to evaluate whether the system was resistant, and for a third time after a recovery period has passed. This approach does not consider that the system at equilibrium exhibits a constant variability which is intrinsic to the system, called normal operating range (NOR) [72], and thus cannot distinguish whether the response of a community to disturbance fits within 'normal' ranges of fluctuation or not, or whether a system that is deemed recovered is in a similar equilibrium to its undisturbed state.

6. Moving forward: Redundancy and adaptive capacity

The concept of resilience employed in the measurement of functional stability—engineering resilience—differs from the ecological concept as it was originally proposed [73]—ecological resilience. The former sees ecosystems as simple, rebounding springs, while the latter includes the possibility of the system shifting to alternative states due to perturbation, and thus is much harder to measure [73]. If ecological resilience is considered as a function in the same way as invasion resistance and stability, then it too can be progressively eroded. While research on this topic has been sparse and has not directly manipulated diversity, evidence of this phenomenon exists [74–78]. For example, mercury-contaminated, heat-shocked soils responded much more slowly to substrate additions than the transiently tylosin-contaminated or control soils [79]. The authors observed a significant decrease in the microbial diversity of the mercury-contaminated soils, which may explain the reduced response following additional disturbances. Mercury constitutes a long-term stress, so the heat-shocked communities were already coping with the original disturbance; however some studies find that even when the soils are allowed to recover from transient perturbations, their response to further disturbances is slower than that of the control soils: in another case, grassland soils which had experienced various forms of perturbation (reseeding, application of sewage-sludge, biocide/nitrogen and lime additions) recovered their ability to decompose more slowly following both copper and transient heat stresses than the unperturbed controls [78].

The concept of ecological resilience can be broken down into three characteristics: 1) the amount of change a system can undergo while retaining the same controls on function and structure; 2) the degree to which the system is capable of self-organization; and 3) the ability to build and increase the capacity for learning and adaptation [80]. Systems in which ecological resilience has been lost are unable to adapt to environmental change beyond a certain threshold, and in response to change shift to alternative stable states, in which the community is characterized by a different set of interactions (Figure 2d). One question that arises from this is whether microbial systems have stable equilibria to begin with. This is unclear, since the detection of alternative stable states requires the measurement of intrinsic variability which, as mentioned in the previous section, is not common practice in microbial ecology.

Another question is whether these irreversible shifts to alternative stable states have any relevance to ecosystem cycles. By analyzing the available literature, we may find mechanisms by which they do: the previously mentioned experiment in which tillage and no-tillage agricultural lands experienced an exchange in practice and the productivity and structure of the nitrite oxidizing communities underwent a catastrophic shift as, in response to an environmental change (tilling), the dominant members of a functional group (nitrite oxidizing bacteria) fundamentally changed from a those belonging to a more efficient genus (*Nitrobacter*) to a less efficient one (*Nitrospira*), leading to a decrease in function. Furthermore, cessation of tillage did not result in the opposite change in community. This may be an example of hysteresis—the phenomenon in which a system fails to return to its original state once perturbation ceases [81]—which is a property of systems that exhibit alternative stable states. In this case, the system may fail to return to its no-tillage state because the *Nitrobacter* community has been eroded beyond its ability to recover, or because the new dominant, generalist group is well suited for a wide range of environmental states, and it cannot easily be outcompeted when the system returns to its original conformation. Regardless of the underlying mechanism, this study provides evidence that a shift in the *identity* of the dominant organisms in a functional group may have an effect on functioning, and that this change may be irreversible. The implications of applying the ecological resilience concept to BEF studies are unknown but potentially very relevant, however to our knowledge, no work explicitly targets the measurement of ecological resilience in microbial systems, and this represents a serious gap in ecological research.

7. Conclusion

The last decade has seen a shift in focus, from a function-focused to a stability and probability-focused perspective. This is to be expected: at a time in which climate is expected to become more unpredictable [3], and biodiversity loss is expected to accelerate [2], it becomes important to be able to guarantee not only that ecosystems will be able to function, but that they will still be able to function in the face of drastic change. As mentioned earlier, the concept of functional redundancy was developed as a way determine which species within a community required the most conservation attention, and was later used to refer to a 'minimum' amount of biodiversity needed to keep the system functioning. As the focus shifts from the from the short- to the long-term effects of redundancy on ecosystem functioning, it becomes clear that the ecological value of redundant species lies in their ability to buffer against environmental change. Microbial communities are excellent model systems to study such buffering, not only due to the extremely high level of functional redundancy found here, but also due to the fact that these systems routinely experience rapid changes which may be catastrophic from a microbial perspective, and yet as a community they are able to maintain function. It seems that, even in the extremely diverse soil microbial system, diversity reductions result in reductions in either long-term or short-term function, or both, although the current gap in knowledge regarding microbial functional responses impairs our ability to understand the mechanics of this reduction as well as our ability to predict when environmental change results in functional change [82].

While the relevance of diversity to resilience and self-organization, and their contribution to the maintenance of function may be elusive and hard to study experimentally, these relationships warrant our attention. Initial studies have already revealed the importance of rare species in restructuring communities. Given current knowledge, it seems that in changing environments, every species matters, even in communities as diverse as the soil microbial community. Future research must delve into whether certain individuals matter more by evaluating the functional response profiles of individuals and communities, and quantifying the effect of changes of community composition on function.

Author details

Stephanie D. Jurburg and Joana Falcão Salles[*]

*Address all correspondence to: falcao.salles@rug.nl

Department of Microbial Ecology, Centre for Life Sciences, University of Groningen, Groningen, The Netherlands

References

[1] Hooper D, Chapin F, Ewel J, Hector A, Inchausti P, Lavorel S, et al. Effects of biodiversity on ecosystem functioning: A consensus of current knowledge. Ecological monographs. 2005;75(1):3–35.

[2] Millenium Ecosystem Assessment. Ecosystems and human well-being: Synthesis. Ecosystems. Washington D.C.: Island Press; 2005.

[3] IPCC 2007. Climate change 2007: Synthesis report. Contribution of Working Groups I, II and III to the Fourth Assessment Report of the Intergovernmnetal Panel on Climate Change. II and III to the Fourth Assessment Report of the Intergovernmental Panel on Climate Change. Intergovernmental Panel on Climate Change, Geneva. Geneva: IPCC; 2007.

[4] Finlay BJ, Maberly SC, Cooper JI. Microbial diversity and ecosystem function. Oikos. 1997;80(2):209–13.

[5] Allison SD, Martiny JBH. Resistance, resilience, and redundancy in microbial communities. Proceedings of the National Academy of Science. 2008;105(Supplement 1): 11512.

[6] Reed HE, Martiny JBH. Testing the functional significance of microbial composition in natural communities. FEMS Microbiology Ecology. 2007 Nov;62(2):161–70.

[7] Keiser AD, Strickland MS, Fierer N, Bradford MA. The effect of resource history on the functioning of soil microbial communities is maintained across time. Biogeosciences. 2011 Feb 23;8(1):1643–67.

[8] Strickland MS, Lauber C, Fierer N, Bradford MA. Testing the functional significance of microbial community composition. Ecology. 2009 Feb;90(2):441–51.

[9] Pagaling E, Strathdee F, Spears BM, Cates ME, Allen RJ, Free A. Community history affects the predictability of microbial ecosystem development. The ISME journal. Nature Publishing Group; 2013;8(1):19–30.

[10] McGuire KL, Treseder KK. Microbial communities and their relevance for ecosystem models: Decomposition as a case study. Soil Biology and Biochemistry. 2010 Apr; 42(4):529–35.

[11] Shade A, Peter H, Allison SD, Baho DL, Berga M, Bürgmann H, et al. Fundamentals of Microbial Community Resistance and Resilience. Frontiers in Microbiology. 2012;3(December):1–19.

[12] Griffiths BS, Philippot L. Insights into the resistance and resilience of the soil microbial community. FEMS Microbiology Reviews. 2012;

[13] Le Roux X, Recous S, Attard E. Soil microbial diversity in grasslands and its importance for grassland functioning and services. In: Lemaire G, Hodgson J, Chabbi A,

editors. Grassland Productivity and Ecosystem Services. Wallingford: CAB international; 2011. p. 158–65.

[14] Cohan FM. What are bacterial species? Annual Reviews in Microbiology. 2002;56(1): 457–87.

[15] Pedrós-Alió C. The rare bacterial biosphere. Annual Review of Marine Science. Annual Reviews; 2012;4:449–66.

[16] Torsvik V, Øvreås L. Microbial diversity and function in soil: From genes to ecosystems. Current opinion in microbiology. 2002;5(3):240–5.

[17] Gans J, Wolinsky M, Dunbar J. Computational improvements reveal great bacterial diversity and high metal toxicity in soil. Science. 2005 Aug 26;309(5739):1387–90.

[18] Van Elsas JD, Jansson JK, Trevors JT, editors. Modern Soil Microbiology. 2nd ed. Boca Raton, Florida: CRC Press; 2006.

[19] Prosser JI, Bohannan BJM, Curtis TP, Ellis RJ, Firestone MK, Freckleton RP, et al. The role of ecological theory in microbial ecology. Nature Reviews Microbiology. 2007;5(5):384–92.

[20] Jeffery S, Gardi C, Jones A, Montanarella L, Marmo L, Miko L, et al. European Atlas of Soil Biodiversity. Luxembourg: Publications Office of the European Union; 2010.

[21] Pimentel D, Wilson C, McCullum C, Huang R, Dwen P, Flack J, et al. Economic and environmental benefits of biodiversity. BioScience. JSTOR; 1997;747–57.

[22] Costanza R, D'Arge R, De Groot R, Farber S, Grasso M, Hannon B, et al. The value of the world's ecosystem services and natural capital. Nature. Londres; 1997;387(6630): 253–60.

[23] Bouskill NJ, Tang J, Riley WJ, Brodie EL. Trait-based representation of biological nitrification: model development, testing, and predicted community composition. Frontiers in microbiology. 2012 Jan;3(October):364.

[24] Prosser JI. Ecosystem processes and interactions in a morass of diversity. FEMS Microbiology Ecology. 2012;

[25] Krause S, Le Roux X, Niklaus P a., Van Bodegom PM, Lennon JT, Bertilsson S, et al. Trait-based approaches for understanding microbial biodiversity and ecosystem functioning. Frontiers in Microbiology. 2014 May 27;5(May):1–10.

[26] Vos M, Wolf AB, Jennings SJ, Kowalchuk GA. Micro-scale determinants of bacterial diversity in soil. FEMS Microbiology Reviews. Wiley Online Library; 2013 Apr 3;1–19.

[27] Foster KR, Bell T. Competition, not cooperation, dominates interactions among culturable microbial species. Current Biology. Elsevier; 2012;22(19):1845–50.

[28] Pereira e Silva M, Schlotter-Hai B, Poly F, Guillaumaud N, Schloter M, van Elsas J, et al. Phylogenetic distance and abundance of thaumarchaeal ammonia oxidizing communities affect potential nitrification rates in agricultural soils. submitted.

[29] Philippot L, Spor A, Hénault C, Bru D, Bizouard F, Jones CM, et al. Loss in microbial diversity affects nitrogen cycling in soil. The ISME journal. 2013 Aug;7(8):1609–19.

[30] Girvan MS, Campbell CD, Killham K, Prosser JI, Glover LA. Bacterial diversity promotes community stability and functional resilience after perturbation. Environmental Microbiology. 2005 Mar;7(3):301–13.

[31] Bell T, Newman JA, Silverman BW, Turner SL, Lilley AK. The contribution of species richness and composition to bacterial services. Nature. 2005;436(7054):1157–60.

[32] Torsvik V, Ovreas L, Thingstad TF. Prokaryotic diversity--magnitude, dynamics, and controlling factors. Science. 2002;296(5570):1064.

[33] Salles JF, Poly F, Schmid B, Le Roux X. Community niche predicts the functioning of denitrifying bacterial assemblages. Ecology. 2009 Dec;90(12):3324–32.

[34] Awasthi A, Singh M, Soni SK, Singh R, Kalra A. Biodiversity acts as insurance of productivity of bacterial communities under abiotic perturbations. The ISME journal. Nature Publishing Group; 2014;

[35] Salles JF, Le Roux X, Poly F. Relating phylogenetic and functional diversity among denitrifiers and quantifying their capacity to predict community functioning. Frontiers in microbiology. Frontiers Media SA; 2012;3.

[36] Jousset A, Schulz W, Scheu S, Eisenhauer N. Intraspecific genotypic richness and relatedness predict the invasibility of microbial communities. The ISME journal. Nature Publishing Group; 2011 Jul;5(7):1108–14.

[37] Wittebolle L, Marzorati M, Clement L, Balloi A, Daffonchio D, Heylen K, et al. Initial community evenness favours functionality under selective stress. Nature. 2009;458(7238):623–6.

[38] Griffiths B, Ritz K, Bardgett R, Cook R, Christensen S, Ekelund F, et al. Ecosystem response of pasture soil communities to fumigation-induced microbial diversity reductions: An examination of the biodiversity–ecosystem function relationship. Oikos. 2000;90(2):279–94.

[39] Van Elsas JD, Chiurazzi M, Mallon CA, Elhottová D, Krištůfek V, Salles JF. Microbial diversity determines the invasion of soil by a bacterial pathogen. PNAS. National Acad Sciences; 2012;109(4):1159–64.

[40] Wertz S, Degrange V, Prosser JI, Poly F, Commeaux C, Freitag T, et al. Maintenance of soil functioning following erosion of microbial diversity. Environmental Microbiology. 2006;8(12):2162–9.

[41] Loreau M. Does functional redundancy exist? Oikos. 2004;3(140).

[42] Walker BH. Biodiversity and ecological redundancy. Conservation Biology. 1992;6(1):18–23.

[43] Lawton JH, Brown VK. Redundancy in ecosystems. In: Schulze ED, Mooney HA, editors. Biodiversity and Ecosystem Function. Berlin, Heidelberg: Springer; 1994. p. 255–70.

[44] Petchey OL, Gaston KJ. Functional diversity (FD), species richness and community composition. Ecology Letters. 2002;5(3):402–11.

[45] Eisenhauer N, Schulz W, Scheu S, Jousset A. Niche dimensionality links biodiversity and invasibility of microbial communities. Functional Ecology. Wiley Online Library; 2013;27(1):282–8.

[46] Pereira e Silva Semenov, A.V., Schmitt, H, van Elsas, J.D., Falcao Salles, J. MC. Microbe-mediated processes as indicators to establish the normal operating range of soil functioning. Biology and Biochemistry. 2012;submitted.

[47] Voigt W, Perner J, Hefin Jones T. Using functional groups to investigate community response to environmental changes: Two grassland case studies. Global Change Biology. 2007 Aug;13(8):1710–21.

[48] Root RB. The Niche Exploitation Pattern of the Blue-Gray Gnatcatcher. Ecological Society of America. 1967;37(4):317–50.

[49] Gitay H, Noble IR. What are functional types and how should we seek them? In: Smith T, Shugart H, FI W, editors. Plant functional types: their relevance to ecosystem properties and global change. Cambridge University Press; 1997.

[50] Lavorel S, Garnier E. Predicting changes in community composition and ecosystem functioning from plant traits: revisiting the Holy Grail. Functional Ecology. Blackwell Science Ltd; 2002 Oct 1;16(5):545–56.

[51] Lennon JT, Jones SE. Microbial seed banks: the ecological and evolutionary implications of dormancy. Nature Reviews Microbiology. Nature Publishing Group; 2011;9(2):119–30.

[52] Lennon JT, Aanderud ZT, Lehmkuhl BK, Schoolmaster DR. Mapping the niche space of soil microorganisms using taxonomy and traits. Ecology. 2012 Aug;93(8):1867–79.

[53] Shade A, Handelsman J. Beyond the Venn diagram: the hunt for a core microbiome. Environmental Microbiology. 2012 Jan;14(1):4–12.

[54] Pimm SL. The complexity and stability of ecosystems. Nature. 1984;307(5949):321–6.

[55] Darwin C. On the origin of species by means of natural selection. Murray, London. 1859;

[56] Tilman D, Lehman CL, Thomson KT. Plant diversity and ecosystem productivity: Theoretical considerations. Proceedings of the National Academy of Sciences. National Acad Sciences; 1997;94(5):1857–61.

[57] Tilman D, Lehman CL, Thomson KT. Plant diversity and ecosystem productivity: Theoretical considerations. Proceedings of the National Academy of Sciences. National Acad Sciences; 1997;94(5):1857–61.

[58] Loreau M. Biodiversity and ecosystem functioning: Recent theoretical advances. Oikos. 2000 Oct;91(1):3–17.

[59] Wohl DL, Arora S, Gladstone JR. Functional redundancy supports biodiversity and ecosystem function in a closed and constant environment. Ecology. 2004; 85(6): 1534-1540.

[60] Naeem S. Species redundancy and ecosystem reliability. Conservation Biology. 1998;12(1):39–45.

[61] Yachi S, Loreau M. Biodiversity and ecosystem productivity in a fluctuating environment: the insurance hypothesis. Proceedings of the National Academy of Sciences. 1999;96(4):1463.

[62] Pereira e Silva MC, Poly F, Guillaumaud N, van Elsas JD, Salles JF, Pereira E Silva MC. Fluctuations in ammonia oxidizing communities across agricultural soils are driven by soil structure and pH. Frontiers in Microbiology. 2012 Jan;3(March):77.

[63] Pereira e Silva MC, Dias ACF, van Elsas JD, Salles JF. Spatial and temporal variation of archaeal, bacterial and fungal communities in agricultural soils. PLOS ONE. Public Library of Science; 2012;7(12):e51554.

[64] Sjöstedt J, Koch-Schmidt P, Pontarp M, Canbäck B, Tunlid A, Lundberg P, et al. Recruitment of members from the rare biosphere of marine bacterioplankton communities after an environmental disturbance. Applied and Environmental Microbiology. Am Soc Microbiol; 2012;78(5):1361–9.

[65] Deng H. A review of diversity-stability relationship of soil microbial community: What do we not know? Journal of Environmental Sciences. The Research Centre for Eco-Environmental Sciences, Chinese Academy of Sciences; 2012 Jun;24(6):1027–35.

[66] Wertz S, Degrange V, Prosser JI, Poly F, Commeaux C, Guillaumaud N, et al. Decline of soil microbial diversity does not influence the resistance and resilience of key soil microbial functional groups following a model disturbance. Environmental Microbiology. 2007;9(9):2211–9.

[67] Balser TC, Firestone MK. Linking microbial community composition and soil processes in a California annual grassland and mixed-conifer forest. Biogeochemistry. 2005 Apr;73(2):395–415.

[68] Attard E, Poly F, Commeaux C, Laurent F, Terada A, Smets BF, et al. Shifts between Nitrospira-and Nitrobacter-like nitrite oxidizers underlie the response of soil poten-

tial nitrite oxidation to changes in tillage practices. Environmental Microbiology. Wiley Online Library; 2010;12(2):315–26.

[69] Lu Y, Slater FR, Mohd-Zaki Z, Pratt S, Batstone DJ. Impact of operating history on mixed culture fermentation microbial ecology and product mixture. Water Science & Technology. IWA Publishing; 2011;64(3):760–5.

[70] Degens BP, Schipper LA, Sparling GP, Duncan LC. Is the microbial community in a soil with reduced catabolic diversity less resistant to stress or disturbance? Soil Biology and Biochemistry. Elsevier; 2001;33(9):1143–53.

[71] Mallon C., Poly F, Le Roux X, Marring I, van Elsas J, Salles JF. Resource pulses can alleviate the biodiversity-invasion relationship in soil microbial communities. Ecology. (submitted).

[72] Jousset a, Schmid B, Scheu S, Eisenhauer N. Genotypic richness and dissimilarity opposingly affect ecosystem functioning. Ecology Letters. 2011 Jun;14(6):537–45.

[73] Kersting K. Normalized ecosystem strain: A system parameter for the analysis of toxic stress in (micro-) ecosystems. Ecological Bulletins. 1984;150–3.

[74] Holling CS. Resilience and stability of ecological systems. Annual Reviews in Ecology, Evolution and Systematics. 1973;1–23.

[75] Philippot L, Cregut M, Chèneby D, Bressan M, Dequiet S, Martin-Laurent F, et al. Effect of primary mild stresses on resilience and resistance of the nitrate reducer community to a subsequent severe stress. FEMS Microbiology Letters. 2008 Aug;285(1): 51–7.

[76] Zhang B, Deng H, Wang H, Yin R, Hallett PD, Griffiths BS, et al. Does microbial habitat or community structure drive the functional stability of microbes to stresses following re-vegetation of a severely degraded soil? Soil Biology and Biochemistry. Elsevier Ltd; 2010 May;42(5):850–9.

[77] Tobor-Kaplon MA, Bloem J, De Ruiter PC. Functional stability of microbial communities from long-term stressed soils to additional disturbance. Environmental Toxicology and Chemistry. Wiley Online Library; 2006;25(8):1993–9.

[78] Chaer G, Fernandes M, Myrold D, Bottomley P. Comparative resistance and resilience of soil microbial communities and enzyme activities in adjacent native forest and agricultural soils. Microbial Ecology. Springer; 2009;58(2):414–24.

[79] Kuan HL, Fenwick C, Glover LA, Griffiths BS, Ritz K. Functional resilience of microbial communities from perturbed upland grassland soils to further persistent or transient stresses. Soil Biology and Biochemistry. 2006;38(8):2300–6.

[80] Müller AK, Westergaard K, Christensen S, Sørensen SJ. The diversity and function of soil microbial communities exposed to different disturbances. Microbial Ecology. Springer; 2002 Jul;44(1):49–58.

[81] Carpenter SR, Brock WA. Adaptive capacity and traps. Ecology and Society. 2008;13(2):40.

[82] Beisner BE, Haydon DT, Cuddington K. Alternative stable states in ecology. Frontiers in Ecology and the Environment. Eco Soc America; 2003;1(7):376–82.

[83] Wallenstein MD, Hall EK. A trait-based framework for predicting when and where microbial adaptation to climate change will affect ecosystem functioning. Biogeochemistry. 2012;109(1):35–47.

Chapter 3

Biodiversity and Ecosystem Functioning in Tropical Habitats — Case Studies and Future Perspectives in Atlantic Rainforest and Cerrado Landscapes

Tathiana G. Sobrinho, Lucas N. Paolucci,
Dalana C. Muscardi, Ana C. Maradini,
Elisangela A. Silva, Ricardo R. C. Solar and
José H. Schoereder

Additional information is available at the end of the chapter

http://dx.doi.org/10.5772/59042

1. Introduction

Currently, environmental changes can be seen as an intrinsic feature of ecosystems, once finding ecosystems that do not suffer of anthropogenic pressures, either direct or indirect, is rare [1]. Such pressures come from the continuous and exponential human population growth, which propels urbanization, activities and processes directly linked to the use of fossil fuels, mining, agriculture and cattle growth. The maintenance of current human population growth implies in the supply of a huge demand for food and technology, resulting in rising pollution and loss of habitats and entire ecosystems [2-5].

Anthropogenic impacts alter the physical environmental characteristics, climate, temperature, soil and water quality, and biogeochemical cycles, interfering directly on the biota [6-11]. The immense resource consumption exerted by the human population demands an ongoing exploitation of natural resources. This causes a constant increase of greenhouse gas emissions and, consequently, the temperature around the globe, also generating an intense conversion of soil use [12,13]. The shift from natural environments into cultivated soils became noteworthy in several countries mainly after the Green Revolution, and together with the frequent use of fertilizers, have changed or destroyed natural habitats, decreasing biodiversity directly or indirectly [9,14]. In addition to these factors, species overexploitation and species invasion

have contributed to the decline of biodiversity [15]. Current data indicates that almost 30% of known species in the World became extinct or are endangered due to anthropogenic pressures.

The loss of biodiversity itself is not the only problem associated with human disturbance, as such habitat changes may have a harmful cascade effects that alter other environmental properties. Every biodiversity loss may be translated into a loss of functional diversity [16], which is related to the characteristics of organisms that allow them to perform different function in the ecosystem [16,17]. These functions, such as seed dispersal [18,19], pest and weed biological control [20], pollination [21], nutrient cycling [22], and the decomposition of organic matter, among others, are essential to ecosystem functioning maintenance. Ecosystem functioning include biogeochemical and ecosystem processes [23], responsible by matter cycling and energy flow, being directly related to resource dynamics and ecosystem stability [17]. Generally, the ecosystem functioning can be estimated as the magnitude and dynamics of ecosystem processes resultant from the interactions within and between different levels of biota [24].

Traditionally, the main parameters used for estimate ecosystem functioning are linked to plant communities, such as primary productivity and biomass stability [25-27]. Processes mediated by other organisms, including arthropods, have seldom been used for such estimates [28,29], regardless their crucial role. Arthropods are very diverse organisms, abundant everywhere, highly successful and spread across the globe. They represent more than 80% of all described biodiversity and, among them, insects are the most abundant [30]. Arthropods perform several functions in ecosystems, at different levels. These organisms inhabit from the underground soil layers to the top of tress, are engaged in several trophic levels, and interfere in the occurrence and distribution of several other organisms through intricate interactions such as predation, competition, herbivory and mutualism. Arthropods perform soil bioturbation [31], pest, weed, parasite and disease control [32-34], pollination [35,36], seed dispersal [18,37], dung and carrion removal [37], and act in decomposition and nutrient cycling [22,38,39]. Thus, it is expected that changes in the diversity of arthropods can also trigger changes in the ecosystem functioning.

Functional features of species may influence how ecosystem functioning will be altered with biodiversity loss [40]. Different species may be redundant in the functions they play in the ecosystem, and in this case species loss would be compensated by another one that performs the same function. Hence, biodiversity loss would not necessarily cause decrease on ecosystem functioning, as well as it would not increase if new species were added (redundancy hypothesis, or null hypothesis). Alternatively, some species may be singular or unique in the functions they play within the ecosystems, and their loss would eventually result in a decrease of ecosystem functioning (linearity hypothesis). Finally, the effect of species loss or gain on ecosystem functioning may be dependent on the conditions (community species composition or soil fertility, for instance) under which these biodiversity alterations occur, and the outcome would be unpredictable (idiosyncrasy hypothesis) [40].

The linearity hypothesis is a pattern frequently reported in studies carried out in temperate regions [41], while in the tropics the null hypothesis seems to be the most common. This difference in the reported patterns may be linked to higher biodiversity levels found in the tropics, suggesting a possible functional redundancy among species, which does not seem to occur in temperate regions. Notably, tropical regions harbor the highest World biodiversity [42,43], once 16 of the 25 biodiversity hotspots are located in these regions [43,44]. Conversely tropical regions also exhibit the highest rates of species loss due to human activities [45,46]. Among the main human alterations that cause biodiversity loss is land use change [47-49] and, according to estimates, it will remain as the main activity during the next 100 years [50]. Several tropical biomes have experienced high biodiversity loss due to land use change and, in Brazil, the Atlantic Forest and the Cerrado (Brazilian savanna) may be highlighted [8,51-53].

The Atlantic Forest originally covered ca. 150 million hectares along the Brazilian coast [52], occurring in tropical and subtropical regions, and including sites with large altitude variation, humidity, temperature and rainfall regimes. Such a variation in abiotic conditions allowed the differentiation of several phytophysiognomies, high endemism and the occurrence of numerous rare species, harboring ca. of 8% of Worlds' biodiversity. However, recent estimates indicate that more than 90% of its vegetation cover has been lost, and the Atlantic Forest is nowadays composed by forest fragments, mostly smaller than 250 ha, immerse in a landscape of different human modified habitats [52]. The effects of habitat loss due to human activities are extensively reported in the literature concerning the Atlantic Forest [8,54-56], but the effects of biodiversity decrease on the processes related to ecosystem functioning still need more consistent information. The Brazilian Cerrado is the second larger biome of the country, occupying originally approximately 22% of its area [57], and stands out by its high biodiversity and endemism. The Cerrado is currently suffering an elevated degree of human exploitation, and nowadays remains less than 30% of its original area [43]. High fragmentation and conversion in pasture or agriculture areas, cause biodiversity loss, soil erosion, arrival and establishment of invasive species, and shifts on fire regimes, carbon cycles and climate [51-53].

In this chapter we aimed to evaluate the relationship between biodiversity and ecosystem functioning in the tropical biomes Atlantic Forest and Cerrado. We report here three case studies that investigate different ecosystem processes modulated by arthropods: litter decomposition, seed dispersal and protection against herbivores. In these studies we seek to understand the relative importance of species richness and of the presence of keystone species on the studied ecosystem functions.

We performed the first case study in a secondary forest fragment, in the Atlantic Forest biome. In this study we test the relationship between litter decomposition and the biodiversity of several functional groups of soil and litter arthropods. We performed the second case study in the Cerrado biome, testing the effect of ants biodiversity that visit extrafloral nectaries on the protection of these plants against herbivores and herbivory.

Lastly, in the third case study, carried out in the same region of the first one, we analyzed the effect of ant biodiversity on seed removal, comparing secondary forests and *Eucalyptus* crops. In this study we test the direct effect of land use change on the relationship between biodiversity and functioning. From the analysis of these three case studies we concluded this chapter presenting some future perspectives of studies on this subject, to solve some knowledge gaps related to the biodiversity and ecosystem functioning relationship in tropical ecosystems.

2. Case study 1

Decomposition is the process that transforms nutrients retained in organic matter into their inorganic form, making them available in the soil to the primary producers [58,59], and is therefore a key supporting process for the functioning of ecosystems. This process is ruled by three main factors: the physicochemical environment, the quality of the decomposing material and the soil and litter fauna [58,60-63]. These factors present different interaction routes [64] and the relative importance of each component changes in different time and spatial scales [65].

The physicochemical environment is related to the climate, or microclimate, mainly humidity and temperature [66,67]. Abiotic conditions may indirectly affect decomposition, altering litter characteristics, or directly, controlling the activity of decomposers [66,68]. Litter quality is usually associated to foliar material degradability [69], as the concentration of some nutrients has been frequently associated to its palatability to organisms [70]. Usually a higher initial nitrogen concentration reflects in a higher organic matter quality to decomposers. Finally, organisms living in litter and soil are crucial for decomposition processes and nutrient release [60,71-73]. These organisms revolve, mix, break and digest the detritus, metabolizing the litter constituents [58]. Among the components of the soil community, fungi and bacteria are the main decomposing agents. Nevertheless, the micro and mesofauna of soil and litter arthropods have an important role in the decomposition process, through fragmentation of organic matter, through the mixing and vertical movement of organic matter [74]. The existence of an abundant and diversified arthropod fauna is expected, then, to favor an enhanced nutrient cycling [75] and a subsequent faster plant growth [76].

The abundant arthropod fauna composing soil and litter communities can be categorized into different guilds or functional groups, according to their activities, which may affect the microbial community by several ways [77]. Fungivores and bacteriovores consume exclusively the microorganisms, decreasing their abundance. Moreover, they can decrease their prey species richness, through an intense predation, or else an increase of this species richness, through the top-down control of the more competitive species, mediating their coexistence. Detritivores, on the other hand, consume part of organic matter together with the film of microorganisms, releasing the broken and partially digested organic matter in their faeces. As a result, besides their negative effects on microorganisms due to predation, detritivores may increase litter fragmentation, resulting in more decomposing surface and higher decomposer abundance an species richness. These organisms interact in complex food webs and therefore diversity and abundance changes of a given functional group or guild may alter abundance, diversity and functioning of another group [78,79]. It is important hence the investigation of the functional groups role on the decomposition process, as different guilds may interfere more than others.

The process of litter decomposition, as well as the intricate relationships among the diverse components of the edaphic fauna associated to the litter, offers an excellent study system of the relationship between biodiversity and ecosystem processes, mainly in tropical environments with their huge diversity. In this study case we verify how soil and litter arthropod biodiversity affects litter decomposition in a tropical habitat. Our hypothesis is that increasing arthropod abundance and species richness cause higher decomposition rates, and that some functional groups may have stronger roles in this process.

2.1. Methods

We carried out this study from July 2008 to February 2009, in a ca. 300 ha secondary forest in Viçosa, Minas Gerais, Southeast Brazil (20°45'S e 42°55'W). The main vegetation is composed by Semidecidual Seasonal Atlantic Forest, located within the domain of the Atlantic Forest [80]. In the study area we set two 75m parallel transects, apart 5 m from each other. Along each transect we delimited 15 1m² squares, 5m distant from each other, in a total of 30 sampling points. We collected approximately 200g of freshly fallen leaves from predominant tree species in each sampling point. These leaves were mixed and oven-dried at 60°C for 72 hours. Dried leaves were weighted and separated in groups of 5g, which were placed into litter bags, measuring 15 x 15 cm, with a mesh of 2 mm [81,63]. In each sampling point we set 15 litter bags and, after 30 days we started to remove them. Litter bags were removed fortnightly along 225 days. At the end of the experiment we took a 20 cm deep soil sample in each sampling point, which were taken to soil analyses. The soil analyses were performed at the Soil Lab analyses of the Federal University of Viçosa, and consisted of organic matter content and macroporosity, variables that could interfere in the decomposition process.

After removal, we placed litter bags in Berlese funnels for 48 hours, to extract the arthropods. After their identification, arthropods were sorted according to their feeding habits: detritivores, fungivores and predators [82-86]. The arthropods that we cannot sort in the above categories, because we could not identify feeding habits, were classified as "other arthropods", and were considered only in the analyses that included all arthropods.

After the arthropod extraction, litter was oven-dried at 60°C for 72 hours and weighted to compare with the initial weight (5g). We considered litter weight loss as the difference between initial and final weight (after 225 days), and we used this as an estimate of decomposition in each sampling point.

2.2. Statistical analyses

To test the hypothesis that more arthropod richness and abundance leads to a higher litter decomposition rate we used a model selection approach [87,88]. The response variable was litter weight loss, and explanatory variables were: total abundance and richness of arthropods, abundance and richness of fungivores, detritivores and predators, as well as macroporosity and soil organic matter. Before structuring the model, we carried out a correlation test among the explanatory variables, using the package "psych", and whenever two variables presented a correlation higher than 0.7 we removed the variable considered biologically less relevant [89]. Variables that presented correlation higher than 0.7 were: total arthropod species richness and predator species richness (0.73) and total arthropod abundance and detritivore abundance (0.94). We opted, then, to remove predator species richness and total arthropod abundance, as the former represents a possible action of organisms distant from the focal process of decomposition and the latter because it is an estimate more general that detritivore abundance.

The procedure of model selection involved the "MuMIn" package [90], that allows the construction of all possible models starting from the global model containing all variables. For each model, the procedure calculates model weight, based on the Akaike Information Criteria– AICc(ω). After doing so, it ranks all models and the best models are those containing lower AICc and higher weight values. We standardized and centralized all explanatory variables [91], using the package "arm" and the models were built with these transformed variables prior to model selection. All models within ΔAICc < 2 bounds were considered to obtain a good evidence of support [87]. In the case of more than one model, we averaged the models to obtain only one final model with averaged model coefficients, including their respective confidence internal. Parameters for which the confidence interval crossed zero were considered non-significant [88]. All analyses were performed under the platform R [92].

2.3. Results and discussion

2.3.1. Litter arthropod fauna

We sampled 2,284 individual and 198 arthropod species, from seven classes: (i) Arachnida, (ii) Malacostraca, (iii) Symphyla, (iv) Chilopoda, (v) Diplopoda, (vi) Entognatha and (vii) Insecta. The class with more orders was Insecta (10 orders), followed by Arachnida (four), Entognatha (three), Malacostraca, Symphyla, Chilopoda and Diplopoda (one order each). The most abundant arthropods in our sampling were Acari and Collembola, which are usually described as more abundant in soil and litter [93]. Besides, high abundance of these two groups had already been reported by [94] and [95], who studied forest fragments in the same region. Oribatid mites were the most respresentative group in all sampling, and these mites have an important role in decomposition process, as most are detritivore [93]. Collembola also presented high abundance and species richness in the samples. These organisms are fungivores and their trophic activity includes both the direct consumption of microorganisms and organic matter fragmentation [96]. Besides, they constitute an important source of food to predatory organisms, being very important in food webs to soil and litter [97].

2.3.2. Arthropod biodiversity and ecosystem functioning

Opposed to what we expected, there was no effect of arthropod species richness and abundance on decomposition rates, both considering total arthropods and when they were sorted by their feeding habits. Although our final model presented soil macroporosity and detritivore species richness as explanatory variables, their 95% confidence interval includes zero, and were considered non-significant (Table 1).

Our results contrast with others, which reported a positive effect of species richness on ecosystem processes [98-103]. The lack of relationship in our study may have occurred due to a high functional redundancy among arthropod species [40,104]. Accordingly, we infer that the studied community is composed by species with similar functions, thus species loss does not cause changes on ecosystem functioning. However, this possible redundancy assumed in this case study does not necessarily exclude another hypothesis to explain the biodiversity-

Response variable	Explanatory variables (parameters)	Estimate	SE	Lower CI	Upper CI
Decomposition	Intercept	1.7027	0.0882	1.5297	1.8756
(litter weight loss)	Macroporosity	0.0977	0.0953	-0.0777	0.2732
	Detritivore species richness	-0.0920	0.0897	-0.2679	0.0838

Table 1. Summary of model averaging results, detailing the explanatory variables present in the final average model. Parameters estimates were obtained from standardized variables.

functioning observed: the linearity hypothesis. Two curves may be generated by these two hypotheses: a linear relationship (Type I curve) in the case of singular species and an asymptotic curve (Type II curve) in the case of redundant species [105]. Therefore, both hypotheses may be explained by the same curve, depending on the scale data was sampled. Linearity, then, would be a component of redundancy curve, but that would only be expected in cases with low diversity. From a given species richness a saturation of the functions would occur, with species playing similar roles. Data obtained in the present study would fit into this latter diversity scale. To test such assumption one can manipulatively reduce arthropod abundance and richness, studying a broad range of species richness, and effectively testing the redundancy hypothesis in tropical environments.

Another possible explanation to the absence of relationship between arthropod diversity and litter decomposition is the similar litter constitution across all sampling points. It is known that the chemical and physical composition of litter has an effect on decomposition rates [61,62,70,106,107]. The manipulation of litter diversity and composition in litter bags may lead to the establishment of different arthropod communities, according to leaf degradability. The manipulation of species richness and composition of plants under decomposition may lead to different responses of arthropod species richness, which might mirror variation in decomposition rates.

Furthermore, in this study we evaluated only the role of arthropod diversity of the soil-litter system. It is known, however, that fungi and bacteria (the microflora) are the decomposers and responsible for organic matter mineralization [58,60,71]. Conversely arthropods act indirectly on decomposition and, even though they facilitate the action of true decomposers, detecting their action on decomposition may be less straightforward.

The absence of relationship between litter decomposition and arthropod species richness and abundance must be evaluated with caution, because assuming functional redundancy among species may be uncertain. Such outcome may lead us to the wrong conclusion that species loss does not affect ecosystem functioning at all, and this early, which may be wrong as discussed above.

Our conclusion is that litter decomposition process in the tropics and other hyper diverse habitats may be more complex than it is for the well known temperate habitats. Studying only one component may not give a precise response, due to the immense assembly of components in complex habitats and their equally complex interactions. Manipulative experiments, microbial activity estimates, as well as manipulation of litter diversity and composition should give us a more precise knowledge of the biodiversity role on ecosystem key processes.

3. Case study 2

Besides the ecosystem processes, as decomposition and nutrient cycling, species interactions are also important in the maintenance of ecosystem stability. The importance of predation and competition in community structuring is well studied [108-110], although mutualism may also have a central role in species distribution in ecosystems [111,112].Ants may establish a wide variety of mutualistic associations with plants [113,114]. Plants may offer resources like shelter, food, or both, that may be used by ants in several ways. The mutualistic interactions between ants and plants vary from diffuse, such as secondary seed dispersal [115] and use of extrafloral nectaries (EFNs) by generalist species [116], to more specialized interactions, such as domatia colonization by *Azteca* sp. [117-119,]. On the other hand, ants may also be beneficial to plants, increasing seed dispersal or reducing herbivory, for example [120].

EFNs, are nectar producing structures associated to plant vegetative organs, as leaves or petioles [121]. Extrafloral nectar is a liquid resource, composed by glucose, sucrose and fructose, and containing sometimes amino acids and proteins [122]. EFN-bearing plants are more visited by ants than plants without them [123], and ants that use extrafloral nectar as a resource may establish a generalist association of protection in exchange for food [114,119]. Therefore, the benefits arising from this interaction may explain its success [124]. The interaction between ants visiting EFNs and the plants has been the subject of several studies, although there are some divergences among the obtained results. Several studies relate advantages for EFN-bearing plants, such as decreasing the herbivory and the abundance of herbivores, or even positive effects on plant fitness [120,121,123,125-127]. However, some studies did not spot beneficial effects of visiting ants [128,131,132], indicating that in some cases ants may not be efficient in reducing herbivory [128]. The outcome of the interaction between ants and plants may depend on feeding habits of the herbivores, due to the interaction between ants and sap-feeding insects. Several authors [123,130-134] suggest that generalist ants feeding on their honeydew protect these insects from predators and competitors over chewing insects [130]. Moreover, plant protection may be related to ant species composition, as different ant species present varied behavior and defensive characteristics [129].

The interaction among ants, EFN-bearing ants and herbivores is very common in the Brazilian Cerrado. This biome is composed by herbs, shrubs and small trees that vary in density, composing different phytophysiognomies [135]. These physiognomies are usually divided in three groups, characterized by fields, savannas and forests [136]. *Qualea grandiflora* (Vochysiacveae) is among the several EFN-bearing plant species of Cerrado, which are medium to large-sized trees that reach 30 meters [137]. It is very studied in the Cerrado as it has a large distribution and abundance, and also because it attracts several ant species to their EFNs, placed at the basis of the petioles, near to leaf insertion [138].

This case study evaluates whether the ants foraging on *Qualea grandiflora* protect these plants against herbivory. We tested the hypotheses that increased ant species richness and abundance (i) decreases herbivore species richness and abundance, and (ii) changes the proportion of herbivore guilds in Cerrado.

3.1. Methods

Sampling was carried out in Panga Ecologic Station (PEE), situated in Minas Gerais, Brazil (19°09'20"-19°11'10" S, 48°23'20"-48°24'35" W). The area is a 409 ha of Cerrado, with several phytophysiognomies [139, 140]. Climate is Aw, tropical with a rainy summer and dry winter [141]. The average temperature during winter is 18°C and during summer is 23°C, and monthly rainfall is 60 mm during winter and 250 mm during summer.

Insects were sampled during January 2013, during the rainy season. We chose 90 individuals of *Qualea grandiflora*, 30 in each phytophysiognomy: *Cerradão* (Forest), *Cerrado Stricto Sensu* (Dense Savanna) and *Campo Cerrado* (Field Savanna). As it is known that ant species richness and abundance may vary with tree density in Cerrado [142], we expect that sampling in three different plant densities would produce a higher range of variation on ant community parameters. We sampled herbivores by beating, using an entomological umbrella of 1 m² [143, 144]. We did 10 beatings in each tree, and all insects were collected. All herbivores were counted, identified up to the family level, and sorted into two groups (guilds): leaf chewing insects and sap-sucking insects. Ants were sampled by pitfall traps, placed in the trunk of trees at 1.5 meters above ground level. In each tree we installed four pitfalls, to maximize ant sampling. Pitfall traps remained open for 48 hours, and ants were identified to the lower level as possible (genus or species). When identification up to species level was not possible, we asserted the individuals into morphospecies. Herbivores eventually collected in the pitfall traps were added to the beating sampling.

3.2. Statistical analyses

To test whether an increase of ant species richness and abundance decreases herbivore species richness and abundance, we carried out an ANCOVA (analysis of covariance), with Poisson distribution, considering phytophysiognomies as covariates. To test the hypothesis that increasing ant species richness and abundance decreases the proportion of leaf chewing insects and increases sap-sucking insects, we carried out an ANCOVA to one of the herbivore guilds, using a binomial distribution, corrected for overdispersion. Only one analysis is needed in this case, as the two response variables (proportion of each guild) are complementary. All analyses were carried out in R platform and models were simplified by removing non-significant variables and obtaining the minimal adequate model [145].

3.3. Results and discussion

We sampled 2,597 ants, from 150 species, 25 genera and seven subfamilies. The most abundant subfamily was Formicinae (1,293 individuals), followed by Myrmicinae (737), Dolichoderinae (426), Pseudomyrmecinae (110), Ectatomminae (22), Ponerinae (eight) and Heteroponerinae (one). The subfamily with the highest species richness was Myrmicinae (55 species), followed by Formicinae (40), Pseudomyrmecinae and Dolichoderinae (20 species each), Ectatomminae (three), Ponerinae (two) and Heteroponerinae (one). We sampled 233 herbivore insects, and Coleoptera was the most abundant order (141 individuals), followed by Hemiptera (75), Lepidoptera (11) and Orthoptera (six). Herbivores were sorted in 97 species, and the order with highest species richness was Coleoptera (62 species), followed by Hemiptera (24), Orthoptera (six) and Lepidoptera (five).

Ant species richness did not affect both herbivores abundance and species richness. As the studied plant species is a mirmecophile, it was not expected to maintain obligatory associations with ant species, being visited by generalist ant species foraging both during the day and the night. Several species in a given community may have a redundant role in some ecological functions, such as predation [146]. Moreover, as mirmecophyles are usually visited by several ant species, species richness may not contribute effectively for herbivore decrease. Therefore, ant species richness visiting EFN-bearing plants may not contribute to herbivore decrease because (i) they may be highly redundant, and/or (ii) they encompass non-aggressive ant species.

Nevertheless, we observed a decrease of herbivore abundance with the increase of abundance of ants in the trees of *Cerradão* and *Campo Cerrado* (χ^2=0.7; p=0.02) (Figure 1). The higher number of ants present in a given site gives a higher probability of encounters between them and herbivores, decreasing the number of herbivores [147]. Some ant species may present aggressive behavior, or efficient recruitment ability, and if the most frequent ants have these attributes, there would be a higher chance of herbivore attack and decreasing.

Conversely, we also observed that herbivore abundance increased with the abundance of ants in *Cerrado Stricto Sensu* (χ^2=1.7; p=0.05) (Figure 1). Another interesting result regards a higher sap-sucker insect abundance in *Cerrado Stricto Sensu*, in comparison with the other two phytophysiognomies (Figure 2). This latter result may explain the positive relationship between ant and herbivore abundance, because most of the herbivores belong to sap-sucking insects, and the positive association between them and ants is well known. As there are more sap-sucking insects in the *Cerrado Stricto Sensu*, ants may be consuming more sugars from the honeydew than from the EFNs [148], possibly leading to a dominance of more aggressive, abundant and frequent ants [149]. Therefore, the association between ants and EFN-bearing plants may shift from mutualistic to antagonistic when sap-sucking insects are present. When there is high resource competition on the plants, ants have a tendency to consume more honeydew than nectar [150], protecting sap-sucking insects and harming the plants.

Although the above scenario may arise from the three trophic level interaction mentioned above, it has been suggested that ants may also repel chewing insects by consuming honeydew, decreasing their abundance and activity on plants [133]. The different responses of herbivores to abundance of ants found in our study suggest that the effect of ants on herbivores is dependent of herbivore feeding habits. Several studies have reported non-obligatory interactions between ants and sap-sucking insects [123,130-134]. In this interaction, ants feed on the honeydew and protect the insects against predators and competitors [130]. Such an interaction may produce an explanation for the above results, because most ants participating in this interaction are generalists, consuming both honeydew and nectar from EFNs. Therefore, both sap-sucking insects and plants may be protected by the ants, as both provide resources to them. In this scenario, leaf chewing herbivores would be repelled or predated by the foraging ants.

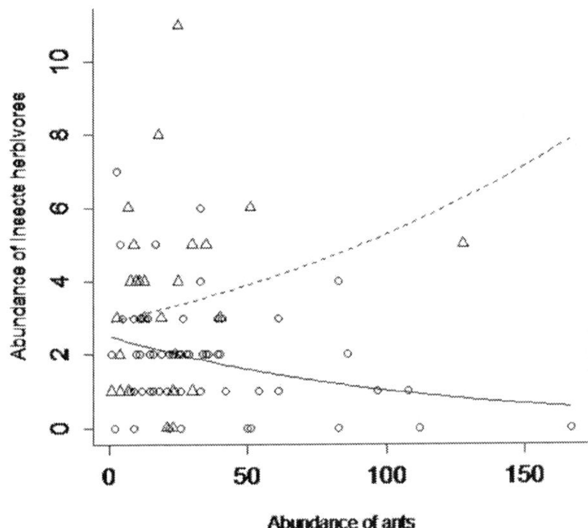

Figure 1. Relationship between abundance of ants and abundance of herbivore insects in the three studied phytophy-siognomies. The continuous curve (circles) represent the decrease of herbivores in *Cerradão* and *Campo Cerrado* (χ^2=0.7; p=0.02), and the dashed line (triangles) the increase of herbivores in *Cerrado Strictu Senso* (χ^2=1.7; p=0.05).

However, ant species distribution and their consequent effect on interactions may be modulated by habitat type and conditions [151]. Habitats with larger resource availability may facilitate the coexistence of ant species [131]. As Cerrado is composed by vegetation types with different tree abundances [136], the relationships among organisms may also vary accordingly. EFN-bearing plants were found to be more frequent in the Forest formations of Cerrado (*Cerradão*) than in other phytophysiognomies, which indicates more extrafloral nectar in this vegetation type [152]. As tree density and species richness influences ant species richness due to higher resource availability to generalist and specialist species [142], mutualistic interactions may be dependent of resource amount and distribution. More heterogeneous habitats may generate diverse resource availability, promoting the found differences among the phytophysiognomies studied here. As there are more resources in the *Cerradão*, ant foraging may be more opportunistic, resulting in a less effective protection against herbivory by chewing insects. Additionally, as there are more resources provided by EFNs, associations between sap-sucking insects and ants may be less effective, decreasing their abundance.

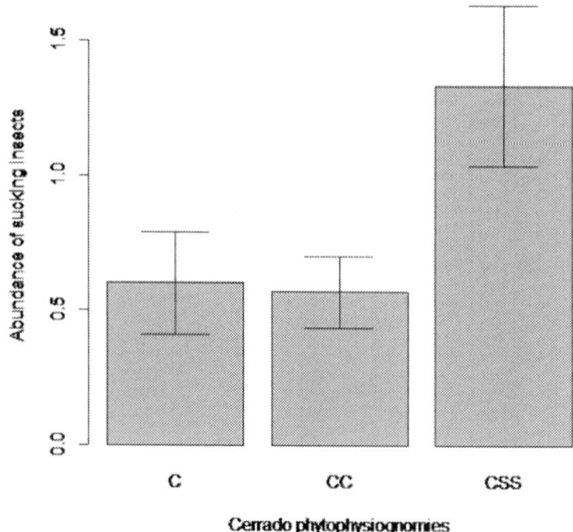

Figure 2. Relationship between SAP-sucking insects in the three phytophysiognomies. C – *Cerradão*, CC – *Campo Cerrado*, CSS – *Cerrado Stricto Sensu*. C and CC did not differed statistically.

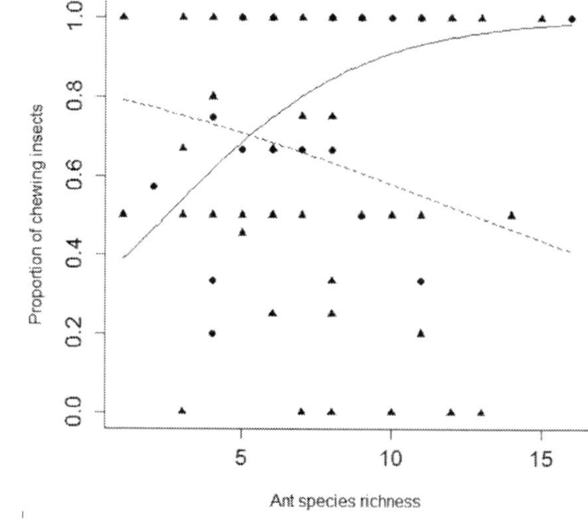

Figure 3. Relationship between the proportion of chewing insects and ant species richness. While ant species richness increased chewing herbivores in *Cerradão* (circles; continuous line), in decreases this proportion in *Cerrado Stricto Sensu* and *Campo Cerrado* (triangles; dashed line). The relationship between sap-sucking insects and ants follows a pattern contrary to the above, as these proportions are complementary.

4. Case study 3

The conversion of pristine environments into human-modified landscapes is rising around the World. Such habitat conversions may culminate in altered environmental conditions, reduction in the availability of resources and decrease in habitat heterogeneity [153]. Consequently, many authors have been warning to the existence of a biodiversity crisis [154,155]. In general the conversion of natural systems introduces newer and simplified ecosystems composed by one or few economically valuable crop species. Whereas habitat loss per se is enough to generate local extinctions [156,157], what is observed is that these habitats are usually substituted by agricultural systems as well. Therefore, most human-modified landscapes are altered by the joint action of these processes, habitat loss and conversion.

Habitat heterogeneity can be defined as the variety and the relative amount of different microhabitats available to organisms, and has been considered over the years a major variable determinant on local species richness and abundance [158-160]. Structurally more complex habitats provide more spatial niches and different types of resource exploitation, thus increasing species diversity [153,161-162], although this relationship may not be always straight [163,164]. Habitat heterogeneity reduction, for instance, can lead to lower resource availability, changes in environmental conditions and eventually species and ecosystem functions losses [165,166].

Eucalyptus crops are one of the economic activities that may lead to the above mentioned biodiversity loss in the Brazilian biomes. This culture was introduced in the country by the beginning of 19th century, and up to 2012 it is estimated to cover 5.105.246 hectares [167]. Once *Eucalyptus* is classically grown as a monoculture in Brazil, habitats are extremely simplified and homogeneous, potentially triggering the mentioned effects on biodiversity.

Several functions may be altered in *Eucalyptus* plantations, potentially due to homogenization, such as litter decomposition [97,168], nutrient cycling [169] and seed dispersal [170]. The latter is usually associated with habitat recovery, as there are several advantages for plants such as avoiding rodent predation [171,172], dispersion for nutrient-rich sites [173], protection against fire [174] and smaller competition with the parental plant [175]. Thereby, such mutualism may play a central role on local plant dynamics [176].

Seed dispersal by animals is considered a diffuse interaction, once can be performed by several generalist frugivores [177,178]. Ants are mainly reported as secondary dispersers, as they take fallen diaspores to their nests. Within the nests conditions are more suitable for the seed, because it is protected from herbivores [179-181], it is a nutrient-rich microhabitat [182,183] and free of competition with the parental plant [175,180,184].

A common observed trend is the reduction in ant species richness with habitat conversion, such as pastures [185], crops [186] and in *Eucalyptus* cultures [187,188]. Species composition is also usually altered by habitat conversion [185,189]. These changes may strongly alter seed dispersal dynamics by ants in fragmented and modified landscapes [185,190], although no mechanism was proposed to explain this correlation. In this study we tested the effect of natural habitat conversion into *Eucalyptus* crops and the consequences for seed removal by ants. We hypothesized that higher ant species richness increases seed removal, and this relationship is more pronounced in natural forest ecosystems than in *Eucalyptus* crops. Furthermore, we hypothesized that ant species composition changes between natural and *Eucalyptus* crops, and this shift is also responsible for supposed differential seed removal rates. Finally, as we observed differential seed removal between studied habitats we tested the presence of keystone species would influence removal rates.

4.1. Methods

4.1.1. Study site

We carried out the study in Viçosa, Minas Gerais (20°45′S, 42°50′W), Brazil, during summer 2010/2011, the rainy season. The pristine vegetation in this region is within the Atlantic Forest Domain, and is classified as Seasonal Semi-deciduous Forest. From the 1930's decade an intense fragmentation process has begun, and the native vegetation was mainly substituted by coffee crops and pastures. Now days, the landscape is highly fragmented, and is composed by several secondary forest fragments, intermingled with pasture, coffee and *Eucalyptus* crop, among others. We arbitrarily chose five forest fragments and five neighboring *Eucalyptus* for our study sites.

4.1.2. Experimental design

We used *Mabea fistulifera* (Euphorbiaceae) seeds, an abundant native myrmecochoricous species with elaiosome. Seeds of this species have a diplochoric dispersion, primarily ballistic and secondarily mainly by ants. Seeds, collected directly from branches of several native trees, were obtained from Forest Seeds Laboratory of the Federal University of Viçosa two months before the experimental set up. Seeds had their elaiosomes preserved and were maintained in cold chamber at 20°C straight after natural dehiscence and kept until their use.

In each of the 10 sampling sites (5 native forests and 5 *Eucalyptus* crops), we set 10 sampling units, which were distributed 10 meters apart from each other. Each sampling unit consisted of one ant sampling point and one seed removal spot, distanced 2 meters from each other. Ant sampling points consisted of unbaited pitfall traps (diameter 8 cm, 12 cm height), buried at soil level, and half filled with a killing solution of water, detergent and salt. Seed removal spots consisted of the provision of 10 seeds of *M. fistulifera*, which were covered by a cage with a mesh of 10 mm, to avoid seed removal by vertebrates [192]. Both pitfall traps and seeds remained in the field for 48 hours. After that, we counted the number of remaining seeds and the number and identity of ant species.

We identified ants to genera using the keys by [192], and when possible to species by comparisons with the reference collection of the Community Ecology Lab/UFV, where voucher specimens were deposited. Species identification was confirmed by a specialist.

4.2. Statistical analyses

To test the relationship between seed removal and ant species richness we used an ANCOVA, in which the response variable was the proportion of removed seeds in each site, and the explanatory variables were ant species richness and environment type (native forests or *Eucalyptus* crops). As the response variable was a ratio, we used a binomial error distribution, corrected for overdispersion when necessary. We performed this analysis in the software R [92] and we did residual analysis to check for model fit and distribution suitability.

To test whether ant species composition changes between studied habitats, we performed a NMDS (Non-metric Multidimensional Scaling), using Bray-Curtis dissimilarity index. We computed species abundances as the number of traps they occurred in each sampling site. The significance of differences was checked through PERMANOVA [193]. This analysis was performed in the package vegan within the software R [92]. We tested if the most frequent ant species act as keystone species [194] in seed removal by an ANOVA, in which we compared seed removal in the presence and in the absence of each species.

We removed one of the *Eucalyptus* areas a priory from all the analyses due to a heavy rain that removed all seeds, and another *Eucalyptus* area from the ANCOVA after the residual analysis as it was considered an outlier, therefore reducing our total sampling units to eight (five native and three *Eucalyptus* crops).

4.3. Results

We sampled 43 ant species, from 25 genera and seven subfamilies. From these, 23 species occurred exclusively in the native forests, five were exclusive from *Eucalyptus* and 15 occurred in both. The most frequent species were *Pheidole radoskowskii* Mayr, 1884 and *Ectatomma muticum* Mayr, 1870, which occurred in 40.24% and 30.49% of pitfall traps, respectively.

As expected, ant species richness was higher in native forest than in *Eucalyptus* crops (χ^2=6.93; p=0.008). Moreover, seed removal rate increased with the number of ant species ($F_{1,6}$=11.01; p=0.021; Fig. 4), however it was higher in the *Eucalyptus* than in the native forest ($F_{1,6}$=8.75; p=0.032). Conversely, species composition did not differ between the two habitats (Fig. 5, PERMANOVA $F_{1,7}$=1.12, p=0.32). Neither the presence of *P. radoskowskii* ($F_{1,80}$=0.87; p=0. 35), more frequent in *Eucalyptus*, nor of *E. muticum* (χ^2=0.94; p=0.33), more frequent in native forests, influenced seed removal.

5. Discussion

Differences in species richness, abundance and composition may affect ecosystem functioning [195]. In this case study we investigated the role of these three biodiversity components on seed removal by ants in native and Eucalyptus forests. Concerning species richness our results confirm the general pattern of reduction in modified habitats. The main causes reported for such pattern include habitat loss, homogenization, and harshness conditions for native species [196]. In comparison with native forest, *Eucalyptus* crops may be homogeneous habitats, which might have contributed to its lower species richness. Ant species richness is strictly related with environmental features such as higher plant species diversity, litter amount and habitat complexity [142,197-200]. From these, plant species diversity and habitat complexity decrease in *Eucalyptus* crops, which may have caused the loss of ant species that did not survive in the modified habitat. We observed the expected positive relationship between seed removal and ant species richness, both in the native forests and *Eucalyptus* crops. Nevertheless, the maximal seed removal at *Eucalyptus* was around 30% while at native forest was about 65%, which may be related to the smaller capacity of *Eucalyptus* crops of harboring species when compared to native forests. This pattern could also be attributed to keystone species (sensu [201]) at native forest, thus promoting the observed higher removal rates. However, the sole effect of potential keystone species did not explain the rates we observed, as seed removal did not change in their absence. Therefore, we have no evidence to consider the existence of some specialist seed remover species inhabiting either of the environments, reinforcing the role of ant species richness in the studied process.

On the other hand we did not find differential species composition between the two habitats types, thus we cannot assign the higher seed removal at native forest due to some keystone species. Moreover, seed removal rates at native forest did not differ when we analyzed the effects of the presence of the most abundant ant species (*E. muticum*). Therefore, we conclude that species richness is the only biodiversity component influencing the ecosystem process in the studied system. The positive relationship between ant species richness and seed removal rate may have important concerns on conservation. The maintenance of natural species richness levels can contribute to a suitable ecosystem functioning due to the role of the seed dispersal for seedling establishment and the community assembly.

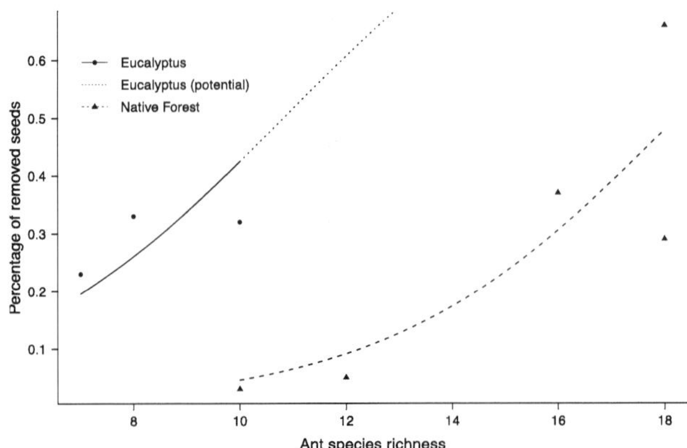

Figure 4. Seed removal rates increased with ant species richness ($F_{1,6}$=11.01; p=0.021), and were higher in *Eucalyptus* crop.

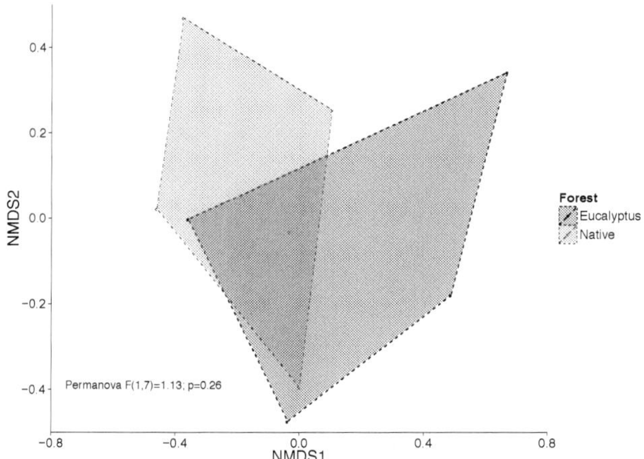

Figure 5. NMDS map of species composition according to treatment (Native or *Eucalyptus* crop). We analyzed significance by using Permanova test, which was non-significant.

6. Conclusions and perspectives

Although a positive relationship between biodiversity and ecosystem functioning is commonly reported [28, 202], we did not find such evidence from all the studies presented here. The main results presented allow us to conclude that the general effect of arthropods on ecosystem functioning is dependent on the studied process and the proximity with their agents. As more direct is the action of arthropods on the ecosystem processes, more detectable are their effects on functioning.

Arthropods contribute indirectly in the process of litter decomposition by modifying the substrate to the decomposers (the microbiota), besides acting through predation in a top-down effect on these microorganisms. Therefore, their indirect effect on litter decomposition may have produced the lack of relationship between their biodiversity and ecosystem functioning. Similarly, in the second case study we observed an absence of ant species richness and plant defense against herbivory. Nevertheless, in this study we noticed that the abundance of ants partially resulted in a decrease of the abundance of herbivore insects. The results obtained in this case study may reflect the plenty of defense mechanisms against herbivory, and ants are only a further mechanism within several others by which plants achieve a better protection. In the third case study we could evaluate a process that directly involved the importance of biodiversity on ecosystem processes, as there is a direct interaction between ants and the seeds they remove, without intermediate agents between them. Therefore, we could notice that the effect of ant biodiversity on the ecosystem process was stronger when compared to the two previous case studies. The above rationale points that a greater proximity between the agent and process turns the relationship stronger and detectable.

Based on the studies presented here, we suggest the following steps to improve the studies of the relationship between biodiversity and ecosystem functioning. Firstly, the control of variables through manipulative approaches should be increased, as confounding variables might decrease the chance of unveiling significant relationships. Secondly, as described above, it should be investigated relationships in which the processes and their agents are more directly connected. As reported elsewhere [28], the effects on productivity decrease with the increase of the number of trophic levels between manipulated (biodiversity) and estimated (ecosystem process) elements. Finally, studies of less complex systems may produce stronger results, once in complex systems several agents may influence concomitantly a given process, decreasing the chance of detecting a relationship between biodiversity and ecosystem functioning. Our second case study is an example, as several agents may influence plant herbivory, besides the presence of a higher ant species richness and abundance. Similar conclusions have been found in a meta-analysis study involving several results obtained from different regions of the World [28]. Hence, the comparison among our results and those obtained by other authors indicate that, despite the high complexity and biodiversity found in tropical regions, the trends reported here are comparable to those found worldwide.

This chapter integrated different case studies relating biodiversity and ecosystem functioning, with varying degrees of proximity between the agents and the processes. Because human activities would certainly continue to produce loss of species, we suggest that future studies relating biodiversity and ecosystem processes consider the linkage among the agents involved in the processes, to improve the understanding of this relationship, as well as the prognosis involving changes in biodiversity.

Acknowledgements

Authors are indebted to Lucas G. Dornelas for providing his data of case study 3. Authors received grants from Conselho Nacional de Desenvolvimento Científico e Tecnológico (CNPq – Brazil), Fundação Capes – Brazil, and Fundação de Amparo à Pesquisa do Estado de Minas Gerais (FAPEMIG).

Author details

Tathiana G. Sobrinho, Lucas N. Paolucci, Dalana C. Muscardi, Ana C. Maradini, Elisangela A. Silva, Ricardo R. C. Solar and José H. Schoereder*

*Address all correspondence to: jschoere@ufv.br

Departamento de Biologia Geral, Universidade Federal de Viçosa. Viçosa, MG, Brazil

References

[1] Barlow J., Gardner TA., Lees AC., Parry L., Peres CA. How pristine are tropical forests? An ecological perspective on the pre-Columbian human footprint in Amazonia and implications for contemporary conservation. Biological Conservation 2012;151(1) 45-49.

[2] Martens P., Rotmans J., Groot D. "Biodiversity: Luxury or Necessity?". Global Environmental Change 2003;13 75-81.

[3] Chapin III FS., Kofinas GP., Folke C. editors. Principles of ecosystem stewardship: resilience-based natural resource management in a changing world. Springer: New York; 2009.

[4] Gardner TA. Monitoring forest biodiversity: improving conservation through ecologically-responsible management. Earthscan; London; 2010.

[5] Tscharntke T., Tylianakis JM,, Rand TA., Didham R., Fahrig L, Batáry P., Bengtsson J., Clough Y., Crist TO., Dormann CF., Ewers RM., Fründ J., Holt RD., Holzschuh A., Klein AM., Kleijn D., Kremen C., Landis DA., Laurance W., Lindenmayer D., Scherber C., Sodhi N., Steffan-Dewenter I., Thies C., Putten WH., Westphal C. Landscape moderation of biodiversity patterns and processes-Eight hypotheses. Biological Reviews 2012;87(3) 661-685.

[6] Fritz H., Saïd S., Renaud PC., Mutake S., Coid C., Monicat F. The Effects of agricultural fields and human settlements on the se of rivers by wildlife in the mid-Zambezi valley, Zimbabwe. Landscape Ecology 2003;18(3) 293-302.

[7] Jiang Z., Liu J., Chen J., Chen Q., Yan X., Xuan J., Zeng J. Responses of summer phytoplankton community to drastic environmental changes in the Changjiang (Yangtze River) estuary during the past 50 years. Water Research 2014;54 1-11.

[8] Colombo AF., Joly AC. Brazilian Atlantic Forest Lato Sensu:the most ancient Brazilian forest, and a biodiversity hotspot,is highly threatened by climate change. Brazilian Journal of Biology 2010;70(3) 697-708.

[9] Cardini U., Bednarz VN., Foster RA., Wild C. Benthic N_2 fixation in coral reefs and the potential effects of human-induced environmental change. Ecology and Evolution 2014;4(9) 1706-1727.

[10] Gruber N. The marine nitrogen cycle: overview and challenges. In: Capone DG., Bronk DA., Mulholland MR., Carpenter EJ. (eds.) Nitrogen in the Marine Environment. Salt lake city: Academic press; 2008. p1-50.

[11] Halpern BS., Walbridge S., Selkoe KA., Kappel CV., Fiorenza M., D'Agrosa C., Bruno JF., Casey KS., Ebert C., Fox HE., Fujita R., Heinemann D., Lenihan HS., Madin EMP., Perry MT., Selig ER., Spalding M., Steneck R., Watson R. A global map of human impact in marine ecosystem. Science 2008;319(5865) 948-952.

[12]   IPCC. Climate Change: The scientific basis. Cambridge: University Press; 2001.

[13]   Grau HL., Aide M. Globalization and land-use transition in Latin America. Ecology and Society 2008;13(2) 16.

[14]   Carr DL., Lopez AC., Bilsborrow RE. The population, agriculture, and environment nexus in Latin America: country-level evidence from the latter half of the twentieth century. Population and Environment 2009;30(6) 222-246.

[15]   Morris RJ. Anthropogenic impacts on tropical forest biodiversity: a network structure and ecosystem functioning perspective. Philosophical Transactions of the Royal Society B: Biological Sciences 2010;365(1558) 3709-3718.

[16]   Tilman D. Functional diversity. Encyclopedia of Biodiversity 2001;3(1) 109-120.

[17]   Díaz S., Cabido M. Vive la différence: plant functional diversity matters to ecosystem processes. Trends in Ecology and Evolution 2001;16(11) 646-655.

[18]   Gorb SN., Gorb EV., Punttila P. Effects of redispersal of seeds by ants on the vegetation pattern in a deciduous forest: A case study. Acta Oecologica 2000;21(4) 293-301.

[19]   Dominguez-Haydar Y., Armbrecht I. Response of ants and their seed removal in rehabilitation areas and forest at El Cerrejón coal mine in Colombia. Restoration Ecology 2011;19(201) 178-184.

[20]   Gordon CE., McGill B., Ibarra-Núñez G., Greenberg R., Perfecto I. Simplification of a coffee foliage-dwelling beetle community under low-shade management. Basic and Applied Ecology 2009;10(3) 246-254.

[21]   Vergara CH., Badano EI. Pollinator diversity increases fruit production in Mexican coffee plantations: the importance of rustic management systems. Agriculture, Ecosystems & Environment 2009;129(1) 117-123.

[22]   Nichols E., Spector S., Louzada J., Larsen T., Amezquitad S., Favila ME. Ecological functions and ecosystem services provided by Scarabaeinae dung beetles. Biological Conservation 2008;141(6) 1461-1474.

[23]   Naeem S., Bunker DE., Hector A., Loreau M., Perrings C. Introduction: the ecological and social implications of changing biodiversity. An overview of a decade of biodiversity and ecosystem functioning research. In Naeem S., Bunker DE., Hector A., Loreau M., Perrings C. (eds.). Biodiversity, Ecosystem Functioning, and Human Wellbeing: An Ecological and Economic Perspective. Oxford: University Press; 2009.

[24]   Naeem S., Chair FS., Chapin III FS., Costanza R., Ehrlich PR., Golley FB., Hooper DU., Lawton JH., O'Neill RV., Mooney HA., Sala OE., Symstad AJ., Tilman D. Biodiversity and Ecosystem Functioning: Maintaining Natural Life Support Processes. Ecological Society of America 1999;4 1-11.

[25]   Naeem S., Thompson LJ., Lawler P., Lawton JH., Woodfin RM. Declining biodiversity can alter the performance of ecosystems. Nature 1994;386 734-737.

[26] Tilman D., Wedin D., Knops J. Productivity and sustainability influenced by biodiversity in grassland ecosystems. Nature 1996;379(6567) 718-720.

[27] Aarssen LW. High productivity in grassland ecosystems: effected by species diversity or productive species? Oikos 1997;80 183-184.

[28] Balvanera P., Pfisterer AB., Buchmann N., Jing-Shen H., Nakashizuka T., Raffaelli D., Schmid B. Quantifying the evidence for biodiversity effects on ecosystem functioning and services. Ecology Letters 2006;9(10) 1146-1156.

[29] Slade EM., Mann DJ., Villanueva JF., Lewis OT. Experimental evidence for the effects of dung beetle functional group richness and composition on ecosystem function in a tropical forest. Journal of Animal Ecology 2007;76(6) 1094-1104.

[30] Borror DJ., Triplehorn CA., Johnson NF. An introduction to the study of insects. 7ed. Connecticut: Cengage Learning; 2004.

[31] Lavelle P., Decaëns T., Aubert M., Barot S., Blouin M., Bureau F., Margerie P., Mora P., Rossi JP. Soil invertebrates and ecosystem services. European Journal of Soil Biology 2006;42(1) 3-15.

[32] Perfecto I. Ants (Hymenoptera: Formicidae) as natural control agents of pests in irrigated maize in Nicaragua. Journal of Economic Entomology 1991;84(1) 65-70.

[33] Perfecto I., Sediles A. Vegetacional diversity, ants (Hymenoptera, Formicidae), and herbivorous pests in a neotropical agroecossystem. Environmental Entomology 1992;21 61-67.

[34] Philpott SM., Armbrecht I. Biodiversity in tropical agroforests and the ecological role of ants and ant diversity in predatory function. Ecological Entomology 2006;31(4) 369-377.

[35] Farwig N., Bailey D., Bochud E., Herrmann JD., Kindler E., Reusser N., Schüepp C., Schmidt-Entling MH. Isolation from forest reduces pollination, seed predation and insect scavenging in Swiss farmland. Landscape Ecology 2009;24(7) 919-927.

[36] Vergara CH., Badano EI. Pollinator diversity increases fruit production in Mexican coffee plantations: the importance of rustic management systems. Agriculture, ecosystems and environment 2009;129(1) 117-123.

[37] Andresen E. Effect of forest fragmentation on dung beetle communities and functional consequences for plant regeneration. Ecography 2003;26(1) 87-97.

[38] Cerdà A., Jurgensen MF. The influence of ants on soil and water losses from an orange orchard in eastern Spain. Journal of Applied Entomology 2008;132(4) 306-314.

[39] Dostal P., Breznova M., Kozlickova V., Herben T., Kovar P. Ant-induced soil modification and its effect on plant below-ground biomass. Pedobiologia 2005;49(2) 127-137.

[40] Naeem S., Loreau M., Inchausti P. Biodiversity and ecosystem functioning: the emergence of a synthetic ecological framework. In: Naeem S., Loreau M., Inchausti P. (eds.) Biodiversity and Ecosystem Functioning: Synthesis and Perspective. Oxford: University Press; 2002. p3-11.

[41] Loreau M., Naeem S., Inchausti P. Biodiversity and ecosystem functioning: synthesis and perspectives. Oxford: University Press; 2002.

[42] Mittermeier RA., Myers N., Thomsen JB., Da Fonseca GA., Olivieri S. Biodiversity Hotspots and Major Tropical Wilderness Areas:approaches to setting conservation priorities. Conservation Biology 1998;12(3) 516-520.

[43] Myers N., Mittermeier RA., Mittermeier CG., Da Fonseca GA., Kent J. Biodiversity hotspots for conservation priorities. Nature 2000;403(6772) 853-858.

[44] Brooks TM., Mittermeier RA., Mittermeier CG., Da Fonseca GA., Rylands AB., Konstant WR., Flick P., Pilgrim J., Oldfield S., Magin G., Hilton-Taylor C. Habitat loss and extinction in the hotspots of biodiversity. Conservation biology 2002;16(4) 909-923.

[45] Hansen MC., Potapov PV., Moore R., Hancher M., Turubanova SA., Tyukavina A., Thau D., Stehman SV., Goetz SJ., Loveland TR., Kommareddy A., Egorov A., Chini L., Justice CO., Townshend JRG. High-resolution global maps of 21st-century forest cover change. Science 2013;342(6160) 850-853.

[46] Putz FE., Zuidema PA., Synnott T., Peña-Claros M., Pinard MA., Sheil D., Vanclay JK., Sist P., Gourlet-Fleury S., Griscom B., Palmer J., Zagt R. Sustaining conservation values in selectively logged tropical forests: the attained and the attainable. Conservation Letters 2012;5(4) 296-303.

[47] Fearnside PM. Global warming and tropical land-use change: greenhouse gas emissions from biomass burning, decomposition and soils in forest conversion, shifting cultivation and secondary vegetation. Climatic Change 2000;46(1-2) 115-158.

[48] Gardner TA., Barlow J., Sodhi NS., Peres CA. A multi-region assessment of tropical forest biodiversity in a human-modified world. Biological Conservation 2010;143(10) 2293-2300.

[49] Gibson L., Lee TM., Koh LP., Brook BW., Gardner TA., Barlow J., Peres CA., Sodhi N.S. Primary forests are irreplaceable for sustaining tropical biodiversity. Nature 2011;478(7369) 378-381.

[50] Sala OE., Chapin III FS., Armesto JJ., Berlow E., Bloomfield J., Dirzo R., Huber-Sanwald E., Huenneke LF., Jackson RB., Kinzig A., Leemans R., Lodge DM., Mooney HA., Oesterheld M., Poff LeRoy N., Sykes MT., Walker BH., Walker M., Wall DH. Global Biodiversity Scenarios for the Year 2010. Science 2000;287(5459) 1770-1774.

[51] Klink CA., Machado RB. Conservation of the Brazilian Cerrado. Conservation Biology 2005;19(3)707-713.

[52] Ribeiro MC., Metzger JP., Martensen AC., Ponzoni FJ., Hirota MM. The Brazilian Atlantic Forest: How much is left, and how is the remaining forest distributed? Implications for conservation. Biological Conservation 2009;142(6) 1141-1153.

[53] Almeida S., Louzada J., Sperber C., Barlow J. Subtle land-use change and tropical biodiversity: dung beetle communities in Cerrado grasslands and exotic pastures. Biotropica 2011;43(6) 704-710.

[54] Chiarello AG. Density and Population Size of Mammals in Remnants of Brazilian Atlantic Forest. Conservation Biology 2000;14(6) 1649-1657.

[55] Filho JAS., Tabarelli M. Bromeliad species of the Atlantic forest of north-east Brazil: losses of critical populations of endemic species. Oryx 2006;40(2) 218-224.

[56] Silva RR., Machado Feitosa RS., Eberhardt F. Reduced ant diversity along a habitat regeneration gradient in the southern Brazilian Atlantic Forest. Forest Ecology an. Management 2007;240(1) 61-69.

[57] Oliveira-Filho AT., Ratter A. Vegetation physiognomies and woody flora of the Cerrado biome. In: P. S. Oliveira PS., Marquis RJ. editors. The Cerrados of Brazil: ecology and natural history of a Neotropical savanna. New York: Columbia University Press; 2002. P91-120.

[58] Hattenschwiler S., Gasser P. Soil animals alter plant litter diversity effects on decomposition. Proceedings of the National Academy of Sciences of the United States of America 2005;102(5) 1519-1524.

[59] Bragazza L., Siffi C., Iacumin P., Gerdol R. Mass loss and nutrient release during litter decay in peatland: The role of microbial adaptability to litter chemistry. Soil Biology and Biochemistry 2007;39(1) 257-267.

[60] González G., Seastedt TR. Soil fauna and plant litter decomposition in tropical and subalpine forests. Ecology 2001;82(4) 955-964.

[61] Xuluc-Tolosa FJ., Vester HFM., Ramírez-Marcial N., Castellanos-Albores J., Lawrence D. Leaf litter decomposition of tree species in three successional phases of tropical dry secondary forest in Campeche, Mexico. Forest Ecology and Management 2003;174(1) 401-412.

[62] Harguindeguy NP., Blundo CM., Gurvich DE., Díaz S., Cuevas E. More than the sum of its parts? Assessing litter heterogeneity effects on the decomposition of litter mixtures through leaf chemistry. Plant and Soil 2008;303(1-2) 151-159.

[63] Castanho CT., Oliveira AA. Relative effect of litter quality, forest type and their interaction on leaf decomposition in south-east Brazilian forests. Journal of Tropical Ecology 2008;24(2) 149-156.

[64] Muscardi DC., Schoereder JH., Sperber CF. Biodiversity and ecosystem functioning: a conceptual model of leaf litter decomposition. In: Oscar Grillo. (ed.). Biodiversity-The Dynamic Balance of the Planet. Rijeka: Intech; 2014. p33-50.

[65] Lavelle P., Blanchart E., Martin S., Martin A. A hierarchical model for decomposition in terrestrial ecosystems: Application to soils in the humid tropics. Biotropica 1993;25(2) 130-150.

[66] Aerts R. Climate, leaf litter chemistry and leaf litter decomposition in terrestrial ecosystems: a triangular relationship. Oikos 1997;79(3) 439-449.

[67] Cotrufo M., Del Gado F., Piermatteo ID. Litter decomposition: concepts, methods and future perspectives. In: Soil Carbon Dinamics. An integrated methodology. Kutsch WL., Bahn M., Heinemeyer A. (eds.). Cambridge: University Press; 2009. p76-90.

[68] Hättenschwiler S., Coq S., Barantal S., Handa IT. Leaf traits and decomposition in tropical rainforests: revisiting some commonly held views and towards a new hypothesis. New Phytologist 2011;189(4) 950-965.

[69] Melillo JM., Aber JD., Muratore JF. Nitrogen and lignin control of hardwood leaf litter decomposition dynamics. Ecology 1982;63(3) 621-626.

[70] Li LJ., Zeng DH., Yu ZY., Fan ZP., Yang D., Liu YX. Impact of litter quality and soil nutrient availability on leaf decomposition rate in a semi-arid grassland of Northeast China. Journal of Arid Environments 2011;75(9) 787-792.

[71] Smith VC., Bradford MA. Do non-additive effects on decomposition in litter-mix experiments result from differences in resource quality between litters? Oikos 2003;102(2) 235-242.

[72] Kurzatkowski D., Martius C., Höfer H., Garcia M., Förster B., Beck L., Vlek P. Litter decomposition, microbial biomass and activity of soil organisms in three agroforestry sites in central Amazonia. Nutrient Cycling in Agroecosystems 2004;69(3) 257-267.

[73] Wickings K., Grandy AS. The oribatid mite Scheloribates moestus (Acari:Oribatida) alters litter chemistry and nutrient cycling during decomposition. Soil Biology and Biochemistry 2011;43(2) 351-358.

[74] Pramanik R., Sarkar K., Joy VC. Efficiency of detritivore soil arthropods in mobilizing nutrients from leaf litter. Tropical Ecology 2001;42(1) 51-58.

[75] Coleman DC., Crossley DA. Fundamentals of Soil Ecology. London: Academic Press; 1996.

[76] Spain AV., Lavelle P., Mariotti A. Stimulation of plant growth by tropical earthworms. Soil Biology and Biochemistry 1992;24(12) 162-1633.

[77] Bardgett RD., Chan KF. Experimental evidence that soil fauna enhance nutrient mineralization and plant nutrient uptake in montane grassland ecosystems. Soil Biology and Biochemistry. Oxford 1999;31(7) 1007-1014.

[78] Hunt HW., Wall DH. Modelling the effects of loss of soil biodiversity on eco-system function. Global Change Biology 2002;8(1) 33-50.

[79] Duffy JE., Cardinale BJ., France KE., McIntyre PB., Thébault E., Loreau M. The functional role of biodiversity in ecosystems: Incorporating trophic complexity. Ecology Letters 2007;10(6) 522-538.

[80] Veloso HP., Rangel Filho ALR., Lima JCA. editors. Classificação da vegetação brasileira, adaptada a um sistema universal. Rio de Janeiro: IBGE; 1991.

[81] Mason CF. Decomposição. São Paulo: Editora da Universidade de São Paulo; 1980.

[82] Moore JC., Walter DE., Hunt HW. Arthropod regulation of micro-and mesobiota in bellow-ground detrital food webs. Annual Reviews of Entomology 1988;33 419-439.

[83] Höfer H., Hanagarth W., Martius C., Beck L., Garcia M. Soil fauna and litter decomposition in primary and secondary forest and a mixed culture system in Amazonia. Manaus: EMBRAPA; 1998.

[84] Delabie JHC., Agosti D., Nascimento IC. Litter ant communities of the Brazilian Atlantic rain forest region. In: Agosti D., Majer JD., Tennant L., Schultz T. (eds). Sampling Ground-dwelling Ants: Case. Studies from the World's Rainforests. Western Australia: Curtin-School of Environmental Biology; 2000.

[85] Correia MEF. Potencial de Utilização dos Atributos das Comunidades de Fauna de Solo e de Grupos Chave de Invertebrados como Bioindicadores do Manejo de Ecossistemas. Seropédica: Embrapa Agrobiologia; 2002.

[86] Marinoni RC, Ganho NG, Monné ML, Mermudes JRM. Hábitos alimentares de Coleoptera (Insecta). Ribeirão Preto: Holos; 2003.

[87] Burnham KP., Anderson DR. Model selection and multimodel inference: a practical information-theoretic approach. New York: Springer; 2002.

[88] Grueber CE., Nakagawa S., Laws RJ., Jamieson IG. Multimodel inference in ecology and evolution: challenges and solutions. Journal of Evolutionary Biology 2011;24(4) 699-711.

[89] Zuur AF., Ieno EN., Elphick CS. A protocol for data exploration to avoid common statistical problems. Methods in Ecology and Evolution 2010;1 3-14.

[90] Barton, K. Package 'MuMIn'. Model selection and model averaging based on information criteria. 2012. R package version 1.7.11

[91] Schielzeth H. Simple means to improve the interpretability of regression coefficients. Methods in Ecology and Evolution 2010;1(2) 103-113.

[92] R Development Core Team. R: A language and environment for statistical computing. R Foundation for Statistical Computing, Vienna: Austria; url http://www.R-project.org; 2014.

[93] Krantz GW., Walter DE. A manual of acarology. 3ed. Texas: Tech University Press; 2009.

[94] Collevatti RG., Schoereder JH. Microclimate ordination and litter arthropod distribution. Journal of the Brasilian Association for the Advanced of Science 1995;47 38-40.

[95] Louzada JNC., Schoereder JH., De Marco Jr P. Litter decomposition in semideciduous forest and *Eucalyptus spp.* crop in Brazil: a comparison. Forest Ecology and Management 1997;94(1) 31-36.

[96] Correia MEF., Andrade AD., Santos GA., Camargo FDO. Formação de serapilheira e ciclagem de nutrientes. Fundamentos da matéria orgânica do solo: ecossistemas tropicais e subtropicais. Porto Alegre: Gênesis; 1999.

[97] Lavelle P. Diversity of soil fauna and ecosystem function. Biology International 1996;33 3-16.

[98] Tilman D. Biodiversity: population versus ecosystem stability. Ecology 1996;77(2) 350-363.

[99] Tilman D., Knops J., Wedin D., Reich P., Ritchie M., Siemann E. The influence of functional diversity and composition on ecosystem processes. Science 1997;277(5330) 1300-1302.

[100] Hooper DU., Vitousek PM. Effects of plant composition and diversity on nutrient cycling. Ecological Monographs 1998;68(1) 121-149.

[101] Symstad AJ., Tilman D. Diversity loss, recruitment limitation, and ecosystem functioning: lessons learned from a removal experiment. Oikos 2001; 92(3) 424-435.

[102] Tilman D., Knops J., Wedin D., Reich P. Plant diversity and composition: effects on productivity and nutrients dynamics of experimental grasslands. In: In: Naeem S., Loreau M., Inchausti P. (eds.) Biodiversity and Ecosystem Functioning: Synthesis and Perspective. Oxford: University Press: 2002. P12-294.

[103] Aarssen LW., Laird RA., Pither J. Is the productivity of vegetation plots higher or lower when there are more species? Variable predictions from interaction of the 'sampling effect' and 'competitive dominance effect' on the habitat templet. Oikos 2003;102(2) 427-432.

[104] Wardle DA., Bonner KI., Nicholson KS. Biodiversity and plant litter: experimental evidence which does not support the view that enhanced species richness improves ecosystem function. Oikos 1997;79(2) 247-258.

[105] Schwartz MW., Brigham CA., Hoeksema JD., Lyons KG., Mills MH., Mantgem PJ. Linking biodiversity to ecosystem function: implications for conservation ecology. Oecologia 2000;122(3) 297-305.

[106] Wang S., Ruan H., Wang B. Effects of soil micro arthropods on plant litter decomposition across an elevation gradient in the Wuyi Mountains. Soil Biology and Biochemistry 2009;41(5) 891-897.

[107] Song F., Fan X., Song R. Review of mixed forest litter decomposition researches. Acta Ecologica Sinica 2010;30(4) 221-225.

[108] Inouye RS., Byers GS., Brown JH. Effects of predation and competition on survivorship, fecundidy, and community structure of desert annuals. Ecology 1980;61(6) 1344-1351.

[109] Menge BA., Sutherland J.P. Community regulation: variation in disturbance, competition and predation in relation to environmental stress and recruitment. American Naturalist 1987;130(5) 730-757.

[110] Savolainen R., Vepsalainen K. A competition hierarchy among boreal ants: Impact on resource partitioning and community structure. Oikos 1988;51(2) 135-155.

[111] Stachowics JJ. Mutualism, facilitation, and the structure of ecological communities positive interactions play a critical, but underappreciated, role in ecological communities by reducing physical or biotic stresses in existing habitats and by creating new habitats on which many species depend. BioScience 2001;51(3) 235-246.

[112] Heil M., McKey D. Protective ant-plant interactions as model systems in ecological and evoluctionary Research. Annual Review of Ecology, Evolution, and Systematics 2003;34 425-453.

[113] Rudgers JA. Enemies of herbivores can shape plant traits: Selection in a facultative ant-plant mutualism. Ecology 2004;85(1) 192-205.

[114] Chamberlain SA., Holland JN. Quantitative synthesis of context dependency in ant-plant protection mutualisms. Ecology 2009;90(9) 2384-2392.

[115] Nathan R., Muller-Landau HC. Spatial patterns of seed dispersal, their determinants and consequences for recruitment. Trends in ecology and evolution 2000;15(7) 278-285.

[116] Schoereder JH., Sobrinho TG., Madureira MS., Ribas CR., Oliveira PS. The arboreal ants community visiting extrafloral nectaries in the Neotropical Cerrado Savanna. Terrestrial Arthropod Reviews 2010;3(1) 3-27.

[117] Davidson DW., McKey D. Evolutionary Ecology of Symbiotic ant-plant relationships. Journal of Hymenoptera Research 1993;2(1) 13-83.

[118] Oliveira PS., Pie MR. Interaction between ants and plants bearing extrafloral nectaries in Cerrado vegetation. Anais da Sociedade Entomologica do Brasil 1998;27(2) 161-176.

[119] Goitía W., Jaffé K. Ant-Plant Associations in different forests in Venezuela. Neotropical Entomology 2009;38(1) 7-31.

[120] Bronstein JL., Alárcon R., Geber M. The evolution of plant-insects mutualisms. New Phytologist 2006;172(3) 412-428.

[121] Rico-Gray V., Oliveira PS. The ecology and evolution of ant-plant interactions. Chicago: The University of Chicago Press; 2007.

[122] Baker DA., Hall JL., Thorpe JL. A study of the extrafloral nectaries of *Ricinus communis*. New Phytologist 1978;81(1) 129-137.

[123] Oliveira PS., Freitas AVL. Ant-plant-herbivore interaction in the neotropical cerrado savanna. Naturwessenschaften 2004;91(12) 557-570.

[124] Del-Claro K., Berto V., Réu W. Effect of herbivore deterrance by ants on the fruit set of an extrafloral nectary plant *Qualea multiflora* (Vochysiaceae). Journal of Tropical Ecology 1996;12(6) 887-892.

[125] Costa FM., Oliveira-Filho AT., Oliveira PS. The role of extrafloral nectaries in *Qualea grandiflora* (Vochysiaceae) in limiting herbivory: An experiment of ant protection in Cerrado vegetation. Ecological Entomology 1992;17(4) 363-365.

[126] Sobrinho TG., Schoereder JH., Rodrigues LL., Collevatti RG. Ant Visitation (Hymenoptera: Formicidae) to Extrafloral Nectaries Increases Seed Set and Seed Viability in the Tropical Weed *Triumfetta semitriloba*. Sociobiology 2002;39(2) 353-368.

[127] Rosumek FB., Silveira FAO., Neves FS., Barbosa NPU., Diniz L., Oki Y., Pezzini F., Fernandes GW., Cornelissen T. Ants on plants: a meta-analysis of the roles of ants as plant biotic defenses. Oecologia 2009;160(3) 537-549.

[128] O'Dowd DJ., Cathpole EA. Ants and extrafloral nectaries: No evidence for plant protection in *Helicrysum spp*. Oecologia 1983;59(2-3) 191-200.

[129] Byk J., Del-Claro K. Nectar-and pollen-gathering *Cephalotes* ants provide no protection against herbivory: A new manipulative experiment to test ant protective capabilities. Acta Ethologica 2010;13(1) 33-38.

[130] Way MJ. Mutualism between ants and honeydew-producing Homoptera. Annual Review of Entomol 1963;8(1) 307-344.

[131] Davidson DW. The role of resource imbalances in the evolutionary ecology of tropical arboreal ants. Biological Journal of the Linnean Society 1997;81(2) 153-181.

[132] Delabie JHC. Trofhobiosis between Formicidae and Hemiptera (Sternoeehyncha and Auchenorryncha): an Overview. Neotropical Entomology 2001;30(4) 501-516.

[133] Grinath JB., Inouye BD., Underwood N., Billick I. The indirect consequences of a mutualism: comparing positive and negative components of the net interaction between honeydew-tending ants and host plants. Journal of Animal Ecology 2012;81(2) 494-502.

[134] Rice KB., Eubanks MD. No Enemies Needed: Cotton Aphids (Hemiptera: Aphididae) Directly Benefit from Red Imported Fire Ant (Hymenoptera: Formicidae) Tending. Florida Entomologist 2013;96(3) 929-932.

[135] Ratter JA., Ribeiro JF., Bridgewater S. The Brazilian cerrado vegetation and threats to its biodiversity. Annals of Botany 1997;80(3) 223-230.

[136] Ribeiro JF., Walter BMT., Sano SM., Almeida SD. Fitofisionomias do bioma Cerrado. Cerrado: ambiente e flora. Brasília: Embrapa; 1998.

[137] Almeida SD., Proença SEB., Sano SM., Ribeiro JF. Cerrado: Espécies Vegetais Úteis. Brasília: Embrapa; 1998.

[138] Oliveira PS., Da Silva AF., Martins AB. Ant foraging on extrafloral nectaries of *Qualea grandiflora* (Vochysiaceae) in Cerrado vegetation: ants as potential antiherbivore agents. Oecologia 1987;74(2) 228-230.

[139] Schiavini I., Araújo GM. Considerações sobre a vegetação da Reserva Ecológica do Panga (Uberlândia). Sociedade e Natureza 1989;1 61-66.

[140] Moreno MIC., Schiavini I. Relação entre vegetação e solo em um gradiente florestal na Estação Ecológica do Panga, Uberlândia (MG). Revista Brasileira de Botânica 2001;24(4) 537-544

[141] Rosa R., Lima SC., Assunção LW. Abordagem preliminar das condições climáticas de Uberlândia (MG). Sociedade e Natureza 1991;3 91-108.

[142] Ribas CR., Schoereder JH., Pic M., Soares SM. Tree heterogeneity, resource availability, and larger scale processes regulating arboreal ant species richness. Austral Ecology 2003;28(3) 305-314.

[143] Gallo D., Nakano O., Silveira Neto S., Carvalho RL., Batista GD., Berti Filho E., Parra JRP., Zucchi RA., Alves SB., Vendramim J. D. Manual de Entomologia Agrícola. 2vol. São Paulo: Agronômica Ceres;1978.

[144] Basset Y. Diversity and abundance of insect herbivores collected on *Castanopsis acuminatissima* (Fagaceae) in New Guinea: Relationship with leaf production and surrounding vegetation. European Journal of Entomology 1999;96(4) 381-391.

[145] Crawley MJ. The R Book. Hoboken: Wiley; 2007

[146] Joner F., Blanco C., Sosinski Jr EE., Muller SC., Pillar VD. Riqueza, redundância funcional e resistência de comunidades campestres sob pastejo. Revista Brasileira de Biociências 2008;5 528-530.

[147] Bentley BR. The protective function of ants visiting the extrafloral nectaries of *Bixa orellana* (Bixaceae). Journal of Ecology 1977;65 27-38.

[148] Katayama N., Hembry DH., Hojo MK., Suzuki N. Why do ants shift their foraging from extrafloral nectar to aphid honeydew?. Ecological Research, 2013;28(5) 919-926.

[149] Bluthgen N., Verhaagh M., Goitía W., Jaffé K., Morawetz W., Barthlott W. How plants shape the ant community in the Amazonian rainforest canopy: the key role of extrafloral nectaries and homopteran honeydew. Oecologia 2000;125(2) 229-240.

[150] Savage AM., Rudgers JA. Non-additive benefit or cost? Disentangling the direct effects that occur when plants bearing extrafloral nectaries and honeydew-producing insects share exotic ants mutualists. Annals of Botany 2013;111 1295-1307.

[151] Perfecto I., Snelling R. Biodiversity and the transformation of a tropical agrosystem: Ants in coffe plantations. Ecological Application 1995;5 1084-1097.

[152] Oliveira OS., Oliveira-Filho AT. Distribution of Extrafloral Nectaries in the Wood Flora of Tropical Communities in Western Brazil. In: Price PW., Lewinsohn TM., Fernandes G W., Benson WW. (eds.) Plant-Animal Interactions: Evolutionary Ecology in Tropical and Temperate Regions: New York: Wiley-Interscience; 1991.

[153] Stein A., Gerstner K., Kreft H. Environmental heterogeneity as a universal driver of species richness across taxa, biomes and spatial scales. Ecology letters 2014;17(7) 866-880.

[154] Hoekstra JM., Boucher TM., Ricketts TH., Roberts C. Confronting a biome crisis: global disparities of habitat loss and protection. Ecology Letters 2005;8(1) 23-29.

[155] Turner WR., Brandon K., Brooks TM., Costanza R., Da Fonseca GA., Portela R. Global conservation of biodiversity and ecosystem services. Bioscience 2007;57(10) 868-873.

[156] Fahrig L. Effects of habitat fragmentation on biodiversity. Annual Review of Ecology, evolution, and Systematics 2003;34 487-515.

[157] Swift TL., Hannon SJ. Critical thresholds associated with habitat loss: a review of the concepts, evidence, and applications. Biological Reviews 2010;85(1) 35-53.

[158] MacArthur RH., MacArthur JW. On bird species diversity. Ecology 1961;42(3) 594-598.

[159] Lawton JH. Plant architecture and the diversity of phytophagous insects. Annual Review of Entomology 1983;28(1) 23-39.

[160] Tews JU., Brose V., Grimm K., Tielbörger MC., SchwagerWM., Jeltsch F. Animal species diversity driven by habitat heterogeneity/diversity: the importance of keystone structures. Journal of Biogeography 2004;31(1) 79-92.

[161] Hutchinson GE. Concluding remarks. Cold Spring Harbor Symposia on Quantitative Biology 1957;22 415-427.

[162] Chase JM., Leibold MA. Ecological Niches: linking classical and contemporary approaches. Chicago: University Press; 2003.

[163] Kadmon R., Allouche O. Integrating the effects of area, isolation, and habitat heterogeneity on species diversity: a unification of island biogeography and niche theory. The American Naturalist 2007;170(3) 443-454.

[164] Allouche O., Kalyuzhny M., Moreno-Rueda G., Pizarro M., Kadmon R. Area–heterogeneity tradeoff and the diversity of ecological communities. Proceedings of the National Academy of Sciences 2012;109(43) 17495-17500.

[165] MacArthur RH. Environmental factors affecting birds pecies diversity. American Naturalist 1964;98 387-396.

[166] Kerr JT., Packer L. Habitat heterogeneity as determine mammalian of species richness in high-energy regions. Nature 1997;385(6613) 252-254.

[167] ABRAF. Anuário estatístico ABRAF 2013 ano base 2012. Brasília: ABRAF; 2013. p148.

[168] Barlow J., Gardner TA., Ferreira LV., Peres CA. Litter fall and decomposition in primary, secondary and plantation forests in the Brazilian Amazon. Forest Ecology and Management 2007;247(1) 91-97.

[169] Parrotta JA. Productivity, nutrient cycling, and succession in single-and mixed-species plantations of Casuarina equisetifolia, Eucalyptus robusta, and Leucaena leucocephala in Puerto Rico. Forest Ecology and Management 1999;124(1) 45-77.

[170] Neuschulz EL., Botzat A., Farwig N. Effects of forest modification on bird community composition and seed removal in a heterogeneous landscape in South Africa. Oikos 2011;120(9) 1371-1379.

[171] Bond W., Slingsby P. Collapse of an ant-plant mutalism: the Argentine ant (Iridomyrmex humilis) and myrmecochorous Proteaceae. Ecology 1984;65(4) 1031-1037.

[172] Auld TD., Denham AJ. The role of ants and mammals in dispersal and post-dispersal seed predation of the shrubs Grevillea (Proteaceae). Plant Ecology 1999;144(2) 201-213.

[173] Davidson DW., Morton SR. Myrmecochory in some plants (F. Chenopodiaceae) of the Australian arid zone. Oecologia 1981;50(3) 357-366.

[174] Beattie AJ. The evolutionary ecology of ant-plant mutualisms. Cambridge: University Press; 1985.

[175] Boyd RS. Ecological benefits of myrmecochory for the endangered chaparral shrub Fremontodendron decumbens (Sterculiaceae). American Journal of Botany 2001;88(2) 234-241.

[176] Christian CE. Consequences of a biological invasion reveal the importance of mutualism for plant communities. Nature 2001;413(6856) 635-639.

[177] Wheelwright NT., Orians GH. Seed dispersal by animals: contrasts with pollen dispersal, problems of terminology, and constraints on coevolution. American Naturalist 1982;119 (3) 402-413.

[178] Blüthgen N., Menzel F., Hovestadt T., Fiala B., Blüthgen N. Specialization, constraints, and conflicting interests in mutualistic networks. Current Biology 2007;17(4) 341-346.

[179] Heithaus ER. Seed predation by rodents on three antdispersed plants. Ecology 1981;62(1) 136-145.

[180] Higashi S., Tsuyuzaki S., Ohara M., Ito F. Adaptive advantages of ant-dispersed seeds in the myrmecochorous plant Trillium Tschonoskii (Liliaceae). Oikos 1989;54 389-394.

[181] Manzaneda AJ., Fedriani JM., Rey PJ. Adaptive advantages of myrmecochory: The predator-avoidance hypothesis tested over a wide geographic range. Ecography 2005;28(5) 583-592.

[182] Culver DC., Beattie AJ. Myrmecochory in Viola: dynamics of seed-ant interactions in some West Virginia species. The journal of Ecology 1978;66(1) 53-72.

[183] Hanzawa FM., Beattie AJ., Culver DC. Directed dispersal: demographic analysis of an ant-seed mutualism. American Naturalist 1988;131(1) 1-13.

[184] Andersen AN. Dispersal distance as a benefit of myrmecochory.Oecologia 1988;75(4) 507-511.

[185] Lomov B., Keith DA., Hochuli DF. Linking ecological function to species composition in ecological restoration: Seed removal by ants in recreated woodland. Austral Ecology 2009;34(7) 751-760.

[186] Armbrecht I., Rivera L., Perfecto I. Reduced diversity and complexity in the leaf-litter ant assemblage of Colombian coffee plantations. Conservation Biology 2005;19(3) 897-907.

[187] Braga DL., Louzada JNC., Zanetti R., Delabie JHC. Avaliação rápida da diversidade de formigas em sistemas de uso do solo no sul da Bahia. Neotropical Entomology 2010;39(4) 464-469.

[188] Suguituru SS., Silva RR., Souza DRD., Munhae CDB., Morini MSDC. Ant community richness and composition across a gradient from Eucalyptus plantations to secondary Atlantic Forest. Biota Neotropica 2011;11(1) 369-376.

[189] Schmidt FA., Ribas CR., Schoereder JH. How predictable is the response of ant assemblages to natural forest recovery? Implications for their use as bioindicators. Ecological Indicators 2013;24 158-166.

[190] Crist TO. Biodiversity, species interactions, and functional roles of ants (Hymenoptera: Formicidae) in fragmented landscapes: a review Myrmecological News 2009;12 3-13.

[191] Silveira JM., Barlow J., Louzada J., Moutinho P. Factors Affecting the Abundance of Leaf-Litter Arthropods in Unburned and Thrice-Burned Seasonally-Dry Amazonian Forests. Plos One 2010;5(9) 12877.

[192] Palacio EE., Fernández F. Claves para las subfamilias y géneros. In: Fernández F. (ed). Introducción a las hormigas de la region Neotropical. Bogotá: Instituto de Investigación de Recursos Biológicos Alexander von Humbolt 2003. P26-398.

[193] Anderson MJ. A new method for non-parametric multivariate analysis of variance. Austral Ecology 2001;26(1) 32-46.

[194] Gove AD., Majer JD., Dunn RR. A keystone ant species promotes seed dispersal in a "diffuse" mutualism. Oecologia 2007;153(3) 687-697.

[195] Loreau M., Naeem S., Inchausti P., Bengtsson J., Grime J. P., Hector, A., Hooper D. U., Huston M. A., Raffaelli D., Schmid B., Tilman D., Wardle DA. Biodiversity and ecosystem functioning: current knowledge and future challenges. Science 2001;294 (5543) 804-808.

[196] Western D. Human-modified ecosystems and future evolution. Proceedings of the National Academy of Sciences 2001;98(10) 5458-5465.

[197] Majer JD, Day JE, Kabbay ED, Perriman WS. Recolonization by ants in bauxite mines rehabilitated by a number of different methods. Journal of Applied Ecology 1984; 21 355-375.

[198] Palmer TM. Spatial habitat heterogeneity influences competition and coexistence in an african acacia ant guild. Ecology 2003;84(11) 2843-2855.

[199] Lassau SA., Cassis G., Flemons PKJ., Wilkie L., Hochuli DF. Using high-resolution multi-spectral imagery to estimate habitat complexity in open-canopy forests: can we predict ant community patterns? Ecography 2005;28(4) 495-504.

[200] Paolucci LN., Solar RRC., Schoereder JH. Litter and associated ant fauna recovery dynamics after a complete clearance. Sociobiology 2010;55(1) 133-144.

[201] Loreau M. Are communities saturated? On the relationship between alpha, beta and gamma diversity. Ecology Letters 2000;3(2) 73-76.

[202] Hooper, DU., Adair EC., Cardinale BJ., Byrnes J E K., Hungate BA., Matulich K L., Gonzalez A., Duffy J E., Gamfeldt L., O'Connor M I. A global synthesis reveals biodiversity loss as a major driver of ecosystem change. Nature 2012;486 105-108.

Chapter 4

Climate Change, Range Shifts and Multitrophic Interactions

Jeffrey A. Harvey and Miriama Malcicka

Additional information is available at the end of the chapter

http://dx.doi.org/10.5772/59269

1. Introduction

Climate change represents one of the most serious threats to biodiversity and ecosystem functioning. The current rate of temperature change, driven primarily by the human combustion of fossil fuels, far exceeds rates that have occurred in at least 10,000 years (lower Pleistocene) and perhaps much longer (IPCC, 2014). That last major climate change event precipitated a mass extinction that led to the sudden demise of many large quadrupeds, including such characteristic species as the woolly mammoth, woolly rhinoceros, mastodon, giant elk, saber-toothed tiger and dire wolf [1]. One of the major differences between landscapes at the time of previous climate change events and the current one is that the biosphere is now dominated by a single species, *Homo sapiens sapiens*, which has profoundly altered and simplified many terrestrial and aquatic ecosystems. Thus, in addition to climate change, natural ecosystems have been altered by other human-induced changes including deforestation, eutrophication, over-harvesting, the introduction of non-native species and various types of pollution. Consequently, species and populations are being challenged by multiple stressors, making it more difficult for them to adapt to rapid shifts in climate regimes. One can strongly argue that we no longer live in the Holocene but in the Anthropocene [2,3].

In a warming world, many species and populations are responding by changing various aspects of their life cycles, such as seasonal growth and phenology patterns, as well as by shifting their ranges pole-wards and/or to higher elevations [4,5,6,7]. The ability of species to shift their distributions is often limited by various eco-physiological constraints. These include the loss of habitat corridors through urban and agricultural expansion, which enable species to disperse over landscapes to other suitable habitat patches. Furthermore, some species lack traits, such as wings, which enable them to easily track changes in biotic conditions. As a result of these factors, we can expect ecological communities to change over time and for this to lead

---

to unpredictable new assemblages which may or may not stabilize as temperatures continue to rise [8].

2. Climate change and multitrophic interactions

It has long been known that species do not exist in isolation in nature. Instead, the survival and persistence of species in food webs and communities is dependent upon an array of interactions with other organisms occurring over highly variable spatial and temporal scales. Indeed, ecologist Daniel Janzen [8] once remarked that 'the ultimate extinction is the extinction of ecological interactions'. More recently, Pimm and Raven [9] argued that for every species of plant that becomes extinct in the tropics, many tens of species dependent on that plant for food or shelter also disappear. Given that different species in food webs may respond differently to climate change, warming has the potential of unravelling and/or destabilizing plant-insect communities, and that these can trickle through to affect other trophic interactions, even involving vertebrates [10]. In this chapter we discuss the effects and potential consequences of warming on trophic interactions involving plants, insect herbivores and specialist natural enemies, focusing on parasitic wasps (or parasitoids). Parasitoids are insects that develop on or in the bodies of other insects, whereas the adults are free living [11]. Hosts attacked by parasitoids are often not much larger than the adult parasitoid, meaning they are under intense selection to allocate and utilize limited host resources for different and often competing fitness functions such as reproduction and survival [12]. It is well established that many parasitoids are often highly specialized in attacking only one or a few species of hosts in nature [11]. Warming-induced changes in the environment may therefore affect various aspects of the biology and ecology of food plants and insect herbivores dependent on them, and this may trickle up the food chain, in particular affecting natural enemies that are more specialized on certain host types. Parasitoids thus make model organisms for examining a range of biotic and abiotic constraints in the environment.

2.1. Outline of the chapter

The chapter will be broken down into separate sections examining the effects of warming on the biology and ecology of the three trophic levels separately, and then move on to integrate these interactions and to provide testable predictions for these processes. Given that insects are ecto-therms, it is by now well established that metabolic rate and the developmental program of insects is closely co-ordinated with changes in temperature. However, it is less well established how changes in temperature, as well as attendant changes in precipitation etc. will affect tightly linked two and three-trophic level interactions. On this basis, the chapter will be broken down thusly:

1. Range shifts in plants and effects to (i) primary [nutrients] (ii) secondary [defensive compounds] metabolites, as well as in plant volatiles under herbivore damage (HIPVs). How will changes in plant quality affect multitrophic interactions?

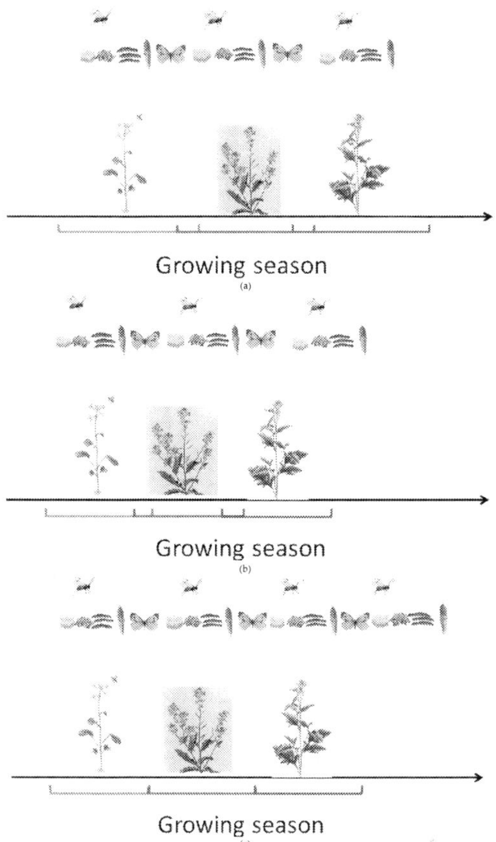

Figure 1. Phenological interactions involving three plant species, a specialist herbivore and its specialist endoparasitoid under normal conditions and under climate warming scenarios in Europe. In (A), *Pieris brassicae* (middle) the large cabbage white butterfly is trivoltine, and different generations lay their eggs on different species of large, short-lived annual brassicaceous (mustard) plants (bottom) that grow in aggregated populations. Different stages of the herbivore – eggs, young larvae, fully grown larvae, pupa and adult are shown. The mustards in turn grow only for 2-3 months during the year and at different times. *Brassica rapa* (wild turnip, left, green line) supports the first generation of *P. brassicae*, *Sinapsis arvensis* (charlock mustard, middle, red line) the second, and *Brassica nigra* (black mustard, right, blue line) the third. In turn, the gregarious endoparasitoid wasp, *Cotesia glomerata* (top) also has three generations where the adult wasps emerge in time to find and parasitize young caterpillars of *P. brassicae*, and emerge from fully grown caterpillars pupating on the food plant. In (B) and (C), warmer temperatures lead to dissociation of the multi-trophic interaction by differentially affecting the seasonal growth and activity patterns of the plants and insects. In (B), the plants are more strongly affected and their growing phenology is shifted to an earlier point in the season. However, the insects respond less to warming and their temporal life cycles become desynchronized with the growth cycles of their plants. There is also a period later in the season when few or no suitable plants are available. In (C), the plants only marginally shift their growing patterns to an earlier time point in the season, whereas the insects have an additional (=4th) generation later in the year. In both climate warming scenarios, there are gaps for the insects when food plants are scarce or absent. This could have profoundly negative consequences on the persistence of this trophic chain.

2. Range shifts and outbreaks in herbivores in response to climate change and effects on plant responses and interactions with their natural enemies; range shifts in parasitoids: tracking their hosts or switching to novel native hosts? Examples of competitive interactions and potential displacement in interspecific competition will be explored.

3. The effects of temperature on insect growth, development and other traits. Given that insects are ectothermic, how will higher temperatures affect their development, and what will the consequences be for biotic interactions?

4. Changes in the seasonal phenology of multitrophic interactions; does warming differentially affect the growth/life cycles of plants, herbivores and their natural enemies, and what are the consequences for the persistence of these interactions?

5. How will warming potentially affect the dynamics of multi-trophic interactions, scaling up to food webs, communities and ecosystems, and will this in turn affect resilience and stability? What are the possible effects on top-down and bottom-up regulation?

An important theme of this chapter is to stress the importance of examining multi-species interactions under climate change scenarios. This entails a close examination of mechanisms as well as the consequences of warming in a community-related context. Thus far, both climate change and invasion ecology have been studied independently or have been largely restricted to the study of bi-trophic interactions (e.g. plant-herbivore) with very few studies going to three (or more) trophic interactions [13]. There are a number of excellent reviews which focus on each of these areas [14,15,16,17] but only a few have begun to address community level-effects, and again these generally do not integrate climate change with range shifts in plants and insects [18,19,20]. Fewer still have explored climate-change related range shifts on plant-herbivore-parasitoid interactions at both reductionist and larger scale perspectives. The main aim of this chapter is therefore to explore and discuss how climate change and related abiotic changes (e.g. precipitation) will affect multi-trophic interactions from both the perspective of mechanisms and processes, scaling up from individuals to communities and ecosystems. We will discuss the potentially cascading effects of climate warming and other parameters related to climate change on ecosystems, suggesting that many will be simplified (Fig 1.), reducing their resilience against other natural and anthropogenic challenges in the environment.

3. Range shifts in plants

Two aspects of warming will affect the structure of plant communities. The first involves native plants which may be adapted to cooler conditions and thus become increasingly stressed as conditions warm. This is especially true for plant species growing at the southern edge of their ranges. The second involves plants which shift their ranges northwards in order to track warming and to exploit more optimal thermal conditions. These plants will increasingly expand into habitats occupied by native plants, leading to potential increases in competition and, at least to some extent, rearrangement in plant community structure. As the climate warms, we may expect physiological and metabolic responses in both native and range expanding plants that will in turn potentially affect the behaviour and performance of higher trophic level consumers associated with them. More specifically, the metabolic changes in plants may be borne out on traits such as primary and secondary metabolism which play a key role in insect nutrition and plant defence responses. Plant volatiles, which are often induced in response to herbivore damage and which potentially have many functions, might also be affected by warming or in response to new selection pressures in range expanding plants. At the same time, range expanding plants may, at least partially, escape from some of their co-evolved enemies, such as herbivores and pathogens, also affecting the costs and benefits of metabolic investment into plant defences. Below, we examine these areas in more detail.

4. Primary and secondary metabolism

Plant tissues contain various concentrations of nutrients, with carbon (C), nitrogen (N), and phosphorus (P) being considered as the most important, as well as amino acids [21]. Although atmospheric CO_2 concentrations are deterministic at various scales, and are rising slowly at about 2-3 ppm annually, temperatures are not. Thus one can question the efficacy of experiments that expose plants (and insects) to extremely high concentrations of CO_2 (e.g. 450 ppm and higher) when these levels are not expected to be reached for several decades. However, there is no way to successfully circumvent this obstacle, and many studies exposing plants to higher CO_2 levels have been conducted. Many have shown that under increased CO_2 regimes attendant with climate change that plants will possibly take up more C at the potential expense of foliar levels of N and P [22,23,24] and that changes in concentrations of other nutrients, such as amino acids, will also occur [25]. (However, these processes are indeed complex – for instance they vary in plants with different metabolic C pathways as well as in different successional stages but a highly detailed discussion of this is not possible here). N is considered a limiting nutrient for insect development [26,27,28,29] and it has been suggested that a reduction in foliar concentrations of N will lead to compensatory feeding in herbivores to ensure optimal levels of this nutrient are acquired [26,30,14,31]. For range expanding plants in a warming world, it is difficult to extrapolate cause-and-effect relationships related to temperature and precipitation, unless atmospheric changes in CO_2 are also taken into account.

Most plants also produce secondary metabolites (or allelochemicals) whose primary function is considered to be defence against antagonists such as pathogens and herbivores [32,33,34,35]. There is a vast array of different types of allelochemicals produced by plants that are based on the phylogeny of a species group, often at the family level [36]. For instance, one can find glucosinolates, alkaloids, iridoid glycosides, furanocoumarins and many other allelochemicals produced by plants in different families. These toxins are often inducible, meaning that they are found in basal levels in intact plant tissues but markedly increase after tissue damage [37, 35]. These compounds have long been known to affect the behaviour and performance of insects that are intimately associated with the plants that produce them. High levels of phytotoxins in plant tissues can impair the development and hence fitness of herbivores, through increased rates of mortality, extended development time and reduced adult size [38,

39,40,41,42,43,44]. These effects often differ between specialist and generalist herbivores. Generalists usually do not possess highly refined mechanisms of dealing with specific secondary plant compounds, but instead rely on general responses, (e.g. P-450 mono-oxygernases) that are effective against a range of different metabolites [45]. However, they are very often less effective than the mechanisms employed by specialists which have strongly co-evolved histories with certain types of plant phytotoxins. For this reason, specialists are frequently assumed to be in a co-evolutionary 'arms race' with plant defences [46] in which increased herbivory leads to increased plant defences which in turn leads to adaptations on the part of the herbivore and so on in a 'Red Queen' type stalemate scenario. In fact, elevated levels of toxins actually stimulate feeding and oviposition behaviour in the many specialist herbivores [35]. Despite this, very well defended plants are often problematical even for specialists [47].

The development of natural enemies such as parasitoid wasps is also known to be affected by host attributes as mediated through the quality of the host diet [41,42]. High levels of plant toxins ingested by the host can also detrimentally affect the growth and survival of immature parasitoids and even hyperparasitoids one trophic level higher [48,49]. One major phylogenetic constraint on parasitoid larvae is that the alimentary tract is not externally connected until they terminate the relationship with the host [50]. Therefore, all host tissues ingested during development are stored in the body pending the voiding of the meconium (or even later as an adult). This includes deleterious materials such as plant toxins that were stored in the body of the herbivore host. If concentrations of these toxins become too high, then it is possible that the developing parasitoid larvae cannot safely store them and will die precociously as a result. It is interesting that the diversity of many parasitoid taxa is higher in temperate than in tropical biomes, a trend that contradicts that shown in most other groups of insects. Bolton [51] hypothesized that tropical vegetation is generally much more toxic than vegetation in the temperate zone, and that parasitoids have not been able to adapt to hosts that ingest and store these toxins in their body tissues (the 'nasty-host-hypothesis').

Several climate-change related effects on plant allelochemistry are possible, and these may trickle their way through trophic webs. Elevated CO_2 regimes will also affect plant allelo-chemistry in potentially different directions depending upon whether a plant's defence chemistry is C-or N-based. Plants with C-based allelochemistry may become more toxic (e.g. better defended) whereas the opposite may occur with plants with N-based allelochemistry [52,53,54,55,56,57]. A second affect will be that warming, along with changes in precipitation, as well as reduced or extended drought periods attendant with climate change will affect metabolic allocation by plants to secondary defense metabolites. Higher temperatures can sometimes lead to a reduction in plant allocation to direct defence [58]. Veteli [59] found that elevated CO_2 and temperature increased plant growth but had opposite effects on the growth rate of an insect herbivore. A third effect will be that range shifting plants will escape from some of their co-evolved herbivore enemies, leading to changes in metabolic allocation towards plant defence [60]. If native specialist and generalist herbivores select for high chemical defences in native plants, then range-expanding plants may invest less in these defences if they are attacked by fewer herbivores (and thus suffer less herbivore damage) in their new range [the 'enemy-release hypothesis' [ERH],[61,62,63] and more towards other functions, such as growth that enable plants to outcompete natives (the 'evolution of increased competitive ability hypothesis' [EICA], [62,64]. A third hypothesis, the 'shifting defence hypothesis' [SDH], posits that co-evolved specialists are adapted to high concentrations of plant allelochemicals, driving selection for a reduction in them to make them less attractive as oviposition sites [65]. In new habitats, these plants escape their specialists but attract more generalists, selecting for an increase in chemical defences. Lastly, many invasive plants bring with them novel secondary metabolites to which the native herbivore fauna is not adapted and which therefore allows the plant to invest metabolic energy to other vital functions (the 'novel weapons hypothesis' [NWH], [66,67]. Thus only a single study has considered climate change-mediated changes in plant communities on natural enemies of the herbivores on native and invasive plants [68]. This revealed that there may be shifts in the intensity of natural enemy-mediated top down trophic cascades versus bottom-up plant-mediated effects on herbivores on natives and invasives. However, it is far too early to draw conclusions as many more studies are needed to tease out potential patterns in communities where shifts in plant composition are occurring as a result of warming.

There is ample evidence in support of these different hypotheses for inter-continental invasions [63,69]. However, intra-continental invasions based on climate-change related range shifts [70,71] are likely to have more subtle effects on plant-consumer interactions because many insects can track their food plants as they move pole-wards [72]. As a result, consistent patterns remain elusive. Engelkes [68] found that range expanding plants in the Netherlands were more toxic to naïve herbivores than related natives, and that the range-expanders also produced higher levels of general defence compounds (phenolics). Fortuna [73] compared allelochemistry and herbivore performance in native and range-expanding populations of warty cabbage (*Bunias orientalis*) and reported that the insects were larger and survived better on the range-expanding than on the native populations. Moreover, the authors found both quantitative and qualitative differences in plant allelochemistry between the native and range expanding plants. It appears that *B. orientalis* is comparatively rare and local in its native range, which is perhaps evidence of biotic resistance amongst coevolved members of the native plant community as well as more top-down control exerted by native co-evolved herbivores and/or pathogens [61,64]. Other studies have found little or no differences in insect performance on native and range-expanding populations in plants [74].

Plants also release chemically-based odours via the production of volatiles that are often induced by stresses such as herbivory [75,76,77]. The precise function of these volatiles is unclear and remains the subject of considerable debate [78,79,80,81]. It is known that these volatiles are used by insects to locate plants on which to oviposit [75,82] or on which natural enemies find their prey or hosts [83,84,85,86]. If insects can drive selection for types and concentrations of volatiles, then changes in top-down pressures as plants are released from their enemies (e.g. herbivores) or potential allies (e.g. parasitoids) may also alter volatile profiles over time. However, thus far the effect of plant volatiles in range expanding plants on native natural enemies has received virtually no attention. Fortuna [87] found that adult females of the large cabbage white butterfly, *Pieris brassicae*, preferred to oviposit on shoots of a native brassicaceous plant, *Sinapis arvensis*, over those of a range expander, *Bunias orientalis*. Herbivore performance was higher on the native plant. However, its major natural enemy, the parasitoid *Cotesia glomerata*, did not distinguish between volatiles of the two plants. This suggests that herbivores and parasitoids may respond to different kinds of volatiles – herbivores clearly on those released by the plant alone and specialist natural enemies to a combination of plant and host-or prey-related odours.

A recent study by Yuan [88] laid out a possible framework for the effects of climate change on plant volatiles, and projected this to the community level. They argued that patterns are likely to be variable and association-specific, because of the immense complexity in the blends that make up a plant's volatile profile. Moreover, other anthropogenic stresses must be factored in, leading to highly unpredictable scenarios across different scales of space and time. They urged that more studies with different systems are necessary to tease out mechanisms that may cascade up to affect higher trophic level consumers.

5. Effects of climate change on insect growth and development

Insects, like all invertebrates, are ectotherms and thus they are highly susceptible to changes in temperature as well as other abiotic processes linked with climate change [14,15]. Many studies have examined how insects across different trophic levels respond to variations in temperature [14]. Rarely have these responses, however, been placed within the context of climate change, perhaps because warming has only been broadly acknowledged in the past 20 years or so by the scientific community. As with plants, insects can respond in two ways to local warming regimes. First, they must adapt behaviourally, morphologically and physiologically to such processes as an increased incidence of heat waves and other attendant stresses such as droughts or higher precipitation regimes or an increase in frost-free periods [15]. Second they can shift their ranges and move pole-wards or to higher elevations (below).

Many studies have reported that the survival, development rate and adult body mass of insect herbivores and their natural enemies are affected by rearing temperatures [89]. Much less attention has been paid to transient periods of high or low temperatures (e.g. combining them under the umbrella of a single experiment, thus creating a more realistic picture of events transpiring in nature) or rainfall. It is by now known that climate warming is likely to generate more extreme weather events at local scales, rather than simply resulting in gradual changes that are measured at large spatial scales across the biosphere. When confronted with these conditions, insects have to adapt or to move to new habitats where conditions are more suitable. In time, the latter will result in range shifts, a phenomenon by now well described in many studies (see next section, below). Those insects that 'remain behind'will exhibit physiological responses to warming and phenomena associated with its such as an increase in heat waves and changes in precipitation.

6. Range shifts in herbivores and their natural enemies

As with other organisms, various insect taxa are adapted to climate windows and have well-defined ranges which coincide with both biotic and abiotic conditions [90]. Climate warming is already known to be driving demographic and geographic responses in insects [6,8,14,91,92]. Range shifts in herbivores and their natural enemies depend on a number of ecological factors that go beyond warming and which are often complex. For specialist herbivores that feed on only one or a few related plant species, a major impediment to movement is the availability of nutritionally suitable plants in their new habitats [6]. Generalist herbivores, on the other hand, may benefit if they are able to feed on a range of unrelated plants in both their native and invasive ranges. For many herbivores this is not a problem if their food plants also track the warming climate or if related plants with similar allelochemistries are also found in the new ranges of the insects. The oak processionary caterpillar (*Thaumetopoea processionea*) has expanded its range dramatically to the north within Europe over the past 30 years, coinciding with the recent warming episode [93]. Suitable oak trees on which the larvae can feed and develop, are found over much of Europe, helping to facilitate its spread. Its close relative, the pine processionary caterpillar (*Thaumetopoea pityocampa*), has also spread northwards as a result of recent warming and is projected to arrive in the Benelux region in the near future [94, 95]. Both species are considered as major health hazards owing to the production of numerous urticating hairs in mid-and late larval instars that contain soluble proteins and which are highly irritating to the skin and mucous membranes of humans [96]. The oak processionary caterpillar is actually more abundant now in many parts of its invasive than in its native range, perhaps because its specialist natural enemies have not effectively tracked its northwards expansion [93]. This is a worrying pattern that, if repeated in many trophic interactions, could facilitate pest outbreaks with large attendant economic costs.

Many other insects are known to be expanding into new habitats as a result of climate warming [7,97,98]. As they do so, they interact with native plants and their associated arthropod communities. The broader ecological outcome of these interactions is open to considerable debate. There is the possibility of community reorganization or reassembly as some species compete with and potentially displace others (see more detailed discussion of this below). Studies examining a suite of ecophysiological processes that underpin the ways in which these interactions work are urgently required. For instance, different species within food webs may each respond differently to changes in abiotic conditions such as temperature and moisture. The diamondback moth (*Plutella xylostella*) is native to Africa and the Mediterranean region, but has been introduced over many parts of the world where it is a serious pest of cabbages and related crops [99]. The moth only began to successfully overwinter in central Europe in the past 20 years, allowing to have two generations per year and to build up numerically faster by mid-summer [99]. It is attacked by several larval endoparasitoids, each of which exhibits differential responses to temperature. Recent warming in central Europe appears to favor thermophilic parasitoids like *Costesia vestalis* and *Dolichogenidea sicaria* over cool-favoring species like *Diadegma semiclausum* [99, J. Harvey, personal observations]. In addition to range expansions, many native insects will benefit from warming as a result of longer growing seasons and more favorable conditions for populations to grow. There is already evidence that some species are experiencing outbreaks as a result of warming as well as range expansions [100].

7. Effects of warming on the seasonal phenology of multitrophic interactions

Jeffs and Lewis [17] examined the potential effects of climate warming on host-parasitoid interactions and developed three primary ways in which parasitoids might respond to warming: (1) by shifting distributions polewards or to higher elevations; (2) altering their phenology; (3) adjusting to persist in their current ranges through phenotypic plasticity or evolutionary adaptations. However, this ignores the potentially negative effects of a failure to respond to warming, or else the consequences of local changes on the survival and persistence of parasitoids. For example, warming is occurring so rapidly in many places that many species or populations may not be able to adapt in sufficient time. The authors fail to discuss the physiological costs of warming and how this might affect the acquisition and metabolic allocation of resources by the larvae (from the host) and the adult (from both host and/or non-host sources). If development of immature stages is negatively affected, this might have profound effects on adult fitness and thus lead to declines and possible extinction. Furthermore, phonological shifts depend on the ability of the parasitoid not only to track the host but the host's foodplant(s). Changes in important abiotic parameters may unravel trophic interactions if the species in these links respond differently to warming in terms of their life cycles. This has already been demonstrated in oak-winter moth interactions and the effects of this are negatively affecting the reproductive success of both migratory and resident insectivorous birds. There is a possibility that parasitoids of winter moths are also being negatively affected by warming. Some plants that are vital for the development and survival of specialist herbivores may also shift their seasonal growth patterns. For instance, interactions involving the large cabbage white butterfly, *Pieris brassicae*, its natural food plants and a specialist gregarious endoparasitoid, *Cotesia glomerata*, are complex in the context of life-history interactions involving the various parties. The herbivore and parasitoid each have 2-3 generations per year, each of which must seek out new food plant species in which to exploit. This is because most of its suitable food plants – brassicaceous species – are short-lived annuals or biennials, whereby different species grow at different times of the growing season [101]. Some species, such as *Brassica nigra*, grow early in the spring, whereas others, including *Sinapis arvensis* and *B. nigra* grow in late spring and summer respectively. The consequences of warming on the phenology of this trophic interaction may critically hinge on how the plants and insects each respond to increasing temperatures, and how this in turn affects the availability and suitability of the resources which they exploit as food. For the herbivore, of course, this means the availability of nutritious shoots on which the caterpillars feed, and for the parasitoid young caterpillars in which the female wasps oviposit clutches of up to 50 eggs. Warming will certainly increase the number of generations the insects have, and there already indications, based on populations in the Mediterranean region that up to 4 generations are indeed possible in more central and northern parts of Europe. However, if plant growth is temporally advanced (something that occurred in 2014) then later generations of the insects may emerge into habitats with little plant food available. There are, however, many possible scenarios, whereby the insects may experience neutral or negative effects of warming on their survival and fitness. Furthermore, this example is hardly likely to be an isolated one; indeed, many trophic interactions involving specialized consumers are under the same constraints.

8. Climate warming in the context of larger ecological scales

Thus far, the relationship between temperature and insect behavior and development have been largely confined to pairwise interactions involving a plant and a herbivorous insect, or even the insect alone when reared on an artificial medium. Slightly more complexity has been achieved by incorporating a predator or a parasitoid into these studies, but the vast majority of them have been focused on optimal rearing conditions for biological control rather than on anthropogenic stresses such as climate change in natural and managed ecosystems. Indeed, in other fields where anthropogenic stresses are involved, such as invasive species or habitats loss, little attention has often been paid to insects in a multitrophic framework [10,13,102]. Climate change certainly represents a serious challenge to insects across vastly different scales of space and time because it will have cascading effects on a wide range of ecological characteristics and processes in habitats. An important challenge is to scale up the results of small-scale studies to see how these play out in communities, ecosystems and biomes. In this context we need to understand how biodiversity over large scales regulates ecosystem-level processes and how warming, by weakening processes and interactions at smaller scales, will affect this regulation.

The traditional approach in examining the relationship between biodiversity plays in ecosystem functioning has been based species interactions and the consequences of such interactions for community structure and function. These interactions can be classified as direct, involving pairwise interactions between species (e.g. predation/parasitism) or indirect, involving mediation by a third party [103]. Therefore, studying how biodiversity influences ecosystem functioning in multitrophic systems (involving mediation by a third party) is important for several reasons: (1) multiple trophic levels represent the core of ecosystems [104]; (2) as multitrophic diversity increases, average ecosystem properties could increase, decrease, stay the same or follow more complex non-linear patterns [105]. Consequently, as the number of species change within a community, the occurrence and significance of (in)direct interactions will also change. These, in turn, may be modified by abiotic factors, which may generate cascading reactions that generate large ecological changes with important ecological consequences [103,106].

Three of the well-documented global changes mediated by human activities are: increasing concentrations of carbon dioxide in the atmosphere; alterations in the biogeochemistry of the global nitrogen cycle; and ongoing land use/land cover change [106]. Human activity is now considered as the prime driver of global environmental change [106,107,108,109]. However, our ability to generate linkages at spatial (landscape) scales relevant to the human enterprise is limited at present [110]. Most importantly, the consequences of biodiversity loss on ecosystem services (e.g. primary and secondary production, nutrient cycling, pest control, pollination, etc.) is poorly understood, as is our knowledge of the effects of warming on ecosystem processes [105]. Therefore, understanding an array of mechanisms that drive the biodiversity–ecosystem functioning relationship is thus difficult to evaluate in multitrophic systems [105], because of the unexpected consequences of warming on species interactions and demographics e.g. when the biology of one species is influenced by the biology of another species [111]; under simultaneously changing landscape characteristics such as habitat availability and landscape structure that affect biodiversity [112,113]; through the loss and/or fragmentation of habitats that drive changes in species abundance with both winners and losers [114].

Multitrophic interactions involve microbes, pathogens, plants, animals and other functional groups that are found in different positions of the food chain and provide vital functions to communities and ecosystems [115]. Here, we focus on organisms inextricably linked to plant-insect interactions. Microorganisms from diverse environments have played an important role in ecosystem sustainability. Since the spatial and temporal stresses of the microbial system may be quite different from those of plants and animals [116], many studies of the ecological responses to global changes have suggested that belowground processes, often mediated by soil microorganisms, are central to the response of ecological systems to global change [117]. Below-ground microbial mediated processes can both immobilize and release nutrients that limit primary production and can influence the long-term response of ecosystems to global change [118]. Investigations into microbial parameters involved in soil quality are increasing [118,119,120] and it has been shown that human activity can directly or indirectly affect the functioning and diversity of the soil community [116,121] and these effects are transferred aboveground where they effect the structure and function of plant-insect communities [19,122,123]. Bardgett [124] described some potential outcomes for soil microbes and carbon exchange that include: (i) increases in soil carbon loss by respiration and in drainage waters as dissolved organic carbon due to stimulation of microbial abundance and activity, and enhanced mineralization of recent and old soil organic carbon [125,126]; (ii) stimulation of microbial biomass and immobilization of soil N, thereby limiting N availability to plants, creating a negative feedback that constrains future increases in plant growth and carbon transfer to soil [127]; (iii) increased plant-microbial competition for N, leading to reduced soil N availability and microbial activity and suppression of microbial decomposition and ultimately increased ecosystem carbon accumulation [117]; (iv) increased growth of mycorrhizal fungi [128,129], which receives carbon in the form of photosynthate directly from the host plant and retains this carbon, controlling its release to the soil microbial community [130]; and changes in root exudation that are known to play a potentially important role through the promotion of methanogenesis and hence carbon loss from soil as methane [131], although the mechanisms involved in this process are poorly understood.

The response of the soil free-living bacterial group (β subclass of the *Proteobacteria*) on simulated multifactorial global change was investigated in grassland vegetation dominated by annual grasses and forbs growing in sandstone-derived soil [118]. The results demonstrate that shifts in community composition were associated with increases in nitrification, but changes in abundance were not, confirming that microbial communities can be consistently altered by global changes and that these changes can have implications for communities and ecosystems.

Alternatively, atmospheric CO_2 on microbial decomposition in peat was found to be greater than when these factors operated alone, creating a stronger positive feedback on carbon loss from soil as DOC (dissolved organic carbon) and respiration [126].

Soil microbes and their activities are inexorably linked to above and below ground communities and therefore it is necessary to understand the effects of climate change on carbon budgets [124]. However, it is important to be aware also of adverse effect of soil microbes on multitrophic interactions, mostly in the case of pathogens. Soil borne pathogens can significantly alter the spatial distribution and reduce the yield and quality of plants [132,133]. Climate change generally has a beneficial effect on pathogen suppression and stress tolerance [115]. Bacteria, fungi and nematodes perform specific functions that help suppress the infection and colonization of plant roots by pathogens in the rhizospehere [115].

Plants play a major role in controlling and mediating directly or indirectly all soil multitrophic interactions [115] and contribute to ecosystems on by providing food, shelter or "ecological islands" for microbes, fungi, insects, and other organisms. However, dead plants also continue to supply the environment with nutrients. Thus, qualitative and quantitative changes in plant physiology, chemistry, tissue composition and signaling pathways under rising CO_2 and temperature regimes may influence the entire cascade of above and below-ground multitrophic interactions [134,135,136,137]. Moreover, the geographical distribution of plants is also affected by changing temperature and precipitation that alter the structure of plant communities at larger scales and which in turn generate new (positive or negative) trophic interactions [138]. The principle effect of changes in the spatial distribution of plants is also influenced by the presence or absence of mutualists (pollinator) and antagonist (pathogen and herbivore) in a changing environment [139] as well as intra-interspecific competitors (taller plants replace lower congeneric species) [140]. On the other hand, plant phenology is temperature-sensitive with strong connections with photoperiod [141]. Accordingly, there is critical springtime photoperiod, which is an important pre-requisite before floral development can begin [141,142]. Net photosynthesis (primary production) typically peaks within the range of normal temperatures [143,144], however increasing temperatures can shift this peak and extending the growing season and consequently accelerate plant growth [145]. This situation can lead to key differences with respect to interactions with insect mutualists or antagonists [16] and subsequently changes in the population dynamics can alter evolutionary trajectories [146], creating "evolutionary noise" with unpredictable consequences for the strength and persistence of trophic interactions. While evidence of adaptive responses via variation among genotypes in response to increasing temperature has been described in a marine parasite *Maritrema novaezealandensis* [147], the extinction of local populations of the angiosperm *Fagus sylvatica* still occurred at the southern range margin despite strong signals of genetic adaptation in this species to climate warming [148,149]. This variation in responses at the species level makes it virtually impossible to predict the effects of climate change on trophic interactions, although available evidence suggests that there will be many more losers than winners [6] and as a result we can expect multitrophic (ecosystem) interactions to be simplified (Fig. 2) and ecosystems therefore to become less stable and resilient [150].

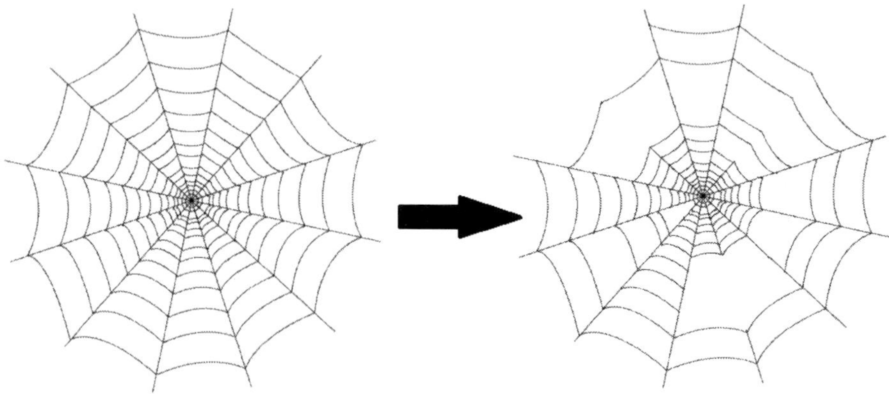

Figure 2. Variation in responses at the species level makes it virtually impossible to predict the effects of climate change on trophic interactions, although available evidence suggests that there will be many more losers than winners and as a result we can expect ecosystem interactions to be simplified and ecosystems therefore to become less stable and resilient.

9. Summary and future directions

Climate change clearly represents a major challenge to biodiversity at all levels of organization. This includes physiological and behaviorial responses of individual species through trophic interactions and beyond to the scale of communities and ecosystems. Given that they are ectotherms, insects will respond to warming in a number of ways. First, they may seek out micro-climates with more optimal temperatures for biological activity; second, they may shift their diel patterns of activity to coincide with more optimal temperature regimes; third they will adjust their broader distributions either moving to higher elevations or polewards. However, these responses also critically depend upon interactions with other species, includ-ing both resources (e.g, plant for herbivores and herbivores for predators) as well as indirect interactions with species two levels or more apart on the food chain.

Thus far, few studies have explored climate change on trophic interactions integrating three or more levels of a food chain. Instead, the main focus thus far has been on descriptive studies or else on two-trophic level interactions, with little attention paid to members higher up the food chain (e.g. parasitoids and hyperparasitoids), especially in a multitrophic framework. Given the paucity of studies in this area, it is virtually impossible to predict larger scale consequences of warming for communities and ecosystems and the services that emerge from them. It is therefore important to scale up the biological effects of warming on individual species of producers (plants) and consumers (herbivores and natural enemies) to trophic chains and from there to communities, in order to better understand the potential consequences on the stability and resilience of ecosystems and the biomes in which they are embedded. This is a daunting challenge, but one that must be tackled if we are to develop predictive power of climate change effects on natural and managed ecosystems. Another important area is to combine warming in an above-and below ground approach integrating trophic interactions occurring amongst insects and other invertebrates in both compartments, an area that is now being increasingly considered in community-based research [151].

Author details

Jeffrey A. Harvey[1,2*] and Miriama Malcicka[2]

*Address all correspondence to: j.harvey@nioo.knaw.nl

1 Netherlands Institute of Ecology, Department of Terrestrial Ecology, Wageningen, The Netherlands

2 VU University Amsterdam, Department of Ecological Sciences, Animal Ecology, Amsterdam, The Netherlands

References

[1] Graham RW, Lundelius EL Jr. Coevolutionary disequilibrium and Pleistocene extinctions. 1984.

[2] Zalasiewicz J, Williams M, Haywood A, Ellis M. The Anthropocene a new epoch of geological time?. Philosophical Transactions of the Royal Society A Mathematical, Physical and Engineering Sciences 2011; 369(1938) 835-841.

[3] Steffen W, Persson Å, Deutsch L, Zalasiewicz J, Williams M, Richardson K, Crumley C. et al. The Anthropocene From global change to planetary stewardship. Ambio 2011; 40(7) 739-761.

[4] Hughes L. Biological consequences of global warming is the signal already apparent? Trends in Ecology & Evolution 2000; 15(2) 56-61.

[5] Post E, Forchhammer MC. Synchronization of animal population dynamics by large-scale climate. Nature 2002; 420(6912) 168-171.

[6] Thomas CD, Cameron A, Green RE, Bakkenes M, Beaumont LJ, Collingham YC, Barend FN Erasmus et al. Extinction risk from climate change. Nature 2004; 427(6970) 145-148.

[7]   Parmesan C. Influences of species, latitudes and methodologies on estimates of phenological response to global warming. Global Change Biology 2007; 13(9) 1860-1872.

[8]   Janzen DH. Tropical dry forests. The Most Endangered Major Tropical Ecosystem, Pp en EO Wilson, Biodiversity; 1988.

[9]   Pimm SL, Raven P. Biodiversity extinction by numbers. Nature 2000; 403(6772) 843-845.

[10]   Cronin JT, Haynes KJ. An invasive plant promotes unstable host-parasitoid patch dynamics. Ecology 2004; 85(10) 2772-2782.

[11]   Godfray HCJ. Parasitoids, behavioral and evolutionary ecology. Princeton University Press; 1994.

[12]   Jervis MA, Ellers J, Harvey JA. Resource acquisition, allocation, and utilization in parasitoid reproductive strategies. Annual Review of Entomology 2008; 53 361-385.

[13]   Harvey JA, Bukovinszky T, van der Putten WH. Interactions between invasive plants and insect herbivores a plea for a multitrophic perspective. Biological Conservation 2010; 143(10) 2251-2259.

[14]   Bale JS, Masters G, Hodkinson I, Awmack C, Bezemer TM, Brown V, Butterfield J, Buse A, Coulson J, Farrar J. Herbivory in global climate change research direct effects of rising temperature on insect herbivores. Global Change Biology 2002; 8(16) 1-16.

[15]   Hance T, Van Baaren J, Vernon P, Boivin G. Impact of extreme temperatures on parasitoids in a climate change perspective. Annual Review of Entomology 2006; 52(1) 107.

[16]   Jamieson MA, Trowbridge AM, Raffa FK, Lindroth RL. Consequences of climate warming and altered precipitation patterns for plant-insect and multitrophic interactions. Plant Physiology 2012; 160 (4) 1719-1727.

[17]   Jeffs CT, Lewis OT. Effects of climate warming on host–parasitoid interactions. Ecological Entomology 2013; 38(3) 209-218.

[18]   Tylianakis JM, Didham RK, Bascompte J, Wardle DA. Global change and species interactions in terrestrial ecosystems. Ecology Letters 2008; 11(12) 1351-1363.

[19]   Van der Putten WH, de Ruiter PC, Bezemer TM, Harvey JA, Wassen M, Wolters V. Trophic interactions in a changing world. Basic and Applied Ecology 2004; 5(6) 487-494.

[20]   Van der Putten WH, Macel M, Visser ME. Predicting species distribution and abundance responses to climate change why it is essential to include biotic interactions across trophic levels. Philosophical Transactions of the Royal Society B Biological Sciences 2010; 365(1549) 2025-2034.

[21]   Epstein E. Mineral nutrition of plants principles and perspectives. 1972.

[22] Ryan MG. Effects of climate change on plant respiration. Ecological Applications 1991; 1(2) 157-167.

[23] Melillo JM, McGuire AD, Kicklighter DW, Moore B, Vorosmarty CJ, Schloss AL. Global climate change and terrestrial net primary production. Nature 1993; 363(6426) 234-240.

[24] Cramer W, Bondeau A, Woodward FI, Prentice IC, Betts RA, Brovkin V, Cox PM et al. Global response of terrestrial ecosystem structure and function to CO2 and climate change results from six dynamic global vegetation models. Global Change Biology 2001; 7(4) 357-373.

[25] Bezemer TM, Jones TH. Plant-insect herbivore interactions in elevated atmospheric CO_2 quantitative analyses and guild effects. Oikos 1998; 212-222.

[26] Mattson Jr, William J. Herbivory in relation to plant nitrogen content. Annual Review of Ecology and Systematics 1980; 119-161.

[27] Chapin III FS, Vitousek PM, Van Cleve K. The nature of nutrient limitation in plant communities. American Naturalist 1986; 48-58.

[28] Lambers H. Rising CO_2, secondary plant metabolism, plant-herbivore interactions and litter decomposition. Vegetatio 1993; 104(1) 263-271.

[29] Elser JJ, Hayakawa K, Urabe J. Nutrient limitation reduces food quality for zooplankton Daphnia response to seston phosphorus enrichment. Ecology 200; 82(3) 898-903.

[30] Lincoln DE. The influence of plant carbon dioxide and nutrient supply on susceptibility to insect herbivores. Vegetatio 1993; 104(1) 273-280.

[31] Berner D, Blanckenhorn WU, Körner C. Grasshoppers cope with low host plant quality by compensatory feeding and food selection N limitation challenged. Oikos 2005; 111(3) 525-533.

[32] Rosenthal GA, Berenbaum MR. Herbivores Their interactions with secondary plant metabolites Ecological and Evolutionary Processes. Vol. 2. Academic Press; 1992.

[33] Bennett RN, Wallsgrove RM. Secondary metabolites in plant defence mechanisms. New Phytologist 1994; 127(4) 617-633.

[34] Karban R, Baldwin IT. Induced responses to herbivory. Chicago University of Chicago Press; 1997.

[35] Schoonhoven LM, van Loon JJA, Dicke M. Insect-Plant Biology, Ed 2. Oxford University Press, Oxford; 2005.

[36] Wink M. Evolution of secondary metabolites from an ecological and molecular phylogenetic perspective. Phytochemistry 2003; 63(1) 3-19.

[37] Bourgaud F, Gravot A, Milesi S, Gontier E. Production of plant secondary metabolites a historical perspective. Plant Science 2001; 161(5) 839-851.

[38] Barbosa P, Krischik VA, Jones CG. Microbial mediation of plant-herbivore interactions. John Wiley & Sons; 1991.

[39] Hunter MD, Linnen CR, Reynolds BC. Effects of endemic densities of canopy herbivores on nutrient dynamics along a gradient in elevation in the southern Appalachians. Pedobiologia 2003; 47(3) 231-244.

[40] Harvey JA, Van Nouhuys S, Biere A. Effects of quantitative variation in allelochemicals in Plantago lanceolata on development of a generalist and a specialist herbivore and their endoparasitoids. Journal of Chemical Ecology 2005; 31(2) 287-302.

[41] Harvey JA. Factors affecting the evolution of development strategies in parasitoid wasps the importance of functional constraints and incorporating complexity. Entomologia Experimentalis et Applicata 2005; 117(1) 1-13.

[42] Ode PJ, Plant chemistry and natural enemy fitness effects on herbivore and natural enemy interactions. Annual Review of Entomology 2006; 51 163-185.

[43] Gols R, Wagenaar R, Bukovinszky T, Van Dam NM, Dicke M, Bullock JM, Harvey JA. Genetic variation in defense chemistry in wild cabbages affects herbivores and their endoparasitoids. Ecology 2008; 89(6) 1616-1626.

[44] Morse DH. Size and development times of herbivorous host and parasitoid on distantly related foodplants. The American Midland Naturalist 2011; 166(2) 252-265.

[45] Schuler MA. The role of cytochrome P450 monooxygenases in plant-insect interactions. Plant Physiology 1996; 112(4) 1411.

[46] Ehrlich PR, Raven PH. Butterflies and Plants A Study in Coevolution. Evolution 1964; 18(4)586-608.

[47] Agrawal AA, Kurashige NS. A role for isothiocyanates in plant resistance against the specialist herbivore Pieris rapae. Journal of Chemical Ecology 2003; 29(6) 1403-1415.

[48] Orr DB, Boethel DJ. Influence of plant antibiosis through four trophic levels. Oecologia 1986; 70(2) 242-249.

[49] Harvey JA, Van Dam NM, Gols R. Interactions over four trophic levels foodplant quality affects development of a hyperparasitoid as mediated through a herbivore and its primary parasitoid. Journal of Animal Ecology 2003; 72 (3) 520-531.

[50] Lambert AM. Effects of prey availability, facultative plant feeding, and plant defenses on a generalist insect predator. Arthropod-Plant Interactions 2007; 1(3) 167-173.

[51] Bolton B, Ficken L. Identification guide to the ant genera of the world. Cambridge Harvard University Press; 1994.

[52] Fajer ED, Bowers MD, Bazzaz FA. The effect of nutrients and enriched CO2 environments on production of carbon-based allelochemicals in Plantago a test of the carbon/nutrient balance hypothesis. American Naturalist 1992; 707-723.

[53] Lincoln DE, Fajer ED, Johnson RH. Plant-insect herbivore interactions in elevated CO2 environments. Trends in Ecology & Evolution 1993; 8(2) 64-68.

[54] Lindroth RL, Kinney KK, Platz CL. Responses of diciduous trees to elevated atmospheric CO_2 productivity, phytochemistry, and insect performance. Ecology 1993; 763-777.

[55] Gleadow RM, Foley WJ, Woodrow IE. Enhanced CO_2 alters the relationship between photosynthesis and defence in cyanogenic Eucalyptus cladocalyx F. Muell. Plant, Cell & Environment 1998; 21(1) 12-22.

[56] Stiling P, Moon DC. Quality or quantity the direct and indirect effects of host plants on herbivores and their natural enemies. Oecologia 2005;142(3) 413-420.

[57] Stiling P, Cornelissen T. How does elevated carbon dioxide (CO_2) affect plant–herbivore interactions? A field experiment and meta-analysis of CO_2-mediated changes on plant chemistry and herbivore performance.Global Change Biology 2007; 13(9) 1823-1842.

[58] Stamp NE, Temple MP, Traugott MS, Wilkens RT. Temperature-allelochemical interactive effects on performance of *Manduca sexta* caterpillars. Entomologia Experimentalis et Applicata 1994; 73(3) 199-210.

[59] Veteli TO, Kuokkanen K, Julkunen-Tiittor, Roininen H, Tahvanainen J. Effects of elevated CO_2 and temperature on plant growth and herbivore defensive chemistry. Global Change Biology 2002; 8(12) 1240-1252.

[60] Müller-Schärer H, Schaffner U, Steinger T. Evolution in invasive plants implications for biological control. Trends in Ecology & Evolution 2004; 19(8) 417-422.

[61] Keane RM, Crawley MJ. Exotic plant invasions and the enemy release hypothesis. Trends in Ecology & Evolution 2002; 17(4) 164-170.

[62] Joshi J, Vrieling K. The enemy release and EICA hypothesis revisited incorporating the fundamental difference between specialist and generalist herbivores. Ecology Letters 2005; 8(7) 704-714.

[63] Liu H, Stiling P, Pemberton RW. Does enemy release matter for invasive plants? Evidence from a comparison of insect herbivore damage among invasive, non-invasive and native congeners. Biological Invasions 2007; 9(7) 773-781.

[64] Blossey B, Notzold R. Evolution of increased competitive ability in invasive nonindigenous plants a hypothesis. Journal of Ecology 1995; 887-889.

[65] Doorduin LJ, Vrieling K. A review of the phytochemical support for the shifting defence hypothesis. Phytochemistry Reviews 2011; 10(1) 99-106.

[66] Callaway RM, Ridenour WM. Novel weapons invasive success and the evolution of increased competitive ability. Frontiers in Ecology and the Environment 2004; 2(8) 436-443.

[67] Barto E, Powell KFR, Cipollini D. How novel are the chemical weapons of garlic mustard in North American forest understories? Biological Invasions 2010; 12(10) 3465-3471.

[68] Engelkes T, Morriën E, Verhoeven KJF, Bezemer TM, Biere A, Harvey JA, McIntyre LM, Tamis WLM, van der Putten WH. Successful range-expanding plants experience less above-ground and below-ground enemy impact. Nature 2008; 456(7224) 946-948.

[69] Wolfe LM, Elzinga JA, Biere A. Increased susceptibility to enemies following introduction in the invasive plant Silene latifolia. Ecology Letters 2004; 7(9) 813-820.

[70] Morriën E, Engelkes T, Macel M, Meisner A, Van der Putten WH. Climate change and invasion by intracontinental range-expanding exotic plants the role of biotic interactions. Annals of Botany 2010; mcq064.

[71] Cunze S, Leiblein MC, Tackenberg O. Range expansion of Ambrosia artemisiifolia in Europe is promoted by climate change. ISRN Ecology; 2013.

[72] Pearson RG, Dawson TP. Predicting the impacts of climate change on the distribution of species are bioclimate envelope models useful? Global Ecology and Biogeography 2003; 12(5) 361-371.

[73] Fortuna TM, Eckert S, Harvey JA, Vet LEM, Müller C, Gols R. Variation in plant defences among populations of a range-expanding plant consequences for trophic interactions. New Phytologist; 2014.

[74] Huberty M, Tielbörger K, Harvey JA, Müller C, Macel M. Chemical Defenses (Glucosinolates) of Native and Invasive Populations of the Range Expanding Invasive Plant Rorippa austriaca. Journal of Chemical Ecology 2014; 40(4) 363-370.

[75] Paré PW, Tumlinson JH. Plant volatiles as a defense against insect herbivores. Plant Physiology 1999; 121(2) 325-332.

[76] Dudareva N, Pichersky E, Gershenzon J. Biochemistry of plant volatiles. Plant Physiology 2004; 135(4) 1893-1902.

[77] Baldwin IT. Plant volatiles. Current Biology 2010; 20(9) 392-397.

[78] Dicke M, Baldwin IT. The evolutionary context for herbivore-induced plant volatiles beyond the 'cry for help'. Trends in Plant Science 2010; 15(3) 167-175.

[79] Hare JD. Ecological role of volatiles produced by plants in response to damage by herbivorous insects. Annual Review of Entomology 2011; 56 161-180.

[80] Kessler A, Heil M. The multiple faces of indirect defences and their agents of natural selection. Functional Ecology 2011; 25(2) 348-357.

[81] Heil M. Herbivore-induced plant volatiles targets, perception and unanswered questions. New Phytologist 2014.

[82] Bruce TJA, Wadhams LJ, Woodcock CM. Insect host location a volatile situation. Trends in Plant Science 2005; 10(6) 269-274.

[83] Turlings TCJ, Ted CJ, James H. Tumlinson, and W. J. Lewis. Exploitation of herbivore-induced plant odors by host-seeking parasitic wasps. Science 1990; 250(4985) 1251-1253.

[84] De Moraes CM, Lewis WJ, Pare PW, Alborn HT, Tumlinson JH. Herbivore-infested plants selectively attract parasitoids. Nature 1998; 393(6685) 570-573.

[85] Wajnberg E, Colazza S. Chemical ecology of insect parasitoids. John Wiley & Sons; 2013.

[86] Gols R. Direct and indirect chemical defences against insects in a multitrophic framework. Plant, Cell & Environment 2014.

[87] Fortuna TM, Woelke JB, Hordijk CA, Jansen JJ, van Dam NM, Vet LEM, Harvey JA. A tritrophic approach to the preference–performance hypothesis involving an exotic and a native plant. Biological Invasions 2013; 15 (11)2387-2401.

[88] Yuan JS, Himanen SJ, Holopainen JK, Chen F, Stewart CN Jr. Smelling global climate change mitigation of function for plant volatile organic compounds. Trends in Ecology & Evolution 2009; 24(6) 323-331.

[89] Tauber MJ, Tauber CA. Insect seasonality diapause maintenance, termination, and postdiapause development. Annual Review of Entomology 1976; 21(1) 81-107.

[90] Chapman RF. The insects structure and function. Cambridge university press; 1998.

[91] Saulich AKh, Musolin DL. Diapause in the seasonal cycle of stink bugs (Heteroptera, Pentatomidae) from the Temperate Zone. Entomological Review 2012; 92(1) 1-26.

[92] Sánchez-Guillén RA, Muñoz J, Rodríguez-Tapia G, Arroyo TPF, Córdoba-Aguilar A. Climate-Induced Range Shifts and Possible Hybridisation Consequences in Insects. PloS one 2013; 8(11) e80531.

[93] Groenen F, Meurisse N. Historical distribution of the oak processionary moth Thaumetopoea processionea in Europe suggests recolonization instead of expansion. Agricultural and Forest Entomology 2012; 14(2) 147-155.

[94] Battisti A, Stastny M, Buffo E, Larsson S. A rapid altitudinal range expansion in the pine processionary moth produced by the 2003 climatic anomaly. Global Change Biology 2006; 12(4) 662-671.

[95] Battisti A, Stastny M, Netherer S, Robinet C, Schopf A, Roques A, Larsson S. Expansion of geographic range in the pine processionary moth caused by increased winter temperatures. Ecological Applications 2005; 15(6) 2084-2096.

[96] Lamy M, Novak F. The oak processionary caterpillar (*Thaumetopoea processionea* L.) an urticating caterpillar related to the pine processionary caterpillar (*Thaumetopoea pityocampa* Schiff.)(Lepidoptera, Thaumetopoeidae). Experientia 1987; 43(4) 456-458.

[97] Carroll AL, Taylor SW, Régnière J, Safranyik L. Effect of climate change on range expansion by the mountain pine beetle in British Columbia. In Pages 223-232 in TL Shore et al.(eds) Mountain Pine Beetle Symposium Challenges and Solutions, Oct. 30-31, 2003. Kelowna BC. Natural Resources Canada, Infromation Report BC-X-399, Victoria; 2003.

[98] Desurmont GA, Harvey JA, van Dam NM, Cristescu SM, Schiestl FP, Cozzolino S, Anderson P et al. Alien interference disruption of infochemical networks by invasive insect herbivores. Plant, Cell & Environment 2014.

[99] Furlong MJ, Wright DJ, Dosdall LM. Diamondback moth ecology and management problems, progress, and prospects. Annual Review of Entomology 2013; 58 517-541.

[100] Jepsen JU, Hagen SB, Ims RA, Yoccoz NG. Climate change and outbreaks of the geometrids Operophtera brumata and Epirrita autumnata in subarctic birch forest evidence of a recent outbreak range expansion. Journal of Animal Ecology 2008; 77(2) 257-264.

[101] Fei M, Gols R, Harvey JA. Seasonal phenology of interactions involving short-lived annual plants, a multivoltine herbivore and its endoparasitoid wasp. Journal of Animal Ecology 2014; 83(1) 234-244.

[102] Bezemer TM, Harvey JA, Cronin JT. Response of native insect communities to invasive plants. Annual Review of Entomology 2014; 59 119-141.

[103] Gange AC, Brown VK. Soil food web components affect plant community structure during early succession. Ecological Research 2002; 17(2) 217-227.

[104] Petchey OL. Integrating methods that investigate how complementarity influences ecosystem functioning. Oikos 2003; 101(2) 323-330.

[105] Thebault E, Loreau M. The relationship between biodiversity and ecosystem functioning in food webs. Ecological Research 2006; 21(1) 17-25.

[106] Vitousek PM. Beyond global warming ecology and global change. Ecology 1994; 75(7) 1861-1876.

[107] Gibson CC, Ostrom E, Ahn T-K. The concept of scale and the human dimensions of global change a survey. Ecological Economics 2000; 32(2) 217-239.

[108] Sutherst RW. Global change and human vulnerability to vector-borne diseases. Clinical Microbiology Reviews 2004; 17(1) 136-173.

[109] Halpern BS, Walbridge S, Selkoe KA, Kappel CV, Micheli F, D'Agrosa C, Bruno JF et al. A global map of human impact on marine ecosystems. Science; 319 (5865) 948-952.

[110] Kremen C. Managing ecosystem services what do we need to know about their ecology? Ecology Letters 2005; 8(5) 468-479.

[111] Schweiger O, Settele J, Kudrna O, Klotz S, Kühn I. Climate change can cause spatial mismatch of trophically interacting species. Ecology 2008; 89(12) 3472-3479.

[112] Warren MS, Hill JK, Thomas JA, Asher J, Fox R, Huntley B, Roy DB et al. Rapid responses of British butterflies to opposing forces of climate and habitat change. Nature 2001; 414(6859) 65-69.

[113] Hill MO, Roy DB, Thompson K. Hemeroby, urbanity and ruderality bioindicators of disturbance and human impact. Journal of Applied Ecology 2002; 39(5) 708-720.

[114] Hewitt CD, Broccoli AJ, Mitchell JFB, Stouffer RJ. A coupled model study of the last glacial maximum Was part of the North Atlantic relatively warm? Geophysical Research Letters 2001; 28(8) 1571-1574.

[115] Chakraborty K, Soovoojeet J, Kar TK. Global dynamics and bifurcation in a stage structured prey–predator fishery model with harvesting. Applied Mathematics and Computation 2012; 218(18) 9271-9290.

[116] Kennedy AC, Smith KL. Soil microbial diversity and the sustainability of agricultural soils. Plant and Soil 1995; 170(1) 75-86.

[117] Hu SJ, Van Bruggen AHC, Grünwald NJ. Dynamics of bacterial populations in relation to carbon availability in a residue-amended soil. Applied Soil Ecology 1999; 13(1) 21-30.

[118] Horz H-P, Barbrook A, Field CB, Bohannan BJM. Ammonia-oxidizing bacteria respond to multifactorial global change. Proceedings of the National Academy of Sciences of the United States of America 2004; 101(42) 15136-15141.

[119] Stewart BA, Hatfield JL. Crops residue management. Lewis Publishers; 1994.

[120] Doran JW, Coleman DC, Bezdicek DF, Stewart BA. Microbial indicators of soil quality. 1994.

[121] Di Castri F, Younes T. Ecosystem function of biological diversity. Biological International Special Issue 1990; 22.

[122] Bezemer TM, van Dam NM. Linking aboveground and belowground interactions via induced plant defenses. Trends in Ecology & Evolution 2005; 20(11) 617-624.

[123] De Deyn, Gerlinde B, Van der Putten WH. Linking aboveground and belowground diversity. Trends in Ecology & Evolution 2005; 20(11) 625-633.

[124] Bardgett RD, Freeman C, Ostle NJ. Microbial contributions to climate change through carbon cycle feedbacks. The ISME Journal 2008; 2(8) 805-814.

[125] Zak JC, Willig MR, Moorhead DL, Wildman HG. Functional diversity of microbial communities a quantitative approach. Soil Biology and Biochemistry 1994; 26(9) 1101-1108.

[126] Freeman C, Ostle N, Kang H. An enzymic'latch'on a global carbon store. Nature 2001; 409(6817) 149-149.

[127] Diaz S, Grime JP, Harris J, McPherson E. Evidence of a feedback mechanism limiting plant response to elevated carbon dioxide. Nature 1993; 364(6438) 616-617.

[128] Klironomos J, Rillig M, Allen M, Zak D, Kubiske M, Pregitzer K. Soil fungal-arthropod responses to *Populus tremuloides* grown under enriched atmospheric CO2 under field conditions. Global Change Biology 1997; 3(6) 473-478.

[129] Staddon PL, Gregersen R, Jakobsen I. The response of two Glomus mycorrhizal fungi and a fine endophyte to elevated atmospheric CO_2, soil warming and drought. Global Change Biology 2004; 10(11) 1909-1921.

[130] Högberg P, Read DJ. Towards a more plant physiological perspective on soil ecology. Trends in Ecology & Evolution 2006; 21(10) 548-554.

[131] Ström L, Mastepanov M, Christensen TR. Species-specific effects of vascular plants on carbon turnover and methane emissions from wetlands. Biogeochemistry 2005; 75(1) 65-82.

[132] Abawi GS, Widmer TL. Impact of soil health management practices on soilborne pathogens, nematodes and root diseases of vegetable crops. Applied Soil Ecology 2000; 15(1) 37-47.

[133] Packer A, Clay K. Soil pathogens and spatial patterns of seedling mortality in a temperate tree. Nature 2000; 404(6775) 278-281.

[134] Hartley S, Harris R, Lester PJ. Quantifying uncertainty in the potential distribution of an invasive species climate and the Argentine ant. Ecology Letters 2006; 9(9) 1068-1079.

[135] Percy KE, Legge AH, Krupa SV. Tropospheric ozone a continuing threat to global forests? Developments in Environmental Science 2003; 3 85-118.

[136] Mcelrone AJ, Reid CD, Hoye KA, Hart E, Jackson RB. Elevated CO_2 reduces disease incidence and severity of a red maple fungal pathogen via changes in host physiology and leaf chemistry.Global Change Biology 2005; 11(10) 1828-1836.

[137] Johnson SN, Staley JT, McLeod FA, Hartley SE. Plant-mediated effects of soil invertebrates and summer drought on above-ground multitrophic interactions. Journal of Ecology 2011; 99(1) 57-65.

[138] Wardle DA, Bonner KI, Barker GM. Linkages between plant litter decomposition, litter quality, and vegetation responses to herbivores. Functional Ecology 2002; 16(5) 585-595.

[139] Colautti RI, Ricciardi A, Grigorovich IA, MacIsaac HJ. Is invasion success explained by the enemy release hypothesis? Ecology Letters 2004; 7(8) 721-733.

[140] Hilbig W. The vegetation of Mongolia. SPB Publishing,Amsterdam, The Netherlands; 1995.

[141] van Asch M, Visser ME. Phenology of forest caterpillars and their host trees the importance of synchrony. Annual Review of Entomology 2007; 52 37-55.

[142] Thomas B, Vince-Prue D. Photoperiodism in plants. Academic Press; 1996.

[143] Kirschbaum MUF. Will changes in soil organic carbon act as a positive or negative feedback on global warming? Biogeochemistry 2000; 48(1) 21-51.

[144] Berggren Å, Björkman C, Bylund H, Ayres MP. The distribution and abundance of animal populations in a climate of uncertainty. Oikos 2009; 118(8) 1121-1126.

[145] Norby RJ, Luo Y. Evaluating ecosystem responses to rising atmospheric CO_2 and global warming in a multi-factor world. New Phytologist 2004; 162(2) 281-293.

[146] Kinnison MT, Hairston NG. Eco-evolutionary conservation biology contemporary evolution and the dynamics of persistence. Functional Ecology 2007; 21(3) 444-454.

[147] Berkhout BW, Lloyd MM, Poulin R, Studer A. Variation among genotypes in responses to increasing temperature in a marine parasite evolutionary potential in the face of global warming? International Journal for Parasitology; 2014.

[148] Jump AS, Hunt JM, Penuelas J. Rapid climate change-related growth decline at the southern range edge of Fagus sylvatica. Global Change Biology 2006; 12(11) 2163-2174.

[149] Jump AS, Hunt JM, Martinez-Izquierdoja JA, Penuelas J. Natural selection and climate change temperature-linked spatial and temporal trends in gene frequency in *Fagus sylvatica*. Molecular Ecology 2006; 15(11) 3469-3480.

[150] Pimm SL. The balance of nature? ecological issues in the conservation of species and communities. University of Chicago Press; 1991.

[151] Soler R, Van der Putten WH, Harvey JA, Vet LEM, Dicke M, Bezemer TM. Root herbivore effects on aboveground multitrophic interactions patterns, processes and mechanisms. Journal of Chemical Ecology 2012; 38(6) 755-767.

Chapter 5

Responding to Change — Criteria and Indicators for Managing the Transformation of Vegetated Landscapes to Maintain or Restore Ecosystem Diversity

Graham A. Yapp and Richard Thackway

Additional information is available at the end of the chapter

http://dx.doi.org/10.5772/58960

1. Introduction

Australia is an old, flat and dry continent that has thin and nutrient poor soils. However, as an island continent with a long separation from other land masses, it has distinctive and highly endemic terrestrial biota — for example, about 92% of flowering plants, 85% of terrestrial mammals and 89% of reptiles [1, 2].

European settlement, beginning only towards the end of the 18th century, brought the introduction of pastoralism and cropping to landscapes where the previous human impact had mainly been the use of fire to assist hunting [3]. Various studies with a focus on the loss of species suggest that, since European settlement in the late 18th Century, Australia has had the world's highest rate of extinctions of mammalian fauna and parallel losses of biodiversity across many of its ecosystems [4, 5].

Concern about the loss of biodiversity is not new and not surprising. The major causes are outlined below, but the context, in almost all cases, has been land use change, inappropriate management of fire regimes, and/or ineffective management of threats imposed by introduced, often highly invasive, plants and animals. Also, historically there was a general unwillingness to learn about the structure, composition and function of Australia's ecosystems and to adapt land management practices that had been learned in Europe and imported to Australian ecosystems.

Biodiversity is fundamentally important to ecosystem function [6, 7]. At local scales, biodiversity dependent ecosystem processes provide key ecosystem services — e.g. pollination, the generation of soils and the recycling of nutrients. At wider landscape or regional scales the role of biodiversity in ecosystem function and the provision of ecosystem services such as

regulation of the hydrological regime and control of soil erosion may be less direct but nevertheless important [8, 9].

With growing concern about the impacts of global warming, the capacity of managers to anticipate, recognise and respond to climatic variability will be a key requirement for maintenance of ecosystem function and ecosystem services that depend on biodiversity.

Vegetation is commonly used as the key descriptor of Australian landscapes and the condition of native vegetation is widely used as an indicator of value for biodiversity and the potential for its protection and maintenance. We describe an approach (VAST) by which the condition of vegetation can be effectively described in terms of three diagnostic components — its structure, composition and regenerative capacity [10]. VAST provides a rationale and method for measurement, monitoring and mapping change in 22 indicators of ecosystem structure, composition and function (i.e. key ecological characteristics) over time. Case studies show that purposeful redirection of the way land is managed is supporting the recovery of biodiversity through the transformation of vegetation condition.

2. Land use in Australia

Almost 64% of Australia's land is being used to meet the food and fibre and domestic needs and to provide export income to support a population that has grown, by current estimates, to exceed 23.5 million [11, 12]. Of the total land area, only a small proportion — less than 4% — is under intensive agricultural and urban use where native vegetation has, in general, been completely removed. About 2% is used for production forestry including plantations. Around 57% of the land area has been modified for agriculture and for pastoral use, with the bulk of this area used for livestock grazing on natural or modified pastures [see Table 1]. In areas that have been cleared for agriculture (primarily cereal cropping) and for grazing on improved pastures, remnant areas of native vegetation provide a much-reduced but very important reservoir for biodiversity. However, grazing has had considerable impact on biodiversity even where it uses natural vegetation in semi-arid and arid landscapes.

On the other side of the ledger, nearly 37% of Australia's land is in national parks, nature reserves and other protected areas (see Table 1). These areas contribute to maintenance of biodiversity as the major component of Australia's National Reserve System (see below). Extensive areas where use is limited or negligible also contribute to biodiversity conservation objectives although, despite being little affected by (or in advanced recovery from) clearing, the impact of fire and feral animals can be significant in these areas.

Changes in land use and land management in Australia are tracked and mapped by cooperative national programs [1, 2, 15, 16]. Related arrangements are also in place for describing and mapping change in vegetative cover [17].

LAND USE	AREA (sq. km)	Proportion (%)
Dryland cropping and horticulture	256 616	3.33
Irrigated cropping and horticulture and intensive animal and plant production	20 146	0.21
Grazing modified and irrigated pastures	730 193	10.50
Grazing native vegetation	3 558 785	46.30
Production and plantation forestry	138 243	1.80
Nature conservation	569 240	7.41
Other protected areas, including Indigenous uses	1 015 359	13.21
Areas of minimal use	1 242 715	16.17
Other (urban, rural residential, waste and mining, water and unknown)	155 850	2.02
Total	7 687 147	100.00

Source: Based on ABARE–BRS [13, 14]

Table 1. Summary of land use in Australia

3. Major impacts on Australian biodiversity

Schedules to Australia's Environmental Protection and Biodiversity Conservation Act (EPBC) [18] list 1298 species of "Threatened Flora" (of which 39 are recorded as "extinct", 139 as critically endangered, 528 as endangered, and 592 as vulnerable [19], and 451 species of threatened fauna of which 54 are recorded as extinct, 1 as extinct in the wild, 54 as critically endangered, 142 as endangered, 193 as vulnerable and 6 as "conservation dependent" [20]. The EPBC also lists 70 Threatened Ecological Communities as vulnerable, endangered or critically endangered [21]. (Note: this is the current number in the EPBC schedule with the most recent entry dated 17 July 2014. The State of the Forests report [1, p94] states "at December 2012, the EPBC Act listed 58 threatened ecological communities, of which 27 are forest communities or contain significant proportions of forest ". For this chapter, we accept the EPBC schedule).

The EPBC Act recognises 21 specific threatening processes. Clearing and replacement of vegetation, along with the introduction of new animal species (including both herbivores and predators), the spread of invasive plant and animal species, changes in fire regimes, manipulation of the distribution and quantity of water resources in both time and space, and changes to the physical, chemical and biological state of the continent's generally ancient and impoverished soils, have all contributed to rapid and extensive loss of biodiversity, including a high rate of species extinction [2, 22]. Widely accepted assessments of the course of climate change point to a further increase in pressures on the biodiversity and ecosystem services of Australian environments [3, 23-25].

3.1. Vegetation clearance

The most dramatic and widespread impact on ecosystem structure and ecological function has been land clearance – removal and/or conversion of vegetative type and cover. Australia's 2013 State of the Forests report [1] notes that clearing or modification for urban settlements and intensive agriculture has affected as much as one third of native vegetation, with some communities reduced to as little as 1% of their original areas and others highly fragmented. Resultant landscapes are a diverse mosaic of fragmented and modified native vegetation and converted and replaced land cover [26, 27]. The loss of habitat is reflected in long lists of threatened plant and animal species.

3.2. Fire in Australian landscapes

Fire is a major driver, responsible for the composition of a great many Australian ecosystems, including most of its forests [28, 29]. Even before European settlement, many Australian ecosystems were fire-generated and fire-dependant [28]. As settlement extended further inland, fire was widely used as a tool for land clearance.

Fire is an important tool for management and can be used in the service of biodiversity objectives but uncontrolled wildfires and inappropriate fire regimes continue to pose an ongoing threat to maintenance of biodiversity. Problems range from fire regimes that are too infrequent and allow the build-up of excessive fuel loads (often contributing to highly damaging wildfires) to burning regimes that are too frequent to allow regeneration, re-establishment and regrowth of affected communities [30, 31]. Suppression and, in many cases, the exclusion of fire has been equally as problematic for ecosystem function as have wildfire or too frequent fire regimes.

Appropriate fire regimes will be increasingly important for management of the expected effects of climate change on fuel loads, temperature, humidity and other drivers. An increase in the frequency, intensity and area affected by fire and a reduction in the period available for safe control burning to manage fuel loads can be expected [1, 24, 25].

3.3. Climate change

There is evidence that anthropogenic climate change is already impacting ecosystem processes and biodiversity in Australia [2, 4, 32, 33]. Continuing, and probably accelerating impacts, will require adaptive management to counter increasing risks of pests and diseases, altered fire regimes and other expected variations of temperature and rainfall and extreme events [5, 34, 35].

3.4. Biogeographic regionalisation based on vegetation

Australia's landscapes have been classified in large homogeneous and distinctive bioregions based on common climate, geology, landform, soil, native vegetation and species information [36]. Currently, there are 89 geographically distinct bioregions which have been further subdivided and mapped into 419 more localized subregions with greater geomorphological homogeneity — see Australia's Bioregions (IBRA) [37].

Large areas of Australia, readily identifiable as distinctive biogeographic regions and subregions, are dominated by two genera – *Eucalyptus* and *Acacia*. The first, *Eucalyptus*, is almost unique to Australia-only seven of the more than 800 species occur (naturally) beyond Australia and only one of these does not occur within it (38, 39). Native forest and woodland forests dominated by eucalypts comprise 92 million hectares; 75% of the native forest area [1, 40].

The second widely dominant genus is *Acacia*. Though the genus is widespread elsewhere, most of the phyllodinous acacias are Australian. As with the characteristic sclerophylly of eucalypts, the phyllodinous adaptation of acacias is recognised as a response to characteristic edaphic and climatic conditions of the Australian land mass complicated by the role of fire and their adaptation to it. Acacias are dominant in almost 10 million hectares — 8% of the forest area [1].

4. The condition of native plant communities (i.e. vegetation condition) as an indicator of ecosystem function

Vegetation has long been used to characterise Australian landscapes. Key roles of vegetation in maintaining ecosystem function in Australia are listed in Table 2. The Vegetation Assets, States and Transitions (VAST) system, as described below, builds on the observation that vegetation is a major driving force in the dynamics of terrestrial ecosystems such that, in Australia, it is often used as a proxy to classify ecosystem type and function.

In the 57% of Australia's land that has been modified to support agriculture and grazing, there is a mix of intact, disturbed and replaced cover types with identifiable vegetation composition, structure and condition classes [17]. This 'mix' largely determines ecosystem function and the ecosystem services that a landscape will support [42]. It also underpins the ability to maintain and enhance, or restore, biodiversity — the 'regenerative capacity' of ecosystems. Accordingly, the VAST system recognises that actions to arrest and/or reverse loss of biodiversity must be based on an understanding of the present condition of vegetation, how that condition was reached and what management changes will be required in order to change the structure, composition and function of the vegetation.

While restoring vegetation structure, species composition and regenerative capacity is an effective step towards improving ecosystem function, it may not always be appropriate to aim for overall recovery of biodiversity. The responses of individual species or species assemblages may change as ecosystem function changes in response to management actions or climate change or other events such as wildfire.

5. Australia's National Reserve System (NRS)

Australia's Commonwealth (i.e. national), state and territory governments have jointly agreed on a strategy for establishment and management of a national reserve system – a "network of public, indigenous and private protected areas over land and inland freshwater" [43]. The objectives for the system are long-term protection of selected ecosystems and the biological diversity they support. The strategy implies a whole-of-landscape approach in recognition that, faced with climate change, the complete, adequate and representative objectives for biodiversity conservation require a "full range of conservation measures applied to other lands across the landscape" [43, p3]

The national system of reserves includes areas representative of all 89 bioregions. Some 54 (61%) of these have greater than 10% areal (but not necessarily ecosystem) representation. Others — mainly those extensively cleared for cultivation or improved pastures — have very small proportions of their areas in the reserve system. Representation is lower at the next level of the classification where 44 of the 419 subregions (10.5%) are unrepresented and only 183 (43%) have more than 10% of their area included in the reserve system.

Currently, the National Reserve System (NRS) covers about 13% of Australia's land area (approximately 100 million hectares) but it has, essentially, been established from what remains of lands with relatively intact vegetation or areas with special biological, historical and/or recreational values. This is particularly the case in over-cleared landscapes that have been converted to intensive agriculture.

In this chapter, we are primarily (but not exclusively) concerned with lands outside the reserve system – i.e. ecosystems or landscapes that are primarily used for agriculture and/or grazing on improved pastures. These show varying degrees of modification. The course of these changes to their original composition, structure and function and the pathway for recovery of these landscapes – i.e. transformation back towards their original state – can be tracked, quantified and displayed by the VAST system here outlined and demonstrated by two case studies. Because the area outside the NRS is very large, reverse transformation of modified landscapes will significantly complement the biodiversity objectives being maintained in reserves. One of the major issues facing public and private land management agencies is how to assess and report the quantum that these landscapes contribute to the national biodiversity account and how their contribution is changing over time. VAST provides information that is relevant to this need. Further, it adds to the available approaches for monitoring responses to climate change and habitat protection at a landscape scale [34].

6. Approaches to biodiversity objectives — Aiming for maintenance or recovery

Two approaches are common – one based on land use and the other on land management.

1.   As outlined above, Australia has a significant proportion of its land in a system of reserves that are (meant to be) providing for the maintenance of biodiversity. More land (and ecological communities) are given a level of protection under various forms of agreements and these add to the total area of in which maintenance of biodiversity is being supported by control over threatening processes. The total area under various forms of protective designation or agreement is about 118 750 sq km – close to 15.5% of Australia's land surface

(excluding Antarctica) [1, 13, 44]. The caveat is that it has not been possible – and never will be possible – to completely control threatening processes such as fire and introduced weeds and animals. Note: similar approaches are being used with a view to maintenance of marine ecosystems and biodiversity, but we are not concerned with marine area protection here. Privately owned and managed reserves comprise only 1% of Australia (795 sq km) [45].

2. Outside urban areas and the National Reserve System, the majority of Australia's land is managed by farmers and graziers. Most of this is extensive cropping and pastoralism — largely due to impoverished soils and the vagaries of climate — but intensification is increasing. Many landowners are well aware of and sympathetic to biodiversity concerns and a number of individual and community initiatives, supported by government programs, are in operation across rural lands. This approach is the main focus of the VAST approach and this chapter.

This effectively means that, for large areas the role of land managers (and citizen scientists and community groups) is critically important to biodiversity maintenance. At local level, much has been done to control invasive weeds and plant trees and shrubs to control land degradation and provide habitat for native species – especially through movements such as LANDCARE. However, a comprehensive study by the National Land and Water Resources Audit [35] included a sub-regional assessment of constraints to community capacity for adaptive management to integrate biodiversity conservation into natural resource management. It concluded that capacity was severely constrained in 14% of Australia's 419 sub-regions and significantly constrained in another 33%. It found that conservation is 'well integrated into production systems' in only 1.5% of subregions [35;p130].

Here we describe the use of a Vegetation Assets, States and Transitions (VAST) concept that is proving an effective way to support adaptive management at site, property and local levels. VAST addresses the need for managers to:

1. obtain information on vegetation condition by assessing the effects that land management practices have had on compositional, structural and functional characteristics; and

2. track and report how these characteristics are responding to adaptive changes in their management practices and, where appropriate, relate these responses to indications of climate change.

7. The VAST transformation approach to assessing change and trend

VAST is a standardised nationally consistent system for monitoring, evaluating and reporting the effects of changes in land use and management and assessing their effects on ecosystem function and the delivery of ecosystem services. It uses key functional, structural and composition characteristics of a site and its vegetation — fire regime, soil structure, hydrology, nutrients and biology, over and understorey vegetation structure and over and understorey species composition. These ecological criteria are hierarchically integrated with 22 indicators that are linked to diagnostic attributes of vegetation condition [17]. A wide range of spatial, temporal and thematic data and information (e.g. environmental histories, relevant time series aerial photography and remote sensing and site-based monitoring) can be used to populate the indicators. The framework is shown in Table 2.

Key functional, structural and composition criteria	Key Indicators	VAST diagnostic components
Fire regime	1. Area /size of fire foot prints	
	2. Interval between fire starts	
Soil hydrology	3. Plant available water holding capacity	
	4. Ground water dynamics	
Soil physical state	5. Effective rooting depth of the soil profile	
	6. Bulk density of the soil through changes to soil structure or soil removal	
Soil nutrient state	7. Nutrient stress – rundown (deficiency) relative to reference soil fertility	Regenerative capacity
	8. Nutrient stress – excess (toxicity) relative to reference soil fertility	
Soil biological state	9. Organisms responsible for maintaining soil porosity and nutrient recycling	
	10. Surface organic matter, soil crusts	
Reproductive potential	11. Reproductive potential of overstorey structuring species	
	12. Reproductive potential of understorey structuring species	
Overstorey structure	13. Overstorey top height (mean) of the plant community	
	14. Overstorey foliage projective cover (mean) of the plant community	
	15. Overstorey structural diversity (i.e. a diversity of age classes) of the stand	Vegetation structure
Understorey structure	16. Understorey top height (mean) of the plant community	
	17. Understorey ground cover (mean) of the plant community	
	18. Understorey structural diversity (i.e. a diversity of age classes) of the plant	
Overstorey composition	19. Densities of overstorey species functional groups	
	22. Richness – the number of indigenous overstorey species relative to the number of exotic species	Species Composition
Understorey composition	21. Densities of understorey species functional groups	
	22. Richness – the number of indigenous understorey species relative to the number of exotic species	

Source: Thackway, 2013 [46]

Table 2. VAST ecological criteria, performance indicators and diagnostic components

For practical purposes, the VAST approach uses information on the condition of individual vegetation sites and patches on the premise that the condition of vegetation can be linked to the structure (height, cover, growth form, strata), regenerative capacity (resilience), and composition (species diversity) of vegetation. These are the three components that are used in VAST as indicators or surrogates to determine the response of the plant community and to derive scores for the diagnostic components used to define change in condition and its trend.

Vegetation condition is described as the difference between observed or measured values for diagnostic attributes or criteria relative to a benchmark or 'reference state'. The identification of reference states is a fundamental underpinning of the VAST concept (for details see reference [47]). In Australia the 'original' natural vegetation is often used as a benchmark because of its assumed relationship to biodiversity. Accordingly, residual patches of native vegetation are generally selected as reference sites. In effect, a reference site is an area that is representative of low impact uses and management of the plant community such that the native vegetation has persisted in a largely unmodified state since first European settlement [47].

VAST uses these plant community reference states and representative sites to enable consistent tracking and reporting of change and trend in key characteristics, performance indicators, and diagnostic components. The observed 'distance' of these diagnostic components from any 'fully natural' benchmark provides relative scores that can be summed to assign condition classes.

Table 3 shows the VAST vegetation condition classes. VAST uses a metric approach to classify and map vegetation 'condition states' that are determined by the effects that land use and/or land management practices have on compositional, structural and functional characteristics of the site. The extent to which these characteristics remain intact or, if changed, have been modified, replaced or altogether removed defines the six VAST vegetation classes.

	Vegetation condition classes		Characteristics of the vegetation
	0	Naturally bare	Areas where native vegetation does not naturally persist, and recently naturally disturbed areas where native vegetation has been entirely removed (i.e. subject to primary succession).
Increasing Vegetation Modification	I	Residual	Native vegetation community structure and composition, with regenerative capacity intact – no significant perturbation from land use/land management practices.
	II	Modified	Native vegetation community structure, composition and regenerative capacity more or less intact, perturbed by land use/ land management practices such as intermittent low intensity grazing.
	III	Transformed	Native vegetation partly removed but community structure, composition and regenerative capacity has been significantly altered by land use / land management practices.
	IV	Largely replaced and degraded	Native vegetation largely replaced by invasive native and/or exotic plant species (commonly areas abandoned or burnt).

V	Replaced and managed for intensive production	Native vegetation completely removed and replaced with intensive agriculture: rain-fed broad acre crops, feed lots, horticulture, irrigation agriculture and long or short rotation forestry. Various types are recognised in the vegetation classification.
VI	Replaced with man-made structures	Settlements and cultural features – e.g. buildings, roads, water reservoirs; gardens, parks and amenity plantings.

Table 3. Vegetation characteristics of VAST condition classes

8. Theoretical underpinning of VAST-a model of temporal change

A land manager who wants to change a landscape will need to understand how managing the 10 VAST criteria can restore a site (or landscape) towards a reference state. This applies whether specifically to recover some species or improve species diversity in general, or to more broadly improve the production of ecosystem services (e.g. the generation of food and fibre and regulation of the hydrological regime while providing habitats and food sources to support increased biodiversity).

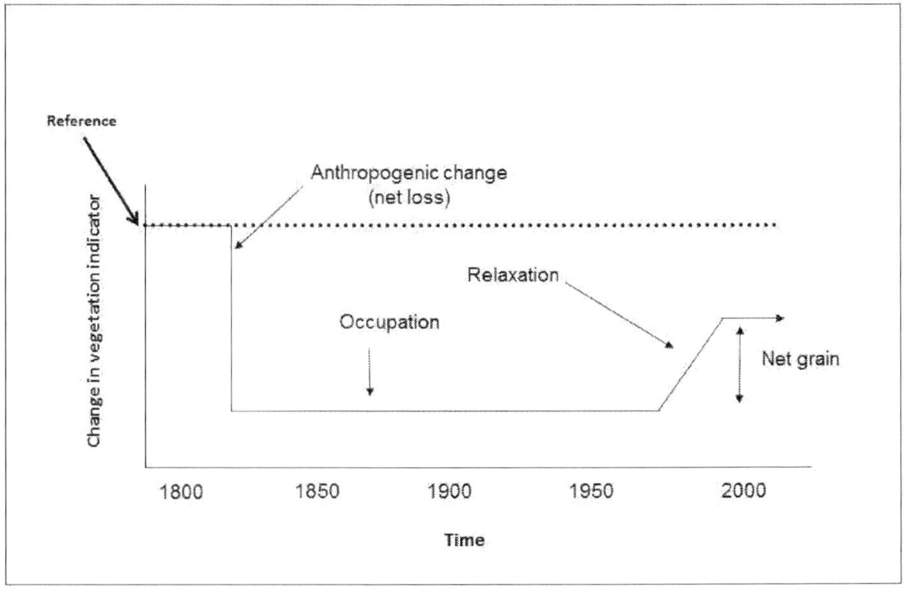

Figure 1. Transformation, occupation and relaxation of a given area of land (based on Hamilton et al. 2008 [48]).

Figure 1 presents the theory underpinning the VAST approach to understanding temporal changes in vegetation indicators in response to changes in land use and management. While a site is being managed using a suite of land management practices it is 'occupied' and there is a net loss in some (or all) of the key indicators relative to the reference state. Where a land manager chooses to relax previous land management practices, allowing reversion of their effect so that the site can recover or be significantly restored, there is a net gain in indicator(s) relative to the reference state.

9. A report card on change — Tracking impacts of land use and management

The VAST system can provide a simple graphical report card showing change and trend relative to a reference state (i.e. natural benchmark). The manager and/or researcher compiles information on how ecosystems within a (selected) landscape have reached their current state (i.e. what happened and what is happening now). This information can be used to model how vegetation — and therefore ecosystems — might be restored within the scope of the objectives of the manager(s) of a landscape. This provides a baseline and rationale for monitoring changes in the vegetation structure, species composition and regenerative capacity of the site following the changes made to land management practices. The results can be used to inform stake-holders of progress towards agreed targets for ecosystem structure and function and/or biodiversity.

One of the most important premises of the VAST system is that, while it emphasises vegetation structure and composition, it essentially operates on and achieves its results through improved understanding and management of the key functional, structural and composition compo-nents at a site relative to the reference state. Within the limits imposed by regional climate and microclimates, the recovery of native vegetation communities largely depends on partially or fully reinstating the soil structure, hydrology, nutrients and biology – often the very criteria most affected by land clearance, cultivation and the introduction of hard-hoofed animals. Fire regime and reproductive potential are also critically important criteria particularly in most grassland and woodland ecosystems many of which, along with some forest ecosystems, are managed for pastoral production.

Figure 2 illustrates these concepts using a Poplar Box (*Eucalyptus populnea*) plant community. At time one (t1) the site is minimally managed. At t1 the site is classified as in an unmodified state (VAST class I). At time two (t2), land management practices have been used to change the vegetation structure and composition characteristics, although the functional characteris-tics of the site are largely unchanged. The condition at t2 is classified as a modified state (VAST class II). At time three (t3) land management practices are intensified to further change the vegetation structure and composition characteristics, although the functional characteristics of the site still are largely unchanged. The condition at t3 is classified as a transformed state (VAST class III). Figure 2 shows the direct relationship between the VAST condition classes of the site and the delivery of ecosystems services [41].

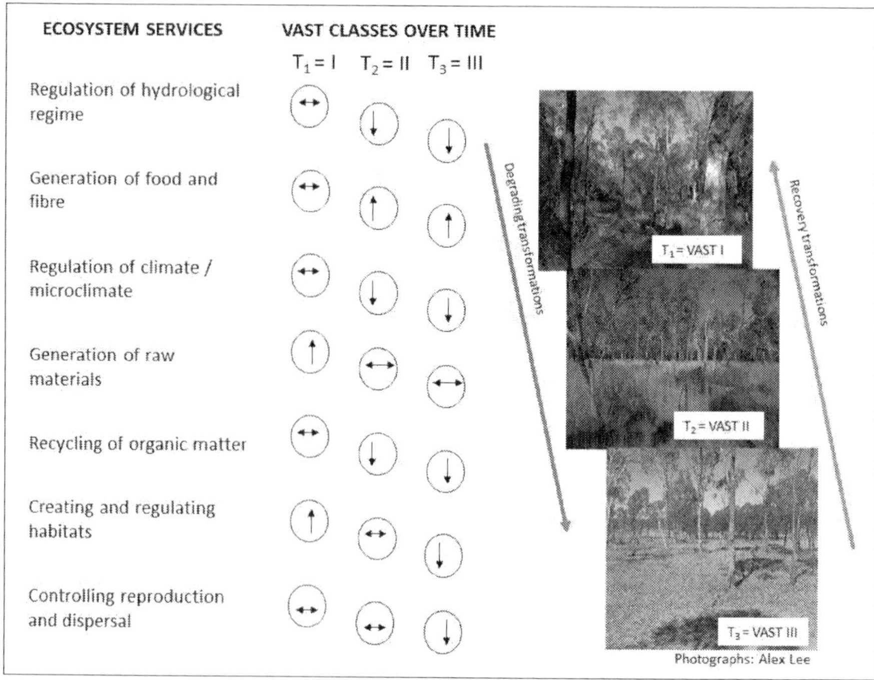

Figure 2. Trajectory of VAST classes and the effect on ecosystem services

10. Application of VAST to the problem of fragmentation

Given the importance attached to the integrity and extent of native vegetation, its fragmentation at local and regional scales poses a number of threats to maintenance of biodiversity. Apart from lack of connectivity for passage between refuges and food sources, there is increasing concern about the risk of declining genetic diversity (inbreeding), especially in the context of climate change [1, 24]. In the VAST system, this is largely encapsulated in indicators of reproductive potential and in overstorey and understorey structure and composition (see Table 2).

McIntyre and Hobbs [22] proposed a framework for conceptualizing the effects of landscape fragmentation and increasing degrees of site modification and understanding their relevance to management. The framework is widely known and understood and assists conservation biologists and natural resource managers to address the full spectrum of human impacts observed across agricultural and fragmented landscapes.

Muetendeudzi and Thackway [49] have adapted this concept to recognise and map 'landscape alteration levels' as an emergent property of vegetation condition at the site level. Figure 3 shows a schematic model characterising landscape alteration levels in terms of the relationship between increasing fragmentation and increasing degrees of modification defined in terms of VAST classes. An appropriate vegetation condition dataset is required for generation of maps of landscape alteration levels. For this, condition classes must be described as homogeneous map units at the site level — i.e. as either mapped polygons or raster based maps [17]. VAST provides a suitable input dataset.

VAST spatial data for fragmentation and degrees of site modification represented are used to classify landscapes into four categories — intact, variegated, fragmented, and relictual — to generate national and regional maps and statistics.. This information can provide a valuable basis for setting conservation priorities in natural management programs [50].

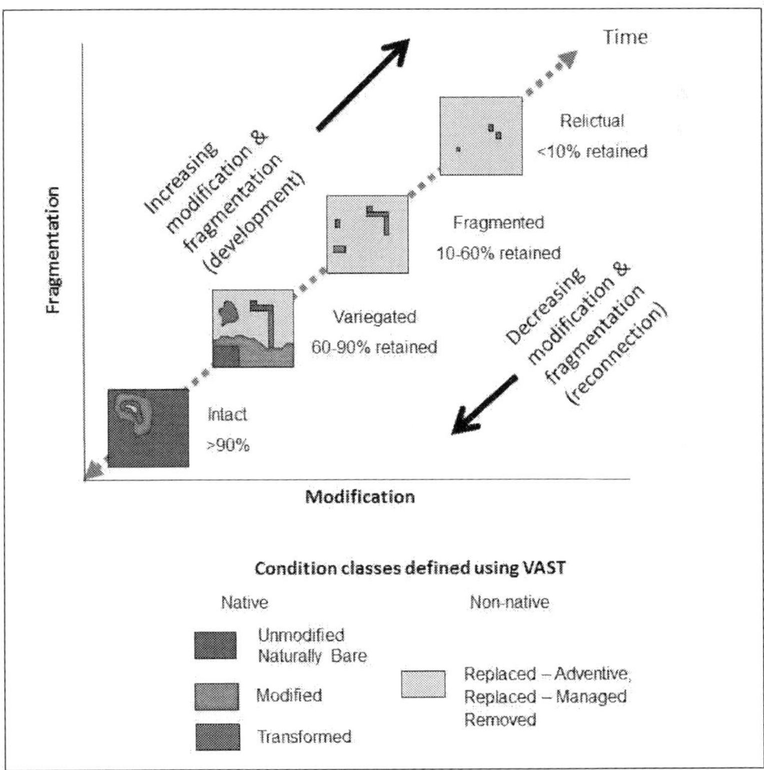

Figure 3. A framework for characterising landscape alteration levels based on increasing fragmentation and increasing degrees of modification defined in terms of VAST classes.

11. Application of the VAST approach to site transformation — Two case studies

Implementation of the approach is shown below in two considerably different case studies — one in an area originally sub-tropical rainforest, the other an area of *Eucalyptus* grassy woodland in a cool temperate climatic environment. The location of the case study sites is shown in Figure 4.

11.1. Rocky Creek Dam, Big Scrub, New South Wales

The Rocky Creek Dam, Big Scrub, New South Wales case study is an area of 25 hectares of hilly terrain located 28°38'8.54"S, 153°20'32.58"E – see Figure 4. This 'transformation site' was originally Lowland Subtropical Rainforest on basalt-derived soils (see reference [47] for full definition of 'transformation site'). The site is located in the South Eastern Queensland bioregion and the Scenic Rim sub-bioregion. In 1910 the transformation site was completely cleared of the native rainforest and converted to agriculture. It was cropped in 1910 and then from 1911-49 it was managed as a continuously grazed pasture supporting dairy cattle.

The transformation site is located near a large remnant area of Lowland Subtropical Rainforest. In 1990, the land manager, Rous Water, defined the reference state for the Rocky Creek Dam transformation site as the nearby Rocky Creek Flora Reserve (172 ha), now part of Nightcap Range National Park [51].

The VAST method (criteria and indicators) was used to assess the link between environmental response and land management activity and change over time. Information on land management history and its progressive effects on key ecological criteria and indicators was derived from published and unpublished sources including interviews with the land manager and discussions with ecologists familiar with the site.

Assessment of the indicators was completed as part of field work in 2010 [52].

11.1.1. Phase 1: 1750-1910

This period coincided with cessation of management by the Goori indigenous people, the area was opened-up for selection as private land, the 'cedar getters' selecting, cutting and removing large trees. During this period the area remained as VAST I (unmodified 80-100%).

11.1.2. Phase 2: 1911-1949

Commencing in 1911 the site was transformed from VAST I unmodified 80-100%, through VAST II modified 60-80%; to VAST III transformed 40-60%.

This phase was characterised by the conversion of the rainforest to pasture and its continuing use for dairying. This included pasture improvement and continuous or set stocking. The aim of land management during this period was to maintain and improve milk production by grazing dairy cows on improved exotic pastures. Regeneration of the rainforest was actively controlled.

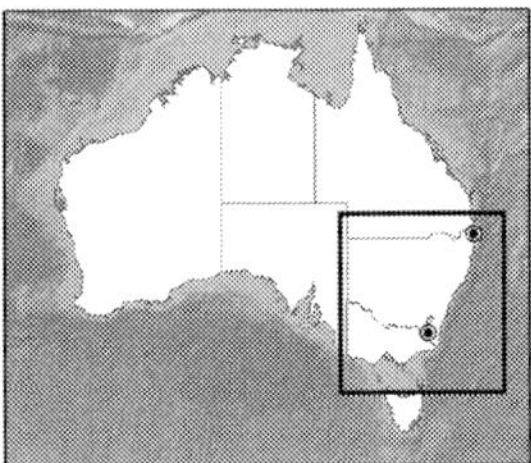

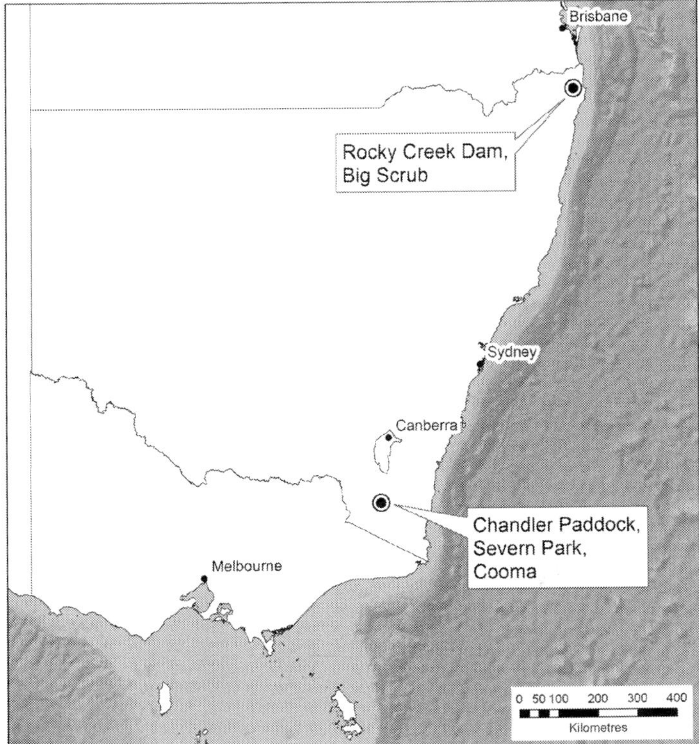

Figure 4. Location of the Case Study areas

During this period the vegetation condition scores were dramatically reduced from 100% to 43%. Initially, with the conversion of the rainforest vegetation, condition scores were reduced from unmodified (80-100%) to transformed (56%) with the major change due to the loss of vegetation structure and species composition. Subsequently ecological function was further diminished due to changes in soil hydrology, soil nutrients, soil biology and reproductive potential of the plant community. At the end of this phase the score was 43%.

11.1.3. Phase 3: 1950-1982

Commencing in 1950 the site was classified as VAST III (transformed 40-60%).

This period was characterised by the transfer of ownership of the land from private to public. Several private properties, including this site, were purchased by Rous Water as part of the Rocky Creek water catchment area within which the Rocky Creek Dam was to be built. The cessation of dairying and the abandonment of pasture management, resulted in a major incursion of lantana (*Lantana montevidensis*). The Dam was built in 1950-52.

During this period birds and flying foxes (*Pteropus spp.*) which had spent time in the adjacent large remnant dropped their faeces into the lantana. The lantana was also ameliorating the soil conditions that had been degraded during the earlier period of dairying.

The land manager commenced experiments on small test plots using assisted regeneration of the rainforest by controlling the regeneration of lantana. This showed that the rainforest could be readily re-established.

At the end of Phase 3 in 1990 the site as a whole was still classified as VAST III (transformed 40-60%).

11.1.4. Phase 4: 1983-2010

This period was characterised by large-scale assisted regeneration which commenced in 1993 and was largely completed in 1999. The process used was to systematically convert small patches covered entirely with lantana, within the site, back towards the original Lowland Subtropical Rainforest composition, structure and condition. The main technique was to slash the lantana to create ground cover mulch and then, as germination of the lantana was observed, use weed killer to halt its re-establishment. The earlier experiments had shown that the rainforest plants geminated later than lantana, thus enabling this system of assisted regeneration to be highly successful.

In 1991 the land manager, Rous Water, commenced detailed monitoring and reporting of the key attributes of species composition and vegetation structure associated with the assisted regeneration.

Since 2001, the site of restored Lowland Subtropical Rainforest, adjacent to the Rocky Creek Dam reservoir, has been managed continuously as open public space.

At the end of Phase 4 in 2010 the site was classified as VAST I (unmodified) with a score of (about) 90%.

Figure 5 shows the change and trend in the transformation of the Rocky Creek Dam transformation site. The symbols depicted in the graph are based on qualitative observations and quantitative measurements (x1, x2, x3) [53, 54].

The figure shows total vegetation status score calibrated to VAST classes:

80-100% of the reference state corresponds to VAST class I– Residual /Unmodified;

60-80% corresponds to VAST class II– Modified;

40-60% corresponds to VAST class III– Transformed;

Scores from 0-39% corresponding to VAST class IV– Replaced and adventive; VAST class V– Replaced and managed; and VAST class VI Replaced and removed have not been differentiated.

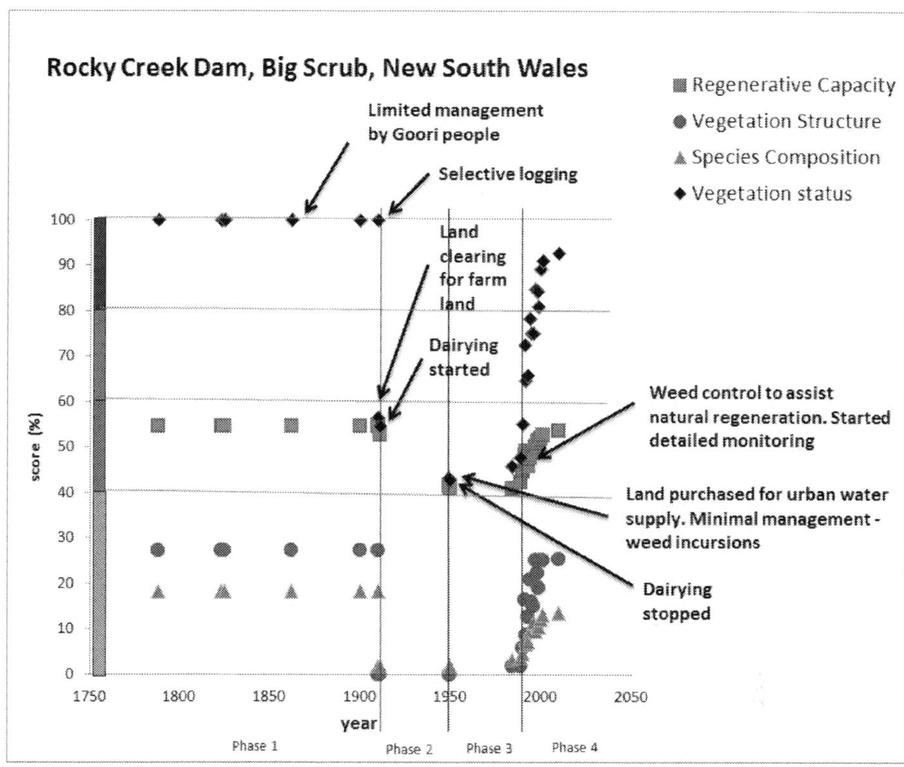

Figure 5. Phases in the transformation of Rocky Creek Dam site.

11.2. Chandler Paddock, Monaro Region, New South Wales, Case study

11.2.1. Introduction

Chandler paddock, Severn Park, Cooma, New South Wales is an area of 107 hectares comprising rolling and hilly terrain; located-36°26'23.4600"S, 148°55'42.0000"E – see Figure 4. The site is located in the South Eastern Highlands bioregion and the Monaro sub-bioregion of the Interim Bioregions of Australia classification [36].

This 'transformation site' was originally *Eucalyptus pauciflora* and *E. rubida* woodland over scattered shrubs, over a *Themeda* and *Poa* grassland with mixed forbs. The soils are derived from granite. In 2013, the land manager, defined the reference state for Chandler paddock transformation site as an extant patch of *Eucalyptus pauciflora* woodland in Carter paddock, also located on Severn Park (Charles Massy pers comm).

The VAST method (criteria and indicators) was used to assess the link between environmental response and land management activity and change over time. Information on land management history and their effects on key ecological indicators and attributes were derived from published and unpublished sources including interviews with the land manager and discussions with ecologists familiar with the site. Assessment of the VAST indicators was completed as part of fieldwork in 2013.

11.2.2. History

Between the 1870s-90s almost all the trees on the upper, mid and lower slopes were ringbarked and killed to promote pasture production. Ridges were not cleared.

11.2.3. Phase 1: 1750-1855

This period coincided with cessation of management by the Ngarigo indigenous people in the mid 1830s. By the end of this period the site was classified as VAST I with a score of 100%.

From 1830, the area was opened-up as grazing land with shepherds continuously moving large flocks of sheep on unfenced natural pastures.

11.2.4. Phase 2: 1855-1875

Grazing on unfenced pastures continued into 1870. Permanent wire fences were erected after 1871 and set or continuous stocking has continued virtually unbroken until July 2014.

Commencing in 1855 almost all trees on the site were killed using technique called ringbarking to promote pasture production. Regeneration of the Eucalyptus trees and understory shrubs was actively controlled. The aim of land management during this period was to improve wool production by grazing merino sheep on native pastures.

By the end of this period the site was classified as VAST I with a score of 95%.

11.2.5. Phase 3: 1876-1929

Commencing in 1876 set or continuous grazing was enabled through the erection of wire fences. Over the next 50 years the intensive stocking regime and the naturally relatively low levels of natural productivity resulted in the relatively rapid diminution of key indicators associated with vegetation structure, species composition and ecological function. During this period the site was transformed from VAST I (unmodified 80-100%), through VAST II (modified 60-80%) to VAST III (transformed 40-60%).

This period was characterised by high stocking rates and active control of regenerating *Eucalyptus* trees and understory shrubs. Wildfire was actively excluded. In 1900-02 rabbits also arrived in the district, adding pressure to already overgrazed pastures coinciding with the end of the Federation Drought in 1902. The result was sheet and rill erosion and loss of the top soil, the development of gully erosion and a change in the dominance of pasture grasses from more to less palatable.

By the end of this period the site was classified as VAST III with a score of 50%.

11.2.6. Phase 4: 1929-1990

A new land manager assumed management of the land on its purchase in 1929. A high density of rabbits infestation was brought under control by 1934. In the 1930s, with the declining productivity of the pasture, the land manager gained access to off-site mountain grazing in summer in the nearby Snowy Mountains. Access to off-site grazing rested the pastures and gave relief to the sheep which were stressed from the damaging spear grass (*Aristida spp*). In 1965 access to summer grazing ceased with the establishment of Kosciuszko National Park. With access to off-site grazing during summer no longer available, the land manager commenced applications of 142 kg per hectare between 1980 and 2008 to improve the productivity of the pasture. Applications of super-phosphate continued until 2008.

In 1980 an attempt to continue to further increase pasture productivity Chandler paddock was ploughed and sown to a mix of exotic pasture species. By the end of this period the site was classified as VAST IV with a score of 41%.

The previous phases of increased pasture production led the land manager to establish a stud merino flock grazing on improved pastures in 1985.

11.2.7. Phase 5: 1991-1999

During this period the land manager observed that the ecosystem had little capacity to recover quickly from drought, had little natural resilience, and that species composition and vegetation structure of the pasture was dependant on high levels of super-phosphate.

Figure 6 shows vegetation structure and species composition scores were the lowest for any of the phases and, by the end of this period the site was classified as VAST IV with a score of 39%.

11.2.8. Phase 6: 2000-2014

In 2000 the land manager ceased operating the merino stud and began managing the pasture using principles of in this paddock with flock sheep.

This period is characterised by the cessation of applications of super-phosphate and use of time-based cell grazing to re-establish the native pasture. This was observed to result in a higher number and greater diversity of native grasses and herbs, and increased amounts of ground cover during all seasons of the year, along with higher levels of soil moisture and increasing levels of biological activity in the soil. In July 2014 the paddock was sub-divided into 7 sub-paddocks so as to fully implement holistic grazing management.

By the end of this period the site was classified as VAST IV with a score of 47%. The aim of the land manager now is to rehabilitate and to maintain the site within the range of scores 60-80% i.e. VAST II.

Figure 6 shows the change and trend in the Chandler paddock, Cooma transformation site. The symbols depicted in the graph are based on qualitative observations obtained using expert elicitation by one of the authors interviewing Charles Massey pers comm.

It is important to note that, since the commencement of the assisted regeneration in 1991, there was a small scale treatment to investigate the feasibility of changing the land management objectives to improve the biodiversity objectives.

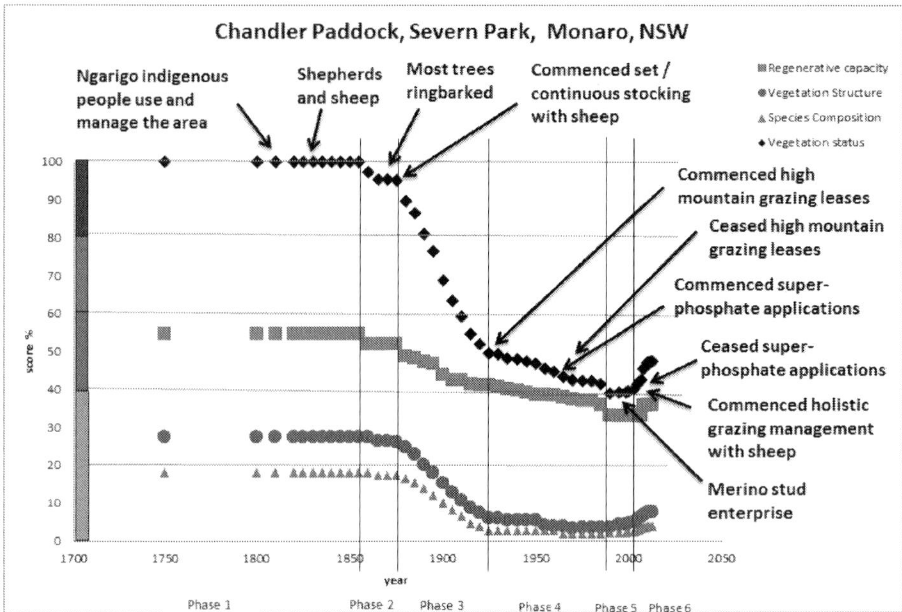

Figure 6. Six phases in the transformation of Chandler paddock, Severn Park, Monaro.

The figure shows total vegetation status score calibrated to VAST classes:

80-100% of the reference state corresponds to VAST class I– Residual /Unmodified;

60-80% corresponds to VAST class II– Modified;

40-60% corresponds to VAST class III– Transformed;

Scores from 0-39% corresponding to VAST class IV– Replaced and adventive; VAST class V– Replaced and managed; and VAST class VI Replaced and removed have not been differentiated.

11.3. Summary of the case studies

Table 2 outlined the ten functional, structural and composition criteria that are the key characteristics for describing condition relative to a 'natural' benchmark. Table 4 shows the relative values for these criteria for three periods — 1910, 1950 and 2010 for the two case study sites. Differences between the values for the two sites in the 2010 records can be partly explained by the interactions between environmental factors — especially wetter sub-tropical vs dryer cool temperate location — and by differences in management objectives. For the Rocky Creek Dam site, the managers objective was full recovery towards the original sub-tropical rainforest with no further incursions of lantana after the exclusion of grazing. For the Chandler Paddock site, the manager's objective was to restore the condition and resilience of a native pasture sufficiently to maintain a sustainable grazing regime with increased productivity.

		Rocky Creek Dam, Big Scrub			Chandler Paddock, Monaro Region		
	Year	1910	1950	2010	1910	1950	2010
	Fire regime	100%	100%	100%	100%	100%	100%
	Soil hydrology	100%	100%	100%	70%	70%	70%
	Soil physical state	100%	95%	100%	70%	70%	70%
	Soil nutrient state	100%	70%	100%	90%	90%	70%
	Soil biological state	100%	10%	100%	70%	70%	70%
Key ecological characteristics (10)	Reproductive potential	100%	75%	90%	65%	25%	15%
	Overstorey structure	0%	0%	87%	40%	10%	0%
	Understorey structure	0%	0%	100%	37%	30%	27%
	Overstorey composition	10%	10%	65%	40%	10%	0%
	Understorey composition	10%	10%	85%	30%	20%	20%
	Average scores	62%	47%	93%	61%	50%	44%

Table 4. Relative value for VAST criteria for 3 snapshots at the 2 case study sites, benchmarked to the reference state.

A benefit of a standardised national system is that it enables a comparison between the 10 criteria for the two case study sites from two different bioregions (Table 4), however it should be noted that such widely spaced temporal snap shots do not reflect short term improvements in condition which has occurred in several criteria between 1999 and 2010 for the Chandler Paddock, Monaro Region case study site (Figure 6).

12. Other applications of the VAST approach

Australia's vegetation has been classified and mapped into 23 Major Vegetation Groups using information from the National Vegetation Information System [55]. The VAST approach has been applied to assess the condition of these vegetation types and compare the pre-European extent with the current vegetation condition for these groups. Figure 7 shows the 23 groups in seven classes — naturally bare and the six VAST classes ranging from Class I (unchanged) to Class VI (removed).

This information can be used for natural resource planning and program development. For example, it provides an indication of which vegetation types may require additional formal public protection in conservation reserves. The information could also be used to identify those vegetation types where existing VAST class I and class II areas are likely to be further modified by progressive adaptation of land use to the impacts of climate change.

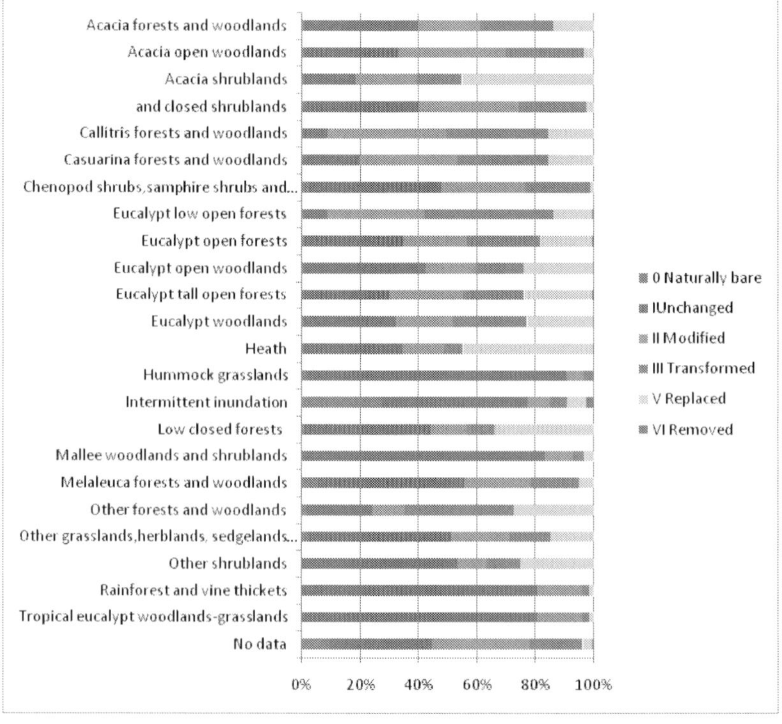

Source ABARES 2011 [55].

Figure 7. Extent of modification of major vegetation groups, as assessed by VAST. Sorted alphabetically

Information derived for Figure 7 has also been used to assess gross modification between native vegetation (VAST classes 0, I, II and III) and non-native vegetation (VAST classes V and VI) to show the relative areal extent of change from original pre-European vegetation cover to the present extent of Major Vegetation Groups. The summary results are shown in Figure 8.

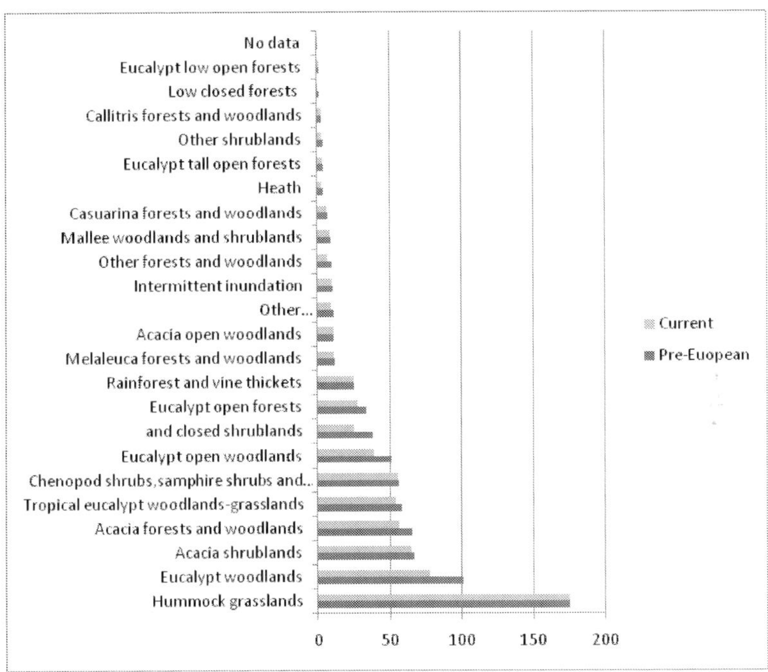

Sources: ABARES [55] and ERIN [56, 57].

Figure 8. Current and pre-1750 extent of Australian major vegetation groups. Sorted by area of the major vegetation groups. Area in millions of hectares.

13. Discussion

We have described the VAST approach as a tool for managers and researchers to understand the condition of a site or landscape relative to a reference, essentially natural, state. The case studies show how intentional changes in land use and/or land management practices can begin to change condition, sometimes rapidly, on a pathway of reverse transformation. The manager's objectives may be manifold, often economic and aesthetic, but very commonly inclusive of a desire to maintain and enhance biodiversity by improving the quality, extent and/or connectivity of suitable habitat.

The collection of data for key indicators, their use to inform strategies and targets for change in key functional, structural and composition components, and monitoring of the effect of these strategies is fundamental to the VAST approach. Recording information on the pathway(s) from "original' condition (as defined by a reference state that, as closely as possible, represents pre-European condition) is an essential step if the VAST approach is to be used for reporting progress.

To date, the VAST 2 system has concentrated on changes that have already been made. The next essential step is to record the actions now being taken and to define their intention-i.e. what these actions were intended to achieve and in what time scale it was expected (or hoped) that the changes would have effect. Then it is necessary to have a schedule and defensible method for monitoring how the site and the broader landscape are affected both by the management actions and by external drivers such as weather events and unplanned fire.

We suggest that these essential steps are more broadly applicable to any activity undertaken with the intention to redress degrading impacts in the landscape — e.g. removal and control of invasive weeds. While farmers and other landowners have primary responsibility for such activities on private lands, the work is often done in conjunction with community and other interest groups. In Australia, involvement in 'greening' and a wide range of other environmental activities, is supported by a very large number of groups involving many thousands of often highly committed individuals. While much has been attempted in the design and implementation of government programs to support environmental action — including the development of indicators and guidelines for monitoring and evaluation — there is much more that can be done to achieve wider and more consistent monitoring and reporting, especially of actions undertaken by community groups independently of government programs.

Pilots of VAST have shown that it is possible for management changes to restore a site to conditions that so closely resemble a reference site as to be judged as 'natural'. However, while at times it is possible to achieve such results — usually with a complete change of land use — some caution is appropriate with regard to sites for which there continues to be a measurable need for 'production' ecosystem services. Attempting to fully restore original composition is not necessarily always the right approach. Parrotta et al [58] emphasise that *the functional argument for biodiversity conservation does not necessarily depend on reinstating previous ecological conditions, although provisioning, cultural, aesthetic and other benefits or services are often enhanced by native biodiversity*.

As noted above, many of Australia's landscapes are now a diverse mosaic of modified native vegetation, 'replaced' vegetation cover types, and fragments of vegetation in 'original' condition. At a landscape scale there is no right or wrong degree of landscape alteration level, i.e. combinations of vegetation modification and fragmentation. However, as public and/or private needs change amid concerns about threats to biodiversity, ecosystem services and economic activity, there is benefit in improving the quality of vegetation condition information that can be used in decision making for natural resource management. Along with this, it is clearly important to have an accounting system that can be used to track progress towards the sustainable use and management of native vegetation across all land use types, with obvious relevance for managing biodiversity.

The VAST system is a flexible tool for identifying what component(s) might to be manipulated to improve vegetation condition; for demonstrating progress toward the desired vegetation condition; and selecting sites which represent least-cost options for future land use changes to improve biodiversity outcomes.

The VAST approach has now been tested and applied to a wide range of sites and landscapes including sub-tropical wet forests, temperate and semi-arid woodlands, grasslands and arid ecosystems. Aside from the two case studies presented here, other examples include:

- regenerating a sub-tropical eucalypt forest on a sand mined dune;

- regenerating a temperate grassy eucalypt woodland to improve its biodiversity and grazing productivity which had become degraded by long-term continuous grazing with sheep;

- regenerating a protected temperate grassland to improve its biodiversity because the area was degraded by heavy grazing by kangaroos;

- regenerating a eucalypt forest to improve its saw log productivity because the area was degraded by heavy and sustained logging pressure;

- regenerating former tropical rainforest on an abandoned dairy farm recently occupied by a dense stand of weeds;

- restoring a former temperate grassy woodland which was converted to improved pasture; and

- regenerating a temperate rainforest after it was logged once.

We suggest that the approach could now be implemented more broadly at both site and landscape scales. We envisage a program for monitoring and reporting by community--based groups to provide consistent information on the trend in biodiversity maintenance that would be especially important as the widely expected impacts of climate change take effect. The elements would be:

1. agreed standards for collection and recording of data

2. standards for description of the location, size, previous and current use and management etc of sites — based on geocoding standards and LUMIS protocols [13]

3. agreed methods for assessing indicators of structure, composition and function — based on the VAST 2 Handbook [47]

4. standards for defining objectives and targets and monitoring progress — based on Monitoring, Evaluation, Reporting and Improvement (MERI) schedules for cooperative government programs [59].

14. Conclusion

The Vegetation Assets, States and Conditions concept provides a reliable approach to identifying the current condition of a site, identifying appropriate actions to transform the site, and recording change in response to both these actions and external events. Being a landscape approach, it enables a user to visualise and evaluate the costs and benefits of trade-offs between economic and environmental objectives at a spatial scale at which the delivery of ecosystem services can meaningfully be influenced and reported. The results also provide a rationale for prioritising investment in restoring ecological function and for assisting stakeholders to prepare for, and adapt, to climate change.

Restoring the composition, structure and resilience of vegetation is a viable and effective pathway for improving ecological function and maintaining or recovering biodiversity. We have presented a range of information products that are designed to improve the general understanding of the spatio-temporal processes and patterns and of vegetation condition and how these are affected positively and negatively by land use and the management of native vegetation. These information products highlight the observation that, contrary to much popular opinion, degraded and highly modified ecosystems can be restored and regenerated and that restored ecosystems can again actively contribute to achieving multiple benefit outcomes at site and landscape scales. These benefits include the maintenance and recovery of biodiversity and sustainable use of vegetation.

The case studies show that compiling information and reporting on standardized indicators will give land managers and other decision makers a proper understanding of the nature of change and trend and enable them to consider future states, plan management interventions, and make adjustments in land management programs. Thereafter it is important and, using the VAST format, it is a relatively straightforward task to monitor and report ecological outcomes in the short, medium and long-term.

Author details

Graham A. Yapp[1] and Richard Thackway[2*]

*Address all correspondence to: r.thackway@uq.edu.au

1 Auricht Projects, Canberra, Australia

2 School of Geography, Planning and Environmental Management, University of Queensland, Australia

References

[1] Montreal Process Implementation Group for Australia and National Forest Inventory Steering Committee. Australia's State of the Forests Report 2013. Canberra: ABARES; 2013. http://www.daff.gov.au/ABARES/forestsaustralia/Documents/sofr2013-web2.pdf (accessed 23 July, 2014).

[2] State of the Environment Committee. Australia state of the environment 2011. Independent report to the Australian Government Minister for Sustainability, Environment, Water, Population and Communities. Canberra: Australian Government; 2011.

[3] Gammage B. The Biggest Estate on Earth. Sydney:Allen and Unwin; 2011.

[4] Australian Bureau of Statistics. 4613.0-Australia's Environment: Issues and Trends, January 2010. http://www.abs.gov.au/AUSSTATS/abs@.nsf/Lookup/ DD486E7A8C8F95A4CA2576C000194E09?opendocument (accessed 23 July 2014).

[5] Steffen W, Burbidge A, Hughes L, Kitching R, Lindenmayer D, Musgrave W, Stafford Smith M, Werner P. Australia's Biodiversity and Climate Change. A report to the Natural Resource Management Ministerial Council commissioned by the Australian Government. Collingwood, Victoria: CSIRO Publishing; 2009. http:// www.climatechange.gov.au/sites/climatechange/files/documents/04 2013/biodiversity-vulnerability-assessment.pdf (accessed 23 July, 2014).

[6] Vihervaara P, Kumpula T, Ruokolainen A, Tanskanen A, Burkhard B. The use of detailed biotope data for linking biodiversity with ecosystem services in Finland. International Journal of Biodiversity Science, Ecosystem Services and Management 2012;8(1-2): 169-185. http://dx.doi.org/10.1080/21513732.2012.686120 (accessed 21 July 2014).

[7] Hooper DU, Chapin FS, Ewel JJ, Hector A, Inchausti P, Lavorel S, Lawton JH, Lodge DM, Loreau M, Naeem S, Schmid B, Setala H, Symstad AJ, Vandermeer J, Wardle DA. Effects of biodiversity on ecosystem functioning: a consensus of current knowledge. Ecological Monographs 2005;75 3-35.

[8] Balvanera P, Pfisterer AB, Buchmann N, He JS, Nakashizuka T, Raffaelli R, Schmid B. Quantifying the evidence for biodiversity effects on ecosystem functioning and services. Ecology Letters 2006;9 1146-1156. doi: 10.1111/j.1461-0248.2006.00963.x (accessed 23 July 2014).

[9] Guariguata MR, Balvanera P. Tropical forest service flows: improving our understanding of the biophysical dimension of ecosystem services. Forest Ecology and Management 2009;258 1835-1829.

[10] Thackway R, Lesslie R. Reporting vegetation condition using the Vegetation Assets, States and Transitions (VAST) framework. Ecological Management and Restoration 2006;7(S1): S53-S62. http://onlinelibrary.wiley.com/doi/10.1111/j. 1442-8903.2006.00292.x/abstract (accessed 23 July 2014).

[11] Australian Bureau of Statistics. Population Clock. http://www.abs.gov.au/ausstats/ abs@.nsf/Web+Pages/Population+Clock?opendocument (accessed 23 July 2014).

[12] Lesslie R, Mewett J. (2013). Land use and management: the Australian context. Canberra: ABARES research report 13.1; 2013. http://data.daff.gov.au/data/warehouse/

9aal/2013/RR13.1lumAc/RR13.1LandUseManageAustContext v1.0.0.pdf (accessed 23 July 2014).

[13] Australian Bureau of Agricultural and Resource Economics and Sciences (ABARES). Land use of Australia, Interim Version5, 2010-2011. Canberra: Australian Government Department of Agriculture; 2014. http://www.daff.gov.au/ABARES/pages/publications/display.aspx?url=http://143.188.17.20/anrdl/DAFFService/display.php?fid=pb_ilua5g9ablu20140708_11a.xml (accessed 28 July 2014).

[14] Australian Bureau of Agricultural and Resource Economics – Bureau of Rural Sciences (ABARE-BRS). Catchment Scale Land Use Mapping for Australia. http://www.daff.gov.au/abares/aclump/pages/land-use/catchment-scale-land-use-reports.aspx (accessed 20 July 2014).

[15] Yapp, GA. Land use: status of information for reporting against indicators under the National Natural Resource Management Monitoring and Evaluation Framework. Canberra: Australian Government National Land and Water Resources Audit; 2008.

[16] Lesslie R, Mewett J, Walcott J. Landscapes in transition: tracking land use change in Australia. Australian Government Department of Agriculture, Fisheries and Forestry: Science and Economic Insights Issue 2.2; 2011. http://data.daff.gov.au/data/warehouse/litlud9abl079/litlud9abll0790111a/SEI2011.2.2LndscpInTrans   LR   1.0.0.pdf (accessed 23 July 2014).

[17] Thackway R, Lesslie R. Describing and mapping human-induced vegetation change in the Australian landscape. Environmental Management 2008;42(4) 572–590.

[18] Australian Government Department of the Environment EPBC Act. Listed Key Threatening Processes. http://www.environment.gov.au/cgi-bin/sprat/public/publicgetkeythreats.pl (accessed 23 July 2014).

[19] Australian Government Department of the Environment. EPBC Act List of Threatened Flora. http://www.environment.gov.au/cgi-bin/sprat/public/publicthreatenedlist.pl?wanted=flora (accessed 23 July 2014).

[20] Australian Government Department of the Environment. EPBC Act List of Threatened Fauna. http://www.environment.gov.au/cgi-bin/sprat/public/publicthreatenedlist.pl?wanted=fauna (accessed 23 July 2014).

[21] Australian Government Department of the Environment EPBC Act. List of Threatened Ecological Communities. http://www.environment.gov.au/cgi-bin/sprat/public/publiclookupcommunities.pl (accessed 21 July 2014).

[22] McIntyre S, Hobbs RJ. (1999) A framework for conceptualising human impacts on landscapes and its relevance to management and research models. Conservation Biology 1999;13(6) 1282–1292. DOI: 10.1046/j.1523-1739.1999.97509.x (accessed 15 June 2014).

[23] Byrne M, Prober S, McLean E, Steane D, Stock W, Potts B, Vaillancourt R. Adaptation to Climate in Widespread Eucalypt Species. Gold Coast: National Climate Change Adaptation Research Facility; 2013.

[24] Doley D. The response of forests to climate change: the role of silviculture in conserving threatened species. Australian Forestry 2010;73 115–125.

[25] Williams RJ, Bradstock RA, Cary GJ, Enright NJ, Gill AM, Liedloff AC, Lucas C, Whelan RJ, Andersen AN, Bowman DJMS, Clarke PJ, Cook GD, Hennessy KJ, York A. Interactions Between Climate Change, Fire Regimes and Biodiversity in Australia: A Preliminary Assessment. Canberra Australian Government Department of Climate Change and Department of the Environment, Water, Heritage and the Arts; 2009. http://climatechange.gov.au/climate-change/publications/interactions-between-climate-change-fire-regimes-and-biodiversity-australia-preliminary-assessment (accessed 21 July 2014).

[26] Thackway R, Lymburner L, Guerschman JP. Dynamic land cover information: bridging the gap between remote sensing and natural resource management. Ecology and Society 2013;18(1) 2. http://dx.doi.org/10.5751/ES-05229-180102 (accessed 23 July 2014).

[27] Lymburner L, Tan P, Mueller N, Thackway R, Thankappan M, Islam A, Lewis A, Randall L and Senarath U. The National Dynamic Land Cover Dataset. Canberra: Geoscience Australia; 2011.

[28] Gill AM. Fire and the Australian flora: a review. Australian Forestry 1975;38 4-25.

[29] Luke H, McArthur AG. Bushfires in Australia. Canberra: Australian Government Publishing Service; 1978.

[30] Russell-Smith J, Yates CP, Whitehead PJ, Smith, R. Craig R, Allan GE, Thackway R, Frakes I, Cridland S, Meyer MCP, Gill AM. Bushfires 'down under': patterns and implications of contemporary Australian landscape burning. International Journal of Wildland Fire 2007;16(4) 361–377.

[31] Gill, AM. How fires affect biodiversity. In: Fire and Biodiversity: the effects and effectiveness of fire management. Biodiversity Series1996; Paper Number 8. Canberra: Department of Environment, Sport and Territories, 1996. http://www.environment.gov.au/archive/biodiversity/publications/series/paper8/paper4.html (accessed 26 July 2014).

[32] House of Representatives Standing Committee on Climate Change, Environment and the Arts. Managing Australia's biodiversity in a changing climate: the way forward. Final report of the inquiry into Australia's biodiversity in a changing climate. Canberra: The Parliament of the Commonwealth of Australia; 2013. http://www.aph.gov.au/parliamentary business/committees/house of representatives committees?url=ccea/ccbio/report/finalreport/index.htm (accessed 24 July 2014).

[33] Reside AE, VanDerWal J, Phillips BL, Shoo LP, Rosauer DF, Anderson BJ, Welbergen JA, Moritz C, Ferrier S, Harwood TD, Williams KJ, Mackey B, Hugh S, Williams YM, Williams SE. Climate change refugia for terrestrial biodiversity: defining areas that promote species persistence and ecosystem resilience in the face of global climate change. Gold Coast: National Climate Change Adaptation Research Facility; 2013. http://www.nccarf.edu.au/publications/climate-change-refugia-terrestrial-biodiversity (accessed 23 July 2014).

[34] Dunlop M, Brown PR. Implications of Climate Change for Australia's National Reserve System—A Preliminary Assessment. Report to the Department of Climate Change, February 2008. Canberra: Department of Climate Change: 2008. http://www.climatechange.gov.au/sites/climatechange/files/documents/03   2013/nrs-report.pdf (accessed 10 July 2014).

[35] Sattler P, Creighton C. Australian Terrestrial Biodiversity Assessment 2002. Canberra: National Land and Water Resources Audit; 2002.

[36] Thackway R, Creswell ID. An interim biogeographic regionalization for Australia: a framework for setting priorities in the National Reserves System Cooperative Program. Canberra: Australian Nature Conservation Agency; 1995.

[37] Australian Government Department of the Environment. Australia's Bioregions (IBRA).   http://www.environment.gov.au/topics/land/national-reserve-system/science-maps-and-data/australias-bioregions-ibra/australias (accessed 20 July 2014).

[38] Bureau of Flora and Fauna. Flora of Australia: Volume 1 Introduction. Canberra: Australian Government Publishing Service; 1981.

[39] Brooker MIH, Slee AV, Connors JR. Eucalypts of Southern Australia. (EUCLID 2nd Edition about eucalypts) Collingwood, Victoria: CSIRO Publishing; 2002. http://www.anbg.gov.au/cpbr/cd-keys/Euclid/sample/html/learn.htm (accessed 24 July 2014).

[40] ABARES. Australia's forests at a glance 2012. Canberra: Australian Government Department of Agriculture, Fisheries and Forestry; 2012. http://www.forestlearning.edu.au/sites/default/files/resources/documents/ForestsAtGlance 2012 v1.0.0.pdf (accessed 20 July 2014).

[41] Yapp G, Walker J, Thackway R. Linking vegetation type and condition to ecosystem goods and services. Ecological Complexity 2010;7(3) 292–301.

[42] Folke C, Carpenter S, Walker B, Scheffer M, Elmqvist T, Gunderson L, Holling CS. Regime shifts, resilience, and biodiversity in ecosystem management. Annual Revue of Ecology, Evolution and Systematics 2004;35 557–581.

[43] National Reserve System Task Group. Australia's Strategy for the National Reserve System 2009-2030. Canberra: Australian Government; 2009. http://www.environ-

ment.gov.au/system/files/pages/643fb071-77c0-49e4-ab2f-220733beb30d/files/
nrsstrat.pdf (accessed 24 July 2014).

[44] Australian Government Department of Environment. Collaborative Australian Protected Area Database-CAPAD 2012. http://www.environment.gov.au/node/34737 (accessed 24 July 2014).

[45] Australian Government Department of the Environment. Ownership of protected areas. http://www.environment.gov.au/topics/land/nrs/about-nrs/ownership (accessed 20 July 2014).

[46] Thackway R. Applying a system for tracking the changes in vegetation condition to Australia's forests. In: Brown AG, Wells KF, Parsons M, Kerruish CM. (eds.) Managing our Forests into the 21st Century: proceedings of National Conference, Institute of Foresters of Australia, 4-7 April 2013, Canberra, Australia, p79-91. http://forestry-conference.org.au/program/program-abstracts/ifa-2013-conf-papers-thackway (accessed 31 July 2014).

[47] Thackway R. VAST-2: Tracking the Transformation of Vegetated Landscapes. Handbook for recording site-based effects of land use and land management on the condition of native plant communities – Version 3. Brisbane: The University of Queensland, Australian Centre for Ecological Analysis and Synthesis, Terrestrial Ecosystem Research Network 2014. http://aceas.org.au/VAST-2_Manual_v3.0.pdf (accessed 28 July 2014).

[48] Hamilton M, Brown M, Nolan, G. Comparing the biodiversity impacts of timber and other building materials. Report prepared for Forests and Wood Products Australia, Project No. PR07.1050 p50. Melbourne: Forests and Wood Products Australia; 2008. http://www.fwpa.com.au/images/resources/PR07_1050_WEB_Report.pdf (accessed 25 July 2014).

[49] Mutendeudzi M, Thackway R. A method for deriving maps of landscape alteration levels from vegetation condition state datasets. Canberra: Bureau of Rural Sciences; 2010. http://data.daff.gov.au/brs/data/warehouse/pe brs90000004186/lalMethTechRpt 20091208 ap14.pdf (accessed 23 July 2014).

[50] Thackway R. Towards a framework for describing and mapping vegetation condition: observations from temperate woodlands. In: Lindenmayer D, Hobbs R, Bennett A. (eds.) Woodland Conservation and Management. Melbourne: CSIRO Publishing; 2010. p261-270.

[51] Woodford R. Converting a dairy farm back to a rainforest water catchment: the Rocky Creek Dam story. Ecological Management and Restoration 2000:1(2) 83-92.

[52] Sanger JC, Kanowski J, Catterall CP, Woodford R. Ecological Management & Restoration 2008:9(2); 143-145.

[53] Sanger JC, Kanowski J, Catterall CP, Woodford R. (unpublished field survey data)

[54] Thackway R. Transformation of Australia's vegetated Landscapes, Big Scrub Rocky Creek Dam, NSW. Brisbane: Australian Centre for Ecological Analysis and Synthesis, University of Queensland; 2012. http://dx.doi.org/10.4227/05/5088F3D486414 (accessed 26 July 2014).

[55] ABARES. Assessing the modification of Australia's native vegetation. Canberra: Australian Government Department of Agriculture 2011. http://data.daff.gov.au/VAST/ (accessed 31 July 2014).

[56] Australian Government Department of the Environment. National Vegetation Information System (NVIS). http://www.environment.gov.au/topics/science-and-research/databases-and-maps/national-vegetation-information-system (accessed 30 July 2014).

[57] ERIN. Australia-Estimated Pre1750 Major Vegetation Groups-NVIS Version 4.1. http://www.environment.gov.au/metadataexplorer/full_metadata.jsp?docId=%7BFC5643D0-E060-4CCC-B67F-59CBE7E14BE5%7D&loggedIn=false (accessed 31 July 2014).

[58] Parrotta JA, Wildburger C, Mansourian S. eds. Understanding Relationships between Biodiversity, Carbon, Forests and People: the Key to Achieving REDD+Objectives. A Global Assessment Report. Prepared by the Global Forest Expert Panel on Biodiversity, Forest Management, and REDD+. Vienna: IUFRO World Series Volume 31; 2012.

[59] Department of Agriculture and Department of Environment. Monitoring, Evaluation, Reporting and Improvement (MERI). Canberra: Australian Government; 2014. http://www.nrm.gov.au/funding/meri/ (accessed 15 July 2014).

Chapter 6

Maintaining Ecosystem Function by Restoring Forest Biodiversity – Reviewing Decision-Support Tools that link Biology, Hydrology and Geochemistry

Yueh-Hsin Lo, Juan A. Blanco, Clive Welham and
Mike Wang

Additional information is available at the end of the chapter

http://dx.doi.org/10.5772/59390

1. Introduction

Land use change in forest ecosystems is a worldwide problem. In many cases, however, the change is only temporary, and after a period of economic activity, the original forest must be reclaimed back to its original (or as close as possible) estate. A typical case is in open-pit mining. In many juridictions there is a legal requirement for the company to engage in restorative activities designed to bring back biodiversity and function to those areas espoiled by mining.

To reclaim a fully functional forest ecosystem, soil, topographic and hydrological properties must foster the biogeochemical and ecological processes required to support a vigorous vegetative community. While significant advances have been made in this regard, each reclaimed forest ecosystem is unique, and there remains considerable uncertainty as to how these interdependent processes will be manifest in any particular instance. One of the most important interdepencies is between water availability and plant uptake. Our understanding of biodiversity and linkage with surface water availability and distribution is limited because this relationship has not been examined within and across scales. The availability and distribution of water can influence ecosystem structure and function at a range of scales and levels of organization through its influences on various processes and feedbacks that can affect both animals and plants. For example, at a landscape level, the distribution of herbivore home ranges and vegetation communities may be influenced by the mosaic created by water sources. At the ecosystem and community levels, ecosystem processes such as nutrient cycling, predator-prey interactions, and interspecific competition may be influenced by the availablility and location of water sources. At a population level, surface and soil water availability and

distribution may influence herbivore and vegetation survivorship through processes such as droughts, pests and diseases.

Water scarcity can be produced by seasonal or annual droughts, but also by difficulties in uptake due to salinity or other contaminants. Naturally saline systems within the boreal forest are infrequent but are widespread across western North America [1]. Typical saline sites can be found in partially-closed catchment areas with inflows of groundwater [2]. Levels of salinity in shallow soils generally increase as the elevation declines toward a basin, reflecting the movement of the salts along the topographic gradient.

Reclaimed landscapes must hold enough water for plant growth but have enough downward movement of water to flush any contaminants from the rooting zone. Excessive percolation is also a concern, however, since drainage from toe slopes can carry dissolved contaminates out of a landform [3]. In northern climates, evapotranspiration is the largest annual loss of water [4], and a primary factor in the movement of salts to the surface. Interactions between vegetation productivity, nutrient budgets and evapotranspiration must therefore be assessed simultaneously [5]. There has been little research, however, on how changes in vegetation and climate affect the energy balance and water movement in northern ecosystems [3].

A broad array of decision support tools is available for projecting forest stand development in reclaimed ecosystems. These include forest ecosystem classification and ecosite manuals, stand establishment keys/guides, competition index methodologies, volume tables, site index curves, soil-site equations, stand density management tools, and growth and yield equations/ tables/decision systems [6]. Many of these tools are empirically based. This means they have significant limitations in their ability to accurately project productivity because in a strict sense, their application should be restricted to the stand conditions from which their underlying relationships were derived [7]. This can be problematic in mine reclamation for two reasons. First, the soil prescriptions that form the basis for reclamation often lack the historical legacy (propagule bank, organic matter, soil structural and biochemical properties, etc.) common to natural sites [8-11]. Secondly, forests are growing under climatic conditions that differ from the historical climate regime, particularly in more northerly regions. Global circulation model projections indicate continued increases in atmospheric and surface temperatures at least through this century, along with associated changes in the precipitation regime. Historical properties are now no longer tenable as the sole basis for deriving empirical growth relationships.

With the widespread increase in computing power, model sophistication and complexity have seen a rapid increase though this has not necessarily resulted in better and more reliable outcomes [12]. One issue with greater complexity is the increased cost and difficulty of obtaining calibration data sets for specific local ecosystems. From the perspective of reclamation, another issue is that most forest stand decision support tools have little or no representation of hydrological processes. In essence, the implicit assumption underlying these models is that the hydrological regime is in an equilibrium condition such that short-term temporal or spatial fluctuations in available moisture are of little consequence to long-term trends in productivity.

One way of resolving these limitations is to build a fully integrated vegetation-hydrology tool that combines a detailed representation of all critical processes within a single computing environment. The resulting 'mega-model' would possess the benefits of a fully integrated and interactive system with appropriate feedbacks and system controls. In such models, linkage between the different components of the model would be in real time, with vegetation growth and development in each time step related to the hydrological processes driving available moisture [13].

Seldom it is feasible or desirable, however, to build a model that includes all processes and scales of interest [14]. There are no a single model can be used in every situation that could arise during planning or management. If such a model is constructed, it will probably have one or more issues of being too generalist, having issues to represent processes at different scales, and continuous tuning and modifications to handle new data and issues [15]. Furthermore, ensuring these models are reliable requires substantial investments of time and energy, and their applicability tends to become overly specialized thereby reducing flexibility and portability (applying the model in different locations or circumstances). Other disadvantages include an increased calibration load associated with the myriad feedback and system controls, validation (establishing the veracity of model structure and architecture) and verification (assessing output accuracy). In this respect, more complex models are not 'better' due simply to the fact they incorporate a greater representation of reality because complexity does not necessarily equate to improved accuracy and precision. This leads to the modelling mantra that models should be only as complex as absolutely necessary.

A compromise (and popular) approach is to construct a 'meta-model' wherein the most suitable forest productivity and hydrological models would be coupled as input-output (I/O) systems. This I/O linkage refers to the idea that output from a given model serves as input to another. Fall [14] provided a list of the benefits and costs of the meta-modeling approach. The most important benefits of the meta-modelling approach are that it can use previous knowledge and expertise generated when developed well document models, but at the same time allow for flexibility to match the meta-model to the local conditions, data availability and other user needs. In addition, different teams can work in different processes and sub-models at the same time, improving the use of time and resources. Such advantage is particularly important when individuals are separated geographically. Another advantage is that by linking different models or sub-models, each of them can be analyzed and validated separately. Such advantage is important to increase understanding of complex interactions between different ecosystem components, and to allow comparisons of different model components. Data flow in the meta-modelling approach is also more flexible, and output from one model can be used as input for several other model components. This allows partial verification as intermediate data can be analyzed and stored, something important in adaptive management and monitoring. Finally, in a meta-modelling framework model sensitivity and scenario analyses are facilitated as they can be performed for different model components.

Not all forest models are applicable to a meta-modelling approach. Hence, the objective of the research presented here was to identify and compare the available forest models already being used in research, and to evaluate their suitability for use as decision-support tools in designing successful restoration plans to bring forest biodiversity and function back to sites disturbed by industrial activities (mining in particular).

2. Review methodology

2.1. Literature review on forest growth models

The review covered papers available at the beginning of 2013 in scientific journals, scientific books, proceedings of scientific workshops and conferences, and technical reports. The literature search was conducted using the term "forest growth model" in combination with each of the following 8 keywords: "climate change, gap model, hydrology, mixedwood, productivity, regeneration, simulation, and succession". 'Hits' were then screened and those pertaining to ecosystems other than temperate and boreal forests were eliminated (tropical and subtropical ecosystems, Mediterranean ecosystems, grasslands, etc.). The data bases consulted were:

- *Canadian Forest Service Bookstore* (list of publication by Canadian authors compiled by the Canadian Forest Service compiled by the Canadian Forest Service)

Available at http://bookstore.cfs.nrcan.gc.ca

- *ISI Web of knowledge* (academic search engine by Thomson Reuters).

Available at http://www.isiwebofknowledge.com

- *Google Scholar* (academic search engine by Google Inc.).

Available at http://scholar.google.ca

- *C.E.M.A. Research library* (collection of reports and publication for the Alberta Oil Sands area compiled by the Cumulative Environmental Management Association)

Available at http://www.cemaonline.ca

- *U.B.C. library* (academic library at the University of British Columbia)

- On-line catalogue available at http://www.library.ubc.ca

- *Register Of Ecological Models* (self-registration tool to compile ecological models by the Kasel University).

Available at http://ecobas.org/www-server/index.html

2.2. Ranking of forest models

Development of a system for ranking models is a challenging and inexact exercise. The criterion used to build the ranking system, for example, as well as the relative weighting attached to each ranking variable are important decisions. First and foremost, we are of the opinion that the scientific peer review process provides the best assurance that model structure and application are sound and have been subjected to expert scrutiny, particularly for those models with multiple entries in the scientific literature. Hence, only models cited in three or more peer-reviewed publications were considered in creating a ranking of model suitability. Models with lesser publications were therefore omitted from further analysis because their low publication rates indicate that they have not yet received a proper assessment of their suitability. The objective of the ranking exercise was to identify the best 3 to 5 models according to five criteria. The ranking was created in two steps:

2.2.1. Initial partial score (0 to 8 points)

Each model was scored initially using four criteria that varied in maximum value between 1.5 to 2.5 points. A proportional approach was used to derive a relative ranking for each model within a given criterion: The model with the best mark received the maximum criterion score, and the remainder were scored in direct proportion to how they compared with the top model. The four criteria used to create the rankings were:

- *Number of publications in the database* (for models with ≥ 3 publications; maximum 2.5 points): An index of model application.

- *Time between first and most recent publications in the database* (in years; maximum 1.5 points): An index of model durability.

- *Time to last publication* (in years; maximum 1.5 points): an index of current activity around the model.

- *Number of citations in Web of Science®* (maximum 2.0 points): an index of the utility and relevance of the work done with the model, as perceived by the scientific community.

A final partial score for a given model was calculated by adding up the score for each criterion.

2.2.2. Full score (0 to 10 points) and shortlisting

An additional criterion was defined as the number of countries, ecosystem types and forest types in which each model had been applied. This criterion was considered a measure of model versatility, and was assigned a potential maximum of 2 points. Given the time-consuming nature of gathering data to calculate the values of this criterion, only models with an accumulated score of 5 or above (out of a maximum score of 8) for the previous four criteria were given scores for model versatility. Following the ranking exercise, a total score (out of a maximum of 10 points) was calculated and those models with 6.5 points or more made the shortlist.

All models can be broadly classified into three categories, depending on their structure and how they are parameterized:

- *Empirical models:* use a bioassay method to estimate tree growth. These models are constructed from historical growth patterns of, for example volume-age curves, height-age cures, yield tables, etc.

- *Process-based models:* simulate the physical processes underlying ecological dynamics.

• *Hybrid models:* combine elements of the above two categories. In this approach, empirical data are used to parameterize one or more of the ecophysiological processes driving in tree growth and ecosystem production.

3. Results and discussion

3.1. Literature review

A total of 466 documents were identified through the literature search. Most of the documents were detected using the general modeling-related keywords: "gap model" (83 documents), "productivity" (89 documents), "regeneration" (86 documents), and "simulation" (33 documents). These 291 documents account for 62.3% of the total. Following the pioneering work of Botkin et al. [16] with development of the JABOWA gap model, the number of models addressing forest productivity has increased steadily over the previous four decades (Figure 1). Several factors have likely contributed to this trend. Most models developed prior to the 1990s were designed to simulate timber production. Since then, forest management has moved from an almost exclusive focus on timber towards an emphasis on the sustainable production of multiple ecosystem goods and services [7]. Subsequent model developments have reflected this change.

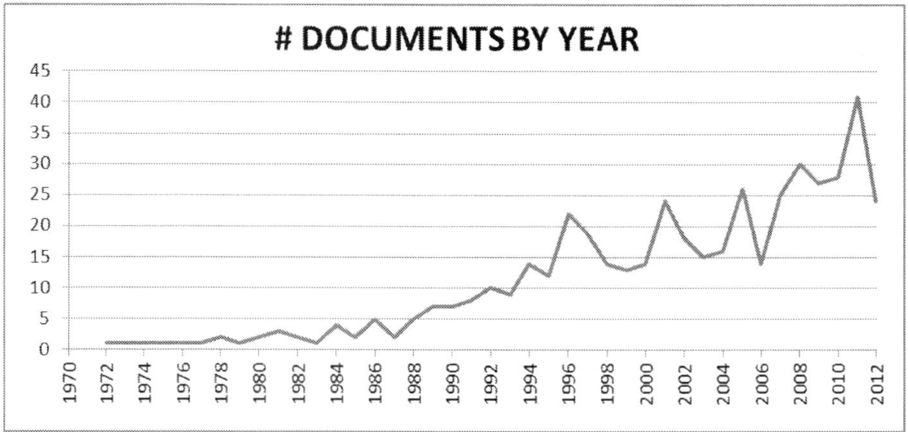

Figure 1. Number of documents published in a given year as derived from the keyword searches.

Climate and climate change have also emerged as a dominant issue in forest management, as government and industry strive to understand its impact on the present and future flow of goods and services. Keywords related to moisture ("climate change", "hydrology") accounted for 96 documents, 20.8% of the total. Of the 96 documents, 33 were related to the "climate change' keyword and 63 under "hydrology". A proportion of the documents under the keyword "hydrology" were also related to how climate change might alter the hydrological cycle. To avoid duplication, those documents were not accounted under the results for "climate change". Hence, under the "climate change" keyword only documents that deal with climate change and anything other than hydrological process are accounted for (mostly temperature-related research).

Tian et al. [17] argue that current forest growth models are seldom "well balanced" in terms of equivalence in the detail with which water, C, and N cycles are represented (see also [18, 19]). For instance, G'DAY [20], PnET-CN [21], and Biome-BGC [22] simulate forest growth and/ or biogeochemical processes in detail but are much less rigorous in their approach to representing forest hydrology. Unbalanced model design is likely to limit the ability of a given model to accurately predict the hydrology and biogeochemistry of forest ecosystems in response to changes in climate, land use, and/or management practices [19, 23].

In recent years, there have been several attempts to better link forest growth with hydrology. Chen and Driscoll [24] demonstrated that incorporating a more detailed hydrologic cycle into the Biome-BGC model improved predictions of seasonal effluent nitrate concentrations. Seely et al. [25] developed a stand-alone hydrology model for forest management applications (ForWaDy), with the explicit objective of minimizing data requirements. This model has been incorporated into the forest ecosystem simulation model, FORECAST [7]. Evidence suggests it provides a robust representation of moisture availability on tree growth, based on the balance between inputs from precipitation and seepage, and outputs by canopy interception, evapo-transpiration, plant uptake, percolation and runoff [26].

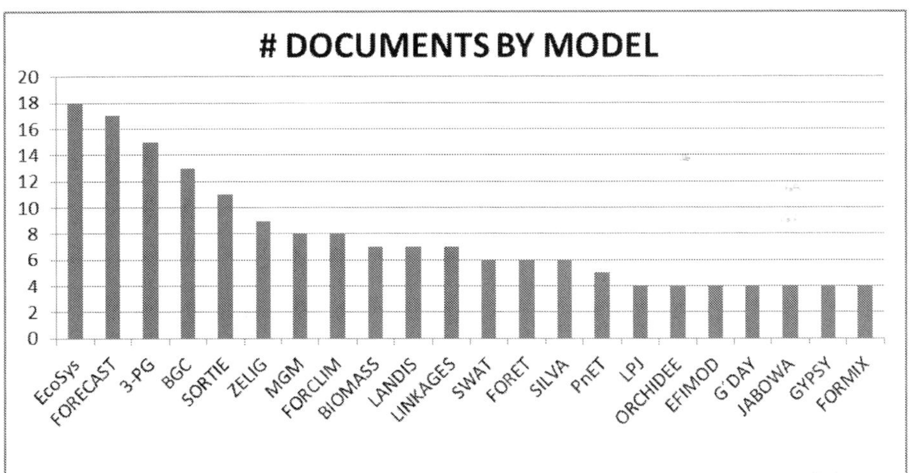

Figure 2. Number of document per model, for models with four or more documents in the database.

Despite the large number of models identified from the initial search, only 22 had 3 or more references (Figure 2). This suggests that many models are developed as one-time tools to explore issues of scientific importance, rather than as decision-support tool in support of management. This can limit the ease with which a model can be applied to situations different from that for which it was originally designed (i.e., the model's portability). It might also constrain the flexibility in model architecture, making it difficult to add management capabilities at a later date [7].

3.2. Model shortlisting

Only five models had more than 10 references: ECOSYS (18), FORECAST (17), 3-PG (15), BGC (13), and SORTIE (11; including both of its versions, SORTIE-ND, SORTIE-BC). It should be noted that the count for BGC is inflated by the fact it has three different variants (TREE-BGC, FOREST-BGC, and BIOME-BGC) which were grouped together for purposes of analysis. Arguably, it may be more appropriate to consider each separately since they are applicable to widely different scales (tree, stand, and biome, respectively). In that case, the ranking for each separate model would be much lower. The models LINKAGES, BIOMASS and CENTURY had relatively few publications but they ranked fairly well in the remaining criteria (Table 1). In the case of ECOSYS, 14 of the 18 references listed the developer (Grant) as the primary author. This is an unusually high number despite the fact the model was first published 14 years ago (Table 1). Publications of the remaining top models in the review encompassed a broad range of authors. This could be an indication that ECOSYS has not received broad acceptance, which could at least partly explain its low citation rate (see below). The common features of the top seven models (in green and yellow colours in Table 1) are that they are subject to ongoing development ($\geq$ 14 years), have been broadly applied in temperate and boreal ecosystems, and been cited within the last 2 years. The one exception is BIOMASS, which has not been cited in the previous 4 years. The top models were also heavily cited, with more 80 citations; ECOSYS was the clear exception, with only 25 citations (Table 1).

On its own, a high citation rate alone does not necessarily mean that a model is indeed being used extensively or is pertinent to the needs of Total. For example, a model that has been cited extensively is CENTURY. This model was originally developed for grasslands and so references to CENTURY were relatively frequent in the agricultural literature. In its current version, the model possesses a crude ability to represent forest growth. Its focus, however, is still mainly on soil processes even though this may be within the context of forest management. To an extent our ranking system was designed to take these factors into account by assigning the publication rate a higher weighting than the citation rate (a maximum score of 2.5 versus 2, respectively).

An important aspect of model suitability to oil sands reclamation is its portability. Portability refers to the ease with which a model can be calibrated, and its algorithms applied to, an ecosystem different from that in which it was originally developed. This is because no tool has been developed specifically for the conditions that characterize oil sands materials. Hence, the higher the number of countries and ecosystems where the model has been successfully applied, the higher its portability. The portability criterion was the final key factor that discriminated among the "shortlisted" models (BGC, FORECAST, 3-PG, and ECOSYS) and the remainder (Figure 2). These four models had more than 10 documents in the database, meaning they have been used in more than 10 different countries and/or ecosystem types, indicating a high portability potential.

3.2.1. ECOSYS

ECOSYS is a process-based, ecosystem-level model. It was originally developed as a soil model for agricultural ecosystems, but since then it has evolved into a complex simulator of the plant-atmosphere-soil system [27]. ECOSYS has a time step of one hour. It has representation of multi-layered canopies and soils. In this model, flows and transformation of growth resources (radiation, water, C, N, and P) are simulated for populations of plants and microorganisms. The model is constructed in order to link at different spatial and biological scales ecological processes that determine the ecophysiology of linked plant and microbial populations.

The model can represent ecosystems scales from homogeneous stands to heterogeneous landscapes, including natural and human-made disturbances. The model estimates ecosystem productivity through an energy balance approach.

Energy flows are simulated between the atmosphere and ground surfaces (snow, soil, litter). For plants, energy flows are simulated between the atmosphere and leaf or stem surfaces [29]. To calculate total exchange energy, energy exchanges between all plant and ground surfaces are added up. Hydrological processes (surface runoff, infiltration macro- and micro-pore flow) are then coupled with surface energy exchange and soil heat transfer [28].

ECOSYS calculates energy exchange in the canopy at an hourly basis, using a two-stage convergence solution to estimate heat and water transfers for the soil-root-canopy system for several plant populations and layers of soil and canopy. In the first stage, a canopy temperature value is calculated for each plant population by closing the canopy energy balance (sensible heat, latent heat flux, net radiation, and change in stored heat).These fluxes are controlled by aerodynamic and canopy stomatal resistances [29].

The simulation of water status effects on energy exchange is based on coupling the uptake of water from the soil through the root to the canopy, with the evaporation of water from the canopy to the atmosphere. This coupling determines the water status of the canopy and hence its conductance to water vapor [27].

Leaf C fixation is determined by carboxylation, which is controlled by irradiance, temperature, and leaf CO_2 concentration, and by diffusion, which is controlled by the atmosphere-leaf CO_2 concentration gradient and leaf conductance. The coupling of carboxylation and diffusion in ECOSYS allows the calculation of a leaf C fixation rate, which is then aggregated to the canopy level. Net C exchange between plants and the atmosphere is the difference between the two. Losses of leaf C are accelerated by reducing C fixation compared to maintenance respiration by reduced availability of growth resources (N, heat, or water). Net CO_2 fixation is calculated for each branch as the difference between gross fixation and the sum of respiration through maintenance, growth, and reproduction [27].

Model	Model	References		Time elapsed		Time since last reference		Citations in Web of Science		PARTIAL	Forest types applied				FINAL
	Type	#	score	years	score	years	score	#	Score	SCORE	# C	# E	Total	score	SCORE
BGC	process	[13]	1.81	23	1.33	2	1.89	178	1.67	6.70	6	14	20	1.48	8.18
FORECAST	hybrid	[17]	2.36	17	0.98	1	2.00	81	0.76	6.10	6	21	27	2.00	8.10
3-PG	process	[15]	2.08	14	0.81	2	1.89	118	1.11	5.89	7	10	17	1.26	7.15
ECOSYS	process	[18]	2.50	14	0.81	1	2.00	25	0.23	5.54	2	11	13	0.96	6.51
LINKAGES	process	[7]	0.97	26	1.50	2	1.89	110	1.03	5.40	3	7	10	0.74	6.14
BIOMASS	process	[7]	0.97	20	1.15	4	1.68	149	1.40	5.21	4	7	11	0.81	6.02
CENTURY	process	[3]	0.42	18	1.04	2	1.89	213	2.00	5.35	4	3	7	0.52	5.87
LANDIS	process	[7]	0.97	12	0.69	1	2.00	132	1.24	4.90	N / A				4.90
ZELIG	process	[9]	1.25	24	1.38	1	2.00	27	0.25	4.89	N / A				4.89
SORTIE	process	[11]	1.53	18	1.04	2	1.89	44	0.41	4.87	N / A				4.87
MGM	hybrid	[8]	1.11	17	0.98	1	2.00	37	0.35	4.44	N / A				4.44
SILVA	process	[6]	0.83	24	1.38	5	1.58	45	0.42	4.22	N / A				4.22
FORCLIM	process	[8]	1.11	18	1.04	4	1.68	35	0.33	4.16	N / A				4.16
SWAT	process	[6]	0.83	6	0.35	2	1.89	105	0.99	4.06	N / A				4.06
PnET	process	[5]	0.69	17	0.98	4	1.68	29	0.27	3.63	N / A				3.63
G'DAY	process	[4]	0.56	17	0.98	3	1.79	28	0.26	3.59	N / A				3.59
LPJ	process	[4]	0.56	11	0.63	1	2.00	18	0.17	3.36	N / A				3.36
EFIMOD	hybrid	[4]	0.56	12	0.69	2	1.89	10	0.09	3.24	N / A				3.24
FORWADY	hybrid	[3]	0.42	14	0.81	2	1.89	4	0.04	3.16	N / A				3.16
PICUS	process	[3]	0.42	12	0.69	3	1.79	20	0.19	3.09	N / A				3.09
ORCHIDEE	process	[4]	0.56	4	0.23	3	1.79	39	0.37	2.94	N / A				2.94
TRIPLEX	hybrid	[3]	0.42	8	0.46	3	1.79	11	0.10	2.77	N / A				2.77
GYPSY	empirical	[4]	0.56	5	0.29	4	1.68	9	0.08	2.61	N / A				2.61
PROGNOSIS	hybrid	[3]	0.42	8	0.46	8	1.26	31	0.29	2.43	N / A				2.43
JABOWA	process	[4]	0.56	22	1.27	19	0.11	25	0.23	2.16	N / A				2.16
CLASS	process	[3]	0.42	4	0.23	9	1.16	9	0.08	1.89	N / A				1.89
FORMIX	process	[4]	0.56	9	0.52	13	0.74	3	0.03	1.84	N / A				1.84
FORET	process	[3]	0.42	17	0.98	19	0.11	32	0.30	1.80	N / A				1.80
FORGRO	process	[3]	0.42	5	0.29	12	0.84	11	0.10	1.65	N / A				1.65
FORECE	process	[3]	0.42	7	0.40	17	0.32	5	0.05	1.18	N / A				1.18

C: Number of different countries where the model has been applied.

E: Number of different ecosystems where the model has been applied.

N / A: Non applicable, for models that did not pass the cut-off score of 5.0, the number of countries and ecosystems was not assessed.

Table 1. Ranking and scores of the models included in the comparative study (with 3 or more documents in the database).

The simulation of nutrient status effects on energy exchange is based on coupling nutrient (N and P) uptake from the soil through the root to the canopy, with nutrient assimilation in the root and canopy. This coupling determines nutrient concentrations in the leaf, which in turn determines leaf carboxylation rates and hence leaf conductance.

Growth respiration is linked to expansive growth of vegetative and reproductive organs at different nodes of each shoot branch, using data on biochemistry of growth and yield to estimate coefficients to partition mobilized C, N and P. Such coefficients also depend on phenological stages. Estimated growth is then allocated to different stem internodes, leaves, and sheaths, changing their lengths, areas and volumes [30, 31]. Then, leaf and stem surfaces (heights and areas) are estimated and used to calculate irradiance interception and aerodynamic conductance. Root and mycrorrizhal axes (both primary and secondary) extensions are driven by growth respiration, mobilizing stored C, N and P [32].

Microbial activity in ECOSYS is represented as a parallel set of substrate-microbe complexes, which includes the rhizosphere, plant residues and animal manure, and native organic matter [32-34]. To simulate microbial growth (facultative and obligated aerobic and anaerobic heterotrophs) at an hourly step, the temperature and water contents of the litter and soil layers are used [34-36]. Temperature and moisture are derived from the energy balance calculations described above.

ECOSYS is a highly complex model with substantial calibration requirements. The strength of its approach is its flexibility, provided by a detailed representation of ecophysiological processes that allow the exploration of the ecological consequences of modifying many different environmental factors. The main weakness of this approach is that validating the accuracy of its simulation algorithms and verifying output are significant challenges, due to the difficulty of finding independent values of many ecophysiological values. In addition, its management capabilities appear limited suggesting that the model is best catagorized as a research tool.

3.2.2. 3-PG

3-PG (the acronym represents Physiological Principles in Predicting Growth) was originally developed to simulate homogeneous, fast-growing plantations such as Eucalyptus [37], but has since been calibrated for other forest types [38]. 3-PG is a monthly time-step model working at stand and population levels. It is a model that includes general ecological processes and therefore needs to be calibrated for each individual species. It is designed for homogeneous forests, particularly even-aged or planted stands.

The model is built around the basic principles that drive ecosystem production. These same principles underlie earlier models such as FOREST-BGC [39] and BIOMASS [40]. The structure of 3-PG is based on two linked sets of calculations [41]: one set estimates biomass and growth values, whereas the other set estimates biomass allocation among different tree components. 3-PG is a conservation-of-mass model.

The model, like most process-based approaches, calculates rates of photosynthesis, transpiration, growth allocation and litter production. 3-PG derives estimates of radiation interception, gross primary production (GPP), net primary production (NPP) and allocation of the resultant carbohydrate pool to component parts of the trees. NPP is calculated as a fixed fraction of gross photosynthesis [42]. GPP is derived by applying a canopy quantum efficiency value to the amount of photosynthetically active radiation absorbed by a stand.

Quantum efficiency (the potential rate of photosynthesis) is a constant fraction of absorbed photosynthetically active radiation, and is constrained by atmospheric vapour pressure deficit. The latter is a function of stomatal conductance, which is influenced by air temperature, frost, water balance and nutrition. Canopy conductance is estimated as a function of leaf area index. The ratio of actual/potential photosynthesis is assumed to decrease in response to a suite of limiting environmental factors. It decreases with reduced availability of water and nutrients, which triggers a higher proportion of photosynthate allocated belowground.

Soil nutritional status (the availability of nutrients such as N and P) is represented by an index, the *fertility rating*, which can assume a value between 0 and 1 [38]. The fraction of production not allocated to roots is partitioned among foliage, stem and branches based on species-specific allometric equations.

3-PG can be used as a stand-level tool, or ground-based forest inventory data can be incorporated into a Geographical Information System (GIS) to simulate forest growth over large areas. 3-PG has a wide range of predicted stand properties that are directly compatible with conventional inventory measurements, including stem density, DBH, basal area, total volume, current and mean annual increment. In addition, the model outputs information pertaining to the underlying biophysical relationships. This means that growth patterns can be linked to specific controls, such as resource deficiencies and climate.

From the perspective of reclamation, a strength of 3-PG is that it appears suitable for predicting tree growth in areas currently devoid of tree cover and has relatively low calibration requirements [38]. Whether it could be reliably calibrated for oil sands materials, however, is unknown. 3-PG can be used to evaluate different management effects of stand density, thinning and fertilization (within the limitations of the fertility rating approach used for simulating nutrient availability).Arguably, the main weakness of 3-PG is its relative simplicity. It does not accommodate stands with complex structure (either in space or in terms of multiple aged trees), multiple species, and it has no understory representation. In addition, representation of soil nutritional status is overly simplified and is considered a static site property (it cannot vary through time). This significantly limits its application to oil sands materials and how soil properties might be expected to change over time.

3.2.3. BGC

BGC is a family of models, designed to accommodate different biological scales (TREE-BGC, FOREST-BGC, and BIOME-BGC). The original model was FOREST-BGC [39], an individual-entity, distance-independent model [42]. The term "entity" is used because STAND-BGC (a derivative of FOREST-BGC; [43]) grows shrubs and grass in addition to trees. Shrubs and grasses are described as per unit area entities, while trees have unique dimensions. All the models have the same core architecture and work on a daily time step, with results typically summarized annually. BIOME-BGC is a biome/ecosystem model, with spatial scales from stand to region.

BGC simulates fluxes and storage of water, carbon, and nitrogen [44-46]. BGC simulates fluxes and storage of water, carbon, and nitrogen [44-46]. The model has been designed to study the interactions between management, disturbances, climate and vegetation ecophysiological features, and their influences in water, nitrogen and carbon flows.

Net primary productivity is calculated as the difference between gross primary productivity (GPP) and autotrophic respiration, where GPP is a function of air temperature, water vapour pressure deficit, soil moisture, CO_2 concentration, LAI, and solar radiation at the top of the canopy. N concentrations in root and leaf, combined with temperature, are used to estimate respiration [47]. Canopy is simulated as one layer with sunlit and shaded foliage. The Farquhar equation is used to calculate photosynthesis [48]. Atmospheric CO_2 and humidity, leaf water and N contents, radiation and air temperature are used to calculate leaf conductance. Then, based on LAI values at leaf level, canopy C and water fluxes are calculated.

BGC is fundamentally driven by daily weather data. Therefore, ecophysiological descriptors of site vegetation, daily weather records and site physical properties are used by the model to simulate plant, soil, and litter variables, as well as water, carbon and nitrogen fluxes between the soil, the vegetation and the atmosphere. Unlike earlier models in the BGC model family (e.g. Forest-BGC, [39]), in Biome-BGC LAI is predicted as a function of the amount of leaf carbon, one of multiple vegetation state variables that are updated daily within the model [22]. Vegetation type is a user-defined, constant set of ecophysiological parameters. However, the model simulates changes in vegetation structure as consequence of disturbance, climate and ecophysiological characteristics of each vegetation type simulated.

The main strength of the model is its application in a broad range of ecosystem types. BGC's structure makes the model a suitable research tool to predict the impact of climate change. Forest-BGC, for example, has been widely used to predict climate change effects on natural disturbance and carbon dynamics [49]. In addition, BIOME-BGC offers a link between input data and GIS databases, which is useful for application of data collected from regional studies. A shortcoming of BGC is that the canopy is homogeneous. Therefore, although leaf area index is proportional to canopy depth, this may not be sufficient to capture water and carbon budgets accurately [39]. Its main drawback is the lack of a management interface, which makes it difficult to consider BGC as a decision-support tool for forest management and land reclamation.

3.2.4. FORECAST

FORECAST is a management-oriented, stand-level forest growth and ecosystem dynamics simulator [50]. The model was originally designed to accommodate a wide variety of harvesting and silvicultural systems in order to compare and contrast their effect on forest productivity, stand dynamics and a series of biophysical indicators of non-timber values. FORECAST-Climate version (see below) calculates climate modifiers on forest productivity on a daily basis. The modifiers are then accumulated across the year to estimate annual biomass production. FORECAST performs many calculations at the stand level but it also disaggregates stand-level productivity across individual stems in relation to age-specific stem size distributions. Top height and DBH are calculated for each stem and used in a taper function to calculate total and individual gross and merchantable volumes, and biomass.

Stand growth and ecosystem dynamics are based on a representation of the rates of key ecological processes regulating the availability of, and competition for, light and nutrient resources. FORECAST calculates biomass productivity (NPP) based on estimates of inherent productivity derived from historical bioassay data (see below) constrained by site-specific nutrient and water availability determined from within the model. The rates of the key ecological processes driving tree and plant growth are calculated from the bioassay data and inputted values for ecosystem variables (decomposition rates, photosynthetic saturation curves, for example) and their relation to nutrient uptake, the capture of light energy, and net primary production. Using this 'internal calibration' (hybrid simulation) approach, the model generates a suite of growth properties for each tree and plant species [50]. These growth properties are retained within the model and used to model subsequent growth as a function of resource availability and competition.

FORECAST's reliance on historical bioassay data serves to reduce calibration requirements while ensuring its projections of productivity are reasonable. Calibration data are assembled that describe the accumulation of biomass (above and below-ground components) in trees and minor vegetation for three chronosequences of stands, representing three different nutritional qualities. Tree biomass and stand self-thinning data can be derived from height, diameter at breast height, and stand density output generated by traditional growth and yield models in conjunction with species-specific biomass allometric equations [51]. To calibrate the nutritional aspects of the model, data describing the concentration of nutrients in the various biomass components are required. FORECAST also requires data on the degree of shading produced by different quantities of foliage and the photosynthetic response of foliage to different light levels. A comparable but simpler set of data for minor vegetation must be provided if the user wishes to represent this ecosystem component (see, for example, [52]). Lastly, data describing the rates of decomposition of various litter types and soil organic matter are required for the model to simulate nutrient cycling. The second aspect of calibration requires running the model in "spin-up" mode to establish initial site conditions. This component is a key feature in the ability of the model to simulate the site conditions characteristic of oil sands reclamation. For a broader discussion on this topic, see [7, 53, 54]).

Stand hydrology and water limitation for tree growth (see [25]) are simulated within the FORECAST-Climate model [55], which on a daily time step provides a mechanistic representation of above and belowground hydrological interactions in forest stands with multiple soil and canopy layers. This facilitates a representation of competition between trees in different canopy layers and minor vegetation for available soil water. In addition, the hydrological model also estimates the influence of drought on litter decomposition rates, and therefore on nutrient mineralization and its availability for vegetation [56]. Hence, as noted above the model tracks the balance between inputs from precipitation and seepage, and outputs by canopy interception, evapotranspiration, plant uptake, percolation and runoff.

FORECAST has been calibrated for the Ft. McMurray region. It has been applied to oil sands reclamation for over almost 15 years, in large part to compare current and alternative reclamation practices and their relationship to indicators of ecosystem function and the achievement of end land-use objectives. In this regard, FORECAST output was used to derive multipliers and nutrient regime classes for the Landscape Capability Classification System [57]; to explore issues associated with peat decomposition rates; the depth and type of the capping material; nitrogen deposition; subsoil organic matter content; species mixes, planting densities, understory dynamics, and dead organic matter dynamics (specifically snags), all within the context of growth and yield [58 – 60]. Recently, FORECAST-Climate was used in a risk analysis of the potential development of water stress in young reclamation plantations consisting of white spruce, trembling aspen, and jack pine established on different ecosites, as a function of soil texture and slope position [61]. In the second phase of this work, the principal objective was an evaluation of the impact of climate and climate change on reclamation success, as compared to the base case analysis (no climate-related impacts) [62]. The potential effect of different climate change scenarios on growth and mortality in reclamation areas was therefore projected using the FORECAST Climate model and associated modelling tools to evaluate their combined impacts on overall ecosystem development in a risk assessment context. A final component of this work consisted of: (1) Model projections of tree regeneration under climate change on actual oil sands reclamation materials, and (2) A comprehensive model analysis of the risks to ecosystem productivity from climate change as a consequence of the impact of moisture stress on tree mortality [55]. Recently, funding was approved for a project to:

a.   Improve the applicability of two established models that have been used to support adaptation decision-making within the context of oil sands reclamation, a state-and-transition simulation model (STSM; [63], and the process based forest ecosystem model, FORECAST-Climate [55, 62].

b.   Develop a decision support tool (DST) by linking the STSM and FORECAST-Climate.

c.   Use the DST to evaluate reclamation best management practices in the oil sands sector in terms of climate-related risk exposure and then inform adaptation and management planning within the context of climate change at both the stand and landscape scale.

Produce a guidance document on how to implement the tools, interpret output, and assess the implications for reclamation principles and practices as reflective of an adaptive decision framework.

4. Conclusions

Over the last four decades, a large number of ecological models that can simulate tree growth and forest hydrology have been developed for temperate and boreal ecosystems. The models best suited for simulating forest growth and hydrology in reclamation are likely to be at the scale of the stand level and in the daily to yearly time scale, as these scales provide sufficient detail to account for the key processes involved in tree growth but can also use operational data from forest management for calibration. In addition, a variety of tools have been developed to assist biodiversity planning in forest management. Among these are statistical models that utilize correlations between forest attributes and the presence of a particular wildlife or plant species or guild to determine habitat suitability [64]. These models have gained popularity because habitat descriptors can be derived from variables commonly available in forestry databases tor through modeling (for example, timber volume, forest age, dominant tree height, and species composition) [65-68]. When properly applied, they can also be used to predict the response of selected species to forest reclamation and to evaluate the efficacy of alternative practices [6, 69, 70].

Few models achieve recognition and use much beyond their development team, and even less have used within an operational setting [7]. Even among the four shortlisted models (ECOSYS, BGC, 3PG, and FORECAST) there is considerable variation in their ultility as decision support tools, particularly within the context of reclamation.

ECOSYS [28] is a complex model, with a strong representation of plant ecophysiological processes. It is a research tool to explore energy and matter fluxes in forest ecosystems. Its calibration requirements are substantial. BGC, particularly its most recent variant BIOME-BGC, is designed to represent the state and fluxes of carbon (C), nitrogen (N), and water (H_2O). The model has been applied to several forest and non-forest ecosystems around the world. The latest versions of the model include options for alternative forest management activities (see Table 4). BGC, however, is mainly a research tool designed to start from equilibrium conditions in a well-established ecosystem [71]. Hence, it is questionable whether the model is suitable for representing the biophysical characteristics of a reclaimed site. BGC also has fairly extensive and elaborate calibration requirements, though not as data-intensive as ECOSYS.

3-PG is a relatively popular forest growth model. It has been used as a research tool in a variety of forest ecosystems around the world. The model has been applied mostly in plantations, especially fast-growing species such as Eucalyptus and subtropical pines. 3-PG has been streamlined in recent years to facilitate its calibration with remote sensing data, therefore making it easy to apply to new sites and over large spatial scales [72]. One conceptual limitation of the model in terms of its application to reclamation is that site quality is represented as a fixed property [49]. This is problematic for two reasons. First, site quality must be known beforehand. This is generally not an issue in established natural forests (though it can be) but it has much greater uncertainty in a peat-based reclaimed system. Secondly, a reclaimed site is expected to transition from nutrient cycling based on the peat/mineral mix to that derived from the dead organic matter deposited by the developing plant community. It is unclear whether this transition will accompany a change in site quality. 3-PG also has no understory representation. Shrubs and herbs can be a key determinant of ecosystem development and productivity [52, 73].

FORECAST is model with a long history of development, but with a strong focus on management applications [50]. With the inclusion of a hydrology submodel (ForWaDy; see [25]), FORECAST now has the capability to simulate climate and climate impacts, and its impact on moisture availability, and C and N fluxes. The calibration requirements of FORECAST are moderate (but they are not trivial) though many parameters can be calibrated with standard inventory data and/or growth and yield tables. Some parameter values are universal and exhibit little variation; for others, the model is relatively insensitive to their variability (see [74], for a sensitivity analysis). Although FORECAST is a stand-alone model, it has been used for landscape-level analysis by linking it to GIS systems that classify the area under study into different ecosystem types [75, 76]. One advantage with FORECAST is that it has already been used extensively in oil sands reclamation (12, and references therein), and so datasets have already been constructed for the dominant tree and understory species. In this respect, FORECAST can be used to simulate complex mixtures of tree and understory species [77].

"A model should be as simple as possible, but no simpler". This is the principle put forward by Albert Einstein (in reference to scientific theories) and is applicable to model construction. Complex models are often required in ecology when the interactions between different ecological factors, both biotic and abiotic, need to be explicitly represented and understood [12]. This is especially important for ecosystems in which there are often no natural analogues, such as reclaimed landscapes [78]. The four shortlisted models provide a good representation of the range of complexity and approaches used to estimate biomass production, nutrient and water cycling. These differences are also reflected in the calibration requirements and calibration load associated with a given model. For example, ECOSYS is fundamentally a 'bottom-up' model in that it integrates ecophysiological processes starting at leaf scale to generate values of biomass production and water consumption at the stand level. BGC, in contrast, is more of a top-down model. FORECAST and 3-PG are somewhere 'in-between', estimating stand productivity with some simplification of the ecophysiological processes that occur at the cellular or leaf levels. The range in modeling approaches is also a reflection of the different origins of each model; FORECAST and 3-PG are forest management models, ECOSYS began as a crop research model, and BGC a forest ecology research model.

Determining the appropriateness of a given model to support biodiversity restoration within the context of reclamation depends on the balance between the accuracy required from the model output, the calibration effort and data available for calibration, model complexity, model flexibility, model robustness, and the capability to assess model performance [51]. Highly complex models such as ECOSYS simulate a large array of ecophysiological processes at fine temporal and spatial scales. Consequently, they require a considerable effort to assemble the data required for calibration. Often, it is necessary to make educated guesses for parameter values that are difficult to measure or which may not exist for the particular circumstances to which the model is to be applied. For obvious reasons, uncertainty in the input data reduces confidence in model output, an issue that becomes more problematic as the calibration requirements increase. Relatively simple models such as 3-PG have low calibration needs which allows for easier portability of the model to new ecosystem types. An overly simplified structure, however, also reduces model applicability (and flexibility) to complex systems and to account for interactions among all the ecosystem compartments. Conversely, robustness refers to a model's capability to produce acceptable estimates of the target variables in the required application. Robustness is not an inherent property of model complexity, and both complex and simple models can be robust, provided that calibration parameters are estimated with low uncertainty, especially for those key parameters for which the model is more sensitive [7].

Recovery of biodiversity in reclaimed sites depends on the timing of reclamation events, the type of forest system reclaimed, and how progressive reclamation impacts the vegetation (understory and stem distribution) relative to what would have been present had the landscape not been mined. Reclamation practices could be targeted toward the habitat requirements of particular wildlife or vegetation species by preferentially reclaiming more favourable ecological sites. Conversely, a broad range of ecological sites is necessary to promote suitable habitats for a diverse range of species on the reclaimed landscape. Such planning needs decision support tools that incorporate the best scientific knowledge available.

Acknowledgements

This work was conducted with funding generously supplied by Total E&P Canada Ltd., Calgary, Alberta, Canada. The opinions expressed herein are solely those of the authors and are not necessarily in accordance with that of any other group or individual.

Author details

Yueh-Hsin Lo[1], Juan A. Blanco[1], Clive Welham[2*] and Mike Wang[3]

*Address all correspondence to: clive.welham@ubc.ca

1 Dep. Ciencias del Medio Natural, Universidad Pública de Navarra, Campus de Arrosadía, Pamplona, Navarra, Spain

2 Dep. Forest Management, University of British Columbia, 2424 Main Mall, Vanocuver, British Columbia, Canada

3 Total E&P Canada Ltd., 2900-240 4th Ave SW, Calgary, Alberta, Canada

References

[1] Schwarz A.G., Wein R.W. 1997. Threatened dry grassland in the continental boreal forest of Wood Buffalo National Park. Canadian Journal of Botany 75: 1363-1370.

[2] Beyen W., Meire P. 2003. Ecohydrology of saline grasslands: consequences for their restoration. Applied Vegetation Science 6: 153-160.

[3] Carey S.K. 2008. Growing season energy and water exchange from an oil sands overburden reclamation soil cover, Fort McMurray, Alberta, Canada. Hydrological Processes 22: 2847-2857.

[4] Devito K., Creed I., Gan T., Mendoza C., Petrone R., Silins U., Smerdon B. 2005a. A framework for broad-scale classificatio of hydrologic response units on the Boreal Plain: is topography the last thing to consider? Hydrological Processes 19: 1705-1714.

[5] Johnson E.A., Miyanishi K. 2008. Creating new landscapes and ecosystems: the Alberta Oil Sands. Annals of New York Academy of Sciences 1134: 120-145.

[6] Welham C., Blanco J.A., Seely B., Bampfylde C. 2012a. Oil sands reclamation and the projected development of wildlife hábitat attributes. In: Vitt. D.H. (Ed.) Reclamation and Restoration of Boreal Ecosystems: attaining sustainable development. Pp 218-258. Cambridge University Press, Cambridge, UK. ISBN 978-1107015715.

[7] Kimmins J.P., Blanco J.A., Seely B., Welham C., Scoullar K. 2010. Forecasting Forest Futures: A Hybrid Modelling Approach to the Assessment of Sustainability of Forest Ecosystems and their Values. Earthscan Ltd. London, UK. 281 pp. ISBN: 978-1-84407-922-3.

[8] Rowland S.M., Prescott C.E., Grayston S.J., Quideau S.A., Bradfield G.E. 2009. Recreating a functioning forest soil in reclaimed oil sands in northern Alberta: an approach for measuring success in ecological restoration. Journal of Environmental Quality 38: 1580-1590.

[9] Naeth, M. A., Chanasyk, D. S. and Burgers, T. D. 2011. Vegetation and soil water interactions on a tailings sand storage facility in the Athabasca oil sands region of Alberta Canada.Phys. Chem. Earth 36: 19-30.

[10] Leatherdale J., Chanasyk D.S., Quideau S. 2012. Soil water regimes of reclaimed upland slopes in the oil sands region of Alberta. Canadian Journal of Soil Science 92: 117-129.

[11] Daly C., Price J., Rezanezhad F., Pouliot R., Rochefort L., Graf M.D. 2012. Initiatives in oil sand reclamation: considerations for building a fen peatland in a post-mined oil sands landscape. In: Restoration and Reclamation of Boreal Ecosystems. Ed. Vitt D., Bhatti J. Pp 179-201. Cambridge University Press, Cambridge, UK.

[12] Kimmins J.P., Blanco J.A., Seely B., Welham C., Scoullar K. 2008. Complexity in Modeling Forest Ecosystems; How Much is Enough? Forest Ecology and Management, 256(10), 1646-1658.

[13] Blanco J.A. 2012. Beyond growth models: Hybrid ecological models in the context of sustainable forest management. Cuadernos de la Sociedad Española de Ciencias Forestales, 34, 11-25.

[14] Fall, A. 2009. A practical approach for comparing management strategies in complex forest ecosystems using meta-modelling toolkits. Sustainable Forest Management

Network, University of Alberta, Edmonton, AB. SFM Network Research Note Series No. 56.

[15] Dixon R, Meldahl R, Ruark G, Warren W (1990).Process modeling of forest growth responses to environmental stress. Timber Press, Portland, OR, USA.

[16] Botkin, D. B., Janak, J. F., and Wallis, J. R.: 1972b, 'Some Ecological Consequences of a Computer Model of Forest Growth', J. Ecol. 60, 849–872.

[17] Tian S., Youssef M.A., Skaggs R.W., Amatya D.M., Chescheir G.M. 2012. DRAIN-MOD-FOREST: integrated modeling of hydrology, soil carbon and nitrogen dynamics, and plant growth for drained forests. Journal of Environmental Quality 41, 764-782.

[18] Tiktak, A., and H.J.M. van Grinsven. 1995. Review of 16 forest-soil-atmosphere models. Ecological Modelling 83:35–53.

[19] Waring, R.H., and S.W. Running. 2007. Forest ecosystems: Analysis at multiple scales. 3rd ed. Academic Press, San Diego, CA.

[20] Comins, H.N., and R.E. McMurtrie. 1993. Long-term response of nutrient-limited forests to CO2 enrichment; equilibrium behavior of plant–soil models. Ecological Applications 3:666–681.

[21] Aber, J.D., S.V. Ollinger, and C.T. Driscoll. 1997. Modeling nitrogen saturation in forest ecosystems in response to land use and atmospheric deposition. Ecological Modelling 101:61–78.

[22] Thornton, P.E., B.E. Law, H.L. Gholz, K.L. Clark, E. Falge, D.S. Ellsworth, A.H. Golstein, R.K. Monson, D. Hollinger, M. Falk, J. Chen, and J.P. Sparks. 2002. Modeling and measuring the effects of disturbance history and climate on carbon and water budgets in evergreen needle leaf forests. Agric. For. Meteorol. 113:185–222.

[23] Wallman, P., M.G.E. Svensson, H. Sverdrup, and S. Belyazid. 2005.ForSAFE—An integrated process-oriented forest model for long-term sustainability assessments. For. Ecol. Manage. 207:19–36.

[24] Chen, L.M., and C.T. Driscoll. 2005. A two-layer model to simulate variations in surface water chemistry draining a northern forest watershed. Water Resour. Res. 41:W09425.

[25] Seely, B.; Arp, P. & Kimmins, J. P. (1997). A forest hydrology submodel for simulating the effect of management and climate change on stand water stress. In Proceedings of Empirical and Process-based models for forest, tree and stand growth simulation, Amaro A. & Tomé M. (ed) Edições Salamandra, Lisboa, Portugal, September 1997.

[26] Seely B., Welham C. 2013. Simulating the impact of climate change on decomposition, nutrient turnover and forest growth rates: a case study from British Columbia,

Canada. Proceedings of the 11th congress of the Spanish Association for Terrestrial Ecology, Pamplona, May 6-10, 2013.

[27] Grant R.F., Black T.A., den Hartog G., Berry J.A., Neumann H.H., Blanken P.D., Yang P.C., Russell C., Nalder I.A. 1999. Diurnal and annual exchanges of mass and energy between an aspen-hazelnut forest and the atmosphere: testing the mathematical model Ecosys with data from the BOREAS experiment. Journal of Geophysical Research 104, 27699-2717.

[28] Grant, R. F. 2004. Modelling topographic effects on net ecosystem productivity of boreal black spruce forests. Tree Physiology 24:1–18.

[29] Grant R.F., Hutyra L.R., De Olivera R.C., Munger J.W., Saleska S.R., Wofsy S.C. 2009. Modeling the carbon balance of Amazonian rain forests: resolving ecological controls on net ecosystem productivity. Ecologicla Monographs 79, 445-463.

[30] Grant, R. F., and J. D. Hesketh, 1992. Canopy structure of maize (*Zea mays* L.) at different populations: Simulation and experimental verifica-tion, Biotronics, 21, 11-24.

[31] Grant, R. F., 1994.Simulation of ecological controls on nitrification, Soil Biol. Biochem., 26, 305-315.

[32] Grant, R. F., 1993a. Simulation model of soil compaction and root growth, I, Model development, Plant Soil, 150, 1-14,

[33] Grant, R. F., 1993b. Rhizodeposition by crop plants and its relationship to microbial activity and nitrogen distribution, Mod. Geol. Biol. Proc., 2, 193-209.

[34] Grant, R. F., and P. Rochette 1994. Soil microbial respiration at different temperatures and water potentials: Theory and mathematical modelling, Soil Sci. Soc. Am. J., 58, 1681-1690.

[35] Grant, R. F. 1997.Changes in soil organic matter under different tillage and rotation: Mathematical modelling in ecosys, Soil Sci. Soc. Am. J., 61, 1159-1174.

[36] Grant, R. F., R. C. Izaurralde, M. Nyborg, S.S. Malhi, E. D. Solberg, and D. Jans-Hammermeister, 1997. Modelling tillage and surface residue effects on soil C storage under current vs. elevated CO2 and temperature in ECOSYS, in Soil Processes and the Carbon Cycle, ed. R. Lal, J. M. Kimble, R. F. Follet, and B. A. Stewart, pp. 527-547 CRC Press, Boca Raton, Fla.

[37] Landsberg, J.J. and R.H. Waring. 1997. A generalized model of forest productivity using simplified concepts of radiation-use efficiency,carbon balance and partitioning. For. Ecol. Manage. 95: 209–228.

[38] Landsberg, J.J., R.H. Waring and N.C. Coops. 2003. Performance of the forest productivity model 3-PG applied to a wide range of forest types. For.Ecol. Manage. 172: 199–214.

[39] Running S.W., Coughlan J.C. 1988. A general model of forest ecosystem processes for regional application. I. Hydrologic balance, canopy gas exchange and primary production processes. Ecol. Model. 42, 125-154.

[40] McMurtrie R.E., Rook D.A., Kelliher F.M. 1990. Modelling the yield of *Pinus radiate* on a site limited by water and nutrition. For. Ecol. Manage. 30, 381-413.

[41] Landsberg J., 2003. Modelling forest ecosystems: state of the art, challenges, and future directions. Canadian Journal of Forest Research 33, 385-397.

[42] Waring, R.H., J.J. Landsberg and M. Williams. 1998. Net primary production of forests: a constant fraction of gross primary production? Tree Physiology 18: 129–134. [44] Milner, K.S., Coble, D.W., McMahan, A.J., Smith, E.L. 2003. FVSBGC: a hybrid of the physiological model STAND-BGC and the forest vegetation simulator. Can. J. For. Res. 33: 466–479.

[43] Milner, K.S., and Coble, D.W. 1995. A mechanistic approach to predicting the growth and yield of stands with complex structures. *In* Proceedings of the Conference: Uneven-Aged Manage- ment: Opportunities, Constraints, and Methodologies, 29 Apr. 1995, Missoula, Mont. *Edited by* K.S. O'Hara. University of Montana, Missoula, Mont. MFCES Misc. Publ. 56. pp. 144– 166.

[44] Thornton, P.E., 1998. Regional ecosystem simulation: combining surface- and satellite-based observations to study linkages between terrestrial energy and mass budgets. Ph.D. dissertation.School of Forestry, University of Montana, Missoula MT, 280 pp.

[45] Kimball, J.S., Thornton, P.E., White, M.A., Running, S.W., 1997. Simulating forest productivity and surface-atmosphere exchange in the BOREAS study region. Tree Physiol. 17, 589 and 599

[46] White, M.A., Thornton, P.E., Running, S.W., Nemani, R.R., 2000. Parameterization and sensitivity analysis of the Biome-BGC terrestrial ecosystem model: net primary production controls. Earth Interactions 4 (3), 1–85

[47] Luo, Z., Sun, O.J., Wang E., Ren H., Xu H. 2010. Modeling productivity in mangrove forests as impacted by effective soil water availability and its sensitivity to climate change using Biome-BGC. Ecosystems 13: 949-965.

[48] Farquhar, G.D., von Caemmerer, S., Berry, J.A., 1980. A biochemical model of photosynthetic CO2 assimilation in leaves of C3 species. Planta 149, 78–90. [53] Blanco J.A., Seely B., Welham C., Kimmins J.P., Seebacher T.M. 2007. Testing the performance of FORECAST, a forest ecosystem model, against 29 years of field data in a *Pseudotsuga menziesii* plantation. Canadian Journal of Forest Research, 37: 1808-1820.

[49] Rodríguez-Suárez J.A., Soto B., Iglesias M.L., Díaz-Fierros F. 2010. Application of the 3PG forest growth model to Eucalyptus globulus plantation in Northwest Spain. European Journal of Forest Research 129, 573-583.

[50] Kimmins J.P., Mailly D., Seely B. 1999. Modelling forest ecosystem net primary production: the hybrid simulation approach used in FORECAST. Ecological Modelling, 122: 195-224.

[51] Blanco J.A., Seely B., Welham C., Kimmins J.P., Seebacher T.M. 2007. Testing the performance of FORECAST, a forest ecosystem model, against 29 years of field data in a Pseudotsuga menziesii plantation. Canadian Journal of Forest Research, 37, 1808-1820.

[52] Bi J., Blanco J.A., Kimmins J.P., Ding Y., Seely B., Welham C. 2007. Yield decline in Chinese Fir plantations: A simulation investigation with implications for model complexity. Canadian Journal of Forest Research, 37, 1615-1630.

[53] Seely, B., Welham, C., and Kimmins, H. 2002. Carbon sequestration in a boreal forest ecosystem: results from the ecosystem simulation model, FORECAST. For. Ecol. Manage. 169: 123-135.

[54] Welham, C., Seely, B., and Kimmins, H. 2002. The utility of the two-pass harvesting system: an analysis using the ecosystem simulation model FORECAST. Can. J. For. Res. 32: 1071-1079.

[55] Welham, C. and B. Seely, 2013. Oil Sands Terrestrial Habitat and Risk Modelling for Disturbance and Reclamation: The Impact of Climate Change on Tree Regeneration and Productivity – Phase III Report. Oil Sands Research and Information Network, University of Alberta, School of Energy and the Environment, Edmonton, Alberta. OSRIN Report No. TR-36. 65 pp.

[56] Dordel J., Seely B., Simard S.W. 2011. Relationships between simulated water stress and mortality and growth rates in underplanted *Toona ciliate* Roem. In subtropical Argentinean plantations. Ecological Modelling 222, 3226-3225.

[57] Welham, C. 2004. Deriving multipliers and nutrient regime classes for the Land Capability Classification System using the ecosystem simulation model, FORECAST. Final report in partial fulfillment of CEMA Contract No. 2003-0007.

[58] Welham, C. 2005 a. Evaluating a prescriptive approach to creating target ecosites using d-ecosites as a test case. Final report in partial fulfillment of CEMA Contract No. 2004-0014.

[59] Welham, C. 2005 b. Evaluating existing prescriptions for creating target ecosites using the ecosystem simulation model, FORECAST: Implications for ecosystem productivity and community composition. Final report in partial fulfillment of CEMA Contract No. 2005-0025

[60] Welham, C. 2006. Evaluating existing prescriptions for creating target ecosites using the ecosystem simulation model, FORECAST: Implications for ecosystem productivity and community composition in reclaimed overburden. Final report in partial fulfillment of CEMA Contract No. 2006-0030.

[61] Welham, C., 2010. *Oil Sands Terrestrial Habitat and Risk Modeling for Disturbance and Reclamation – Phase I Report*. OSRIN Report No. TR-8. 109 pp.

[62] Welham, C. and B. Seely, 2011. Oil Sands Terrestrial Habitat and Risk Modelling for Disturbance and Reclamation – Phase II Report. Oil Sands Research and Information Network, University of Alberta, School of Energy and the Environment, Edmonton, Alberta. OSRIN Report No. TR-15. 93 pp. http://hdl.handle.net/10402/era.24547

[63] Frid, L. and Daniel, C. 2012. Development of a State-and-Transition Simulation Model in Support of Reclamation Planning. Report Prepared for the Reclamation Working Group of the Cumulative Environmental Management Association. Fort McMurray, AB.

[64] Edenius, L., and Mikusinski, G. 2006. Utility of habitat suitability models as bioviersity assessment tools in forest management. Scandinavian Journal of Forest Research 21:62-72.

[65] Allen, A.W. 1983a. Habitat suitability index models: Beaver. Washington, DC: U.S. Fish and Wildlife Service, Department of the Interior. Biological Report FWS/OBC-82/10.30, revised.

[66] Allen, A.W. 1983b. Habitat suitability index models: Fisher. Washington, DC: U.S. Fish and Wildlife Service, Department of the Interior. Biological Report FWS/OBC-82/10.45.

[67] Eccles, T.R., Green, J.A., Thompson, C., Searing, G.F. 1986. Slave River Hydro Project Mammal Studies – Volume 1. Final report. Prepared for the Slave River Hydro Project Study Group by LGL Ltd., Calgary, AB. Canada.

[68] Jalkotsky, P., Van Egmond, T.D., Eccles, T.R., Berger, R. 1990. Wildlife Habitat Evaluation, Assess,emt and Mapping for the OSLO Wildlife Study Area. Prepared for the OSLO project by the DEltan Environmental Management Group Ltd., Calgary, AB. Canada.

[69] Kliskey, A.D., Lofroth, E.C., Thompson, W.A., Brown, S., Schreier, H. 1999. Simulating and evaluating alternative resource-use strategies using GIS-based habitat suitability indices. Landscape and Urban Planning 45: 163-175.

[70] Welham C., Seely B., Blanco J.A. 2012. Projected patterns of C storage in upland forests reclaimed after oil sands mining. In: Vitt. D.H., Bhatti J.S. (Ed.) Reclamation and Restoration of Boreal Ecosystems: attaining sustainable development. Cambridge University Press, Cambridge, UK. Pp 218-258. ISBN 978-1107015715.

[71] Thornton P.E., Rosenbloom N.A. 2005. Ecosystem model spin-up: estimating steady state conditions in a coupled terrestrial carbon and nitrogen cycle model. Ecological Modelling 189, 25-48.

[72] Waring R.H., Coops, N.C., Landsberg J.J. 2010. Improving predictions of forest growth using the 3-PGS model with observations made by remote sensing. Forest Ecology and Management 259, 1722-1729.

[73] Welham C., Van Rees K., Seely B., Kimmins H. 2007. Projected long-term productivity in Saskatchewan hybrid poplar plantations: weed competition and fertilizer effects. Canadian Journal of Forest Research 37, 356-370.

[74] Seely, B., Kimmins, J.P., Welham, C., Scoullar, K.A., 1999. Ecosystem management models: defining stand-level sustainability, exploring stand-level stewardship. Journal of Forestry 97(6):4-10.

[75] Flanders D., Sheppard S.R.J., Blanco J.A. 2009. The Potential for Local Bioenergy in Low-Carbon Community Planning. Smart Growth on the ground: Prince George. Foundation research Bullletin #4. Smart growth BC, Vancouver, BC, Canada. 9 p.

[76] Blanco J.A., Dubois D., Littlejohn D., Flanders D., Robinson P., Moshofsky M., Welham C. 2014. Soil Organic Matter: a sustainability indicator for wildfire control and bioenergy production in the urban/forest interface. Soil Science Society of America Journal, 78(S1): S105-S117.

[77] Seely B., Hawkins C., Blanco J.A., Welham C., Kimmins J.P. 2008. Evaluation of an ecosystem-based approach to mixedwood modelling. Forest Chronicle, 84: 181-193.

[78] Lo Y.-H., Blanco J.A., Kimmins J.P. 2010. A word of caution when projecting future shifts of tree species ranges. The Forestry Chronicle, 86(3), 312-316.

Chapter 7

Microbial Assembly in Agroecosystems — From the Small Arise the Big

Dennis Goss de Souza, Lucas William Mendes,
Acácio Aparecido Navarrete and Siu Mui Tsai

Additional information is available at the end of the chapter

http://dx.doi.org/10.5772/58992

1. Introduction

Since the dawn of classical microbiology, scientists have applied efforts to unravel the ecological patterns of occurrence, distribution and function of microorganisms in several ecosystems, including soil. In the last years, the famous Baas-Becking affirmation "Everything is everywhere, but, the environment selects", have been largely used as central question, in microbial biogeographic studies and also in theoretical niche-based theories [1].

Despite we known the importance of soil "small"-microorganisms to maintain the dynamic balance and resilience of the "big"-ecosystems, little is known about their patterns of assembly and its relationship with the functions resilience due the conversion of natural areas, such as tropical forests, to agriculture. Advances arising from genomics era have enabled microbial ecologists to access the ecological dimension of genetic and functional biodiversity, through genomic sequencing techniques, at scale and depth never seen before [2].

However, a topic that remains unclear is how to analyze and interpret these patterns of biodiversity generated by millions (or even billions) of genetic and functional data, resulting in robust and concise answers about ecological issues, among them: which is/are the effect(s) of conversion of forest to agriculture on microbial ecological patterns? Moreover, how to integrate this dimension of genetic and functional biodiversity, with the dimension expressed by metabolic products and ecological relations of microorganisms, and with a third and not less important, environmental dimension, which can modulate the patterns of occurrence and distribution of microorganisms in several ecosystems around the globe? Elucidating these dimensions, through metacommunity ecology and biogeography may allow us to unravel the black box of microbial assembly and functionality in the agroecosystems, and give answers to Baas-Becking affirmation supporters and opponents.

The dimension 1, has been massively analyzed through large scale sequencing of nucleic acids. Recently, metagenomic libraries from soil microbial DNA are being used as template [3], in order to evaluate the effects of the conversion of forest into agriculture [4]. The dimension 2 has been assessed by biochemical assays of production/consumption of microbial-mediated greenhouse gases and microbial enzyme activities in soil [5–8]. The Dimension 3 is evaluated by observation or data collection and analysis of environmental factors, including soil physicochemical, climatic and geographical attributes and their relation with microbial molecular parameters [9–12]. A multidimensional approach, linking these dimensions that modulate the distribution and abundance of microorganisms in the ecosystems is obtained by multivariate pairwise correlations among the parameters evaluated in both the three dimensions, generating an integrative view in systems biology (Figure 1).

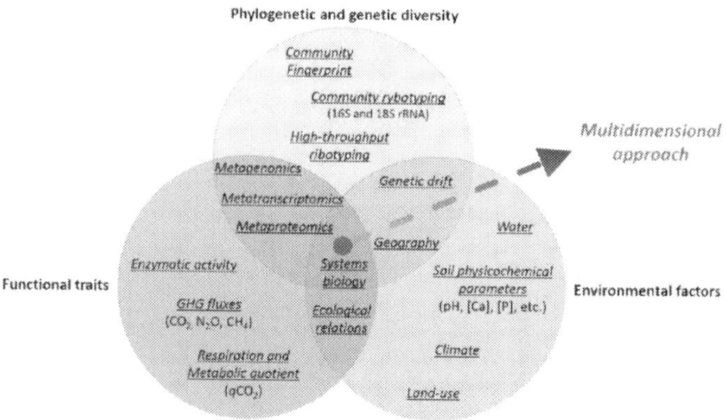

Figure 1. Dimensions to unravel the patterns of distribution and abundance of soil microorganisms in agroecosystems.

The aim of this chapter is to provide to readers some conceptual and practical bases for analysis and interpretation of microbial metacommunity assembly (structure), functions and their linkage, with applications in agroecosystems conservation. To achieve that, we consider results from the recent advances in high throughput next-generation sequencing (NGS) and bioinformatics that allow us to assess deeply, both the taxonomic, the phylogenetic and the genetic microbial biodiversity, establishing a novel border in microbial metacommunity ecology. We argue that metacommunity ecology and biogeography may be used as cornerstones to microbial ecology studies, helping us to elucidate tricking questions regarding microbial distribution and ecological relationships, from the local community level to the global level.

2. Molecular advances in microbial ecology

The rapid development of molecular biology techniques at the end of the twentieth century and their successful application to the study of microbial ecology has changed our view of the

assess structure and function of microorganisms. In recent years, advances in the field of molecular microbial ecology, in which are included the NGS techniques [13], have revealed a far unknown microbial biodiversity that was not detected previously by classical microbiology.

The NGS tools have decreased the relative costs of sequencing and increased massively the capacity of data production and quality. Their advances have contributed significantly to the understanding of the structure and function of soil microbial communities. Several molecular methods have been used to investigate the microbial diversity and changes in the microbial community structure in a wide range of environments (e.g. Shotgun metagenomics).

The studies in microbial ecology have been improved with the development and advance of the sequencing technologies. DNA sequencing is the process of determining the order of nucleotides that constitute a DNA molecule. The method determines the order of the four bases, i.e. adenine (A), guanine (G), cytosine (C), and thymine (T), in a strand of DNA. The DNA sequencing provides a mean of identifying organisms by comparing to databases. From a known and identified species, a molecular marker (e.g. 16S rRNA gene) is sequenced and deposited in publicly available databases for future comparison. DNA sequencing is suitable for sequence individual genes, molecular markers, larger genetic regions, full chromosomes or the entire genome. The sequencing approach is a powerful tool for the study of microbial communities inhabiting soil and could be useful to predict changes in soil properties and quality, as well as to understand the community assembly in these environments. The assessment of the microbial diversity will be advanced by the development of new technologies that answer some key questions about the "who, what, when, where, why and how" of microbial communities [14].

The rapid development of molecular biology techniques at the end of the twentieth century and their successful application to the study of microbial ecology has changed our view of the assess structure and function of microorganisms. In recent years, advances in the field of molecular microbial ecology, in which are included the Next Generation Sequencing (NGS) techniques [13], have revealed a far unknown microbial biodiversity that was not detected previously by classical microbiology.

The advance in sequencing technologies from Sanger to 454 pyrosequencing and Illumina has opened new possibilities in microbial community analysis by making it possible to collect millions of sequences, spanning hundreds of samples. The increase in the number of sequences per run from parallel pyrosequencing technologies such as the Roche 454 GS FLX (5×10^5) to Illumina GAIIx (1×10^8) is of the order of 1,000-fold and greater than the increase in the number of sequences per run from Sanger (1×10^3 through 1×10^4) to 454 [15]. In addition, the use of barcode strategies allows the analysis of thousands of samples in a single run. With the advance of such technologies the read length has increased, although they are far shorter than the desirable length or the read length obtained from traditional Sanger sequencing (~1000 bp) [16]. The 454 pyrosequencing was the first next-generation sequencing technology available as a commercial product [17] and can be considered the cornerstone of the sequencing revolution. The development of the pyrosequencing method allowed an advance of metagenome studies by increasing the number of reads and decreasing costs per sequence, enabling a deep phylogenetic community analysis.

The NGS tools have decreased the relative costs of sequencing and increased massively the capacity of data production and quality. Their advances have contributed significantly to the understanding of the structure and function of microbial communities. Several molecular methods are used to investigate the microbial diversity and changes in the microbial community structure in a wide range of environments. The use of metagenomics in the studies of microbial communities has enabled researchers to have an overview not only of the diversity, but also the functional traits, which are also an important approach to link the microbial community structure to functions. The rapid advance of sequence technologies allied to bioinformatics tools are increasing the possibility of massive studies on microbial ecology for a deep comprehension of the composition and function that soil microorganisms play in a wide range of ecosystems. The new information available will be useful for a better understanding of microbial assembly, at both phylogenetic and functional aspect of a community.

3. The metacommunity concept in microbial ecology

The first and simplest concept defines *metacommunity* as a set of communities that interact each other exchanging individuals of multiple species, linked by dispersal [18]. Different species interact each other via ecological relations, at the *community* (local level). There might be events of immigration, dispersal, besides other, that modulate the exchange of individuals from local communities to a broader range of communities, culminating with species evolution [19]. The use of different terms and different perspectives is a concerning question in metacommunity ecology. To reach scales of organization and set populations and communities dynamics within metacommunities, we use the terms and definitions conceptualized by [20] and applied by [21]. In order to assess the metacommunity assembly we regard the four central theories in Metacommunity Ecology, namely: (I) patch-dynamic, (II) species-sorting, (III) mass effects and (IV) neutral theory [22].

Metacommunity studies have been applied to ecology [19,23]. The patters of community distribution can vary in regional scale across environments, between environments at the same region and inside a specific environment [24]. Thus, a multidimensional approach is needed to have a comprehensive picture. To achieve that, four paradigms of metacommunities can be reached (Figure 2):

i.   Patch-dynamic (stochastic+deterministic) – as found to the neutral model, it assumes that the habitat quality is constant across different arrays (microbial cores) of the landscape. In this model, both stochastic and deterministic extinctions are affected by interspecific relations e counterbalanced by dispersion [20].

The approaches undergoing this paradigm are based on two different versions, based in occupancy formalisms, in which the patches are occupied or vacant by certain populations at

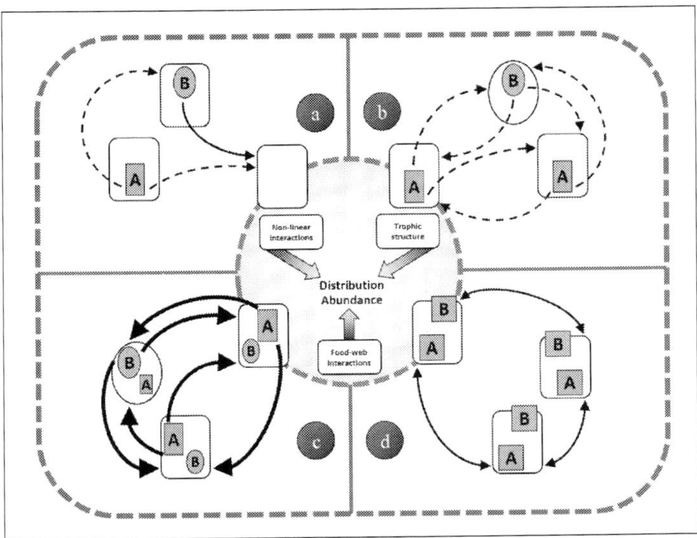

Figure 2. Representation of four metacommunity hypothetical situations at the local scale (dashed brown circle) and at the regional scale (dashed blue frame). Schematic situations for two competing species with populations A and B. (a) Patch-dynamic paradigm, (b) species-sorting paradigm, (c) mass-effects paradigm and (d) neutral paradigm. Adapted from [20].

their equilibrium. Both versions assume that the local and the regional population dynamics have a time gap, which means that the effects of changes in extinction-colonization patterns in the local community take certain period to affect the regional metacommunity dynamics.

In the first version for this paradigm, only regional coexistence is considered to influence the metacommunity patterns. It assumes that species that coexist compete for niche resources, but there is no interactions between species that influence local community dynamics, since local communities are not considered in the model. In the second version, given a homogeneous environment where a set of species co-occur in equilibrium, the regional coexistence is possible due a trade-off between competition and dispersal or fecundity and dispersal [25] (Figure 1a). A limitation of patch-dynamic paradigm is that a set of local communities or patches are assumed identical.

ii.   Species-sorting (deterministic) – based on the traditional theory of niche segregation of species that co-occur in certain environment [26]. The theory infers about changes in communities across environmental gradients [27].

In this case, the role of environmental parameters such as soil fertility and plant cover species, soil organic matter content, besides other, acquire fundamental importance in modulate the patterns of distribution and abundance of microorganisms [21]. The species-sorting paradigm infers that local patches differ in some attributes and the result of local species interactions depends on the environmental abiotic factors [28].

This paradigm assumes that different habitats patches are heterogeneous and that rates of dispersal are moderate, which means that all species have similar probability to reach all patches of the metacommunity (Figure 1b). Thus, it is expected to occur a species-sorting through niche partitioning, since there species are differently adapted to particular conditions, defined by environmental gradients.

iii.   Mass-effects (deterministic) – assumes that a certain population can vary in regional and local scales. This population can be affected quantitatively by dispersion. This model of mass effects due dispersion requires that different arrays of habitat have different conditions in certain moment, and that these relations should sufficiently tightly related. Thus, dispersion results in a sink/source dynamic between populations in different arrays at the landscape [29].

Dispersal has a great role in mass-effects paradigm. In one hand, an increase in immigration rates might enhance the abundance of certain populations in a local community, in detriment of neighbor communities from the metacommunity. In other hand an increase in emigration rates can decrease the rates of birth of local populations apart from the abundance expected in a close metacommunity (Figure 1c). Considering competing species of microorganisms in a certain environment where the total community has a constant abundance, a fluctuation in local population abundances may occur by chance [29].

iv.   Neutral Model (stochastic) – one thing that all the previous paradigms have in common is the assumption that species in local communities differ from each other in their capacities of competition for niche occupation. The dynamics of a metacommunity depends on the trade-offs resulting from the assembly of several co-occurring populations.

Neutral paradigm emerges as a null hypothesis for microbial assembly. Thus, in its models, the persistence in a certain community is the result of random processes of immigration and extinction (Figure 1d). The species have equal competition capacity [22]. Neural theory is the simplest way to characterize the complexity of a set of populations in a local community. To asses that, we only need a θ number of potential species in a given community, and a m immigration rate parameter [30].

A classical approach to evaluate neutral paradigm was described by [22], through a re-interpretation of Ewens' sampling distribution, that was initially developed for genetics studies [31]. The model undergoing this approach is based on zero-sum dynamics of a metacommunity. Indices deriving from this view are being also used for local communities [32], with recent applications in studies related to the patterns of microbial assembly in agroecosystems and rhizosphere [4,21]

A more comprehensive picture of the application of these paradigms in microbial ecology studies can be reached by both theoretical (e.g. Classic Metapopulation – neutral) and numerical (e.g. Zero-Sum Model – neutral, Broken-Stick Model – niche-based) models. The models describe the organization of microorganisms into communities, at the local level [4], and along metacommunities, at the regional level [21] (e.g. Rates of dispersal and immigration coefficient).

4. Application of metacommunity models to unravel microbial structure and functions in agroecosystems

Although we know a lot about plant and animal distribution, demography and functions, these patterns remain abstruse, when it comes to microorganisms. Knowledge about microbial assembly and functions, due conversion of pristine forests into agricultural systems, is vital to understand the possible ecological consequences.

Biogeography and metacommunity ecology studies have made possible to investigate the mechanisms leading to microbial diversity generation and maintenance in these ecosystems, such as emergence of new species, extinction, dispersal and ecological interactions [20] in several levels of complexity and scale ranges. A framework to investigate microbial patterns is needed, with references to that found to macroorganisms and the establishment of possible exceptions regarding microbial assembly and functional niche occupation. This comprise the knowledge whether microbial assembly differ among environments and space, besides the effects that modulate this variation. Moreover, a biogeographic multiscale approach can help us to unravel if spatial variation is due to punctual environmental factors, such as land-use and seasonality [11] or evolutionary selecting events [33].

As mentioned in the previous section, different species inside a local community and even communities along a metacommunity, use to have different patterns of assembly through space and time, due different ability to compete, occupy niches, and disperse along the ecosystems gradient. Thus, besides the application of metacommunity paradigms to describe microbial assembly and niche occupancy, we can also argue about the limitations and barriers to dispersal that make some species behavior to differ in a biogeographic scale (Figure 3).

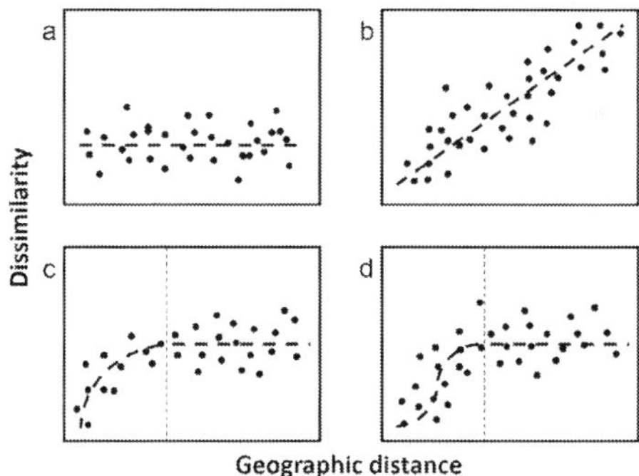

Figure 3. Relationship between geographic distance and microbial community dissimilarity. (a) According to the assumption "Everything is everywhere", all communities appear to be similar to each other, apart from distance. (b) Communities have a continuous decay of dissimilarity over geographic distance. (c) Communities have autocorrelation until a threshold (vertical dashed line), in which the limit of spatial correlation is reached. (d) Communities have a lag before autocorrelation begins (vertical dashed line), what means that in low distances, we are sampling systematically the same community. Adapted from [34].

An early conceptual groundwork for microbial biogeography can be found in Candolle province and habitat definitions for plants [35]. Bringing that into microbial boundaries, a province could be any area, in which the microbial structure reflects historical evolutionary events. Thus, the limits of a single province should vary greatly in size and are inwardly linked to the resolution and the taxa in study [33]. Areas of soybean cultivation, hundreds of miles distant each other, might be considered particular provinces, considering the general structure of bacterial communities that inhabit their rhizosphere and surrounding soil. Although, those areas may also be treated as members of a single province, taking into account that members of the bacterial class Alphaproteobacteria, are able to colonize the rhizosphere and nodulate these plants, apart from distance, due a high level of conserved genes related to this mechanism and a large number of strategies of signaling to a broader range of environments and plant species [36].

4.1. Local, regional and global factors affect soil microbial community structure

Understanding controls over the distribution of soil microbial communities is a fundamental step toward describing soil ecosystems, understanding their functional capabilities, and predicting their responses to environmental change. However, the complexity of these communities and their interactions with environmental characteristics have made generalizations difficult. Recently, high throughput sequencing technologies have facilitated the investigation of soil bacterial communities at local [37], regional [12], and global scales [38].

Microbial groups related to environmental characteristics has been recognized as the most important mechanism controlling soil microbial communities [39] with chemical soil factors identified as a master variable explaining significant portions of the variation in soil microbial diversity and community structure at local [40,41], regional [42–44] and global [45,46] scales.

However, while environmental factors have been identified as exerting primary control on soil microbial distribution, on average approximately 50% of the variation in microbial diversity and structure remains unexplained [39]. Additionally, very few examinations have been made of how controls on soil microbial communities operate simultaneously at multiple scales to contrast local and regional drivers of microbial diversity and community structure.

The factors that control soil microbial community composition are much debated. It has been suggested that while local scale variation in soil microbial communities can be explained by plant identity, substrate hotspots and soil chemical factors [47–49], at regional and continental scales, additional factors, such as climate, topography, and soil pH, become more important [50–52]. However, pH has been shown to shape soil microbial communities over distances < 1 m^2 [53], as well as at field and continental scales [38,41]. Microbial communities have also been shown to be influenced by vegetation type, land use, soil nutrient status, and soil organic matter quality and quantity at landscape scales [51,52,54].

The relationship among soil microbial communities and landscape factors, soil factors and plant communities at different spatial scales is relatively lacking, despite their importance for ecosystem functioning. This lack of understanding of the factors that explain variation in microbial communities at larger spatial scales is surprising given their functional importance in regulating ecosystem processes, such as carbon and nitrogen cycling [49,55] and the resistance of nutrient cycles to climate change-related disturbances [56]. In this sense, no studies have simultaneously tested the importance of a range of abiotic factors, including climate and soil properties, and biotic factors, such as vegetation composition, across a wide range of spatial scales. According to [56], this represents a major gap in knowledge given the potential for both abiotic and biotic factors to explain variation in soil microbial communities.

4.2. Case study: Niche-based theory explains microbial assembly in soybean rhizosphere

The rhizosphere is the immediate surroundings of the plant root, the portion of soil under influence of root exudates. The rhizosphere is considered a hot spot of microbial species, and the communities inhabiting this environment are shaped by the nutrients released by the plant, such as exudates, border cells and mucilage. Studies on rhizosphere microbiome increased in the last year, mainly because this microbiota can have profound effects on the growth, nutrition and health of plants in agroecosystems [for rhizosphere microbiome review see [57]].

In an experimental research, [4] studied the process of microbial selection in the rhizosphere from bulk soil reservoir under agricultural management of soybean in Amazon forest soils. They used a shotgun metagenomics approach to investigate the taxonomic and functional diversities of microbial communities and to test the validity of neutral and niche theories to explain the community assembly in the rhizosphere. The species rank abundance distribution generated by metagenomics was fitted to five theoretical models of assembly. The neutral theory predicts that rank abundance distribution will be consistent with ZSM model [22] and niche-based fits the pre-emption, broken stick, log-normal and Zipf-Mandelbrot models [58–60].

The authors collected samples of bulk and rhizosphere soil of soybean harvested in agricultural fields in order to evaluate which microbial groups and functional genes are selected in the rhizosphere when compared to the bulk soil. At first, they showed that there is a selection process in the rhizosphere, where the species abundance fitted the log-normal distribution model, which is an indicator of the occurrence of niche-based process. The niche theory predicts that changes in species community composition are related to changes in environmental variables, since species have unique properties that allow them to exploit unique niches available [61]. Thus, the root exudates may select organisms to inhabit the rhizosphere environment.

With the sequencing data, the authors also could show what groups of organisms are selected in the rhizosphere and what function they are playing. In this study [4], they showed that there was a selection process at both taxonomic and functional levels operating in the assembly of the rhizospheric community, with different community structure compared to the bulk soil community. The phyla Actinobacteria, Acidobacteria, Chloroflexi, Cyanobacteria, Chlamydiae, Tenericutes, Deferribacteres, Chlorobi, Verrucomicrobia and Aquificae were selected in the rhizosphere. In addition, the functional analysis indicated that functions related to the metabolism of nutrients, such as nitrogen, phosphorus, potassium and iron were more abundant in rhizosphere than the bulk soil (Figure 4).

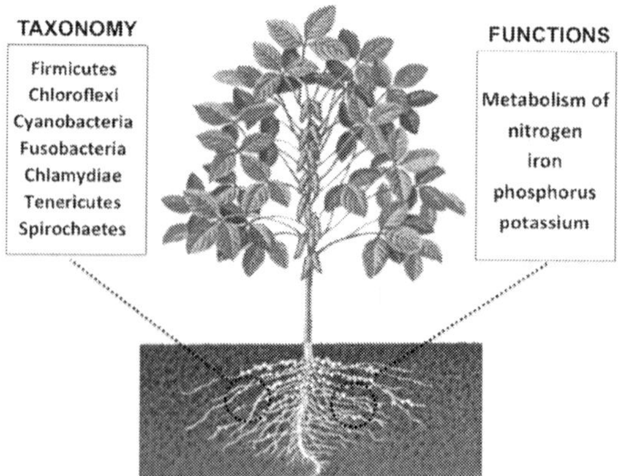

Figure 4. Taxonomical and functional groups selected in soybean rhizosphere, following the niche-based mechanisms.

The phyla indicate in the figure 4 were selected in the rhizosphere and are playing important functions related to the metabolism of some important nutrients to the plant. The community selection in rhizosphere is influenced by exudates released from the roots, which create different niches to be exploited by the soil microorganisms. The roots deposits consist mainly of carbon, and secondary metabolites such as antimicrobial compounds and flavonoids. Other soil parameters also are affected by the root system, such as pH, moisture, oxygen pressure and nutrient availability. In this study, the authors used a community assembly approach to understand the microbial selection process in rhizosphere, and they have shown that soybean selects a specific microbial community inhabiting the rhizosphere based on functional traits, which may be related to benefits to the plant, as growth promotion and nutrition. The microbial community assembly in the rhizosphere follows largely the niche-based mechanisms, showing that variations in the rhizosphere promoted by roots exudates shape the microbial community structure.

5. Concluding remarks

In this chapter, we have discussed the considerations of the applications of metacommunity theories and biogeography for land-use management and agroecosystems conservation. The contribution of these models to explain the patterns of structure, abundance and functional traits at local and regional scales were emphasized here. We settled some bases for the application of metacommunity models, regarding community assembly and microbial functions in agroecosystems, including recent results from our group and several theoretical and experimental studies available in the literature.

Studies of microbial assembly and its linkage to the functional resilience in the agroecosystems are very important for microbial ecologists. Comparative studies in different agroecosystems and regions of the globe are needed, to stablish a huge conceptual view about the patterns of microbial distribution and ecological relationships. Based on these several studies, we can argue about soil quality and find global bioindicators of soil health as well as endemic microbial populations with local and regional importance to maintain the ecosystems equilibrium and guarantee the biodiversity, acting as niche holders.

Metacommunity and biogeography concepts emerge as important tools to evaluate bioindicators of soil quality and functional resilience, since both can be applied to a broader range of environments, from the microcosm scale up to the landscape or regional scale, independently of the type of soil, management or species to be reached.

Acknowledgements

We wish to thank Fundação de Amparo à Pesquisa do Estado de São Paulo (FAPESP 2008/58114-3, 2011/51749-6) and Conselho Nacional de Desenvolvimento Científico e Tecnológico (CNPq 485801/2011-6) for funding.

Author details

Dennis Goss de Souza[1,2], Lucas William Mendes[1], Acácio Aparecido Navarrete[1] and Siu Mui Tsai[1*]

*Address all correspondence to: tsai@cena.usp.br

1 Cellular and Molecular Biology Laboratory, Center for Nuclear Energy in Agriculture (CENA), University of São Paulo (USP), Piracicaba, Brazil

2 Department of Land, Air and Water Resources, College of Agricultural and Environmental Sciences, University of California, Davis (UCDavis), Davis, USA

References

[1] De Wit R, Bouvier T. "Everything is everywhere, but, the environment selects"; what did Baas Becking and Beijerinck really say? Environ Microbiol 2006;8:755–8.

[2] Metzker ML. Sequencing technologies-the next generation. Nat Rev Genet 2010;11:31–46.

[3] Daniel R. The metagenomics of soil. Nat Rev Microbiol 2005;3:470–8.

[4] Mendes LW, Kuramae EE, Navarrete AA, van Veen JA, Tsai SM. Taxonomical and functional microbial community selection in soybean rhizosphere. ISME J 2014:1–11.

[5] Tabatabai MA. Soil enzymes. In: Weaver RW, Angle JS, Bottomley PS, editors. Methods soil Anal. Part 2, Madison: SSSA Book Series; 1994, p. 775–833.

[6] Yang Z, Liu S-Q, Zheng D, Feng S. Effects of cadium, zinc and lead on soil enzyme activities. J Environ Sci 2006;18:1135–41.

[7] Sainju UM, Stevens WB, Caesar-Tonthat T, Liebig M a. Soil greenhouse gas emissions affected by irrigation, tillage, crop rotation, and nitrogen fertilization. J Environ Qual 2012;41:1774–86.

[8] Morales SE, Cosart T, Holben WE. Bacterial gene abundances as indicators of greenhouse gas emission in soils. ISME J 2010;4:799–808.

[9] Drenovsky RE, Steenwerth KL, Jackson LE, Scow KM. Land use and climatic factors structure regional patterns in soil microbial communities. Glob Ecol Biogeogr a J Macroecology 2010;19:27–39.

[10] Bell C, McIntyre N, Cox S, Tissue D, Zak J. Soil microbial responses to temporal variations of moisture and temperature in a chihuahuan desert grassland. Microb Ecol 2008;56:153–67.

[11] Lauber CL, Ramirez KS, Aanderud Z, Lennon J, Fierer N. Temporal variability in soil microbial communities across land-use types. ISME J 2013;7:1641–50.

[12] Dequiedt S, Thioulouse J, Jolivet C, Saby NP a, Lelievre M, Maron P-A, et al. Biogeographical patterns of soil bacterial communities. Environ Microbiol Rep 2009;1:251–5.

[13] Handelsman J, Smalla K. Techniques: conversations with silent majority. Curr Opin Microbiol 2003;6:271–3.

[14] Knight R, Jansson J, Field D, Fierer N, Desai N, Fuhrman JA, et al. Unlocking the potential of metagenomics through replicated experimental design. Nat Biotechnol 2012;30:513–20.

[15] Caporaso JG, Lauber CL, Walters WA, Berg-Lyons D, Lozupone CA, Turnbaugh PJ, et al. Global patterns of 16S rRNA diversity at a depth of millions of sequences per sample. Proc Natl Acad Sci U S A 2011;108:4516–22.

[16] Luo C, Tsementzi D, Kyrpides N, Read T, Konstantinidis KT. Direct comparisons of Illumina vs. Roche 454 sequencing technologies on the same microbial community DNA sample. PLoS One 2012;7:1–12.

[17] Margulies M, Egholm M, Altman WE, Attiya S, Bader JS, Bemben LA, et al. Genome sequencing in microfabricated high-density picolitre reactors. Nature 2005;437:376–80.

[18] Wilson DS. Complex interactions in metacommunities, with implications for biodiversity and higher levels of selection. Ecology 1992;73:1984–2000.

[19] Loeuille N, Leibold MA. Effects of local negative feedbacks on the evolution of species within metacommunities. Ecol Lett 2014;17:563–73.

[20] Leibold MA, Holyoak M, Mouquet N, Amarasekare P, Chase JM, Hoopes MF, et al. The metacommunity concept: a framework for multi-scale community ecology. Ecol Lett 2004;7:601–13.

[21] Barberán A, Casamayor EO. Global phylogenetic community structure and β-diversity patterns in surface bacterioplankton metacommunities. Aquat Microb Ecol 2010;59:1–10.

[22] Hubbell SP. The unified neutral theory of biodiversity and biogeography. Princeton: Princeton University Press; 2001.

[23] Siqueira T, Bini LM, Roque FO, Cottenie K. A metacommunity framework for enhancing the effectiveness of biological monitoring strategies. PLoS One 2012;7:1–12.

[24] Hovatter SR, Dejelo C, Case AL, Blackwood CB. Metacommunity organization of soil microorganisms depends on habitat defined by presence of Lobelia siphilitica plants. Ecology 2011;92:57–65.

[25] Yu DW, Wilson HB. The competition-colonization trade-off is dead; long live the competition-colonization trade-off. Am Nat 2001;158:49–63.

[26] Pianka ER. Latitudinal gradients in species diversity: a review of concepts. Am Nat 1966;100:33–46.

[27] Whittaker RH. Classification of natural communities. Bot Rev 1962;28:1–239.

[28] Chase JM, Leibold MA. Ecological niches: linking classical and contemporary approaches. Chicago: University of Chicago Press; 2003.

[29] Mouquet N, Loreau M. Community patterns in source-sink metacommunities. Am Nat 2003;162:544–57.

[30] Alonso D, McKane AJ. Sampling Hubbell's neutral theory of biodiversity. Ecol Lett 2004;7:901–10.

[31] Tabare S, Ewens WJ. Multivariate Ewens distribution. In: Johnson NL, Kotz S, Balakrishman N, editors. Discret. Multivar. Distrib., New York: Wiley; 1997, p. 232–46.

[32] Volkov I, Banavar JR, Hubbell SP, Maritan A. Neutral theory and relative species abundance in ecology. Lett to Nat 2003;424:1035–7.

[33] Martiny JBH, Bohannan BJM, Brown JH, Colwell RK, Fuhrman JA, Green JL, et al. Microbial biogeography: putting microorganisms on the map. Nat Rev Microbiol 2006;4:102–12.

[34] Robeson MS, King AJ, Freeman KR, Birky CW, Martin AP, Schmidt SK. Soil rotifer communities are extremely diverse globally but spatially autocorrelated locally. Proc Natl Acad Sci U S A 2011;108:4406–10.

[35] Nelson G. From Candolle to Croizat: comments on the history of biogeography. J Hist Biol 1978;11:269–305.

[36] Pini F, Galardini M, Bazzicalupo M, Mengoni A. Plant-bacteria association and symbiosis: are there common genomic traits in alphaproteobacteria? Genes (Basel) 2011;2:1017–32.

[37] Acosta-Martínez V, Dowd S, Sun Y, Allen V. Tag-encoded pyrosequencing analysis of bacterial diversity in a single soil type as affected by management and land use. Soil Biol Biochem 2008;40:2762–70.

[38] Lauber CL, Hamady M, Knight R, Fierer N. Pyrosequencing-based assessment of soil pH as a predictor of soil bacterial community structure at the continental scale. Appl Environ Microbiol 2009;75:5111–20.

[39] Hanson CA, Fuhrman JA, Horner-devine MC, Martiny JBH. Beyond biogeographic patterns: processes shaping the microbial landscape. Nat Rev Microbiol 2012;10:497–506.

[40] Navarrete AA, Kuramae EE, de Hollander M, Pijl AS, van Veen JA, Tsai SM. Acidobacterial community responses to agricultural management of soybean in Amazon forest soils. FEMS Microbiol Ecol 2013;83:607–21.

[41] Rousk J, Baath E, Brookes PC, Lauber CL, Lozupone C, Caporaso JG, et al. Soil bacterial and fungal communities across a pH gradient in an arable soil. ISME J 2010:1–12.

[42] Jesus EC, Marsh TL, Tiedje JM, Moreira FM. Changes in land use alter the structure of bacterial communities in Western Amazon soils. ISME J 2009;3:1004–11.

[43] Navarrete AA, Taketani RG, Mendes W, Cannavan FDS, Moreira FM de S, Tsai SM. Land-use systems affect archaeal community structure and functional diversity in western amazon soils. Rev Bras Ciência Do Solo 2011;35:1527–40.

[44] Rodrigues JLM, Pellizari VH, Mueller R, Baek K, Jesus EDC, Paula FS, et al. Conversion of the Amazon rainforest to agriculture results in biotic homogenization of soil bacterial communities. Proc Natl Acad Sci U S A 2013;110:988–93.

[45] Chu H, Fierer N, Lauber CL, Caporaso JG, Knight R, Grogan P. Soil bacterial diversity in the Arctic is not fundamentally different from that found in other biomes. Environ Microbiol 2010;12:2998–3006.

[46] Fierer N, Jackson RB. The diversity and biogeography of soil bacterial communities. Proc Natl Acad Sci U S A 2006;103:626–31.

[47] Bezemer TM, Fountain MT, Barea JM, Christensen S, Dekker SC, Duyts H, et al. Divergent composition but similar function of soil food webs of individual plants: plant species and community effects. Ecology 2010;91:3027–36.

[48] Orwin KH, Buckland SM, Johnson D, Turner BL, Smart S, Oakley S, et al. Linkages of plant traits to soil properties and the functioning of temperate grassland. J Ecol 2010;98:1074–83.

[49] Thomson BC, Ostle N, McNamara N, Bailey MJ, Whiteley AS, Griffiths RI. Vegetation affects the relative abundances of dominant soil bacterial taxa and soil respiration rates in an upland grassland soil. Microb Ecol 2010;59:335–43.

[50] Ettema CH, Wardle DA. Spatial soil ecology. Trends Ecol Evol 2002;17:177–83.

[51] Lauber CL, Strickland MS, Bradford MA, Fierer N. The influence of soil properties on the structure of bacterial and fungal communities across land-use types. Soil Biol Biochem 2008;40:2407–15.

[52] Fierer N, Strickland MS, Liptzin D, Bradford MA, Cleveland CC. Global patterns in belowground communities. Ecol Lett 2009;12:1238–49.

[53] Baker KL, Langenheder S, Nicol GW, Ricketts D, Killham K, Campbell CD, et al. Environmental and spatial characterisation of bacterial community composition in soil to inform sampling strategies. Soil Biol Biochem 2009;41:2292–8.

[54] Jangid K, Williams M a., Franzluebbers AJ, Sanderlin JS, Reeves JH, Jenkins MB, et al. Relative impacts of land-use, management intensity and fertilization upon soil microbial community structure in agricultural systems. Soil Biol Biochem 2008;40:2843–53.

[55] Balser TC, Firestone MK. Linking microbial community composition and soil processes in a California annual grassland and mixed-conifer forest. Biogeochemistry 2005;73:395–415.

[56] De Vries FT, Liiri ME, Bjørnlund L, Bowker MA, Christensen S, Setälä HM, et al. Land use alters the resistance and resilience of soil food webs to drought. Nat Clim Chang 2012;2:276–80.

[57] Philippot L, Raaijmakers JM, Lemanceau P, Putten WH Van Der. Going back to the roots: the microbial ecology of the rhizosphere. Nat Rev Microbiol 2013;11:789–99.

[58] Motomura I. On the statistical treatment of communities. Zool Mag 1932;44:379–83.

[59] Macarthur RH. On the relative abundance of bird species. Zoölogy 1957;43:293–5.

[60] McGill BJ, Etienne RS, Gray JS, Alonso D, Anderson MJ, Benecha HK, et al. Species abundance distributions: moving beyond single prediction theories to integration within an ecological framework. Ecol Lett 2007;10:995–1015.

[61] Jongman RHG, ter Braak CJ, van Tongeren OFR. Data analysis in community and landscape ecology. Cambridge: Cambridge University Press; 1995.

Chapter 8

Impact of Organic Farming on Biodiversity

Martina Bavec and Franc Bavec

Additional information is available at the end of the chapter

http://dx.doi.org/10.5772/58974

1. Introduction

In last several decades agriculture has been oriented towards industrial and extremely intensive farming practices, aimed at ensuring enough food for the human population, a goal that was not achieved. These types of farming practices also caused several negative environmental impacts such as decreasing biodiversity, including the farm bird index, where a decline has been observed in Slovenia since 2008. Many farms intensified their activities and became highly mechanized, whilst those unable to do so became increasingly marginalized and were sometimes forced to abandon their land, causing equally devastating consequences for biodiversity [1]. Today, it is globally imperative that the growing demand for food be met in a manner that is socially equitable and ecologically sustainable over the long term. It is possible to design farming systems that are equally productive and that maintain or enhance the provisioning of ecosystem services (i.e., biodiversity, soil quality, nutrient management, water-holding capacity, control of weeds, diseases and pests, pollination services, carbon sequestration, energy efficiency and reducing global warming potential, as well as resistance and resilience to climate change and crop productivity) and thus agroecosystem resilience and sustainability [2].

Organic agriculture refers to a farming system that enhances soil fertility by maximizing the efficient use of local resources, while foregoing the use of agrochemicals, genetically modified organisms and the many synthetic compounds used as food additives. The high quality of organic food and its added value relies on a number of farming practices based on ecological cycles, and aims at minimizing the environmental impact of the food industry, preserving the long term sustainability of soil and reducing to a minimum the use of non-renewable resources [3].

Organic farming practices have been promoted as reducing the environmental impacts of agriculture. The results of meta-analysis of studies that compare the environmental impacts

of organic and conventional farming in Europe show that organic farming practices generally have positive impacts on the environment per unit of area, but not necessarily per product unit. Significant differences between the two farming systems include soil organic matter content, nitrogen leaching, nitrous oxide emissions per unit of field area, energy use and land use. Most of the studies that compared biodiversity in organic and conventional farming demonstrated lower environmental impacts from organic farming [4]. Furthermore, organic farming appears to perform better than conventional farming and also provides other important environmental advantages such as halting the use of harmful chemicals and their spread in the environment and along the trophic chain, and reducing water use [3]. A life cycle analysis approach calculating the ecological footprint of different productions systems confirmed, respectively, 8.5 and 5.9 times better environmental performance of organic farming practices, compared to their conventional counterparts in winter wheat and spelt production [5].

Biodiversity loss and the degradation of ecosystems have important implications for the environment and are costly for society as a whole [6]. In Europe, loss of plant biodiversity is primarily reflected in the decline of many species of plants and in the disappearing of local and old plant varieties. In 2011, the European Parliament adopted the European Union (EU) Biodiversity Strategy to 2020 with aim of preventing biodiversity loss and the degradation of ecosystems. The strategy includes combating invasive alien species that jeopardize biodiversity and aims also at enhancing the positive contribution of European agriculture, forest and fishery sectors to biodiversity conservation and sustainable use, and to increase by 2020 the EUs contribution to drawing attention to global biodiversity loss [1]. The World Trade Organization notes a crop variety loss of 75% during the past 100 years, even of 90% in the EU. Only 17% of species and habitats assessed under the Habitats Directive have been deemed to be in good status and the degradation and loss of natural capital is jeopardizing efforts for attaining the EUs biodiversity and climate change objectives [7, 4], which did not reach its 2010 biodiversity target [1].

Organic farming is a production method that preserves or even enrich biodiversity at the field level, at the farm level and in the ecosystem as per its regulatory demands, where the objectives of organic farming in EU regulation 834/2007 is noted thus: that organic farming shall pursue to establish a sustainable management system for agriculture with respect to nature's systems and cycles, and sustain and enhance the health of soil, water, plants and animals and the balance between them, and to contribute to a high level of biological diversity [5]. Organic farming systems generally harbour larger floral and faunal biodiversity, more so than conventional systems; however, when properly managed, the latter can also improve biodiversity. Importantly, the landscape surrounding farmed land also appears to have the potential to enhance biodiversity in agricultural areas [3]. However, the benefits of organic farming to biodiversity in agriculture landscapes are still being discussed.

Agrobiodiversity is an important aspect of biodiversity that is directly influenced by different production methods, especially at the field level. It can also supply several ecosystem services to agriculture, thus reducing environmental externalities and the need for off-farm inputs. Organic farming is considered an environmentally-friendly agriculture practice and a holistic approach encompassing several demands and bans from a regulatory point of view [8], and receives primarily from European countries additional agri-environmental payments for ecosystem services, including biodiversity. In some countries, payments are available as single biodiversity measures (i.e., hedgerows, insectary strips, crop rotation, or the retention of semi-natural areas) in agri-environmental programmes that are also aimed at conventional agriculture.

2. Aim and methodology

The aim of this paper is to establish whether organic farming fulfils the promise of protecting biodiversity better than conventional farming, based on the review of recent publications emphasizing the importance of precisely quantifying the effect of organic vs. conventional farming. Additional to an extensive review, data from the University of Maribor regarding the effects of different production systems on the earthworm population [9] and the biodiversity of weed species from field experiments in the north east of Slovenia [10] were compared with other findings.

The reader is kindly referred to previously mentioned sources [5, 9, 10, 36] for a detailed description of differences between farming systems. For a better general understanding, some details are explained. Earthworms were collected in October 2009, 2010 and 2011 using a mustard aqueous solution as a non-toxic irritant that drove deep burrowing earthworm species to the surface [11]. After measurements were taken, earthworms were returned back to the soil. Analyses were carried out using the Statgraphics Centurion XV statistical program [12]. The biodiversity of weed species [9] was measured using two methods: (i) above-ground weed population sampling; (ii) seedbank sampling. The size of the weed seedbank was determined within the 0 to 0.2 m soil layer of each plot using the green-house emergence method [13]. The in situ number of the above-ground weed population per m^2 was measured at the end of June or at the beginning of July 2009, 2010 and 2011, after mechanical control and the use of herbicides. Weeds were counted in four 0.25 m^2 quadrates randomly located in the centre of each plot, parallel to the working direction of machinery. The weed species were determined when a 2/3 population was at the stages of 2 to 3 true leaves and 1/3 was at the stages of 4 to 5 true leaves. Species diversity was calculated for both seedbank and weed communities using H' [14].

3. Results and discussion

3.1. Overall data

Results of several research studies and published scientific articles showed that organic farming benefits to the environment, including biodiversity. Comparison of biodiversity in organic and conventional farms has shown that organic farming generally had positive impacts on many species [15]. Results of meta-analyses that compared biodiversity in organic and conventional farms found that organic farms generally have 30% higher species richness and 50% higher abundance of organisms than conventional farms. However, there are wide variations between different studies, which have to be discussed; for example, 16% of studies found a negative effect of organic farming on species richness. Additionally, it was also found that the effect of organic farming on species richness was larger for intensively managed landscapes than for diverse landscapes with many non-crop biotopes [16]. In 327 out of 396 relevant results [17], a higher degree of biodiversity in organic farming was found when compared to conventional farming. In 56 papers (14 %), no difference was verified and in 13 contributions (3%), organic farming yielded less biodiversity (seven of them for soil invertebrates). Significantly, the positive effect of organic farming on biodiversity compared to conventional farming was noticed in 80% of cases; in 16%, differences were unclear and less biodiversity was found in 4% of comparisons (Table 1).

Author	Number of comparisons[1]	Number of biodiversity indicators[2]	Significantly positive effect – more biodiversity	No significant differences – unclear, indifferent	Significantly negative effect – less biodiversity
Rahmann [17]	343	10	327	56	13
Hole et al. [15]	76	9	66	25	8
Pfiffner [18]	44	7	49	5	1
Sum			442	86	22
Share (%)			80	16	4

[1] Multiple citations of used studies are possible due to different conclusions for different species or multiple answers; [2] biodiversity indicators i.e., flora, weeds, soil biota, earthworms, pollinators, birds, etc.

Table 1. Impact of organic farming on biodiversity based on the literature review.

On average, organic farming increased species richness by about 30%. This result has been robust over the past 30 years of published studies. Organic farming had a greater effect on biodiversity as the percentage of arable fields of the landscape increased, that is, it is higher in intensively farmed regions [19]. Thus, it may be concluded that organic farming produces more biodiversity. Research gaps still exist for the understanding of functional biodiversity and ecosystem impact, which comprise soil biota, landscape (ecosystem and habitat) and genetic biodiversity on agricultural land in natural habitats [17]. The majority of current studies are from Northern and Western Europe and North American agriculture practices, while other regions with large areas of organic farming have been poorly investigated. Comparison between paired organic and conventional fields in India assessed a wide range of taxa (plants, soil microbes, earthworms, butterflies, dragonflies and other arthropods, reptiles, molluscs, amphibians/frogs and birds) trough different methods that showed similar trends. Habitat area, composition and management of organic fields were likely to favour higher levels of biodiversity by supporting higher numbers of species, dominance and abundance across most taxa. Organic fields are systems that are less dependent on external inputs to restore and rejuvenate the environment, resulting in higher biodiversity that promotes higher sustainable production on a long-term basis [20]. The effects of time since conversion to organic farming on species richness and abundance have been poorly researched. Plant and butterfly species richness was 20% higher on organic farms and butterfly abundance was about 60% higher, compared to conventional farms. Time since conversion to organic farming affected butterfly abundance gradually over a 25-year period, resulting in a 100% increase; however, no effect was found on plant or butterfly species richness, indicating that the main effect took place immediately after the conversion to organic farming [21].

Three recent multiregional studies from Europe have also demonstrated the negative effects of both agricultural intensification (increased use of synthetic fertilizers and pesticides combined with the reduced use of diversified farming system techniques) and landscape simplification on components of biodiversity [2]. The EU Biodiversity Strategy to 2020 also focuses on sustainable farming and forestry as the focus of one of six targets in the form of improving the integration of biodiversity conservation into key policies for agriculture and forestry. Combined, these two sectors include almost 72% of land in the EU and play a major role in Europe's biodiversity [1].

Crop rotation brings biodiversity in the time scale. It is mandatory on organic farms and is stated as a method to maintain and increase the fertility and biological activity of the soil, and means the prevention of damage caused by pests, diseases and weeds [8]. Due to more diverse crop rotation and the use of green manure and intercroppings on organic farms, there is also greater biodiversity. Furthermore, using domestic populations of seed varieties preserves biodiversity, but the production of alternative crops (rare, underutilized, disregarded, neglected or new) increase biodiversity at the filed level [22].

3.2. Weeds biodiversity

The biodiversity of weed communities in agro-ecosystems provides several valuable ecological functions [23]. Conventional and integrated production systems tend to be similar in both intensity of management and within-field biodiversity, but organic production tends to support greater density, species number and biological diversity in comparison with other investigated production systems [24]. At the field level, species richness was the greatest on organic farms where there was a greater abundance of weeds [24-27, 31; organic production system had the highest biodiversity of weed species [28-31]. Organic agricultural practices yielded more weed species in root crops, red clover/grass mixtures and in winter triticale. Weed species richness was reduced in red clover/grass stands, while root crops and spring barley undersown with red clover and grasses decreased weed species diversity, which is also important for achieving higher yields in an organic production system. The species composition and in particular the quantitative structure of weeds were affected more by crop species and cultivation regime, compared to different agriculture practices (organic vs. integrated). Weed communities of crops grown using organic and integrated farming systems were more similar in terms of species composition than quantitative structure [30].

The maintenance of a diverse weed community is one step towards the sustainability of an agro-ecosystem through improved nutrient cycling and pest control, improved soil chemical and physical properties, and the reduction of soil erosion. An important aspect in the evaluation of the environmental impact of production systems is the biodiversity index for weed species (Table 2). Using the Shannon-Weaver diversity index for weeds of different production systems (conventional, integrated, organic) growing white cabbage and red beet showed that the biodiversity index was significantly higher in organic systems (0.86 in organic vs. 0.66 in conventional systems for cabbage and 0.81 in organic vs. 0.59 in conventional for red beet). Using ecological footprint calculation for the evaluation of different production systems showed that organic farming had the lowest impact on the environment. In the case of white cabbage and red beet production, ratio in ecological footprint between organic and conventional production was 1 to 3.5 [10].

The emerged weed flora is more affected by recent agrochemical inputs than the seedbank, which is buffered by the persistence of weed seeds in the soil. The seedbank is more strongly influenced by soil characteristics, such as the percentage organic carbon and percentage total nitrogen than by management [26]. The same weed species were in the seedbank and at field counted as germinated weeds, totalling 29 weed species in the survey (Table 2). The accumulated number of observed species pooled over fields was highest in the organic production of white cabbage and red beet, with 29 and 28 species, respectively. Within the conventional crop rotations, 18 species were observed in the cabbage field and 17 in the red beets field, while 20 and 19 were observed in the integrated crop rotation for cabbage and red beets. The differences in the number of weed species between conventional and integrated fields for cabbage were not significantly different; however, the difference when comparing organic and conventional fields was significantly different for both vegetables. For red beet, differences among all production systems were significant, which is contrary to the findings of [30], where weed communities of crops grown under organic and integrated farming systems were similar with regard to species composition but not quantitative structure. Different farming practices (described as organic, integrated and conventional) appeared to exert selection pressure on the species composition of the seedbank, building up different communities under the three farming systems over time [26]. These effects were scale dependent. At a within-field scale, species richness was greatest in organic farms, where there was a greater abundance of weeds; this was similar to our results and those of many others [24-31]. These results suggest that weed species diversity can be promoted by using organic cropping practices [31].

| Production system | Weeds in white cabbage | | Weeds in red beet | | Earthworm population (no./ 0.25m^2) |
	H'	O	H'	O	
Control	0.38d	14c	0.32d	13c	11.58b
Conventional	0.66c	18b	0.59c	17b	11.25b
Organic	0.86a	29a	0.81a	28a	22.41a
Integrated	0.74b	20b	0.64b	19b	13.00b
Biodynamic	-	-	-	-	24.00a

[a-d] Mean values followed by different letters within a column are significantly different (Duncan, α=0.05)

Table 2. Shannon-Weaver diversity index (H') and the frequency of occurrence (O) of weed species from the 30 species present in white cabbage and red beet in different production systems [10],3 and the influence on earthworm population [9, 36].

3.3. Earthworms population

Organic farming systems are generally associated with increased biological activity and increased below-ground biodiversity. The main impacts on biological fertility do not result from the systems per se, but are related to the amount and quality of the soil organic matter that is used in the farming system, as well as the disruptions of soil habitat using different tillage tools. Even within the constraints of organic farming practices, it is possible for farmers to make changes to management practices using less tillage, which will tend to improved soil biological quality [32]. An important part of soil biodiversity is arbuscular mycorrhizal fungi, which can provide several benefits to plants and ecosystems. Organic farming enhances arbuscular mycorrhizal fungi, communities of which are similar in organically managed fields and in semi-natural species-rich grasslands; however, significantly less communities are found in conventionally managed fields. Their richness increased significantly over time since conversion to organic agriculture [33]. Soil microorganisms and other parts of soil biota including earthworms are also important drivers of soil fertility. Organic farming is based on the principle of the maintenance and enhancement of soil life and natural soil fertility, soil stability and soil biodiversity for preventing and combating soil compaction and soil erosion, and for the nourishing of plants primarily through the soil ecosystem [8]. Furthermore, our research results investigating the number and mass of earthworms as an indicator of soil biodiversity confirmed the effects of different production systems (conventional, integrated, organic, biodynamic) on the population of earthworms following the harvesting of different crops [9].

The studied production systems significantly influenced total earthworm population (Table 2) and small earthworms [36]. Both were shown to be higher in number in the biodynamic and organic production systems compared to the control, conventional and integrated production systems. When compared to control plots, as well as those managed without fertilizers and plant protection agents, there were roughly 2.7 and 2.5 times more small earthworms in biodynamic and organic production systems, respectively. In the same manner, the total earthworm population in the biodynamic production system was 207% and in the organic production system, 193% of this was counted for the control treatments. Similarly, the beneficial effect of organic farming on earthworms has been emphasized by other investiga-tions [34, 35]. The abundance of earthworms, as well as their total body mass, was affected by plant species occurring in crop rotation. Oil pumpkins were revealed to have a beneficial effect on earthworms. There was also a significant production system and plant species interaction concerning the population of small earthworms [36]. In addition to a production system, tillage is also a major driver for altering communities of earthworms and microorganisms in arable soils. The use of reduced tillage provides an approach for eco-intensification by enhancing inherent soil biota functions in organic arable farming [37].

3.4. Some other ecosystem services connected to biodiversity

Biodiversity, as one of the most important ecosystem services of organic farming, is firmly connected to biocontrol and pollination services [2]. While the field of organic crop production has increased globally, the potential interactions between pest management in organic and conventionally managed systems have to date received little attention [38]. Organic agriculture improves biodiversity at the field level, but potential interactions with the surrounding landscape and the potential effects on ecosystem services are less well known. Predation of aphids was the highest in organic fields in mixed landscapes and lower in more uniform surroundings. The results of comparing 153 farms from five countries showed that organic agriculture improved the biodiversity of plants and birds in all landscapes, but only in more diverse surroundings did it improve the potential for biological control. Contradictory results showed the necessity for taking into consideration production methods (organic vs. conventional) and regional landscape complexity for developing agri-environmental schemes for the future [39]. Organic farming is one of the most successful agri environmental schemes, as humans benefit from high quality food and farmers from higher prices for their products; additionally, this approach often successfully protects biodiversity. Based on the assessment of 30 triticale fields (15 organic vs. 15 conventional) and the comparison of five conventional fields that were treated with insecticides and 10 non-treated conventional fields, it was found out that organic fields had five times higher plant species richness and about 20 times higher pollinator species richness compared to conventional fields. In contrast, the abundance of cereal aphids was five times lower in organic fields, while predator abundances were three times higher and predator-prey ratios 20 times higher in organic fields, indicating a significantly higher potential for biological pest control in organic fields [40]. Aphid density was also significantly lower in organic wheat fields compared to conventional fields, based on the assessment of 216 wheat fields during a two-year study [41]. Another positive impact of crop genetic diversity where wheat is concerned was found on below (collembola) and above-ground arthropod (spiders and predatory carbides) diversity at the field scale, which may be the result of a wider variety of food resources or more complex crop architecture. Increasing crop genetic diversity can therefore be an easy-to-implement scheme for benefiting farmland biodiversity [42].

Despite decades of European policy to ban harmful pesticides, the negative effects of pesticides on wild plant and animal species are nonetheless present and can be observed through losses pertaining to biodiversity. Chemical pesticides minimize opportunities for biological pest control. If there is an aim for biodiversity to be restored in Europe, opportunities should be created for crop production utilizing biodiversity-based ecosystem services such as biological pest control; what is needed is a Europe-wide shift towards farming employing the minimal use of pesticides over large areas, not only on organic farming areas [43]. Insecticide treatment in conventional fields had only a short-term effect on aphid densities, while later in the season, aphid abundances were even higher and predator abundances lower in treated compared to untreated conventional fields. Preventative insecticide application in conventional fields has only short-term effects on aphid densities but long-term negative effects on biological pest control. Therefore, conventional farmers should restrict insecticide applications to situations where thresholds for pest densities have been reached. Organic farming increases biodiversity, including important functional groups like plants, pollinators and predators, which in turn enhance natural pest control [40].

Biodiversity supplies multiple ecosystem services to agriculture. In addition to the potential for biological pest control, pollination problems are a topic now also being addressed in EU agriculture policy [43]. Declines in insect-pollinated plants and their pollinators have been reported as a result of agricultural intensification [44]. Reducing farming intensity with conventionally managed leys does not seem to be as effective as organic farming for delivering crop pollination services [45]. The abundance of pollinators was more than 100 times higher on organic fields. Plant and pollinator species richness, as well as predator abundances and predator-prey ratios, were higher at field edges compared to field centres, highlighting the importance of field edges for ecosystem services [40].

Pollination systems within intensive grassland communities may be different from those in arable systems. Results from comparing plant community composition among 10 pairs of organic and conventional dairy farms indicate that organic management increases plant richness in field centres, but that landscape complexity exerts a strong influence on both organic and conventional field edges. Insect-pollinated forb richness showed positive relationships to landscape complexity, reflecting what has been documented for bees and other pollinators [44].

Hedges provide important nesting, feeding and sheltering sites for birds in agricultural areas, while organic farming also enhances the environments of farmland birds [15, 18, 46]. However, little is known about how the interaction of (the amount of) hedges and variables pertaining to the organic management of the landscape scale affects birds. Birds were surveyed in the fields and in the adjoining hedges on conventional and organic winter wheat fields and meadows. More bird species occurred in organic than in conventional fields, regardless of land-use type. Hedge length had a much stronger effect on bird richness than organic farming practice. The interaction of landscape complexity and hedge length was found to be connected. Hedge length enhanced bird richness only in the case of simple landscapes. In more complex landscapes, the local effect of hedge length levelled off, because bird richness was high even without local hedges. Adding hedges or introducing organic farming practices should be primarily promoted in simple landscapes, where it particularly makes a difference for biodiversity [46].

The effect of organic farming differs depending on the scale of uptake of a particular landscape. The local effect of organic farming was found to be consistently strong, with higher diversity in borders adjoining organic fields, most likely due to the lack of herbicides used on organically managed farmland. In addition to the proportion of semi-natural habitat, which is important for farmland biodiversity, the management practice of cropland can also influence diversity in semi-natural habitats. Forb richness, which was evaluated as an agri-environmental indicator for biodiversity was also higher within borders situated in landscapes with a high proportion of organic land, irrespective of local management; this was possibly as a result of the dispersal of primarily annual plant species from the organically managed fields into the borders (mass effect). Farming practice at a local and a landscape scale can independently influence plant species richness, indicating that organic farming can also influence diversity at larger spatial scales, as well as outside organically managed land [47]. Organic farming enhances species richness and the abundance of many common taxa, but its effects are often species specific, as well as trait or context dependant. Landscape enhances or reduces the positive effects of organic farming, or acts through interactions where the surrounding landscape affects biodiversity differently on organic and conventional farms [48].

Around the world, small farms are those that practice high-diversity agriculture. Small farmers often choose to cultivate several varieties of the same crop; additionally and perhaps more importantly, different farmers in a given locality often cultivate different varieties. On the other hand large farms usually sow a single variety over a wide area [49]. Small farms may in this way have an indirect, positive effect on biodiversity, since these farms normally have smaller land parcels and thus more field edges, which are relatively species-rich. Although the average organic farm is bigger in the EU than its conventional counterparts [50] and in some cases is "conventionalized", organic farming is nonetheless generally viewed as small farms. The world's majority of food is produced by smallholder farmers who grow over 70% of all our food. Organic farming on small farms leads to an increase in food production and to greater benefits for the ecosystem by improving soil organic matter, reducing erosion and increasing biodiversity. At the same time, organic farming also allows farmers to receive higher prices for their value-added produce and provide them with opportunities to export to markets niche [51]. The report of a study focusing on farming systems in Africa showed that it is possible to set broad priorities for agricultural intensification based on the organic principles of health, ecology, fairness and caring for the earth. Ecological principles and technologies can be used to support farmers in obtaining food security and improving their livelihoods without destroying the local indigenous biodiversity [52].

3.5. Agri-environmental payments and farmers' attitudes towards biodiversity

Agricultural intensification has caused significant declines in biodiversity, while the profound intensification of European agricultural practices in the past number of decades continues. This is due to decreasing crop diversity, simplification of cropping methods, the use of fertilizers and pesticides and the homogenization of landscapes, all of which have negative effects on biodiversity in agricultural areas. Agricultural management practices can have a substantial positive impact on the conservation of the EUs wild flora and fauna. Agri-environmental schemes including organic farming are thought to benefit biodiversity. Agri-environmental payments are part of Common agriculture policy, which promotes the multifunctional role of farming as a provider of food products and a steward of diverse landscapes, as well as the cultural and natural heritage of rural areas. Furthermore, in the future, according to the EU regulation 1305/2013, each member state has to introduce agri-environmental measures for enhancing biodiversity and the preservation of high nature value farming and forestry systems [53]. Ecosystem services payments must be based on a standardized and transparent assessment of the goods and services provided. This is especially relevant in the context of EU agri-environmental programmes, but also for organic-food companies that foster environmental services on their contractor farms [54].

Agri-environmental schemes have been introduced to minimize the effects of agricultural intensification and enhance farmland biodiversity, but evaluations have produced inconsistent results [47]. Biodiversity is in different countries supported by different measures (i.e., strips and hedges, crop rotation, autochthone varieties, Nature 2000 measures), as is organic farming, which enhances the species richness and abundance of above and below soil taxa [15-20]. Traditional farming contributes to the safeguarding of certain natural or semi-natural habitats. Many valuable habitats and the presence of species have a direct interdependence with agriculture (e.g., many bird species nest and feed on farmland). Two major changes have contributed to upsetting the delicate balance between agriculture and biodiversity: (i) specialization and intensification of certain production methods (such as the use of more chemicals and heavy machinery); (ii) marginalization or abandonment of traditional land management, a key factor in preserving certain habitats and site-specific biodiversity. In some EU member states, land abandonment and the withdrawal of traditional management may become a threat to biodiversity on farmland. Therefore, preventing these processes is a key action for halting the loss of biodiversity. The Common agricultural policy addresses the preservation of habitats and biodiversity by specific rural development measures targeted at the preservation of habitats and biodiversity (agri-environmental and Nature 2000 payments), as well as requirements included in the scope of cross compliance for birds and habitats [55].

Agri-environmental payments to farmers for the conversion from conventional to organic farming or remaining inorganic should encourage them to participate in schemes, thereby responding to the increasing demand by society for the use of environmentally-friendly farm practices and also for high standards of animal welfare, as is the case in organic farming. In order to increase synergy in biodiversity and the benefits delivered by the organic farming, other measures should also be promoted and supported among organic farmers in order to cover larger areas or other protected areas, e.g., Nature 2000 [53].

In agricultural landscapes, farmers have a large impact on biodiversity through the management decisions and agricultural practices that are used on their farms. Farmers' perceptions of biodiversity and its different values influence their willingness to apply biodiversity-friendly farming practices. Organic and conventional farmers' perceptions of the different values of biodiversity were analysed across three European countries. Farmers' perceptions of biodiversity were strongly connected to their everyday lives and linked to farming practices. In addition to recognizing the importance of variety, species and habitat diversity, farmers also acknowledged wider landscape processes and attached value to the complexity of ecological systems. It was found that organic farmers tended to have a more complex and philosophical approach to biodiversity, with little differences being observed between these farmers; conventional farmers, on the other hand, exhibited more differences among themselves. Furthermore, ethical and social values were important for all farmers, but economic value was more important for conventional farmers, which has an impact on their behaviour [56].

Based on a survey among organic and conventional farmers, it was concluded that they had similar attitudes to farming results and to the environment; however, organic farmers were better informed about environmental issues and carried out more environmentally-friendly practices and behaviours. More biodiversity was found on environmentally-friendly orientated farms and less on high production-orientated farms. Organic farmers with more positive attitudes to the environment and who were better informed about environmental topics had higher biodiversity on their farms compared to others. Although there were disparities between attitudes and actual behaviours in relation to the environment among organic farmers sharing similar attitudes to conventional farmers, they were more prepared to inform themselves about and carry out environmentally-friendly farming. Results of the comparison study showed that biodiversity benefitted more from organic farming and environmentally-oriented farmers, and that there is an important link between farmers' environmental attitudes and knowledge and the beneficial effects of organic farming on biodiversity [57].

Farmers strongly acknowledged ethical and social biodiversity values. This suggests that soft policy tools can also foster biodiverse-sensitive farming methods that are complementary to mainstream monetary incentives [55]. As farmers receive a majority of agri-environmental payments, they can be more involved in data generation and conservation management. Farm size is very important in terms of the amount of payments that are provided per hectare and for improving biodiversity on a bigger scale. A standardized model for measuring on-farm biodiversity does not yet exist in practice. Performance indicators should be focused on and farmers should be included in generating this information. A framework is needed for assessment of the results and for management measures that can be employed on farms. Another requirement is ease of application, which encompasses the simplicity of gathering input data and its clarity to those farmers who will apply it [54]. Conservation-oriented thinking and better environmental education among farmers should be encouraged for those who already participate in an agri-environmental scheme and even more so amongst new-comers. In this way, the benefits of the agri-environmental schemes for the environment can be maximized [57].

An open source farm assessment system was prepared for the assessment of biodiversity including biotopes, species, biotope connectivity and the influence of land use. Interviews with the test farmers showed that the assessment methods can be implemented on farms and that they were understood by farmers [54].

4. Conclusions

The analysed data showed that in the past decades, the specialization and intensification of agriculture production methods have had negative effects on biodiversity. The future holds the challenge of designing more sustainable farming systems that are productive and maintain or enhance the provision of ecosystem services, including biodiversity. The significantly positive effect of organic farming on biodiversity compared to conventional farming was noticed in 80% of cases; in 16%, differences were unclear and less biodiversity was found in 4% of comparisons [15, 17, 18], where seven to 10 biodiversity indicators were taken into account. Small farms in particular may have an indirect positive effect on biodiversity. These farms generally have smaller land parcels and thus more field edges, which are relatively species-rich.

We can conclude that the benefits of organic farming on biodiversity are as follows:

i.   Organic farming increased species richness by about 30% and had a greater effect on biodiversity, as the percentage of the landscape consisting of arable fields increased. It was found that organic fields had up to five times higher plant species richness compared to conventional fields. For example, plant and butterfly species richness was up to 20% higher on organic farms and butterfly abundance was about 60% higher. After the conversion from conventional to organic farming abundance of butterflies was increased for 100%. Organic farming enhanced arbuscular mycorrhizal fungi and its communities. This was similar in organically managed fields and in semi-natural species rich grasslands, but significantly fewer communities were found in conventionally managed fields. Their richness increased significantly over time from the point of a conversion to organic agriculture.

ii.   The occurrence of weed species was significantly higher in the organic production of white cabbage and red beet compared to integrated and conventional production. The biodiversity index was significantly higher in organic production compared to the conventional method, 0.86 vs. 0.66 for cabbages and 0.81 vs. 0.59 for red beets. Conventional and integrated production systems tended to be similar both in terms of the intensity of management and regarding within-field biodiversity; however, organic production tended to support greater density, species number and biological diversity compared to other investigated production systems.

Earthworms were more abundant on organically managed fields. In organic and biodynamic farming plots, the number of earthworms was on average two times higher compared to integrated, conventional and control plots.

iii.   Biodiversity as one of the most important ecosystem services of organic farming is firmly connected to biocontrol and pollination services, which are enhanced when using no or less chemicals. The abundance of cereal aphids was five times lower in organic fields, while predator abundances were 20 times higher in organic fields, indicating a significantly higher potential for biological pest control in organic fields. Organic fields had 20 times higher pollinator species richness compared to conventional fields. Pollinators and predator abundance was higher at field edges compared to field centres, highlighting the importance of field edges for ecosystem services. Edges provide important nesting, feeding and sheltering sites for birds in agricultural areas. Thus, organic farming enhances farmland birds.

Overall, organic agriculture appears to perform better than conventional farming and provides important environmental advantages such as halting the use of harmful chemicals and their spread in the environment and along the trophic chain, reducing water use, as well as reducing carbon and ecological footprints. As we have underscored, organic farming fulfils the promise to protect biodiversity better than conventional farming. However, in the European commission document, The EU Biodiversity Strategy to 2020 [1], organic farming is not even mentioned, while in the European Parliament resolution regarding the strategy [6], organic farming is mentioned only once in the context of a call for a strengthening of Pillar II and for drastic improvements to the environmental focus of that pillar, and to the effectiveness of its agri-environmental measures. Supporting farmers to convert their properties to organic land and to maintain organic farming within the scope of agri-environmental schemes as a part of Common agriculture policy can have a significant impact on biodiversity as a result of management decisions farmers apply to their agricultural land.

Acknowledgements

The results presented in this paper are in part an output of the national research project J4-9532: "The quality of food dependent on the agricultural production method", funded by the Ministry of Higher Education, Science and Technology of the Republic of Slovenia and Slovenian Research Agency.

Author details

Martina Bavec* and Franc Bavec

*Address all correspondence to: martina.bavec@um.si

University of Maribor, Faculty of Agriculture and Life Sciences, Hoče/Maribor, Slovenia

References

[1] European Commission. The EU Biodiversity Strategy to 2020. 2011. http://ec.euro-pa.eu/environment   /nature/info/pubs/docs/brochures/2020%20Biod%20brochure%20final%20lowres.pdf (Accessed 1 April 2014).

[2] Kremen, C., Miles, A. Ecosystem services in biologically diversified versus conventional farming systems: benefits, externalities, and trade-offs. Ecology and Society. 2012;17 http://dx.doi.org/10.5751/ES-05035-170440 (Accessed 1 April 2014).

[3] Gomiero, T., Pimentel, D., Paoletti, M.G. Environmental Impact of Different Agricultural Management Practices: Conventional vs. Organic Agriculture. Critical Reviews in Plant Sciences. 2011:30 95-124.

[4] Tuomisto, H.L., Hodge, I.D., Riordan, P., Macdonald, D.W. Does organic farming reduce environmental impacts? - A meta-analysis of European research. Journal of Environmental Management. 2012;112 309-320.

[5] Bavec, M., Naradoslawsky, M., Bavec, F., Turinek, M. Ecological impact of wheat and spelt production under industrial and alternative farming systems. Renewable Agriculture and Food Systems, 2012;27 242-250.

[6] Our life insurance, our natural capital: an EU biodiversity strategy to 2020. European Parliament resolution of 20 April 2012 on our life insurance, our natural capital: an EU biodiversity strategy to 2020 (2011/2307(INI)). http://consilium.europa.eu/media/1379139/st18862.en11.pdf (Accessed 20 March 2014).

[7] EC Proposal for a Decision of the European Parliament and of the Council on a Genaral Union Environment Action Programme to 2020 "Living well, within the limits of our planet" /* COM/2012/0710 final - 2012/0337 (COD). http://www.europarl.euro-

pa.eu/sides/getDoc.do?type= REPORT&reference=A7-2013-0166&language=EN (Accessed 20 March 2014).

[8] Council Regulation (EC) No 834/2007 of 28 June 2007 on organic production and labelling of organic products and repealing Regulation (EEC) No 2092/91. http://eurlex.europa.eu/LexUriServ/LexUriServ.do?uri=OJ:L:2007:189:0001:0023:EN:PDF (Accessed 1 April 2014).

[9] Turinek, M. Comparability of the biodynamic production system regarding agronomic, environmental and quality parameters. Ph.D. thesis. University of Maribor; 2011.

[10] Štraus, S. Potential indicators for sustainability assessment of food production on the field level. Ph.D. thesis. University of Maribor; 2012.

[11] Lawrence, A.P., Bowers, M.A. A test of the 'hot' mustard extraction method of sampling earthworms. Soil Biology and Biochemistry. 2002;34 549-552.

[12] Statgraphics® Centurion XV. User Manual. Herndon, StatPoint, Inc. 2005, 287.

[13] Teasdale, J.R., Mangum, R.W., Radhakrishnan, J., Cavigelli, M.A. Weed seedbank dynamics in three organic farming crop rotations. Agronomy Journal. 2004;96 1429-1435.

[14] Magurran, A.E. Ecological diversity and its measurement. Princeton University Press, Princeton. 1988, 179.

[15] Hole, D.G., Perkins, A.J., Wilson, J.D., Alexander, I.H., Grice, F., Evans, A.D. Does organic farming benefit biodiversity? Biological Conservation. 2005;122 113-130.

[16] Bengtsson, J., Ahnstrom, J., Weibull, A. The effects of organic agriculture on biodiversity and abundance: a meta-analysis. Journal of Applied Ecology. 2005;42 261-269.

[17] Rahman, R. Biodiversity and organic farming: What do we know? Agriculture and Forestry Research. 2011:61 189-208.

[18] Pfiffner, L. Which farming methods enhance faunal diversity? Agrarforschung. 1996:3 527-530.

[19] Tuck, S.L., Winqvist, C., Mota, F., Ahnstrom, J., Turnbull, L.A., Bengtsson, J. Land-use intensity and the effects of organic farming on biodiversity: a hierarchical meta-analysis. Journal of Applied Ecology, 2014;51 746-755.

[20] Padmavathy, A., Poyyamoli, G. Biodiversity comparison between paired organic and conventional fields in Puducherry, India. Pakistan Journal of Biological Sciences. 2013;16 1675-1686.

[21] Jonason, D., Andersson, G.K.S., Ockinger, E., Rundlof, M., Smith, H.G., Bengtsson, J. Assessing the effect of the time since transition to organic farming on plants and butterflies. Journal of Applied Ecology. 2011:48 543-550.

[22] Bavec, F., Bavec, M. Organic production and use of alternative crops. Boca Raton, New York, London : Taylor & Francis: CRC Press. 2006.

[23] Marshall, E.J.P., Brown, V.K., Boatman, N.D., Lutman, P.J.W., Squire, G.R., Ward, L.K. The role of weeds in supporting biological diversity within crop fields. Weed Research. 2003;43 77-89.

[24] Kleijen, D., Baquero, R.A., Clough, Y., Diaz, M., De Esteban, J., Fernandez, D.G., Herzog, F., Holzschuh, A., Johl, R., Knop, E., Kruess, A., Marshall, E.J.P., Steffan-Dewenter, I., Tscharntke, T., Verhulst, J., West, T.M., Yela, J.L. Mixed biodiversity benefits of agri-environment schemes in five European countries. Ecol. Lett. 2006;9 243-254.

[25] Hyvonen, T., Ketoja, E., Salonen, J., Jalli, H., Tiainen, J. Weed species diversity and community composition in organic and conventional cropping of spring cereals. Agric. Ecosyst. Environ. 2003;97 131-149.

[26] Hawes, C., Squire, G.R., Hallett, P.D., Watson, C.A., Young, M. Arable plant communities as indicators of farming practice. Agric. Ecosyst. Environ. 2010;138 17-26.

[27] Pacini, C., Wossink, A., Giesen, G., Vazzana, C., Huirne, R. Evaluation of sustainability of organic, integrated and conventional farming systems: a farm and field-scale analysis. Agric. Ecosyst. Environ. 2003;95 273-288.

[28] Pacini, C., Wossink, A., Giesen, G., Vazzana, C., Omodei-Zorini, L., Huirne, R. The EU's Agenda 2000 reform in the sustainability of organic farming in Tuscany: ecological-economic modeling at field and farm level. Agric. Syst. 2004;80 171-197.

[29] Wortman, E.S., Lindquist, J.L., Haar, M.J., Francis, C.A. Increased weed diversity, density and above-ground biomass in long-term organic crop rotations. Renewable Agriculture and Food Systems. 2010;25 281-295.

[30] Jastrzebska, M., Jastrzebski, W.P. Holdynski, C. Kostrzewska, M.K. Weed species diversity in organic and integrated farming system. Acta Agrobotanica. 2013;66 113-124.

[31] Edesi, L., Jarvan, M., Adamson, A., Lauringson, E., Kuht, J. Weed species diversity and community composition in conventional and organic farming: a five-year experiment. Zemdirbyste-Agriculture. 2012;99 339-346.

[32] Stockdale, E.A., Watson, C.A. Biological indicators of soil quality in organic farming systems. Renewable agriculture and food systems. 2009;24 308-318.

[33] Verbruggen, E., Roling, WFM, Gamper, H.A., Kowalchuk, G.A., Verhoef, H.A., van der Heijden, M.G.A. Positive effects of organic farming on below-ground mutualists: large-scale comparison of mycorrhizal fungal communities in agricultural soils. New Phytologist. 2010;186 968-979.

[34] Riley, H., Pommeresche, R., Eltun, R., Hansen, S., Korsaeth, A. Soil structure, organic matter and earthworm activity in a comparison of cropping systems with contrasting

tillage, rotations, fertilizer levels and manure use. Agriculture, Ecosystems and Environment. 2008;124 275-284.

[35] Irmler, U. Changes in earthworm populations during conversion from conventional to organic farming. Agriculture, Ecosystems and Environment. 2010;135 194-198.

[36] Bavec M., Prašnički M., Grobelnik Mlakar S., Turinek M., Robačer M., Bavec F. Influence of different production systems on body mass and number of earthworms. V: 46th Croatian and 6th International Symposium on Agriculture, Opatija, Croatia, February 14-18, 2011. Pospišil, Milan (ed.). Proceedings. Zagreb: University of Zagreb, Faculty of Agriculture. 2011; 61-65.

[37] Kuntz, M., Berner, A., Gattinger, A., Scholberg, J.M., Maeder, P., Pfiffner, L. Influence of reduced tillage on earthworm and microbial communities under organic arable farming. Pedobiologia. 2013;56 251-260.

[38] Bianchi, F.J.J.A., Ives, A.R., Schellhorn, N.A. Interactions between conventional and organic farming for biocontrol services across the landscape. Ecological Applications. 2013;23 1531-1543.

[39] Winqvist, C., Bengtsson, J., Aavik, T., Berendse, F., Clement, L.W., Eggers, S., Fischer, C., Flohre, A., Geiger, F., Liira, J. Mixed effects of organic farming and landscape complexity on farmland biodiversity and biological control potential across Europe. Journal of Applied Ecology. 2011;48 570-579.

[40] Krauss, J., Gallenberger, I., Steffan-Dewenter, I. Decreased Functional Diversity and Biological Pest Control in Conventional Compared to Organic Crop Fields. Plos One. 2011;6.

[41] Gosme, M., de Villemandy, M., Bazot, M. Jeuffroy, M.H. Local and neighbourhood effects of organic and conventional wheat management on aphids, weeds, and foliar diseases. Agriculture Ecosystems and Environment. 2012;161 121-129.

[42] Chateil, C., Goldringer, I., Tarallo, L., Kerbiriou, C., Le Viol, I., Ponge, J.F., Salmon, S., Gachet, S., Porcher, E. Crop genetic diversity benefits farmland biodiversity in cultivated fields. Agriculture Ecosystems and Environment. 2013;171 25-32.

[43] Geiger, F., Bengtsson, J., Berendse, F., Weisser, W.W., Emmerson, M., Morales, M.B., Ceryngier, P.,; Liira, J., Tscharntke, T., Winqvist, C. Persistent negative effects of pesticides on biodiversity and biological control potential on European farmland. Basic and Applied Ecology. 2010;1 97-105.

[44] Power, E.F., Kelly, D.L., Stout, J.C. Organic Farming and Landscape Structure: Effects on Insect-Pollinated Plant Diversity in Intensively Managed Grasslands. Plos One. 2012;7.

[45] Andersson, G.K.S., Ekroos, J., Stjernman, M., Rundlof, M., Smith, H.G. Effects of farming intensity, crop rotation and landscape heterogeneity on field bean pollination. Agriculture Ecosystems and Environment. 2014;184 145-148.

[46] Batary, P., Matthiesen, T., Tscharntke, T. Landscape-moderated importance of hedges in conserving farmland bird diversity of organic vs. conventional croplands and grasslands. Biological Conservation. 2010;143 2020-2027.

[47] Rundlof, M., Edlund, M., Smith, H.G. Organic farming at local and landscape scales benefits plant diversity. Ecography. 2010;33 514-522.

[48] Winqvist, C., Ahnstrom, J., Bengtsson, J. Effects of organic farming on biodiversity and ecosystem services: taking landscape complexity into account. Year in ecology and conservation biology (ed. Ostfeld, RS; Schlesinger, WH) Annals of the New York Academy of Sciences. 2012;1249 191-213.

[49] Boyce, J.K. A Future for Small Farms? Biodiversity and Sustainable Agriculture. Working paper series. 2004: No 86. http://scholarworks.umass.edu/cgi/viewcontent.cgi?article=1071&context=peri_workingpapers (accessed 24 April 2014).

[50] European Union Agriculture and rural development. Facts and figures on organic agriculture in the European Union. 2013. http://ec.europa.eu/agriculture/markets-and-prices/more-reports/pdf/organic-2013_en.pdf (accessed 24 April 2014).

[51] IFOAM press release. Highlights the Plight of Smallholder Farmers on Earth Day. http://www.ifoam.org/ sites/default/files/pr_earth_day_0.pdf (accessed 25 April 2014).

[52] Berhan, T. Egziabher, G., Edwards, S. Knowledge of Africa's rich plant biodiversity as a basis for the ecological intensification of agriculture and protecting biodiversity. Acta Universitatis Upsaliensis Symbolae Botanicae Upsalienses. 2011;35 229-250.

[53] Regulation (EU) No 1305/2013 of the European Parliament and of the Council of 17 December 2013 on support for rural development by the European Agricultural Fund for Rural Development (EAFRD) and repealing Council Regulation (EC) No 1698/2005. http://ec.europa.eu/digital-agenda/en/news/regulation-eu-no-13052013-european-parliament-and-council (accessed 24 July 2014).

[54] von Haaren, C. Kempa, D; Vogel, K; Ruter, S. Assessing biodiversity on the farm scale as basis for ecosystem service payments. Journal of environmental management. 2012;113 40-50.

[55] Agriculture and biodiversity. 2014. http://ec.europa.eu/agriculture/envir/biodiv/index_en.htm (accessed 24 July 2014).

[56] Kelemen, E., Nguyen, G., Gomiero, T., Kovacs, E·, Choisis, J.P., Choisis, N., Paoletti, M.G., Podmaniczky, L., Ryschawy, J., Sarthou, J.P. Farmers' perceptions of biodiversity: Lessons from a discourse-based deliberative valuation study. Land use policy. 2013;35 318-328.

[57] Power, E.F., Kelly, D.L., Stout, J.C. Impacts of organic and conventional dairy farmer attitude, behaviour and knowledge on farm biodiversity in Ireland. Journal for nature conservation. 2013;21 272-278.

Evidence and Facts

Hydromorphology and Biodiversity in Headwaters — An Eco-Faunistic Substrate Preference Assessment in Forest Springs of the German Subdued Mountains

Martin Reiss and Peter Chifflard

Additional information is available at the end of the chapter

http://dx.doi.org/10.5772/59072

1. Introduction

Springs are autochthonous inland freshwater ecosystems, which occur where groundwater reaches the surface [1-2]. From a limnological point of view springs are divided into two subtypes: the springhead (eucrenal) and the springbrook (hypocrenal), because of a differentiation in their species composition caused by differences of structural and environmental parameters [3]. That is only part of the reality for the hypocrenal when springs connected with flowing surface waters and be integrated into the upper part of a stream system (headwater). Regarding the common limnological spring types based on hydromorphological properties (rheocrene spring: fast flowing or falling water occurrence; helocrene spring: diffuse or laminar flowing water occurrence; limnocrene spring: water occurrence in a still water pool), springs can also occur in still surface waters without run-off [4-5]. This is of importance for the understanding and interpretation of species presence and biodiversity of springs, because depending on the spring type it is a lotic or a lentic aquatic ecosystem with an appropriate flow velocity as a hydromorphological factor (lotic: 0.1 to 1 m/s; lentic: 0.001 to 0.01 m/s) [6]. Furthermore, it should be emphasized that springs are ecotones with boundary or transition areas between different habitats [7]. The species composition is influenced by interacting with other different species communities and can be characterized as taxa rich regarding the whole habitat (crenon) [8]. Beside typically aquatic spring species (crenocenosis) other aquatic fauna elements occur from groundwater (stygobionts) and related surface waters (brook/river biota or still-water biota). Also semi-aquatic and terrestrial fauna are an integrated part in spring ecotones with specific transition zones as fauna elements (semi-aquatic: Fauna hygropetrica, Fauna liminaria; terrestrial: hydrophilic terrestrial fauna) [9-10]. Springs in the German subdued mountains are commonly cold stenothermic habitats, which means the mean annual

water temperature is about the local mean air temperature (8-12° C) [5] without higher annual amplitudes for the springhead (2°C) and moderate low annual amplitudes for the springbrook (5°C) [11]. This abiotic peculiarity of a more or less isotherm setting means that in spring ecosystems relatively constant environmental conditions are proclaimed [12]. However, there are other important key factors or filters [13], especially geochemical parameters (e.g. pH value, nutrient content) that influences the occurrence and distribution of species in springs. In this case, the spatial dimension or scale is taken into consideration. The spring area size is usually small (a few square meters), but structures and functions of the spring ecosystem are an integral component of the landscape and manifold linked with other landscape elements. Based on the concept that a water body is strongly influenced by landform and land use within the surrounding catchment at multiple scales [14], the term *springscape* illustrates the relationship and spatial embedding of ecological structures and functions regarding biodiversity [15]. Most ecological studies of spring species and communities focus on the distribution within the entire spring area as the habitat, e.g. to characterize the strength of binding to the spring habitat (stenotypy) [10]. Undisturbed forest springs of the mid-latitudes in Europe have a predominantly mosaic hydromorphological structure that suggests a potential differentiation of the colonization of substrates as microhabitats. It follows that the eucrenal itself is not a discrete spatial entity at the micro scale, because it is made of different substrate types that build heterogeneous mosaic-like structures or patches [16]. It is possible to subdivide the spring level at the nano scale, because invertebrates and other organisms inhabit the substrata. However, the fauna-microhabitat-relationship of springheads (eucrenal) has not been studied sufficiently [17], so this research wants to fill that gap to quantitatively describe and qualitatively assess substrate preferences of invertebrates in springheads as an ecotone.

2. The ecohydrological importance of substrates as microhabitats in springheads — State of research and open questions

Substrate is a complex variable of the physical environment and itself a basic material usually to build out heterogeneous patches in aquatic ecosystems [13]. Substrate is an important ecohydrological component that influences the occurrence, adaptation, survival and reproduction of the springhead fauna (Figure 1). Catchment properties like land use pattern (e.g. forest type), parent rock material of soil genesis, slope position and slope inclination as well as hydrological structures like spring type (flow regime), surface roughness, vegetation / forest structures and soil texture determine substrate types and their composition. There are inorganic or mineral and organic substrate types with separate corresponding nomenclature and classification. Table 1 show a classification based on size categories of mineral particles. Organic substrate types vary greatly in size and a systematic classification by size class does not seem very practical. However, a consistent and therefore comparable nomenclature in freshwater ecology is helpful for interpreting structures and functions of microhabitats. Here, especially for the river bed assessment within the implementation of the European Water Framework Directive a standardized designation for organic substrates as organic microhabitats exists [33] and provides the basis for adaptation to the conditions of springheads (Table 2).

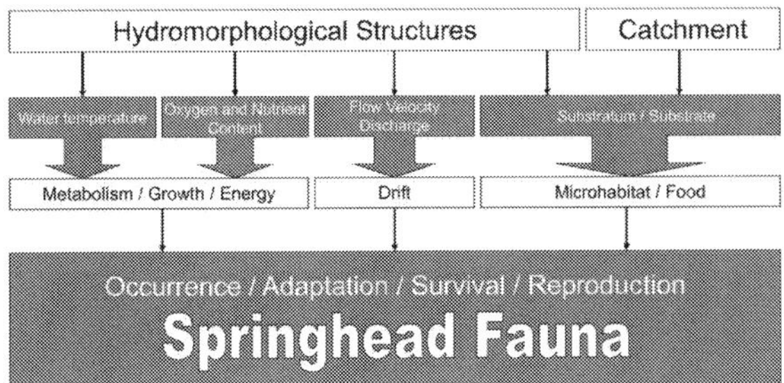

Figure 1. Influence of hydromorphological properties on the springhead fauna. Modified after [32].

Mineral Substrate		Description	Particle Size (Equivalent Diameter)
Megalithal		Upper Side / Top of blocks or in-situ rock	> 40 cm
Macrolithal		Head-sized boulder with a variable proportion of smaller grain sizes	20 cm – 40 cm
Mesolithal		Fist-sized cobbles / stones with a variable proportion of smaller grain sizes	6,3 cm – 20 cm
Microlithal		Pebbles and course gravel with a variable proportion of smaller grain sizes	2,0 cm – 6,3 cm
Akal		Gravel (fine to middle grained gravel) with a variable proportion of smaller grain sizes	0,2 mm – 2,0 cm
Psammal	Psammopelal	Sand (fine to coarse grained sand) with a variable proportion of smaller grain sizes	0,063 mm – 0,2 mm
Argyllal		Fine sediment (clay, silt) more or less solidified	< 0,063 mm
Pelal		Fine sediment (clay, silt) mixed with organic matter (loose material)	< 0,063 mm

Table 1. Nomenclature and classification of inorganic / mineral substrate types in springheads. See [1] and [33].

Structures and functions of substrates in running waters with a distinct hydrological flow regime are very well investigated [13, 18-19] and a special discipline has evolved: The river bottom ecology [20]. The research results from fluvial ecosystems cannot be transferred automatically, because of the special environmental conditions of springheads (e.g. cold stenothermic, oligotrophic, mostly low flow velocity) and their small sized ecotone character-istics. However, the bottom substrate in fluvial ecosystems like brooks and rivers is often one of the most significant factors affecting the species composition of the benthic fauna in the substratum [21-26]. Some studies consider the mapping of substrate types in springs for a hydromorphological based water typology [27-29], but without a given classification scheme and a method instruction for the assessment of substrate type coverage within a field survey. Referring to an estimation procedure by [30] a first combining example for a coverage classification and a description for an ecological assessment procedure for springheads and springbrooks is given by [31]. There are five classes based on the aggregation of levels of

Organic Substrate	Description
Emergent macrophytes	Spring-fed herbaceous macrophytes
Submerged macrophytes	Subaqueous herbaceous macrophytes (partly above the water or completely under water)
Moss cushions	Contiguous patches or layers of mosses
Fine roots	Floated living fine roots of the riparian area
Xylal	Dead wood, non-living tree trunks, branches and/or roots
CPOM	Course particular organic material (e.g. leaf litter)
Coniferous litter	Only needle litter of coniferous trees or shrubs
FPOM	Fine particular organic material
Algae	Filamentous algae, algal tufts

Table 2. Nomenclature of organic substrate types in springheads. See [10] and [33].

coverage: 0 – absent; 1 – low level (10-20 % coverage); 2 – medium level (30-40 % coverage); 3 – strong level (50-60 % coverage) and 4 – continuous level (> 70 % coverage). The mapping of the substrate as a potential microhabitat, combined with a simultaneous integrated field sampling of the invertebrate fauna based on the coverage of substrate types as a water type specific method has so far been developed only for brooks (rhithral) and rivers (potamal), e.g. [33-34]. Studies on the role of substratum for species richness in springs are executed, but with different levels of detail of the research question with respect to the substrate preference of species. A study in the limestone Alps of Austria shows that different substrate types emphasized the differences of species composition and abundance in springs on carbonate substrata [35]. The microhabitat preferences of benthic invertebrates and especially for Oligochaeta has been studied and illustrated by the example of karst springs in the Krakow-Czestochowa Upland in Southern Poland [17]. Here, the substrate type was found to be the main discriminatory factor with regard to the fauna density. Orthocladiinae (subfamily of non-biting midges), Cyprididae (ostracods), Turbellaria (planarians) and only one Oligochaeta (*Nais communis*) were more abundant in coarse mineral substratum whereas Chironominae (non-biting midges), Limnephilidae (northern caddisflies), Bythinellinae (prosobranch spring snails) and most Oligochaeta (subclass of earthworms) were more numerous in fine mineral substratum. Another outcome of this research is, that from a higher substrate heterogeneity results a higher biodiversity. In a study with the aim to find environment variables, which represent the species composition of fauna in certain assemblages of Danish springs, the result is that higher substrate heterogeneity increases the biodiversity especially in helocrene springheads [36]. The correlation between substrate type diversity and richness in species are also confirmed by investigations in Canadian springs [37] and in springs of the USA [38] for North America. Even [39] can also show a clear relationship between the species composition of insects in springs of the Sacra catchment (Adamello Brenta Regional Park, Italy) and the grain size of mineral substrate. The occurrence of certain substrate types determines fauna assemblages as a key factor beside physical-chemical parameters in a study in perennial limestone springs in Northwest Switzerland [40]. In a different geological setting of perennial siliceous sandstone springs in the Nationalpark Pfälzerwald (Southwest Germany) also the

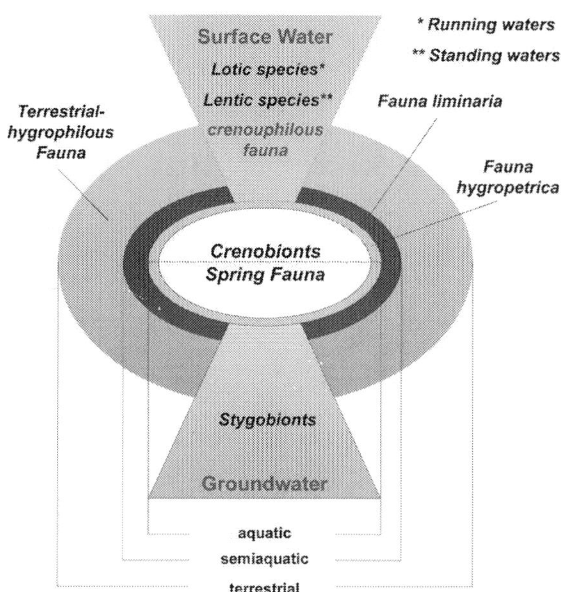

Figure 2. The eucrenal of spring ecosystems as an ecotone with its related transition fauna zones. See [10].

substratum leads to main separations in the species composition of aquatic invertebrates [41]. For alpine limestone springs in the Schütt catchment in Kärnten, Austria the role of habitat structure on the community composition was studied with a particular focus on the spring-dwelling animals colonizing the aquatic and the adjacent aquatic-terrestrial transition zones [42]. Here, microhabitat composition and the concomitance of lotic and lentic areas in the springheads furthered a high species diversity and abundance without an influence of the altitude of the investigated springs. In certain dominant microhabitats taxa specific substrate preferences were detected. Ephemeroptera (mayflies) prefer micro- and mesolithal, the caddisfly *Crunoecia irrorata*, however found mainly in CPOM. The study shows also a certain distribution of taxa according to the different spring ecotone zones, at which crenobionte species mainly occur in semi-aquatic areas (Fauna hygropetrica and Fauna liminaria) and only a few crenobionte taxa exclusively in the aquatic environment. This finding underlines the importance to investigate springheads as an ecotone and to include all transition zones from aquatic to terrestrial areas within the methodological concept of eucrenal studies (Figure 2).

For diatoms, only the grain size of mineral substrates has an influence on the colonization of certain species, because a significant correlation with different microhabitats in springs cannot be determined [43]. There are also studies that achieve no or an unclear relationship between substrate occurrence and species diversity in the results. [44] deduce a mixture of substrate specific microhabitat types and general spring types from empirical field data: Mineral dominated springbrooks, helocrene springs, moss cushions, limnocrene springs. For all these subtypes of springs specific inhabited taxa of Crustaceans (Crustacae) and insects are found, with the exception of helocrene springs. A statistically significant correlation between these habitat types and species diversity cannot be described. For the latter result, it should be noted that it is not useful to aggregate data of fauna assemblages at different spatial scales (e.g. cumulate microhabitat and habitat scale) to run statistical analysis to differentiate fauna communities, because the hierarchical levels of spatial scales must be considered [15, 45]. By using multivariate statistical ordination methods to analyze fauna composition in the eucrenal of springs in Northwest Switzerland (Swiss Plateau and the Jura Mountains) on the habitat scale the most important identification criteria are the spring type, substrate and discharge intensity [46]. A further regional specific faunistic relevant differentiation of spring types is possible on the basis of the criteria substrate (microhabitat scale). Especially for the structural separation into fine and coarse mineral substrates and particularly specific organic substrates such as CPOM and emergent macrophytes a faunistic relevance is detectable. Even in studies of karst springs in the Wye catchment in the Peak District National Park (Derbyshire) in England no dominant relationship between the occurrence of different microhabitats and the species composition of invertebrates was found [47-48]. The results obtained from the springs and springbrooks examined that discharge variability has a greater influence on macroinvertebrate community composition than the distribution and diversity of substrate types. A separate data analysis according to the areas springhead and springbrook would show a more significant influence of microhabitats to differentiate fauna communities of the eucrenal and hypocrenal. The characteristics of a springbrook (hypocrenal) are that here, significantly more lotic and crenoxene taxa are to be expected caused by a higher velocity flow than in the springhead (eucrenal) with a higher proportion of crenobionts within the fauna community [49]. The springhead should be seen as an autonomous ecotone with a complex of microhabitats, so that sampling and analyzing methods has to be performed using tools adapted to every microhabitat type [50].

In summary, the review of the state of research about fauna-microhabitat-relationships and an eco-faunistic substrate preference assessment to analyze research deficits shows that some structural hydromorphologically based water type subdivisions of springs using a variety of substrate types already exist, e.g. to differentiate existing spring typology approaches. The integrative joint consideration of the function and the ecological significance of the substratum as a hydromorphological element and as a microhabitat for invertebrates of springheads are lacking in assessment methods and analyses. In eco-faunistic studies that interpret the fauna-microhabitat-relationship in the eucrenal of springs, combined quantitative and qualitative investigations and analysis of the substrate preferences of invertebrate taxa regarding the springhead as an ecotone are still missing. Thereby faunistic research focuses mostly on the aquatic taxa only, rarely on terrestrial organisms. The scientific deficits described are the motivation for new research about fauna-habitat-relationships in springs. The results of the prospective study presented here were conducted in order to answer the following main research questions:

1. Is there a substrate preference for specific taxa considering the ecotone characteristics of springs? (Quantitative Structural Analysis);

2. Which functions of microhabitat types of springs could be characterized with the inves‐
tigation of the substrate preference of specific taxa? (Qualitative Functional Analysis);

3. How strong is the relationship (or correlation) between microhabitat diversity (substrate
type richness) and biodiversity? (Structure-Function Synthesis).

3. Study area and methods

The selection of study sites (Figure 3) was based on two main criteria. First, only forest springs
have been investigated in order to ensure a wide range of possible selection of different
substrate types. Therefore, certain categories of protected areas were deliberately chosen in
forest landscapes as forest reserve, national park, and core zone of the biosphere reserve or
nature reserve to identify an equally wide variety of hydromorphological structures. Different
land uses or management strategies in forests implicated anthropogenic influences such as
artificial water control structures and were included consciously. Second, study sites were
selected that were as little as possible or not studied to close regional gaps in knowledge for
species inventory and for locational eco-faunistic characterization. The study sites are located
in the central area of the German subdued mountains. Table 3 gives an overview to their natural
physical geographic characteristics.

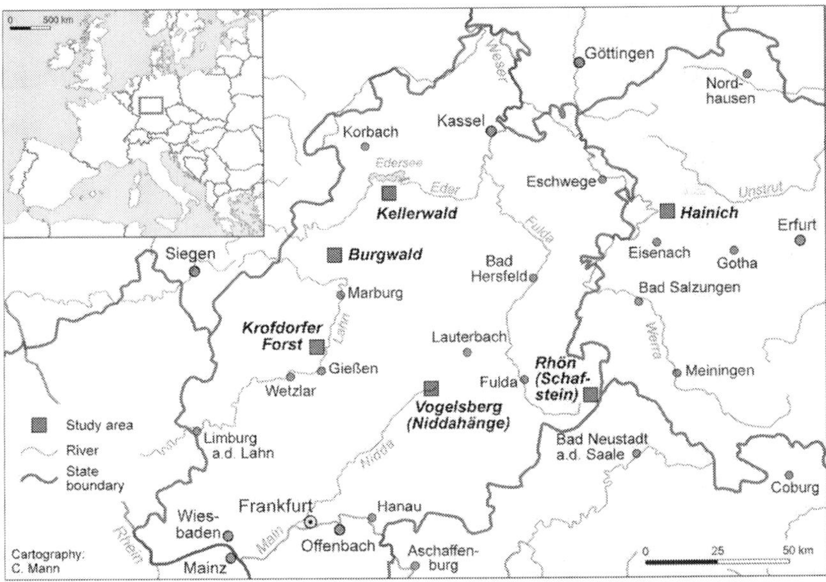

Figure 3. Study Area in Central Germany.

Study site	Altitude and Climate	Groundwater Body	Main Forest Communities	No. of investigated springs
Niddahänge (Vogelsberg) *Forest Reserve*	Up to 670 m a.s.l., 1200-1300 mm annual mean precipitation, 5-6° C annual mean air temperature	Volcanic rocks (Miocene)	Beech Forest, Alder Swamp Forest, Sycamore-Ash-Forest	24
Schafstein (Rhön) *Core Zone Biosphere Reserve*	Up to 830 m a.s.l., 1100-1200 mm annual mean precipitation, 5° C annual mean air temperature		Beech Forest, Birch-Rowan-Forest, Linden-Wych Elm-Forest	9
Hainich *National Park*	Up to 500 m a.s.l.,	Limestone (Middle Triassic)	Beech Forest	11
Burgwald *Nature Reserve* (partial)	Up to 440 m a.s.l., 600-700 mm annual mean precipitation, 7-8° C annual mean air temperature	Sandstone (Lower Triassic)	Beech Forest, Pine and Spruce Forest	30
Kellerwald *National Park*	Up to 630 m a.s.l., 600-800 mm annual mean precipitation, 6-8° C annual mean air temperature	Greywacke, Clay Shale (Lower Carbonifereous)	Beech Forest	40
Krofdorfer Forst *FFH* Site*	Up to 400 m a.s.l., 600-700 mm annual mean precipitation, 8-9° C annual mean air temperature	Greywacke, Clay Shale (Upper-Devonian)	Beech Forest, Spruce Forest	38
Total no. of springs				**152**

Table 3. Natural physiogeographic characteristics of the study sites. * FFH: European Habitat Directive (Flora-Fauna Directive).

In this study, a total number of 152 springs are surveyed and analyzed. Related to the natural area classification of the Federal Republic of Germany [51] the study sites can be grouped in 4 different landscapes (thick lined frames in Table 3) regarding geological subsoil characteristics (*Groundwater Body* in Table 3). On the landscape scale the study sites can be aggregated into two main groups concerning chemical groundwater criteria: 1) study sites with siliceous springs (Niddahänge, Schafstein, Burgwald, Kellerwald and Krofdorfer Forst); 2) study site with limestone springs (Hainich).

The investigation approach based on a hierarchical spatial framework for spring habitats to aid the illustration and understanding of functional, structural and process relationships on different scales [15]. The springscape (Figure 4) is a theoretical concept concerning specific geographical dimensions as levels of habitat filters that operate to influence species distribution and abundance within the landscape [52]. This implies that hierarchically nested environmental factors like substrate type influences the assemblage of species at progressively more localized spatial scales (e.g. at the microhabitat scale) [13].

Every substrate type is arranged at the microhabitat scale (or nano scale) and can be seen as the smallest habitat unit as a relatively homogeneous minor area where species occur. It is

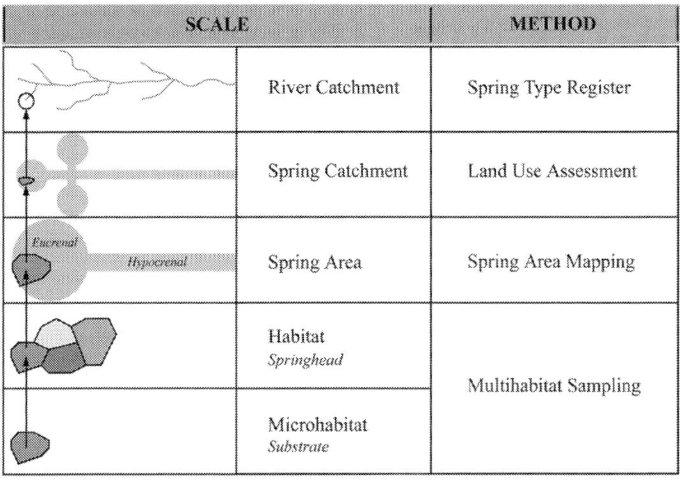

SCALE		METHOD
	River Catchment	Spring Type Register
	Spring Catchment	Land Use Assessment
Eucrenal *Hypocrenal*	Spring Area	Spring Area Mapping
	Habitat *Springhead*	Multihabitat Sampling
	Microhabitat *Substrate*	

Figure 4. The springscape: A hierarchical spatial system of springs. See [15].

similar to the habitat scale of the patch dynamic concept [16]. These substrate types form a mosaic-structured complex, which determine the entire substratum within the ecotone of a springhead on the habitat scale. The arrangement of substrate types corresponds to the patch scale of the patch dynamic concept [16]. Springhead (eucrenal) and springbrook (hypocrenal) consolidated the spring area at the meso scale within the stream system [31]. Several spring areas are part of headwater catchments, which are taken together in a higher-level system of a river catchment. Spring and river catchments are part of the landscape scale of the patch dynamics concept [16]. Finally, such stream systems can be a part of major, continent-scale river basins. For the microhabitat scale a new method to detect substrate types within springhead ecosystems and to sample the invertebrate fauna of each substrate type within an ecotone approach was developed. It is a multi-habitat sampling technique with a 2-layer approach (Figure 5).

The principle is similar to the AQEM/STAR approach to assess the riverbed of river segments [33,53], but with basic changes in the procedure considering essential springhead environmental characteristics. The inorganic and organic layers are considered individually in a 2-layer approach by taking the area of the whole springhead habitat as a reference surface (5-10 square meters). The appraisement of substrate type coverage was documented in a record sheet. The number of sub-samples taken in each layer corresponds to the fraction of the substrate types of the reference surface that layer has, with one sample taken per 5 percent coverage. On the example of the substrate type microlithal (coarse gravel in Figure 5) a coverage ratio of 40 percentages was estimated, 8 separate samples of fauna collections have to be performed. For each sample, a substrate specific sampling technique (e.g., sampling by net, collecting with tweezers) is performed for 2 minutes over a 10 cm by 10 cm reference area. A specific handheld net sampler was used with a mesh width of 100 µm. For taxonomic

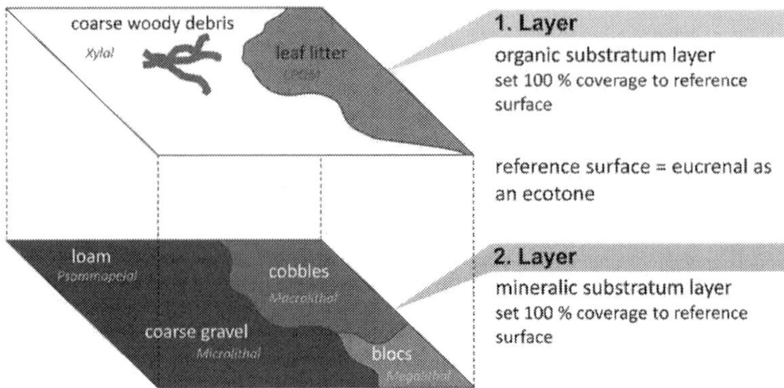

Figure 5. The 2-Layer approach for a multihabitat sampling technique for springheads. See [10].

determination invertebrates were preserved in ethanol alcohol (90 %) and stored in small (6 ml) Wheaton polyethylene jars. The samples are archived in the laboratory of the Biospeleological Register maintained by the Hesse Federation for Cave and Karst Research [54] and are available for genetic research by the Bavarian State Collection of Zoology in Munich. Some taxonomic groups were passed on to specialist taxonomists for detail determination: Dr. Peter Martin (Kiel, Germany) for Halacaridae and Hydrachnellae of the order Acari (water mites); Dr. Axel Schönhofer (Mainz, Germany) for Opiliones (harvestmen); Christoph Bückle (Tübingen, Germany) for Auchenorrhyncha (cicadas) and Andreas Allspach (Frankfurt / Main, Germany) for Trichoniscidae (woodlice, isopods). Mapping and sampling were taken once a time for 152 springs in 2008. In 2009 a control sample in 4 representative helocrene springs carried out to identify possible changes in substrate coverage. As a descriptive statistics method the relative frequency (f_i) was calculated to compare the habitat type occurrence of the different substrate types for the quantitative structural analysis (Equation 1):

$$f_i = \frac{n_i}{N}$$

Equation 1. Calculation of the relative frequency (f_i) of a taxon within a substrate type (=substrate preference); n_i: absolute frequency of a taxon within a substrate type; N: total number of samples of a substrate type.

The SIMPER analysis (*similarity percentages*) was executed (Equation 2) to test the validity of aggregated microhabitat types with specific taxa as statistical descriptors by ranking similarity in fauna community pattern [55-56].

$$Sjk = \sum_{i=}^{p} / Sjk(i)$$

Equation 2. Calculation of the SIMPER analysis. S: Group within a pair of samples j, k; i: *i*th term of S_{jk}; l Sjk(i): Bray-Curtis coefficient, see [57].

Therefor a similarity coefficient with a standard deviation regarding the abundances of taxa was calculated. Most commonly occurring taxa with high abundances are the best descriptors to identify ecological relevance (or validity) in microhabitat types. The qualitative functional analysis of diet types was performed using existing feeding type valence values [58-59] and new established values for water mites in cooperation with Dr. Peter Martin [10]. To calculate a metric for the biodiversity of the invertebrate fauna the Shannon Index was used as a basis for the Structure-Function Synthesis [60]. In addition, the Evenness Index was performed as a structure metric to analyze the statistical distribution of the Shannon Index [61]. The interpretation of the relationship between structure and function within the context of analyzing hydromorphological structures and biodiversity the Pearson correlation coefficient was applied for statistical calculation. A multivariate statistical method was applied using a principal component analysis (PCA) to characterize variables to differentiate springheads.

A modeling of aggregated microhabitat types was performed using a new and specially developed three-step decision scheme to subdivide hydromorphological based habitat types for springheads (Figure 6).

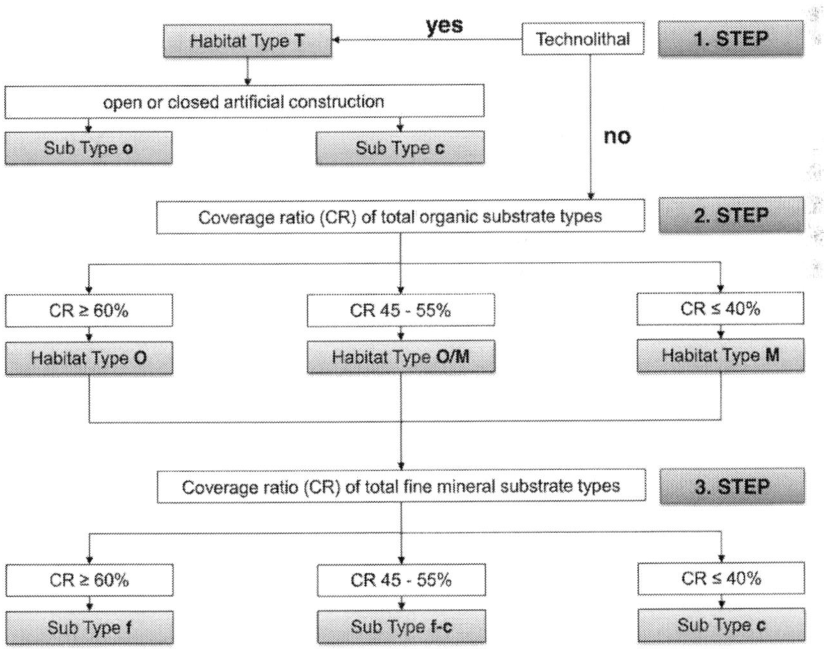

Figure 6. Decision Scheme for modelling microhabitat types of springheads. See [10].

4. Results

In this study 11.663 individuals (single organisms) in 639 sampling jars were sampled and determined, which corresponds to an overage value of 76 individuals per springhead. Arthropoda accounted for the largest share (81%), followed by Mollusca (11 %), Annelida (4 %), Nemathelminthes (2 %) and Plathelminthes (1 %) regarding the phylum of invertebrates. The dominant class is Insects (51 %), followed by Arachnida (13 %), Crustacea (12 %), Gastropoda (8 %), Clitellata (4 %), Entognatha (4 %), Bivalvia (3 %), Nematoda (1 %), Turbellaria (1 %) and Others (1 %). The major group within the order is Diptera (24 %), followed by Coleoptera (15 %), Trichoptera (8 %), Araneae (7 %), Plecoptera (5 %), Stylommatophora (5 %), Oligochaeta (5 %), Amphipoda (4 %), Veneroida (3 %), Acari (3 %), Isopoda (3 %), Cyclopoida (3 %), Hemiptera (3 %), Neotaenioglossa (3 %), Harpacticoida (2 %), Basammotophora (2 %), Seriata (2 %), Opiliones (1 %), Hymenoptera (1 %), Gordiida (1 %), Diplopoda (1 %), Lepidoptera (1 %) and Others (1 %; 11 further groups). A detailed presentation of results regarding the taxonomic rank of families is given by [10]. The taxonomic ranks of Genus and Species are considered in the results of the substrate and microhabitat preferences. The species composition of the six different study areas is very similar, regarding a cluster analysis and a nonmetric multidimensional scaling. Using mean abundances a separation is possible corresponding to a differentiation based on natural physiogeographic characteristics (Table 3). The overage value of springhead specific genus and species is about 28 percentages for all study areas.

4.1. Results of the quantitative structural analysis

Relative presence within all investigated springs means the percentages of the occurrence of a substrate type in 152 springs as the statistical main unit (Example: Psammopelal is present in 61 springs of 152 total springs, that results 40 %). The relative coverage ratio was calculated as the mean value of the related substrate type of all studied springheads. The results in Table 4 showing the presence and the coverage of the investigated substrate types of all 152 studied springs.

The most present substrate type is psammopelal with a ratio of 40 percentages. This mineral substrate type represents 51 percentages of the coverage in comparison to all substrate types. The second common substrate type is coarse particular organic matter (CPOM) with 21 percentages presence and 37 percentages coverage. The most common coarse mineral substrate type is microlithal with 18 percentages presence and 20 percentages coverage. Further representative organic substrate types are xylal (coarse woody debris), emergent macrophytes and moss cushions. Artificial substrates are of minor importance. Most of the springheads can be characterized as structurally undisturbed and non-degraded habitats. Nevertheless, the results of the found substrate types represent diverse microstructures related to forestland cover.

The results of the aggregated microhabitat types performed using the decision scheme (Figure 6) is documented in Table 5. In comparison to the findings of the substrate types (Table 4) it is to ascertain that the most common habitat types of all studied springs are organic dominated (74 %). Here, the dominant microhabitat type is the organic-dominated, fine-material-

Substrate Type	Presence (relative)		Coverage (relative)	
Argyllal	2 %	Fine Sediment	2 %	Fine Sediment
Psammal	4 %		2 %	
Psammopelal	40 %	47 %	51 %	55 %
Akal	10 %	Coarse Sediment	4 %	Coarse Sediment
Microlithal	18 %		20 %	
Mesolithal	10 %		5 %	
Macrolithal	7 %	49 %	4 %	38 %
Megalithal	5 %		5 %	
Open Construction	2 %	Technolithal	3 %	Technolithal
Closed Construction	2 %	4 %	4 %	7 %
Emergent macrophytes	15 %	Organic Matter	20 %	Organic Matter
Submerged macrophytes	1 %		1 %	
Moss cushions	19 %		10 %	
Xylal	22 %	100 %	12 %	85 %
CPOM	21 %		37 %	
Coniferous litter	4 %		3 %	
FPOM	17 %		2 %	
Algae	1 %		0 %*	
Without organic substrates			15 %	15 %

Table 4. Presence and coverage of subtrate types within the investigated springheads. * 0,4 % rounded down=0 %. First layer: Mineral and artificial substrate types (100%); Second layer: Organic substrate types and coverage without substrates (100%).

abounded microhabitat type (43 %), which represents the importance of fine mineral substrate. Exclusively mineral habitat types are less representative (11 %). However, their presence and importance as habitat types had not been sufficiently documented by previously existing mapping methods, because of the non-regarding of overlapped substrates. Here the application of the 2-layer approach is a benefit for ecological characterization and classification.

Habitat Type	Percentages	Microhabitat Type (HT)	Percentages
Organic dominated	74 %	Organic-dominated, fine-material-abounded HT (O_f)	43%
		Organic-dominated, coarse-material-abounded HT (O_c)	24%
		Organic-dominated, fine- to coarse-material-abounded HT ($O_{f\text{-}c}$)	7%
Mineral dominated	11 %	Mineral-dominated, fine-material-abounded HT (M_f)	6%
		Mineral-dominated, coarse-material-abounded HT (M_c)	3%
		Mineral-dominated, fine- to coarse-material-abounded HT ($M_{f\text{-}c}$)	1%
Mixed Type	7 %	Mixed type (organic/mineral), fine-material-abounded HT (O/M_f)	3%
		Mixed type (organic/mineral), coarse-material-abounded HT (O/M_c)	4%
		Mixed type (organic/mineral), fine- to coarse-material-abounded HT ($O/M_{f\text{-}c}$)	0%
Artificial (Technolithal)	7 %	Technolithal with open construction (T_o)	3%
		Technolithal with closed construction (T_c)	4%
Special Type	1 %	Special Type (S)	1%
Total	100 %	Total	100 %

Table 5. Ecohydrological microhabitat types for springheads within the investigated springheads.

Relative Frequency (f_i)		Preference Classification
≥ 50%	strong	++
25 – 49 %	common	+
< 25 %	rare	-
0 %	absent	

Table 6. Assessment Scheme to classify the substrate preference. See [62].

http://dx.doi.org/10.5772/59072

| Fauna Area | Taxon | Mineralic Substrates | | | | | | | | Organic Substrates | | | | | |
| | | Fine Sediment | | | Coarse Sediment | | | | | | | | | | |
		Arg	Psa	Psp	Aka	Mic	Mes	Mac	Meg	eMp	sMp	Moss	Xyl	CPOM	CoL
aquatic	Agabus sp.		-	-									-	+	-
	Arr font.					-	-			-			-	+	
	Bezzia sp.		-	++	-										
	Byt com			+	-	-	-					-		-	
	Byt dun		-	-	-	-	-	-	-		-		-	+	
	Cord bid			+		++								+	
	Cren alp		-	-	-	-	-	-					-	+	
	Galba tr			-				-	-				-	+	
	Gamm fos	-	-	-	-	-	-	-	-				-	+	
	Gams pul		-	+	-	-			-				-	+	
	Habrol con				-	-							++	++	
	Helop sp.											++	+	+	
	Hydrov pla			++		+									
	Hygrob nor			++											
	Leuctra sp.		-	+	-	-		-	-				-	+	
	Loboh web			++											
	Nemoura sp.			-	-	-	-	-					-	+	-
	Niph aqu			-	-	-		-					-	++	
	Niph schell		-	-	-	-	-						-	+	
	Partn steinm		-	+									-	-	
	Pisidium sp.	-	-	++	-	-	-	-	-			-	-	+	
	Polyc fel			+					-		-			+	
	Proton sp.				-	-			-	-			+	-	
	Protz squ squ			+										++	
	Seric sp.	-	-	-	-	+	-	-	-			-	-	-	
	Sold chap			++											
	Sold mon			++											
	Sperchon sp.			++		-									

| Fauna Area | Taxon | Mineralic Substrates | | | | | | | | Organic Substrates | | | | | |
| | | Fine Sediment | | | | Coarse Sediment | | | | | | | | | |
		Arg	Psa	Psp	Aka	Mic	Mes	Mac	Meg	eMp	sMp	Moss	Xyl	CPOM	CoL
	Velia sp.					+	-	-				-		+	
hygp	*Anac* sp.		-	-	-	-	-			-		-	+	+	-
	Cruno irr	-	-	-	-	-	-	-				-	-	+	
	Dixa sp.		-		-	-	-			-		-	-	+	
ln	*Carych* sp.		-	+			-					-	-	+	
	Carych trid		-	+			-					-	-	+	
terrestrial-hygropilous	*Cicad vir*									++					
	Discus rot		-		-	-	-	-					+	-	
	Eisen tetr		-			-	-	-				-	+	+	
	Ligid hyp									-		++	-	-	
	Monac inc					-		-		++		-	-	-	
	Oligol trid									++					
	Oniscus as						-					-	++	-	
	Paran quadrip								+				++		
	Polydesmus sp.												++	+	
	Trichoniscus sp.					-	-	-				+	+	+	
terrestrial	*Bryo pter*									++					
	Eucon fulv						-						-	++	
	Euconulus sp.		-					-	-				+	+	
	Ixodes sp.											++			
	Leiob blackw									++				+	
	Lithobius sp.									+		-	-	-	
	Neob carc											-	+	++	
	Neob sim											++	+		
	Stenod hols									++					
	Stenod laev									++					

Table 7. Substrate preference. Fauna Area: *hygp* – hygropetric; *ln* – liminaria. See assessment scheme in Table 6. Abbreviation: Arg: Argyllal; Psa: Psammal; Psp: Psammopelal; Aka: Akal; Mic: Microlithal; Mes: Mesolithal; Mac: Macrolithal; Meg: Megalithal; eMp: Emergent Macrophytes; sMp: Submerged Macrophytes; Moss: Moss cushions; Xyl: Xylal; CPOM: Coarse particular organic matter; CoL: Coniferous litter.

The classification of the substrate preference of the found taxa was performed using the assessment scheme showing in Table 6. Taxa with a relative frequency of 25 and more percentages are classified as good descriptors for substrate type preference. The results of the substrate preference analysis are shown in Table 7. Generally, we found 30 taxa with a substrate preference for CPOM, 17 taxa with a substrate preference for psammopelal, 12 taxa with a substrate preference for xylal, 8 taxa with a substrate preference for emergent macrophytes, 5 taxa with a substrate preference for moss cushions, 4 taxa with a substrate preference for microlithal and 1 taxa with a substrate preference for megalithal. For all other substrate types we cannot found a substrate preference. These results represent not only the quantity of the methodological approach, because of the more intensive sampling of fauna in more representative substrate types. The results are also characterizing qualitative aspects like choosing a specific food source. In usually oligotrophic springheads organic matter is an important substrate as a food basis. That means, the most representative substrate type psammopelal (mineral substrate) is not the substrate type with the most fauna preference value for spring related invertebrates. Taxa are more present in organic substrates like CPOM or xylal. In FPOM and algae no taxa were found.

The results of the microhabitat type preference are documented in Table 8. The SIMPER test shows a differentiation in the microhabitat preference for the most abundant taxa. We analyzed an excerpt of the most representative organic and mineral microhabitat types (Table 5). Although organic substrates clearly dominate (74 % mean coverage), also a substrate preference of mineral substrates (11 % mean coverage) can be recognized. It is also interesting that not only aquatic taxa contribute to the characterization of the faunal relevance of these aggregated microhabitats. We found also hygropetric fauna (*Anacaena* sp., *Crunoecia irrorata*) and terrestrial-hygropilous fauna (*Trichoniscus* sp.) in the species pool.

Already two taxa (*Pisidium* sp., *Anacaena* sp.) describe almost half (46 %) the contribution to the substrate preference of the organic-dominated, fine-material-abounded microhabitat type (O_f). The organic-dominated, coarse-material-abounded microhabitat type is signified by 4 taxa (*Sericostoma* sp., *Crunoecia irrorata*, *Anacaena* sp., *Trichoniscus* sp.) with more than the half of there contribution (52 %). Here, the caddisfly *Sericostoma* sp. seems to be a good taxon to differentiate organic dominated microhabitats with coarse mineral abounded substrates, because of the preferential occurrence in such substrate types (substrate preference: microlithal). Therefor a precise quantitative analysis of the substrate preference (Table 7) is implicitly necessary to interpret SIMPER test results. The faunistic relevance of the less representative mineral dominated microhabitats (M_f, M_c) is partial uncertain. The most dominant taxon is the stonefly *Leuctra* sp., which characterizes fine mineral substrates (psammopelal) and the organic substrate type CPOM. A similar uncertainty can be observed for the water scavenger beetle *Anacaena* sp. as the second most representative taxa for the mineral-dominated, coarse-material-abounded microhabitat type. This taxon occurs mostly in organic substrates like CPOM and xylal. However, other faunistic findings are very plausible to interpret microhabitat preferences. For example, the pea clam *Pisidium* sp. prefers organic and fine mineral substrates and determines the mineral-dominated, fine-material-abounded microhabitat type. Although, spring taxa are normally not very abundant, it is possible to statistically validate modeled microhabitat types within a SIMPER analysis and to differentiate microhabitat preferences by taxon related contributions.

Taxon	Substrate Preference	Contribution (SIMPER) in % (Microhabitat Preference)			
		O_f	O_c	M_f	M_c
Bythinella compressa	mineral		3		
Leuctra sp.	mineral	3	2		30
Sericostoma sp.	mineral	3	15		15
Pisidium sp.	mineral / organic	24	8	13	11
Anacaena sp.	organic	22	12		17
Bythinella dunkeri	organic	2	6	6	3
Crenobia alpina	organic		5		7
Crunoecia irrorata	organic	6	15	19	3
Dixa sp.	organic	6	5	13	
Eiseniella tetraedra	organic	3			3
Galba truncatula	organic	4		11	2
Gammarus fossarum	organic		6	10	
Nemoura sp.	organic	13	3	19	
Trichoniscus sp.	organic	5	10		

Table 8. Results for the contribution of the SIMPER analysis. O_f: organic-dominated, fine-material-abounded; O_c: organic-dominated, coarse-material-abounded; M_f: mineral-dominated, fine-material-abounded; M_c: mineral-dominated, coarse-material-abounded.

It is to summarize that there is a significant substrate preference of certain taxa within separate fauna areas of the spring ecotone. A quantitative determination of indicator taxa of aquatic, hygropetric, liminarian, terrestrial-hygropilous and terrestrial fauna areas can be given as a basis for an eco-faunistic substrate preference assessment in forest springs of the German subdued mountains.

4.2. Results of the qualitative functional analysis

The qualitative function of substrates as microhabitats is related to the life strategy of an animal, which means the question about the use of a substrate type by a specific taxon. Can we qualitatively validate a specific quantitative assessed substrate preference by regarding autecological information about a taxon? Life strategies are diverse to characterize (movement type, diet type), however, they can all lead to a certain adaptation to the habitat [63]. Therefore, a suitable variable to analyze microhabitat functions is to typify the feeding group of a taxon. It allows the classification whether the taxon occurs for a direct food intake (substrate as food basis), indirectly for food intake (e.g. predators follows taxa with direct food intake) or another reason is to describe. The result of the qualitative functional analysis is summarized in Table

9 and Table 10. Most aquatic insects, especially stoneflies (Plecoptera), caddies flies (Trichoptera) and mayflies (Ephemeroptera) are almost exclusively present as larvae. The aquatic and hygropetric beetles were only found as imago. For most of the taxa the microhabitat function can be interpreted as the area of food intake or the substrate itself is the food source. The latter means, e.g. shredder organisms occurred dominantly in CPOM (coarse particular organic matter), because leaf litter is the original food basis. Interesting is the fact, that CPOM is the most dominantly organic substrate type and the preferential substrate type while FPOM is not representative. Here, we can assume that fine particular organic matter is transported downwards into parts of the springbrook or the epirhithral of the headwater, because of a dominant activity of shredders in the springheads. Therefore, barely collectors were found which filters or catch FPOM. Here, we have to confirm the River Continuum Concept with respect to headwaters and the declaration that shredders play a major role [64]. An importance of emergent macrophytes is conspicuous for the fauna areas of the terrestrial-hygrophilous and terrestrial zones. Microhabitat function is also food intake, but here, non-aquatic plant suckers occur. Other functional feeding groups are also existent, e.g. xylophages on coarse woody debris (xylal) or detritus and/or sediment feeder in fine mineral substrates, while psammopelal is the dominant mineral substrate type. Predators also occur, partial there are the major feeding group regarding the equivalence values of feeding groups [10], as in the aquatic and terrestrial fauna area. The substrate type itself has no direct significance as a food basis, because predators using microhabitats as hunting grounds. Therefore, it is indirectly of importance that specific taxa from other functional feeding groups showing a distinctive substrate preference, because predators reproduce a similar substrate preference, as the prey seeks for special microhabitats. We can classify corresponding substrate preferences for CPOM or psammopelal considering predators. A similar conclusion can be made for parasites like the spring specific taxa group of most water mites, certainly with a possible specific host preference within certain microhabitats. Another function of substrates can be deduced without analyzing the trophic state of taxa. Microhabitats are refuge areas for different organisms within the whole ecotone. Aquatic taxa like the pea clam (*Pisidium* sp.) or the terrestrial non-spring specific ticks (*Ixodes* sp.) find an area to retreat suboptimal environmental conditions. Pea clams burrowed actively into fine wet sediment (psammopelal) to survive times without discharge in the springhead, while ticks waiting in more or less bodily immobilization for host organisms. For some taxa a certain interpretation about their diet type is not really possible, because autecological information is lacking. For example, we found biting midges of *Bezzia* sp. larvae (Ceratopogonidae) with a high abundance and a specific substrate preference for fine mineral sediment (psammopelal). The Taxon is not specified in common functional feeding group reference lists [58-59]. Adult animals are plant and bloodsuckers, so that for the larvae the aquatic environment of fine sediment is a refuge area or a nursery ground. The larvae also survive droughts in springheads in wet fine sediment [65], so that this taxon needs more attention as a substrate preference indicator for temporary springs. For the two marine mite species of *Soldanellonyx* we did not found any information about diet type, what makes an interpretation of the microhabitat function impossible.

Fauna	Area	Taxon	Substrate Preference	Diet Type (Feeding Group)[i]	Microhabitate Function
aquatic		*Agabus* sp.	CPOM	Predator (9)	Hunting Ground
		Arr font.	CPOM	Predator (7), Parasite (3)	Hunting Ground
		Bezzia sp.	Psammopelal	Not specified; Bezzia are plant and blood suckers (host insects)	Refuge Area for larvae?
		Byt com	Psammopelal	Grazer (7), Sediment/Detritus Feeder (3)	Area of food intake; Substrate as food source
		Byt dun	CPOM	Grazer (10)	Area of food intake; Substrate as food source
		Cord bid	Microlithal, Psammopelal, CPOM	Predator (10)	Hunting Ground
		Cren alp	CPOM	Predator (10)	Hunting Ground
		Galba tr	CPOM	Sediment/Detritus Feeder (4), Grazer (3), Shredder (3)	Area of food intake; Substrate as food source
		Gamm fos	CPOM	Shredder (7), Sediment/Detritus Feeder (2), Grazer (1)	Area of food intake; Substrate as food source
		Gams pul	Psammopelal, CPOM	Shredder (7), Sediment/Detritus Feeder (2), Grazer (1)	Area of food intake; Substrate as food source
		Habrol con	CPOM	Grazer (7), Sediment/Detritus Feeder (3)	Area of food intake; Substrate as food source
		Helop sp.	Moss, Xylal, CPOM	not specified	Larvae are predators (= Hunting Ground)
		Hydrov pla	Psammopelal, Microlithal	Predator (7), Parasite (3)	Hunting Ground
		Hygrob nor	Psammopelal	Predator (7), Parasite (3)	Hunting Ground
		Leuctra sp.	Psammopelal, CPOM	Shredder (4), Sediment/Detritus Feeder (4), Grazer (2)	Area of food intake; Substrate as food source
		Loboh web	Psammopelal	not specified	Opiliones are predators (= Hunting Ground)
		Nemoura sp.	CPOM	Shredder (6), Sediment/Detritus Feeder (4)	Area of food intake; Substrate as food source
		Niph aqu	CPOM	Sediment/Detritus Feeder (10)	Area of food intake; Substrate as food source

Fauna	Area	Taxon	Substrate Preference	Diet Type (Feeding Group)[i]	Microhabitate Function
		Niph schell	CPOM	Sediment/Detritus Feeder (10)	Area of food intake; Substrate as food source
		Partn steinm	Psammopelal	Predator (7), Parasite (3)	Hunting Ground
		Pisidium sp.	Psammopelal, CPOM	Filtering Collectors	Area of food intake; Refuge Area (dry period)
		Polyc fel	Psammopelal, CPOM	Predator (10)	Hunting Ground
		Proton sp.	Xylal	Shredder (6), Sediment/Detritus Feeder (2), Grazer (2)	Area of food intake; Substrate as food source
		Protz squ squ	CPOM, Psammopelal	Predator (7), Parasite (3)	Hunting Ground
		Seric sp.	Microlithal	Shredder (7), Sediment/Detritus Feeder (1), Grazer (1), Predator	Area of food intake; Substrate as food source; Hunting Ground
		Sold chap	Psammopelal	Not specified	No interpretation possible (Food: Bacteria, Algae; Plant suckers, Predators)
		Sold mon	Psammopelal	Not specified	No interpretation possible (Food: Bacteria, Algae; Plant suckers, Predators)
		Sperchon sp.	Psammopelal	Predator (7), Parasite (3)	Hunting Ground
		Velia sp.	Microlithal	Predator (10)	Hunting Ground

Table 9. Diet types and microhabitat functions of the investigated springheads for aquatic taxa. * see Table 5; [i] see [58-59]; (*) clear preference, but without value.

Fauna	Area	Taxon	Substrate Preference	Diet Type (Feeding Group)[i]	Microhabitate Function
	hygropetric	*Anac* sp.	Xylal, CPOM	Sediment/Detritus Feeder (4), Grazer (4), Shredder (2)	Area of food intake; Substrate as food source
		Cruno irr	CPOM	Xylophage (5), Shredder (3), Predator (2)	Area of food intake; Substrate as food source; Hunting Ground
		Dixa sp.	CPOM	Filtering Collectors (7), Sediment/Detritus Feeder (3)	Area of food intake; Substrate as food source
	liminaria	*Carych* sp.	Psammopelal, CPOM	Shredder (10)	Area of food intake; Substrate as food source

Fauna Area	Taxon	Substrate Preference	Diet Type (Feeding Group)[1]	Microhabitate Function
terrestrial-hygrophilous	*Carych trid*	Psammopelal, CPOM	Shredder (10)	Area of food intake; Substrate as food source
	Cicad vir	Emergent Macrophytes	Plant Sucker (10)	Area of food intake; Substrate as food source
	Discus rot	Xylal	Shredder (*), Sediment/Detritus Feeder (*)	Area of food intake; Substrate as food source
	Eisen tetr	Xylal, CPOM	Sediment/Detritus Feeder (10)	Area of food intake; Substrate as food source
	Ligid hyp	Emergent Macrophytes	Shredder (6), Xylophage (2), Sediment/Detritus Feeder (2)	Area of food intake; Substrate as food source
	Monac inc	Emergent Macrophytes	Xylophage (*), Shredder (*)	Area of food intake; Substrate as food source
	Oligol trid	Emergent Macrophytes	Predator (10)	Hunting Ground
	Oniscus as	Xylal	Shredder (6), Xylophage (2), Sediment/Detritus Feeder (2)	Area of food intake; Substrate as food source
	Paran quadrip	Xylal, Megalithal	Predator (10)	Hunting Ground
	Polydesmus sp.	Xylal, CPOM	Shredder (7), Xylophage (3)	Area of food intake; Substrate as food source
	Trichoniscus sp.	Moss, Xylal, CPOM	Shredder (8), Sediment/Detritus Feeder (2)	Area of food intake; Substrate as food source
terrestrial	*Bryo pter*	Emergent Macrophytes	Plant Sucker (10)	Area of food intake; Substrate as food source (only ferns)
	Eucon fulv	CPOM	Shredder (*), Xylophage (?)	Area of food intake; Substrate as food source
	Euconulus sp.	Xylal	Shredder (*), Xylophage (?)	Area of food intake; Substrate as food source
	Ixodes sp.	Moss	Parasite (10)	Refuge Area
	Leiob blackw	Emergent Macrophytes, CPOM	Predator (10)	Hunting Ground
	Lithobius sp.	Emergent Macrophytes	Predator (10)	Hunting Ground
	Neob carc	CPOM, Xylal	Predator (10)	Hunting Ground

Fauna Area	Taxon	Substrate Preference	Diet Type (Feeding Group)[i]	Microhabitate Function
	Neob sim	Moss, Xylal	Predator (10)	Hunting Ground
	Stenod hols	Emergent Macrophytes	Plant Sucker (10)	Area of food intake; Substrate as food source
	Stenod laev	Emergent Macrophytes	Plant Sucker (10)	Area of food intake; Substrate as food source

Table 10. Diet types and microhabitat functions of the investigated springheads for the other taxa. ˣ see Table 5; ⁱ see [58-59]; (*) clear preference, but without value.

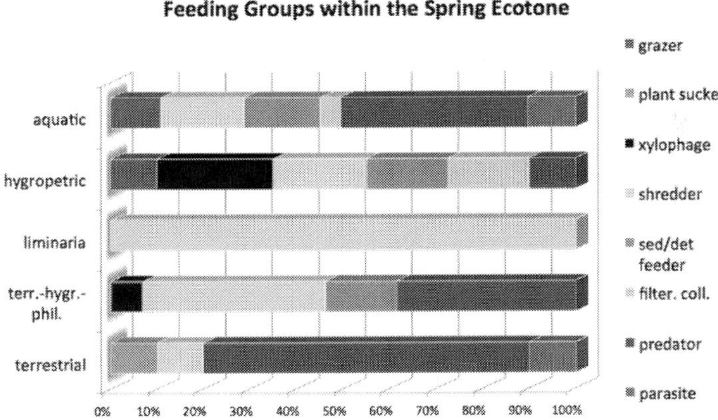

Figure 7. Feeding groups within the spring ecotone.

In general, there is a heterogeneous feeding group composition within the aquatic and hygropetric fauna areas of the spring ecotone (Figure 7), although only three different taxa could be indexed for the hygropetric fauna, but with diverse feeding type valence values. In contrast, for the Fauna liminaria just one feeding group (shredder) is dominant, because only the small air-breathing snail *Carychium* with only one main feeding type valence value is indicated. The terrestrial-hygrophilous and terrestrial fauna is also characterized by a heterogeneous feeding group arrangement. Here, shredders and predators are of similar importance in comparison to the aquatic fauna area, but with different taxa and substrate preferences. That means, also the adjacent non-aquatic spring areas showing a high diversity concerning their trophic state. That underlines a basic necessity of sampling and indicating terrestrial invertebrates in spring ecotones. Thereby, we can interpret trophic functions within hydromorphological structures with the result, that for terrestrial non-aquatic spring invertebrates similar functions of microhabitats can be ascertained, but in comparison to the aquatic spring invertebrates within different hydromorphological structures (substrate types).

It is to summarize that we can identify specific trophic functions of different microhabitat types. Aquatic and terrestrial spring invertebrates using specific substrates as a food basis, so that the substrate type is the area of food intake. Otherwise microhabitats were used as hunting grounds and refuge areas.

4.3. Results of the structure-function synthesis

There is an important relationship between the diversity of substrates and species diversity. The statistical correlation (R^2=0,88) between substrate diversity and biodiversity is highly significant (Figure 8). It is remarkable that the trend of the two curves (substrate diversity, Shannon-Index) is very similar, i.e. an increase in the substrate diversity leads to an almost identical increase in the Shannon index as an indicator value for biodiversity. The evenness values are between 0,7 (study areas: KW, VB) and 0,9 (study area: H) and emphasize the good quality of the results with a normal distribution of the fauna data (evenness values for the study areas RH: 8,0; BW and KR: 8,3). A further univariate analysis of the Shannon-Index with other location parameters and a correlation between these parameters and the occurrence of spring related taxa showing that substrate diversity is a key parameter determining biodiversity in springheads (Figure 9 and Figure 10).

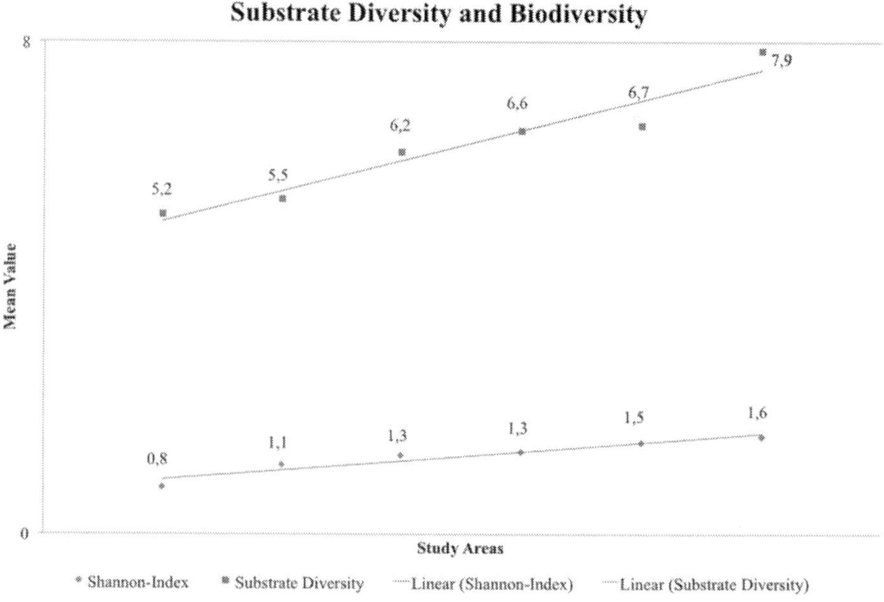

Figure 8. Substrate Diversity and Biodiversity. Study Areas: BW: Burgwald, H: Hainich, KR: Krofdorfer Forst, RH: Rhön (Schaftstein), VB: Vogelsberg (Niddahänge).

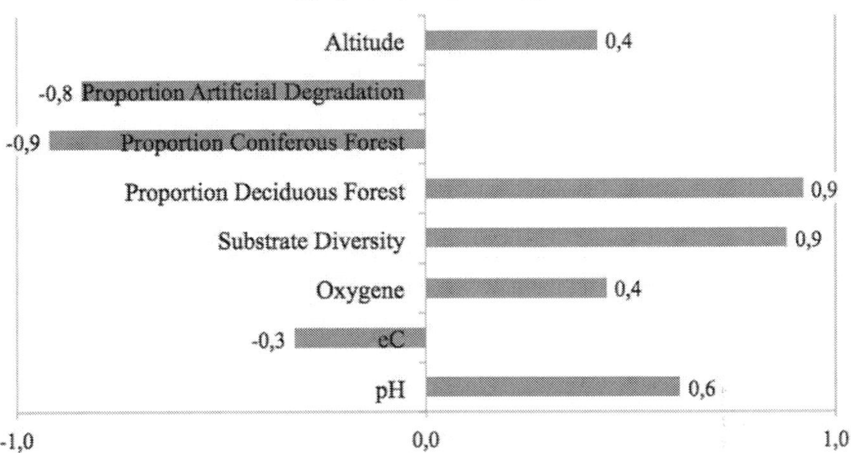

Figure 9. Univariate Correlation Shannon-Index and other location parameters.

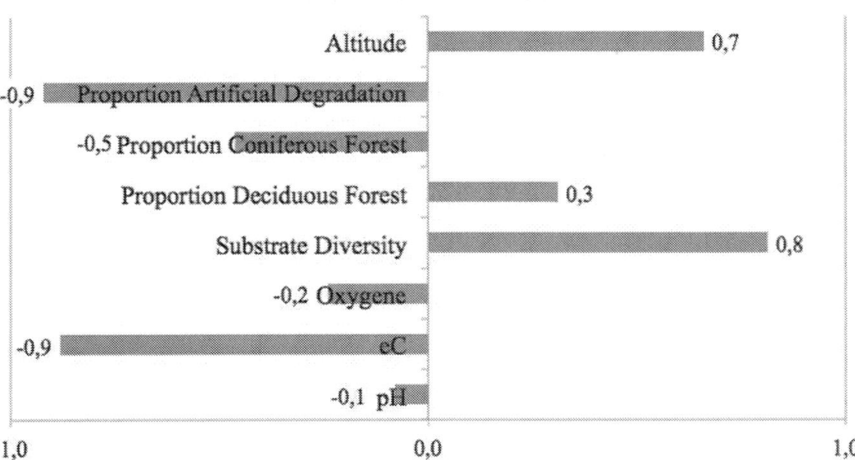

Figure 10. Univariate Correlation spring fauna (occurrence) and other location parameters.

The substrate diversity is one of the most important discriminatory factors for biodiversity in springheads besides forest cover type and pH (Figure 9). It is also an essential key driver for the occurrence of the spring fauna (crenobionts), which means taxa with a very strong and exclusive relationship to the eucrenal (Figure 10).

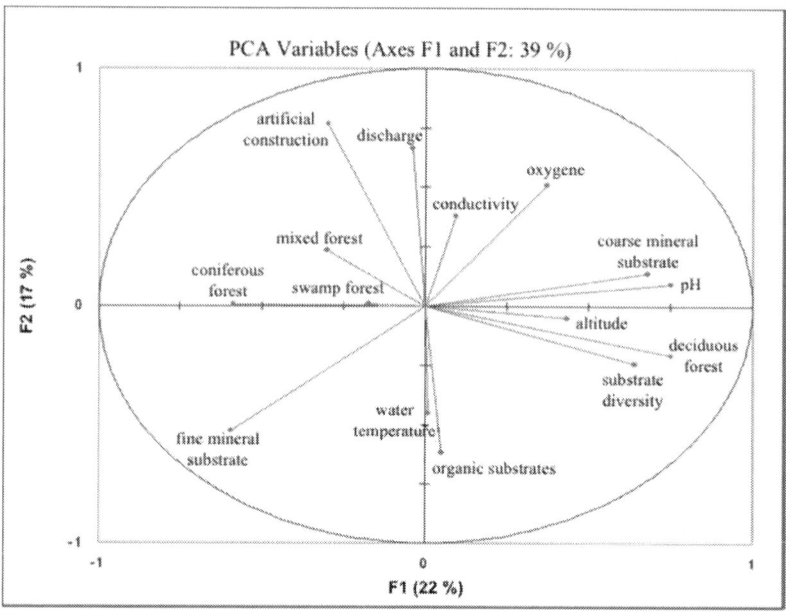

Figure 11. Principal component analysis.

The importance of substrate diversity as a key parameter determining biodiversity in springheads is also confirmed by a statistical multivariate analysis (Figure 11).

Indices	Undisturbed	Artificial	Relative Tendency
Shannon-Index	2	1	↘48% Decrease
No. of Species (mean)	8	3	↘57% Decrease
No. of Individuals (mean)	41	11	↘73% Decrease

Table 11. Biodiversity of undisturbed and artificial degraded springsheads. Data rounded off to whole numbers.

The artificial degradation of springheads with open or closed technical constructions (spring tapping and/or piping) is an immense stressor for fauna species in the eucrenal (Table 11). This can be shown strongly on the detailed quantitative analysis; not only the Shannon-Index and the number of species decrease significantly, especially the number of individuals' decreases sharply. The loss of biodiversity is significantly caused by spring tapping and is a consequence of the loss of substrate richness and microhabitat diversity. Here, the hydromorphological structure (substrate type diversity) is an important ecosystem service to preserve and develop biodiversity in springheads.

It is to summarize that there is a very strong relationship between microhabitat and substrate type diversity and biodiversity. The substrate diversity is one of the most significant discriminatory factors for biodiversity in springheads. The degradation of hydromorphological structures causes a substantial loss of species and abundance of species. Nature conservation strategies for spring ecotones have to consider the importance of substrate type richness and heterogeneity to protect and develop biodiversity in springheads.

---

5. Further research

The results of this study provide numerous starting points for further research. One of the most pressing issues is the question about the relevance of dynamics and resulting changes in the occurrence and coverage of substrate types on species presence and species composition. The investigated springs of the study areas are predominantly helocrene springs with low amplitudes of the annual discharge. Further research in karst springs with episodic and temporally very high discharge may abruptly change hydromorphological structures. Are these disturbances significant also to detect in a temporally or more long-term variation in species composition? How stable are such more or less dynamic hydromorphological springheads or is there a tendency to equilibrium conditions after substrates changing events? Hereby, it is to verify the transferability of the methodological approach of the research. In addition, generally long-term monitoring of hydromorphological and environmental monitoring is still lacking in studying spring ecology. A representative selection of already fewer objects of investigated springheads would determine a good approach to analyze long-term changes and trends in the future. Particularly, research is needed about the impact of land use change, which will require future projections of probably modifications of the occurrence and mosaic structures of substrates. What is the influence of a potential forest conversion to microhabitat heterogeneity and biodiversity in forest springheads? Regarding land use pattern comparison research is necessary to study substrate preferences of the spring fauna in non-forest areas, e.g. extensive wetland, grassland and springs in flood plains. Can we observe shifts of substrate preferences of known spring species caused by the absent of substrate types or can we characterize an absent of these known taxa or can we find complete different taxa? Beside the empirical study from field surveys also habitat modeling is crucial to answer those questions. Therefore, more detailed experimental research to strengthen the knowledge of autecological conditions, especially for non-aquatic spring species, but also for aquatic fauna with a recent not specified classification of the feeding type is needed. Especially, there is no robust information about the indexing of feeding groups for species of the Family Halacaridae (marine mites), which are a consistent part of springhead communities in the Meiobenthos (Mesofauna). Considering the background of future climate change conditions the importance of microhabitats like fine mineral substrates as refuge area ("moist islands") caused by decreasing time periods of drought not only for aquatic organisms with adaption strategies, but also for terrestrial-hygrophilous of the adjacent areas of springheads should be investigated. Also applied research for ecological assessment procedures is an essential issue and would benefits the practical orientated outcome of this basic research in spring ecology. For the protection and management of springs it is useful to implement the quantitative results of the substrate preference in existing or new metrics to characterize the ecological quality of these freshwater habitats.

6. Conclusion

Springs are considered as unknown habitats, most notable the relationship between invertebrates and hydromorphological structures. Research about the ecological importance of substrates for the inhabitation of species and consequences for biodiversity is still necessary to improve the knowledge about the relationship between structures and functions in springheads. This is needed if effective protection strategies and ecologically worthwhile nature conservation shall stand on a scientifically founded basis. Therefore, a first and operable mapping, sampling and assessment method was developed and can be used for further research and methodologically advances and modifications. Mainly, the theoretically background of the 2-layer approach is meaningful to assess also biased, not representative substrate types. Nevertheless, it is practicable to classify and verify ecological valid microhabitat types within representative substrate types for springheads. Here, we use a common limnological substrate type nomenclature, similar used for running waters, to compare the results with other water types or segments of brooks and rivers (rhithral, potamal). A quantitative approach to categorize substrate preferences is possible and can use as a basis to characterize the importance of mineral and organic substrate types in spring ecosystems. For specific invertebrate taxa a significant substrate preference is notable. Therefore, springheads were analyzed regarding their ecotone characteristics. Springheads are both, firstly an interface between the subterranean groundwater and the surface freshwater, secondly an embedded aquatic ecosystem with transition zones to terrestrial ecosystems. Hence, the whole importance of substrate heterogeneity and complexity in relation to biodiversity can be illustrated, although springheads are small sized inland water ecosystems or sometimes classified within small water bodies. The results of the found fauna reflecting the ecotone and a separate consideration of the substrate preference by fauna areas like the aquatic, hygropetric, liminaria and adjacent terrestrial fauna zone can be conducted. A taxa specific substrate preference considering the ecotone characteristics of springs can be determined. A qualitative functional analysis was done concerning each categorization of the feeding group (diet type) of the specific taxa. Thereby, an interpretation of microhabitats functions shows, that most of the taxa are present, because the substrate itself is the food basis or the place of food intake, especially for shredders, but also as a hunting ground for predators or a refuge area to survive non-optimal environmental conditions. To conclude the structure-function synthesis we can significantly prove a strong relationship between the diversity of substrates and species diversity. An increasing diversity of substrate types leads to a higher biodiversity. Hydromorphological degradation results in the distinctive decrease of invertebrate species and their abundances, especially caused by technical spring tapping. Substrate respectively substrate diversity is an important discriminatory factor to classify springhead ecosystems and their invertebrate fauna. It shows mainly the susceptibility and the need of nature conservation of these special habitats.

Nomenclature

We used abbreviations for taxa names in tables as listed below. (common name mentioned as far as applicable).

Abbreviation	Taxon	Common Name
Agabus sp.	*Agabus* sp.	Aquatic Beetle
Arr font.	*Arrenurus fontinalis*	Water Mite
Bezzia sp.	*Bezzia* sp.	Biting Midge
Byt com	*Bythinella compressa*	Spring Snail (Rhoen Spring Snail)
Byt dun	*Bythinella dunkeri*	Spring Snail (Dunkers Spring Snail)
Cord bid	*Cordulegaster bidentata*	Dragonfly (Sombre Goldenring)
Cren alp	*Crenobia alpina*	Triclad (Turbellaria)
Galba tr	*Galba truncatula*	Freshwater Snail
Gamm fos	*Gammarus fossarum*	Scud (Amphipod Crustacean)
Gams pul	*Gammarus pulex*	Scud (Amphipod Crustacean)
Habrol con	*Habroleptoides confusa*	Mayfly
Helop sp.	*Helophorus* sp.	Scavenger Beetle
Hydrov pla	*Hydrovolzia placophora*	Water Mite
Hygrob nor	*Hygrobates norvegicus*	Water Mite
Leuctra sp.	*Leuctra* sp.	Stonefly
Loboh web	*Lobohalacarus weberi*	Marine Mite
Nemoura sp.	*Nemoura* sp.	Stonefly
Niph aqu	*Niphargus aquilex*	Groundwater Amphipod (Crustacean)
Niph schell	*Niphargus schellenbergi*	Groundwater Amphipod (Crustacean)
Partn steinm	*Partnunia steinmanni*	Water Mite
Pisidium sp.	*Pisidium* sp.	Pea Clam

Abbreviation	Taxon	Common Name
Polyc fel	*Polycelis felina*	Planaria
Proton sp.	*Protonemura* sp.	Stonefly
Protz squ squ	*Protzia squamosa squamosa*	Water Mite
Seric sp.	*Sericostoma* sp.	Caddisfly
Sold chap	*Soldanellonyx chappuisi*	Marine Mite
Sold mon	*Soldanellonyx monardi*	Marine Mite
Sperchon sp.	*Sperchon* sp.	Water Mite
Velia sp.	*Velia* sp.	Water Strider
Anac sp.	*Anacaena* sp.	Water Beetle
Cruno irr	*Crunoecia irrorata*	Caddiesfly
Dixa sp.	*Dixa* sp.	Meniscus Midge
Carych sp.	*Carychium* sp.	Hollow-shelled Snails (Ellobiidae)
Carych trid	*Carychium tridentatum*	Herald Snail
Cicad vir	*Cicadella viridis*	Leafhopper (Cicada)
Discus rot	*Discus rotundatus*	Rotund Disc
Eisen tetr	*Eiseniella tetraedra*	Square Tail Worm (Earthworms)
Ligid hyp	*Ligidium hypnorum*	Woodlouse
Monac inc	*Monachoides incarnatus*	Land Snail ("Incarnadine Snail")
Oligol trid	*Oligolophus tridens*	Harvestman (Arachnids)
Oniscus as	*Oniscus asellus*	Woodlouse
Paran quadrip	*Paranemastoma quadripunctatum*	Harvestman (Arachnids)
Polydesmus sp.	*Polydesmus* sp.	Flat-backed Millipede
Trichoniscus sp.	*Trichoniscus* sp.	Woodlouse
Bryo pter	*Bryocoris pteridis*	Bug
Eucon fulv	*Euconulus fulvus*	Hive Snail (Land Snail)
Euconulus sp.	*Euconulus* sp.	Hive Snail (Land Snail)
Ixodes sp.	*Ixodes* sp.	Tick
Leiob blackw	*Leiobunum blackwalli*	Harvestman (Arachnids)
Lithobius sp.	*Lithobius* sp.	Stone Centipede
Neob carc	*Neobisium carcinoides*	Pseudoscorpion
Neob sim	*Neobisium simile*	Pseudoscorpion
Stenod hols	*Stenodema holsata*	Bug
Stenod laev	*Stenodema laevigata*	Bug

Acknowledgements

The results presented here are the main outcomes of the Ph.D.-thesis of the first author. He would like to thank the two reviewers Prof. Dr. Christian Opp (University of Marburg, Germany) and PD Dr. Hans Jürgen Hahn (University of Koblenz-Landau, Institute for Groundwater Ecology, Germany) for their helpful and motivating comments and advisory activities. The Hesse Federation for Cave and Karst Research supported the study with material resources. Thank you Stefan Zaenker (Chairman of the Hesse Federation for Cave and Karst Research in Fulda, Germany).

Author details

Martin Reiss* and Peter Chifflard

*Address all correspondence to: reissm@geo.uni-marburg.de

University of Marburg, Faculty of Geography, Marburg, Germany

References

[1] Schönborn, W. Lehrbuch der Limnologie. Stuttgart: E. Schweizerbart'sche Verlags-buchhandlung; 2013.

[2] Kresic, N. Types and Classification of Springs. In: Kresic, N., Stevanovic, Z. (eds.) Groundwater Hydrology of Springs. Oxford: Elsevier; 2010. p31-86.

[3] Illies, J., Botosaneanu, L. Problèmes et methodes de la classification et de la zonation ecologique des eaux courantes, considerées surtout du point de vue faunistique. Stuttgart: E. Schweizerbart'sche Verlagsbuchhandlung; 1963.

[4] Steinmann, P. Praktikum der Süßwasserbiologie. I. Teil: Die Organismen des fließenden Wassers. Berlin: Borntraeger; 1915.

[5] Thienemann, A. Die Gewässer Mitteleuropas. Eine hydrobiologische Charakteristik ihrer Haupttypen. Stuttgart: E. Schweizerbart'sche Verlagsbuchhandlung; 1924.

[6] Wetzel, R.G. Limnology: Lakes and River Ecosystems. San Diego: Academic press; 2001.

[7] Cantonati, M., Gerecke, R., Bertuzzi, E. Springs of the Alps – sensitive ecosystems to environmental change: from biodiversity assessment to long-term studies. Hydrobiologia 2006;562 59-96

[8] Cantonati, M., Füreder, L., Gerecke, R., Jüttner, I., & Cox, E. J. Crenic habitats, hotspots for freshwater biodiversity conservation: toward an understanding of their ecology. Freshwater Science 2012; 31(2) 463-480

[9] Fischer, J., Fischer, F., Schnabel, S. Wagner, R., Bohle, H.-W. Die Quellenfauna der hessischen Mittelgebirgsregion. Besiedlungsstruktur, Anpassungsmechanismen und Habitatbindung der Makroinvertebraten am Beispiel von Quellen aus dem Rheinischen Schiefergebirge und der osthessischen Buntsandsteinlandschaft. In: Botosaneanu, L. (ed.) Studies in Crenobiology. The biology of springs and springbrooks. Leiden: Backhuys Publishers; 1998. p183-199.

[10] Reiss, M. Substratpräferenz und Mikrohabitat-Fauna-Beziehung im Eukrenal von Quellgewässern. PhD thesis. University of Marburg; 2011

[11] Ward, J.V. Ecology of alpine streams. Freshwater Biology 1994; 32 277-294

[12] Odum, E.P. Fundamentals of Ecology. Philadelphia: Saunders.

[13] Allan, J.D., Castillo, M.M. Stream Ecology. Structure and Function of Running Waters. Dordrecht: Springer; 2007

[14] Allan, J.D. Landscapes and Riverscapes: The Influence of Land Use on Stream Ecosystems. Annual Review of Ecology 2004; 35 257-284

[15] Reiss, M. An integrative hierarchical spatial framework for spring habitats. Journal of Landscape Ecology 2013,6(2) 65-77.

[16] White, P.S., Pickett, S.T.A. Natural disturbance and patch dynamics: an introduction. In: Pickett, S.T.A., White, P.S. (eds.) The Ecology of Natural Disturbance and Patch Dynamics. Elsevier: San Diego; 1985. p1-16.

[17] Dumnicka, E., Galas, J., Koperski. Benthic Invertebrates in Karst Springs: Does Substratum or Location Define Communities? International Review of Hydrobiology 2007; 92 452-464.

[18] Hildrew, A.G. Physical habitat and the benthic ecology of streams and rivers. In: Bretschko, G., Helešic, J. (eds.) Advances in River Bottom Ecology. Leiden: Backhuys Prublishers; 1998. p13-22.

[19] Hynes, H.B.N. The Ecology of Running Waters. Liverpool: Liverpool University Press; 1970

[20] Bretschko, G., Helešic, J. (eds.) Advances in River Bottom Ecology. Leiden: Backhuys Prublishers; 1998.

[21] Brainwood, M., Burgin, S., Byrne, M. The role of geomorphology in substratum patch selection by freshwater mussels in the Hawkesbury-Nepean River (New South Wales) Australia. Aquatic Conservation: Marine and Freshwater Ecosystems 2008; 18 1285-1301.

[22] Urbanaič, G., Toman, M.J., Krušnik, C. Microhabitat type selection of caddiesfly larvae (Insecta: Trichoptera) in a shallow lowland stream. Hydrobiologia 2005; 541 1-12.

[23] Lamouroux, N., Dolédec, S., Gayraud, S. Biological traits of stream macroinvertebrate communities: effects of microhabitat, reach and basin filter. Journal of the North American Benthological Society 2004; 23 449-466.

[24] Boyero, L. The effect of substrate texture on colonization by stream macroinvertebrates. Annales de Limnologie-International Journal of Limnology 2003; 39 211-218.

[25] Angradi, T.R. Fine sediment and macroinvertebrate assemblages in Appalachian streams: a field experiment with biomonitoring applications. Journal of the North American Benthological Society 1999; 18 49-66.

[26] Erman, N.A., Erman, D.C. The response of stream macroinvertebrates to substrate size and heterogeneity. Hydrobiologia 1984; 108 75-82.

[27] Büttner, G., Fetz, R., Hotzy, R., Römheld, J. Aktionsprogramm Quellen in Bayern – 1.Teil Bayerischer Quelltypenatlas. Augsburg: Bayerisches Landesamt für Umwelt. 2008

[28] Spitale, D. Assessing the ecomorphology of mountain springs: suggestions from a survey in the south-eastern Alps. In: Cantonati, M., Bertuzzi, E., Spitale, D. (eds.). Trento: Museo Tridentino di Scienze Naturali; 2007. p31-39.

[29] Howein, H., Schröder, H. Geomorphologische Untersuchungen. In: Gerecke, R., Franz, H. (eds.) Quellen im Nationalpark Berchtesgaden. Lebensgemeinschaften als Indikatoren des Klimawandels. Berchtesgaden: Nationalparkverwaltung Berchtesgaden; 2006. p71-86.

[30] Mühlenberg, M. Freilandökologie. Heidelberg: Quelle & Meyer; 1993.

[31] Reiss, M., Opp, C. Ein Erfassung- und Bewertungsverfahren der Gewässerstrukturgüte von Quellen und Quellbächen. In: Opp, C. (ed.). Wasserressourcen – Nutzung und Schutz. Marburg: Marburger Geographische Gesellschaft; 2004. p155-189.

[32] Koop, J.H.E. Gewässerstruktur und Biodiversität. In: Gewässermorphologisches Kolloquium, 05 May 2002, Koblenz, Germany. Koblenz: Bundesanstalt für Gewässerkunde; 2003.

[33] Hering, D., Moog, O., Sandin, L., Verdonschot, P. F. Overview and application of the AQEM assessment system. Hydrobiologia 2004; 516 1-20.

[34] Barbour, M.T., Gerritsen, J., Snyder, B.D., Stribling, J.B. Rapid Bioassessment Protocols for Use in Streams and Wadeable Rivers: Periphyton, Benthic Macroinvertebrates and Fish. Washington; U.S. Environmental Protection Agency, Office of Water; 1999.

[35] Weigand, E. Limnologisch-faunistische Charakterisierung von Karstquellen, Quellbächen und unterirdischen Gewässern nach Choriotopen und biozönotischen Ge-

wässerregionen. Unpublished Research Report, Leonstein: Nationalpark Kalkalpen; 1998.

[36] Lindegaard, C., Brodersen, K.P., Wiberg-Larsen, P., Skriver, J. Multivariate analyses of macrofaunal communities in Danish springs and springbrooks. In: Botosaneanu, L. (Ed.) Studies in Crenobiology. The biology of springs and springbrooks. Leiden: Backhuys Publishers; 1998. p201-219.

[37] Williams, D.D., Williams, N. E. Canadian Springs: postglacial development of the invertebrate fauna. In: Batzer, D.P., Rader, R.B., Wissinger, S.A. (Eds.) Invertebrates in Freshwater Wetlands of North America. Ecology and Management. New York: Wiley & Sons; 1999; p447-467.

[38] Gaskin, B., Bass, D. Macroinvertebrates Collected From Seven Oklahoma Springs. In: Proceedings of the Oklahoma Academy of Science 2000;80 17-23.

[39] Bonettini, A. M., Cantonati, M. Macroinvertebrate assemblages of springs of the River Sarca catchment (Adamello-Brenta Natural Park, Trentino, Italy). Crunoecia 1996;5 71–78.

[40] von Fumetti, S., Nagel, P., Scheifhacken, N., Baltes, B. Factors governing macrozoobenthic assemblages in perennial springs in north-western Switzerland. In: Hydrobiologia, 2006;568 467-475.

[41] Hahn, H.-J. Studies on Classifying of Undisturbed Springs in Southwestern Germany by Macrobenthic Communities. In: Limnologica 2000;30 247-259.

[42] Staudacher, K., Füreder, L. Habitat Complexity and Invertebrates in Selected Alpine Springs (Schütt, Carinthia, Austria). In: International Review of Hydrobiology 2007;92 465-479.

[43] Cantonati, M., Spitale, D. The role of environmental variables in structuring epiphytic and epilithic diatom assemblages in springs and streams of Dolomiti Bellunesi National Park (south-eastern Alps). Fundamental and Applied Limnology 2009;174 117-133.

[44] Ilmonen, J., Paasivirta, L. Benthic macrocrustacean and insect assemblages in relation to spring habitat characteristics: patterns in abundance and diversity. Hydrobiologia 2005;533 99-113.

[45] Morrison, M.L., Marcot, B.G., Mannan, R.W. Wildlife-habitat relationships. Concepts and Applications. Washington: Island Press; 2006.

[46] Zollhöfer, J.M., Brunke, M., Gonser, T. A typology of springs in Switzerland by integrating habitat variables and fauna. In: Archiv für Hydrobiologie 2000;Supplement 121 349-376.

[47] Smith, H. The Hydro-Ecology of Limestone Springs in the Wye Valley, Derbyshire. In: Water and Environment Journal 2002;16 253-259.

[48] Smith, H., Wood, P.J., Gunn, J. The influence of habitat structure and flow permanence on invertebrate communities in karst spring systems. In: Hydrobiologia 2003;510 53-66.

[49] Gerecke, R., Meisch, C., Stoch, F. Acri, F., Franz, H. Eucrenon/Hypocrenon ecotone and spring typology in the Alps of Berchtesgaden (Upper Bavaria, Germany). A study of microcrustacea (Crustacea: Copepoda, Ostracoda) and water mites (Acari: Halacaridae, Hydrachnellae). In: Botosaneanu, L. (Ed.) Studies in Crenobiology. The biology of springs and springbrooks. Leiden: Backhuys Publishers; 1998. p167-182.

[50] Guidicelli, J., Bournaud, M. Invertebrate Biodiversity in land-inland water ecotonal habitats. In: Lachavanne, J.-B., Juge, R. (Eds.) Biodiversity in land-inland water ecotones. Paris: United Nations Educational, Scientific and Cultural Organization (UNESCO); 1997. p143-160.

[51] Ssymank, A., Hauke, U., Rückriem, C., Schröder, E. Das europäische Schutzgebietssystem NATURA 2000. BfN-Handbuch zur Umsetzung der Fauna-Flora-Habitat-Richtlinie und der Vogelschutz-Richtlinie. Bad Godesberg: Bundesamt für Naturschutz; 1998.

[52] Poff, N.L. Landscape filters and species traits: towards mechanistic understanding and prediction in stream ecology. Journal of the North American Benthological Society 1997;16(2) 391-409.

[53] Cheshmedjiev, S., Soufi, R., Vidinova, Y., Tyufekchieva, V., Yaneva, I., Uzunov, Y., Varadinova, E. Multi-habitat sampling method for benthic macroinvertebrate communities in different river types in Bulgaria. Water Research and Management 2011; 1(3) 55-58.

[54] Reiss, M., Steiner, H., Zaenker, S. The Biospeleological Register of the Hesse Federation for Cave and Karst Research (Germany). Cave and Karst Science 2009;35(1) 25-34.

[55] Clarke, K.R. Non-parametric multivariate analyses of changes in community structure. Australian Journal of Ecology 1993;18 117-143.

[56] Clarke, K.R., Warwick, R.M. Changes in marine communities: an approach to statistical analysis and interpretation. 2nd Edition. Plymouth: Plymouth Marine Laboratory.

[57] Bray J.R., Curtis J.T. An ordination of the upland forest communities of southern Wisconsin. Ecological Monographs 1957;27 325–349.

[58] Moog, O., editor. Fauna Aquatica Austriaca. Wien: Wasserwirtschaftskataster, Bundesministerium für Land-und Forstwirtschaft, Umwelt-und Wasserwirtschaft; 2002.

[59] Schmedtje, U., Colling, M. Ökologische Typisierung der aquatischen Makrofauna. München: Bayerisches Landesamt für Wasserwirtschaft; 1996.

[60] Shannon, C.E., Weaver, W. The Mathematical Theory of Communication. Urbana: University of Illinois Press; 1949.

[61] Pielou, E.C. The Measurement of Diversity in different Types of Biological Collections. Journal of Theoretical Biology 1966;13 131-144.

[62] Braukmann, U. Zoozönologische und saprobiologische Beiträge zu einer allgemeinen regionalen Bachtypologie. Stuttgart: Schweizerbart'sche Verlagsbuchhandlung; 1987.

[63] Poff, N. L., Ward, J. V. Physical habitat template of lotic systems: recovery in the context of historical pattern of spatiotemporal heterogeneity. Environmental Management 1990;14.5 629-645.

[64] Vannote, R.L., Minshall, G. W., Cummins, K. W., Sedell, J. R., Cushing, C. E. The river continuum concept. Canadian journal of fisheries and aquatic sciences 1980;37(1) 130-137.

[65] Hennig, W. Die Larvenformen der Dipteren. 2. Teil. Berlin: Akademie-Verlag; 1950.

Chapter 10

Biodiversity and Conservation of Temporary Ponds — Assessment of the Conservation Status of "Veiga de Ponteliñares", NW Spain (Natura 2000 Network), Using Freshwater Invertebrates

Amaia Pérez-Bilbao, Cesar João Benetti and Josefina Garrido

Additional information is available at the end of the chapter

http://dx.doi.org/10.5772/59104

1. Introduction

Freshwater biodiversity provides a broad variety of valuable goods and services for human societies, some of them irreplaceable [1]. Globally, the biodiversity of freshwater ecosystems is rapidly deteriorating as a result of human activities [2]. It is possible that in future decades human pressure on water resources will further endanger aquatic biodiversity present in these systems [3]. The need to protect these ecosystems and many others led to the creation of the Natura 2000 network in Europe. This network is the most important conservation and management tool in the European Union. It was established under the Habitats Directive (92/43/EEC) and the Birds Directive (79/409/EEC), and its main objective is to ensure the long-term conservation of the most important European species and habitats in a sustainable way with human activities. It is formed by Special Areas of Conservation (SAC), which are protected areas established with the purpose of conservation of habitat types and/or species included in the Habitats and Birds directives. In Spain, there are 1,448 SAC covering a total of 23.17% of the territory. Only 11.65% of the Autonomous Community of Galicia (North-western Spain) is protected (59 SAC), in spite of having a great variety of freshwater ecosystems.

Wetlands are sites of high biodiversity and productivity [4], but these ecosystems have undergone a serious decline worldwide [5,6]. Among stagnant water bodies, ponds constitute essential freshwater ecosystems for biodiversity conservation. Due to their heterogeneity and the varied network of habitats they provide, they often support higher diversity than more

permanent and large freshwater habitats and act as stepping stones for the dispersal of species [7]. Among these freshwater ecosystems are temporary ponds, which are endangered due to their small size and shallowness [8]. Small changes in hydrological regimes can greatly impact the ecological regime of temporary ponds [9] and it is expected that the reduction in rainfall brought about by climate change will affect their hydrology [10]. Temporary ponds have been neglected for years and are affected by human activities, such as agriculture, urbanization, etc. The inclusion of the Mediterranean temporary ponds (code 3170*) as a priority habitat for conservation in the Habitats Directive (EEC, 1992) highlights the importance of these ecosystems and the necessity to conserve them.

Temporary ponds support a great diversity of freshwater fauna, including invertebrates. Under this assumption, the purpose of this chapter was to analyze the importance of the invertebrate fauna in maintaining the ecological balance of temporary ponds and assess the effectiveness of protecting areas in the conservation of their biological values. So, we wonder if SAC are efficient in wildlife conservation and whether the freshwater invertebrates are good indicators of the environmental quality of temporary ponds. To do this, we studied the invertebrate fauna in different ponds within the SAC "Veiga de Panteliñares" (North-Western Spain).

2. Definition and characteristics of temporary ponds

According to the Ramsar Convention, temporary ponds are usually small (< 10 ha) and shallow wetlands which are characterized by alternating of flooded and dry phases, and whose hydrology is largely autonomous (Figure 1). They occupy depressions, often endorheic, which are flooded for a sufficiently long period to allow the development of hydromorphic soils and wetland-dependent aquatic or amphibious vegetation and fauna communities. However, equally importantly, temporary ponds dry out for long enough periods to prevent the development of the more widespread plant and animal communities characteristic of more permanent wetlands.

Temporary ponds are habitats with a predictable annual dry phase of 3-8 months, usually during summer and autumn [11]. According to [12], temporary ponds can be classified as intermittent (with a seasonal cyclic pattern of dryness and flooding) or episodic (unpredictably flooded). These habitats must undergo a periodic cycling of flooding and drought for a correct functioning [13]. They are usually located in shallow areas with impermeable ground and present a small catchment area [9]. Water volume depends on the balance between inputs (precipitation, surface runoff, melting snow and inflows of groundwater) and outputs (evapotranspiratiom, infiltration and overflow) [14]. One of the main characteristics of temporary ponds is their isolation. If they were connected to more permanent habitats, this would probably cause the colonization of species typical of permanent habitats and the disappearance of those typical of temporary habitats due to competition and predation.

Figure 1. Temporary pond in the SAC Serra do Careón (NW Spain).

Extreme fluctuations affect the physical and chemical characteristics of the water. Nutrients such as nitrate or phosphate appear usually in low concentrations, but vary throughout the different hydrological stages of the pond. The same happens with pH, dissolved oxygen or salinity. The latter present higher values in the last part of the wet phase because the ion concentration increases when the pond is drying. In general, dissolved oxygen concentration is low and organisms have developed several adaptations to survive, for example, swimming near to the water surface or having more haemoglobin [9]. Mineralization is low with electric conductivity reaching values of 0.05-0.3 mS/cm in the maximum flooding period [15].

These environments occur in many parts of the world, but are well represented in arid, semi-arid and Mediterranean areas. Mediterranean temporary ponds constitute one type of temporary ponds and are considered a priority habitat type in Europe (code 3170*). According to the *Interpretation Manual of European Union Habitats* (EUR27, July 2007) these are very shallow temporary ponds (a few centimetres deep) which exist only in winter or late spring with a flora mainly composed of Mediterranean therophytic and geophytic species belonging to the alliances *Isoetion, Nanocyperion flavescentis, Preslion cervinae, Agrostion salmanticae, Heleochloion* and *Lythrion tribracteati*.

3. Biodiversity of temporary ponds

Temporary ponds constitute an important habitat for the breeding, feeding and migration of amphibians, reptiles, invertebrates, birds and mammals [16]. Because of their relatively isolated status in comparison to permanent water bodies, their unpredictable date of flooding and their small size and shallow conditions, biodiversity in these ponds is high [9]. In addition, several studies have probed the importance of temporary ponds for rare and endangered invertebrate species [8,17,18].

According to [19], hydroperiod is one of the main factors affecting the composition and structure of aquatic assemblages. As a result, temporary ponds support biological communities different to those that inhabit permanent habitats. Organisms living in temporary waters have to adapt to temporary drought conditions to survive, sometimes being exclusive to these ecosystems [20]. They develop morphological adaptations, life cycles and dispersion mechanisms which make them survive in dry seasons. Based on the four groups proposed by [21], aquatic organisms have two strategies to survive drought: to pass the dry phase via resistant life stages or to actively migrate when water disappears. For example the aquatic beetle *Berosus signaticollis* (Charpentier) remains embedded in the sediment to complete its life cycle [19], the crustacean *Tanymastix stagnalis* Daday resists the drought with resting eggs [22] or the damselfly *Lestes dryas* Kirby and the dragonfly *Sympetrum sanguineum* (Muller) complete their life cycles before the drying season [23].

Invertebrates of temporary ponds usually exhibit traits of r-selected species, especially great dispersal ability, rapid growth, short life-span, small size, and opportunistic/generalistic feeding [12]. Insects and crustaceans constitute the larger invertebrate groups in these ponds [19]. Among aquatic invertebrates, several groups or species can be considered typical of temporary ponds, like the large branchiopods *Lepidurus apus* (L.) and the genus *Triops*; cladocerans in the genus *Daphnia*; the fairy shrimp *Tanymastix stagnalis*; odonates in the genera *Lestes* or *Sympetrum*; hemipterans in the genera *Gerris*, *Notonecta*, *Sigara* or *Hesperocorixa*; aquatic beetles in the genera *Agabus*, *Graptodytes*, *Berosus*, *Helophorus* or *Hydroporus*, among others [15,24,25].

Temporary ponds are especially favorable habitats for amphibians (*Bufo*, *Hyla*, *Rana or Triturus*) to feed and breed (Figure 2). Larvae can feed on the abundant phyto and zooplankton, and aquatic vegetation is ideal for egg laying. Many of these ponds are fishless, thus reducing predation pressure that has a great impact on larvae [26]. Reproductive success depends on the hydroperiod length, because if the pond dries out too soon the offspring can die [9]. In addition, temporary freshwater bodies are very important not only for waterfowl and migratory birds but also for other bird species that inhabit temporary ponds and their surroundings [13].

These ponds also support rich and diverse plant communities, especially in the Mediterranean region, hosting rare and endangered species [9]. Species composition depends on the flooding length, the type of substrate and water depth. In general these species are able to produce seeds in a short period of time to complete their life cycles [13]. We can typically find species of the

Figure 2. Frogs in a temporary pond in A Serra da Capela (NW Spain).

genera *Isoetes, Callitriche, Ranunculus, Eryngium* or *Juncus*. Regarding phytoplankton, a great number of different species belonging to Dinophyceae, Clorophyceae, Euglenophyceae, Zygnematophyceae or Cryptophyceae are usually found in these habitats. Phytoplankton and periphyton constitute the basis of the food webs in temporary ponds [27].

4. Threats

Temporary ponds constitute endangered habitats due to their shallowness and temporality, and probably in future decades their situation will get worse. In general stagnant water bodies have been considered for a long time as unproductive areas inhabited by disease-transmitting insects. Besides, the need for new farmlands has caused the reduction of these habitats [28]. The lack of information on the natural values of temporary ponds and appropriate management measures results in their deterioration or even disappearance. For example, in Spain it was estimated that in the late twentieth century more than 60% of wetlands have disappeared [29].

According to [13], threats to temporary ponds, especially Mediterranean ones, are mainly related to invasive species, pollution, changes in the hydrological functioning and climate change. These threats can be resumed as follows:

- Invasive species: allochthonous flora and fauna species can displace autochthonous ones. The introduction of invasive species of fishes, crabs or reptiles that predate crustaceans, insects and amphibians can produce the disappearance of the latter ones which are not used to inhabit ponds with this type of predators.

- Pollution: fertilizers and pesticides affect the water chemistry of the ponds. Many species are sensitive to pesticides and disappear in polluted habitats. On the other hand, fertilizers cause the eutrophication of the water, which can modify flora and fauna assemblages.

- Changes in the hydrological functioning: these changes are all related to human activities, such as urbanization, over-exploitation of aquifers, draining for new farmlands or dredging for watering cattle.

- Climate change effects: temporary ponds depend on flooding and drought phases for a correct functioning of the habitat. Changes in rainfall and temperature due to climate change can greatly affect these ecosystems. Global warming is also transforming permanent water bodies into temporary ones.

In relation to the latter, in the twenty-first century ponds will have to face global challenges [30]. The expected reduction of humid years and of rainfall globally may lead to a decrease in the probability of survival of populations of characteristic pond species [31]. Thus, changes in the hydrology will be a key factor to investigate. In this sense, long-term monitoring can provide particularly rich information, especially in the context of global change, and many protected areas have now set up systems to monitor their biodiversity [30].

5. Case study

5.1. Introduction

As mentioned before, hydroperiod is the main stressor in temporary habitats. Due to the instability of the system, temporary ponds are very diverse ecosystems. Information about the relative biodiversity value of different water body types is a vital pre-requisite for many strategic conservation goals [32], including sustainable catchment management as required by the EC Water Framework Directive (2000/60/EC). In this sense, the composition and abundance of benthic invertebrates is one of the most important criteria to be considered. So, the knowledge about the aquatic invertebrates inhabiting the different temporary pond types is essential for developing management strategies.

The use of different groups as indicators for monitoring population trends in other groups of aquatic invertebrates and for identifying high biodiversity areas at a regional scale has been suggested. For example, aquatic Coleoptera are generally considered a suitable group to assess the environmental and conservation value of wetland sites and habitats [33-36,28,37]. Aquatic Hemiptera are also considered potential bioindicators of water quality [38], so they can be used in terms of regional or global conservation planning of freshwater biodiversity [39].

In Spain, studies focused on temporary ponds have been mainly carried out in the Mediterranean region [e.g. 19,40-45]. In Galicia (NW Spain) several studies in stagnant waters have been carried out during the last years [e. g. 25]. We can conclude that the Mediterranean area of this region has a great biodiversity of aquatic organisms, but there are no studies dealing with the biology and ecology of temporary ponds.

The main objectives of this study were (a) to analyze the composition and structure of the invertebrate fauna assemblages in three temporary ponds using data obtained in studies carried out in two periods of time; and (b) to assess the effectiveness of the SAC on the conservation of their biological values by studying the change in these assemblages.

5.2. Study area

The study was carried out in "A Veiga de Panteliñares", located in the Autonomous Region of Galicia (North-western Spain) (Figure 3). Although this area is protected under the Natura 2000 network as a SAC and is included in the Biosphere Reserve Área de Allariz, it is one of the smallest protected areas in Galicia, with a surface of 130 Ha. It is formed by alluvial water meadows with temporary flooding along the banks of the Limia River. These meadows represent a small part of what it was in this area until the middle of the 20th century, when these fields were dried including the Antela Lake. This shallow lake, which it was considered the biggest in the Iberian Peninsula, was 6 km long from northeast to southeast and 4 km wide, with a depth of 3 m in the rainy season and less than 1 m in the dry season. The Limia River valley is a highly humanized area with many crops, blasting companies and poultry and pig farms.

The SAC is included in the Mediterranean Region. The climate of the study area is warm temperate, with a mild temperature [46]. Dry summers cause the ponds to dry out, for between two and four months, depending on the year.

Three temporary ponds were sampled in the study area: Veiga da Pencha, A Telleira and A Veiga (Figures 4-6). Table 1 shows the pond codes, which are used in the paper, and the location in UTM coordinates. These ponds are Mediterranean temporary ponds, which is a priority habitat for conservation (code 3170) included in the Annex I of the Habitats Directive and are classified as intermittent waters according to [12].

Pond	Code	UTM X	UTM Y
A Veiga	AV	29T 595.518	4.655.117
Telleira	TE	29T 595.966	4.655.513
Veiga da Pencha	VP	29T 594.548	4.654.969

Table 1. Names of the studied ponds with their codes and UTM coordinates.

Veiga da Pencha (VP) is a temporary pond (15 m maximum width x 100 m maximum length) formed at the end of a stream that was probably connected to the Limia River before the construction of a road on the river side. It is not independent of the stream until late spring

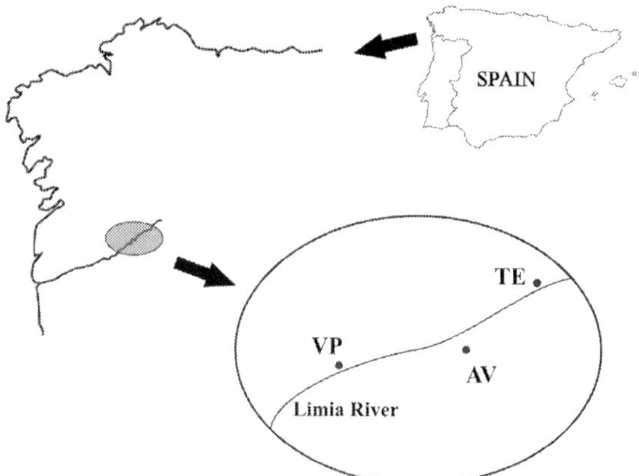

Figure 3. Map indicating the location of the studied ponds within the catchment of the Limia River. VP: Veiga da Pencha; AV: A Veiga; TE: A Telleira.

and dries completely at the end of the summer. The pond is surrounded by willows and birches.

Figure 4. A Veiga da Pencha pond.

A Telleira (TE) is the smallest sampled pond (4 m width x 7 m length). It is located at the end of a temporary stream connected to the Limia River. The pond is formed due to the temporality of the stream. The size of the pond decreases during the dry season but maintains its water volume longer than the other ponds probably due to a freatic contact with the Limia River.

Figure 5. A Telleira pond.

A Veiga (AV) is located in the southern limit of the SAC. It is an irregular and elongated (10 m maximum width x 52 m length) temporary pond. The extremes of the pond dry out rapidly at the end of the rainy season and the central part of the pond dries at the end of the summer.

The studied ponds have important riparian vegetation that consists mainly of grasses and autochthonous deciduous trees, as well as a significant macrophyte cover. The riparian vegetation is mainly composed of *Alnus glutinosa* (L.) Gaertn, *Erica* sp., *Eryngium viviparum* J. Gay, *Fraxinus excelsior* L., *Quercus robur* L. and *Salix atrocinerea* Brot. The main macrophyte species are *Agrostis* sp., *Callitriche palustris* L., *Cyperus* sp., *Damasonium* sp., *Ranunculus peltatus* Schrank and *Scirpus lacustris* (L.) Palla.

Figure 6. A Veiga pond.

5.3. Material and methods

5.3.1. Sampling

The three ponds were sampled in two periods, 2001-2002 (before being included in the Natura 2000 network) and 2007-2011 (after being included in the Natura 2000 network). During 2001 and 2002 samples were taken seasonally, and during the period 2007-2011 surveys were carried out in spring and summer of every year. In total, 24 samples were processed. Fauna was collected using an entomological net (500 μm mesh, 30 cm diameter and 60 cm deep) across a 10m transect running parallel to the margin. This semi-quantitative method allows for direct comparisons across sites or time because sampling effort can be assumed equivalent. The material was preserved in 99% ethanol, and sorted out and identified at the laboratory. After being studied, the specimens were conserved in 70% ethanol and deposited in the scientific collection of the Aquatic Entomology Laboratory at Vigo University.

5.3.2. Data analysis

The structure of the assemblage was assessed using different diversity indices: Richness (S); Rarefied richness (ES); Abundance (N) and the Shannon-Wiener diversity index (H'). Rarefied richness is the expected number of taxa for a given number of randomly sampled individuals and facilitates comparison of areas in which densities may differ [48]. It calculates the number of expected species from each sample if all samples were reduced to a standard size of n individuals. The values of this index were calculated for 200 individuals ES (200). The Shannon-Wiener index measures the average degree of uncertainty in predicting to which species will belong an individual randomly chosen from a collection [49]. It assumes that all the individuals are randomly selected and that all species are represented in the sample. The greater the index value, the greater the diversity in the habitat. According to [49], the usual range for this index is 1.5-3.5.

Similarity relationships among invertebrate assemblages in all samples were determined by the Bray-Curtis coefficient and graphically presented using non-metric Multi-Dimensional Scaling (NMDS) mapped in two dimensions. Then, we grouped the samples according to two factors: Pond (A Veiga, Telleira and Veiga da Pencha), and the two periods (2001-2002 and 2007-2011). An analysis of similarity (ANOSIM) was used to test whether the established groups based on biotic data differed significantly.

To investigate the groups' consistency the SIMilarity PERcentages-species contributions (SIMPER) analysis was used to obtain differences between all pairs of groups and the contribution of each species for the groups. SIMPER examines the contribution of each species to the average Bray-Curtis dissimilarity between groups of samples and also determines the contribution to similarity within a group [47]. Statistical analyses were carried out with PRIMER version 6.

5.4. Results

5.4.1. Diversity indices

Table 2 shows the minimum, maximum and mean (including standard deviation) diversity indices (total abundance, richness, rarefied richness and Shannon-Wiener diversity) for each sample.

	TE			AV			VP		
	Minimum	Maximum	Mean ± SD	Minimum	Maximum	Mean ± SD	Minimum	Maximum	Mean ± SD
S	13	35	23.4 ± 8.02	10	61	31.75 ± 18.59	21	77	41 ± 20.38
ES (200)	10.58	22.47	15.78 ± 5	10	25	18.93 ± 5.62	16,77	34,26	24.04 ± 5.36
N	246	5469	1512 ± 2258	68	18149	3185 ± 6381	110	16349	3405 ± 5290
H'(log^2)	1.43	3.15	2.46 ± 0.71	1,86	4	2.97 ± 0.74	2,19	4,58	3.41 ± 0.74

Table 2. Minimum, maximum and mean values of the different diversity indices.

The total abundance was of 70,512 specimens. The most abundant were insects (34,600) and crustaceans (32,582). Among insects Coleoptera was the most abundant group (15,497 specimens), followed by Diptera (7,760), Hemiptera (3,681) and Odonata (3,119). The most abundant crustaceans were Cladocera (23,712), Copepoda (5,599) and Ostracoda (2,501).

The greatest abundances were observed in AV in April 2007, with 18,149 specimens collected, and in VP in the same sampling (16,349 specimens). On the contrary, the lowest values were obtained in AV in December 2001 with only 68 individuals captured and in VP in May 2002 (110 specimens). Mean abundance was significantly higher in VP (3,405 specimens) and AV (3,185) than in TE (1,512).

Regarding species, the most abundant were the water beetles *Graptodytes flavipes* (Olivier) (5,216 specimens), *Anacaena lutescens* (Stephens) (1,580), *Haliplus lineatocollis* (Marsham) (1,114) and *Hydrochus angustatus* Germar (1,027). Several species were represented by only one specimen, among which we highlight the crustacean *Lepidurus apus*, the beetle *Helophorus bameuli* Angus and the dragonfly *Boyeria irene* (Fonscolombe).

In total, 169 invertebrate taxa were collected in the three studied ponds (Table 3). The most representative group were insects (145 taxa), followed by gastropods (6 taxa) and crustaceans (5 taxa). Among insects, we highlight Coleoptera (94 species), Hemiptera (20 species) and Odonata (14 species). The other faunal groups recorded were Cnidaria, Bivalvia, Hirudinea, Oligochaeta, Arachnida and Collembola.

Taxa	TE	AV	VP
Cnidaria			
Hydridae			
Hydra sp.		X	
Bivalvia			
Sphaeriidae		X	X
Pisidium sp.		X	X
Gastropoda			
Ancylidae		X	X
Lymnaeidae			X
Myxas sp.		X	X
Radix sp.			X
Physidae	X		
Planorbidae		X	X
Hirudinea			
Erpobdellidae		X	X
Glossiphonidae			X
Haemopidae			
Haemopis sanguisuga (L., 1758)			X
Hirudinidae		X	X
Oligochaeta			
Enchytraeidae		X	
Lumbricidae		X	
Lumbriculidae		X	X
Naididae		X	

Taxa	TE	AV	VP
Crustacea			
Asellidae	X	X	X
Cladocera		X	X
Copepoda		X	X
Ostracoda		X	X
Triopsidae			
Lepidurus apus (L., 1758)			X
Arachnida			
Hydracarina	X	X	X
Collembola		X	X
Insecta			
Coleoptera			
Gyrinidae			
Gyrinus substriatus Stephens, 1829	X	X	X
Haliplidae			
Haliplus guttatus Aubé, 1836			X
Haliplus heydeni Wehncke, 1875	X	X	X
Haliplus lineatocollis (Marsham, 1802)	X	X	X
Haliplus sp.		X	X
Peltodytes rotundatus (Aubé, 1836)		X	
Noteridae			
Noterus laevis Sturm, 1834	X	X	X
Dytiscidae			
Agabus bipustulatus (L., 1767)	X	X	X
Agabus brunneus (Fabricius, 1798)			X
Agabus didymus (Olivier, 1795)	X		X
Agabus labiatus (Brahm, 1791)		X	
Agabus paludosus (Fabricius, 1801)			X
Agabus sp.		X	X
Bidessus goudotii (Laporte, 1835)	X	X	X
Colymbetes fuscus (L., 1758)	X	X	X
Cybister lateralimarginalis (De Geer, 1774)			X
Dytiscus marginalis L., 1758	X		X
Dytiscus pisanus Laporte, 1835		X	
Dytiscus semisulcatus O.F. Müller, 1776		X	X
Dytiscus sp.			X
Graptodytes bilineatus (Sturm, 1835)			X
Graptodytes castilianus Fery, 1995	X	X	X
Graptodytes flavipes (Olivier, 1795)	X	X	X
Graptodytes fractus (Sharp, 1882)			X
Graptodytes ignotus (Mulsant & Rey, 1861)	X		X

Taxa	TE	AV	VP
Graptodytes varius (Aubé, 1838)			X
Hydroporus gyllenhalii Schiødte, 1841	X	X	X
Hydroporus planus (Fabricius, 1781)			X
Hydroporus pubescens (Gyllenhal, 1808)		X	X
Hydroporus sp.		X	X
Hydroporus vagepictus Fairmaire & Laboulbène, 1854	X	X	X
Hydroporus vespertinus Fery & Hendrich, 1988		X	X
Hydrovatus clypealis Sharp, 1876		X	X
Hygrotus inaequalis (Fabricius, 1776)	X	X	X
Hyphydrus aubei Ganglbauer, 1892		X	
Ilybius meridionalis Aubé, 1837			X
Ilybius montanus (Stephens, 1828)	X		X
Laccophilus hyalinus (De Geer, 1774)	X	X	
Laccophilus minutus (L., 1758)	X	X	X
Laccophilus sp.			X
Liopterus atriceps Sharp, 1882	X	X	X
Metaporus meridionalis (Aubé, 1838)			X
Rhantus hispanicus Sharp, 1882		X	X
Rhantus suturalis (McLeay, 1825)			X
Stictonectes lepidus (Olivier, 1795)	X		
Hydrophilidae			
Anacaena globulus (Paykull, 1798)			X
Anacaena limbata (Fabricius, 1792)	X		X
Anacaena lutescens (Stephens, 1829)	X	X	X
Anacaena sp.			X
Berosus affinis Brullé, 1835	X		X
Berosus signaticollis (Charpentier, 1825)	X	X	X
Berosus sp.	X	X	X
Enochrus fuscipennis (Thomson, 1884)	X		X
Enochrus nigritus (Sharp, 1872)	X	X	X
Enochrus sp.		X	X
Helochares punctatus Sharp, 1869	X	X	X
Hydrobius convexus Brullé, 1835			X
Hydrobius fuscipes (L., 1758)	X	X	X
Hydrobius sp.		X	X
Hydrophilus pistaceus Laporte, 1840			X
Laccobius ytenensis Sharp, 1910	X		
Limnoxenus niger (Gmelin, 1790)	X	X	X
Paracymus scutellaris (Rosenhauer, 1856)	X		X
Hydrochidae			
Hydrochus angustatus Germar, 1824	X	X	X

Taxa	TE	AV	VP
Hydrochus flavipennis Küster, 1852	X	X	X
Hydrochus nitidicollis Mulsant, 1844	X	X	X
Helophoridae			
Helophorus alternans Gené, 1836		X	X
Helophorus bameuli Angus, 1987		X	
Helophorus flavipes Fabricius, 1792		X	X
Helophorus maritimus Rey, 1885		X	
Helophorus minutus Fabricius, 1775		X	X
Helophorus seidlitzii Kuwert, 1885			X
Helophorus sp.	X		X
Hydraenidae			
Aulacochthebius exaratus (Mulsant, 1844)		X	X
Hydraena brachymera D'Orchymont, 1936	X	X	
Hydraena sp.		X	X
Hydraena marcosae Aguilera,Hernando & Ribera, 1997	X		X
Hydraena rugosa Mulsant, 1844	X	X	X
Hydraena testacea Curtis, 1830	X	X	X
Hydrena exasperata D'Orchymont, 1935	X	X	
Limnebius gerhardti Heyden, 1870	X	X	
Limnebius lusitanus Balfour-Browne, 1979		X	X
Limnebius sp.			X
Limnebius truncatellus (Thunberg, 1794)		X	
Ochthebius sp.	X	X	X
Ochthebius viridis fallaciosus Ganglbauer, 1901			X
Dryopidae			
Dryops luridus (Erichson, 1847)	X	X	X
Dryops sp.		X	X
Dryops striatellus (Fairmaire & Brisout, 1859)		X	
Elmidae			
Oulimnius rivularis (Rosenhauer, 1856)	X		X
Oulimnius sp.		X	X
Scirtidae			
Helodes sp.			X
Hydrocyphon sp.		X	X
Diptera			
Ceratopogonidae		X	X
Chironomidae	X	X	X
Culicidae			
Anopheles sp.	X	X	X
Culex sp.		X	X
Dixidae			

Taxa	TE	AV	VP
Dixella sp.		X	X
Dolichopodidae			X
Empididae		X	X
Limoniidae		X	
Rhagionidae		X	
Sciomyzidae			X
Tabanidae		X	X
Ephemeroptera			
Baetidae		X	X
Siphlonuridae			
Siphlonurus sp.		X	X
Hemiptera			
Corixidae			
Corixa iberica Jansson, 1981		X	X
Hesperocorixa moesta (Fieber, 1848)		X	X
Hesperocorixa sahlbergi (Fieber, 1848)	X	X	X
Hesperocorixa sp.		X	
Sigara janssoni Lucas, 1983	X	X	X
Sigara limitata (Fieber, 1848)			X
Sigara scotti (Douglas & Scott, 1868)			X
Sigara sp.			X
Gerridae			
Gerris gibbifer Schummel, 1832			X
Gerris lacustris (L., 1758)			X
Gerris sp.		X	X
Gerris thoracicus Schummel, 1832		X	X
Hydrometridae			
Hydrometra stagnorum (L., 1758)			X
Naucoridae			
Naucoris maculatus Fabricius, 1798	X	X	X
Nepidae			
Nepa cinerea L., 1758	X	X	X
Ranatra linearis (L., 1758)		X	X
Notonectidae			
Notonecta glauca Poisson, 1758			X
Notonecta meridionalis Poisson, 1926		X	X
Notonecta obliqua Thunberg, 1787			X
Notonecta sp.		X	X
Pleidae			
Plea minutissima Leach, 1817	X	X	X
Vellidae			X

Taxa	TE	AV	VP
Lepidoptera			
Crambidae			
Elophila sp.			X
Odonata			
Aeshnidae			X
Boyeria irene (Fonscolombe, 1838)	X		
Coenagrionidae	X	X	X
Pyrrhosoma nymphula (Sulzer, 1776)		X	
Cordulegasteridae			
Cordulegaster sp.			X
Cordullidae		X	X
Lestidae			
Lestes barbarus (Fabricius, 1798)	X	X	
Lestes dryas Kirby, 1890			X
Lestes sp.		X	X
Lestes viridis (Vander Linden, 1825)		X	
Libellulidae		X	X
Sympetrum sanguineum (Muller, 1764)			X
Sympetrum vulgatum (L., 1758)	X	X	X
Platycnemididae			
Platycnemis sp.			X
Plecoptera			
Nemouridae			X
Trichoptera			
Limnephilidae		X	X
Limnephilus sp.			X

Table 3. List of the identified taxa. The crosses indicate the presence of the taxa in the ponds.

The greatest richness values were observed in VP in April 2007 (77 taxa recorded) and in July 2008 (64 taxa). On the other hand, the lowest values were observed in AV in December 2001 (10 taxa) and in TE in September 2002 (13 taxa). The mean richness value was 41 in VP, 31.75 in AV and 23.4 in TE, showing a similar pattern to that observed for the abundance. Considering the accumulated values, the richest pond was VP with 142 taxa, followed by AV (110 taxa) and TE (59 taxa).

In this study, the highest values of rarefied richness correspond to VP in July 2007 (34.26) and July 2008 (31.16). The lowest values were observed in AV in December 2001 (10) and in TE in the same sample (10.58). The mean rarefied richness was 24.04 in VP, 18.93 in AV and 15.78 in TE.

The Shannon-Wiener diversity index ($H'(\log^2)$) revealed that, in general, the studied ponds presented high diversity values. The highest values were observed in VP in July 2007 (4.58) and in July 2008 (4.27), and the lowest values were recorded in TE in December 2001 (1.43) and in AV in April 2007 (1.86). Among ponds, the mean values were 3.41 (VP), 2.97 (AV) and 2.46 (TE).

5.4.2. Assemblage composition

The NMDS ordination shows the spatial distribution of the samples and the grouping according to faunal similarity. The stress obtained with the ordination was 0.05, which ensures good consistency of results. NMDS allowed us to group the species according to two factors: Pond and Period (Figure 7). In the latter we can see a clear separation into two groups: 2001-2002 and 2007-2011. According to the ANOSIM, the assemblage composition of invertebrates shows significant differences in faunal composition related to the two periods (R=0.572; p=0.001). Regarding Pond factor, this analysis shows significant differences between TE and the two other ponds (TE-AV: R=0.225; p=0.003; TE-VP: R=0.250; p=0.002), but does not show significant differences between AV and VP (R=0.063; p=0.17).

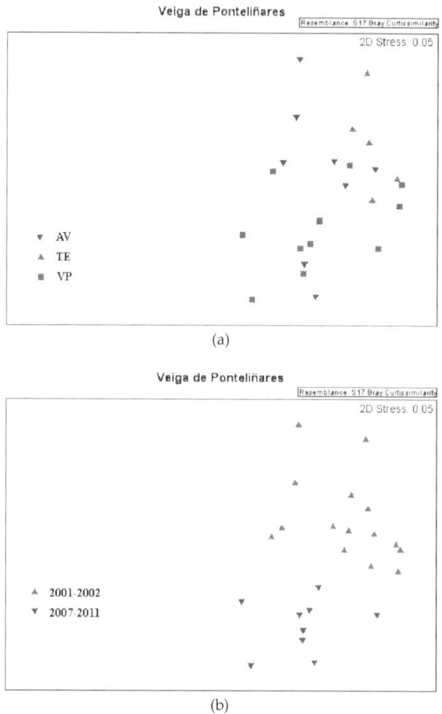

Figure 7. Results of the NMDS analysis: (a) Pond factor, and (b) Period factor. In graph (a) AV is represented with blue triangles, TE with red triangles and VP with green squares. In graph (b) the period 2001-2002 is represented with red triangles and period 2007-2011 with blue ones.

The contribution of the taxa to each group (ponds and periods) according to the SIMPER analysis is given in the Table 4. Regarding ponds, the dissimilarity between groups ranged from 84.26% (AV-VP) to 91.55% (TE-AV) and between periods, the dissimilarity was 95.55%. The faunal groups that most contributed to group similarity were insects and crustaceans, for both pond factor and period factor. Among insects, the taxa that most contributed were the beetles *Graptodytes flavipes*, *Hydroporus vagepictus* Fairmaire & Laboulbène, *Haliplus lineatocollis*, *Hydrochus angustatus* and the dipterans *Culex* sp. and Chironomidae. The most contributive crustaceans were Cladocera, Copepoda and Ostracoda, especially important for 2007-2011 group characterization.

Factor	Number of samples	Most contributive taxa	Contribution to the group characterization
AV	8	*Hydroporus vagepictus*, *Culex* sp., *Hydrovatus clypealis*, *Hydroporus vespertinus*	57.95% (*Hydroporus vagepictus* contributed with 31.01%)
TE	5	*Graptodytes flavipes*, *Haliplus lineatocollis*, *Hydraena marcosae*, *Hydraena testacea*	83.22% (*Graptodytes flavipes* contributed with 48.76%)
VP	11	*Graptodytes flavipes*, *Hydrochus angustatus*, Chironomidae, *Anacaena lutescens*, Copepoda, Sphaeriidae	40.74% (*Graptodytes flavipes* contributed with 8.14%)
2001-2002	15	*Graptodytes flavipes*, *Hydroporus vagepictus*, *Haliplus lineatocollis*, *Hydrochus angustatus*	56.17% (*Graptodytes flavipes* contributed with 27.35%)
2007-2011	9	Cladocera, Chironomidae, Copepoda, Ostracoda	60.32% (Cladocera contributed with 20.95%)

Table 4. Results of the SIMPER analysis separated by ponds and by periods showing the most contributive taxa in each group.

5.5. Discussion

Invertebrate richness in temporary waters can be considered similar to that found in natural and artificial permanent ponds [50-52]. Other permanent systems, such as mountain peatlands [53], show lower values. Our temporary ponds had a rich invertebrate fauna with similar or higher richness than other temporary systems [54,19,55,42]. When permanent and temporary waters are compared, temporary ones are richer, probably due to the trophic state of the pond, a factor associated with water permanence [56].

In this study we found some interesting and rare species, as already highlighted in other temporary pond studies in different parts of the world, in which the importance of this habitat for rare and often endemic species has been emphasized [e. g. 57,19,58]. On the other hand, it is often assumed that there is little overlap between invertebrate species found in temporary and permanent ponds [12]. However, the results of this study show that many of the species recorded in temporary ponds can also be found in permanent ones, in agreement with other studies [59,8,60].

Despite their ephemeral nature and small size, temporary ponds have a great importance for biodiversity maintenance in particular as regards some rare invertebrate species [12]. According to [61], both common and rare species rely on a variety of pond types within each region for their continued survival. According to [19], the faunal composition of temperate temporary aquatic ecosystems includes a remarkable number of uncommon species, species associated with permanent environments and species that frequently or exclusively inhabit these environments due to their biological adaptations. Similar results were observed in our study.

The main groups observed in the studied ponds were insects and crustaceans. Many of them rely on temporary ponds for reproduction (e.g. beetles, dipterans, dragonflies) and others spend their entire life cycle in these ponds as in the case of branchiopods. According to [22], several species of this group of crustaceans are only found in this type of ponds because of the lack of predatory fish. Branchiopods found in these habitats include Notostraca (tadpole shrimps), Cladocera (water fleas), Anostraca (fairy shrimps) and Conchostraca (clam shrimps).

One interesting crustacean species was collected during the surveys, the tadpole shrimp *Lepidurus apus*. This crustacean is rare in the Iberian Peninsula and lives in small temporary waters, such as flooded roadsides and ditches widenings and backwaters, always with aquatic vegetation, and is typical of less mineralized, dystrophic and in general clear waters [22]. The *Lepidurus* genus is present in all continents except Antarctica [62]. This world-wide distribution is due to their antiquity, but possibly also to their passive transport: geographical barriers are more effective for non-passively distributed animals. From an ecological point of view notostracans, like most branchiopods, are restricted to temporary pools [63]. The ephemerality of these extreme habitats may have been selected for the development of resistant stages (dried eggs or cysts) and some unusual reproductive strategies. In this sense, [64] noted that temporary wetlands flagship taxa (e.g. anostracans) rarely co-occur together and therefore suggested that each wetland harbouring them should be given conservation priority.

Among insects, Coleoptera and Hemiptera species dominated the temporary pond assemblages. This result agrees with other studies in temporary ponds [65,19,55,61]. Adults in both groups are mobile and can leave the pond when it dries out due to their excellent dispersal capabilities [21,66]. So, the high abundance of these taxa may be explained by the arrival of dispersers, moving from dry ponds to other ponds while dispersing to more permanent habitats to survive during dry periods [21,67,66,20]. On the contrary, non-mobile invertebrates require adaptations to survive the dry period.

Regarding Coleoptera, several interesting rare species barely recorded in the Iberian Peninsula were found in the studied ponds, like *Agabus labiatus* (Brahm), *Liopterus atriceps* Sharp, *Graptodytes bilineatus* (Sturm), *G. flavipes* or *Hydroporus vagepictus*. According to [68], *G. flavipes* is a rare species, but when it occurs it is usually in high abundance, a fact also observed in our study. *H. vagepictus* is an Iberian endemic beetle which has a wide ecological range and is present in running and stagnant waters [69]. The capture of the aquatic beetle *Hydraena rugosa* Mulsant must be highlighted. This species is little cited in the Iberian Peninsula but when found it is usually collected in high abundance [70].

According to the SIMPER analysis, different beetles were the species that contributed most to the NMDS groups characterization, especially *Haliplus lineatocollis* and *Hydrochus angustatus*. *H. lineatocollis* occurs in different habitats, from temporary ponds to rivers [e. g. 19,71]. A similar pattern was observed for *H. angustatus*, present in running waters [72] and coastal lagoons [73]. This result confirms that these ponds also harbour species with a wide ecological range.

In this study, several interesting Hemiptera species scarcely recorded in the Iberian Peninsula were collected, such as *Hesperocorixa moesta* (Fieber) and *Sigara limitata* (Fieber). These insects are also common inhabitants of temporary ponds [19,55,74,42]. Two Odonata species typical of temporary habitats were recorded, *Lestes dryas* Kirby and *Sympetrum sanguineum* (Muller). These two species can complete their life cycles before the habitat dries out. [23] also found them in Irish turloughs before the drying season. Regarding Thichoptera, the only ones recorded were Limnephilidae. They are common in temporary habitats, and according to [75] they emerge before the dry period and remain inactive in vegetation until the pond refills in autumn, and the eggs may be laid.

Endemic species constitute priority species to conserve, although in many cases are common in their distribution area. In the SAC Veiga de Panteliñares several endemisms were captured during the surveys, the water beetles *Hydroporus vagepictus*, *H. vespertinus* Fery & Hendrich, *Graptodytes castilianus* Fery, *Hydraena marcosae* Aguilera, Hernando & Ribera, *Limnebius lusitanus* Balfour-Browne, *Helophorus bameuli* Angus and *H. seidlitzii* Kuwert, and the water boatman *Sigara janssoni* Lucas. So, their presence in these temporary ponds is of particular interest as it increases their natural value.

Temporary ponds are fluctuating and unstable habitats whose faunal composition varies greatly from one year to another. In this study we have detected significant changes in their invertebrate composition. The NMDS analysis segregated the studied ponds in two groups corresponding with the two periods (2001-2002 and 2007-2011) and the ANOSIM analysis showed significant dissimilarity between the two periods (R=0.572; P=0.001). The results show that these ponds present a remarkable variability in time. Inter-annual variation may be due to variation in climate conditions, as a consequence of the wide differences in rainfall, and in the differences in the lenght of the hydroperiod observed among years. According to [76], historical events, such as very dry years, may affect the invertebrate community composition as much as site-specific abiotic differences among ponds. High variation among years has already been noted by other authors such as [44] in a study conducted in Doñana (southern Spain).

5.6. Conclusion

The use of invertebrates as monitoring tools in freshwater management programmes depends on the ability to discriminate natural (spatial and temporal) variation in community structure from alterations caused by anthropogenic disturbance [77,78]. In this sense, temporary ponds in the SAC Veiga de Panteliñares are a good example. Despite the large inter-annual variability, these ponds can result in sustaining a rich and abundant invertebrate fauna. The high species richness recorded in this study is likely to reflect the high ecological quality of these ponds, which are located in a protected area and are minimally impaired by anthropogenic activity. In addition, the occurrence of rare and uncommon species suggests that effective conservation of aquatic biodiversity in the SAC requires a regional approach of management.

Our study emphasizes the importance of drying out of ponds for the occurrence of rare and uncommon species forasmuch several of these species were found during this study. This finding highlights the contribution of temporary ponds to regional biodiversity and their high conservation value. It is important to preserve these systems, especially in Europe, where the number of temporary ponds is probably a mere fraction of what they probably were in the past [79].

The high richness found in our study does not correspond to a single pond, but to a system of temporary ponds. According to [80] ponds located near each other allow movement and dispersal of species among them. In this kind of systems, the high connectivity and non-fragmentation area are very important factors to conserve their invertebrate biodiversity [81,82]. Some studies have claimed that for the conservation of biodiversity of temporary pond species it is more important to preserve an area with a system of temporary ponds than the preservation of isolated ponds [83-86,9,61]. In this sense, this study highlights the importance of "Veiga de Ponteliñares"in the preservation of biodiversity at a regional scale. In this SAC, there are different small ponds and the connectivity between them favours the dispersion and preservation of species and makes it a hot spot of aquatic biodiversity.

Finally, it would be interesting to expand the SAC trying to include all the ponds formed along the banks of the Limia River. Other possible management measure would be to restore the numerous ponds formed by the sand extraction which are located where once was the Antela Lake. The aquatic fauna will colonized the new habitats rapidly due to the proximity to the Limia River and the other ponds. Thus, the restoration of the Antela Lake would be fundamental for the maintenance of the biodiversity in an area full of farms and crops that polluted the aquatic ecosystems of the region causing the loss of the Galician fauna and flora. In Spain there are clear examples of the restoration of shallow lakes dried during the past century. We can highlight the cases of Cospeito and Caque (Galicia), La Nava and Pedraza (Palencia), and Cañizar (Teruel). These wetlands have been restored through the development of various projects financed by Spanish and European public administrations.

Acknowledgements

This study was financially supported by the Galician Government through different projects ("*Study of the entomological fauna of Veiga de Ponteliñares (Ourense), SAC proposed by Galicia for the Natura 2000 network*"; "*Study and assessment of the diversity of aquatic invertebrates in Galician standing waters included in the Natura 2000 network*" and "*Assessment of the effectiveness of the wetlands included in the Galician Natura 2000 network as priority conservation areas for invertebrates*").

Author details

Amaia Pérez-Bilbao, Cesar João Benetti* and Josefina Garrido

*Address all correspondence to: cjbenetti@uvigo.es

Department of Ecology and Animal Biology, University of Vigo, Spain

References

[1] Dudgeon, D.; Arthington, A.H.; Gessner, M.O.; Kawabata, Z.I.; Knowler, D.J.; Lévêque, C.; Naiman, R.J.; Prieur-Richard, A.H.; Soto, D.; Stiassny, M.L.J.; Sullivan, C.A. Freshwater biodiversity: importance, threats, status and conservation challenges. Biological Reviews of the Cambridge Philosophical Society 2006;81 163-182.

[2] Dahl, J.; Johnson, R.K.; Sandin, L. Detection of organic pollution of streams in southern Sweden using benthic macroinvertebrates. Hydrobiologia 2004;516 161-172.

[3] Strayer, D.L. Challenges for freshwater invertebrate conservation. Journal of the North American Benthological Society 2006;25 271-287.

[4] Mitsch, W.J.; Gosselink, J.G. Wetlands. New York: John Wiley and Sons; 2000.

[5] Shine, C.; Klemm, C. Wetlands, Water and the Law: Using law to Advance Wetland Conservation and Wise Use. Gland: IUCN; 1999.

[6] Stenert, C.; Maltchik, L. Influence of area, altitude and hydroperiod on macroinvertebrate communities in southern Brazil wetlands. Marine and Freshwater Research 2007;58 993-1001.

[7] Ewald, N.; Nicolet, P.; Oertli, B.; Della Bella, V.; Rhazi, L.; Reymond, A-S.; Minssieux, E.; Saber, E.; Rhazi, M.; Biggs, J.; Bressi, N.; Cereghino, R.; Grillas, P.; Kalettka, T.; Hull, A.; Scher, O.; Serrano, L. A preliminary assessment of Important Areas for Ponds (IAPs) in the Mediterranean and Alpine Arc, Technical Report. Geneva: EPCN; 2010.

[8] Collinson, N.H.; Biggs, J.; Corfield, A.; Hodson, M.J.; Walker, D.; Whitfield, M.; Williams, P.J. Temporary and permanent ponds: an assessment of the effects of drying out on the conservation value of aquatic macroinvertebrate communities. Biological Conservation 1995;74 125-133.

[9] Zacharias, I.; Dimitrou, E.; Dekker, A.; Dorsman, E. Overview of temporary ponds in the Mediterranean region: threats, management and conservation issues. Journal of Environmental Biology 2007;28 1-9.

[10] Williams, P.J.; Biggs, J.; Barr, C.J.; Cummins, C.P.; Gillespie, M.K.; Rich, T.C.G.; Baker, A.; Baker, J.; Beesley, J.; Corfield, A.; Dobson, D.; Culling, A.S.; Fox, G.; Howard,

D.C.; Luursema, K.; Rich, M.; Samson, D.; Scott, W.A.; White, R.; Whitfield, M. Lowland Pond Survey 1996. London: Department of the Environment, Trade and the Regions; 1998.

[11] Ward, J.V. Aquatic insect ecology, 1. Biology and habitat. New York: John Wiley; 1992.

[12] Williams, D.D. Temporary ponds and their invertebrate communities. Aquatic Conservation: Marine and Freshwater Ecosystems 1997;7 105-117.

[13] Ruiz, E. Management of Natura 2000 habitats. 3170 *Mediterranean temporary ponds. Brussels: European Commission; 2008.

[14] Keeley, J.E.; Zedler, P.H. Characterization and global distribution of vernal pools. Ecology, conservation and management of Vernal Pool Ecosystems, Proceedings from a Conference, Section I: Past and Present Distribution and Physical and Biological Considerations. Sacramento: California Native Plant Society; 1996.

[15] Camacho, A.; Borja, C.; Valero-Garcés, B.; Sahuquillo, M.; Cirujano, S.; Soria, J.M.; Rico, E.; De la Hera, A.; Santamans, A.C.; García de Domingo, A.; Chicote, A.; Gosálvez, R.U. 31 Aguas continentales retenidas. Ecosistemas leníticos de interior, In: VV.AA., Bases ecológicas preliminares para la conservación de los tipos de hábitat de interés comunitario en España, Madrid: Ministerio de Medio Ambiente, y Medio Rural y Marino; 2009.

[16] Marsh, D.M.; Trenham, P.C. Metapopulation dynamics and amphibian conservation. Conservation Biology 2001;15 40-49.

[17] Standen, V. Quantifying macroinvertebrate taxon richness and abundance in open and forested pool complexes in the Sutherland Flows. Aquatic Conservation: Marine and Freshwater Ecosystems 1999;9 209-217.

[18] Della Bella, V.; Bazzanti, M.; Chiarotti, F. Macroinvertebrate diversity and conservation status of Mediterranean ponds in Italy: water permanence and mesohabitat influence. Aquatic Conservation: Marine and Freshwater Ecosystems 2005;15 583-600.

[19] Boix, D.; Sala, J.; Moreno-Amichi, R. The faunal composition of Espolla pond (NE Iberian Peninsula): the neglected biodiversity of temporary waters. Wetlands 2001;21 577-592.

[20] Williams, D.D. The biology of temporary waters. Oxford: Oxford University Press; 2006.

[21] Wiggins, G.B.; Mackay, R.J.; Smith, I.M. Evolutionary and ecological strategies of animals in annual temporary pools. Archive für Hydrobiologie 1980;58 97-206.

[22] Alonso, M. Crustacea, Branchiopoda, In: Fauna Ibérica, vol. 7, Ramos et al. (eds.) Madrid: Museo Nacional de Ciencias Naturales, CSIC; 1996.

[23] Porst, G.; Naughton, O.; Gill, L.; Johnston, P.; Irvine, K. Adaptation, phenology and disturbance of macroinvertebrates in temporary water bodies. Hydrobiologia 2012;696 47-62.

[24] Pérez-Bilbao, A.; Benetti, C.J.; Garrido, J. Nuevas aportaciones al conocimiento de los heterópteros acuáticos (Heteroptera: Gerromorpha y Nepomorpha) en humedales de Galicia (N.O. España). Boletín de la Asociación Española de Entomología 2012;36 87-107.

[25] Pérez-Bilbao, A.; Benetti, C.J.; Garrido, J. Aquatic Coleoptera assemblages in protected wetlands of North-western Spain. Journal of Limnology 2014;73 81-91.

[26] Babbit, K. The relative importance of wetland size and hydroperiod for amphibians in southern New Hampshire, USA. Wetland Ecology and Management 2005;13 269-279.

[27] Moyá, G.; Conforti, V. Cyanobacteria and microalgae communities in temporary ponds, In: Fraga I Arguimbau P. (ed.) International Conference on Mediterranean Temporary Ponds, Proceedings & Abstracts, 5-8 May 2009, Maó, Menorca: Consell Insular de Menorca; 2009.

[28] Pérez-Bilbao, A.; Garrido, J. Evaluación del estado de conservación de una zona LIC (Gándaras de Budiño, Red Natura 2000) usando los coleópteros acuáticos como indicadores. Limnetica 2009;28 11-22.

[29] Casado, S.; Montes, C. Guía de los lagos y humedales de España. Madrid: J.M. Reyero; 1995.

[30] Oertli, B.; Céréghino, R.; Hull, A.; Miracle, R. Pond conservation: from science to practice. Hydrobiologia 2009;634 1-9.

[31] Rhazi, L.; Grillas, P.; Rhazi, M.;. Aznar, J-C. Ten-year dynamics of vegetation in a Mediterranean temporary pool in western Morocco. Hydrobiologia 2009;634 185-194.

[32] Williams, P.; Whitfield, M.; Biggs, J.; Bray, S.; Fox, G.; Nicolet, P.; Sear, D. Comparative biodiversity of rivers, streams, ditches and ponds in an agricultural landscape in Southern England. Biological Conservation 2003;115 329-341.

[33] Foster, G.N.; Foster, A.P.; Eyre, M.D.; Bilton, D.T. Classification of water beetle assemblages in arable fenland and ranking of sites in relation to conservation value. Freshwater Biology 1990;22 343-354.

[34] Sánchez-Fernández, D.; Abellán, P.; Velasco, J.; Millán, A. Selecting areas to protect the biodiversity of aquatic ecosystems in a semiarid Mediterranean region using water beetles. Aquatic Conservation: Marine and Freshwater Ecosystems 2004;14 465-479.

[35] Sánchez-Fernández, D.; Abellán, P.; Mellado, A.; Velasco, J.; Millán, A. Are water beetles good indicators of biodiversity in Mediterranean aquatic ecosystems? The

case of the Segura River basin (SE Spain). Biodiversity Conservation 2006;15 4507-4520.

[36] Bilton, D.T.; McAbendroth, L.C.; Bedford, A.; Ramsay, P.M. How wide to cast the net? Cross-taxon congruence of species richness, community similarity and indicator taxa in ponds. Freshwater Biology 2006;51 578-590.

[37] Guareschi, S.; Gutiérrez-Cánovas, C.; Picazo, F.; Sánchez-Fernández, D.; Abellán, P.; Velasco, J.; Millán, A. Aquatic macroinvertebrate biodiversity: patterns and surrogates in mountainous Spanish national parks. Aquatic Conservation: Marine and Freshwater Ecosystems 2012;22 598-615.

[38] Jansson, A. Micronectae (Heteroptera, Corixidae) as indicators of water quality in two lakes in southern Finland. Annales Zoologici Fennici 1977;14 118-124.

[39] Polhemus, J.T.; Polhemus, D.A. Global diversity of true bugs (Heteroptera; Insecta) in freshwater. Hydrobiologia 2008;595 379-391.

[40] Boix, D.; Sala. J.; Quintana, X.D.; Moreno-Amichi, R. Succession of the animal community in a Mediterranean temporary pond. Journal of the North American Benthological Society 2004;23 29-49.

[41] Fernández, A.I.; Viedma, O.; Sánchez-Carrillo, S.; Alvarez-Cobelas, M.; Angeler, D.G. Local and landscape effects on temporary pond zooplankton egg banks: conservation implications. Biodiversity Conservation 2009; 18 2373-2386.

[42] Florencio, M.; Serrano, L.; Gómez-Rodríguez, C.; Millán, A.; Díaz-Paniagua, C. Inter- and intra-annual variations of macroinvertebrate assemblages are related to the hydroperiod in Mediterranean temporary ponds. Hydrobiologia 2009;634 167-183.

[43] Florencio, M.; Díaz-Paniagua, C.; Gomez-Mestre, I.; Serrano, L. Sampling macroinvertebrates in a temporary pond: comparing the suitability of two techniques to detect richness, spatial segregation and diel activity. Hydrobiologia 2012;689 121-130.

[44] Díaz-Paniagua, C.; Fernández-Zamudio, R.; Florencio, M.; García-Murillo, P.; Gómez-Rodríguez, C.; Portheault, A.; Serrano, L.; Siljeström, P. Temporary ponds from Doñana National Park: a system of natural habitats for the preservation of aquatic flora and fauna. Limnetica 2010;29 41-58.

[45] Garmendia, A.; Pedrola-Monfort, J. Simulation model comparing the hydroperiod of temporary ponds with different shapes. Limnetica 2010;29 145-152.

[46] Kottek, M.; Grieser, J.; Beck, C.; Rudolf, B.; Rubel, F. World Map of the Köppen-Geiger climate classification updated. Meteorologiche Zeitschrift 2006;15 259-263.

[47] Clarke, K.R.; Warwick, R.M. Change in Marine Communities: An Approach to Statistical Analysis and Interpretation, 2nd Edition, Primer-e Ltd, Plymouth: Plymouth Marine Laboratory; 2001.

[48] McCabe, D.J.; Gotelli, N.J. Effects of disturbance frequency, intensity, and area on assemblages of stream macroinvertebrates. Oecologia 2000;124 270-279.

[49] Magurran, A. E. Ecological diversity and its measurement, New Jersey: Princeton University Press; 1988.

[50] Hillman, T.J. Billabongs, In De Deckker, P.; Williams, W.D. (ed.) Limnology in Australia, Dr W. Junk Publishers, Dordrecht, The Netherlands; 1986.

[51] Friday, L.E. The diversity of macroinvertebrate and macrophyte communities in ponds. Freshwater Biology 1987;18 87-104.

[52] Jeffries, M. The ecology and conservation value of forestry ponds in Scotland, United Kingdom. Biological Conservation 1991;58 191-211.

[53] Erman, D.C.; Erman, N.A. Macroinvertebrate composition and production in some Sierra Nevada minerotrophic peatlands. Ecology 1975;56 591-603.

[54] Brooks, R.T. Annual and seasonal variation and the effects of hydroperiod on benthic macroinvertebrates of seasonal forest ("vernal") ponds in Central Massachusetts, USA. Wetlands 2000;20 707-715.

[55] Nicolet, P.; Biggs, J.; Fox, G.; Hodson, M.J.; Reynolds, C.; Whitfield, M.; Williams, P. The wetland plant and macroinvertebrate assemblages of temporary ponds in England and Wales. Biological Conservation 2004;120 261-278.

[56] Balla, S.A.; Davis, J.A. Seasonal variation in the macroinvertebrate fauna of differing water regime and nutrient status on the Swan Coastal Plain, Western Australia. Hydrobiologia 1995;299 147-161.

[57] Brendonck, L.; Williams, W.D. Biodiversity in wetland of dry regions (Drylands), In: Gopal, B.; Junk, W.J.; Davis, J.A. (ed.), Biodiversity of Wetlands: Assessment, Function and Conservation, vol. 1; 2000.

[58] Hillmann, T.J.; Quinn, G.P. Temporal change in macroinvertebrate assemblages following experimental flooding in permanent and temporary wetlands in an Australian floodplain forest. River Research and Applications 2002;18(2) 137-154.

[59] Jeffries, M. Measuring Talling's element of chance in pond populations'. Freshwater Biology 1989;21 383-393.

[60] Bazzanti, M.; Seminara, M.; Baldoni, S.; Stella, A. Macroinvertebrates and environmental factors of some temporary and permanent ponds in Italy. Verhhandlungen International Verein Limnologie 2000;27 936-941.

[61] Bilton, D.T.; McAbendroth, L.C.; Nicolet, P; Bedford, A.; Rundle, S. D.; Foggo, A.; Ramsay, P.M. Ecology and conservation status of temporary and fluctuating ponds in two areas of Southern England. Aquatic Conservation: Marine and Freshwater Ecosystems 2009;19 134-146.

[62] Brtek, J.; Thiéry, A. The geographic distribution of the European branchiopods (Anostraca, Notostraca, Spinicaudata, Levicaudata). Hydrobiologia 1995;298 263-280.

[63] Kerfoot, W.C.; Lynch, M. Branchiopod communities: association with planctivorous fish in space and time, In Kerfoot, W.C.; Sih, A. (ed.) Predation: Direct and Indirect Impacts on Aquatic Communities, Hanover: University Press of New England; 1987.

[64] De Roeck, E.R.; Vanschoenwinkel, B.J.; Day, J.A.; Xu, Y.; Raitt, L.; Brendonck, L. Conservation status of large branchiopods in the Western Cape, South Africa. Wetlands 2007;27 162-173.

[65] Eyre, M.D.; Carr, R.; McBlane, R.P.; Foster, G.N. The effects of varying site-water duration on the distribution of water beetle assemblages, adults and larvae (Coleoptera: Haliplidae, Dytiscidae, Hydrophilidae). Archiv für Hydrobiologie 1992;124 281–291.

[66] Bilton, D.T.; Freeland, J.R.; Okamura, B. Dispersal in freshwater invertebrates. Annual Review of Ecology and Systematics 2001;32 159-181.

[67] Higgins, M.J. & Merrit, R.W. Invertebrate seasonal patterns and trophic relationships, In Batzer, D.; Rader, R.B.; Wissinger, S.A. (ed.), Invertebrates in Freshwater Wetlands of North America. Wiley, New York; 1999.

[68] Ribera, I.; Isart, J.; Régil J.A. Coleópteros acuáticos de los Estanys de Capmany (Girona): Hydradephaga. Scientia gerundensis 1994;20 17-34.

[69] Ribera, I. Biogeography and conservation of Iberian water beetles. Biological Conservation 2000;92 131-150.

[70] Valladares, L.F.; Garrido, J.; García-Criado, F. The assemblages of aquatic Coleoptera from shallow lakes in the northern Iberian Meseta: Influence of environmental variables. European Journal of Entomology 2002;99 289-298.

[71] Sarr, A.; Benetti, C.J.; Fernández-Díaz, M.; Garrido, J. The microhabitat preference of water beetles in four rivers in Ourense province, Northwest Spain. Limnetica 2013;32(1) 1-10.

[72] Fernández-Díaz, M.; Benetti, C.J.; Garrido, J. Influence of iron and nitrate concentration in water on aquatic coleoptera community structure: application to the Avia river (Ourense, NW. Spain). Limnetica 2008;27(2) 285-298.

[73] Garrido, J.; Munilla, I. Aquatic Coleoptera and Hemiptera assemblages in three coastal lagoons of the NW Iberian Peninsula: assessment of conservation value and response to environmental factors. Aquatic Conservation: Marine and Freshwater Ecosystems 2008;18 557-569.

[74] Culioli, J.L.; Foata, J.; Mori. C.; Orsini, A.; Marchand, B. Temporal succession of the macroinvertebrate fauna in a Corsican temporary pond. Vie et Milieu 2006;56 215-221.

[75] Bratton, J.H. Seasonal pools: an overlooked invertebrate habitat. British Wildlife 1990;2(1) 22-29.

[76] Boulton, A.J.; Lake, P.S. The ecology of two intermittent streams in Victoria, Australia. II. Comparisons of faunal composition between habitats, rivers and years. Freshwater Biology 1992;27 99-121.

[77] White, J. & Irvine, K. The use of littoral mesohabitats and their macroinvertebrate assemblages in the ecological assessment of lakes. Aquatic Conservation: Marine and Freshwater Ecosystems 2003;13 331-351.

[78] Trigal, C.; García-Criado, F.; Fernández Aláez, C. Macroinvertebrate communities of mediterranean ponds (North Iberian Plateau): importance of natural and human-induced variability. Freshwater Biology 2007; 52 2042-2055.

[79] Williams, P.; Biggs, J.; Fox, G.; Nicolet, P.; Whitfield, M. History, origins and importance of temporary ponds. Freshwater Forum 2001;17 7-15.

[80] Fortuna, M.A.; Gómez-Rodríguez, C.; Bascompte, J. Spatial network structure and amphibian persistence in stochastic environments. Proceedings of the Royal Society B: Biological Sciences 2006;273 1429-1434.

[81] Briers, R.A.; Biggs, J. Spatial patterns in pond invertebrate communities: separating environmental and distance effects. Aquatic Conservation: Marine and Freshwater Ecosystems 2005;15 549-557.

[82] Van de Meutter, F.; Stoks, R.; De Meester, L. Lotic dispersal of lentic macroinvertebrates. Ecography 2006;29 223-230.

[83] Oertli, B., Joye, D.A.; Castella, E.; Juge, R.; Cambin, D.; Lachavanne, J.B. Does size matter? The relationship between pond area and biodiversity. Biological Conservation 2002;104 59-70.

[84] Beja, P.; Alcazar, R. Conservation of Mediterranean temporary ponds under agricultural intensification: an evaluation using amphibians. Biological Conservation 2003;114 317-326.

[85] Cheylan, M. Amphibians, In: Grillas, P.; Gauthier, P.; Yavercovski, N.; Perennou, C. (ed.) Mediterranean Temporary Pools; volume 1-Issues relating to conservation, functioning and management, Arles: Station Biologique de la Tour du Valat; 2004.

[86] Angelibert, S.; Marty, P.; Céréghino, R.; Giani, N. Seasonal variations in the physical and chemical characteristics of ponds: implications for biodiversity conservation. Aquatic Conservation: Marine and Freshwater Ecosystems 2004;14 439-456.

Chapter 11

A First Approach to Assess the Impact of Bottom Trawling Over Vulnerable Marine Ecosystems on the High Seas of the Southwest Atlantic

J. Portela, J. Cristobo, P. Ríos, J. Acosta, S. Parra,
J.L. del Río, E. Tel, V. Polonio, A. Muñoz,
T. Patrocinio, R. Vilela, M. Barba and P. Marín

Additional information is available at the end of the chapter

http://dx.doi.org/10.5772/59268

1. Introduction

The Southwest Atlantic (SW Atlantic), corresponding to FAO Statistical Area 41, includes a total continental shelf area of approximately 1.96 million km^2 of which a large portion lies off the Argentine coast (the Patagonian Shelf) and extends beyond Exclusive Economic Zones (EEZs) in the region [1-3]. This area is therefore integrated in the Southeast South American Shelf Large Marine Ecosystem (SSASLME) [4,5]. Currently, this region is the only worldwide significant area for high seas (HS) fisheries not covered by any Regional Fisheries Management Organisation (RFMO) [3].

The Patagonian Shelf (PS) hosts some of the most important fisheries in the world, targeting cephalopods (*Illex argentinus* [Castellanos, 1960] and *Doryteuthis gahi* [D'Orbigny, 1835]), and hakes (*Merluccius hubbsi* [Marini, 1933] *Merluccius australis* [Hutton, 1872]) [3,6-14]. Most of the exploited demersal stocks on the HS are straddling stocks, including Argentine shortfin squid (*I. argentinus*), Argentine hake (*M. hubbsi*) and southern blue whiting (*Micromesistius australis* [Norman, 1937]) [15].

Several authors [2,3,16-23] have studied the potential disturbance of the seabed by bottom otter trawls and the possible negative effects on the structure of benthic communities. In recent years, several resolutions of the United Nations General Assembly [24-28] on sustainable fisheries made a call to States and RFMOs to identify vulnerable marine ecosystems (VMEs)

and determine whether bottom fishing activities would cause a significant adverse impact on such ecosystems.

Sensitive species such as deep-water corals and deep-water sponges are found throughout the world oceans. Thus, the importance of habitat-structuring organisms is not restricted to shallow water, but also to shelf-break, hydrothermal vents, seamounts, and even the once considered constant and uniform deep-sea basins. Deep-water corals are vulnerable organisms occurring in the upper bathyal zones throughout the world and threatened by human activities, particularly fishing and oil exploration [29-31]. Fishing has a significant adverse impact (SAI) on deep-water coral communities in all oceans [32-35], particularly in the Northeast and Northwest Atlantic [36-40], Northeast Pacific [41,42], and Southwest Pacific [43-46]. In the SW Atlantic, the HS are one of the areas where deep-sea science has, to date, not been very active.

Protection of VMEs is a significant element of the management framework for bottom fisheries in high seas areas of the world ocean and its identification for selecting suitable protection areas is a challenge that conventional fisheries science cannot alone solve satisfactorily. Instead, it requires a multidisciplinary approach [21,22,47]. From the point of view of management of bottom fisheries and the governance of high seas areas, the situation in the PS poses an added problem as there is no any RFMO in force [2]. In its 2014 report [48], the Global Ocean Commission (GOC) recognises that continued scientific research is necessary to assess the cumulative impacts of human activities on the high seas so that informed decisions can be made about reversing the degradation of the global ocean.

Submarine canyons are unique habitats in terms of complexity, instability, material processing, and hydrodynamics. They may support diverse assemblages of larger epibenthos [49]. Inside canyons, abundance and diversity of the macrofauna depend, to some extent, on the physical disturbance regime and on the rate and quantity of organic matter deposited. In the study area, canyons and submarine mounts were shown to be hot spots of benthic biodiversity of species and ecosystems.

Benthos refers to the community of organisms which live on, in, or near the seabed, also known as the benthic zone. Megabenthos or macrobenthos comprises the more visible, benthic organisms exceeding 1 mm in size and large enough to be determined on photographs [50,51]. Megabenthos is a key issue of environmental studies, as it represents a major fraction of the deep-sea benthic biomass and plays a key role in deep-sea ecosystems [52]. Tracey et al. (2007) in [53] reported linear and radial annual growth rates of 20 mm and 0.2 mm, respectively, for some genera of the ISIDIDAE Family (Lamouroux, 1812), which is presumably evidence of the high vulnerability of these taxa to direct or indirect mechanical impact produced by the sediment removal, re-suspension, etc. caused by bottom fishing activities.

Some of these organisms form complex 3D structures protruding from the seabed, allowing for the settlement of sessile species needing consolidated substrata to settle and develop (sponges, other cnidarians), and providing shelter and food for a wide range of vagile fauna (crustaceans, echinoderms, molluscs, and some fish).

2. Materials and methods

In accordance with the aforementioned UNGA resolutions [24-28] and the FAO deepwater guidelines [54], the Spanish Institute of Oceanography (Instituto Español de Oceanografía [IEO]) conducted from October 2007 to April 2010 a series of 13 multidisciplinary research cruises on the HS of the SW Atlantic, to identify VMEs and to assess the potential interactions with fishing activities. This paper presents the results of the five first cruises, consistently with UNGA resolutions (paragraphs 80 and 83 to 87 of resolution 61/105 (2007) and paragraphs 117 and 119 to 127 of resolution 64/72 (2010) in [27,28], which support making publicly available information on interactions between bottom fisheries and VMEs in the HS.

The use of spatial management tools to preserve the marine biodiversity of species inhabiting the HS has been broadly discussed in recent years [55]. To make such spatial management possible, our immediate objectives are: assessing specific biodiversity (mainly describing new species to science); describing the different habitats, ecosystems and deep-sea geomorphological features identified; and analysing their interactions and relationships to protect the full range of potentially different habitats.

The explored area during the five cruises conducted between October 2007 and April 2008 (Table 1) was located on the southern part of the HS of the SW Atlantic, to the east of the Argentinian EEZ 200 miles limit and between 44° 40'S and 47° 51'S up to the 1500 m depth contour (Figure 1). The rest of the study area (up to 42°S) was surveyed during the eight following cruises (October 2008-April 2010), but the analysis of the information concerning VMEs collected during those last cruises, is still ongoing.

Cruise name	Start	End	Total days
Patagonia 11/07	28/10/2007	20/11/2007	24
Patagonia 12/07	24/11/2007	21/12/2007	28
Patagonia 01/08	08/01/2008	30/01/2008	23
Patagonia 02/08	30/01/2008	11/03/2008	41
Atlantis 2008	12/03/2008	15/04/2008	40

Table 1. Cruises carried out by R/V "Miguel Oliver".

In the right image of Figure 1 a non coloured area in the shelf can be roughly appreciated around 45°30'S and between 60°00'W-60°40'W, for which it was not possible to collect multibeam bathymetry data (no data) due to bad sea state conditions. The exploration of this area was carried out during one of the cruises conducted in 2009. Nevertheless, this type of data is not relevant for the present study, for which several trawl and CTD stations allowed the collection of pertinent information. The blue lines in the left image of Figure 1 corresponding to the 600, 1000 and 1500 m depth contours.

Key concepts for definition of VMEs were applied according to the FAO International Guidelines for the Management of Deep-Sea Fisheries in the High Seas [54]. These guidelines classify marine ecosystems as vulnerable based on several criteria: (1) uniqueness or rarity; (2) functional significance of the habitat; (3) fragility; (4) life-history traits of component species that make recovery difficult; and (5) structural complexity.

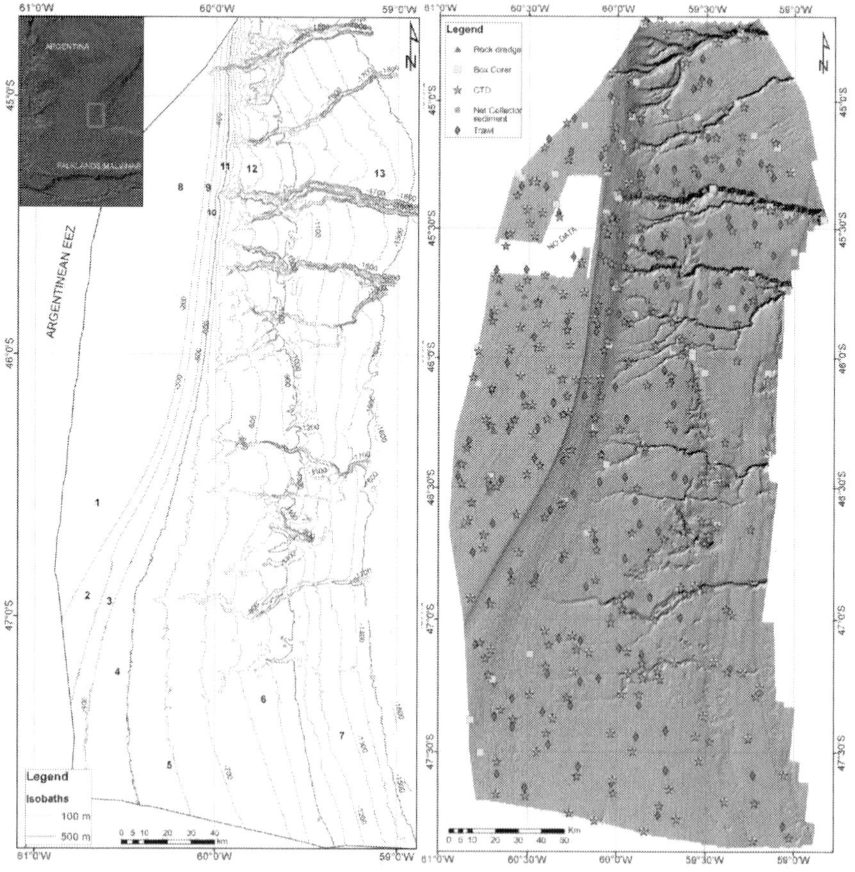

Figure 1. Study area and positioning of the stations carried out during the research cruises onboard the R/V "Miguel Oliver".

For an adequate identification of VMEs, the two approaches in operation since 2008 by the NAFO Scientific Committee and the NAFO Working Group on Ecosystem Approach to

Fisheries Management (WGEAFM) were applied in this study [56,57]: (1) the examination of cumulative catch data by ranking the biomass of VME taxa in each trawl from lowest to highest and then plotting the increase in cumulative biomass with each additional trawl; and (2) the use of Geographical Information System (GIS) to map the density of vulnerable species and groups' by-catch [58].

The study area included part of the outer shelf and upper and middle slope of the PS and was divided into thirteen depth strata in order to obtain a higher resolution in the description of vulnerable organisms. The research cruises involved five scientific disciplines: cartography, geology, benthos, fisheries, and hydrography.

This study used data from three main sources: i) Information from the five research cruises (geological, echosounder and oceanographic data; benthos and fish samples; fishery catch data [cpue]); ii) Data from commercial fishing activity collected by onboard scientific observers from 1989 to 2007 (fishery footprint); and iii) Commercial information on historical landings and effort data (provided by the Spanish fishing sector), as well as catch data for the main commercial species during the period 2000-2007 (logbooks filled in by captains of the fishing vessels, and provided by the Spanish General Secretariat for Fisheries [SGP]).

Geophysical and geological data were collected following internationally accepted standards and protocols for habitat mapping [20,59]. Full sea floor coverage using swath bathymetry provided a very high resolution of sea floor morphology. The backscatter data from multibeam echosounder together with high resolution seismic reflection profiles made available valuable data on the seabed sediments types. These data provided the geomorphological and acoustic basis to design a ground-truth planning strategy allowing for precise habitat mapping. Navigation during the surveys was via differential GPS Simrad GN33 using satellite corrections integrated into an inertial-aided Seapath 200 system. Swath-bathymetric data were acquired using a hull-mounted Kongsberg-Simrad EM 302 multibeam echosounder (288 individual beams, angular coverage up to 150°) operating at a frequency of 30 kHz. To correct the multibeam bathymetry, we carried out systematic casts of direct sound velocity profiles on the water column with an Applied Microsystems SV Plus equipment. Data processing included the removal of anomalies and the necessary sound velocity corrections using the Kongsberg–Simrad swath bathymetric software package NEPTUNE. Valid data were gridded at 50×50 m cell size resolution on a SUN workstation. The seismic parametric system Topas 18 produced very high resolution seismic profiles along all ship tracks. Sub-bottom penetration varied, according to the lithology, between 150 and 250 m. Morphometrical data were obtained using ArcGis (ESRI) and Fledermaus software (Interactive Visualization Systems [IVS]) to provide final 3D images of the seafloor morphology.

Samples of benthic fauna analysed in this study were collected with the Lofoten bottom trawl gear itself. Benthic fauna samples were sorted on board and preserved (70% ethanol or 4% buffered formaldehyde-seawater solution) for further identification analysis. Even if the bottom-trawl by-catch collected information did not allow for a detailed habitat mapping of VMEs, it provided a valuable indication of VME presence/absence that can be used to propose conservation measures, such as candidate areas for bottom fishery closures [23].

Sediment samples were collected using net collectors attached to the Lofoten fishing gear (Atlantis 2008 cruise) and with an USNEL type box-corer (BC) (maximum breakthrough of 60 cm; effective sampling area of 0.25 m² [50×50 cm]). A few samples were taken using a Bouma type box-corer (effective sampling area of 0.0175 m² [10×17.5 cm]). Both gears are designed to take undisturbed samples from the top of the seabed, and are suitable for almost every type of sediment. Sediment temperature and redox profiles (Eh) were immediately performed for the box-corer sample after each station. In the laboratory, the granulometrical analysis of the sediment was carried out by dry sorting the coarse fraction (>62 µm) and the sedimentation of the fine fraction (<62 µm). The organic matter content was assessed after calcinating (at 500°C for 24 h) and drying the sediment sample.

The hydrographical conditions in the studied area during the Atlantis 2008 cruise were characterised by means of a Seabird-25 CTD probe (SBE-25), equipped with oximeter, fluorometer and PAR detector. The survey schedule was optimized by systematically deploying the CTD at fishing stations below 500 m, but not always at greater depths. At each cast, the CTD was deployed to 5 m depth and stabilised for approximately 3 min. Once stable, the CTD was brought back to the surface and started profiling at a constant speed of 1 m·s⁻¹. The SBE-25 worked in auto-contained mode at a frequency of 8 scans·s⁻¹ and the downloaded data were converted into physical units and pre-processed by using the SeaBird software (SeaSave/SBE DataProcesing-win32) with standard calibration values. Quality control and post-processing was performed with MATLAB.

Atlantis 2008 stratified bottom trawl survey enabled the assessing of the biomass and bathymetric distribution of the main commercial and most abundant fishery stocks by means of the swept area method. The survey used a stratified random design with strata boundaries definedby latitude and depth ranges, depth strata 1-7 located south of parallel 45°S and depth strata 8-13 sited north of the referred parallel (Table 2). Scheduled fishing stations (hauls of 30 min) were performed using a Lofoten bottom trawl net fitted with a rockhopper mix train with bobbins and rubber separators, suitable for deep-water fishing over irregular bottoms. Mean trawl speed was of 3.2 knots and trawl direction followed the bathymetric profile in the upper slope, but was variable in the outer shelf and middle slope.

Data recorded by scientific onboard observers from 1989 to 2007 between latitude 42°S and 48°S were used for mapping only the Spanish fishery footprint, since fishing data of other fleets were unavailable to us. The IEO observers' program placed one observer per selected vessel to cover 12% to 15% of the whole fleet. Table 3 summarize the activities (number of hauls year⁻¹) of the IEO observers on the HS of the SW Atlantic, were Divisions 42 and 46 correspond to the areas roughly around parallels 42°S and 46°S.

Data used for each fishing haul corresponded to the middle tow position, since it offers more relevant information than the initial or final positions. All middle tow positions were imported into ArcGIS 9.3 mapping software to plot all trawl tows as straight lines between the reported start and end positions. They were then exported to a grid of 5'×10' min blocks, and any block including at least two tows was retained for mapping the bottom trawl footprint.

Strata	Depth range (m)	Surface (mn²⁾	No. of grids (~5 mn²)	No. of scheduled hauls	No. of hauls made	
					Valid	Null
1	<200	1148	219	12	12	
2	201-300	272	51	3	4	
3	301-400	381	71	4	3	
4	401-500	518	119	7	7	
5	501-700	1513	318	18	18	
6	701-1000	1952	349	20	20	3
7	1001-1500	2007	435	24	2	5
8	<200	1394	254	14	15	
9	201-300	111	24	2	2	
10	301-400	121	21	2	2	
11	401-500	78	26	2	2	
12	501-1000	933	170	10	12	
13	1001-1500	2507	515	29	26	5
Total		12933	2571	147	125	13

Table 2. Scheme of hauls by depth stratum, and main characteristics (ATLANTIS 2008 cruise).

Year	Division 46	Division 42	Total
1989	756	734	1490
1990	411	222	633
1991	152	28	180
1992	561	9	570
1993	515	0	515
1994	469	0	469
1995	186	0	186
1996	310	21	331
1997	811	35	846
1998	709	0	709
1999	384	4	388
2000	590	44	634
2001	673	111	784
2002	452	142	594
2003	191	0	191
2004	472	0	472
2005	561	1	562
2006	477	0	477
2007	333	0	333
Total	9013	1351	10,364

Table 3. Number of hauls/year and division recorded by scientific observers.

Proper identification of the areas where VMEs are present followed the methodology used by the NAFO in its Regulatory Area [60]. Threshold catches, defined as catch levels of significant concentrations of invertebrates to be considered as possible VME areas, were assessed by analysing the cumulative biomass frequencies. Cumulative catch curve method was chosen to calculate the threshold catch. The cumulative frequency was plotted for all capture sets where taxa, considered as vulnerable by the International Guidelines for the Management of Fisheries [54] and by the Convention for the Protection of the marine Environment of the North-East Atlantic (OSPAR), were identified. The threshold selection for each taxon was made on the basis of minimum/maximum catch, density and morphological characteristics. Once a location of significant concentrations of vulnerable organisms was defined (key location), a 2 nm radius buffer zone around it was drawn to provide a safe margin of error on site.

The Random Forest algorithm for classification (RF) was used to predict the potential distribution of vulnerable benthic species by rating environmental conditions on the basis of previous observations.

RF is a non-parametric statistical method for data analysis that makes no distributional assumptions about the predictor or response variables [61], showing high prediction accuracy classifying rocky benthic communities [62] and beating other methods commonly used for ecological prediction [61,63]; The algorithm calculate the suitability of a given habitat for a given species based on known affinities with habitat characteristics, stored as raster maps, and called independent ecogeographical variables (EGV). According to HSI values, a map of species' expected distribution is produced, a value ranging from 0 to 1 showing the probability that the habitat of a given location is suitable for the species occurrence [64]. Thus, for a particular location, high HSI values mean high chances of the species' occurrence. To perform this mapping, presence/absence data from different vulnerable benthic organisms found in the study area were used as dependent variables of the different EGV.

Gathering accurate sampling presence/absence data is a critical part of the study, since the absence of a species in a given location can be due to several reasons: the species is present but is not observed, the species is absent even though the habitat is suitable, or the species is absent because of the unsuitability of the habitat. Only the last reason is considered as a "true absence" [65,66]. As presence data were aggregated into one single group named "vulnerable organisms", the resulting HSI predicted the potential habitat of any of the considered vulnerable organisms in the HS of the SW Atlantic under study.

The RF method offers the possibility to calculate an accurate unbiased estimator, using Out-Of-Bag (OOB) observations as an internal validation data set [67], computed from the resulting confusion matrix [68]. Accuracy is the proportion of the total number of predictions that were correct and this accuracy indicator is offered to the user as a measure of the model's predictive performance. It is determined using the equation:

$$ACC_{OOB} = \frac{TS + TU}{N}$$

Where TS is the number of truly suitable locations, i.e. suitable locations correctly classified by the model; TU is the number of truly unsuitable locations, in other words unsuitable locations that have been correctly classified; and N is the total number of observations.

Data was analyzed using the R statistical software [69] and the "Random Forest" package [70] and predictions were exported to a shapefile format using the "maptools" package [71]. GIS visualization of results was performed using the ESRI ArcMap 10.0 software.

Selected environmental variables involved in the study included depth, slope, sea bottom temperature, substrate characteristics and topographic position (Table 4). The topographic position was sorted into six categories: shelf (1), outcrop areas on the shelf (2), high slope (3), low slope (4), abyssal flats (5), and canyons (6).

	Variable	Type of variable	Range (min- max)
Hydrography	Sea bottom temperature	Continuous	1.70°C – 6.14°C
Topography	Bathymetry	Continuous	110.1m – 1848.6m
	Slope	Continuous	0° – 14.792°
Substrate characteristics	Q50	Continuous	2 – 3.59
	Coarse sand fraction	Continuous	0.17% – 11.2%
	Fine sand fraction	Continuous	57.2% – 97.28%
	Mud fraction	Continuous	2.17% – 41.31%
Topographic position	Seabed morphology	Discrete	1 – 6

Table 4. Summary of the environmental variables used for the Habitat Suitability Index (HSI) modelling of vulnerable organisms.

CTD stations' sea bottom temperature data were interpolated for the whole area using the local polynomial interpolation function (LPI) implemented in the ArcGIS 10.0 software. Slope was derived from the bathymetry high resolution data, and after studying the semivariogram, substrate characteristics were interpolated from granulometrical measures for the whole area using a universal kriging interpolator (Unpublished).

All the explanatory data were extracted for the presence/absence data locations, subsequently exported and analysed with the R statistical software using the BIOMOD package [72]. Several presence/absence models were performed: Generalized Additive Models [73,74], Multivariate Adaptative Regression Splines [74,75], Generalized Boosting Models [74,76] and Random Forest model (RF) [67,74].

3. Results

3.1. Geomorphology

Geomorphological and geophysical data from the five research cruises revealed that the outer shelf was mantled by 15 m high sand ridges, and was 60 to 67 m deeper than the maximum 120 m lowering of sea-level during the last glacially induced regression. This difference in depth indicates that the PS had experienced subsidence in the Holocene. These ridges are relict and were probably constructed during the post-glacial transgression by the north flowing Falkland (Malvinas) Current, since they are resting on shell layers of <35,000 to 11,000 years old [77].

The upper continental slope descends from the shelf break, located at depths from 200 to 750 m, and is scarred by iceberg plough marks whose orientation and morphology suggest that icebergs carried northwards by the Falkland (Malvinas) Current were probably responsible for this erosion during the last glaciations [78].

Scattered over the study area (south of 45°S) we found pockmarks, carbonate mounds formed by deep-water corals, northwards furrows, areas of smooth topography and sediment waves indicating that deposition on this part of the middle slope is controlled by bottom currents [79].

Seven submarine canyons were identified on the middle slope surveyed (Figure 2). Canyons 1 to 6 were cut by turbidity currents, whereas canyon 0 resulted from the combined effect of turbidity currents and coalescence of pockmarks (formed by the expulsion of thermogenic gas). These gas and fluid seepages contributed to the formation of canyons and to the partial detachment of blocks from the canyon walls. Thus, the thermogenic gas responsible for the formation of the identified pockmarks on the middle slope could be deep-seated, probably related to the Falkland Rift Basin, north of the Falkland (Malvinas) Islands [80,81].

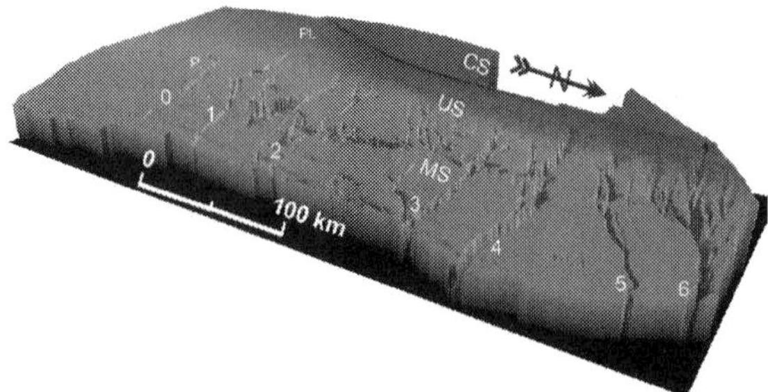

Figure 2. Colour shaded 3D bathymetric map of a segment of the Patagonian Argentinian margin compiled from multibeam backscatter data. Arabic numbers identify submarine canyons discussed in text. CS=Continental shelf; US=Upper continental slope; MS=Middle continental slope; P=Pockmark; PL=Iceberg plough marks.

The association of gas seepage with deep-water corals has been reported by [82] in pockmarks off Brazil. If such association also occurs on the Patagonian margin, those communities may be quite widespread in our study area.

3.2. Benthic communities

Bathelia candida (Moseley, 1881) was found to be one of the main reef builder species in the study area, providing habitat for diverse associated fauna of sponges, crustaceans, echinoderms, molluscs, and other cnidarians. The benthic megafauna caught during the cruises included invertebrates as well as Phyla Chordata and Hemichordata. Phyla Cnidaria and Porifera were dominant in terms of biomass (46% and 30%, respectively [Figure 3A]). The high abundance of Cnidaria is remarkable, since 33.7% of the biomass of this phylum corresponded to the Class Octocorallia, including significant groups such as gorgonians (sea fans), alcyonaceans (leather corals) and pennatulaceans (sea pens). In addition, the VMEs dominated by suspensivore and/or filter feeding organisms are habitats with high biodiversity and many resources.

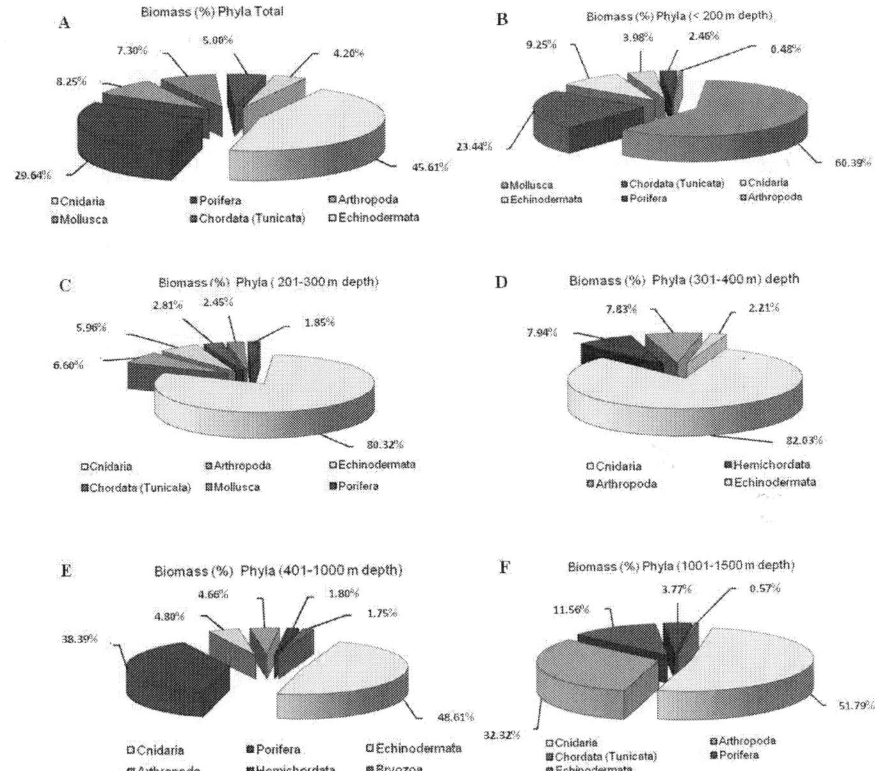

Figure 3. Biomass per Phyla in total strata (A) and by stratum < 200 (B), 201-300 (C), 301-400 (D), 401-1000 (E), 1001-1500 m depth (F).

A large part of the benthic samples contained erect sponges, octocorals, colonial scleractinians, calcified antipatharians, and hydrozoans (Family STYLASTERIDAE), all of them slow-growing organisms considered as vulnerable by the UN and the OSPAR standards (see Table 7).

Bathymetric strata differences clearly arise by comparing the composition of the sampled benthic megafauna (Figures 3B-F):

Strata 1 and 8 (<200 m) showed a low catch of benthos (17,209 and 41,202 g. respectively), both in number and diversity. We observed a strong dominance of pectinid molluscs of the Genus *Zygochlamys* (Ihering, 1907) (60.39% of the biomass [Figure 3B]), mainly *Z. patagonica* (King & Broderip, 1832), followed by those of the Genus *Chlamys* (Röding, 1798). Vulnerable organisms were practically unrepresented in these shallower strata, probably due to the bottom trawling activities for years by bottom trawlers from international fleets.

Strata 2 and 9 (201-300 m) recorded the lowest catch in terms of biomass (2121 and 1576 g, respectively). In these strata, detritivorous and opportunistic species were predominant, and the presence of vulnerable organisms was negligible again. Compared to strata 1 and 8, we observed an increase of the benthic cnidarians' biomass values, dominated by gorgonians from Family PRIMNOIDAE (Milne Edwards, 1857) (Octocorallia; Gorgonacea) (80.32% in biomass, [Figure 3C]).

Strata 3 and 10 (301-400 m) were hardly sampled due to the reduced number of valid hauls (3 in stratum 3 and 2 in stratum 10). The low benthic biomass and the negligible presence of vulnerable organisms (Figure 3D) could be attributed to bottom fishing activities, as above-mentioned for strata 1 and 8.

Strata 4, 11, 5, 6, and 12 of intermediate depths (401-1000 m, [Figure 3E]) recorded high biomass and numbers of octocorals, sponges, colonial scleractinians (*Bathelia candida*), and large hydrocorals. Octocorals included colonies of various genera belonging to families PRIMNOI-DAE and ISIDIDAE. As aforementioned, the increase and proliferation of these species create complex 3D structures providing the ideal habitat for a wide range of organisms. In those strata, the large amount of filter feeders and suspensivore sessile organisms is an indication of the presence of unaltered, complex and structured ecosystems. In the future, ROV and other submersible camera systems could confirm these assumptions.

Strata 7 and 13 (1001-1500 m, [Figure 3F]) were the most difficult ones for trawling. Numerous tows failed to produce valid results. In these strata, the highest proportion of animals was of benthopelagic crustaceans, usually making diel migrations, even though they were normally present on the seafloor. Benthic cnidarians were dominated by octocorals of the Order PENNATULACEA (Verrill, 1865), with a wide bathymetric distribution, adapted to live on soft substrates.

3.3. Sediments

Sediment data obtained during Patagonia 1207, Patagonia 0108, and Atlantis 2008 cruises showed that fine sands were generally predominant throughout the study area, with low contents of organic matter and sediment sorting varying from poor to moderately good. In more detail, the bathymetric sedimentary classification would be as follows:

Depths <200 m: fine sand (mean diameter=210 µm) with low organic matter content (mean value=1.14 %), moderately sorted.

Depths from 201 to 400 m: fine sand (mean diameter ranging from 150 to 189 µm) with low organic matter content (mean value=1.06%), moderately well sorted.

Depths from 401 to 700 m: very fine sand (mean diameter from 110 to 120 µm) recording the highest organic matter content (mean value ranging from 2.23% to 2.35%) and also the highest percentage (up to 44.50%) of silt and clay (<62 µm). Sorting was poor to moderate.

Depths from 701 to 1500 m: fine sand sediments similar to those of the shallowest stratum (mean diameter 160 to 190 µm), with low organic matter contents (mean value ranging from 1.43% to 1.68%). Moderately sorted.

Depths >1501 m: the deepest stratum, located in the bottom of submarine channels and canyons, was characterised by the presence of heterogeneous sediments mainly composed of fine sand (mean diameter=200 µm), with low organic content (mean value=1.68%) and poor sorting. This stratum showed the highest percentage (up to 39.5%) of coarse particles (>500 µm).

3.4. Fishery footprint

The statistical analysis of the bottom trawl footprint plot generated with the georeferenced fishery data obtained by the IEO scientific observers (between 1989 and 2007, 9013 fishing operations) showed that most of the commercial hauls of the Spanish fishing fleet in the study area (99.85%) took place at depths below 300 m (Figure 4).

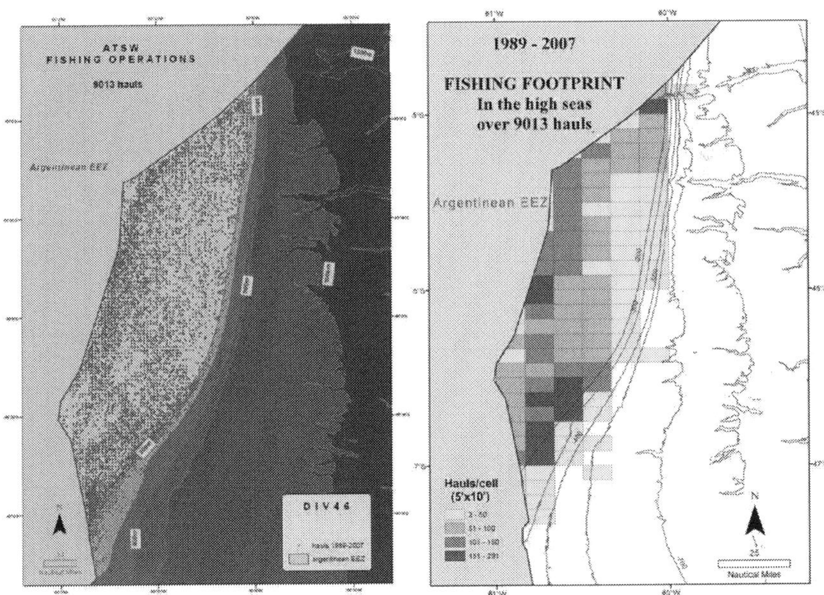

Figure 4. Location of commercial hauls and fishery footprint (5'×10') of the Spanish bottom trawl fleet on the HS of the SW Atlantic (1989-2007).

4. Multivariate analysis

4.1. Model selection

Predictive accuracy of the models was evaluated through multiple cross-validation procedures, splitting the original data three times into two random subsets for calibration (80% data) and evaluation (20% data). The mean area under the receiver operating characteristic (ROC) curve (AUC) obtained from the three repetitions served to assess the predictive performance index of the model. AUC ranks from 0.5 to 1, null accuracy or perfect accuracy of the model, respectively [83]. Table 5 shows the best predictive performance score of the RF model, which was subsequently chosen for vulnerable species modelling.

Model	Mean cross validation score
RF	0.876
GBM	0.825
MARS	0.804
GAM	0.778

Table 5. Validation of the predictive performance of the four candidate presence/absence models tested (RF: Random Forest; GBM: Generalized Boosting Model; MARS: Multivariate Adaptative Regression Splines; GAM: Generalized Additive Model).

4.2. Variable influence

The Mean Decrease Gini method, implemented in the BIOMOD package, was used to measure the importance of the dependent variable. This exploration tool shows graphically the total decrease in node impurities from splitting on the variable, averaged over all trees. Thus, for classification, the node impurity is measured by the Gini index [72]. The higher the value in the X axis, the higher importance the indicated variable will have on the classification of the dependent variable. Figure 5 show that the topographic position is the main variable affecting the distribution of vulnerable organisms in the HS of the PS, followed by the slope and the sea bottom temperature. Comparatively, the sea floor granulometry has a negligible effect on the distribution of the vulnerable organisms.

In addition to this, it is possible to visualize how each environmental variable, independently from any other, influences the response variable using partial dependence plots [73], which graphically represents the relationships between each predictor variable and the predicted occurrence probabilities of the vulnerable organisms obtained from the RF model. Figure 6 show that bathymetry has a positive effect between 500 and 1000 m depth. Regarding the topographic position, the highest interactions with the presence of vulnerable organisms were observed in canyons (6), followed by abyssal flats (5) and the slope (3 and 4). On the shelf, only outcrop areas (2) were positively correlated with the dependent variable.

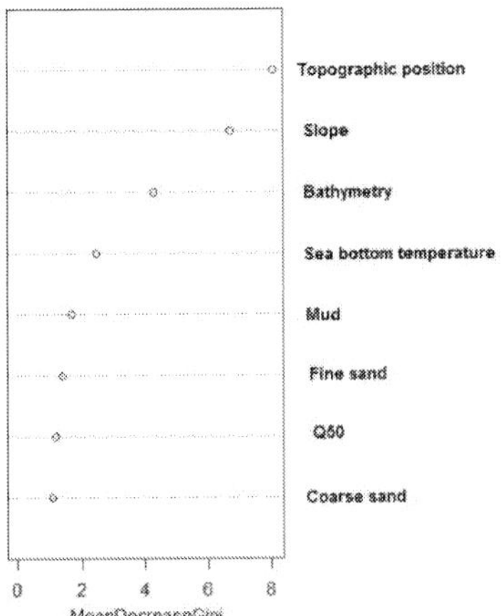

Figure 5. Mean Decrease Gini for each explanatory variable in the RF model. Higher values in the X axis indicate higher influence of the environmental variable on the occurrence of benthic vulnerable organisms.

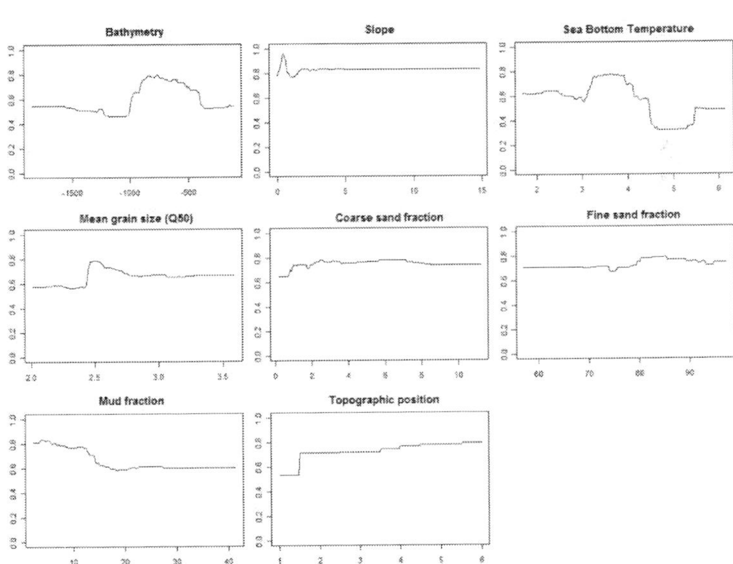

Figure 6. Partial dependence plots showing quantitative influence from each environmental variable on the occurrence of benthic vulnerable organisms predicted probability.

4.3. HSI mapping

Table 6 shows the presence data of benthic vulnerable organisms from the 169 sampled locations. Predicted values were plotted to produce a habitat suitability map showing survey sampling stations with presence/absence data and the vulnerable organisms' probability of occurrence (Figure 7).

Organism	Presence
Alcyonacea	15
Bathelia candida	23
Demospongiae	22
Gorgonacea	24
Hexactinellidae	14
Hydrozoa	41
Pennatulacea	7
Rhodalidae	9
Stylasteridae	25
Total VO	76

Table 6. Summary of the presence sampling data of vulnerable organisms (VO).

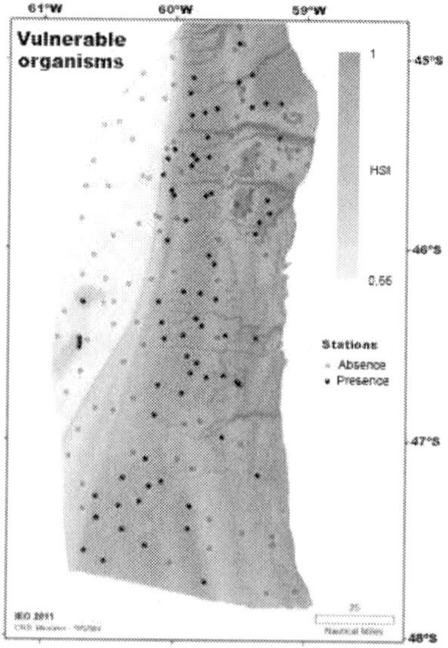

Figure 7. HSI map of benthic vulnerable organisms. Higher probability of occurrence is shown in darker tones. Survey sampling stations are overlapped, showing presence (black dot) or absence (circle) of such organisms.

5. Conclusions

Multibeam acoustic data showed that the upper slope and uppermost middle slope were scarred by iceberg plough marks. The middle slope surveyed was entrenched by seven submarine canyons [78]. Pockmarks and other seismic and morphologic evidence of gas/fluids seepage were pervasive throughout the entire survey area and more intense in the southern middle part [80]. Water coral communities associated with those pockmarks could be quite extensive in the study area.

The highest benthic biodiversity was found between 800 and 1500 m depth. Biodiversity was higher along the continental margin (per an equal number of individuals, and in terms of abundance) than biodiversity found along the continental shelf. Our results have confirmed the existence of close ecological relationships between Patagonian deep-sea fauna and Antarctic fauna of shallow waters. Benthic megafauna collected included invertebrates, chordates, and hemichordates. There was a clear dominance in biomass and diversity of the Phyla Porifera and Cnidaria. Most species of these groups are considered as vulnerable according to UN and OSPAR criteria: sponges, octocorals, colony scleractinians, anthipatarians, calcified hydrozoans (Family STYLASTERIDAE), and erect bryozoans (Table 7).

Shallow waters (<400 m) are the strata having sustained most of the fishing pressure for almost 50 years. Below 400 m we recorded the lowest biomass, abundance and diversity values, most likely due to this fishing pressure. In these strata we noted the presence of sparse organisms with erected growth, a high dominance of pectinid mollusks (*Zygochlamys patagonica*), and minor presence of species considered as indicators of VMEs.

Intermediate depths (401-1000 m) showed an important increase in number and biomass of vulnerable organisms, with outstanding numbers, densities and biomasses of octocorals, sponges, colony scleractinians (*Bathelia candida*) and big hydrocorals (*Errina* spp., *Cheiloporidion pulvinatum* [Cairns, 1983], *Sporadopora* sp., and *Stylaster densicaulis* [Moseley, 1879]). Also remarkable was the presence of sponges of the Family CLADORHIZIDAE (Dendy, 1922), a group of a great zoological importance because they are carnivorous and have developed a trophic adaptation to live in the ocean's depths.

In deeper strata (1001-1500 m) we found more anomuran crustaceans of the Family LITHODIDAE (Samouelle, 1819) (mainly *Paralomis formosa* [Henderson, 1888]). Amongst benthic cnidarians, the pennatulid octocorals (Order PENNATULACEA) were the most abundant.

The model accuracy is acceptable (0.876). Although the modelling' accuracy values were higher when considering each organism, this was an expected fact due to the different environmental preferences of the studied organisms. However, HSI mapping is a useful conservation management tool enabling an initial observation of how environmental conditions control the spatial distribution of vulnerable organisms in the study area. The research will proceed further when data from all 13 survey cruises undertaken in the area are analysed.

The main environmental conditions affecting presence of vulnerable organisms seems to be connected to the topographic position, slope and bathymetry. Sea bed granulometry appeared

Porifera Grant, 1836	*Chondrocladia* sp.
Class Hexactinellida Schmidt, 1870	*Euchelipluma* sp.
Rossella antarctica Carter, 1872	*Mycale (Oxymycale) acerata* Kirkpatrick, 1907
Class Demospongiae Sollas, 1885	*Mycale (Carmia) gaussiana* Hentschel, 1914
Tetilla leptoderma Sollas, 1886	*Isodictya kerguelenensis* Ridley & Dendy, 1886
Cynachyra sp	*Latrunculia* sp.
Geodia sp.	Axinellidae indet.
Polymastia sp.	*Haliclona (Haliclona)* sp.
Radiella sp.	*Haliclona (Gellius)* sp.
Tentorium sp.	Dictyoceratida indet.
Stylocordyla cf. *stipitata*	**Cnidaria Hatschek, 1888**
Timea sp.	**Class Hydrozoa Owen, 1843**
Lithistidindet.	*Errina* sp.
Iophon sp.	*Stylaster* cf. *densicaulis*
Clathria sp.	**Class Anthozoa Ehrenberg, 1831**
Raspailia sp.	*Alcyonium* sp.
Inflatella sp.	*Anthomastus* sp.
Pyloderma latrunculioides Ridley & Dendy, 1886	*Paragorgia* sp.
Desmacidon, sp.	*Primnoella* sp.
Hymedesmia (Hymedesmia) sp.	Isididae indet.
Hymedesmia (Stylopus) sp.	*Anthoptilum* sp.
Phorbas sp.	*Halipteris* sp.
Myxilla (Myxilla) mollis Ridley & Dendy, 1886	*Epizoanthus* sp.
Myxilla (Burtonanchora) lissostyla Burton, 1938	*Actinostola crassicornis* Hertwig, 1882
Tedania (Tedaniopsis) charcoti Topsent, 1907	*Bathelia candida* Moseley, 1881
Tedania (Tedaniopsis) oxeata Topsent, 1916	*Caryophyllia* sp.
Tedania (Tedaniopsis) massa Ridley & Dendy, 1886	*Desmophyllum* sp.
Tedania (Trachytedania) sp.	*Flabellum* sp.
Asbestopluma sp.	

Table 7. Cold-water corals and deep-water sponges concentrations: list of most common species collected in the campaigns of the Atlantis Project in 2007 and 2008.

to have a negligible effect on the presence of vulnerable organisms, contradicting published research results on this subject in other geographical areas, where substrate characteristics determine to a large extent the presence or absence of a particular benthic species [84-86].

Our study only calculated the general trends of the granulometrical parameters, while bathymetry, slope and topographic position were variables derived from high resolution data, strongly correlated with the response variable. Therefore, local conditions are the main factors ruling the potentiality of a habitat to host benthic vulnerable organisms in the HS of the PS.

The use of the Random Forest model offers both higher classification accuracy and determination of variable importance, and more stability where small perturbations of the data exist [76]. RF is a predictive classification and algorithm that does not make any distributional assumptions about the predictor or the response variables. It also handles situations in which the number of predictor variables exceeds the number of observations, offering a powerful non parametrical alternative for ecological modelling [64].

The vulnerable species groups, communities and habitats described here are mainly distributed beyond the 500 m depth contour. The presence of organisms considered as vulnerable is almost negligible in the fishing area. This fact is almost certainly due to bottom trawling operations of international fleets taking place in the study area for nearly 50 years. Also, the fishing grounds are far away from the geographical location of the main geomorphological features such as canyons, trenches, gas and fluid seepages observed in the middle slope, and identified as potential sites for VMEs.

The fishery footprint plot shows that the historical activity of the Spanish bottom trawler fleet has been located in the shallowest depth strata, at depths not generally exceeding 300 m. On this basis we think that the adverse impacts of current bottom fishing activities on VMEs are negligible or small. However, the displacement of the fishing fleet to target deep sea species at greater depths (were the existence of VMEs has been observed) could have a negative impact on those ecosystems. With this in mind and following the FAO deep-water guidelines, the potential threat of such a fishing strategy should be assessed.

Apart from Spanish fishing fleet, other bottom trawling fleets from different nations (former Soviet Union, Poland, GDR, Bulgaria, etc) have been operating intensively in the SW Atlantic (including our study area) from mid 60's until mid 80's, both over the continental shelf and slope [87-92]. Even if no data were made available to us for assessing the eventual negative impact of these fleets on VMEs, some experiences in other geographical areas such as the North Atlantic, Southwest and East Pacific, seamounts off Tasmania, and waters off New Zealand [31,45,92], have shown that high fishing pressure exerted by a large number of bottom trawlers over a long period of time could relevantly affect these VMEs. We therefore think that probably the almost 50 years of intensive bottom trawling in this SW Atlantic area by the abovementioned fleets could have contributed to the low presence of VMEs in the study area at depths lower than 500 m.

Acknowledgements

We wish to thank the crew of the R/V "Miguel Oliver" (owned by the Spanish General Secretariat for Fisheries [SGP]) and her captain, for the professionalism and the courtesy extended towards us during the research cruises. We are also very grateful to all those involved in the five research campaigns, namely the scientific and technical personnel who made this work possible, known as the Atlantis Group: J. M. Cabanas, J. Gago, B. Almón, E. Elvira, P. Jiménez, A. Fontán, C. Alcalá, and V. López, among others. Our thanks also to the Spanish Secretaría General del Mar (SGM, General Secretariat for the Sea), owner of the research ship, for giving us the opportunity to conduct this research.

Author details

J. Portela[1*], J. Cristobo[2], P. Ríos[2], J. Acosta[3], S. Parra[4], J.L. del Río[1], E. Tel[3], V. Polonio[2], A. Muñoz[5], T. Patrocinio[1], R. Vilela[1], M. Barba[1] and P. Marín[5]

*Address all correspondence to: julio.portela@vi.ieo.es

1 Spanish Institute of Oceanography (IEO), Vigo Oceanographic Center. Fisheries Dept. Vigo, Spain

2 Spanish Institute of Oceanography, Gijón Oceanographic Center. Benthos Dept. Gijón, Spain

3 Spanish Institute of Oceanography, Headquarters Madrid. Geology Dept. Madrid, Spain

4 Spanish Institute of Oceanography, A Coruña Oceanographic Center. Epibenthos Dept. A Coruña, Spain

5 Multidisciplinary Mapping Group (TRAGSATEC-SGP), Spanish General Secretariat for Fisheries (SGP). Cartography Dept. Madrid, Spain

References

[1] FAO. Review of the state of world marine fishery resources. FAO Fisheries Technical Paper 457. Rome, FAO. 2005. 235 pp.

[2] Bensch, A., Gianni, M., Gréboval, D., Sanders, J.S., Hjort, A. Worldwide review of bottom fisheries in the high seas. FAO Fisheries and Aquaculture Technical Paper. No. 522. Rome, FAO. 2008. 145p.

[3] Portela, J.M., Pierce, G.J., del Río, J.L., Sacau, M., Patrocinio, T., Vilela, R. Preliminary description of the overlap between squid fisheries and VMEs on the high seas of the PatagonianShelf. Fisheries Research, 106 (2010) 229–238.

[4] Bakun, A. The California current, Benguela current, and Southwestern Atlantic Shelf ecosystems: A comparative approach to identifying factors regulating biomass yields', in K. Sherman, L.M. Alexander & B.D. Gold (Eds.), Large Marine Ecosystems. Stress, Mitigation, and Sustainability, Washington, DC, AAAS Press, 1993, pp 199-221.

[5] Bisbal, G.A. The Southeast South American shelf large marine ecosystem: Evolution and components. Marine Policy, Vol. 19, No. 1, pp. 21-38, 1995.

[6] Rodhouse, P.G., Symon, C., Hatfield E.M.C. Early life cycle of cephalopods in relation to the major oceanographic features of the southwest Atlantic Ocean. Marine Ecology Progress Series, Vol. 89: 183-195, 1992.

[7] Ivanovic, M.L., Brunetti, N.E. Food and feeding of *Illex argentinus*. Antarctic Science, 1994, 6 (2): 185-193.

[8] Arkhipkin, A.I. Intrapopulation structure of winter-spawned Argentine shortfin squid, *Illex argentinus* (Cephalopoda, Ommastrephidae), during its feeding period over the Patagonian Shelf. Fishery Bulletin, 2000, 98: 1-13.

[9] Laptikhovsky, V. Diurnal feeding rhythm of the short-fin squid, *Illex argentinus* (Cephalopoda, Ommastrephidae), in the Falkland waters. Fisheries Research, 2002, 1373 (2002) 1-5.

[10] Arkhipkin, A.I., Middleton, D.A., Portela, J.M., Bellido, J.M. Alternative usage of common feeding grounds by large predators: the case of two hakes (*Merluccius hubbsi* and *M. australis*) in the southwest Atlantic. Aquatic living resources, 2003. 16(6), 487-500.

[11] Barton, A.J., Agnew, D.J., Purchase, L.V. The Southwest Atlantic; achievements of bilateral management and the case for a multilateral arrangement. In A.I.L. Payne, C.M. O'Brien & S.I. Rogers (Eds.), Management of Shared Fish Stocks, Blackwell, 2004, pp 202 – 222. [Proceedings of the Symposium on International Approaches to Management of Shared Stocks – problems and future directions. Centre for Environment, Fisheries and Aquaculture Science (CEFAS), Lowestoft 10-12 July 2002].

[12] Agnew, D.J., Hill, S.L., Beddington, J.R., Purchase, L.V., Wakeford, R.C. Sustainability and Management of Southwest Atlantic Squid Fisheries. In: Proceedings of the World Conference on the Scientific and Technical Bases for the Sustainability of Fisheries, November 26–30, 2001. University of Miami, USA. Bulletin of Marine Science 2005, 76 (2) 579–594.

[13] Waluda, C.M., Grifftiths, H.J., Rodhouse, P.G. Remotely sensed spatial dynamics of the *Illex argentinus* fishery, Southwest Atlantic. Fisheries Research 91 (2008) 196-202.

[14] Pierce, J.G., Portela, J. Fisheries Production and Market Demand. In: J. Iglesias et al. (eds.) Cephalopod Culture, DOI 10.1007/978-94-017-8648-5_3, © Springer Science +Business Media Dordrecht 2014.

[15] Maguire, J.J., Sissenwine, M., Csirke, J., Grainger, R., Garcia, S. The state of world highly migratory, straddling and other high seas fishery resources and associated species. FAO Fisheries Technical Paper 495. Rome, FAO, 2006. 84 pp.

[16] Kaiser M.J. Significance of Bottom-Fishing Disturbance. Conservation Biology, pages 1230–1235. Volume 12, No. 6, December 1998.

[17] Kaiser, M.J., Edwards, D.B., Armstrong, P.J., Radford, K., Lough, N.E.L., Flatt, R.P., Jones H.L. Changes in megafaunal benthic communities in different habitats after trawling disturbance. ICES (International Council for the Exploration of the Sea) Journal of Marine Science, 1998, 55: 353–362.

[18] Collie, J.S., Escanero, G.A., Valentine, P.C. Photographic evaluation of the impacts of bottom fishing on benthic epifauna. ICES Journal of Marine Science 2000, 57: 987–1001. 2000a.

[19] Collie, J.S., Hall, S.J., Kaiser, M.J., Poiner I.R. A quantitative analysis of fishing impacts on shelf-sea benthos. Journal of animal ecology (2000) 69, 785-798. 2000b.

[20] Coggan, R., Populus, J., White, J., Sheehan, K., Fitzpatrick, F. & Piel, S. (Eds.). Review of Standards and Protocols for Seabed Habitat Mapping, 2007. MESH project document.

[21] Durán Muñoz, P., Murillo, F.J., Serrano, A., Sayago-Gil, M., Parra, S., Díaz Del Río, V., Sacau, M., Patrocinio, T. & Cristobo, J. A Case Study of available methodology for the identification of Vulnerable Ecosystems/Habitats in bottom deep-sea fisheries: Possibilities to apply this method in the NAFO Regulatory Area in order to select Marine Protected Areas. NAFO SCR Doc. 08/06 Serial No. N5491, 2008. 20 pp.

[22] Durán Muñoz, Sayago-Gil, M., Cristobo, J., Parra, S., Serrano, A., Díaz Del Río, V., Patrocinio, T., Murillo, F. J., Palomino, D., & Fernández-Salas, L.M. Seabed mapping for selecting cold-water coral protection areas on Hatton Bank, Northeast Atlantic. – ICES Journal of Marine Science. 2009, 66: 2013–2025.

[23] Murillo, F. J., Durán Muñoz, P., Altuna, A. & Serrano, A. Distribution of deep-water corals of the Flemish Cap, Flemish Pass, and the Grand Banks of Newfoundland (Northwest Atlantic Ocean): interaction with fishing activities. ICES Journal of Marine Science. 2010, 68 (2): 319-332.

[24] UNGA. Resolution 58/14. Sustainable fisheries, including through the 1995 Agreement for the Implementation of the Provisions of the United Nations Convention on the Law of the Sea of 10 December 1982 relating to the Conservation and Management of Straddling Fish Stocks and Highly.Migratory Fish Stocks, and related instru-

ments. 2003. Available from: http://daccess-dds-ny.un.org/doc/UNDOC/GEN/ N03/453/75/PDF/N0345375.pdf? OpenElement, 12pp.

[25] UNGA. Resolution 59/25. Sustainable fisheries, including through the 1995 Agreement for the Implementation of the Provisions of the United Nations Convention on the Law of the Sea of 10 December 1982 relating to the Conservation and Management of Straddling Fish Stocks and Highly Migratory Fish Stocks, and related instruments. UNGA A/RES/59/25. 2004. Available from: http://www.un.org/Depts/los/ general_assembly/general_assembly_reports.htm, 16pp.

[26] UNGA. Resolution 60/31. Sustainable fisheries, including through the 1995 Agreement for the Implementation of the Provisions of the United Nations Convention on the Law of the Sea of 10 December 1982 relating to the Conservation and Management of Straddling Fish Stocks and Highly Migratory Fish Stocks, and related instruments. UNGA A/RES/61/105. 2005. Available from: http://daccess-dds-ny.un.org/doc/ UNDOC/GEN/N05/489/40/PDF/N0548940.pdf? OpenElement

[27] UNGA. Resolution 61/105. Sustainable fisheries, including through the 1995 Agreement for the Implementation of the Provisions of the United Nations Convention on the Law of the Sea of 10 December 1982 relating to the Conservation and Management of Straddling Fish Stocks and Highly Migratory Fish Stocks, and related instruments. UNGA A/RES/61/105. 2007. Available from: http://www.un.org/Depts/los/ general_assembly/general_assembly_reports.htm, 21pp.

[28] UNGA. Resolution 64/72. Sustainable fisheries, including through the 1995 Agreement for the Implementation of the Provisions of the United Nations Convention on the Law of the Sea of 10 December 1982 relating to the Conservation and Management of Straddling Fish Stocks and Highly Migratory Fish Stocks, and related instruments. UNGA A/RES/64/72. 2010. Available as General Assembly document A/64/L. 29. Available from: http://daccess-dds-ny.un.org/doc/UNDOC/GEN/ N09/466/15/PDF/N0946615.pdf? OpenElement, 26pp.

[29] Roberts, J.M., Harvey, S.M., Lamont, P.A., Gage, J.D. & Humphery, J.D. Seafloor photography, environmental assessment and evidence for deep-water trawling on the continental margin west of the Hebrides. Hydrobiologia 2000, 441:173-83.

[30] Rogers, A.D. The biology of *Lophelia pertusa* (Linnaeus 1758) and other deep-water reef forming corals and impacts from human activities. International Review of Hydrobiology 1999, 84:315-406.

[31] Trush, S.F., Dayton, P.K. Disturbance to marine habitats by trawling and dredging: Implications for marine biodiversity. Annual Review of Ecology and Systematics 2002, 33: 449-473.

[32] Hiddink, J. G., Jennings, S., Kaiser, M.J. Indicators of the ecological impact of bottom-trawl disturbance on seabed communities. Ecosystems (2006) 9 (7): 1190-1199. DOI: 10.1007/s10021-005-0164-9

[33] Hiddink, J. G., Jennings, S., Kaiser, M.J. Assessing and predicting the relative ecological impacts of disturbance on habitats with different sensitivities. Journal of Applied Ecology (2007) 44 (2): 405-413.

[34] Nellemann, C., Hain, S., Alder, J. (Eds.). February 2008. In Dead Water – Merging of climate change with pollution, over-harvest, and infestations in the world's fishing grounds. United Nations Environment Programme, GRID-Arendal, Norway, www.grida.no.

[35] Weaver, P.P.E., Benn, A., Arana, P.M., Ardron, J.A., Bailey, D.M., Baker, K., Billett, D.S.M., Clark, M.R., Davies, A.J., Durán Muñoz, P., Fuller, S.D., Gianni, M., Grehan, A.J., Guinotte, J., Kenny, A., Koslow, J.A., Morato, T., Penney, A.J., Perez, J.A.A., Priede, I.G., Rogers, A.D., Santos, R.S., Watling, L. The impact of deep-sea fisheries and implementation of the UNGA Resolutions 61/105 and 64/72. Report of an international scientific workshop, National Oceanography Centre, Southampton, (2011) 45 pp. Available from http://hdl.handle.net/10013/epic.37995.

[36] Hall-Spencer, J., Allain, V., Fosså, J.H. Trawling damage to Northeast Atlantic ancient coral reefs. Proceedings of the Royal Society of London, 2002, Series B: Biological Sciences 269(1490): 507–511.

[37] Fosså, J., Mortensen, P., Furevik, D. The deep-water coral *Lophelia pertusa* in Norwegian waters: distribution and fishery impacts. Hydrobiologia (2002) 471: 1–12.

[38] Wheeler, A.J., Bett, B.J., Billett, D.S., Masson, D.G., Mayor, D. The impact of demersal trawling on northeast Atlantic coral habitats: The case of the Darwin mounds, United Kingdom. In: Barnes, P.W., Thomas, J.P. (Eds.) Benthic habitats and the Effects of fishing, American Fisheries Society Symposium 41, 2005, Bethesda, Maryland, USA, pp. 807–817.

[39] Mortensen, P.B., Buhl-Mortensen, L. Gordon Jr, D.C. Fader, G.B.J., McKeown, D.L., Fenton, D.G. Effects of Fisheries on Deep-water Gorgonian Corals in the Northeast Channel, Nova Scotia (Canada). American Fisheries Society Symposium 2005, 41: 369-382.

[40] Edinger, E., Baker, K., Devillers, R., Wareham, V.E. Coldwater Corals off Newfoundland and Labrador: Distribution and Fisheries Impacts. WWF-Canada, Toronto, Canada, 2007, 41pp.

[41] Stone, R.P. Coral habitat in the Aleutian Islands of Alaska: depth distribution, fine-scale species associations, and fisheries interactions. Coral Reefs, 2006, 25: 229–238.

[42] Krieger, K.J. Coral (Primnoa) impacted by fishing gear in the Gulf of Alaska. In: Willison. J.H.M., Gass, S.E., Kenchington, E.l.R., et al. (Eds.) Proceedings of the first International Symposium on Deep-Sea Corals, Ecology Action Centre and Nova Scotia museum, Halifax, Nova Scotia, Canada, 2001, pp. 106–116.

[43] Koslow, J.A., Gowlett-Holmes, K. The seamount fauna off southern Tasmania: benthic communities, their conservation and impacts of trawling: Final Report to Environ-

ment Australia and the Fisheries Research and Development Corporation, 1998. Rep. FRDC Project 95/058, CSIRO, Hobart, Tasmania, Australia, 104pp.

[44] Koslow, J.A., Gowlett-Holmes, K., Lowry, J.K., O'Hara, T., Poore, G.C.B., Williams, A. Seamount benthic macrofauna off southern Tasmania: community structure and impacts of trawling. Marine Ecology Progress Series 2001, 213: 111–125.

[45] Althaus, F., Williams, A., Schlacher, T.A., Kloser, R.J., Green, M.A., Barker, B.A., Bax, N.J., Brodie, P., Schlacher-Hoenlinger, M.A. Impacts of bottom trawling on deep-coral ecosystems of seamounts are long-lasting. Marine Ecology Progress Series 2009, 397: 279-294.

[46] Clark, M.R., Rowden, A.A. Effect of deepwater trawling on the macro-invertebrate assemblages of seamounts on the Chatham Rise, New Zealand. Deep-Sea Research 2009, Part I 56: 1540–1554.

[47] Jones, C.D., Lockhart, S.J. Detecting Vulnerable Marine Ecosystems in the Southern Ocean using research trawls and underwater imagery. Marine Policy 2011, 35 (2011) 732–736

[48] Global Ocean Commission. From Decline to Recovery-A Rescue Package for the Global Ocean. Global Ocean Commission Report 2014, 92 pages. Available from: http://www.globaloceancommission.org/news/the-global-ocean-commissions-final-report-is-now-available/#more-11563

[49] [49]Schlacher, T.A., Schlacher-Hoenlinger, M.A., Williams, A., Althaus, F., Hooper, J.N.A., Kloser, R. Richness and distribution of sponge megabenthos in continental margin canyons off southeastern Australia. Marine Ecology Progress Series 2007, 340, 73–88.

[50] Grassle, J. F. Slow recolonization of deep-sea sediment. Nature 1977, 265:618-619.

[51] Rex, M.A. Community structure in the deep-sea benthos. Annual Review of Ecological Systematics 1981, 12: 331-353.

[52] Smith, C.R., Hamilton, S.C. Epibenthic megafauna of a bathyal basin off southern California: patterns of abundance, biomass, and dispersion. Deep-Sea Research 1983, 30, 907-928.

[53] Tracey, M.D., Neil, H., Marriott, P., Andrews, A.H., Cailliet, M.G., Sanchez J.A. Age and growth of two genera of deep-sea bamboo corals (Family Isididae) in New Zealand waters. Bulletin of Marine Science, 2007. 81 (3), 393-408.

[54] FAO (2009). International Guidelines for the Management of Deep-sea Fisheries in the High Seas. Rome/Roma, FAO. 2009. 73p.

[55] Harris, P.T., Whiteway, T. High seas marine protected areas: Benthic environmental conservation priorities from a GIS analysis of global ocean biophysical data. Ocean & Coastal Management 2009, 52: 22–38.

[56] WGEAFM. Report of the NAFO SC Working Group on Ecosystem Approach to Fisheries Management (WGEAFM). NAFO Scientific Council Meeting June 2008. Serial Number 5511, NAFO SCS Doc. 08/10, 70pp. 2008a

[57] WGEAFM. Report of the NAFO SC Working Group on Ecosystem Approach to Fisheries Management (WGEAFM). Response to Fisheries Commission Request 9.a. NAFO Scientific Council Meeting October 2008, Serial Number 5592, NAFO SCS Doc. 08/24, 19pp. 2008b

[58] Rogers, A.D., Gianni, M. The Implementation of UNGA Resolutions 61/105 and 64/72 in the Management of Deep-Sea Fisheries on the High Seas. Report prepared for the Deep-Sea Conservation Coalition. International Programme on the State of the Ocean, London, United Kingdom, 2010, 97pp.

[59] ICES. Report of the Working Group on Marine Habitat Mapping (WGMHM), 21-24 April 2009, The National Institute of Aquatic Resources, Charlottenlund Castle, Copenhagen, Denmark. ICES CM 2009/MHC: 07. 78 pp.

[60] NAFO. Report of the NAFO SC Working Group on Ecosystem Approach to Fisheries Management (WGEAFM). Response to Fisheries Commission Request 9.a. NAFO SCS Document, 2008, 08/24, Serial No. N5592. 19 pp.

[61] Cutler, D.R., Edwards, T.C., Beard, K.H., Cutler, A., Hess, K.T., Gibson, J., Lawler, J.J. Random forests for classification in ecology. Ecology 2007, 88, 2783–2792.

[62] Crisci, C., Ghattas,B., Perera, G. A review of supervised machine learning algorithms and their applications to ecological data. Ecological Modelling 2012, 240, 113-122.

[63] Prasad, A.M., Iverson, L.R., Liaw, A. Newer classification and regression tree techniques: bagging and random forest for ecological prediction. Ecosystems 2006. 9: 181-199.

[64] Hirzel, A.H., Hausser, J., Chessel, D., Perrin, N. Ecological-niche factor analysis: How to compute habitat-suitability maps without absence data? Ecology, 83(7), 2002, pp. 2027-2036.

[65] Kuhnert, P.M., Martin, T.G., Mengersen, K., Possingham, H.P. Assessing the impacts of grazing levels on bird density in woodland habitat: a Bayesian approach using expert opinion. Environmetrics, 2005, 16: 717–747.

[66] Zuur, A.F., Ieno, E.N., Meesters, E.H.W.G. A Beginner's Guide to R. Springer, 2009, New York.

[67] Breiman, L. Random Forests. (S. R. E, Ed.). Machine Learning 2001. 45(1), 5-32. Springer.

[68] Kohavi R., Provost, F. Glossary of terms. Machine Learning 1998. 30, (2-3): 271-274.

[69] R Development Core Team. R: A language and environment for statistical comput‐ ing. R Foundation for Statistical Computing 2011, Vienna, Austria. ISBN 3-900051-07-0, URL http://www.R-project.org/.

[70] Liaw, A., Wiener, M. Classification and regression by Random Forests. R News 2002, 2/3:18–22. URL: http://CRAN.R-project.org/doc/Rnews/

[71] Bivand, R., Lewin-Koh, N. Maptools: Tools for reading and handling spatial objects. R package version 0.8-23, 2013. http://CRAN.R-project.org/package=maptools

[72] Thuiller, W. BIOMOD – optimizing predictions of species distributions and projec‐ ting potential future shifts under global change. Global Change Biology, 2003, 9: 1353–1362.

[73] Hastie, T.J., Tibshirani, R.J., Friedman, J.H. The elements of statistical learning: Data mining inference, and prediction. Springer Series in Statistics, 2001. Springer, New York.

[74] Elith J., Graham C.H., Anderson R.P., Dudik M., Ferrier S., Guisan A., Hijmans R.J., Huettmann F., Leathwick J.R., Lehmann A., Li J., Lohmann L.G., Loiselle B.A., Man‐ ion G., Moritz C., Nakamura M., Nakazawa Y., Overton J.M., Peterson A.T., Phillips S.J., Richardson K., Scachetti-Pereira R., Schapire R.E., Soberon J., Williams S., Wisz M.S., Zimmermann N.E. Novel methods improve prediction of species' distributions from occurrence data. Ecography, 2006, 29(2), 129-151.

[75] Friedman, J.H. Multivariate Adaptive Regression Splines. Annals of Statistics, 1991, 19, 1.

[76] Friedman J.H., Hastie, T., Tibshirani, R. Additive logistic regression: a statistical view of boosting. Annals of Statistics, 2000, 28, 337-407.

[77] Fray, C., Ewing, M. Pleistocene sedimentation and fauna of the Argentine shelf. I. Wisconsin sea level as indicated in the Argentine continental shelf sediments. Pro‐ ceedings of the Academy of Sciences of Philadelphia, 1963. 115, 113-116.

[78] Lastras, G., Acosta, J., Muñoz, A., Canals, M. Submarine canyon formation and evo‐ lution in the Argentine Continental Margin between 44°30′S and 48°S. Geomorpholo‐ gy, 2011, vol 128, pp 116-136.

[79] Muñoz, A., Cristobo, J., Ríos P., Druet, M., Polonio, V., Uchupi, E., Acosta, J., Atlantis Group. Sediment drifts and cold-water coral reefs in the Patagonian upper and mid‐ dle continental slope. Marine and Petroleum Geology 36 (2012) 70-82. DOI: 10.1016/ j.marpetgeo.2012.05.008.

[80] Richards, P., Duncan, I., Phipps, C., Pickering, G., Grzywacs, J., Hoult, R., Merritt, J. Exploring for fan delta sandstones in the offshore Falklands basins. Journal of Petro‐ leum Geology, 2006, 29(3), 199-214.

[81] Muñoz, A., Acosta, J., Cristobo, J.M., Druet, M., Uchupi, E., Atlantis Group. Geomorphology and shallow structure of a segment of the Atlantic Patagonian margin. Earth-Science Reviews 121 (2013) 73-95.

[82] Sumida, P.Y.G., Yoshinaga, M.Y., Madureira, L.A.S.-P., Hovland, M. Seabed pockmarks associated with deepwater corals off SE Brazilian continental slope, Santos Basin. Marine Geology, 2004. 205, 159-167.

[83] Fawcett, T. ROC Graphs: Notes and Practical Considerations for Researchers. Re-CALL 2004,31, 1-38.

[84] Glockzin, M., Gogina, M., Zettler, M.L. Beyond salty reins. Modeling benthic species spatial response to their physical environment in the Pomeranian Bay (Southern Baltic Sea). Baltic Coastal Zone 2009, 13, 79-95

[85] Brown, C.J., Collier, J.S. Mapping benthic habitat in regions of gradational substrata: An automated approach utilising geophysical, geological, and biological relationships. Estuarine, Coastal and Shelf Science 2008, 78, 203-214.

[86] Franklin E.C., Ault J.S., Smith S.G., Luo J., Meester G.A., Diaz G.A., Chiappone M., Swanson D.W., Miller S.L., Bohnsack J.A. Benthic habitat mapping in the Tortugas region, Florida. Marine Geodesy 2003, 26: 19-34.

[87] Roemmich, G.L. Impact of the Law of the Sea Treaty on the Soviet Fishing Industry. Theses and Major Papers, 1983. Paper 153.

[88] Beddington, J.R., Brault, S., Gulland, J. The Fisheries around the Falklands. IIED/IUCN Marine Resources Assessment Group. Centre for Environmental Technology. Imperial College 1985. London.

[89] Csirke, J. The Patagonian fishery resources and the offshore fisheries in the South-West Atlantic. FAO Fisheries Technical Paper 286.-75 pp., 35 figs. Rome: Food and Agriculture Organization of the United Nations 1987. ISBN 92-5-102564-9.

[90] Willetts, P. Fishing in the South-West Atlantic (London: South Atlantic Council Occasional Papers, No. 4, March 1988), available from www.staff.city.ac.uk/p.willetts/SAC/OCPAPERS.HTM.

[91] Jacobson, D., Weidner, D. Argentine-Soviet Fishery Relations Reviewed, 1966-88. Marine Fisheries Review 1989, 51(2):55-68 [Int. Fish. Rep. 88-108].

[92] Anon. Effects of trawling and dredging on seafloor habitat. Committee on Ecosystem Effects of Fishing: Phase 1 Effects of bottom trawling on seafloor habitats, National Research Council. Ocean Studies Board. Division on Earth and Life Studies. National Academic Press 2002. ISBN: 0-309-50815-0, 136 pages.

Chapter 12

Agrorural Ecosystem Effects on the Macroinvertebrate Assemblages of a Tropical River

Bert Kohlmann, Alejandra Arroyo,
Monika Springer and Danny Vásquez

Additional information is available at the end of the chapter

http://dx.doi.org/10.5772/59073

1. Introduction

Costa Rica is an ideal reference point for global tropical ecology. It has an abundance of tropical forests, wetlands, rivers, estuaries, and active volcanoes. It supports one of the highest known species density (number of species per unit area) [1, 2] on the planet and possesses about 4 % of the world's total species diversity [3]. Because of its tropical setting, it also serves as an important location for agricultural production, including cultivars such as coffee, bananas, palm hearts, and pineapples. The country has also attracted more ecotourists and adventure travelers per square kilometer than any other country in the world [4].

The agrorural frontier on the Caribbean side of Costa Rica started to spread during the 1970s, especially in its northeastern area. Migrations of land-poor people from the Pacific and mountain areas of the country started to colonize the land that the government had made available [5, 6]. These waves of immigrants tended to establish themselves along river systems. In this way, towns, small to medium-scale family farming, ranching, and plantation agriculture began to base themselves along the main river systems. It was during this time that the human settlements originated along the Dos Novillos River [7].

Residual waters produced by all of the aforementioned human activities are at present discharged into the river systems in the Costa Rican Caribbean area. Households not situated in the neighborhood of rivers will use septic tanks; homesteads situated along riverbanks will discharge their effluents directly into the rivers. Other activities like the production of residual waters from dairy farms, pigsties, banana packing plants, plantations' excess fertilization, etc. will drain eventually into a river. The Dos Novillos River is no exception. Water sewage systems in this part of Costa Rica are almost non-existent.

Aquatic biomonitoring in Latin America started at the end of the last century, commencing in Colombia [8-10] and then spreading to other Latin American countries. Revisions of the use of aquatic biomonitoring indices in Latin America are given in de la Lanza Espino *et al.* [8], Prat *et al.* [9], and Springer [10]. In the case of Costa Rica, several studies based on the importance of macroinvertebrates for biomonitoring water quality and community structure, and function in banana and conventional and organic rice systems [14-22], as well as macroinvertebrate field-guides [23, 24] have been published. Some studies of macroinvertebrate assemblage structures also have been reported for rivers on the southern Caribbean coast of Costa Rica [25-34].

Although ecosystem studies have been used in Costa Rica for evaluating possible impacts caused by crop activities on river water quality at specific points, almost nothing is known about how a mix of other human activities can impact macroinvertebrate biodiversity and ecosystem structure along the length of a river. The aim of this study was to use macroinvertebrate biodiversity under the influence of different human activities along the length of a river in order to describe their impact on community structure and function in tropical agrorural environments.

2. Materials and methods

2.1. Study area

This study was conducted at the Dos Novillos River (Figure 1, Table 1) in the province of Limón, Costa Rica. Samples were collected two times a month from January 2005 to March 2005 and monthly from April 2005 to January 2006. The Dos Novillos River drains from the Central Cordillera at an elevation of approximately 2380 masl towards the Caribbean lowlands of the province of Limón. This river is part of the 2950 km² Parismina River watershed in a premontane wet forest and tropical moist forest region [35]. The underlying geology is represented by quaternary sedimentary and volcanic rocks under the influence of nearby volcanoes, with a flat to undulating topography and poorly drained alluvial soils susceptible to flooding [36]. Banana plantations have been developed on the lower reaches of this watershed. The study area is characterized by a humid tropical climate with a mean temperature of 25.8 °C, an average annual relative humidity of 87 %, and an annual precipitation average of 3460 mm ± 750 mm without a pronounced dry season. The sampling area runs in a straight line through the towns and localities of La Argentina, Pocora, and EARTH University in the Province of Limón. Each fore mentioned landmark is separated by approximately 4 km. The total sampling area runs along a length of 13.6 km, across an area with a mixture of a premontane wet forest, pastureland, small town, riparian tropical moist forest, and banana agricultural areas.

Six sampling sites were located along the Dos Novillos River (Figs. 1-7; Table 1) where macroinvertebrates were sampled. The first sampling site (Figure 2) (site 1, "Don Eladio") served as a reference site, being part of the rhithral region, located upstream of the first anthropogenic disturbance (pastureland). This site is surrounded by tropical rain forest;

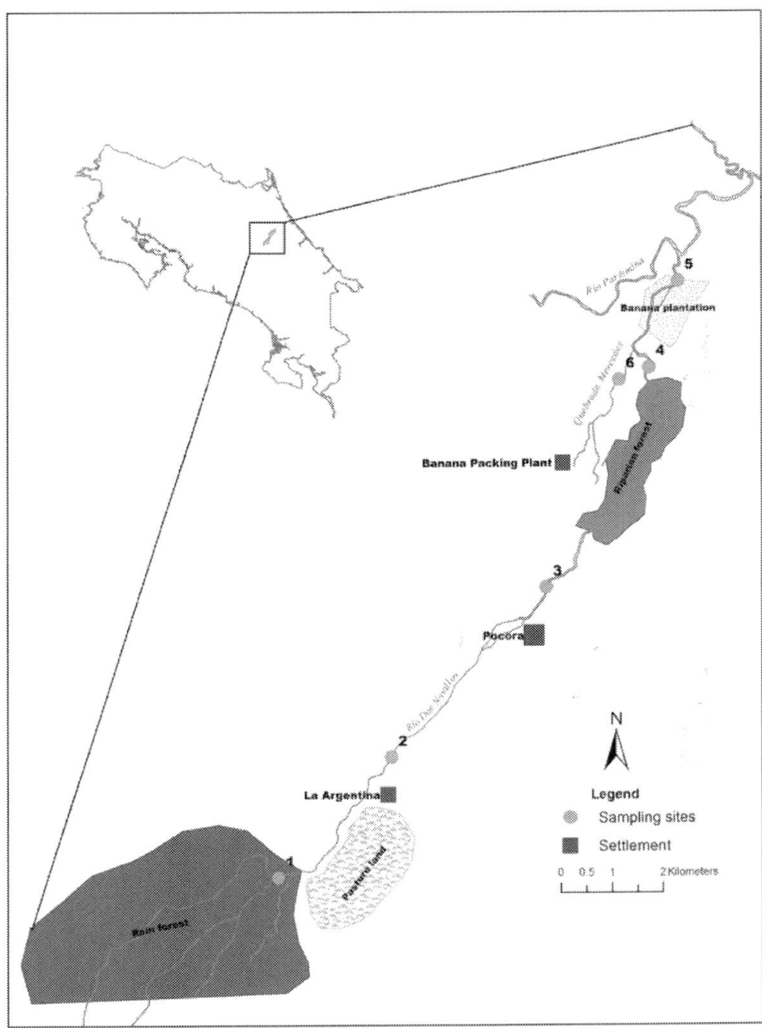

Figure 1. Sampling site location and areas of major potential anthropogenic disturbance at the Dos Novillos River, Guácimo, Limón, Costa Rica [modified from [37].

therefore, natural good water conditions were expected, as well as high taxa richness and an assemblage composition dominated by pollution-sensitive organisms. Site 2, "La Argentina" (Figure 3), was located approximately 5.5 km downstream from site 1. This site was selected

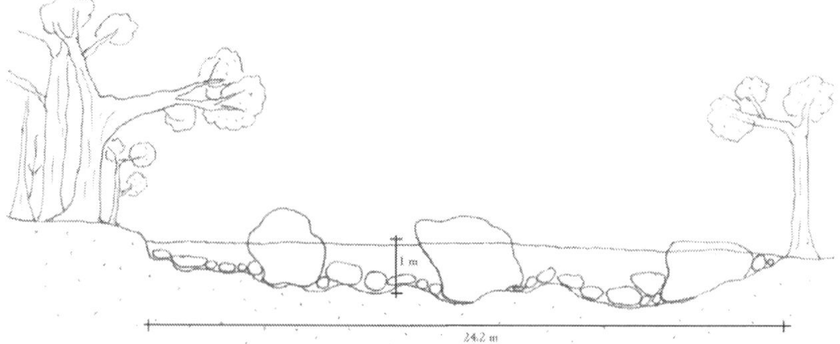

Figure 2. The first site (Don Eladio) was established as a control (reference) site and is located several miles upstream of La Argentina, Pocora, inside natural tropical forest. The site is characterized by a variety of current conditions, with fast flowing riffles, medium laminar flow and pools. The substrate is composed mostly of medium-sized rocks, many of them covered with moss, though large boulders also stand, where large numbers of insects live in the splash area.

to examine the possible extent that small livestock farming might have on river water quality. Site 3, "Chiquitín" (Figure 4), was located approximately 8 km downstream from site 1. High anthropogenic influence was expected at this site because the houses situated at the riverfront discharge their grey and black waters directly into the river. Site 4, "Puente La Hamaca" (Figure 5), was located within the property of EARTH University, approximately 2 km downstream

Figure 3. Site 2 (La Argentina) is located downstream of La Argentina. It is a place dominated by rocks of all sizes, although big boulders are less numerous than site 1. Current conditions are similar as in Site 1.

from site 3. As the intervening river length between sites 3 and 4 runs through forest areas, site 4 was selected in order to examine if water quality was improved by a forest filtering processes. Site 5, "Desembocadura" (Figure 6), is within the EARTH University campus and is located approximately 500 m upstream from the confluence of the Dos Novillos and Parismina Rivers. Site 5 was selected to analyze the impact of banana plantations on river water quality. Site 6, "Quebrada Mercedes" (Figure 7), is one tributary of the Dos Novillos River flowing through a forested area within the EARTH University campus, approximately 2 km downstream from site 3. Site 6 was chosen to examine if the intermittent discharge of a small drain of water used to wash bananas in a packing plant had any effect on the stream. Table 1 indicates the exact location, depth, width, and current conditions for each site.

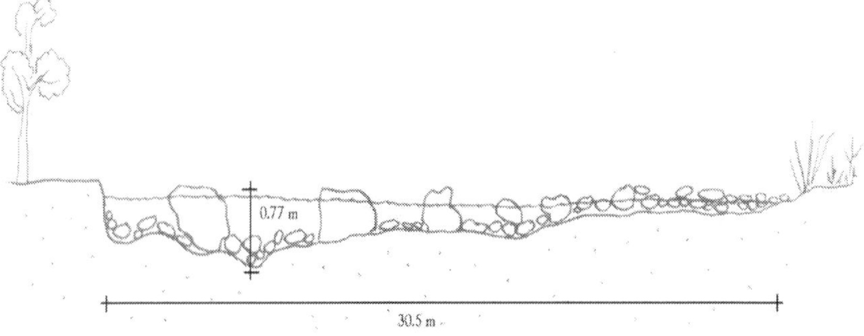

Figure 4. This site (Chiquitín) is located in the center of the town of Pocora. It is at this point where the channel becomes wider and current is more laminar; the substrate is also rocky, but there is no presence of large boulders.

Figure 5. Site 4 (La Hamaca) is located within the EARTH University campus, specifically at the suspension bridge. The diversity of current conditions is similar to the other sites, although the rock size is much smaller as at the other three upstream sites. The gallery forest borders the channel at this point.

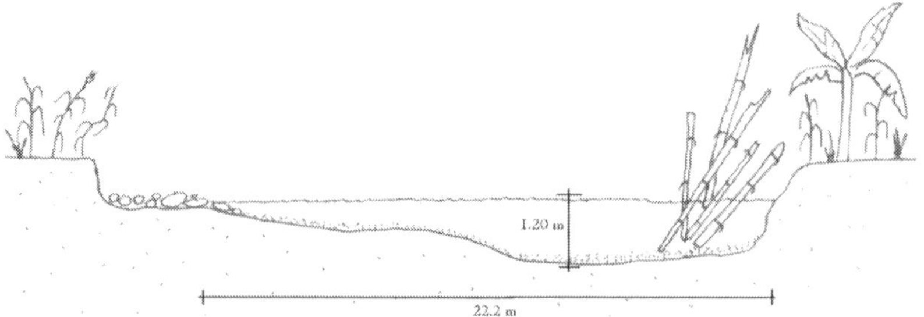

Figure 6. Site 5 (Desembocadura) corresponds to the mouth of the River Dos Novillos with the Parismina. The substrate consists almost entirely of sand. Big boulders are absent, small rocks are scarce and current flow is weak; tall grasses, banana plants, bamboo, and a few trees dominate vegetation along the channel. At this site, small airplanes were observed flying over the river while spraying pesticides on the surrounding banana plantations.

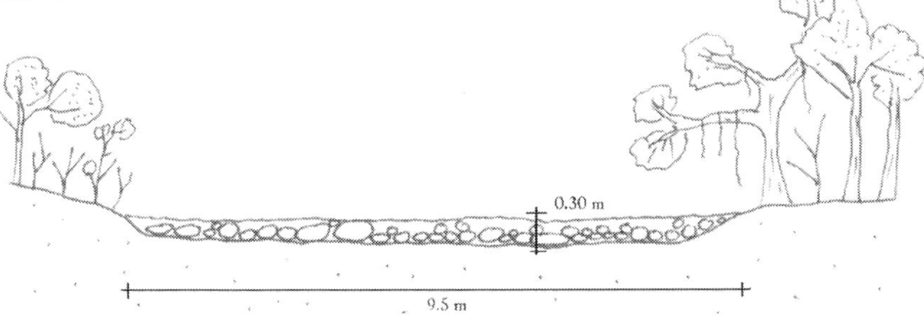

Figure 7. Shallow waters and a moderate current characterize the tributary Quebrada Mercedes (site 6). Small rocks are the predominant substrate and current is moderate and laminar with some faster flowing riffle areas.

Site number	1	2	3	4	5	6
and name	Don Eladio	La Argentina	Chiquitín	La Hamaca	Desembocadura	Quebrada Mercedes
Longitude (N)	10° 07′ 09.7″	10° 09′ 14.3″	10° 10′ 40.8″	10° 13′ 00.9″	10° 14′	10° 12′
Latitude (W)	83° 39′ 15.2″	83° 37′ 24.7″	83° 36′ 10.6″	83° 35′ 18.4″	83° 34′	83° 35′
Altitude (m)	441	187	90	51	40	44
Width (m)	24.2	17.5	30.5	24	22.2	9.5
Depth (m)	0.3-1.3	0.25-0.8	0.25-0.8	0.2-0.86	0.2 – 1.2	0.15-0.3
Current (m/s)	1.67	1.94	0.9	1.05	3.2	1.9
River bottom	Medium-sized rocks	Medium-sized rocks	Small rocks	Small rocks	Sand	Small rocks

Table 1. Geographical and physical characteristics of each sampling site at the Dos Novillos River (1-5) and the Mercedes Stream (6), Guácimo, Province of Limón, Costa Rica.

2.2. Sampling

A plastic strainer with a diameter of 20 cm and 0.5 mm mesh size, and tweezers were used for directly collecting macroinvertebrates. The main criterion for this semi-quantitative collecting method is time; there were no defined sampling areas. All types of microhabitats present at a particular site were examined equally for the macroinvertebrates. Collected organisms were fixed immediately in 70 % ethanol at the time of sampling. Exact details of the sampling methodology can be found in Stein *et al.* [37].

A presampling was carried out in order to determine sampling time [37]. Out of the achieved results, an accumulated taxa curve was elaborated, and 120 min were determined to be a representative sampling time per site. Sampling took place during early morning and under normal current situations in order to avoid the negative effects of flooding and high water conditions.

The government of Costa Rica has officially suggested this method under the water quality monitoring regulation [38]. This method is coupled with the use of a modified Biological Monitoring Working Party index for Costa Rica (BMWP′-CR), a biotic index utilized to define different levels of water quality. Each family of macroinvertebrates has a sensitivity value ranging from 1 to 10, reflecting tolerance to pollution based on the knowledge of distribution and abundance. The values for each family are then summed up independently from abundance and generic or species diversity. Sensitivity scores higher than 120 points indicate undisturbed aquatic ecosystems, while low values indicate serious contamination (mostly organic) of the environment [13, 23, 38,].

2.3. Data analysis

The analyzed data comprised the values of the physical-chemical water quality variables (Table 2) and macroinvertebrate abundances (Table 3) during a collecting period of 13 months (January 2005 to January 2006).

The following physical-chemical variables were measured: pH, temperature, O_2, O_2 saturation, suspended solids, turbidity, conductivity, NO_3^-, NH_4^+, PO_4^+, BOD, and COD. The water samples also were analyzed for the following agrochemicals associated with banana production using gas chromatography-MS and liquid chromatography-PDA: Chlorpyrifos, Diazinon, Dimethoate, Edifenfos, Etoprofos, Fenamifos, Malathion, Parathion-methyl, Parathion-ehtyl, Terbufos, Difenoconazol, Propiconazol, Imazalil, Ametrine, Atrazine, Hexazinone, Terbutylazine, Bromacil, Bitertanol, Chlorothalonil, and Thiabendazole. However, no traces of them could be detected in the river water samples. This does not come as a surprise because in order to monitor pesticides very frequent sampling would be required to detect peak concentrations during pesticide application periods [39], whereas low concentrations are very difficult to detect.

Following the suggestion of Ramírez and Gutiérrez-Fonseca [40], this study is also undertaking an ecosystem process analysis of the functional feeding groups (FGG) of the aquatic macroinvertebrates. This sort of analysis is based on two key aspects of macroinvertebrates: morphological characteristics related to the obtainment of food resources (*e.g.*, mouthparts and related structures) and behavioural mechanisms (*e.g.*, feeding behaviour). FGG is a very useful tool that provides valuable information on ecosystem functioning, facilitating stream ecosystem comparisons, and avoiding the traps of gut content analysis, which is more appropriate for assigning trophic guilds [40].

In this ecosystem study, the use of different parameters of the structure and composition of macroinvertebrate assemblages are presented: total and relative abundances, taxa richness, and functional feeding groups. Also, correlations of different genera and functional feeding groups with environmental variables were analyzed.

2.4. Statistical analysis

The model comparisons between physical-chemical variables, macroinvertebrate abundances, and the BMWP′-CR index at different collecting sites was done by performing an analysis of variance (ANOVA; $\alpha=0.05$). For all three cases, the proposed hypothesis is to test the existence of significant variable differences between sites. Abundances were square root transformed in order to comply with error normality. Evaluation of the best model (homocedastic or heterocedastic) for each variable was performed using the Akaike information criterion (AIC), which is one of the benchmarks of mixed models based on penalized likelihood [41, 42]. When the model detected significant differences, a DGC (Di Rienzo – González – Casanoves) statistical test was performed for the comparison of means [43].

On the other hand, taxonomic groupings and FFG relative frequencies analyses were done using a Chi-square test in order to assay for statistically significant differences between sites. The Chi-square analysis is testing for independence between the sites and the studied variables. Any p value below 0.05 shows the existence of an association between the site and the studied variable.

PLS regression is a technique that combines Principal Component Analysis and Linear Regression [44]. It is applied when it is desired to predict a set of dependent variables (y), in this case the abundance of macroinvertebrate genera and FFG abundances and the BMWP index values, from a set of predictor variables (x), in this case physical-chemical variables. To represent the results obtained from the PLS analysis, a Triplot graph was superimposed on a Biplot graph [45], thus correlating all variables. Then, the observations appear ordered in a Triplot graph (sites), depending on the values of the dependent variables (macroinvertebrate and FGG abundances and BMWP′-CR index) and their correlation with the predictor variables (physical-chemical water-quality variables). For the macroinvertebrate genera PLS analysis, out of the 127 collected taxa (of which, 123 could be identified to the genera level and their different developmental stages: larva, pupa, adult), the multimetric analysis included only 58 taxa, which were chosen using a PCA (Principal Component Analysis). The rest were characterized for repeating the same information. These 58 taxa, composed of 15 688 individuals, showed high projection values on the first two principal components. All statistical analyses were done using the InfoStat program [46].

In order to correlate the abundance of macroinvertebrates, FFG, and the BMWP′-CR index values of different sites, these variables were correlated with physical-chemical variables using the Spearman rank correlation coefficient. In the present case, the hypothesis tries to establish if one variable can be effectively substituted by another one, due to the existence of a significant correlation. The Spearman correlation coefficient was selected, versus Pearson, because its use is recommended in the case of having a small sample.

Finally, in order to evince the consistency of the sites′ congruences arranged in one plane unto physicochemical variables and FFG and macroinvertebrate genera abundances, a Generalized Procrustes Analysis was performed. This analysis is used for harmonizing multivariate configurations obtained on the same set of observations with different types of variables or time points [47]. Alignment is performed through a series of steps including normalization, rotation, reflection, and scaling of data to obtain a consensus array between groups of variables. This series of steps should maintain the distances between individuals from the individual configurations and minimize the distance between similar points [44]. The result of this multivariate method is to present a graph that displays the configurations arrived at by each variable type and the consensus configuration. A percentage consensus analysis was also undertaken. A high consensus indicates that any group of variables characterizes the different sites in the same way; therefore, using any group of variables is indistinct for site characterization.

3. Results

3.1. Physical-chemical parameters

Table 2 shows the results of the physical-chemical analysis (ANOVA, DGC-test, $p > 0,05$). All sites presented neutral pH and high dissolved oxygen levels and saturation. Temperatures varied in a statistically significant way, site 1 being the coolest place and site 5 the warmest. Conductivity was quite low at all sites, but statistically significantly lower in sites 5 and 6;

whereas, turbidity was statistically significantly higher in site 6. NO_3 was statistically significantly higher in sites 5 and 6, and lowest in sites 1, 2, and 3.

Variables		Site 1 Don Eladio	Site 2 Argentina	Site 3 Chiquitín	Site 4 Hamaca	Site 5 Desembocadura	Site 6 Q. Mercedes
pH		7.0±0.5	7.1±0.6	7.1±0.7	7.1±0.7	7.0±0.7	7.2±0.7
Dissolved O_2	mg/L	7.2±1.8	7.4±1.8	7.5±1.2	7.7±1.6	7.00±2.1	7.0±1.8
O_2 Saturation	%	80.3±21.3	88.5±22.4	86.8±20.8	90.5±21.8	86.7±27.5	82.3±24.8
Temperature	°C	21.1±0.6c	23.6±1.3b	24.7±1.4b	24.3±0.8b	25.9±1.5a	24.0±1.0b
Conductivity	µS/cm	46.5±4.3b	49.3±4.4b	46.3±7.5b	43.3±17.5b	58.4±7.2a	56.0±14.5a
Turbidity	NTU	1.3±2.3b	0.8±0.5b	1.0±0.6b	1.5±1.2b	1.4±1.2b	3.4±2.6a
BOD	ppm	16.0±8.1	14.7±10.9	16.3±8.8	11.6±7.7	14.3±9.2	12.2±9.6
COD	ppm	14.9±10.7	10.2±8.0	12.8±7.2	10.4±6.3	15.4±14.9	14.9±9.7
NO_3	ppm	0.05±0.04c	0.03±0.02c	0.06±0.04c	0.10±0.05b	0.13±0.04a	0.15±0.07a
NH_4	ppm	0.06±0.14	0.04±0.06	0.03±0.05	0.06±0.10	0.07±0.17	0.05±0.10
PO_4	ppm	0.08±0.05	0.07±0.06	0.08±0.05	0.09±0.05	0.09±0.05	0.09±0.06
Susp. Solids	mg/l	46.2±27.5	38.3±22.3	28.5±27.2	39.7±29.2	42.1±22.8	42.2±31.8

Table 2. Mean values of physical-chemical variables with their standard deviations used for the PLS analysis at the six collecting sites (January 2005-January 2006). Means in the same row with the same lettering are not significantly different (ANOVA, DGC-test, homocedastic model for the physical-chemical variables, p>0, 05).

3.2. Diversity and composition of macroinvertebrate assemblages

The study collected a total of 17 163 specimens, distributed in the following 6 classes (number of total amount of specimens in parentheses): Clitellata (2), Turbellaria (10), Gastropoda (403), Arachnida (21), Malacostraca (40), and Insecta (16 706) with the following orders: Ephemeroptera (6299), Coleoptera (3150), Trichoptera (3010), Diptera (2868), Plecoptera (683), Odonata (435), Hemiptera (101), Megaloptera (91), Lepidoptera (64), Blattodea (2) (Table 3). The total abundance analysis (Figure 8) shows that Ephemeroptera, Coleoptera, Trichoptera, and Diptera were the most abundant groups, comprising 89.2 % of the collected macroinvertebrates.

Class/Order	Family	Genus	FFG	Site 1 Don Eladio	Site 2 La Argent.	Site 3 Chiquitín	Site 4 Hamaca	Site 5 Desembocadura	Site 6 Q.Merced	Total
Blattodea	Blaberidae	Epilampra		0	1	0	0	0	1	2
Coleoptera	Elmidae	Austrolimnius	CG	3	1	1	0	1	0	6
Coleoptera	Elmidae	Austrolimnius-at	CG	6	15	5	0	1	0	27
Coleoptera	Elmidae	Cylloepust	GC	0	0	0	3	4	3	10

Class/Order	Family	Genus	FFG	Site 1 Don Eladio	Site 2 La Argent.	Site 3 Chiquitín	Site 4 Hamaca	Site 5 Desembocadura	Site 6 Q.Merced	Total
Coleoptera	Elmidae	Cylloepus-a†	CG	13	16	10	2	5	0	46
Coleoptera	Elmidae	Heterelmis†	CG	46	50	26	42	21	3	188
Coleoptera	Elmidae	Heterelmis-a†	CG	206	87	6	3	0	17	319
Coleoptera	Elmidae	Disersus	CG	2	0	0	0	0	0	2
Coleoptera	Elmidae	Hexanchorus†	CG	105	142	10	40	2	2	301
Coleoptera	Elmidae	Hexanchorus-a†	CG	567	205	41	44	1	5	863
Coleoptera	Elmidae	Macrelmis†	CG	26	199	53	108	19	15	420
Coleoptera	Elmidae	Macrelmis-a†	CG	55	95	19	53	10	8	240
Coleoptera	Elmidae	Microcylloepus	CG	4	5	5	3	4	1	22
Coleoptera	Elmidae	Microcylloepus-a†	CG	17	53	1	5	8	6	90
Coleoptera	Elmidae	Neocylloepus	CG	0	0	0	0	0	2	2
Coleoptera	Elmidae	Neoelmis	CG	1	1	3	2	5	0	12
Coleoptera	Elmidae	Neoelmis-a†	CG	1	5	5	2	1	0	14
Coleoptera	Elmidae	Onychelmis	CG	0	0	0	0	1	0	1
Coleoptera	Elmidae	Phanocerus†	CG	29	27	4	20	10	6	96
Coleoptera	Elmidae	Phanocerus-a†	CG	32	9	0	13	10	6	70
Coleoptera	Elmidae	Pseudodisersus†	CG	1	1	0	0	0	0	2
Coleoptera	Elmidae	Stenhelmoides	CG	0	0	0	1	0	0	1
Coleoptera	Elmidae	Stenhelmoides-a	CG	0	0	1	0	0	0	1
Coleoptera	Gyrinidae	Gyretes-a	Pr	0	0	0	0	1	0	1
Coleoptera	Lutrochidae	Lutrochus	Sh	0	1	0	0	0	0	1
Coleoptera	Lutrochidae	Lutrochus-a	Sh	0	2	0	6	0	0	8
Coleoptera	Psephenidae	Psephenops-a	Sc	0	4	0	0	0	0	4
Coleoptera	Psephenidae	Psephenus†	Sc	52	128	87	67	13	12	359
Coleoptera	Ptilodactylidae	Anchytarsus†	Sh	4	15	10	6	2	7	44
Diptera	Blephariceridae	Paltostoma†	Sc	107	1	0	1	0	0	109
Diptera	Blephariceridae	Paltostoma-p†	Sc	2	3	0	0	0	0	5
Diptera	Chironomidae	Chironomus	CG	0	0	0	0	0	2	2
Diptera	Empididae	Hemerodromia†	Pr	1	8	4	7	2	1	23
Diptera	Psychodidae	Maruina†	Sc	263	274	142	24	0	0	703
Diptera	Psychodidae	Maruina-p†	Sc	0	6	0	0	0	0	6
Diptera	Simuliidae	Simulium†	Ft	36	588	442	244	412	130	1852
Diptera	Simuliidae	Simulium-p	Ft	0	5	6	0	0	0	11
Diptera	Tabanidae	Chrysops	Pr	0	0	0	0	0	1	
Diptera	Tipulidae	Hexatoma†	Pr	5	2	3	23	26	82	141
Diptera	Tipulidae	Hexatoma-p	Pr	0	0	3	0	0	2	5

Class/Order	Family	Genus	FFG	Site 1 Don Eladio	Site 2 La Argent.	Site 3 Chiquitín	Site 4 Hamaca	Site 5 Desembocadura	Site 6 Q.Merced	Total
Ephemeroptera	Baetidae	Americabaetis†	CG	3	4	13	197	512	65	794
Ephemeroptera	Baetidae	Baetodes†	Sc	116	246	127	62	17	28	596
Ephemeroptera	Baetidae	Camelobaetidius†	CG	91	179	198	84	21	8	581
Ephemeroptera	Baetidae	Cloeodes	CG	20	6	50	2	9	1	88
Ephemeroptera	Baetidae	Mayobaetis†	CG	119	96	0	0	0	0	215
Ephemeroptera	Caenidae	Caenis†	CG	0	0	0	0	2	4	6
Ephemeroptera	Heptageniidae	Stenonema	Sc	1	1	0	3	1	22	28
Ephemeroptera	Leptohyphidae	Asioplax†	CG	4	2	2	2	3	1	14
Ephemeroptera	Leptohyphidae	Epiphrades†	CG	0	1	0	11	54	17	83
Ephemeroptera	Leptohyphidae	Haplohyphes	CG	0	1	0	0	0	0	1
Ephemeroptera	Leptohyphidae	Leptohyphes†	CG	166	422	242	634	322	87	1873
Ephemeroptera	Leptohyphidae	Tricorythodes†	CG	143	44	74	74	80	64	479
Ephemeroptera	Leptohyphidae	Vacuperinus	CG	0	2	54	48	171	5	280
Ephemeroptera	Leptophlebiidae	Farrodes†	CG	51	40	51	123	223	135	623
Ephemeroptera	Leptophlebiidae	Hydrosmilodon	CG	0	1	0	0	0	0	1
Ephemeroptera	Leptophlebiidae	Terpides	CG	2	1	1	3	96	5	108
Ephemeroptera	Leptophlebiidae	Thraulodes†	CG	106	206	92	90	9	26	529
Hemiptera	Hebridae	Hebrus†	Pr	1	6	1	1	1	0	10
Hemiptera	Mesoveliidae	Mesovelia	Pr	1	6	4	0	1	2	14
Hemiptera	Naucoridae	Cryphocricos	Pr	0	1	0	0	0	4	5
Hemiptera	Naucoridae	Limnocoris	Pr	0	0	1	0	2	0	3
Hemiptera	Naucoridae	Limnocoris-a	Pr	0	0	0	1	1	0	2
Hemiptera	Ochteridae	Ochterus†	Pr	5	8	0	0	0	0	13
Hemiptera	Veliidae	Rhagovelia	Pr	0	17	2	5	24	6	54
Lepidoptera	Crambidae	Petrophila†	Sc	11	25	15	10	2	1	64
Megaloptera	Corydalidae	Chloronia†	Pr	2	0	0	0	2	0	4
Megaloptera	Corydalidae	Corydalus	Pr	13	27	5	36	4	2	87
Odonata	Calopterygidae	Hetaerina	Pr	9	9	6	18	35	11	88
Odonata	Coenagrionidae	Argia	Pr	21	19	39	22	44	22	167
Odonata	Coenagrionidae	Nehalenia†	Pr	0	0	0	0	2	1	3
Odonata	Corduliidae	Neocordulia	Pr	0	0	0	1	0	0	1
Odonata	Gomphidae	Agriogomphus	Pr	0	0	0	0	2	7	9
Odonata	Gomphidae	Desmogomphus†	Pr	4	2	0	0	0	0	6
Odonata	Gomphidae	Epigomphus	Pr	1	1	0	0	1	0	3
Odonata	Gomphidae	Erpetogomphus	Pr	1	1	2	1	4	24	33
Odonata	Gomphidae	Perigomphus	Pr	0	1	0	0	0	0	1

Class/Order	Family	Genus	FFG	Site 1 Don Eladio	Site 2 La Argent.	Site 3 Chiquitín	Site 4 Hamaca	Site 5 Desembocadura	Site 6 Q.Merced	Total
Odonata	Gomphidae	Phyllogomphoides	Pr	0	0	0	0	2	0	2
Odonata	Libellulidae	Brechmorhoga†	Pr	9	3	1	5	1	1	20
Odonata	Libellulidae	Dythemis†	Pr	0	0	0	2	15	4	21
Odonata	Libellulidae	Macrothemis†	Pr	0	7	0	0	3	0	10
Odonata	Libellulidae	Miathyria	Pr	0	0	0	0	1	0	1
Odonata	Megapodagrionidae	Heteragrion†	Pr	10	3	3	0	3	14	33
Odonata	Perilestidae	Perrisolestes†	Pr	0	0	0	0	1	1	2
Odonata	Platystictidae	Palaemnema	Pr	0	0	1	1	0	28	30
Odonata	Polythoridae	Cora†	Pr	1	3	1	0	0	0	5
Plecoptera	Perlidae	Anacroneuria†	Pr	226	258	24	22	5	148	683
Trichoptera	Anomalopsychidae	Contulma	Sc	1	0	0	0	0	0	1
Trichoptera	Calamoceratidae	Phylloicus	Sh	5	0	0	0	0	0	5
Trichoptera	Helicopsychidae	Cochliopsyche†	Sc	1	0	1	0	0	0	2
Trichoptera	Helicopsychidae	Helicopsyche	Sc	9	0	0	0	0	0	9
Trichoptera	Hydrobiosidae	Atopsyche	Pr	2	25	3	0	0	0	30
Trichoptera	Hydrobiosidae	Atopsyche-p†	Pr	0	1	0	1	0	0	2
Trichoptera	Hydropsychidae	Leptonema	Ft	11	43	13	124	37	153	381
Trichoptera	Hydropsychidae	Macronema	Ft	1	0	0	1	2	83	87
Trichoptera	Hydropsychidae	Smicridea	Ft	163	229	120	239	205	50	1006
Trichoptera	Hydropsychidae	Smicridea-p	Ft	0	1	0	0	0	0	1
Trichoptera	Hydroptilidae	Alisotrichia	Pc	0	0	0	4	0	0	4
Trichoptera	Hydroptilidae	Anchitrichia†	Pc	0	0	0	4	0	0	4
Trichoptera	Hydroptilidae	Anchitrichia-p	Pc	1	0	0	0	0	0	1
Trichoptera	Hydroptilidae	Brysopterix†	Pc	265	13	0	0	0	0	278
Trichoptera	Hydroptilidae	Brysopterix-p†	Pc	18	19	0	0	0	0	31
Trichoptera	Hydroptilidae	Leucotrichia	Pc	0	9	32	0	0	0	72
Trichoptera	Hydroptilidae	Leucotrichia-p	Pc	0	0	6	0	0	0	6
Trichoptera	Hydroptilidae	Ochrotrichia	Pc	411	2	0	0	0	0	413
Trichoptera	Hydroptilidae	Ochrotrichia-p	Pc	39	0	0	0	0	0	39
Trichoptera	Hydroptilidae	Oxyethira†	Pc	4	0	2	0	0	0	6
Trichoptera	Hydroptilidae	Oxyethira-p	Pc	2	0	0	0	0	0	2
Trichoptera	Hydroptilidae	Rhyacopsyche-p	Pc	0	73	12	1	0	0	86
Trichoptera	Leptoceridae	Atanatolica†	CG	20	5	0	0	0	0	25
Trichoptera	Leptoceridae	Atanatolica-p†	CG	1	0	0	0	0	0	1
Trichoptera	Leptoceridae	Nectopsyche	Sh	8	3	8	5	14	4	42
Trichoptera	Leptoceridae	Nectopsyche-p	Sh	9	5	5	0	0	5	24

Class/Order	Family	Genus	FFG	Site 1 Don Eladio	Site 2 La Argent.	Site 3 Chiquitín	Site 4 Hamaca	Site 5 Desembocadura	Site 6 Q.Merced	Total
Trichoptera	Leptoceridae	Oecetis	Pr	2	3	0	0	2	0	7
Trichoptera	Leptoceridae	Triplectides+	CG	2	0	0	0	0	2	4
Trichoptera	Philopotamidae	Chimarra+	Ft	22	116	79	130	15	89	451
Trichoptera	Polycentropodidae	Polycentropus+	Pr	5	2	2	0	0	0	9
Trombidiformes	Various families	Various genera	Pr	12	2	2	3	0	2	21
Decapoda	Atyidae	Atya	CG	3	2	1	1	1	1	9
Decapoda	Palaemonidae	Macrobrachium	Pr	7	6	3	4	3	2	25
Decapoda	Pseudothelphusidae	Pseudothelphusa	CG	2	1	1	1	1	0	6
Gastropoda	Ampullaridae	Pomacea	Sc	0	0	0	0	0	1	1
Gastropoda	Ancylidae	Gundlachia	Ft	0	0	2	0	0	0	2
Gastropoda	Hydrobiidae	Aroapyrgus	Sc	10	8	20	24	4	311	377
Gastropoda	Physidae	Gen. indet.	Sc	0	0	12	0	0	0	12
Gastropoda	Thiaridae	Melanoides	Sc	0	0	1	1	9	0	11
Lumbriculida	Lumbriculidae	Gen. indet.	CG	0	0	2	0	0	0	2
Tricladida	Planariidae	Gen. indet.	Pr	0	4	4	2	0	0	10
Total				3757	4170	2227	2722	2528	1759	17163

Table 3. Total number of individuals collected (abundance) per genera and its different life-forms along the studied sites (a=adult, p=pupa, no sign=larva, + taxa considered for the genera PLS analysis). Functional feeding group (FFG) categories: CG=Collector-Gatherers, Ft=Filterers, Pc=Piercers, Pr=Predators, Sc=Scrapers, and Sh=Shredders.

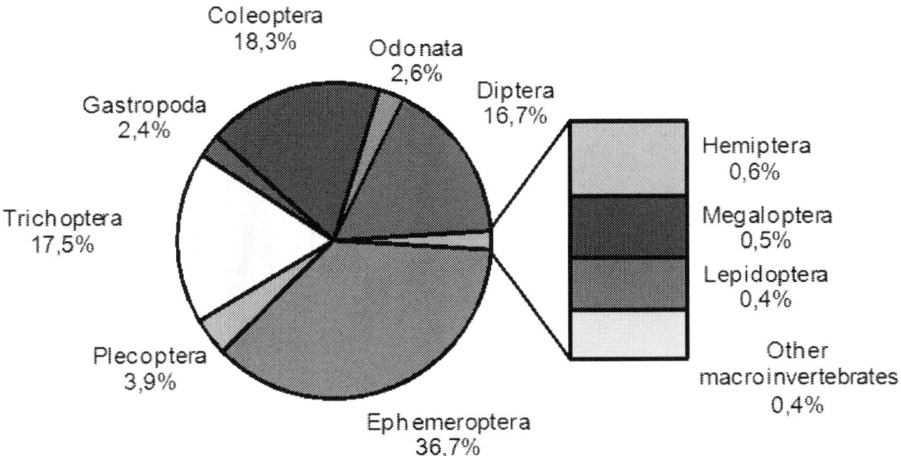

Figure 8. Graph indicating the percentage taxonomic composition of the total study sample of all six collecting sites.

3.3. Comparison between sampling sites with different agrorural influence

If we consider the mean total number of collected individuals per site, clear abundance differences are evinced (Figure 9). The greatest mean total numbers, with almost 300 individuals collected per sampling date, are found in the two least impacted sites (ANOVA, DGC-test; p>0.05),the reference (site 1) and the livestock-pasture site (site 2), whereas the sites under more intense human influence present statistically significantly reduced mean total values (around 150 individuals collected per month).

Table 4 presents a family, genera, and EPT richness analysis of the different collecting sites, as well as the mean BMWP-CR values and resulting water quality. The total family and genera richness does not show significant frequency differences between sampling sites, although the highest number of genera were found at the first two sites, with over 60 genera. The site with the lowest taxonomic richness, both on family and genus level, was site 5, close to the river mouth. The EPT taxa richness was highest at site 1 (14 EPT taxa), decreased downstream towards only nine EPT taxa at sites 5 and 6, although no statistically significant frequency differences were found using the Chi-square test. The BMWP'-CR index shows statistically significant differences (ANOVA, DGC-test; p>0.05), where the index values diminish according to the following site groupings: (sites 1-2)-(site 6)-(sites 3-4)-(site 5). The water quality, indicated by the mean BMWP'-CR index falls into the categories "good quality" (sites 1-2) and "regular quality" (sites 3-6).

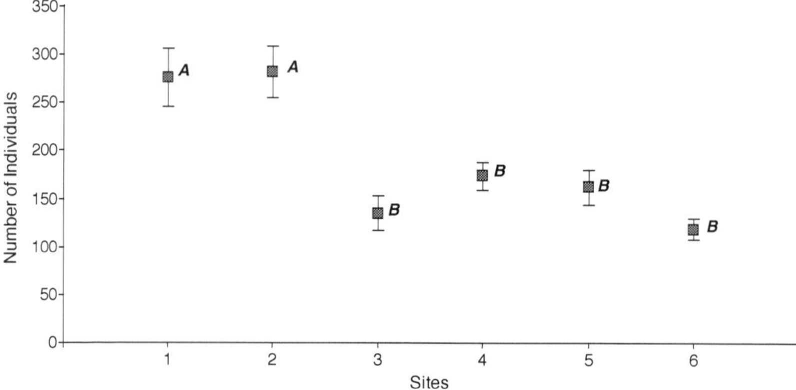

Figure 9. Comparison of the mean of the total number of collected individuals in the six sampling sites. Means with the same letters are not significantly different (ANOVA, DGC-test; p>0.05).

On the other hand, a very clear change in the structure of macroinvertebrate assemblages can be observed along the different sampling sites of the Dos Novillos River (Figure 10). At the first site, the undisturbed sampling point, beetles are significantly more abundant than at the other sites (Chi-square, p<0.0001), which are all under the influence of human impact. The last

Sites	Site 1 Don Eladio	Site 2 La Argentina	Site 3 Chiquitín	Site 4 La Hamaca	Site 5 Desembocadura	Site 6 Q. Mercedes
Families	37	39	40	36	33	35
Genera	63	67	53	52	45	53
EPT	14	11	11	10	9	9
BMWP'-CR	113.8±20.6[a]	119.8±16.3[a]	77.1±12.0[c]	83.9±8.4[c]	68.8±15.2[d]	93.6±13.2[b]
Water quality	good	good	regular	regular	regular	regular

Table 4. Total number of families, genera richness, EPT richness (Ephemeroptera-Plecoptera-Trichoptera), and mean values for the BMWP'-CR index with resulting biological water quality in the different collecting sites. No statistically significant frequency differences were found using the Chi – square test for families, genera and EPT richness (ANOVA, DGC-test, heterocedastic model for the BMWP analysis, p>0, 05).

site (site 6) shows a significantly greater abundance of Gastropoda, Odonata, and Plecoptera (Chi-square, p<0.0001). Here the banana packing plant discharges effluents, carrying small banana pieces and other suspended organic material, into the water of the stream. Also, a tendency of an increase in mayfly (Ephemeroptera) abundance can be observed towards sites 4 and 5, while caddisfly abundance is relatively steady throughout all sampling sites.

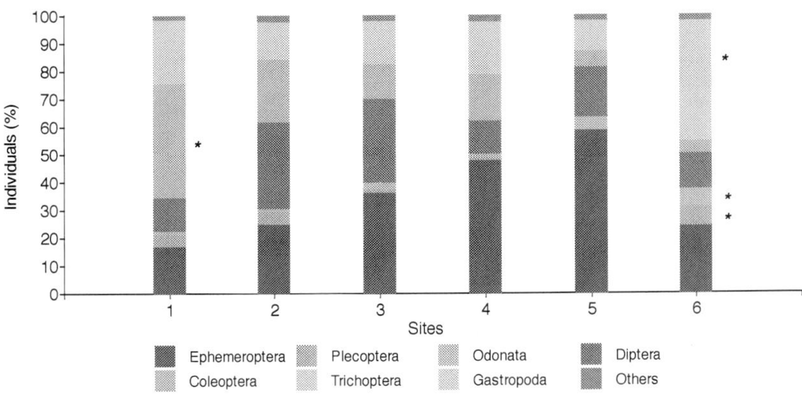

Figure 10. Percentage abundances of the seven most common taxonomic groups at the six different sampling sites. Groups marked with an asterisk (*) have statistically significant frequency differences (Chi-square; p<0.0001).

3.4. Functional feeding group analysis

The FFG analysis is presented in Figure 11 (Chi-square; p<0.0001). Collector-gatherers are the dominant group at all sites along the main river, with around 50% of all individuals collected; at Quebrada Mercedes (site 6), they are the second largest group after the filter-feeders. Filter-feeders show also high percentages at each site, with exception of site 1 (reference site), where they have statistically significant lower relative frequencies. One the other hand, piercers have a statistically significant greater relative frequency at this site, compared to the rest of the sampling sites. Finally, site 6 has a statistically significant lower relative frequency of collector-gatherers and greater relative frequencies of predators, scrapers, and shredders.

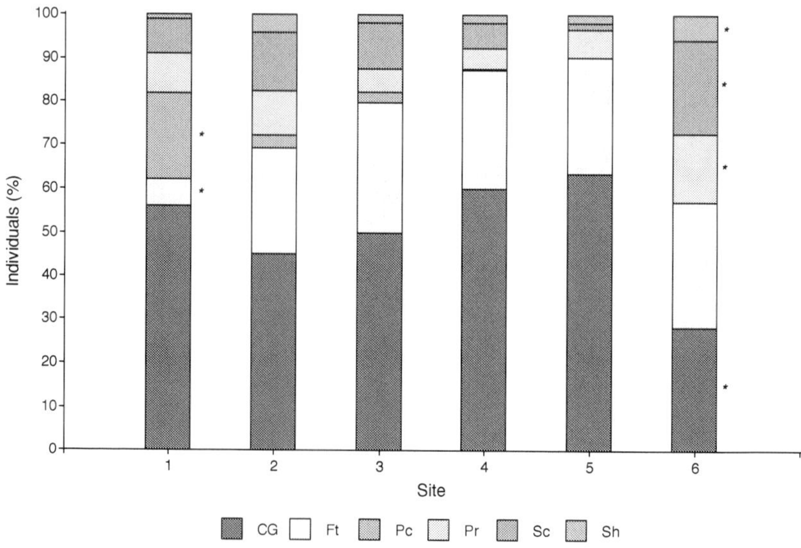

Figure 11. Percentage abundances of the functional feeding groups (FFG) at the six different sampling sites. Groups marked with an asterisk (*) have statistically significant frequency differences (Chi-square; p<0.0001). CG=Collector-Gatherers, Ft=Filterers, Pc=Piercers, Pr=Predators, Sc=Scrapers, and Sh=Shredders.

The FFG abundance PLS (Figure 12) presents high variation explanatory values, with factor 1 explaining 39.1 % of the total variance and factor 2 explaining 25.4 %. The variables' temperature, as well as the piercer and predator abundances, mainly separates the different sites on the horizontal projection, as does the shredder abundance on the vertical projection. The FFG most closely allied to the reference site, is piercer abundance. On the other hand, the most closely allied variables to the agriculturally impacted areas, sites 5 and 6, are the variables turbidity, conductivity, and NO_3.

3.5. Relationship between macroinvertebrate assemblages, physical-chemical parameters and sampling sites

The macroinvertebrate abundance partial least square analysis (Figure 13) presents high variation explanatory values, with factor 1 explaining 42.0 % of the total variance and factor 2 explaining 29.9%. The variable NO_3, and to a lesser degree conductivity and turbidity, separate mainly the different sites on the horizontal projection, as does suspended solids on the vertical

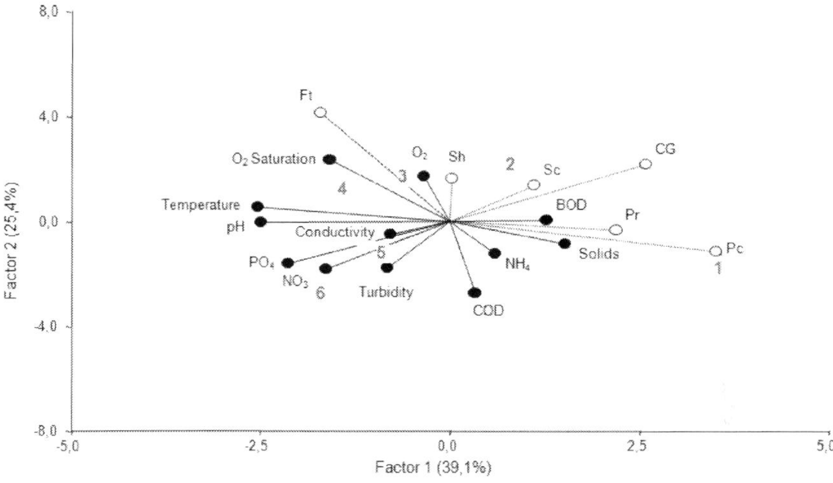

Figure 12. Site ordination (1-6) with a triplot PLS analysis using physical-chemical values as independent variables and functional feeding group abundances as dependent variables. CG=Collector-Gatherers, Ft=Filterers, Pc=Piercers, Pr=Predators, Sc=Scrapers, and Sh=Shredders.

one. The genera most closely allied to the reference site are: *Brysopterix*, *Heterelmis*, and *Paltostoma*. On the other hand, the most closely allied genera to the agriculturally impacted areas, sites 5 and 6, are larvae of the genera *Caenis*, *Dythemis*, *and Hexatoma*. Interestingly, the Spearman rank analysis has resulted in a plethora of correlations. The strongest physical-chemical/biological correlations are presented in Table 5.

Variable (1)	Variable (2)	Spearman rank	p-value
pH	Heterelmis	-0.99	0.0003
NO$_3$	Maruina	-0.99	0.0003
Temperature	Phanocerus	-0.97	0.0012
Turbidity	Atopsyche	-0.94	0.0051
Turbidity	*Austrolimnius*-a	-0.93	0.0077
NO$_3$	Farrodes	0.93	0.0077
O$_2$ Saturation	Hemerodromia	0.93	0.0077
NO$_3$	Piercers	-0.93	0.0077
PO$_4$	Collector-Gatherers	-0.93	0.0080
NO$_3$	Cora	-0.93	0 0080

Variable (1)	Variable (2)	Spearman rank	p-value
Turbidity	Cora	-0.93	0.0080
PO_4	*Cylloepus*-a	-0.93	0.0080
PO_4	*Macrelmis*-a	-0.93	0.0080
PO_4	pH	0.93	0.0080
NO_3	PO_4	0.93	0.0080

Table 5. Statistically significant associations of the Spearman rank correlation coefficient between physical-chemical, genera, and functional feeding groups (a=adults, p=pupae, genera with no lettering=larvae).

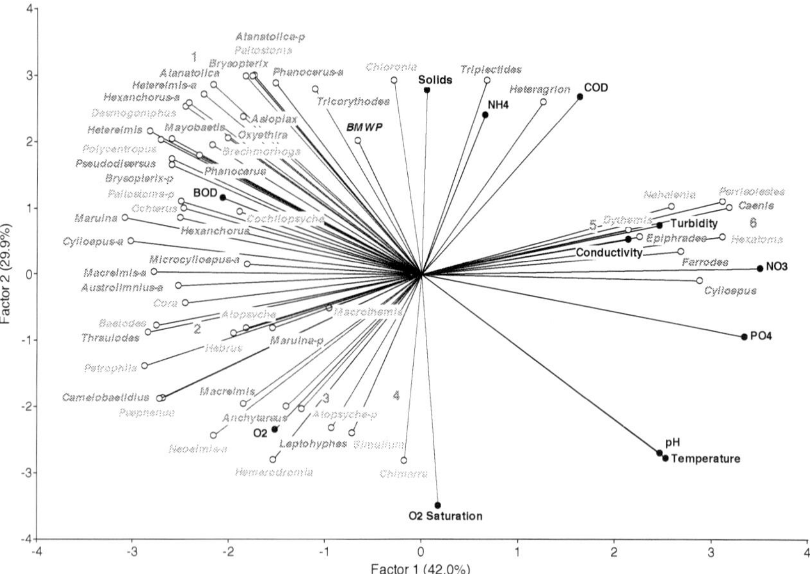

Figure 13. Site ordination (1-6) with a triplot PLS analysis, using physical-chemical values as independent variables, and macroinvertebrate generic abundance and the BMWP index as dependent variables (FFG categories: purple=collector-gatherers, yellow=filterers, green=piercers, grey=predators, red=scrapers, blue=shredders).

The Procrustes analysis produced a good site ordination consensus based on macroinvertebrate abundances, FFG, and the physical-chemical variables. The first axis explains 45.8 % of the variance and the second axis explains 33.8 % (Figure 14). The proportional consensus percentages are also good, ranging from 76 % to 92.4 %, with a mean value of 82.9% (Table 6). One can conclude that the site ordination has a good congruence with the biological and physical-chemical data sets.

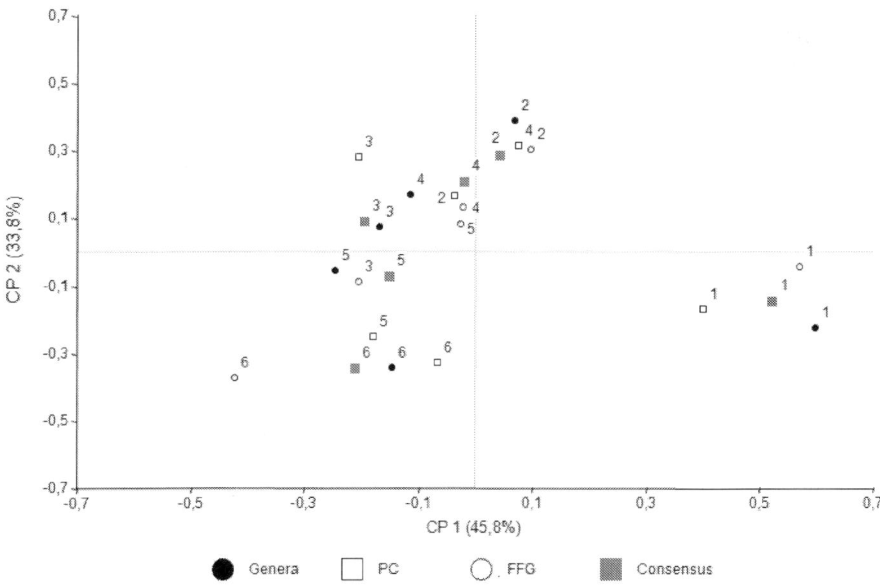

Figure 14. Site ordination configuration congruence according to physical-chemical values (PC), macroinvertebrate and functional feeding group (FFG) abundances (Genera) using a Procrustes analysis.

Variable	Proportional consensus (%)
Genera	92, 4
FFG	78, 3
PC	76, 0
Mean	82, 9

Table 6. Proportional consensus percentages as displayed by the configurations generated between macroinvertebrate generic abundances, physical-chemical values (PC), and functional feeding groups (FFG) with the generalized Procrustes ordination.

4. Discussion

An important consideration for the present study is the ability to distinguish between natural variability and human impacts [49]. In the present case, all sites are located along the same river, spanning only a small distance of 10 km (Figure 1). Moreover, elevation, stream size, and surface geology are relatively similar (Table 1) in order to properly assess human impacts and reduce natural gradients.

4.1. Diversity and composition of macroinvertebrate assemblages

The present study recorded the existence of 16 macroinvertebrate orders. In an analysis undertaken by Castillo *et al.* [17] very near to the present collecting localities, the authors reported the existence of 15-16 macroinvertebrate orders in their reference sites and 12-16 orders in the banana plantation sites. The results of the present study fall within the order range for similar studies in nearby regions. The present analysis resulted in 53 collected families (Table 3). This number compares well with the number of families collected by Lorion and Kennedy [27] and O'Callaghan and Kelly-Quinn [58] in Costa Rican and Honduran neotropical rivers, who reported 56 and 60 families, respectively. The present study also identified a total of 98 genera (Table 3). Montoya Moreno *et al.* [59] reported 69 genera for the Negro River in Colombia; whereas, Sánchez Argüello *et al.* [59] reported 96 genera in their study in Panama. Considering that some groups, such as water mites, were not identified to genus level in the present study, the total number of genera present in the Dos Novillos river and its tributaries is likely to be over 100 genera, reflecting a very high taxa richness.

It is interesting to compare the dominance results obtained in this study with analyses made in neighbouring areas under similar ecological conditions. Ramírez *et al.* [33] sampled the Carbón and Gandoca Rivers and found the following dominance gradient for total abundances: Ephemeroptera-Diptera-Trichoptera-Odonata. Castillo *et al.* [17], sampling on sites very near to the present ones, found the following order dominance in their reference sites: Ephemeroptera-Trichoptera-Coleoptera and the following one in banana plantation sites: Ephemeroptera-Diptera-Coleoptera-Gastropoda-Trichoptera. Lorion and Kennedy [27] studied several streams in the Sixaola River Valley. Considering the total number of individuals, they found a diminishing abundance gradient as follows: Ephemeroptera-Diptera-Coleoptera-Odonata. Gutiérrez-Fonseca and Ramírez [50] have reported at La Selva Biological Station, in a 15-year study, the following dominance sequence in unpolluted streams: Diptera-Trichoptera-Odonata. The present study (Figure 8) has found the following abundance dominance order: Ephemeroptera-Coleoptera-Trichoptera-Diptera. It would seem from these results that dominance sequences might vary depending on several factors dependent on the collecting site, such as substrate, current, and water quality, but also can be a result of different sampling device and mesh size [37]. However, Ephemeroptera would appear to be the most constant dominant group in most lowland rivers and streams in Costa Rica.

4.2. Comparison between sampling sites with different agrorural influence

The analysis of abundance differences demonstrates that less impacted sites clearly show statistically significantly higher abundances than sites under stronger human influence (Figure 9). Similarly, Paaby *et al.* [28] and Lorion and Kennedy [27] detected greater abundances in forested areas versus pastures under neotropical conditions; whereas, the study by Ramírez *et al.* [33] detects greater abundances in Costa Rican tropical riffle habitats than in other habitats.

Taxonomic composition also varies in a clear way along the sampling sites under the influence of different ecological impacts (Figure 10). The reference site, arguably not under the influence of human impact, presented a significantly greater relative abundance of Coleoptera. This is due to the high amount of individuals collected from the riffle beetle family (Elmidae), which are especially diverse and abundant in well oxygenized rivers and streams in forested areas (Springer, *unpubl*). A study of rivers in the Guanacaste area in Costa Rica (Kohlmann, *unpubl.*) has also found very high numbers of Coleoptera in unpolluted rivers. At the other end of the scale, the banana packing plant discharge, carrying banana debris and showing statistically significantly higher values of conductivity, turbidity, and NO_3, presents statistically significantly high numbers of Gastropoda, Odonata, and Plecoptera. This is very interesting because stoneflies (Plecoptera) have always been considered as good indicators of oxygenated, clean, and cool running waters [51-54]. Additionally, the triplot (Figure 12) and Spearman rank correlation analyses show a strong negative correlation between stonefly (*Anacroneuria*) abundance and temperature values (Spearman rank correlation value=-0.89, p=0.0476). Gutiérrez and Springer [55] reported the widespread species *A. holzenthali* from coffee plantations in Costa Rica. Tomanova and Tedesco [56], as well as Thorp and Covich [57] also indicated that stonefly presence is not necessarily a sure sign of water cleanliness. The results from these analyses seem to support this claim.

The EPT taxa richness (Table 4) showed a steady decline along the collecting sites, following an increasing anthropogenic impact trend, similar to what Lorion and Kennedy [27] reported following a forest-forest buffer-pasture gradient. The number of families per site (Table 4) did not vary much (33-40) along the collecting transect in this study (Table 4). An analysis by Castillo *et al*. [17] reported a family number variation going from 39 to 47 families. Similarly, their family numbers did not vary much between their reference (46-47) and their banana plantation collecting sites (39-46). Some families (Table 3) were restricted to only a single site (site number in parentheses), like (1): Anomalopsychidae, Calamoceratidae; (3): Ancylidae, Lumbriculidae, Physidae; (4): Corduliidae; (5): Gyrinidae; (6): Ampullaridae, Chironomidae, and Tabanidae. Certain genera (Table 3) appeared only once in a specific site: (1): *Chloronia, Contulma, Dythemis, Hemerodromia, Ochterus, Palaemnema,* and *Polycentropus*; (2): *Hebrus, Leptohyphes, Mayobaetis,* and *Psephenops*; (3) *Stenonema*; (4): *Anchitrichia* and *Neocylloepus*; (5): *Haplohyphes*; and (6): *Chironomus, Chrysops*. These taxa would seem to be closely associated with the ecological conditions of each site, *e.g.* Ampullaridae, Chironomidae, and Tabanidae in a banana packing-plant effluent with a high organic waste discharge versus Anomalopsychidae and Calamoceratidae in a forested undisturbed condition. Interestingly, the number of families and genera do not show the same trend among sampling sites. The reference and slightly disturbed sites (site 1 and 2, respectively) show a higher number of genera (over 60), in comparison to the more influenced sites (with around 40 to 50 genera), even though the number of families were similar at the latter, and even higher in one case (site 3). These results suggest that family richness is not necessarily an adequate indicator for biomonitoring, and generic identification is necessary to achieve results that are more reliable.

The BMWP index in its different variations has been popularly employed in Latin America. These studies usually have found this index to be satisfactory for reflecting water quality [10,

12, 22, 60-63], especially the Costa Rican adaptation [58]. Sánchez Argüello *et al.* [61] undertook in their Panamanian study a comparison between the Colombian and Costa Rican adaptations of the BMWP index and found the latter to be more unforgiving in its water quality evaluation. Rizo-Patrón *et al.* [22] undertook an analysis using the BMWP index modified for Costa Rica, studying the environmental impact caused by conventional and organic-irrigated rice fields on the macroinvertebrate communities. Their BMWP'-CR results show that the index values were greater in the organically irrigated rice fields. On the other hand, Fenoglio *et al.* [64] recommend, from their experience in Nicaragua, the use of the Indice Biotico Esteso [65] because of its ease of use and low cost. However, these studies have mostly assessed the comparative performance of the various indices; no attempts were made to correlate the BMWP index to specific physical-chemical variables.

The results of the BMWP'-CR index for the present analysis (Table 2) do indeed show a discriminating capacity of the index following a diminishing environmental quality site trend, especially under agricultural impact conditions (site 5), but it also shows a tendency of reporting a higher value when in river waters with high organic pollution (site 6). Interestingly, there is a statistically significant negative correlation of the BMWP'-CR index with temperature, not with pollution variables, as one could expect from the general assumption that the BMWP index reflects organic water pollution quality. These results generate some doubts about the reliability of the BMWP'-CR index as an environmentally representative tool, as the following studies indicate. Sermeño Chicas *et al.* [66] tried to implement the BMWP-CR index in El Salvador where rivers showed consistently high organic pollution conditions that were not reflected by the BMWP index [67]. In their analysis of selected macroinvertebrate-based biotic indices in Honduras, O'Callaghan and Kelly-Quinn [58] found that a BMWP-CR-based version of the ASPT index performed much better than the aforementioned index. Without doubt, more studies will be necessary in order to adjust the biotic indices used for aquatic biomonitoring in Costa Rica and Central America according to the different ecoregions.

4.3. Functional feeding group analysis

FFG relative abundances also change significantly depending on the human impact conditions on the quality of river water. It would seem that under undisturbed conditions filterers' relative abundances tend to be minimal, their increase at disturbed sites might be a result of higher dissolved organic matter. In this study, under conditions of high organic pollution, shredders, scrapers, and predators tend to have maximal relative abundances while collector-gatherers tend to have minimal values. Finally, filter feeders seem to react positively to high concentrations of dissolved O_2 (Table 5), which is positively correlated with fast flowing waters, a condition that also favors the feeding mechanism of filterers.

The taxonomic grouping triplot analysis (Figure 13) suggests a correlation between the reference site and the piercers, and it would appear that the first axis is characterized by a piercers' abundance gradient, diminishing from the reference community (site 1) to the high organic waste discharge sites (site 6). In the present study, piercers are mainly represented by the caddisfly family Hydroptilidae, which is especially abundant in the splash zone of big rocks in riffle areas, which were characteristic for the reference sampling site. The second axis appears to be characterized by filterers, arranging the different agrorural ecosystems along a diminishing abundance gradient.

The FFG triplot analysis (Figure 12) supports a strong agrorural ecosystem ordination process mediated by the abundance of piercers along the main axis as suggested in the previous triplot analysis; however, as this analysis benefits from having more information (127 taxa versus 58 taxa), it also evinces the importance of predators and shredders as relevant ecosystem ordinating biological variables. The strong negative correlation between the piercer's abundance and NO_3 and PO_4 values (Table 5) stresses again the importance of the piercers as an ecosystem characterizing variable, although the presence of suitable microhabitats might be another important factor to consider.

4.4. Relationship between macroinvertebrate assemblages, physical-chemical parameters and sampling sites

The taxonomic grouping triplot analysis (Figure 13) and the Spearman rank correlation analysis (Table 5) show the existence of several genera and FFG that are highly correlated with physical-chemical variables and that possibly could be used as surrogates for these variables. The larvae of *Maruina* showed one of the strongest correlations with NO_3, although other genera like *Farrodes* and *Cora,* and the piercers´ functional feeding group were also significantly correlated with this chemical variable. Species of the genus *Caenis* have been found regularly in organically enriched streams [48]. Of the statistically significant genus list (Table 5), only *Heterelmis* and *Farrodes* already had been cited before as good quality bioindicators for toxicity and pollution-sensitivity testing by Castillo *et al.* [17] and Rizo-Patrón *et al.* [22], respectively, in a similar type of analysis.

Finally, the Procrustes analysis allows an assessment of the goodness of fit of the taxonomic, FFG, and physical-chemical analyses (Figure 14, Table 6). The analysis resulted in a good match of the collecting sites (landmarks) with the values derived from the three blocks of variables, indicating that any one of the three is describing in a similar way the ecology of each study site. The consensus values indicated in Table 6 show a very good concordance in this study between the different variable blocs and the consensus values, where generic abundances seem to generate a better ecological ordination. An environmental impact gradient also becomes very apparent on the first ordination axis, ranging from the undisturbed reference site inside the forest on the right of the graph to the banana packing-plant effluent on the left.

5. Conclusions

The present study clearly shows that tropical river macroinvertebrate diversity changes and at the same time characterizes and defines different river ecosystem conditions under various agrorural impacts. Changes in taxonomic composition and functional feeding group structure are very indicative of ecosystem function.

Ephemeroptera seem to be, in general, a rather constant and numerous group, present in the great majority of collections in neotropical rivers. However, high relative abundances of Coleoptera, especially from the family Elmidae, seem to be indicative of unpolluted conditions in tropical rivers; whereas, high relative abundances of Gastropoda, Odonata, and Plecoptera show up in sites with relatively high (plant-derived) organic pollution, although with well-oxygenized waters and forested stream margins. Addtionally, high numbers of individuals were found in unpolluted or slightly polluted sites; whereas, lower abundances were found in sites under human impact (town and fruit packing plant discharges, agricultural plantations).

Especially illustrative is the change in structure of functional feeding groups. Piercers showed the highest relative abundance in unpolluted sites and seem to be especially sensitive to human impacts because they quickly disappear under altered conditions, most probably reacting to the decrease of their microhabitat and food items. Filterers have the lowest relative abundance under unpolluted conditions and quickly become relatively more abundant under human impact conditions, most probably reflecting an increase in particles in the water column. At the other end of the spectrum, under high (plant-derived) organic pollution, shredders and scrapers show their highest relative abundance concomitantly with an increase in particulate organic matter, and probably as a response to it. Predators here show their highest relative abundance as well and at the same time, collector-gatherers here show their lowest relative abundance. It would seem then that predation, as well as scraping and shredding, increases significantly in river areas with high (plant-derived) organic pollution.

The biomonitoring analysis presented in this chapter used an adaptation of the BMWP index to Costa Rica. In all cases, the method revealed or indicated the existence of an anthropogenic/agricultural impact gradient going from the unpolluted site to the most perturbed locality. ANOVA tests evidenced the fact that the BMWP index has enough sensitivity and discriminating power to detect changes in macroinvertebrate biodiversity, which can be translated into statistically significant differences between sampling sites. The method determined the existence of changes in macroinvertebrate communities associated with agricultural areas, even when analytical methods could not detect the presence of pesticides in river water. A previous analysis of the BMWP score by Pinder and Farr [68, 69] found that it was significantly negatively correlated only with dissolved organic carbon. The present study detects a strong negative correlation between the BMWP score and temperature (Figure 12, Table 5). Due to its simplicity, speed of use, efficiency, and cheapness, this method shall undoubtedly continue to be a very popular one in the future. The BMWP is considered an extremely successful index according to Spellerberg [70]. However, there are some doubts about its suitability as a tool for detecting organic pollution in some regions of Latin America, especially if one considers that in this case the index was more sensitive to the impact of a banana plantation than the one caused by relatively high (plant-derived) organic pollution. The PLS/Procrustes analysis seemed, on this occasion, to be a more suitable method for describing and evaluating anthropogenic/agricultural environmental impacts.

The use of multivariate ordination for environmental studies is becoming more and more common. In particular, the PLS/Procrustes analyses represent a very powerful combination tool as they not only perform a site ordination but different taxa and environmental variables can be correlated at the same time. This makes this method extremely useful for taxa, physical-chemical, and FFG variables' correlations as specific environmental variable surrogates. Castillo *et al.* [17] and Rizo-Patrón *et al.* [22] found similar promising results for the study of agricultural ecosystem analyses. Castillo *et al.* [17] indicated, based on their results, that multivariate analyses are more sensitive in distinguishing pesticide effects than toxicity tests. Therefore, multivariate analyses should be incorporated as an approach for future ecosystem/ biodiversity analyses.

Acknowledgements

We would like to thank Mildred Linkimer for her help in the design of some figures; as well as Roxana Araya for her help in the composition of tables and lists. Finally yet importantly, we are also indebted to Jane Yeomans and Kent McLeod for critically revising the text and to Pablo Gutiérrez for useful suggestions in organizing the results. This material is based upon work supported by the Department of Energy [National Nuclear Security Administration] under Award Number DE FG02-04ER 63856.

Author details

Bert Kohlmann[1]*, Alejandra Arroyo[1], Monika Springer[2] and Danny Vásquez[3]

*Address all correspondence to: bkohlman@earth.ac.cr

1 EARTH University, San José, Costa Rica

2 School of Biology, University of Costa Rica, San José, Costa Rica

3 Private Consultant, San José, Costa Rica

References

[1] Valerio CE. Costa Rica. Ambiente y Sociedad. Santo Domingo de Heredia: Instituto Nacional de Biodiversidad; 1999.

[2] Obando V. Biodiversidad de Costa Rica. Estado del Conocimiento y Gestión. Santo Domingo de Heredia: Instituto Nacional de Biodiversidad; 2002.

[3] Jiménez J. Training parataxonomists and curators to help conservation: the biodiversity inventory. In: Bisonette, JA. and Krausman, PR. (eds.). Integrating people and wildlife for a sustainable future: proceeding of the First International Wildlife Congress. Bethesda: The Wildlife Society; 1995. p165-167.

[4]   Kohlmann B, Mitsch WJ, Hansen DO. Ecological Management and Sustainable Development in the Humid Tropics of Costa Rica. Ecological Engineering 2008; 34(4): 254-266.

[5]   Jones JR. Colonization and Environment: Land Settlement Projects in Central America. Tokio: The United Nations University; 1990.

[6]   Seligson M. Peasants of Costa Rica and the Development of Agrarian Capitalism. Madison: University of Wisconsin Press; 1980.

[7]   de Oñoro MT. El Asentamiento Neguev. Atlantic Zone Programme. Inf. Técnico No. 162, Programme Paper No. 7. Turrialba: Centro Agronómico Tropical de Investigación y Enseñanza; Universidad Agrícola de Wageningen; Ministerio de Agricultura y Ganadería; 1990.

[8]   Roldán G., Builes J., Trujillo CM., Suárez A. Efectos de la Contaminación Industrial y Doméstica sobre la Fauna Béntica del Río Medellín. Actualidades Biológicas 1973; 2(4): 54-60.

[9]   Roldán G. Los Macroinvertebrados y su Valor como Indicadores de la Calidad del Agua. Revista de la Academia Colombiana de Ciencias Exactas, Físicas y Naturales 1999; 23(88): 375-387.

[10] Roldán G. Bioindicación de la Calidad del Agua en Colombia. Antioquia: Universidad de Antioquia; 2003.

[11] de la Lanza Espino G. Criterios Generales para la Elección de Bioindicadores. In: de la Lanza Espino G., Hernández Pulido H., Carbajal Pérez JL. (eds.). Organismos Indicadores de la Calidad del Agua y de la Contaminación (Bioindicadores). México, D.F.: Plaza y Valdés, S.A de C.V.; 2000. p17-41.

[12] Prat N., Ríos B., Acosta R., Rieradevall M. Los Macroinvertebrados como Indicadores de la Calidad de las Aguas. In: Domínguez E., Fernández HR. (eds.). Macroinvertebrados Bentónicos Sudamericanos: Sistemática y Biología. Tucumán: Fundación Miguel Lillo; 2009. P631-654.

[13] Springer M. Biomonitoreo Acuático. In: Springer M., Ramírez A., Hanson P. (eds.). Macroinvertebrados de Agua Dulce de Costa Rica I. Revista de Biología Tropical 2012; 58(4): 53-59.

[14] Castillo LE. Pesticide impact of intensive banana production on aquatic ecosystems in Costa Rica. PhD thesis. Stockholm University: Stockholm; 2000.

[15] Castillo LE., De la Cruz E., Ruepert C. Ecotoxicology and Pesticides in Tropical Aquatic Ecosystems of Central America. Environ. Toxicol. Chem. 1997; 16: 41-51.

[16] Castillo LE., Ruepert C., Pinnock M. Acute toxicity and pesticide residues in surface waters in a pineapple growing area of Costa Rica. 10th International Symposium on Toxicity Assessment, Quebec City. 2001.

[17] Castillo E., Martínez E., Ruepert C., Savage C., Gilek M., Pinnock M., Solís E. Water Quality and Macroinvertebrate Community Response Following Pesticide Applications in a Banana Plantation, Limón, Costa Rica. Science of the Total Environment 2006; 367: 418-432.

[18] Martínez E. Utilización de organismos acuáticos macro-bentónicos en la determinación de la calidad de las aguas naturales en los arrozales de Bagatzí, Guanacaste. MSc thesis. Universidad Nacional: Heredia; 1998.

[19] Monge P., Partanen T., Wesseling C., Bravo V., Ruepert C.B.I. Assessment of Pesticide Exposure in the Agricultural Population of Costa Rica. Ann. Occup. Hyg. 2005; 49(5): 375-384.

[20] Fernández L., Springer M. El Efecto del Beneficiado del Café sobre los Insectos Acuáticos en Tres Ríos del Valle Central (Alajuela) de Costa Rica. Revista de Biología Tropical 2008; 56(4): 237-256.

[21] Rizo-Patrón, F. Estudio de los arrozales del proyecto Tamarindo: agroquímicos y macroinvertebrados bentónicos en relación al Parque Nacional Palo Verde, Guanacaste, Costa Rica. MSc thesis. Universidad Nacional: Heredia; 2003.

[22] Rizo-Patrón F., Kumar A., McCoy Colton MB., Springer M., Trama FA. Macroinvertebrate Communities as Bioindicators of Water Quality in Conventional and Organic Irrigated Rice Fields in Guanacaste, Costa Rica. Ecological Indicators 2013; 29: 68-78.

[23] Springer M., Vásquez D., Castro A., Kohlmann B. Field-guide for water quality bioindicators. San José: EARTH-UCR; 2007.

[24] Vásquez D., Springer M., Castro A., Kohlmann B. Bioindicadores de la Calidad del Agua: Cuenca del Río Tempisque. San José: Universidad EARTH; 2010.

[25] Boyero L., Bosch J. Spatial and Temporal Variation of Macroinvertebrate Drift in Two Neotropical Streams. Biotropica 2002; 34(4): 567-574.

[26] Boyero L., Bosch J. Multiscale Spatial Variation of Stone Recolonization by Macroinvertebrates in a Costa Rican Stream. Journal of Tropical Ecology 2004; 20(1): 85-95.

[27] Lorion CM., Kennedy BP. Relationship between Deforestation, Riparian Forest Buffers and Benthic Macroinvertebrates in Neotropical Headwater Streams. Freshwater Biology 2009; 54(1): 165-180.

[28] Paaby P., Ramírez A., Pringle CM. The Benthic Macroinvertebrate Community in Caribbean Costa Rican Streams and the Effect of Two Sampling Methods. Rev. Biol. Trop. 1998; 46(Supl. 6): 185-199.

[29] Pringle C.M., Ramírez A. Use of Both Benthic and Drift Sampling Techniques to Assess Tropical Stream Invertebrate Communities along an Altitudinal Gradient, Costa Rica. Freshwater Biology 1998; 39(2): 359-373.

[30] Ramírez A., Pringle CM. Invertebrate Drift and Benthic Community Dynamics in a Lowland Neotropical Stream, Costa Rica. Hydrobiologia 1998a; 11(1-3): 19-26.

[31] Ramírez A., Pringle CM. Structure and Production of a Benthic Insect Assemblage in a Neotropical Stream. J. N. Am. Benthol. Soc. 1998b; 17(4): 443-463.

[32] Ramírez A., Pringle CM. Spatial and Temporal Patterns of Invertebrate Drift in Streams Draining in a Neotropical Landscape. Freshwater Biology 2001; 46(1): 47-62.

[33] Ramírez A., Paaby P., Pringle M., Agüero G. Effect of Habitat Type on Benthic Macroinvertebrates in Two Lowland Tropical Streams, Costa Rica. Rev. Biol. Trop. 1998; 46(Supl. 6): 201-213.

[34] Rosemond AD., Pringle CM., Ramírez A. Macroconsumer Effects on Insect Detritivores and Detritus Processing in a Tropical Stream. Freshwater Biology 1998; 39(3): 515-523.

[35] Mitsch WJ., Tejada J., Nahlik A., Kohlmann B., Bernal B. Hernández C. Tropical Wetlands for Climate Change Research, Water Quality Management and Conservation Education on a University Campus in Costa Rica. Ecological Engineering 2008; 34(4): 276-288.

[36] Vásquez Morera A. Soils. In: Janzen DH. (ed.). Costa Rican Natural History. Chicago: The University of Chicago Press; 1983.

[37] Stein H., Springer M., Kohlmann B. Comparison of Two Sampling Methods for Biomonitoring Using Aquatic Macroinvertebrates in the Dos Novillos River, Costa Rica. In: Kohlmann B., and Mitsch MJ. (eds.). Ecological Engineering 2008; 34(4): 267-275.

[38] MINAE (Ministerio de Ambiente y Energía, CR). Reglamento para la Evaluación y Clasificación de la Calidad de Cuerpos de Agua Superficiales. Decreto Ejecutivo Nº 33903-MINAE-S. Alcance No. 8 a La Gaceta No. 78, lunes 17 de septiembre del 2007. San José, CR. Available from: http://historico.gaceta.go.cr/pub/2007/09/17/COMP_17_09_2007.html

[39] Liess M., Brown C., Dohmen P., Duquesne S., Hart A., Heimbach F., Krueger J., Lagadic L., Maund S., Reinert W., Streloke M., Tarazona JV. (eds.). Effects of pesticides in the field. Le Croisic: EU and SETAC Europe Workshop; 2003.

[40] Ramírez A., Gutiérrez-Fonseca PE. Functional Feeding Groups of Aquatic Insect Families in Latin America: A Critical Analysis and Review of Existing Literature. Rev. Biol. Trop. 2014; 62(Suppl. 2): 155-167.

[41] Akaike H. Information theory and an extension of the maximum likelihood principle. Tsahkadsor, Armenian SSR. 2nd International Symposium on Information Theory; 1973.

[42] Arroyo AT. Modelos mixtos en el análisis de *loci* de caracteres cuantitativos y su aplicación en mejoramiento vegetal. PhD thesis. Universidad Nacional de Córdoba: Córdoba; 2008.

[43] Di Rienzo JA., Guzmán AW., Casanoves F. A Multiple Comparisons Method Based on the Distribution of the Root Node Distance of a Binary Tree. Journal of Agricultural, Biological and Environmental Statistics 2002; 7(2): 1-14.

[44] Balzarini MG., González L., Tablada M., Casanoves F., Di Rienzo JA., Robledo CW. Infostat. Manual del Usuario, Córdoba: Editorial Brujas; 2008.

[45] Gabriel KR. Biplot Display of Multivariate Matrices with Application to Principal Components Analysis. Biometrika, 1971; 58: 453-467

[46] Di Rienzo JA., Casanoves F., Balzarini MG., González L., Tablada M., Robledo C.W. InfoStat versión 2012. Universidad Nacional de Córdoba: Grupo InfoStat, FCA; 2012. Available from: http://www.infostat.com.ar

[47] Gower JC. Generalized Procrustes Analysis. Psychometrika 1975; 40:33-51.

[48] Hilsenhoff WL. An Improved Biotic Index of Organic Stream Pollution. Great Lakes Entomologist 1987; 20: 31-39.

[49] Moya, N, Hughes, R.M., Domínguez, E., Gibon, F.M., Goitia, E., Oberdorff, T. 2011. Macroinvertebrate-Based Multimetric Predictive Models for Evaluating the Human Impact on Biotic Conditions of Bolivian Streams. Ecol. Indic. 11: 840-847.

[50] Gutiérrez-Fonseca PE., Ramírez A. Long-term patterns of aquatic macroinvertbrate assemblages in lowland neotropical streams. San José. Poster presented at the 50th Anniversary Meeting of the Association of Tropical Biology and Conservation and Organization of Tropical Studies, 2013. Available at: https://atbc.confex.com/atbc/2013/webprogram/Paper1287.html

[51] Baur WH. Gewässergüte Bestimmen und Beurteilen. Berlin: Parey Buchverlag; 1998.

[52] Stewart KW., Stark B. Plecoptera. In: Merritt RW., Cummins KR., Berg MB.(eds.). An Introduction to the Aquatic Insects of North America. Dubuque: Kendall/Hunt; 2008. p311-384.

[53] Gutiérrez-Fonseca PE. Ecología, reproducción, taxonomía y distribución de *Anacroneuria* spp. Klapálek, 1909 (Insecta: Plecoptera: Perlidae) en Costa Rica. BSc thesis.Universidad de Costa Rica, San José; 2009.

[54] Salcedo S., Trama FA. Manual de Identificación de Macroinvertebrados Acuáticos de la Microcuenca San Alberto, Provincia de Oxapampa, Perú. Lima: CONCYTEC/FONDECYT; 2014.

[55] Gutiérrez-Fonseca PE., Springer M. Description of the final instar nymphs of seven species from *Anacroneuria* Klapalék (Plecoptera: Perlidae) in Costa Rica, and first record for an additional genus in Central America. Zootaxa 2011; 2965: 16-38.

[56] Tomanova S., Tedesco PA. Tamaño Corporal, Tolerancia Biológica y Potencial de Bioindicación de la Calidad del Agua de *Anacroneuria* spp. (Plecoptera: Perlidae) en América del Sur. Revista de Biología Tropical 2007; 55(1): 67-81.

[57] Thorp JH., Covich AP. Ecology and Classification of North American Freshwater Invertebrates. Academic Press; 2010.

[58] O'Callaghan P., Kelly-Quinn M. Performance of Selected Macroinvertebrate-Based Biotic Indices for Rivers Draining the Merendón Mountains Region of Honduras. Cuadernos de Investigación UNED 2013; 5(1): 45-54.

[59] Montoya Moreno Y., Acosta García Y., Zuluaga Zuluaga E., García A. Evaluación de la Biodiversidad de Insectos Acuáticos y de Calidad Fisicoquímica y Biológica del Río Negro (Antioquia-Colombia). Revista Universidad Católica de Oriente 2007; 23: 71-84.

[60] Sánchez Argüello R., Cornejo A., Boyero L., Santos Murgas A. Evaluación de la Calidad del Agua en la Cuenca del Río Capira, Panamá. Tecnociencia 2010; 12(2): 57-70.

[61] Miserendino ML., Pizzolón LA. Rapid Assessment of River Water Quality Using Macroinvertebrates: A Family Level Biotic Index for the Patagonic Andean Zone. Acta Limnologica Brasiliensia 1999; 11: 137-148.

[62] Mafla MH. Guía para Evaluaciones Ecológicas Rápidas con Indicadores Biológicos en Ríos de Tamaño Medio Talamanca – Costa Rica. Macroinvertebrados (BMWP – CR – Biological Monitoring Working Party) y Hábitat (SVAP – Stream Visual Assessment Protocol). Turrialba: Centro Agronómico Tropical de Investigación y Enseñanza (CATIE); 2005.

[63] Macchi PA. Calidad del agua en ecosistemas fluviales utilizando macroinvertebrados bentónicos. Cuenca del arroyo Pocahullo, San Martín de los Andes. BSc thesis. Universidad Nacional del Comuahue, Neuquén; 2007.

[64] Fenoglio S., Badino G., Bona F. Benthic Macroinvertebrate Communities as Indicators of River Environment Quality: An Experience in Nicaragua. Rev. Biol. Trop. 2002; 50(3-4): 1125-1131.

[65] Ghetti RE. Manuale di Applicazione dell'Indice Biotico Esteso (I.B.E.). Trento: Provincia Autonoma di Trento Press; 1997

[66] Sermeño Chicas JM., Serrano Cervantes L., Springer M., Paniagua Cienfuegos MR., Pérez D., Rivas Flores AW., Menjívar Rosa, RA., Bonilla de Torres BL., Carranza Estrada FA., Flores Tensos JM., Gonzáles C. de los A., Gutiérrez Fonseca PE., Hernández Martínez MA., Monterrosa Urías AJ., Arias de Linares AY. Determinación de la Calidad Ambiental de las Aguas de los Ríos de El Salvador, Utilizando Invertebrados Acuáticos: Índice Biológico a Nivel de Familias de Invertebrados Acuáticos en El Salvador (IBF-SC-2010). San Salvador: Editorial Universitaria (UES); 2010.

[67] Hilsenhoff WL. Rapid Field Assessment of Organic Pollution with a Family-Level Biotic Index. Journal of the North American Benthological Society 1988; 7: 65-68.

[68] Pinder LCV., Farr IS. Biological Surveillance of Water Quality 2. Temporal and Spatial Variation in the Macroinvertbrate Fauna of the River Frome, a Dorset Chalk
Stream. Archives of Hydrobiology 1987a; 109: 321-331.

[69] Pinder LCV., Farr IS. Biological Surveillance of Water Quality 3. The Influence of Organic Enrichment on the Macroinvertebrate Fauna of a Small Chalk Stream. Archives
of Hydrobiology 1987b; 109: 619-637.

[70] Spellerberg IF. Monitoring Ecological Change. Cambridge: Cambridge University
Press; 2005.

Chapter 13

Ecohidrology and Nutrient Fluxes in Forest Ecosystems of Southern Chile

Carlos E Oyarzún and Pedro Hervé-Fernandez

Additional information is available at the end of the chapter

http://dx.doi.org/10.5772/59016

1. Introduction

Nitrogen (N) cycling in terrestrial ecosystems is a global environmental concern. The N cycle is a complex interplay where biotic and abiotic processes interact to transform and transfer N in an ecosystem. In general, one can simplify by classifying terrestrial N cycles all over the world in two groups: 'tight' N cycles and 'open' N cycles. The 'tight' N cycle is characterized by its high efficiency in producing bioavailable N and retaining it in the plant-soil system. The 'open' N cycle, on the other hand, is then considered to be less efficient, showing significant loss of N towards aquatic ecosystems and the atmosphere. The latter losses might lead to adverse effects on stream water and air quality, contributing as such to 'global change' [1].

The movement of nutrients between ecosystems is called geochemical cycling or external cycling. Two important input processes to forests are atmospheric deposition and mineral weathering [2]. The atmospheric input to forests consists of dry, and wet deposition. Aerosol and gases can by deposited directly from the air to plant and soil surfaces during rainless periods by dry deposition. Wet deposition is defined as the input of atmospheric compounds to the earth's surface by rain, hail, snow and/or occult deposition that occurs via fogs and clouds, which can be important in mountainous regions [3]. During rain events, dry deposition is washed off from plant parts and, together with wet deposition, reaches the forest floor as throughfall and stem flow. A second input process is the weathering of soil minerals as a result of chemical dissolution. In combination with atmospheric deposition, mineral weathering is the only long-term source of base cations for terrestrial ecosystems [2].

The temperate climate region of southern Chile still reflects undisturbed, pre-industrial environmental conditions [4]. This is in strong contrast with land use, which has been altered significantly over the last decades and centuries. Only fragments of the original forest vegetation remain unaltered, and are located in the Coastal and Andes mountain ranges (CMR

and AMR, respectively). Exotic tree plantations and agricultural areas dominate the central valley of southern Chile [5]. These characteristics make this region an ideal study area to investigate human impacts on biogeochemical nutrient cycling. Temperate forests in Chile are not yet affected by elevated N deposition, as is the case for forests in Europe or northeastern North America [6]. However, anthropogenic activities such as transport, industry and agriculture have been increasing in central and southern Chile. These activities can substantially alter the atmospheric N load and enhance N input on forest ecosystems in Chile [5].

Several biogeochemical studies have been carried out most in humid temperate forest ecosystems between 40° and 43° S in southern Chile [i.e. 7; 8; 9]. The annual mean temperature is 5 to 12° C and precipitation ranges from 2000 to 7000 mm in the AMR [3]. Data from [5] reported that mean annual N composition of the rainwater in the CMR and AMR ranges (41°-43° S), varied between < 30 – 43 NO_3^--N µg L^{-1} and 9.8 – 26.2 NO_3^--N µg L^{-1}. Similarly, NH_4^+-N concentrations were < 50 NH_4^+-N µg L^{-1} and between 39.5 – 45.4 NH_4^+-N µg L^{-1} for CMR and AMR, respectively. Forests in the CMR, are located immediately near the ocean and are unique in this sense that external input of major elements are almost exclusively due to marine aerosols. Since trees canopy act as efficient filters, forests can capture large amounts of atmospheric deposition, especially occult deposition (i.e: fog and cloud). Normally, mountain forest ecosystems are very efficient in trapping nutrients, especially N and cations from clouds and fogs [10; 11; 4].

Stream nutrient loads are heavily dependent on catchment vegetation. Alteration of canopies and the soil under it, have a significant impact on nitrogen (NO_3^--N; NH_4^+-N; DON and TDN) and phosphorus (PO_4^{3+}-P and TDP) reaching the stream. Human disturbances have a direct impact on biological communities and may lead to land degradation, causing a change in ecosystem services and livelihood support. Temperate rain forest ecosystems of southern Chile have efficient mechanisms of retention for essential nutrients, especially NH_4^+ and NO_3^- [7, 3]. [6] described that the dominant form of N leaching was dissolved organic nitrogen (DON) for unpolluted forests of southern Chile. Other studies in the area had reported that conversion from native forests to exotic fast-growing plantations is likely to decrease N retention on catchments [12].

1.1. Native temperate rainforests of southern Chile

Native temperate rainforests of southern Chile represent an important global reserve of temperate forest with an extraordinary genetic, phytogeographic and ecological significance [13] with a worldwide high conservation priority [14]. These forests cover an area of 13.5 million ha. and are isolated by physical and climatic barriers, resulting in high endemism in plants and animals: 28 of 82 genera of woody plants (34%) are endemic to the region, along with 50% of vines, 53% of hemiparasites and 45% of vertebrates [15]. Some taxa are derived from ancient elements in southern Gondwana. Some relict tree species of conifers have the longest recorded lifespan, reaching an age of up to 3,600 years, constituting an excellent historical document for studies in reconstruction of climatic variability [16]. Most of the Valdivian eco-region is also considered as part of the world's 25 hotspots for biodiversity conservation and some of its forest types are included among the last frontier forests in the planet. These forests support fundamental ecological functions, which provide a range of ecosystem services and goods such as conservation of biological diversity, maintenance of soil fertility, and timber and non-timber products [17]. Also they contribute to maintain fresh water supply, which in turn supports the availability of drinkable water for cities [18].

Native forests in the Valdivian eco-region (36° S through 48° S) have suffered anthropical disturbances due to inadequate logging practices, and to agricultural land or exotic fast growing plantations conversion. Rapid conversion to forest plantations between 1975 and 2000 resulted in deforestation rates of 4.5% per year within an area of 578,000 ha in the Maule region (38° S), facilitated through afforestation incentives [19]. Another important cause of deforestation has been human-set fires, with an annual average of 13,000 ha burned in the period 1995–2005 and a high interannual variability associated to rainfall variation [20]. Anthropogenic land cover change in the central depression of southern Chile (40°-42° S) is the most evident process of deforestation and agricultural expansion. A large fraction of the *Nothofagus* forests in that region has been cleared for agriculture during the last century [21]. Patches of second-growth forest cover vast areas of the regional landscape, leaving only scattered stands as a result from intensified agriculture activity. Direct effects of past land use may occur via long-term (> 50 yr) physical alteration of the rhizosphere caused by historic practices. Soil compaction is an enduring consequence of cultivation, grazing, and logging that can cause increased bulk density and reduced pore space [1]. These changes may affect the abundance of aerobic and anaerobic microorganisms and subsequently reduce the cycling of several elements, including N.

1.2. Eucalyptus plantation forests

In south-central Chile (35-40° S), the native vegetation has been converted to agricultural uses, primarily plantation forestry, which has resulted in a landscape dominated by industrial forestry plantations. The amount of land in the region classified as plantation forestry has increased by 55 % between 1998 and 2008 (116–179 thousands ha; [22]. As in other parts of Chile, over 20,000 ha of those new plantations have replaced native forests in the region [19, 23], mainly located in the CMR. The growth of exotic species in non-native environments has uncertain ecohydrological consequences [24]. Therefore, there is much concern about their water consumption. Several authors have concluded that the consequences of exotic fast growing plantations are: (i) the decrease of discharge due to higher evapotranspiration [25, 26]; and (ii) changes in the soil hydrological properties, such as infiltration rates [27] and soil hydrophobicity [28].

2. Objectives

In small headwater catchments located at the Costal mountain range (CMR), in southern Chile (40° S), concentrations and fluxes of NO_3^--N, NH_4^+-N, DON, TDN, TDP and base cations (Ca^{2+}, Mg^{2+}, Na^+ and K^+) in bulk precipitation, throughfall and catchment discharge water were measured. The main objective of this study was to compare how hydrological variability affects catchment nutrient load responses with different land cover of native forests and exotic plantation of *Eucalyptus spp.*, in order to evaluate possible effects of land use

3. Material and methods

3.1. Description of the study sites

We selected five catchments with different land cover: (a) one with old-growth native ever-green rainforest (ONE), (b) one with native deciduous *Nothofagus obliqua* forest (ND), (c) one with secondary native evergreen forest (NE), (d) one covered with exotic fast growing *Eucalyptus nitens* (FEP) and (e) one with fast-growing exotic cover of *Eucalyptus globulus* (EG), located at CMR (40°S), near the city of Valdivia, Chile. All five catchments are located inland from the Pacific coast. The ONE catchment has an area of 2.8 ha at 336 m a.s.l. and is 20 km from the coast. The ND catchment has an area of 10.1 ha at 71-125 m a.s.l., and is 23.0 km from the coast. The NE catchment has an area of 3.1 ha at 227-275 m a.s.l. and is 2.0 km from the coast. The FEP catchment has an area of 54.8 ha and is 18 km from the coast, and the EG catchment has an area of 5.6 ha at 250-297 m a.s.l., and is 2.6 km from the coast.

3.2. Forest cover

In the catchment covered by old-growth native evergreen rainforest (ONE) the main canopy species are *Eucryphia cordifolia* Cav., *Aextoxicon punctatum* Ruiz et Pav. and *Laureliopsis philippiana* (Looser) Schodde. This last shows the highest density (718 tree ha^{-1}) and basal area (37.2 m^2 ha^{-1}) (Figure 1). The understorey is dominated by *Amormyrtus luma, Amomyrtus meli, Drimys winteri* and *Myrceugenia planipes*. The attributes of the old-growth native rainforests in the study area includes: increase in the proportion of successional species, the promotion of better growth rates to reach large diameters, the development of a rich understory and new regeneration cohorts, the increase of vertical structure, the development of increased wildlife habitat, and the presence of dead wood in the system (snags, and coarse woody debris) [29].

The main canopy species in the mixed ND catchment is the deciduous species *Nothofagus obliqua* (Mirb.) Oerst. reaching heights of 35 m, which covers 63.3 % of the catchment. Also, 13.8 and 7.9 percent is covered by native secondary forests of *Gevuina avellana* and *Astrocedrus chilensis* planted in 1983 and 1982, respectively, and 15.0 percent is covered by the fast-growing *Eucalyptus sp.* plantation. Understorey trees include *Luma apiculata, Podocarpus salignus, Aextoxicon punctatum, Amomyrtus meli, Gevuina avellana* and the exotic tree *Acacia melanoxylon*. Shrubs that reach heights over 3 m are mainly *Chusquea quila* Kunth with a 95% canopy cover.

In the NE catchment, the vegetation cover is characterized as a second growth native evergreen forest, dominated by *Myrtaceae* spp., *Amomyrus luma* (Mol.) Legr. et. Kaus (29%), *Amomyrtus meli* (Phil.) Legr. et. Kaus (25%), *Laureliopsis phillipiana* (Mol.) Mol. (14%), *Myrceugenia pla-nipes* (Hook. et Arn.) Berg. (13%), *Dyasaphillum diacanthoides* (Less.) Cabrera (7%), *Gevuina avellana* (Molina) Molina (6%), *Lomatia ferrugina* (Cav.) R. Br., *Persea lingue* (Ruiz et Pav.) Nees ex Koop. and *Myrceugenia exucca* (DC.) Berg. (2% each) and *Aextoxicon punctatum* (1%). This catchment is also used as a source of wood by local residents and as an occasional grazing ground for animals during the winter.

The FEP catchment is covered with *Eucalyptus nitens* of 4 and 14 yr-old. However, this catchment has had already 5 *E. nitens* rotations; density of 2911 tree ha^{-1} and the basal area is 131.9 m^2 ha^{-1}. In FEP, the highest density was observed in the diameter 20-25 and 25-30 cm (Figure 2). The total density ranges between 2911-2733 tree ha^{-1} in the sites. Basal area ranges between 131.9 and 144.4 m^2 ha^{-1}, and the mean height of the trees was 25.4 m. The riparian vegetation of the catchment with *Eucalyptus nitens* plantation has a large proportion of small trees and shrubs with a diameter distribution between 5-10 cm (Figure 3). The main tree species is *Luma apiculata* with 2180 tree ha^{-1} and the shrub *Aristotelia chilensis* with 815 tree ha^{-1} (Figure 3).

In EG catchment, the vegetation cover is composed of 80% exotic plantation of *Eucalyptus globulus* and 20% native evergreen remnant as a buffer zone. This is composed of *Berberis darwini* (Hooker) and *Ovidia pillopillo* (Gray) Hohen ex Meissn. (both with 29% cover), *Eucriphya cordifolia* Cav. (25.8%), *Lomatia ferruguinea* (Cav.) R. Br. (9.7%), *Dasyphyllum diacanthoides* (Less.) Cabrera and *Raphitamnus spinosus* (Juss.) Mold. (both with 3.2%). Originally this catchment was a native evergreen forest. However it was cleared (35 years ago) with fire to open areas for grazing animals, and in some areas, for the extraction of wood. Recently (9 years ago) the grassland was replaced by exotic trees (*Eucalyptus globulus*). Local residents use the forest as a source of wood and also allow animals to graze on the grass as well as on tree shoots.

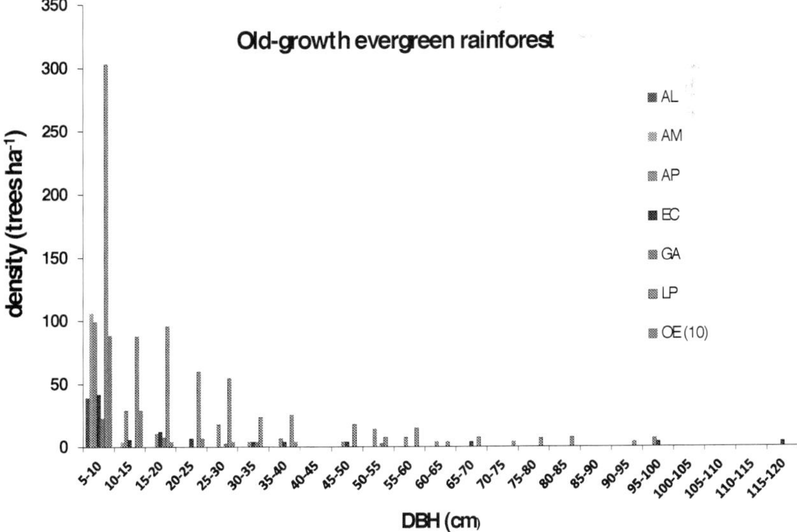

Figure 1. Diameter distributions by species (AL=*Amomyrtus luma*, AM=*Amomyrtus meli*, AP=*Aextoxicon punctatum*, EC=*Eucryphia cordifolia*, GA=*Gevuina avellana*, LP=*Laureliopsis philippiana*, OE=other species) in the catchment with native old-growth evergreen rainforests.

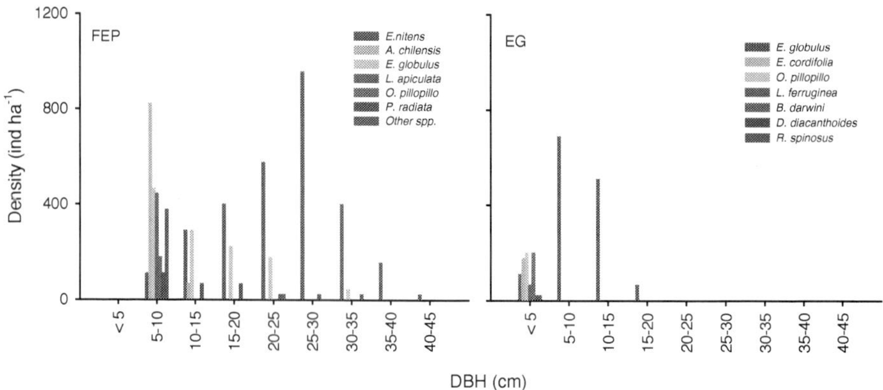

Figure 2. Diametric classes of species found on FEP and EG catchments.

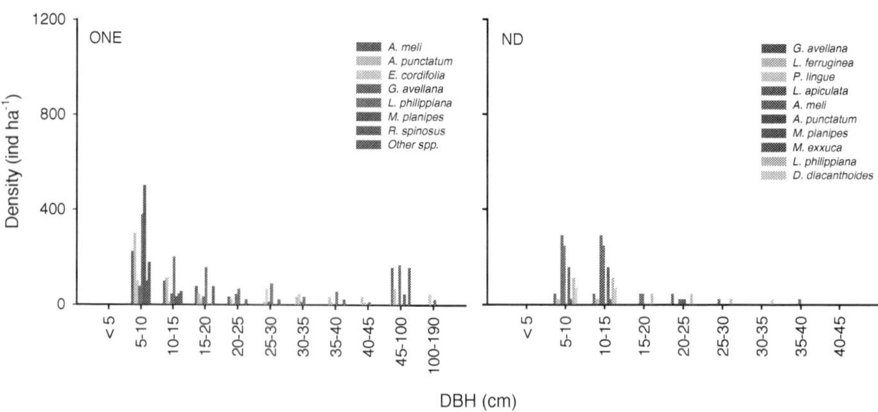

Figure 3. Diametric classes and density of trees for ONE and ND. Note that since ONE had the oldest trees, the last two classes comprise trees within 45 to 100, and 100 to 190 cm diameter at breast high.

3.3. Soils and climate

Climate in the area of study, is rainy temperate. In the meteorological station Isla Teja (25 m a.s.l.), 10 to 20 km from the study sites, the mean annual temperature is 12.0 °C (January mean is 17 °C and July mean is 7.6 °C) and the mean annual precipitation is 2,280 mm. Rainfall is concentrated during winter (May–August, 62 %) and decreases strongly in the summer (January–March, 9 %). Soils in the study area are red clayish derivatives from ancient volcanic ashes, deposited over a metamorphic geological substratum, dominated by micaceous schist and quartz lenses. The soils are shallow (< 1.0 m depth) in EG and NE catchments, and predominantly deep (> 1.0 m) in ND catchment. Soils in the EG catchment are characterized by poor infiltration rates, and in the NE and ND catchments by high infiltration rates [27].

Soils at ONE and FEP catchments have approximately the same texture in the bottom of the 1 meter depth soil profile, however the top layers (0 to 15; and 15 to 30) have consistently 10% more clay, and 1% less sand in FEP compared to ONE soil profiles. In the FEP catchment, clay content ranges between 37.2 – 45.1 %, organic matter content ranges between 1.8 – 17.1%, inorganic-N (NO_3^--N and NH_4^+-N) ranges between 9.8 – 21.0 mg kg^{-1}, Ca^{2+}between 0.19 – 0.23 cmol kg^{-1} and Mg^{2+}ranges between 0.09 – 0.16 cmol kg^{-1}. While, ONE soil clay content ranges between 31.1 – 37.3 % and organic matter content ranges between 5.9 – 17.8 %, inorganic-N ranges between 11.2 – 57.4 mg kg^{-1}, Ca^{2+}ranges between 0.23 – 1.32 cmol kg^{-1} and Mg^{2+}ranges between 0.10 – 0.71 cmol kg^{-1}.

4. Methods

Bulk precipitation was sampled using four plastic rain collectors attached to a 2.5-liter bottle. Bulk precipitation collectors (surface area 200 cm^2,) were installed in open areas (no trees were within 20 m of the sampling point), located between a distance of 100 – 500 m. Throughfall water was collected, using 2-4 collectors (surface area 254 cm^2) were installed inside each type forest. All collectors were installed 1.2 m above the forest floor and installed inside opaque tubes in order to avoid light penetration that could promote algae growth. Throughfall collectors had a thin mesh at the beginning of the neck of the funnel, in order to prevent insects and leaves entering the collection bottles, and designed with a plastic ring in order to exclude bird droppings [30]. Soil water was sampled at two different depths (0.3, 0.6 m) with low-tension porous-cup lysimeters (max 60 kPa of tension was applied) (Soil Moisture equipment corp.).

Discharge from each catchment was constantly measured by a pressure transducer paired with a baro diver (Schlumberger Water Services). Water samples were taken directly from the streams with an ISCO-6712 automatic sampler in each catchment. Stream samples were composed by two 250 mL aliquots taken each 30 minutes (1 h compound sample per bottle). Samples were filtered through a borosilicate glass filter (Whatman) of 0.45 μm. NO_3^--N (NO_3^--N+NO_2^--N) was determined by the cadmium reduction method, where NO_2^--N was always below detection limits. NH_4^+-N was determined with the phenate method (blue indophenol), detection limit (DL) was < 2 μg L^{-1}, for nitrite, nitrate and ammonia. Dissolved Inorganic Nitrogen (DIN) was calculated as follows: DIN=NO_3^--N+NO_2^--N$^+$$NH_4^+$-N. Total dissolved nitrogen (TDN) was determined by the sodium hydroxide and persulfate digestion method (DL < 15 μg L^{-1}). Organic nitrogen (DON) was calculated by subtracting (DON=TDN-DIN) concentration from TDN. Total dissolved phosphorous (TDP) was measured by the sodium hydroxide and persulfate digestion method (DL < 3 μg L^{-1}) at LIMNOLAB (Limnology Laboratory, Universidad Austral de Chile). Ca^{2+}and Mg^{2+}($\pm$ 0.05 mg L^{-1}) were analyzed by AAS, while Na^{+}and K^{+}($\pm$ 0.05 mg L^{-1}) by AES in the Forestry Nutrition and Soil Laboratory, Universidad Austral de Chile.

Canopy enrichment factors were calculated as the ratio between throughfall and bulk precipitation from different forest covers (throughfall / bulk precipitation). Fluxes were calculated using discharge and rainfall volumes. While nutrient retention (R) was calculated as follows:

$$\text{Retention} = (\text{Input} - \text{Output}) / \text{Input}$$
$$\text{Where,} \rightarrow R > 0, + \text{Retention}$$
$$R = 0, \text{Equilibrium}$$
$$R < 0, - \text{Retention}$$

5. Results and discussion

5.1. Throughfall enrichment factors

Canopy enrichment factors are presented in Figure 4. ND and ONE forests showed the highest enrichment and variability, whereas the EG plantation showed the lowest. The nutrient which presented the lowest annual enrichment in all throughfall samples was NO$_3^-$-N ranging from-0.8 for EG, through 1.5 for FEP. The highest enrichment was DON (10.3 times) for ONE and TDP (10.7 times) for ND forests. This enrichment is due to two processes: the washing off of the unquantified N input by dry deposition, on the one hand, and the N uptake from wet, dry particulate and gaseous deposition by leaves, twigs, stem surfaces, and lichens, on the other hand [31]. The old-growth evergreen forests (like ONE catchment) are multi-stratified and have an understory of high diversity, resulting in a complex and diverse structure and species composition. Also, [32] reported that DIN and DON concentrations were higher in throughfall than in bulk precipitation, particularly for nitrate, in a native *Nothofagus obliqua* forest and a *Pinus radiata* plantation, located near of the study sites. [8] observed 3.7 times throughfall enrichment for NO$_3^-$-N, in an evergreen *Nothofagus betuloides* forest (9.8 μg L^{-1} and 36.5 μg L^{-1} for bulk precipitation and throughfall, respectively) and a 1.7 throughfall enrichment under a deciduous *Nothofagus pumilio* forest (26.2 μg L^{-1} and 43.5 μg L^{-1} for bulk precipitation and throughfall, respectively) at cordillera de los Andes (40° S, 1120 m a.s.l.). However, NH$_4^+$-N was retained by canopies. Data from forested sites in the USA and Europe [33] showed that net canopy exchange of N (throughfall plus stemflow minus bulk precipitation) was negative for NH$_4$+and NO$_3$-at all sites, indicating that canopies were clearly sinks for inorganic N.

5.2. Annual nutrient fluxes

TDN annual retention and net annual fluxes (in kg N ha^{-1} yr^{-1}) was 0.58 (1.43); 0.90 (9.31) and-4.79 (-7.14) for NE, ND and EG forests, respectively. TDP annual retention and net annual fluxes (in kg P ha^{-1} yr^{-1}) were 0.70 (0.08); 0.96 (0.06) and -1.44 (0.4) for NE, ND and EG, respectively (Figure 4). Studies in watersheds in the United States [34] reported that thin or porous soils and high infiltration rates have less capacity to retain N. However, in our study, catchments with high infiltration rates, such as NE and ND showed greater N retention than soils with very low infiltration rates, such as EG. In our study, the differences in DIN retention were evident between native forests and *Eucalyptus* plantation, as also has been described previously by [12]. However, [35] observed using land cover, watershed area and precipitation as predictors for water quality (nitrate, ammonia, DON, TDP and electric conductivity) for local models explained 79.5% of the variance.

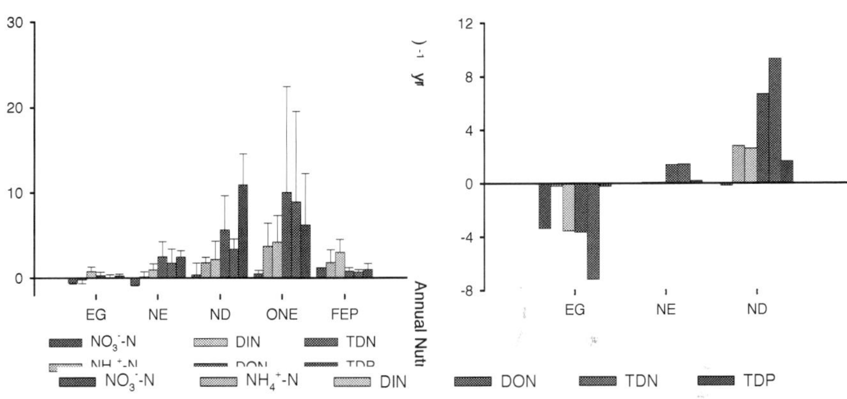

Figure 4. Throughfall enrichment factors for the five catchments (left) and annual nutrient fluxes for three catchments (right). EG=*Eucalyptus globulus* plantation, NE=native secondary evergreen ND=native deciduous, ONE=native old-growth evergreen, FEP=*Eucalyptus nitens* plantation.

5.3. Nutrient concentration in stream water

Nitrogen and phosphorous concentrations in stream water are variable in forest ecosystems of southern Chile (see Table 1). In general, the highest values of TDN and TDP concentrations are in *Fitzroya cuppressoides* forest (176.5 µg N L^{-1}) located in Coastal mountain range and in *Nothofagus pumilio* forest (67.3 µg P L^{-1}) located in Andean mountain range. The lowest values were found in an evergreen forest (36.8 µg N L^{-1}), located in Coastal mountain range and in *Fitzroya cuppresoides* forest (4.6 µg P L^{-1}), and located in the Coastal mountain range. Concentrations of inorganic N were smaller in the evergreen forest (33.2 µg L^{-1}) and in *E. nitens* plantation (33.6 µg L^{-1}) compared to organic N (94.4 and 67.0 µg L^{-1}, respectively), in agreement with previous research in southern Chile [6; 3] demonstrating that dissolved organic nitrogen is responsible for the majority of nitrogen losses from unpolluted forest ecosystems.

Type of forest	Forest description	Location	TDN	TDP	References
Native deciduous	*Nothofagus pumilio*	AMR	nd	67.3	[8]
Native deciduous	*Nothofagus betuloides*	AMR	nd	9.2	[8]
Native deciduous	*Nothofagus betuloides*	AMR	62	nd	[3]
Native deciduous	*N.nervosa-N. obliqua*	AMR	73.3	44	Unpublished
Native evergreen	Evergreen forest	AMR	157	18	Unpublished
Native evergreen	Evergreen forest	AMR	67.3	37.4	Unpublished
Native evergreen	*S. conspicua - L. philippiana*	AMR	109	4.9	Unpublished
Native conifer	*Fitzroya cuppressoides*	CMR	177	4.6	[9]
Native evergreen	Evergreen forest	CMR	36.8	24.1	[12]
Native evergreen	Evergreen forest	CMR	127	11.1	Unpublished
Native deciduous	*Nothofagus dombeyi*	CMR	153	nd	[32]
Exotic monoculture	*Eucalyptus spp.*	CMR	94.8	30.1	[12]
Exotic monoculture	*Eucalyptus nitens*	CMR	100	11	Unpublished
		AMR average	85.6	30.1	
		CMR average	115	16.2	

Table 1. Mean concentrations (μg L^{-1}) of TDN and TDP in stream water for different forest ecosystems under a low-deposition climate, southern Chile. At the end of the table 1, is the average for each location: Andean mountain range (AMR) and Coastal mountain range (CMR).

5.4. Relationships between discharge and nutrient concentrations

Nutrient exportation is related to hydrology, since water transports chemical compounds and particles. The relations of TDN and TDP with catchment discharge were positive for all nutrients except DIN, which showed a negative relation with discharge, during wet season (Figure 5). This negative relation is due to the dilution of nitrate with rainfall water which has higher concentrations of NH_4^+-N.

For dry season, the fitted models showed relatively high adjusted r^2 values for the *E. nitens* covered catchment for TDN and TDP (0,952 and 0,826, respectively; both with p < 0.05). However, the old growth covered catchment showed much lower values for TDN and TDP (0.317 and 0.519, respectively). Nevertheless, only TDP was significant. Dry season event DIN exportation was best fitted with a linear model. However, the fit was poor and not significant for both catchments. During wet season, the adjusted r2 values were higher for *E. nitens* covered catchment than the old growth covered catchments (Table 2). On figure 5, is clearly seen that during dry season TDN, TDP and DIN increase rapidly as discharge increases in *E.*

nitens covered catchment (FEP). However this is not observed for the old growth covered catchment (ONE). However, during wet season TDP shows greater increase in concentrations in ONE, rather than FEP. TDN and DIN shows the same behaviour in both catchments.

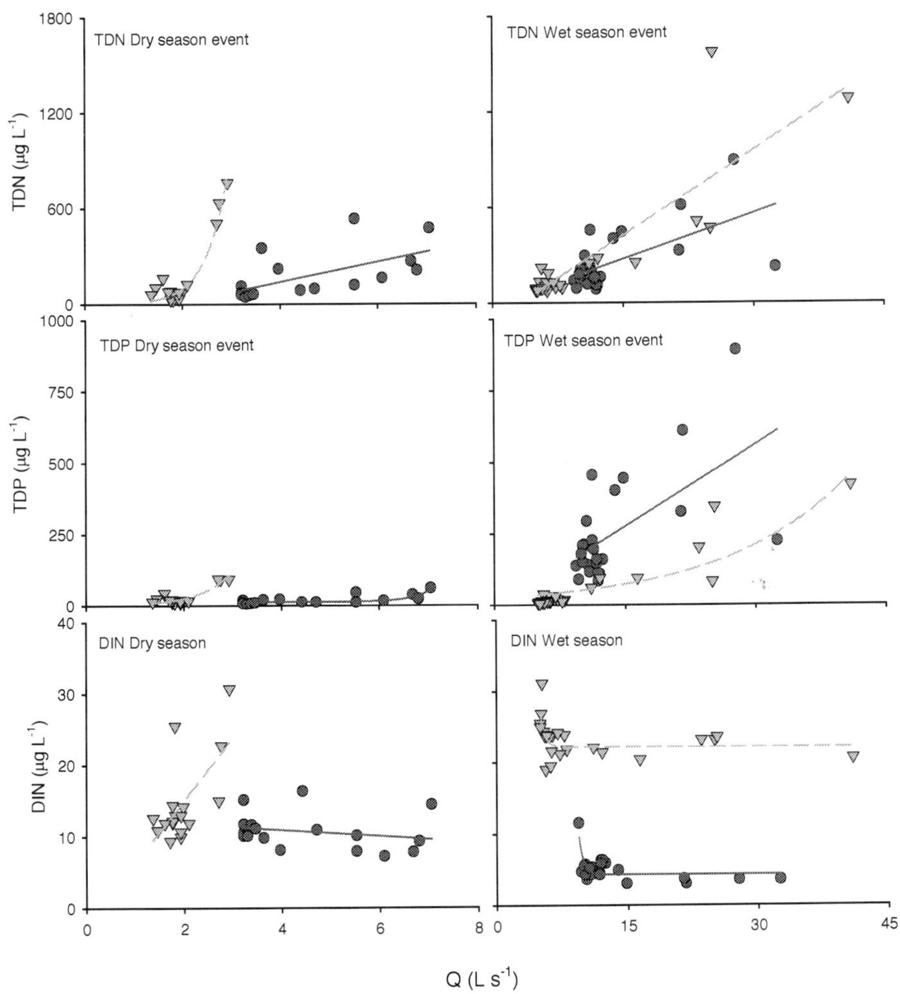

Figure 5. Total dissolved nitrogen (TDN), Total dissolved phosphorus (TDP) and Dissolved inorganic nitrogen (DIN) concentrations during one dry and wet season events (for the period March – November 2013), for the catchments covered with old growth native evergreen (ONE, in dark red circles) and catchment covered with *Eucalyptus nitens* (FEP, inverted orange triangles).

	Dry season event			Wet season event		
Catchment	TDN	TDP	DIN	TDN	TDP	DIN
ONE	0,317 (L)	0,519 (3EG)	0,170 (L)	0,331 (L)	0,331 (L)	0,05 (L)
FEP	0,952 (1EG)	0,826 (2EG)	0,04 (L)	0,728 (L)	0,765 (2EG)	0,388 (L)

Table 2. Adjusted r^2 values after fitting linear (L, f=y0+a x); single parameter exponential growth (1EG, f=e^{(a x)}); 2 parameter exponential growth (2EG, f=a e^{(b x)}) and 3 parameter exponential growth (3EG, f=y0+a e^{(b*x)}) models.

	Dry season event		Wet season event	
Catchment	Ca^{2+}	Mg^{2+}	Ca^{2+}	Mg^{2+}
ONE	nd	nd	0,554 (L)	0,184 (L)
FEP	nd	nd	0,026 (L)	0,857 (ED)

Table 3. Ca^{2+} and Mg^{2+} vs. discharge during events for each catchment. Adjusted r^2 values after fitting linear (L, f=y0+a x) and exponential decay (ED, f=a e^{(-b x)}) models.

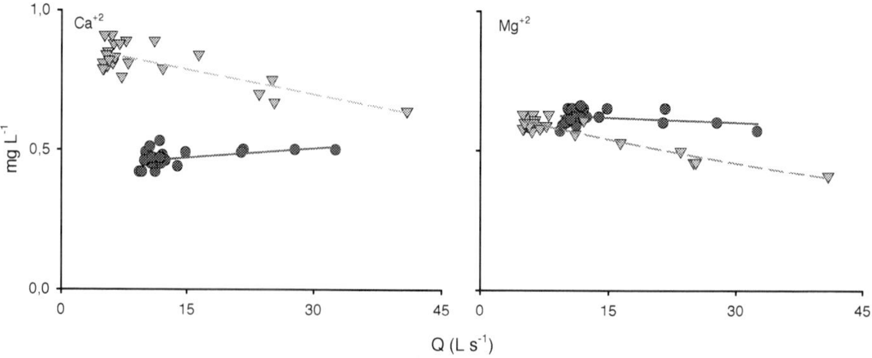

Figure 6. Ca^{2+} and Mg^{2+} concentrations vs discharge for the wet season event. Dark red dots and continuous line stands for old growth evergreen covered catchment (ONE), while inverted orange triangles and segmented line stand for *Eucalyptus nitens* covered catchment (FEP).

Typically, products of mineral weathering (e.g. Ca^{2+} and Mg^{2+}) decline in concentration when the discharge increases caused by rainfall (stream water dilutes). This was observed during wet season event, and only in FEP, for both cations. ONE showed an increase in concentration for Ca^{2+} and a slightly reduced concentration for Mg^{2+}.

We observed negative correlations between stream discharge and base cations concentrations (Figure 6). Typically, products of mineral weathering (e.g. Ca^{2+} and Mg^{2+}) decline in concentration when the discharge increases caused by rainfall (stream water dilutes). [36] reported inverse relationship between stream discharge and concentrations of Ca^{2+} and Mg^{2+}. However, [37] reported that during storms, both positive and negative relationships were observed between stream discharge and Ca^{2+} and Mg^{2+} concentrations and in some storms an initial increase in concentration was followed by dilution. On the other hand, [38] reported in an undisturbed old-growth Chilean forest that Ca^{2+} concentration demonstrated dilution when stream discharge increase and enhanced hydrological access occurred only for H^+. According to [39], mica schists, present in the geological substrate at the coastal mountain range, are rich in micas and minerals and contain high levels or iron and magnesium. Hence, concentration levels of magnesium in stream water probably are influenced by the geological substrate. However, the dilution and increase in concentration (on FEP and ONE, respectively) is mostly due to the dilution of stream water discharge with throughfall.

6. Conclusions

We conclude that the mixed-deciduous (ND) and old-growth evergreen (ONE) forests show the highest canopy enrichment for throughfall, while the *Eucalyptus* plantations (FEP and EG) showed the minimum enrichment. The highest enrichment was DON (10.3 times) for ONE; and TDP (10.7 times) for ND catchment. In general, the differences in enrichment are attributed to high LAI (Leaf Area Index) values in both native forests: the old-growth evergreen forests are multi-stratified and have an understory of high diversity, and particularly in the mixed-deciduous forest the presence of a thick layer of bamboo (*Chusquea quila*), which covered the soil. Our results differing from forested sites in North America and Europe which indicates that the canopies are generally acting as sinks for inorganic-N [33]. Also [40] have reported that NO_3^--N concentrations decreased in stemflow and throughfall relative to precipitation in old-growth forest in North America. However, in a data compilation from 126 European sites with high deposition climate in Scandinavia, Netherlands and Germany, [41] reported that inputs are enhanced by up to 3-5 times in throughfall through addition of dry deposition. On the other hand, our results show that the highest canopy enrichment was DON (dissolved organic nitrogen) especially in both native evergreen and deciduous forests. Also, DON was the most important nutrient fluxes in the native forested catchments, according to the literature [6] that reported that the dominant form of N leaching is dissolved organic nitrogen (DON) in unpolluted forests of southern Chile.

Annual retention of TDN in native deciduous and evergreen forests was 0.90 and 0.58, and TDP retention was 0.96 and 0.70, respectively. While the exotic *Eucalyptus* plantation there was a net release or loss of 4.79 and 1.44 for TDN and TDP, respectively. Studies in watersheds in the United States [34, 42] reported that thin or porous soils and high infiltration rates have less capacity to retain N. However, in our study, catchments with high infiltration rates, such as evergreen and deciduous forests showed greater N retention than soils with very low infiltration rates, such as *Eucalyptus globulus* plantation. Our results suggests that in native forests, rainfall water was infiltrating and percolating (subsurface flow) exporting less N in contrast to *Eucalyptus* plantation in which as soil has less porosity and infiltration rates due to land use history. The *Eucalyptus* plantation catchment was cleared (35 years ago) with fire to open areas for grazing animals, and in some areas, for the extraction of wood, and recently (9 years ago) the grassland was replaced by exotic trees (*Eucalyptus globulus*).

Nutrients (TDN and TDP) shows the same behavior in both catchments, their concentration tends to increase as catchment discharge increases. DIN however, showed a different behavior for dry and wet season events. In the native old growth evergreen forest (ONE), DIN lower its concentrations as discharge increased, however in *E. nitens* covered catchment (FEP) increased its concentration. The latter is mostly due to the dilution or the increase of NO_3^--N in stream discharge. However, during wet season both catchments showed the same DIN exportation behavior, though FEP had twice as much DIN when compared to ONE.

We are aware that modelling help to unravel and understanding hydrological processes and therefore nutrient exportation occurring within soil catchments. However there are many things to take in to account for, like biota (trees and microorganisms). However, discharge appeared to be a good predictor for TDN and TDP, for both events shown here. This was only seen in FEP, and not in ONE. DIN on the other hand showed poor model fitting. This means that there is still one or several unknowns on the control of DIN exportation during events.

The studies of events provide us with a much detailed perspective of what's happening within the catchment as an ecosystem, either pristine or heavily intervened. The reality is that ecosystems are going to keep "developing", each time with more and more relation to rural and city population. These pristine environments are in great danger and have to be protected from the inhabitants and other anthropic pressures, mostly cattle and land cover change to agricultural lands and exotic species.

Pristine study sites are recognized by being scarce and require a lot of efforts (monetary, time and struggle). In Chile, we have the luxury to have such areas near by some cities, nevertheless it will require more effort to keep it as pristine as possible. The prize for keeping this areas are many, from biodiversity hotspots to be able to unravel some of the black boxes that still exists regarding nutrient exportation and what are the effects of land cover change.

We would like also to address that soil use/cover change history, also plays an important role in N and P retention. Therefore before planting or doing forestry and agricultural activities, soil should be treated in order to enhance nutrient and water retention capabilities.

Acknowledgements

This research was supported by the Fondecyt Project 1120188 (Fondo Nacional de Ciencias). We would like to thank the different owners of the research sites, Mr. Armin Alba, CEFOR (Universidad Austral de Chile), Forestal ANCHILE and Llancahue community for providing the facilities and for collaborating in the monitoring and field work.

Author details

Carlos E Oyarzún[1*] and Pedro Hervé-Fernandez[2]

*Address all correspondence to: coyarzun@uach.cl

1 Instituto de Ciencias Ambientales y Evolutivas, Facultad de Ciencias, Universidad Austral de Chile, Valdivia, Chile

2 Laboratory of Hydrology and Water Management, Faculty Bioscience Engineering, University of Ghent, Belgium

References

[1] Huygens D, Boeckx P. Terrestrial nitrogen cycling in southern Chile: looking back and forward. In: Verhoest N, Boeckx P, Oyarzún C, Rodoy R (eds), Ecological advances on Chilean temperate rainforests. Academia Press, Belgium; 2009. p89-101.

[2] Staelens J, Oyarzún C, Almonacid L, Padilla E, Verheyen K. Aboveground nutrient cycling in temperate forest ecosystems of southern Chile. In: Verhoest N, Boeckx P, Oyarzún C, Godoy R (eds), Ecological advances on Chilean temperate rainforests. Academia Press, Belgium; 2009. p103-116.

[3] Oyarzun CE, Godoy R, Deschrijver A, Staelens J, Lust N. Water chemistry and nutrient budgets in an undisturbed evergreen rainforest of southern Chile. Biogeochemistry 2004; 71: 107-123.

[4] Weathers K, Likens G. Clouds in southern Chile: an important source of nitrogen to nitrogen limited ecosystems? Environmental Science and Technology 1997; 31: 210-213.

[5] Godoy R, Paulino L, Oyarzún C, Boeckx P. Atmospheric N deposition in central and southern Chile. An overview. Gayana Botánica 2003; 60 (1): 47-54.

[6] Perakis S, Hedin L. Nitrogen loss from unpolluted South American forest mainly via dissolved organic compounds. Nature 2002; 415: 416-419.

[7] Huygens D, Boeckx P, Templer P, Paulino L, Van Cleemput O, Oyarzún C, Müller Ch, Godoy R. Mechanisms for retention of bioavailable nitrogen in volcanic rainforest soils. Nature Geosciences 2008; 1 (8): 543-548. (Online publication DOI 10.1038/ngeo252)

[8] Godoy R, Oyarzún C, Gerding V. Precipitation chemistry in deciduous and evergreen *Nothofagus* forest of southern Chile under a low-deposition climate. Basic and Applied Ecology 2001; 2: 65-72.

[9] Oyarzun C, Godoy R, Sepulveda A. Water and nutrient fluxes in a cool temperate rainforest at the Cordillera de la Costa in southern Chile. Hydrological Processes 1998; 12: 1067-1077.

[10] Weathers K, Lovett G, Likens G, Caraco N. Cloudwater input of nitrogen to forest ecosystems in southern Chile: forms, fluxes, and sources. Ecosystems 2000; 3: 590-595.

[11] Weathers K. The importance of cloud and fog in the maintenance of ecosystems. Trends in Ecology and Evolution 1999; 14: 214-215.

[12] Oyarzún C, Godoy R, Aracena A, Rutherford P, Deschrijver A. Effects of land use conversion from native forest to exotic plantations on nitrogen and phosphorus retention in catchments of southern Chile. Water, Air and Soil Pollution 2007; 179: 341-350

[13] Armesto JJ, Smith-Ramírez C, Carmona M, Celis-Diez JL, Díaz I, Gaxiola A, Gutiérrez AG, Nuñez-Avila M, Pérez C, Rozzi R. Old-growth temperate rain forests of South America: conservation, plant-animal interactions, and baseline biogeochemical processes. In: Wirth Ch, Gleixner G & Heiman M (eds). Old-growth Forests. Function, Fate and Value. Ecological Studies 207, Springer Berlin, Germany; 2009. p367-390.

[14] Olson D, Dinerstein E. The global 200: a representation approach to conserving the earth's most biologically valuable ecoregions. Conservation Biology 1998; 12: 502-515.

[15] Armesto J, Aravena JC, Villagrán C, Pérez C, Parker GG. Bosques templados de la cordillera de la costa. In: Armesto, J.J., Villagrán, C., Arroyo, M.K. (eds.), Ecología de los Bosques Nativos de Chile. Editorial Universitaria, Santiago, Chile; 1996. p199–213.

[16] Lara A, Villalba R. A 3,620-year temperature record from Fitzroya cuppressoides tree rings in southern South America. Science 1993; 260: 1104-1106.

[17] Nahuelhual L, Donoso P, Lara A, Nuñez D, Oyarzún C, Neira E. Valuing ecosystem services of Chilean temperate rainforest. Environment, Development and Sustainability 2007; 9: 481-499.

[18] Nuñez D, Nahuelhual L, Oyarzun C. Forests and water: the value of native temperate forests in supplying water for human consumption. Ecological Economics 2006; 58: 606-616.

[19] Echeverría C, Coomes D, Salas J, Rey Benayas JM, Lara A, Newton A. Rapid deforestation and fragmentation of Chilean temperate forests. Biology Conservation 2006; 130: 481-494.

[20] Lara A, Reyes R, Urrutia R. Bosques Nativos. In: Instituto de Asuntos Públicos, Universidad de Chile (eds.), Informe País: Estado del Medio Ambiente en Chile, Santiago, Chile; 2006. p107–139.

[21] Godoy R, Paulino L, Valenzuela E, Oyarzun C, Huygens D, Boeckx P. Temperate ecosystems of Chile: characteristic biogeochemical cycles and disturbance regimes. In: Verhoest N, Boeckx P, Oyarzún C, Rodoy R (eds), Ecological advances on Chilean temperate rainforests. Academia Press, Belgium; 2009. p31-39.

[22] CONAF-CONAMA. Catastro de uso del suelo y vegetación: monitoreo y actualización región de los Ríos. Gobierno de Chile. Ministerio de Agricultura; 2008

[23] Armesto JJ, Manuschevich D, Mora A, Smith-Ramirez C, Rozzi A, Abarzúa A, Marquet P. From the Holocene to the Anthropocene: a historical framework for land cover change in southwestern South America in the past 15,000 years. Land Use Policy 2010; 27: 148-160.

[24] Cannel MGR. Environmental impacts of forest monocultures: water use, acidification, wildlife conservation, and carbon storage. New Forests 1999; 17: 239-262.

[25] Brown A, Zhang L, McMahon T, Western A, Vertessy R. A review of paired catchment studies for determining changes in water yield resulting from alterations in vegetation. Journal of Hydrology 2005; 310: 28-61.

[26] Iroumé A, Huber A, Schulz K. Summer flows in experimental catchments with different forest covers, Chile. Journal of Hydrology 2005; 300: 300–313.

[27] Oyarzun C, Frene Ch, Lacrampe G, Huber A, Hervé P. Propiedades hidrológicas del suelo y exportación de sedimentos en dos microcuencas de la Cordillera de la Costa en el sur de Chile con diferente cobertura vegetal. Bosque 2011; 32 (1): 10-19.

[28] Ferreira AJD, Coelho COA, Walsh RPD, Shakesby RA, Ceballos A, Doerr SH. Hydrological implications of soil water-repellency in Eucalyptus globulus forests, north central Portugal. Journal of Hydrology 2000; 231-232: 165-177.

[29] Donoso P, Frene C, Flores M, Moorman MC, Oyarzún CE, Zavaleta JC. Balancing water supply and old-growth forest conservation in the lowlands south-central Chile through adaptative co-management. Landscape Ecology 2014; 29: 245-260. DOI 10.1007/s10980-013-9969-7.

[30] Kleemola S, Soderman G. Manual for Integrated Monitoring. International Co-operative programme on integrated monitoring on air pollution effects. Environmental Report 5. Environment Data Centre, National Board of Waters and the Environment. Helsinki; 1998.

[31] Staelens J, Godoy R, Oyarzun C, Thibo K, Verheyen K. Nitrogen fluxes in throughfall and litterfall in two *Nothofagus* forests in southern Chile. Gayana Botánica 2005; 62: 63-71

[32] Oyarzun CE, Godoy R, Staelens J, Aracena C, Proschle J. Nitrogen fluxes in a *Nothofagus obliqua* forest and a *Pinus radiata* plantation in the central Valley of southern Chile. Gayana Botánica 2005; 62: 88-97.

[33] Lovett GM. Atmospheric deposition and canopy interactions of nitrogen. In: Johnsonn DW, Lindberg S (eds), Atmospheric Deposition and Forest Nutrient Cycling. Springer Verlag, New York; 1992. P152-166.

[34] Lajtha K, Seely B, Valiela I. Retention and leaching losses of atmospherically-derived nitrogen in the aggrading coastal watershed of Waquoit Bay, MA. Biogeochemistry 1995; 28: 33–54.

[35] Cuevas JG, Soto D, Arismendi I, Pino M, Lara A, Oyarzun C. Relating land cover to stream properties in southern Chile watersheds: trade-off between geographic scale, sample size, and explicative power. Biogeochemistry 2006; 81: 313-329.

[36] Reynolds B, Hornung M, Hughes S. Some factors controlling variations in chemistry of an upland stream in Mid Wales. Cambria 1983; 10: 130-145.

[37] Chapman PJ, Reynolds B, Weather HS. Sources and control of calcium and magnesium in storm runoff: the role of groundwater and ion exchange reactions along water flowpaths. Hydrology and Earth System Sciences 1997; 1(3): 671-685.

[38] Salmon C, Water MT, Hedin L, Brown M. Hydrological controls on chemical export from and undisturbed old-growth Chilean forest. Journal of Hydrology 2001; 253: 69-80.

[39] Uyttendaele GYP, Iroume A. The solute budget of a forest catchment and solute fluxes within a Pinus radiata and a secondary native forest site, southern Chile. Hydrological Processes 2002; 16: 2521-2536.

[40] Edmonds R, Thomas T, Blew R. Biogeochemistry of an old-growth forested watershed, Olympic National Park, Washington. Water Resources Bulletin 1995; 31: 409–419.

[41] Dise NB, Matzner E, Gundersen P. Synthesis of nitrogen pools and fluxes from European forest ecosystems. Water, Air & Soil Pollution 1998; 105: 143–154.

[42] Campbell JL, Hornbeck JW, Mitchell MJ, Adams MB, Castro MS, Driscoll CT, Kahl JS, Kochenderfer JN, Likens GE, Lynch JA, Murdoch PS, Nelson SJ, Shanley JB. Input-Output budgets of inorganic nitrogen for 24 forest watersheds in the northeastern United States: A review. Water, Air & Soil Pollution 2004; 151: 373–396.

Ecological Flexibility of the Top Predator in an Island Ecosystem — The Iriomote Cat Changes Feeding Patterns in Relation to Prey Availability

Shinichi Watanabe

Additional information is available at the end of the chapter

http://dx.doi.org/10.5772/59502

1. Introduction

It has long been recognized that islands contain fewer species than comparable pieces of mainland. It is also well-established that the number of species on islands decreases as island area declines. The most successfully established islands species will be those that combine low extinction rates with high immigration rates and where it is generally more difficult for animals without high dispersal ability at higher trophic levels to live on small islands [1]. In particular, carnivorous mammals at the top of terrestrial trophic chain find it most difficult to establish themselves on islands. Extreme examples are considerably large body sized and complete carnivorous species in the family Felidae.

In the region from south-east Asia to the Ryukyu Archipelago in southern Japan, there are thousands of islands of various sizes [2]. In this region, it is well-documented that biodiversity is particularly high among many taxa [3]. The biodiversity of mammalian fauna in this region was summarized in [4]. Among Felidae, seven species are distributed in this region, most of which are distributed only on the continental islands of Java, Sumatra, Borneo and Taiwan [4]. In particular, the leopard cat *Prionailurus bengalensis*, the most widespread species of East Asian wildcats, is an exception to this rule, occurring on several small islands as well as larger islands and the Asian continent [2, 4].

An example of an extreme case is the Iriomote cat *Prionailurus bengalensis iriomotensis* (Figure 1), which lives on the smallest island (284 km², Figure 2) of the Ryukyu Archipelago. The Iriomote cat is unique among the family Felidae, particularly in terms of its food habits [5-8]. Felidae are known as the most successfully evolved and developed predators specialized in feeding on mammalian prey [9, 10]. In contrast, the Iriomote cat preys upon a variety of animals

such as birds, reptiles, amphibians and insects, in addition to mammals [5-8]. The cat shows functional responses according to the availability of various alternative sources of prey [7]. Its principle prey changes seasonally, as the population density of potential prey items change. Moreover, regional differences in the cat's diet have also been reported [5, 11]. The Iriomote cat's diet is more diversified in habitats in which several vegetative environments are included and more similar in habitats where vegetative environments are uniform [11]. For terrestrial vertebrates on the island, distribution of each species is strongly influenced by various topographic and vegetative environmental factors, and distribution patterns vary depending on the type of species [12]. The cat diet changes flexibly in relation to seasonal and regional differences in prey availability [6, 7, 11]. Thus, it is likely that the preferred habitats for this species will also vary depending on seasons and regions.

Figure 1. An Iriomote cat *Prionailurus bengalensis iriomotensis* taken by photo-trap (Mammal Ecology Laboratory, University of the Ryukyus).

Most animals selectively use environments with a good quality of food patches [13, 14]. It is therefore likely that predators specializing in a particular food type that occurs in specific habitats will be habitat specialists, while predators feeding on a range of different food types will be habitat generalists. Variation in prey availability, i.e., the density and distribution of prey animals in an environment, leads to various predator responses [15-18]. For example, predators specialized in catching particular prey types often produce numerical responses to prey availability, so that the density of predators fluctuates alongside prey density [15-17]. In contrast, non-specialized predators often produce functional responses to prey availability, allowing these predators to switch prey types in relation to the availability of alternative resources [18].

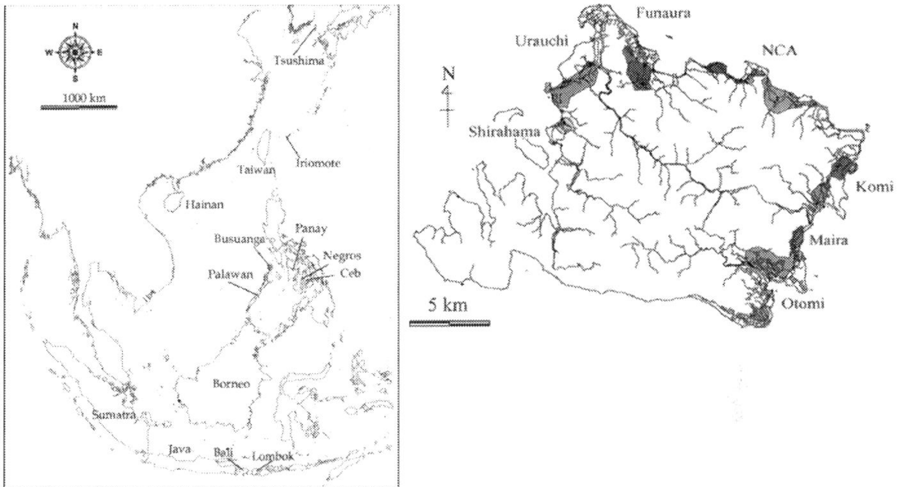

Figure 2. Map of East Asia showing islands with wild felid populations (left) according to [2] and Iriomote Island (Ryukyu Archipelago, Japan, right) showing the locations of seven study sites where radio-tracking were conducted. Each dark area shows the home range of each studied cat, estimated by the 75% harmonic mean method.

In the case of carnivores, food habits are well-documented at interspecific levels [e.g., 19, 20]. Each felid species takes only a few different mammal prey items, while other carnivores eat various food types. The Felidae family is highly specialized in preying on mammals in terms of having developed morphological and behavioural traits [21, 22]. Thus, Felidae are considered typical specialists in terms of food and habitat. Their hunting behaviour is specialized for preying on mammalian prey items [21]. Hence, they often show a numerical response to the density of a particular prey species [e.g., 18, 23].

The Iriomote cat, however, preys on various types of animals. Its diet flexibly changes in relation to seasonal and regional differences of prey availability [5-8; see also the results in the present study]. It is therefore likely that the Iriomote cat makes use of a variety of habitats and movement patterns in response to spatio-temporal variations in prey availability.

A comprehensive and accurate analysis of habitat use and selection, particularly when dealing with large home ranges and high habitat diversity across the geographic range of an animal, must encompass multiple spatial scales [24]. A number of studies have been conducted on the habitat use of the Iriomote cat, but these have only investigated the habitat selection on a univariate scale. Sakaguchi [5] emphasized that the Iriomote prefers to use lowlands (< 50 m above sea level) and avoids highlands, yet other factors potentially affecting their habitat use have not been investigated. In the present study, I will therefore quantify seasonal and regional variations in the habitat use and movement patterns of the Iriomote cat using detailed microhabitat measures at point locations and detailing the movement tracks used by the cat. From the results, I will then discuss their feeding strategies in terms of seasonal and regional variations in prey availability. Furthermore, I will also present the possible reasons for the presence of this species on Iriomote Island.

Study site	Tracked period		Tracked days	No. of cats		No. of location
				Male	Female	
Otomi	Jul. 12. 1999	Jan. 19. 2001	79	2	1	545
Maira	Jun. 22. 2000	Dec. 25. 2001	33		2	238
Komi	Jun. 04. 1999	Dec. 25. 2001	254	1	2	1211
NCA	Nov. 20. 2001	Mar. 07. 2002	62	3		340
Funaura	Jun. 18. 2000	Feb. 15. 2001	69	1	1	457
Urauchi	Jun. 16. 2000	Aug. 28. 2000	13	1		66
Shirahama	Oct. 12. 1998	Nov. 06. 1999	155	1	1	1229
	Total		665	9	7	4086

Table 1. Summary of the radio-tracking survey in seven study sites on Iriomote Island.

2. Methods

2.1. Field survey

A field survey was conducted on Iriomote Island (24°20′N, 123°49′E) in the Ryukyu Archipelago, southern Japan (Figure 2). The island largely consists of highly folded mountains with the highest peak (Mt. Komi) being 469 m above sea level. Its vegetation is mostly natural subtropical evergreen broadleaved forest. Most of the island is protected as a Japanese national park and contains good examples of the natural subtropical forests (see [7] for more information about the island).

Cats were trapped using box traps during the following six periods, May to July and October in 1999, January, June and November in 2000 and November in 2001. For the captures, box-traps equipped with radio-alarm systems were used. Captured cats were immediately brought to a laboratory and anesthetized by a professional veterinarian with an intramuscular injection of ketamine hydrochloride and xylazine (ketamine hydrochloride 10 mg/kg body weight). The animals were weighed and measured; their age-classes were estimated according to tooth wear, body weight and the delivery history of females [5, 25]. The cats were fitted with 40 g radio collars (Advanced Telemetry Systems Inc., Isanti, Minnesota, USA., or alternatively, collars hand-made in my laboratory).

I captured and radio-tracked a total of 16 adult cats (nine males and seven females) (Table 1) in seven separate study sites (Figure 2). All females were parous when captured, as per evidence of previous suckling marks on their nipples [5, 25]. Each cat remained in the same study site throughout the study period.

The movements of radio-collared cats were continuously monitored for seven-to 10-day periods, either by car or on foot. Locations used by the cats were taken at intervals of at least one hour using the triangulation software Loas version 2.1 (Ecological Software Solutions) from two-to four-points, marked with a handheld global positioning system (GPS; model Garmin GPS II Plus) on roads or trails. I then determined the universal transverse mercator (UTM; zone 51, WGS84 Datum) coordinates of cats' locations using a geographic information system (GIS) and the IDRISI Kilimanjaro version 14 software (Clark Labs, Clark University, Worcester MA, USA).

2.2. Scat analysis

Seasonal and local variations on diet compositions among the study sites were examined via scat analysis [6-8]. Scats collected in each study site (Figure 2) were used for the analysis. Diet composition and principal prey items were compared among the sites. I calculated the frequency of occurrences for each prey taxon: mammals, birds, reptiles, amphibians, insects, crustaceans and others; this was done for each season and site.

2.3. Data analysis

2.3.1. Environmental measurements

I measured nine environmental variables related to topographic and vegetative characteristics within habitats to determine regional differences in environments among the study sites, as well as the most important factors that influenced the habitat preferences of the Iriomote cat. IDRISI [26], an integrated GIS and remote sensing software, was used for the analysis and display of digital geospatial information. IDRISI is a PC grid-based system that offers tools for researchers and scientists engaged in analysing earth system dynamics for effective and responsible decision making regarding environmental management [26].

Three topographic variables, elevation (El), slope (Sl) and the presence of drainage (Dr), were derived from digital elevation models (Digital Map 50 m Mesh Elevation, published by the Japan Geographical Survey Institute). The topographic data contained elevation values of one meter precision at the centre of grids by latitudinal 1.5 second and longitudinal 2.25 second; the ground length was roughly 50 m. The elevation data were geometrically-corrected in the UTM coordinate as a raster image showing a 50 m x 50 m grid. Sl and Dr were derived from the image using the Surface and Runoff operations of IDRISI. Elevation data generally contain depressions that hinder flow routing; these were removed and then I calculated the accumu‐ lation of rainfall units per pixel based on the elevation image. Drainage networks can produce a setting for discovering a threshold on the accumulation of runoff [26]. In the present study, the threshold (50 pixels) was able to detect permanent stable water and as such, streams and rivers in the study area were set. These three variables were subdivided into a 10 m x 10 m grid; images of El and Sl were averaged among neighbouring 5 x 5 pixels using the Filter

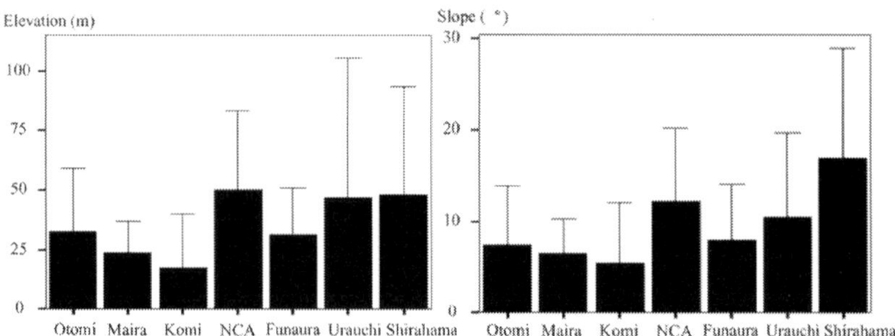

Figure 3. Mean values (+*SD*) of elevation and slope within the home ranges of the Iriomote cat in seven study sites.

operation of IDRISI. Thus, the distance (10 m) and planimetry (100 m²) accuracies in this study were limited by the grid size.

Vegetative variables were derived from a digital vegetation map [27], in which vegetation was classified as 29 categories within the study area. Many categories had a few patches, while some categories were very similar to others. I combined similar habitat types and broadly classified these in the following five categories; natural forest (NF) including subtropical, evergreen and broad-leaved forests; secondary forest (SF), including pine and artificial forests; coastal vegetation (CV), including mangroves and vegetation along shorelines; rice fields and swamps (RS); croplands and postures (CP). The vegetation classified into five categories was also geometrically-correlated in the same UTM coordinate as the raster image showing a 10 m x 10 m grid.

Since the Iriomote cat prefers to use the boundaries of forests and open lands [28, 29], I presumed that the prey availability of some species was high along forest edges, which influenced their food habits. Thus, a variable (FE: presence of forest edges) was derived from the 10 m x 10 m vegetation grid data using the Buffer operation of IDRISI. Forest edge was defined as zones within 50 m toward forests (NF and SF) from other vegetation types.

Variable				Study site						
	Otomi	Maira	Komi	NCA	Funaura	Urauchi	Shirahama	G	d.f.	P
DR	4.9	6.0	5.9	6.7	5.4	7.0	5.4	190	6	<0.0001
FE	18.9	38.6	12.1	12.5	14.2	9.3	14.6	3957	6	<0.0001

Table 2. Percentages of drainages (Dr) and forest edges (FE) in the home ranges of the Iriomote cat in each study site.

2.3.2. Habitat types in home ranges

I estimated the home range of each radio-collared cat from the coordinates of radio-tracking locations using home range estimation software Biotas version 1.0 (Ecological Software Solutions). The home range (HR) was defined as the area enclosed within the 75% utilization contour with the harmonic mean [30]. I overlaid HRs of all radio-collared cats in each study site (Figure 2), in which environments concerning the above nine variables were compared among sites. The differences between the continuous variables (El and Sl, see Figure 3) among study sites were subjected to a Kruskal-Wallis test for those that were significantly different, while the seven other categorical variables (Table 2, Figure 4) were subjected to a likelihood-ratio test (G-test) for independence to test for differences.

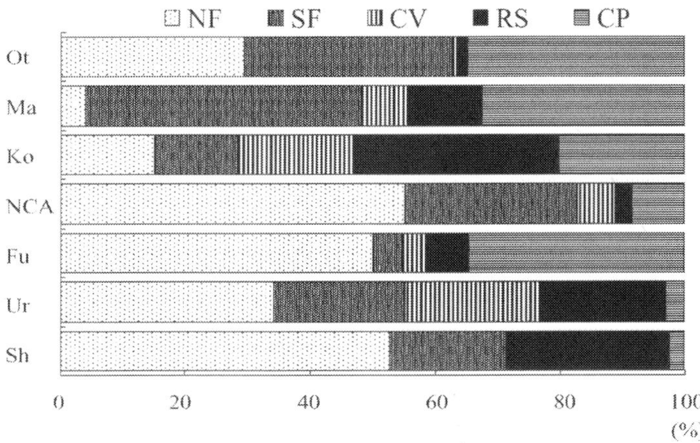

Figure 4. Compositions of vegetation types (NF: natural forest; SF: secondary forest; CV: coastal vegetation; RS: rice fields and swamps; CP: croplands and postures) within the home ranges of the Iriomote cat in seven study sites (Ot: Otomi; Ma: Maira; Ko: Komi; NCA: northern coastal area; Fu: Funaura; Ur: Urauchi; Sh: Shirahama).

2.3.3. Habitat preferences on cat location

To determine the most important topographic and vegetative characteristics influencing the animals' habitat use, I measured the above nine variables in areas within a 20 m radius of cat location sites fixed by radio-triangulation. To determine seasonal differences in habitat use, the location sites of six cats that were studied during all seasons (spring: April to June; summer: July to September; autumn: November to December; winter: January to March) were compared among seasons with regard to the above environmental characteristics.

To determine the habitat preference of radio-collared cats at each site, environmental characteristics in areas within a 20 m radius at cat location sites were compared with those at random locations. As many random locations as cat locations were chosen from HRs in each study site, using IDRISI. Mean values of El and Sl and the proportions of occurrence of other seven categorical variables were compared between areas within a 30 m radius at cat locations and random locations.

As a first step, these statistical differences were tested using a Mann-Whitney U-test for El and Sl, and using a G-test for others. Variables remaining after univariate testing were entered into a logistic regression function following forward stepwise procedures. In the stepwise regression, a forward procedure using the likelihood-ratio statistic was employed, which included a variable in the model at $P=0.05$ level, which was removed if said variable's significance fell below 0.10. The percentage of sites classified correctly (radio tracking location vs. random location) and coefficients of determination (Nagelkerke R^2) determined by the final logistic models, which were indications of the influence of the logistic regression [31], were calculated in each regression.

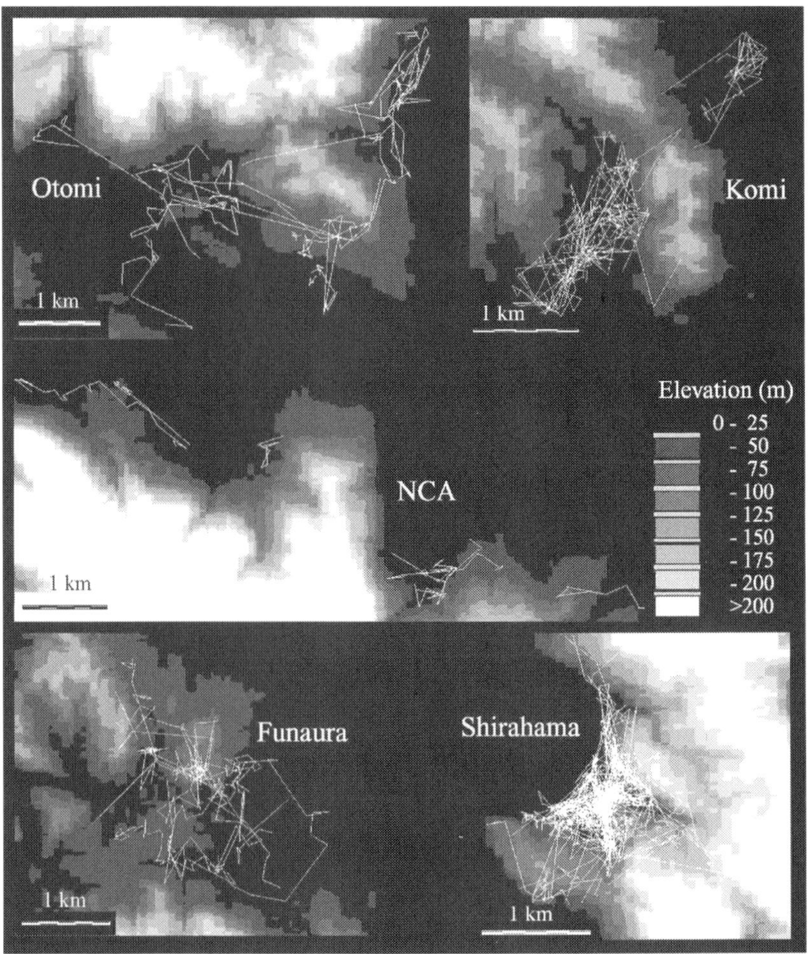

Figure 5. Movement tracks of the Iriomote cat in five study sites shown in digital elevation models. The movement tracks were based on continuous radio-tracking with locations taken every one-to two-hours.

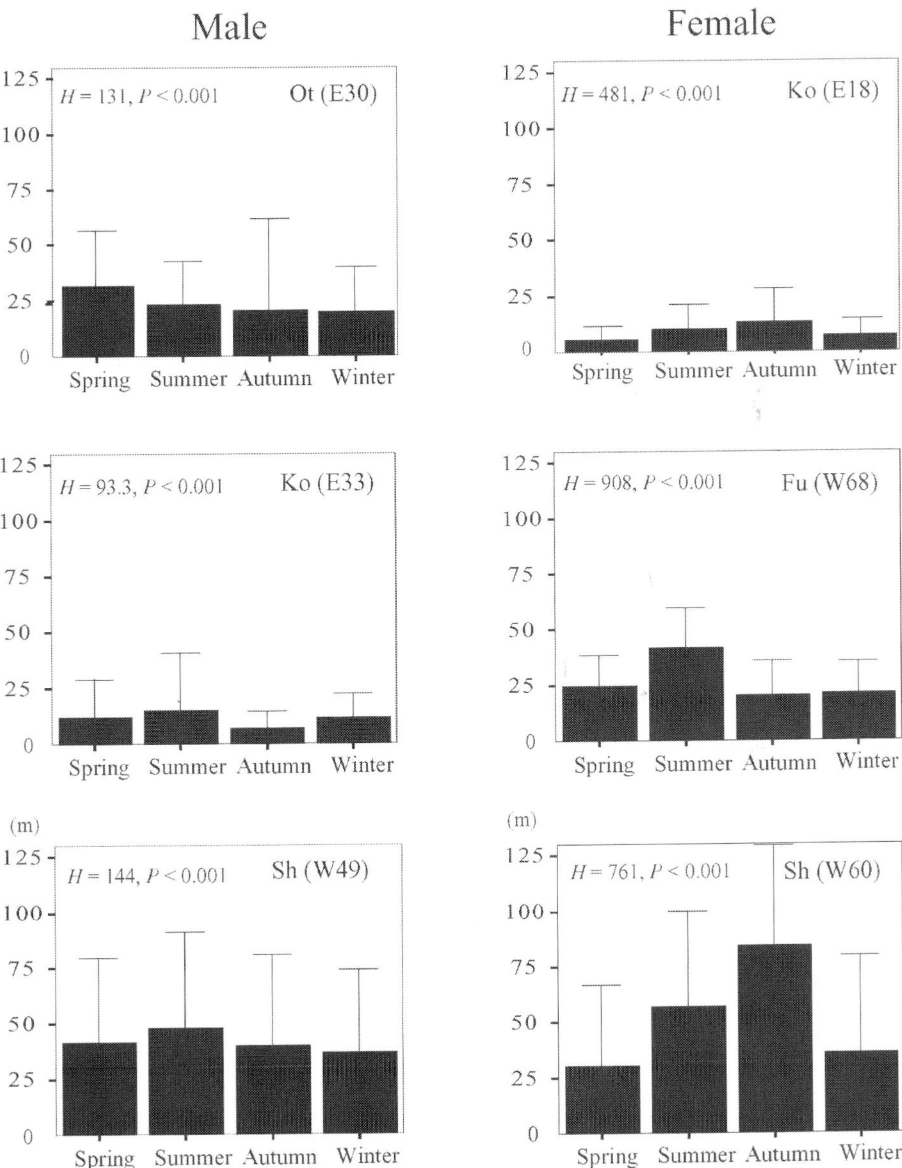

Figure 6. Seasonal elevation changes (mean+*SD*) of the radio-tracking locations of six Iriomote cats in four study sites (Ot: Otomi; Ko: Komi; Fu: Funaura; Sh: Shirahama).

2.3.4. Movement pattern

The movement tracks of radio-collared cats were derived from fulfilled radio-tracking locations (Figure 5) following the procedure in [32]. Data were collected at a rate of at least one hour intervals for more than 24 hours. However, if cats moved less than 100 m and rested at a fixed location, the data were accepted. The movement of each cat was characterized by calculating the daily movement distance (DMD: the sum of straight line distances between consecutive locations during 24-hour tracking sessions).

To determine the most important environmental characteristics influencing the animals' movement patterns, I measured the most important predictors derived from the above analyses of the movement tracks of radio-collared cats (MT) and compared them to the same variables for random tracks (RT), which were created by a Monte Carlo simulation using a random walk program in Biotas. A random walk is the most basic process for creating a spatially unpredictable data set. The process was operated with a point pattern using the same number of cat locations and straight line distance of MT during a consecutive tracking session without any specific direction. This process was able to "walk" in any direction within the home range. These statistical differences were tested using a Mann-Whitney U-test.

All statistical analyses were carried out using SPSS 11.5 for Windows (SPSS Inc., Illinois, USA). Statistical differences were accepted as significant when $P < 0.05$.

3. Results

3.1. Environmental characteristics among study sites

I measured nine environmental variables within the home ranges of the radio-collared cats and compared them among study sites (Figure 3). Continuous variables (El and Sl) were significantly different among study sites (one-way ANOVA: F=2389 for El and 3468 for Sl, $d.f.$=6, both $P < 0.0001$). Mean values of El were highest in Shirahama, followed by Urauchi and NCA. Those of Otomi, Maira, Funaura and Komi were relatively low. A similar pattern emerged for the case of slopes (Figure 3).

For categorical variables, percentages of drainages (Dr) and forest edges (FE) were also statistically different among sites ($P < 0.001$). Dr was the highest in Urauchi, followed by NCA, Maira and Komi. However, these values varied slightly, ranging from 4.9 to 7.0%, whereas FE varied largely depending on study sites. FE was particularly abundant in Maira, followed by Otomi. These values were lower in other sites.

Composition of vegetation type in each study site is illustrated in Figure 4 and differed significantly among sites (G=66826, $d.f.$=24, $P < 0.0001$). In Otomi, NCA, Funaura and Shirahama, vegetative environments were relatively uniform and mostly occupied by one or two vegetation types; NF and CP in Otomi, NF in NCA, NF and CP in Funaura and NF and SW in Shirahama. On the other hand, vegetative environments were more complex in Maira, Komi and Urauchi, as these were occupied by five vegetation types.

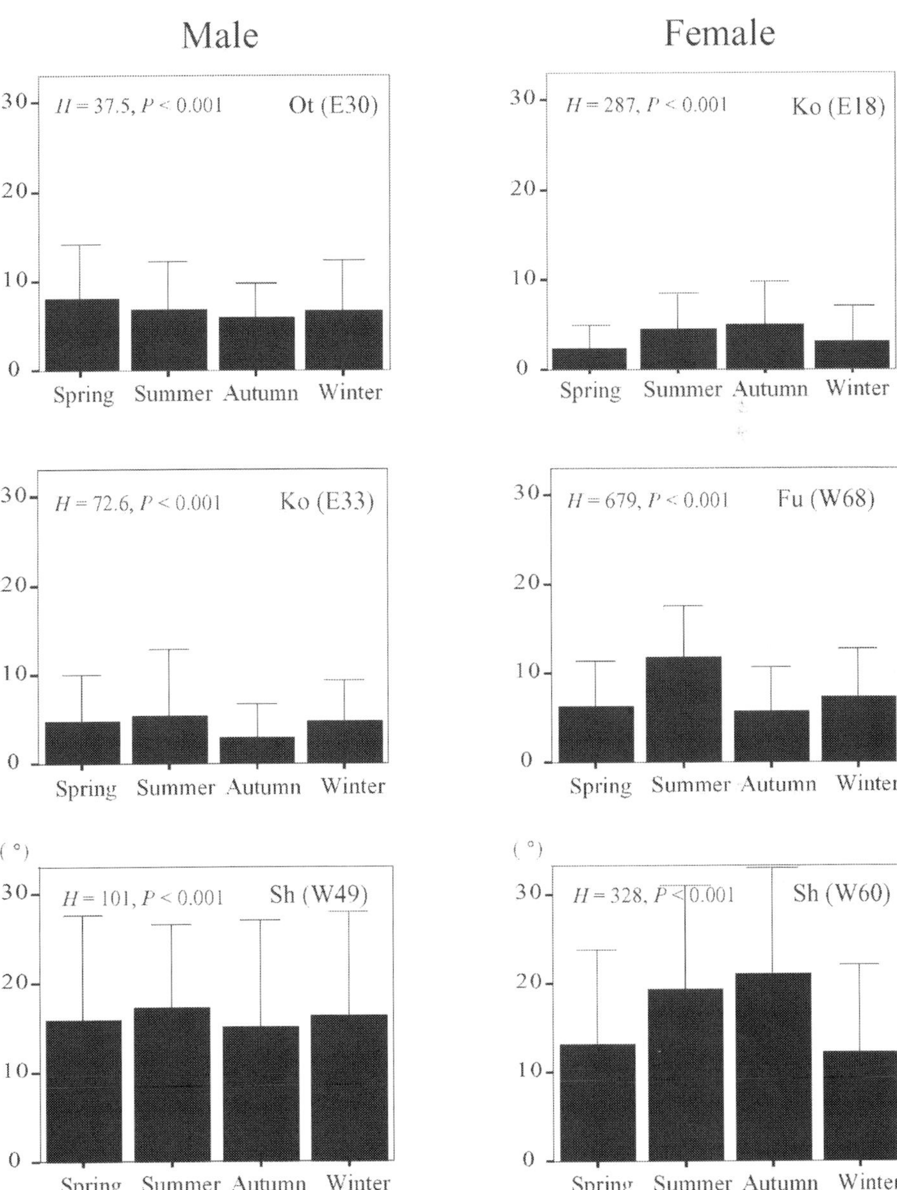

Figure 7. Seasonal slope change (mean+*SD*) of the radio-tracking locations of six Iriomote cats in four study sites (Ot: Otomi; Ko: Komi; Fu: Funaura; Sh: Shirahama).

3.2. Habitat use

3.2.1. Seasonal differences

I measured the seasonal differences for the major environmental characteristics of six radio-collared cats and the results are shown in Figures 6 to 8. For each cat, El and Sl at cat location sites significantly differed among seasons (Kruskal-Wallis test, $P < 0.001$: Figures 6 and 7). Vegetation types at cat locations were also significantly different among seasons for each cat (*G*-test, $P < 0.001$: Figure 8). Seasonal variations for both topographic variables in males were relatively low and varied comparatively more among females. In particular, a female cat in Shirahama used a lowland habitat during winter and spring, and used a higher habitat during summer and autumn (Figures 6 and 7). The vegetative compositions of used habitats varied relatively among females, more so than among males. Although the environments of cat location sites varied seasonally, environmental conditions essentially differed depending on study sites (Figure 8).

Study site	Logistic coefficient of covariate							Correct ratio (%)	Nagelkerke (R^2)	χ^2	P
	El	Sl	FE	SF	CV	RS	Intercept				
Otomi	-0.030		0.463		-2.073		1.13	83.3	0.366	21.9	0.0051
Maira	-0.035		0.821				0.65	85.0	0.339	19.2	0.0138
Komi	-0.040			0.745		1.488	0.50	78.9	0.405	42.7	<0.0001
NCA	-0.037						2.25	97.4	0.688	16.1	0.0414
Funaura	-0.023				-2.385	0.932	1.00	76.4	0.220	23.0	0.0033
Urauchi	-0.015						0.71	81.8	0.202	24.7	0.0017
Shirahama	-0.015	0.036		-1.483		1.041	0.75	83.0	0.502	18.1	0.0208

Table 3. Results of final logistic regression models regarding environmental variables (El: elevation; Sl: slope; FE: forest edges; SF: secondary forest; CV: coastal vegetation, rice fields and swamps) predicting radio-tracking locations vs. random locations in each study site.

3.2.2. Regional differences

I measured habitat variables at radio-tracking locations and random locations in each of the study sites; radio-tracking locations were significantly influenced by one to four environmental variables depending on study sites. These environmental variables were employed in forward stepwise procedures. By doing so, I obtained predictive models for cat location sites that were correctly classified at 76.4 to 97.4% (Table 3).

Elevation (El) was the only predictor selected in the models in all study sites, which indicates that the cat preferred lowland and avoided areas at higher altitudes in all study sites. In NCA and Urauchi, El was the only important predictor of cat locations. The cat preferred lowland, regardless of vegetative types. In four of five study sites (other than Funaura), El was the most important predictor of cat locations. Slope (Sl) was only chosen in the model in Shirahama that indicates the cat preferred sloping lands.

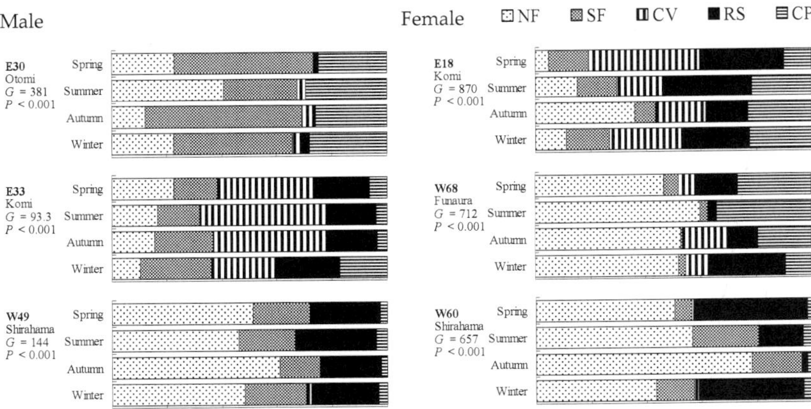

Figure 8. Seasonal vegetative changes for the radio-tracking locations of six Iriomote cats in four study sites (Ot: Otomi; Ko: Komi; Fu: Funaura; Sh: Shirahama).

Although vegetative variables also employed in the models of the five study sites, the influence varied depending on the sites. According to each predictive model, besides lowland areas, the cat preferred areas near forest edges in Maira and Otomi; preferred swamps in Komi, Funaura, and Shirahama; preferred secondary forests in Komi but it was avoided in Shirahama; the cat avoided coastal vegetation in Otomi and Funaura.

3.3. Movement pattern

I analysed the movement patterns of 11 cats (five males and six females, see Table 5). The total tracked time and distance were 3012 h and 359.6 km, respectively. DMD of each individual was calculated and is shown in Table 4. The DMD was slightly longer for males (3.36±1.04 km: x ±SD, N=5) than females (3.02±0.80 km, N=6), but lacked the same statistical support (Mann-Whitney U-test, U=14, P=0.93). Despite slight differences between the sexes, the DMD varied largely among study sites and was the longest in Funaura (4.61 km, N=2), followed by Otomi (3.06 km, N=1), Komi (3.04 km, N=2), Maira (3.03 km, N=2), NCA (2.80 km, N=2) and Shirahama (2.26 km, N=2), in this order. The DMD was positively related with home range size but lacked statistical supports (r=0.522, N=5, P=0.182, for males; r=0.575, N=6, P=0.116, for females).

Correlations of HR sizes and DMD against mean values of elevation and slopes in HR are shown in Figure 9. For male cats, DMD and HR were negatively closely related with mean values of slope, though not related with mean values of elevation. Meanwhile, there was not significant correlation between elevation and slope ($P > 0.05$).

The values of elevation and slope on movement tracks of the cats were compared with random tracks created by the random walk simulation (Table 5). Both variables were significantly lower on movement tracks than those on random tracks in all individuals (Mann-Whitney U-test, $P < 0.001$).

Study site	Cat name	Sex	Distance (m)	Track hour (h)	DMD (km)	HR (km^2)	El (m)	Sl (°)
Otomi	E30	M	32880	229	3.45	3.35	47.5	9.60
Maira	E36	F	9674	90	2.58	0.70	32.4	8.51
	E39	F	7750	53	3.48	0.49	15.6	5.66
Komi	E18	F	52196	458	2.73	0.66	28.4	8.07
	E32	F	20419	146	3.36	1.55	37.6	9.79
NCA	W86	M	2898	28	2.46	2.79	67.3	12.2
	W89	M	7936	61	3.14	2.63	66.3	12.3
Funaura	W61	M	10536	50	5.09	2.65	47.7	10.4
	W68	F	46518	270	4.13	1.75	24.9	6.92
Shirahama	W49	M	140572	1257	2.68	0.55	93.3	16.2
	W60	F	28253	369	1.84	0.76	75.9	16.0

Table 4. Movement characteristics of each individual Iriomote cat. DMD (daily movement distance), HR (home range size, estimated by 75% harmonic mean method), mean values of elevation (El) and slope (Sl) in HR.

Study site		N	Elevation (m: mean± SD)		U	P	Slope (°: mean± SD)		U	P
			Random	Movement			Random	Movement		
Otomi	E30	3036	67.9±49.0	25.5±21.8	2004615	<0.00001	12.6±8.72	6.53±5.17	2660791	<0.00001
Maira	E36	882	56.9±51.1	22.1±15.9	229749	<0.00001	12.0±8.50	6.44±4.94	209086	<0.00001
	E39	694	26.9±15.0	10.8±7.08	67732	<0.00001	7.44±6.04	4.82±3.13	166544	<0.00001
Komi	E18	4669	24.8±29.9	10.8±12.8	7670419	<0.00001	7.39±7.39	4.36±4.86	7626407	<0.00001
	E32	1788	37.1±37.4	9.41±13.9	553259	<0.00001	9.17±8.26	3.58±4.51	725216	<0.00001
NCA	W86	268	48.2±37.4	16.0±12.8	15260	<0.00001	10.2±6.17	5.08±3.65	18011	<0.00001
	W89	719	49.3±38.7	21.6±15.8	145632	<0.00001	11.2±6.92	7.80±5.28	183296	<0.00001
Funaura	W61	966	38.4±30.2	28.4±15.6	399199	<0.00001	10.1±7.72	6.54±4.52	311399	<0.00001
	W68	4191	44.2±46.1	28.7±22.1	6874122	<0.00001	9.33±7.11	8.68±6.55	8301671	0.00047
Shirahama	W49	12720	114 ±80.4	38.8±39.0	33984231	<0.00001	17.6±10.2	15.3±11.1	69846145	<0.00001
	W60	2503	107 ±62.1	43.2±42.9	1248043	<0.00001	19.5±10.7	16.4±12.4	2583308	<0.00001

Table 5. Elevation and slope ranges for the movement tracks of the Iriomote cat and for random tracks created by random walk simulations.

The movement tracks that were analysed are shown on the digital elevation model in Figure 5. When the cats went on long-distance walks, they avoided higher lands and selected flat routes to another area. In Otomi, Komi and Funaura, where flat lands largely occur, the movement tracks were distributed relatively uniform, whereas they were concentrated in flat lands in NCA and Shirahama, where these types of area are extremely limited.

3.4. Regional differences of the diet compositions

I analysed the contents of 805 scats collected within HRs of radio-collared cats: 70 scats in Otomi, 182 scats in Maira, 166 scats in Komi, 169 scats in NCA, 81 scats in Funaura, 45 scats in Urauchi and 92 scats in Shirahama. The result of the scat analysis is summarized in Figure 10.

Principal prey groups were different among sites; reptiles and birds in Otomi, reptiles and amphibians in Maira, birds and amphibians in Komi, birds and reptiles in NCA, birds, reptiles and amphibians in Funaura, mammals and birds in Urauchi, and birds in Shirahama, were most frequently preyed upon, respectively.

The cats' diets were relatively diversified with high frequencies of several prey groups in Otomi, Maira, Komi, NCA and Funaura; however, their diets were narrowed to mainly birds in Urauchi and Shirahama (Figure10).

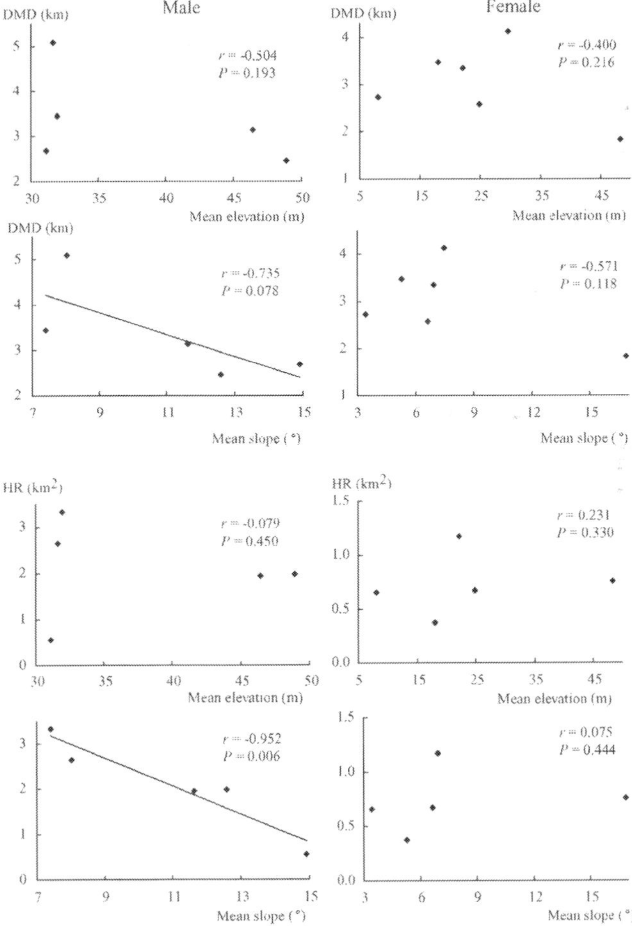

Figure 9. Correlations for daily movement distance (DMD) and home range size (HR) against mean values of elevation and slope in HRs.

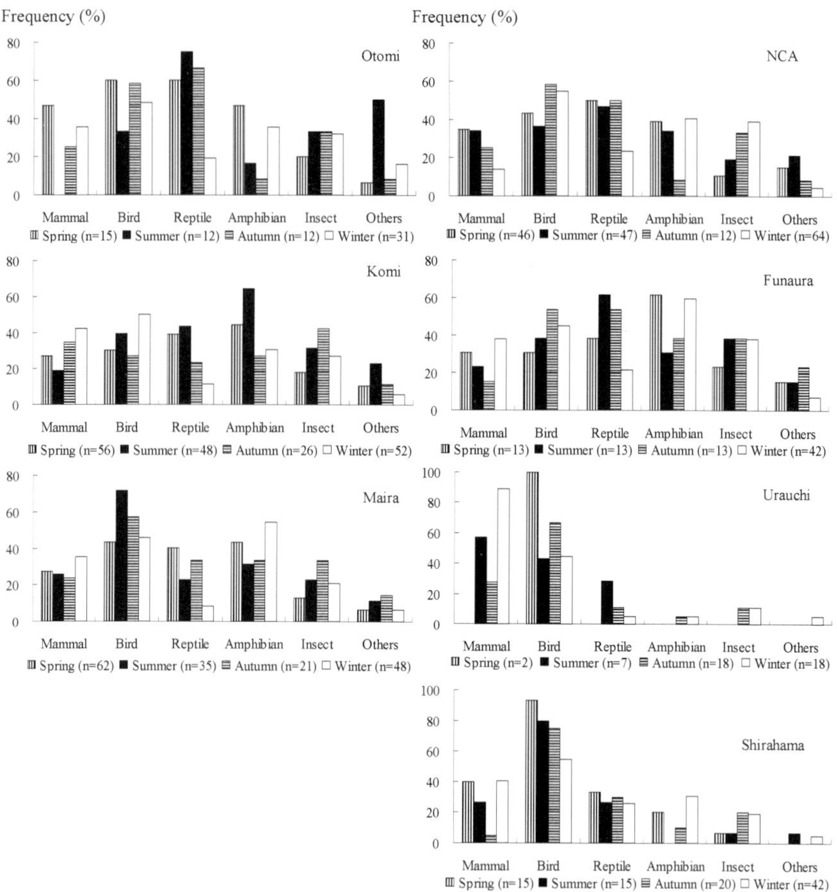

Figure 10. Seasonal and regional patterns of frequency occurrences (%) of prey groups found in 805 scats of the Iriomote cat.

4. Discussion

4.1. Feeding strategy of the Iriomote cat

The results of the present study showed seasonal variance in the habitat use of the Iriomote cat. Several studies have shown seasonal effects on the movements and habitat use of felids at various scales [33-39]. Deep snow and severe winter weather were shown to restrict home range size and habitat use [33, 36], as well as the movement patterns [37] of felids. Severe drought during the dry season was similarly shown to restrict movement patterns and the habitat use of some felids [38, 39]. In contrast with the study areas noted above, the climate of the Iriomote is extremely warm and humid throughout the year [40]. Thus, I believe that the seasonal variations in the habitat use of the Iriomote cat in the present study must be related to other factors.

In the present study, there were different levels of seasonal variations between sexes. Male cats had small seasonal variations; meanwhile, female cats used lowland habitats during winter and spring. The different seasonal patterns of habitat use between sexes may be connected to the gender differences in breeding cost. Female Iriomote cats breed and raise their young, while males are not involved in the raising of offspring [25, 41]. The breeding cycle of the cat is not seasonally restricted but a mating peak is recognized between February and April, and females deliver litters between April and June [41]. Accordingly, females need suitable habitats for breeding dens and for nursing their young [25, 41]. Three females were parous when they were captured. Thus, the females preferred and used lowland habitats during breeding; meanwhile, the habitat use of males were not as related to the breeding cycle.

Some studies of carnivores have suggested that suitable structures and sites for breeding dens are essential and limited within the animals' habitats. The distribution of breeding sites affects the habitat use of breeding females [42-45]. The breeding habitat use of Iberian lynxes *Lynx pardinus* is more strongly influenced by distribution of natal hollow trees than by prey availability and breeding females use old growth forests during breeding season [45]. It has also been reported that a female Iriomote cat used a hollow tree for breeding [46]. On Iriomote Island, hollow trees that are large enough in size for the breeding purposes of Iriomote cats were identified among several tree species, mostly *Quercus miyagii, Castanopsis sieboldii* and *Machilus thunbergii* (Watanabe, unpub. data); these only occurred in the NF vegetative category in the present study. Contrarily, females used lowland habitats with a low proportion of NF during the breeding season, suggesting that the availability of den sites was not the only important factor for the habitat use of breeding Iriomote cats. They may also use other structures or sites for breeding and as such, it is likely that an increase in food requirements for nursing influences the habitat use of female cats during breeding.

During nursing periods, females with young concentrated their movements near the den [47]. In the present study, the proportions of rice fields and swamps in home ranges of three females were higher than those of other vegetation types during the period. There were an abundance of birds and frogs in the habitat type during this period [7, 12, 40, 48]. Thus, the females intensively used the habitat to prey on abundant food sources for nursing their young.

In addition to seasonal differences, I observed significant alterations of habitat use among the Iriomote cats in different study sites. The habitat use of several widespread felids such as bobcats, lynxes, leopards and tigers have been studied in broad geographic ranges across several climatic zones [39]; the habitat use of each species varies somewhat according to region. This is likely the result of different climates, vegetation structures, or the principal prey species present in completely different environments. For leopard cats *Prionailurus bengalensis*, home range sizes differ among broad regions in Thailand [49-51], in Borneo [52], on Tsushima Island

[53] and on Iriomote [5, 47, 54]. On Iriomote, the vegetative and topographic conditions significantly differed among the study sites, which were only several kilometres away from one another. The narrow-regional differences among environments strongly affected the habitat use, movement pattern and diet of the Iriomote cats.

Bekoff et al. [19] assumed that prey distribution strongly affects the habitat use of predators. Generally, vegetation types have a larger effect on the prey distribution of carnivores [e.g., 55-60]. The distribution of principal prey items of the Iriomote cat is also chiefly determined by vegetative environment [7]. The habitat use of the Iriomote cat varied in relation to the regional differences of vegetative environments. Preferred vegetation types differed among regions. In addition, diet composition of the Iriomote cat differed between regions. Therefore, it is my conclusion that the cats changed their use of habitat in order to adapt to differences in prey distribution.

Although most environmental variables affected the habitat use of the Iriomote cat, only elevation showed strong and similar effects within all study sites. All radio-collared cats preferred using lowland habitats, mainly with elevation of less than 50 m, while they hardly used highlands with elevation more than 200 m. In particular, the effect of elevation was more highly correlated with the habitat use of the cats in rough terrains (NCA, Urauchi and Shirahama). Sometimes, elevation contributed as a factor for determining vegetation type and affected the mammalian fauna [e.g., 61-63]. However, this effect was the result of climatic changes that required a vertical interval of several thousand metres. In this study, a vertical interval of only 50 m limited the habitat use and movement of the Iriomote cat. Thus, the small difference of elevation did not cause climatic changes.

Sakaguchi [5] also reported that the Iriomote cat avoided highlands and suggested this to be due to scarce prey resources in the highlands. However, some principal prey items of the Iriomote cat such as reptiles and insects are more abundant in montane forests than in lowland habitats [7]. Thus, I believe that highlands encompass abundant prey items at a similar level of that found in lowland habitats and that prey availability is not a principal factor affecting the avoidance of highlands.

The habitat uses of predators are likely determined by how efficiently they seek and acquire food. This is because predators perform optimum feeding to maximize the efficiency of their energy acquirements [13]. In this regard, prey availability will be an important factor for habitat use. However, at the same time, if there is a high cost for acquiring food, the results will be inefficient. Schoener [64] assumed that energy acquirement efficiency fluctuated according to the relationship between food availability and the cost demanded for acquiring the food.

According to the feeding patterns of the Iriomote cat, the animal is considered to be an opportunistic mobile predator [5]; as such, the feeding cost can be represented as the movement cost expended when seeking prey. Several studies have suggested that movement cost is highly correlated with topographic condition, particularly slope [e.g. 65-67]. Walking speed decreases as the slope on walking routes increase [65] and the energy requirements of humans [66] and goats [67] during walking are much higher on a slope than on a flat surface. In the analysis results of movement tracks of the Iriomote cat, it was shown in all study sites that the animal preferred to move on areas with a lower slope and lower elevation. Thus, I hypothesize that the avoidance of highland areas by the Iriomote cat is a consequence of increasing movement cost based on the optimal feeding strategy: the cat performs habitat use to maximize the energy acquirement efficiency, depending on the cost and benefit of feeding.

Several studies of felids have suggested that the availability of suitable foraging spots in an area is limited [9, 68] and that felids flexibly shift their home range uses in response to prey availability. In studies of bobcats *Lynx rufus*, it has been suggested that prey availability affects home range sizes and large home ranges were reported in habitats with limited prey resources [33]. A cat in an area with scarce prey resources needs to expand its home range to acquire essential hunting spots and prey. However, in the present study, home range sizes tended to contract as the slope in home range increased. If a cat enlarged the home range in hilly habitats where the movement cost was particularly high, the cat acquired more food, but spent much more energy seeking out said prey than it acquired energy.

Corbett [69] reported that feral cats with larger ranges were more likely to use mobile than stationary or ambush hunting strategies, while cats with smaller ranges used stationary or ambush strategies more often. In this study, the movements of cats in hilly areas (NCA and Shirahama) were concentrated in limited lowlands, whereas cats with wide lowland habitats (Otomi, Komi and Funaura) utilized their home ranges uniformly. Thus, it is likely that cats in areas with wide lowland habitats used mobile opportunistic hunting strategies, whereas cats in areas with hilly lands were more inclined to using stationary or ambush hunting strategies in order to raise their feeding efficiency. Consequently, the cats kept small home ranges in hilly areas with high movement cost.

4.2. Evolution of flexible habits of the Iriomote cat

Consumers can be roughly classified as either specialists with a narrow diet range and generalists with a broad diet range. Discussions on the general or specialist characteristics of predators are common in ecological literature [70]. The distribution of diet widths, i.e., the range of food types eaten by an animal, differs among the various types of consumers [71]. In the case of carnivores, the topic is well-documented at interspecific levels [e.g., 19, 20]. Each felid species takes only a few different prey items of mammals, while other carnivores eat various food types. The family Felidae is highly specialized for preying on mammals in terms of developing their morphological and behavioural traits [21, 22]. Thus, Felidae is considered a typical food specialist. These patterns appear to be in diets of *P. bengalensis* in that they feed mostly on mice and rats [39, 51, 72]. However, the Iriomote cat is also considered as being a generalist, because they prey on various types of animals besides mammals [6-8]. What then generalizes their diet?

From an ecological perspective, the diet width of an animal is chiefly determined by the functional limitations of their feeding ability, that is, how many food types in its habitat the animal can consume [71]. For instance, all felids cannot digest vegetable matter due to physiological limitations, though other carnivorous families such as some mustelids and all ursids can. In addition, solitary felids generally do not hunt animals bigger than themselves, due to their morphological and behavioural limitations; however, felid species do hunt in groups, as lions and canids often do [22, 73]. In the diet of the Iriomote cat, *Sus scorfa* (wild pig), the largest animals on Iriomote Island, are hardly eaten. In addition, the Iriomote cat mostly preys on ground-living animals, while arboreal species are infrequently preyed upon [7]. This is likely due to limitations of their feeding ability. However, the Iriomote cat preys on nearly a hundred prey items belonging to wide range of taxonomical groups [7]. Thus, it is possible that they have developed a high feeding technique ability to prey on most types of ground-living animals.

The ancestral species of the Iriomote cat is common to other species of the genus *Prionailurus* [74], which presumably fed on small mammals living in continental environments, in which small mammals coexisted. As such, it is assumed that the Iriomote cat acquired a more diverse diet through further development of feeding functions to include a larger variety of animals.

To improve feeding ability, species are likely to develop morphological or behavioural functions for hunting [22, 73]. However, there is no notable morphological distinction between the Iriomote cat and other closely related species [74]. In addition, it has been reported that the hunting methods of the Iriomote cat have not been well-developed in relation to each of its prey types and is much more primitive than those of other small felids [75]. Furthermore, if a species develop its feeding patterns in relation to feeding on particular food types, they are likely to specialize in a narrow range of food types [76]. According to the dietary studies of *P. bengalensis* in other regions, in addition to mammals, they also eat birds, reptiles, amphibians and insects, though at low frequencies [2, 39, 51]. This suggests that the cats are also capable of feeding on these types of prey but that these prey types may be in the minority in environments, or are avoided by the cats. Therefore, it is likely that the Iriomote cat's well-developed hunting techniques are related to each of its various prey items.

It is more likely that regional differences of potential food resources influence diet widths of *P. bengalensis*. The insular fauna on Iriomote Island is characterized by a geological history that caused the absence of native non-volant small mammals [2], by the humid-subtropical climate that leads to the high abundance of floor-dwelling amphibians [40] and by the island's geographically suitable location for migrant birds that has led to the drastic seasonal changes in the abundance of and species composition of the avifauna [7]. Thus, small vertebrates are remarkably abundant on the island in spite of the scarcity of small mammalian fauna, which may be responsible for large differences in terms of the potential food resources of cats. However, these small vertebrates eaten by the Iriomote cat are commonly also eaten in other regions by other carnivores such as mustelids, viverrids and herpestids [21].

Absolute limits of diet widths are primarily defined by the cats' feeding abilities, but very few animals actually eat all of the different food types they are capable of consuming, thereby exhibiting their fundamental niche. This fundamental niche of a species in the absence of competitors from other species may be restricted to a realized niche in the presence of competitors [71]. In other regions where wild felid populations are present, several species of other carnivores coexist. Interspecific differences of morphological traits and diets among sympatric carnivore species have been reported as evidence for interspecific competition [e.g., 77-79]. Thus, it is likely that sympatric carnivore species compete with felids in the habitats, causing the diet widths of cats to be narrowed. There is no strong competitor for the cat on

Iriomote. Thus, its broad diet width range is possibly as evolutionary consequences of ecological release [e.g.70], due to the absence of competitors.

However, even when a competitor species is absent in a habitat, the species present will select food types with good quality in order to maximize reward intake [13, 80]. As such, its diet width will be narrower than its fundamental niche. At the one extreme, if the qualities of all food types are uniform and scarce, the predator may employ a generalist strategy that will tend to exhibit a broad diet, i.e., it will hunt and eat many of the food items that it comes into contact with. At the other extreme, if a food type with remarkably high quality is abundant in a habitat, the predator may employ a specialist strategy, have a narrow diet and ignore many of the prey items it comes across, preferring to search for a few specific types of food. In general, animals exhibit strategies ranging across a continuum between these two extremes in relation to food condition in their habitats [13, 80]. In addition, strategies will be more generalized when the food condition in a habitat is scarcer [14].

Food quality is generally determined by its energy content and its feeding efficiency, i.e., the ease of predation [13]. For small felids, food types with good quality are the most common prey item, for example, small terrestrial mammals such as rodents. However, these food types are scarce on Iriomote Island [2, 7]. If preferred food types with good qualities decreased in a habitat, the predator has two choices. First, it might maintain the selectivity of its diet and migrate to another suitable habitat; alternatively, it risks starving by staying in the same habitat. Second, the predator changes the selectivity of its diet and adapts to environmental changes by preying on other food types in the same habitat. In general, such flexible adaptations to environmental changes in the second choice are more difficult for specialized species [14]. Thus, it is likely that most felid species will opt for the first choice during food scarcity.

In general, island habitats are restricted in terms of food resources compared to those on continents. For animals living in continental habitats, if food conditions worsen, they may escape from starvation by migrating to other habitats with more resources. Indeed, it has been reported that the home ranges of *P. bengalensis* on the continent shifts seasonally, seeing them move to other habitats or enlarging their habitat in relation to prey distribution [51]. However, where animals in insular habitats are confined to the same habitats all the time, food conditions can vary. Thus, animals in insular habitats likely need some ecological flexibility in order to adapt to environmental variations. Consequently, it is likely that animals that have adapted to island habitats often have peculiar habits when compared to those found in continental sites [81]. For example, some insectivorous lizards adapted to insular environments have expanded their diets to include nectar, pollen and fruit [82]. To respond to flexibly as it concerns different food types, it may be better for feeding patterns to be unspecialized when it comes to particular food types. This is why the feeding behaviour of the Iriomote cat is more primitive and undeveloped when it comes to particular prey. It is also likely that the opportunistic feeding pattern of the Iriomote cat is suited for flexibly responding to variations in food availability.

Prionailurus bengalensis is considered a habitat generalist when compared to other small felids. While other species in the genus inhabit narrow habitat types [39], the habitat of *P. bengalensis* varies [2]. Such flexible habitat use may also play a role allowing the felid population to be present in such a small island. However, the results of the present study showed that the

Iriomote cat does not randomly forage for food in its habitat. They might instead have learned about food availability in relation to density and the distribution of prey from short-term experience, and as a result, adopt the most efficient feeding tactics. These flexible feeding patterns, as well as their diversified diets, are uncommon among Felidae. It is likely that the Iriomote cat optimizes most of habitat types on the island.

Therefore, the wide range of food habits of the Iriomote cat results significantly from the peculiar prey fauna of Iriomote Island [7, 40], the lack of competitors on the island [2] and the limited environment due to small island effects [7]. Moreover, I believe that the potential environmental adaptability, i.e., the fundamental niche of *P. bengalensis* appears only on Iriomote Island, which is therefore an essential area of study for the behavioural evolution of Felidae.

5. Conclusion

The leopard cat, *Prionailurus bengalensis*, is one of most widespread felids distributed through-out Asia. Although there are thousands of islands of various sizes within the range of distribution of the species, the species lives on several small islands as well as larger islands and the Asian continent. Iriomote Island (284 km²) of the Ryukyu Archipelago in southern Japan, is the smallest habitat of this species, on which the Iriomote cat *Prionailurus bengalensis iriomotensis*, a subspecies of leopard cats, lives.

On Iriomote Island, there are no autochthonous terrestrial small mammals such as rodents, which are generally principal prey of wild felids. Thus, it is likely that there are unique characteristics in the biodiversity of the island and in the ecology of this particular cat as the top predator in the ecosystem. In the present study, I investigated the characteristics of the ecology of the Iriomote cat concerning food habits, habitat use and movement patterns.

I conducted radio-tracking surveys in seven study sites. I examined a location fixed by the radio-tracking in each study site in terms of the cats' habitat preferences related to nine topographic and vegetative variables by using a geographic information system (GIS). Then, the seasonal and regional patterns of the cats' habitat use were examined in terms of feeding patterns.

The results showed that all studied cats selectively used their habitats. Cat locations were significantly influenced by six to eight environmental variables, depending on the study sites. To determine the most important topographic and vegetative factors influencing their habitat use, I attempted a logistic regression function following a forward stepwise approach and with environmental determinants in each study site. Suitable habitats evaluated from logistic regression models more or less differed among study sites. For the comparison of habitat use among study sites, elevation was the only variable to significantly relate to the cats' habitat use in all study sites, while the effects of other variables varied depending on the particular study site. In the results of assessing the prey availability of the cat, distribution and the abundance of their principal prey species were chiefly influenced by vegetative environments.

The compositions of vegetative types differed among study sites. Thus, prey distribution and abundance also varied according to site, which potentially influenced the regional differences of suitable habitats.

The diet of the Iriomote cat was examined in terms of the seasonal and regional differences of each prey type by analysing 805 Iriomote cat scat contents collected from various environments in the seven study sites. The results showed that the cat seasonally shifted principal prey items as it concerned food availability. They preyed on items that were abundant in environments. In addition to seasonal differences, the diet compositions also differed among study sites. Thus, it is likely that the cat feeds on abundant prey types depending on the regional differences of environments, as well as seasonal differences.

To determine the most important topographic characteristics influencing Iriomote cats' movement patterns, I measured the elevation and slope of the movement tracks of radio-collared cats and compared them to the same variables on random tracks created by Monte Carlo simulations. The results showed that values of elevation and slope were significantly lower in all individuals on movement tracks than those on random tracks. This suggests that the cat moves from one area to another to avoiding steep paths.

Predators maximize the energy acquirement efficiencies that fluctuate within the relationship between prey abundance and the demanded cost for acquiring prey. For mobile predators such as the Iriomote cat, demanded costs for feeding are significantly determined by costs required for foraging. Thus, highly folded habitats will increase feeding costs. In addition, the cat may avoid areas at high elevation, irrespective of prey abundance.

From the above results, I have concluded that the broad range of food niches of the Iriomote cat likely resulted from making the best possible use of fauna on the small subtropical island. Furthermore, the cat has adapted to the island-wide environment in order to change its principal prey items and feeding patterns in relation to the spatial and temporal variations of food availability. Most small felids may potentially have such flexibility in their ecology. However, this might only be the case in the uniquely biodiverse environment of Iriomote Island.

Acknowledgements

This study would not have been possible without the support of numerous individuals and institutions. I would especially like to acknowledge Professors M. Izawa, M. Tsuchiya, A. Hagiwara and Dr N. Nakanishi at the University of the Ryukyus, as well as Professor T. Doi at Nagasaki University, Dr M. Okamura and Dr N. Sakaguchi at the Ministry of the Environment and Dr A. Takahashi at the National Institute of Polar Research for their valuable comments relating to this study. I am also grateful to the Kumamoto Regional Forestry Office of the Forestry Agency and the Iriomote Wildlife Center of the Japan Ministry of the Environment (IWCC) for supporting this research. The scat censuses of the Iriomote cat were conducted as a part of endangered species conservation projects administered by the Kumamoto

Regional Forestry Office. Field work was carried out using IWCC facilities. I also thank the members of the Ecology Laboratory, Faculty of Science, University of the Ryukyus. I also appreciate the kindness of individuals at the IWCC and the many inhabitants of Iriomote Island, Ms C. Matsumoto and Mr M. Matsumoto in particular, for their assistance during the field work. This study was partially funded by the Inui Memorial Trust for Research on Animal Science, Fujiwara Natural History Foundation, Pro Natura Foundation Japan and the 21st Century COE Program of the University of the Ryukyus.

Author details

Shinichi Watanabe*

Address all correspondence to: watanabe@ma.fuma.fukuyama-u.ac.jp

Department of Marine Bio-Science, Fukuyama University, Japan

References

[1] MacArthur RH, Wilson EO. The theory of island biogeography. New Jersey: Princeton University Press; 1967.

[2] Watanabe S. Factors affecting the distribution of the leopard cat *Prionailurus bengalensis* on East Asian islands. Mammal Study 2009; 34(4): 201-207.

[3] Darlington PJ. Zoogeography: the geographical distribution of animals. New York: John Wiley and Sons; 1957.

[4] Corbet GB, Hill JE. The mammals of the Indomalayan region. Natural history museum publications. London: Oxford University Press; 1992.

[5] Sakaguchi N. Ecological aspects and social system of the Iriomote Cat *Felis iriomotensis* (Carnivore; Felidae) PhD thesis. Kyushu University; 1994.

[6] Sakaguchi N, Ono Y. Seasonal change in the food habits of the Iriomote cat *Felis iriomotensis*. Ecological Research 1994; 9(2): 167-174.

[7] Watanabe S. Ecological flexibility of the top predator in an island ecosystem: food habit of the Iriomote cat. In: Ali M. (ed.) Diversity of Ecosystems. Rijieka: InTech; 2012. p465-484. Available from: http://www.intechopen.com/books/diversity-of-ecosystems/ecological-flexibility-of-the-top-predator-in-anisland-ecosystem-food-habit-of-the-iriomote-cat

[8] Watanabe S, Nakanishi N, Izawa M. Habitat and prey resource overlap between the Iriomote cat *Prionailurus iriomotensis* and introduced feral cat *Felis catus* based on assessment of scat content and distribution. Mammal Study 2003; 28: 47-56.

[9] Kleiman DC, Eisenberg JF Comparisons of canid and felids social systems from an evolutionary perspective. Animal Behaviour 1975; 21(4): 637-659.

[10] Everett DD. (ed.) Cats of the world: Biology, conservation, and management. Washington DC: National Wildlife Federation; 1982. p353-374.

[11] Sakaguchi N, Murata S, Nishihira M. Regional difference of food items of the Iriomote cat, *Felis iriomotensis*, based on the study of feces contents. Island Studies in Okinawa1990; 8: 1-13. (in Japanese with English summary)

[12] Watanabe S, Izawa M. A technique for the estimation of distribution and the evaluation of habitat suitability for terrestrial vertebrate fauna on Iriomote Island. Proceedings of the Geographic Information Systems Association 2002; 11: 345-348. [in Japanese with English abstract]

[13] MacArthur R, Pianka E. On the optimal use of a patchy environment. American Naturalist 1966; 100(916): 603-609.

[14] Stephens DW, Krebs JR. Foraging Theory. New Jersey: Princeton University Press; 1986.

[15] Krebs CJ, Myers JH. Population cycles in small mammals. Advances in Ecological Research 1974; 8: 267-399.

[16] Hansson L, Henttonen H. Gradients in density variations of small rodents: the importance of latitude and snow cover. Oecologia 1985; 67: 394-402.

[17] Hansson L, Henttonen H. Rodent dynamics as community processes. Trends in Ecology and Evolution, 1988; 3(8): 195-200.

[18] Krebs, CJ. Population cycles revisited. Journal of Mammalogy 1996; 77(1): 8-24.

[19] Bekoff M, Daniels TJ, Gittelman JL. Life history patterns and the comparative social ecology of carnivores. Annual review of ecology and systematics 1984; 15: 191–232.

[20] Gittleman JL. Carnivore life histories: are-analysis in light of new models. Symposium of the zoological Society London 1993; 65: 65-88.

[21] Kruuk H. Interactions between Felidae and their prey species: a review. In: Miller SD, Everett DD. (ed.) Cats of the world: Biology, conservation, and management. Washington DC: National Wildlife Federation; 1982. p353-374.

[22] Kleiman DC, Eisenberg JF Comparisons of canid and felids social systems from an evolutionary perspective. Animal Behaviour 1975; 21(4): 637-659.

[23] Andersson M, Erlinge S. Influence of predation on rodent populations. Oikos 1977; 29(3): 12-16.

[24] Kolowski J, Woolf A. Microhabitat use by bobcats in southern Illinois. Journal of Wildlife Management 2002; 66: 822-832.

[25] Okamura M. A study on the reproduction and social systems of the Iriomote cat, *Felis iriomotensis*. PhD thesis. Kyushu University; 2002.

[26] Eastman JR. IDRISI Kilimanjaro Tutorial. [CD-ROM] Worcester, MA: Clark University; 2003.

[27] Environment Agency, Japan. Natural Environment Information GIS. [CD-ROM] Tokyo: Nature Conservation Bureau; 1999 (in Japanese).

[28] Sakaguchi N. Habitat of the Iriomote cat and changes in its response to prey availability. In: (eds.) Proceedings of the International Symposium on Wildlife Conservation, Tokyo, Japan, 21-25 August 1990, Tsukuba and Yokohama; 1991.

[29] Watanabe S, Nakanishi N, Sakagichi N, Doi T, Izawa M. Temporal changes of land covers analyzed by satellite remote sensing and habitat states of the Iriomote cat *Prionailurus iriomotensis*. Bulletin of the International Association for Landscape Ecology-Japan 2002; 7(2): 25-34. (in Japanese with English abstract)

[30] Dixon KR, Chapman, JA. Harmonic mean measure of animal activity areas. Ecology 1980; 61: 1040-1044.

[31] Freund RJ, Wilson WJ. Regression analysis: statistical modeling of a response variable. New York: Academic Press; 1998.

[32] Watanabe S, Nakanishi N, Izawa M, Doi T. Three-dimensional analysis of home range utilization by the Iriomote cat *Felis iriomotensis* using digital elevation models (DEM) (Perspectives on the mammalian home range study by telemetry).Japanese Journal of Ecology 2002; 52(2): 259-263. (in Japanese)

[33] Bailey T N. Social organization in a bobcat population. Journal of Wildlife Management 1974; 38: 435-446.

[34] Koehler GM, Hornocker MG. Influences of seasons on bobcats in Idaho. Journal of Wildlife Management 1989; 53: 197-202.

[35] Palomares F, Delibes M, Revilla E, Calzada J, Fedriani, JM. Spatial ecology of Iberian lynx and abundance of European rabbits in southwestern Spain. Wildlife Monographs 2001;148: 1-36.

[36] Kolowski JM, Woolf A. Microhabitat use by bobcats in southern Illinois. Journal of Wildlife Management 2002; 66: 822-832.

[37] McCord CM. Selection of winter habitat by bobcats on the Quabbin Reservation, Massachusetts. Journal of Mammalogy 1974; 55: 428-437.

[38] Bailey TN. Social organization in a bobcat population. Journal of Wildlife Management 1974; 38: 435-446.

[39] Sunquist M, Sunquist F. Wild cats of the World. Chicago and London: The University of Chicago Press; 2002.

[40] Watanabe S, Nakanishi N, Izawa M. Seasonal abundance of the floor-dwelling frog fauna on Iriomote Island of the Ryukyu Archipelago, Japan. Journal of Tropical Ecology 2005; 21: 85-91.

[41] Okamura M, Doi T, Sakaguchi N, Izawa M. Annual reproductive cycle of the Iriomote cat *Felis iriomotensis*. Mammal Study 2000; 25: 75-85.

[42] Brainerd SM, Helldin JO, Lindstrom ER, Rolstad E, Rolstad J, Storch I. Pine marten (*Martes martes*) selection of resting and denning sites in Scandinavian managed forests. Annales Zoologici Fennici 1995; 32: 151-157.

[43] Buskirk SW, Yiqing M, Zhaowen J. Winter habitat ecology of sables (*Martes zibellina*) in relation to forest management in China. Ecological Applications 1996; 6: 318-325.

[44] Lariviere S. Messier F. Denning ecology of the striped skunk in the Canadian prairies: implications for waterfowl nest predation. Ecological Society 1998; 35: 207-213.

[45] Fernándes N, Palomares F. The selection of breeding dens by the endangered Iberian lynx (*Lynx pardinus*): implications for its conservation. Biological Conservation 2000; 94: 51-61.

[46] Okamura M, Sakaguchi N, Izawa M, Doi T. A record on a breeding site of the Iriomote cat, *Felis iriomotensis*. Island Studies in Okinawa 1995; 13: 1-6. (in Japanese with English abstract)

[47] Schmidt K, Nakanishi N, Okamura M, Doi T, Izawa M. 2003. Movements and use of home range in the Iriomote cat (*Prionailurus bengalensis iriomotensis*). Journal of Zoology 2003; 261: 273-283.

[48] Watanabe S, Izawa M. Species composition and size structure of frogs preyed by the Iriomote cat *Prionailurus bengalensis iriomotensis*. Mammal Study 2005; 30(2): 151-155.

[49] Grassman LI. Movements and diet of the leopard cat *Prionailurus bengalensis* in a seasonal evergreen forest in south-central Thailand. Acta Theriologica 2000; 45: 421-426.

[50] Grassman LI, Tewes ME, Silvy NJ, Kreetiyutanont K. Spatial organization and diet of the leopard cat *Prionailurus bengalensis* in north-central Thailand. Journal of Zoology 2005; 266: 45-54.

[51] Rabinowitz A. Notes on the behavior and movements of leopard cats, *Felis bengalensis*, in a dry tropical forest mosaic in Thailand. Biotropica 1990; 22(4): 397-403.

[52] Rajaratnam R, Sunquist M, Rajaratnam L, Ambu L. Diet and habitat selection of the leopard cat (*Prionailurus bengalensis borneoensis*) in an agricultural landscape in Sabah, Malaysian Borneo. Journal of Tropical Ecology 2007; 23: 209-217.

[53] Izawa M, Doi T, Ono Y. Ecological study on the two species of Felidae in Japan. In: Maruyama N, Bobek B, Ono Y, Regelin W, Bartos L, Ratcliffe PR. (ed.) Wildlife Con-

servation: present trends and perspectives for the 21st century. Tokyo: Sankyo; 1991. p141-143.

[54] Nakanishi N, Okamura M, Watanabe S, Izawa M, Doi T. The effect of habitat on home range size in the Iriomote Cat *Prionailurus bengalensis iriomotensis*. Mammal Study 2005; 30(1): 1-10.

[55] Gese EM, Ruff RL, Crabtree RL. Foraging ecology of coyotes (*Canis latrans*): the influence of extrinsic factors and a dominance hierarchy. *Canadian Journal of Zoology* 1996; 74: 769-783.

[56] Joshi AR, Smith JJD, Cuthbert FJ. Influence of food distribution and predation pressure on spacing behavior in palm civets. Journal of Mammalogy 1995; 76: 1205-1212.

[57] Litvaitis JA, Sherburne JA, Bissonette JA. Bobcat habitat use and home range size in relation to prey density. Journal of Wildlife Management 1986; 50: 110-117.

[58] Powell RA. Effects of scale on habitat selection and foraging behavior of fishers in winter. Journal of Mammalogy 1994; 75: 349-356.

[59] Revilla E, Palomares F, Delibes M. Defining key habitats for low density populations of Eurasian badgers in Mediterranean environments. Biological Conservation 2000; 95: 269-277.

[60] Waller JS, Mace RD. Grizzly bear habitat selection in the Swan Mountains, Montana. Journal of Wildlife Management 1997; 61: 1032-1039.

[61] Heaney LR. Small mammal diversity along elevational gradients in the Philippines: an assessment of patterns and hypotheses. Global ecology and biogeography 2001; 10: 15-39.

[62] Li JS, Song YL, Zeng ZG. 2003. Elevational gradients of small mammal diversity on the northern slopes of Mt. Qilian, China. Global ecology and biogeography 2003; 12: 449-460.

[63] Rickart EA. Elevational diversity gradients, biogeography, and the structure of montane mammal communities in the intermountain region of North America. Global Ecology and Biogeography 2001; 10: 77-100.

[64] Schoener T. Theory of feeding strategies. Annual review of ecology and systematics 1971; 2: 369-404.

[65] Tobler W. Three presentations on geographical analysis and modeling: Technical Report 1993; 93 (1). National Center for Geographic Information and Analysis, Santa Barbara University of California.

[66] Terrier P, Aminian K, Schutz Y. Can accelerometry accurately predict the energy cost of uphill/downhill walking? Ergonomics 2001; 44: 48-62.

[67] Lachica M, Prieto SO. The energy costs of walking on the level and on negative and positive slopes in the Granadina goat (*Capra hircus*). British Journal of Nutrition 1997; 77: 73-81.

[68] Hornocker MG. An analysis of mountain lion predation upon mule deer and elk in the Idaho Primitive area. Wildlife Monograph 1970; 21: 1-39.

[69] Corbett LK. Feeding ecology and social organization of wildcats (*Felis silvestris*) and domestic cats (*Felis catus*) in Scotland. PhD thesis. University of Aberdeen; 1979.

[70] Futuyma DJ, Moreno G. The evolution of ecological specialization. Annual Review of Ecology and Systematics 1988; 19: 207-233.

[71] Begon M, Harper J, Townsend C. Ecology: individuals, populations and communities, 3rd ed. Oxford: Blackwell Science Ltd; 1996.

[72] Inoue T. The food habit of Tsushima leopard cat, *Felis bengalensis* sp., analyzed from their scats. Journal of the Mammalian Society of Japan 1972; 5: 155-169. (in Japanese with English abstract)

[73] Gittleman JL. Carnivore Behavior, Ecology, and Evolution Volumes 1. New York: Cornell University Press; 1989.

[74] Leyhausen P, Pfleiderer M. The systematic status of the Iriomote cat (*Prionailurus iriomotensis Imaizumi 1967) and the subspecies of the leopard cat (Prionailurus bengalensis* Kerr 1792). Journal of Zoological Systematics and Evolutionary Research 1999; 37: 121-131.

[75] Yasuma S. 1979. Food habits and feeding behaviour of the Iriomote cat. PhD thesis. Tokyo University; 1979.

[76] Krebs JR, Davies NB (eds.) Behavioural Ecology: an evolutionary approach (3rd edition). Oxford: Blackwell Scientific Publications; 1991.

[77] Dayan T, Simberloff D. Character displacement, sexual dimorphism and morphological variation among British and Irish mustelids. Ecology 1994; 75: 1063-1073.

[78] Dayan T, Simberloff D, Tchernov E, Yom-Tov Y. Inter-and intraspecific character displacement in mustelids. Ecology 1989; 70: 1526-1539.

[79] Virgós E, Llorente M, Cortés Y. Geographic variation in genet (*Genetta genetta*) diet: a literature review. Mammal Review 1999; 29: 119-128.

[80] Charnov EL. Optimal foraging: attack strategy of a mantid. American Naturalist 1976; 110: 141-151.

[81] Lack D. Island biology, illustrated by the land birds of Jamaica. Berkeley: University of California Press; 1976.

[82] Olesen JM, Valido A. Lizards as pollinators and seed dispersers: an island phenomenon. TRENDS in Ecology and Evolution 2003; 18: 177-181.

Chapter 15

Grasses (Poaceae) of Easter Island — Native and Introduced Species Diversity

Víctor L. Finot, Clodomiro Marticorena,
Alicia Marticorena, Gloria Rojas and Juan A. Barrera

Additional information is available at the end of the chapter

http://dx.doi.org/10.5772/59154

1. Introduction

Rapa Nui (Easter Island, Isla de Pascua), also known as Te Pito O Te Henua, is a small oceanic island of volcanic origin discovered by the Dutch explorer Jakob Roggeveen in April 1722. It has belonged to Chile since 1888 and is administratively part of the Region of Valparaíso, Province Isla de Pascua. At around 163.6 km², it is the largest island of the Chilean insular territory, situated in Polynesia, *ca.* 3,700 km from continental Chile in the Pacific Ocean (27°7′S, 109°22′W). Rapa Nui is considered the most remote inhabited island in the world, with a population of nearly 5,800 inhabitants. Approximately 43.5 % of the island territory is under the protection of the National System of Wild Protected Areas of the State of Chile (SNASPE). The Rapa Nui National Park, administered by the National Forest Corporation of Chile (CONAF) was created on 16th January, 1935 and declared a World Heritage Site by UNESCO in 1995 to protect the Rapa Nui culture, and especially the 887 statues known as *moai* [1].

The climate is warm and sub-tropical. The flora of Rapa Nui is extremely poor compared to other oceanic tropical islands [2]. Approximately 40 % of the flora is indigenous. Nearly 23 % of the vascular flora is represented by endemic species, and some 20 species of the native flora, 10 of which are endemic, have disappeared or are endangered, principally due to invasive plants, fire, overgrazing and agriculture, among other factors [3]. Nearly 90 % of the territory corresponds to herbaceous vegetation, with species of Poaceae (grasses) as the principal component, most of them alien [4]. There are, however, very dense little forests composed of species that have come with human beings since the island was colonized. Wetlands are located chiefly in the craters of volcanoes, the largest in Rano Kau (Fig. 1) and others in Ranu Raraku and Ranu Aroi. The flora of these wetlands consists of *Schoenoplectus californicus, Persicaria acuminata, Cyperus eragrostis, Cyperus polystachyos* and *Sorghum halepense*. It has been suggested

that the original vegetation of the island was represented by palm-dominated forests that have since disappeared and been replaced by a large number of introduced species that became naturalized. Pollen analyses of lake sediment cores showed a replacement of forests dominated by the palm species *Paschalococcus disperta* by grass-dominated vegetation communities [5]. Deforestation would have occurred either in AD 1000-1200 or 600 years later. It has been suggested that the deforestation of the island occurred due to intense human activity, including clearing, introduction of the Polynesian rat, fire and agriculture, among other factors [7, 8]. However, the existence of vegetation dominated by trees, as well as the proposed ecological disaster, still needs to be proven conclusively [5].

Grasslands present on the island can be divided into two types: 1. Very low grasslands, with high species diversity, overgrazed by livestock, especially horses. 2. Higher grasslands composed almost exclusively of *Melinis minutiflora*.

Figure 1. Crater of the volcano Rano Kau. Photo: G. Rojas.

The number of species reported for the island is not consistent in the scientific literature. Most of the differences in number of species occur, probably, because the authors include or exclude cultivated plants, and due to synonyms and nomenclatural changes.

Castro *et al.* [6] report only 40 species of monocots for Rapa Nui from a total of 121 vascular plants; these authors did not specify how many species of Poaceae there are; on the basis of 121 vascular species, Poaceae represent *ca.* 44 % of the entire vascular flora of the island. Previously, Skottsberg [9] indicated 44 species from 18 families of vascular plants, including eight genera and 12 species of Pteridophyta, and 28 genera and 32 species of Spermatophyta; Poaceae comprises 10 species. Two decades before, Fuentes reported *ca.* 124 species from 104 genera and 48 families [10]; five species reported by Fuentes are non-vascular plants and Poaceae numbers 19 species. Zizka [12] reported 100 wild angiosperms. Zizka [13, 14] reported 46 species of Poaceae. Dubois *et al.* [3] found 21 species of grasses. It has been suggested that more than 370 species of vascular plants have been introduced by humans to Rapa Nui, of which some 180 became naturalized [3].

The aim of this chapter is to provide a synopsis of the diversity of the family Poaceae (Gramineae) in Rapa Nui, to provide a catalogue of all species of Poaceae and to analyse the completeness of the inventory, to analyse the taxonomic distribution, life cycle, photosynthetic pathway, and phytogeographical origin of Poaceae in Rapa Nui and to compare the diversity of Poaceae in Rapa Nui with those of other oceanic islands. To date, the most complete list of grass species published on the flora of Rapa Nui comprises 46 species [14]. Our data indicate that in Rapa Nui, the family Poaceae comprises 50 species and one infraspecific taxon, from 37 genera and seven subfamilies. Recent taxonomic treatments have followed to update the nomenclature and the classification of the species.

2. History of the botanical expeditions to Rapa Nui

The first botanical collections made in Rapa Nui were those of Johann Reinhold Forster and his son Georg, during Cook's second voyage aboard the "Resolution", who sighted Rapa Nui on 13th March, 1774, and the next day landed at Hanga Roa [15]. Both Cook and Forster made similar and interesting comments, mainly on crop species such as sugarcane, potatoes and bananas. They also mentioned *Sophora toromiro* (Fabaceae) as the only native shrub species growing on the island, which was scarce, and with hard and heavy wood [16]. Georg Forster [15] cited 9 species on the island in the *Florulae insularum Australium Prodomus* which three belong to the grass family Poaceae: *Saccharum officinarum, Panicum filiforme* and *Avena filiformis (=Lachnagrostis filiformis).*

During a voyage of the Russian ship "Rurik", commanded by Captain Kotzebue, and with Adelbert von Chamisso, a naturalist, on board, it passed by Rapa Nui on a short visit; apparently, a small botanical collection was made [18]. Between 1825 and 1828, Captain Beechey's journey of exploration in the "Blossom" visited several locations in Chile, of which an important record is retained in the publications of Hooker and Arnott (1830 and 1832). The ship arrived at Rapa Nui on 16th November, 1825, but there are no records of plants [19].

Endlicher [20] in *Bermerkungen über die Flora der Südseeinseln*, lists numerous species on the islands they visited. For Rapa Nui, he mentions 11 species previously studied by Chamisso and Forster, but no specimens were deposited in herbarium. In this work, five species of Poaceae were recorded: *Paspalum filiforme, Agrostis conspicua, Deyeuxia chamissonis* (=*Lachnagrostis filiformis*), *Deyeuxia forsteri* (=*Lachnagrostis filiformis*) and *Lepturus repens*.

Savatier, during the campaign of the "Magicienne", reached Rapa Nui in August, 1877. Plants collected on the island are kept in the herbarium of the Museum of Paris, but no list was published. In 1885, Hemsley, in his *Report of the present stage of knowledge of various insular floras* [21], listed species of the island that were present in the work of Endlicher (except *Centaurea apula*), and also mentioned that there were other widely distributed plants that had been collected, such as the already known *Sophora tetraptera* and *Sesuvium portulacastrum*. The zoologist Alexander Agassiz, as a member of the Albatross expedition to the Tropical Eastern Pacific, visited Rapa Nui in 1904, making an important collection of plant specimens that he sent to Cambridge, Gothenburg and Washington.

In 1911, Francisco Fuentes was commissioned by the Chilean government to conduct a study on Rapa Nui. As a result of this work, Fuentes published his *Reseña botánica sobre la isla de Pascua* [10], where he mentions 135 species, of which 40 % are native or naturalized, and of these 25 are typically tropical. Grasses are represented by 19 species and 14 genera. He mentions that grasses cover the entire surface of the island, forming a steppe-like vegetation consisting mainly of *Paspalum orbiculare* (=*P. scrobiculatum* var. *orbiculare*), *Sporobolus indicus*, *Eragrostis diandra* (probably *E. tenuifolia* or *E.atrovirens*), *Andropogon halepensis* (=*Sorghum halepense*) and *Panicum sanguinale* (=*Digitaria sanguinalis*).

Between 5th October, 1916, and 26th, September, 1917, Skottsberg and his wife made a major Swedish expedition to explore and study the Juan Fernández Archipelago and Rapa Nui, which culminated in the publication of *The natural history of Juan Fernández and Easter Island* [17]. On 15th June, they arrived at La Pérouse Bay [19]. They collected 30 species that were probably indigenous or naturalized and four species that were semi-naturalized, but certainly introduced by the first inhabitants because of their usefulness, and 24 species accidentally introduced. This research increased by 23 the species mentioned by Fuentes, in which nine species of grasses are given. In 1927, the names of the plants collected in 1918 by Gusinde were also provided.

The Franco-Belgian mission exploring Rapa Nui from 29th July, 1934, to 3rd January, 1935, collected 61 species; the few that were not reported previously are obviously introduced, and known angiosperms number 142 [16]. The list mentions nine species of grasses, some new with respect to the Skottsberg list, and others that were missing, and notes that there are numerous plants without flowers to identify.

Subsequently, several researchers have conducted the collection of plants, which have been deposited in various herbaria, especially in Chile in the herbarium of the National Museum of Natural History (SGO) and the herbarium of the University of Concepción (CONC) [14]. In this work, 46 grass species have been reported.

3. Material and methods

Specimens were collected and preserved in the herbarium of the National Museum of Natural History at Santiago (SGO). Specimens were identified and photographs were taken using a Zeiss Stemi 2000 C stereomicroscope equipped with an Axiocam ERc5s camera. Images were processed with the software Zen 2011.

A database of the species of grasses of Rapa Nui was constructed, based on the databases of two important Chilean herbaria: CONC (Herbarium of the University of Concepción) and SGO (Herbarium of the National Museum of Natural History, Santiago). Specimens deposited in these herbaria and those collected for this project were included. The database contains the following fields: 1. Genus; 2. Species; 3. Common names in Rapa Nui; 4. Origin (native/introduced/endemic); 4. Geographical origin; 5. Photosynthetic pathway; 6. Life cycle; 7. Subfamily; 8. Tribe; 9. Collector's name; 10. Collector's number; 11. Latitude; 12. Longitude; 13. Altitude; 14. Locality; 15. Date of collection (year); 16. Date (year) of first registration; 17. Herbarium; 18. Herbarium number; 19. Bibliographic citations. A total of 369 specimens were included.

A checklist is provided, including Latin name, origin (endemic, native, introduced), homeland, life cycle (annual, perennial, annual or perennial), photosynthetic pathway (C3/C4) and classification (subfamily, tribe); however, the biogeographic status is sometimes difficult or impossible to establish. Meyer's secondarization index was calculated as the number of native species/number of naturalized species [37].

The diversity of grasses of Rapa Nui was compared with the diversity of grasses of other oceanic islands (Galápagos, Pitcairn, Marquesas, Juan Fernández and Hawaii), using the regional diversity index (D). This index was calculated as D=S/log A, where S is the number of species in the region and A is the area in square kilometres [22]. The floristic affinity between these islands was compared by cluster analysis of presence-absence data for 349 species, using Jaccard coefficient as the similarity measure, UPGMA algorithm and the statistical software Infostat [23]. Species composition was taken from the literature [24-28]. Species accumulation curve and estimated richness was calculated using the software Estimates 8.0 [29].

4. Results

Our database for the island included a total of 369 specimens collected over 12 decades (1900-2013), representing 51 species, 37 genera, 11 tribes and seven subfamilies (Table 1 and 2), that is, approximately 10 % of the total Chilean (continental and insular) grass flora (523 species and 57 infraspecific taxa) [30]. The proportion of species relative to the number of genera is 1.36, similar to the proportion determined by Fuentes [10] for the entire flora of the island (135 species / 104 genera=1.29). Most of the genera are represented only by one or two species, *Paspalum* (five species), *Digitaria* (three species) and *Setaria* (three species) being the most diverse. Details of some Poaceae of Rapa Nui are illustrated in Figs. 1-3.

Only two species (3.9 %) of the family Poaceae are endemic to Rapa Nui (*Rytidosperma paschalis* and *Paspalum forsterianum*). Eight or nine species are most probably native (15 %) and at least 42 (81 %) are introduced (Table 1). Native species are distributed in seven genera (1.29) and alien species in 30 genera (1.4). Among native Poaceae, the genera *Dichelachne* and *Paspalum* include two native species each, whereas the rest of the genera include only one species (*Axonopus, Bromus, Digitaria, Lachnagrostis* and *Piptochaetium*). Among alien Poaceae, the most diverse genera are *Paspalum* (three spp.), *Setaria* (three spp.), *Cenchrus* (two spp.), *Digitaria* (two spp.), *Sorghum* (two spp.) and *Vulpia* (one sp. and one var.).

Although there are relatively few botanical specimens of Poaceae collected in Rapa Nui, the species cited in the botanical literature are well represented in Chilean herbaria. Moreover, only one species was collected for the first time in 2013, suggesting that the inventory of species is fairly comprehensive. The first herbarium specimens entered into our database correspond to those made by Alexander Agassiz, who in 1904 collected 16 specimens representing 12 different species, about 20 % of the currently known diversity of Poaceae in Rapa Nui. By the middle of the 20th century, with the botanical expeditions made by Fuentes, Skottsberg, Balfour, Williamson & Co., Drapkin and the Mission Franco-Belge, the number of known species reached nearly 50 % of the currently known species number (Fig. 5). An important increase in the number of known species occurred after the botanical trips made by Michel Etienne, who published 24 wild and two cultivated species of Poaceae in Rapa Nui [4] and by Georg Zizka, who reported 46 species of grasses, the most comprehensive list until today [13, 14]. In the decade 1981-1990, a total of 149 specimens of Poaceae were collected, most of them by Zizka. In general, the herbarium collections of Poaceae from Rapa Nui are limited. Our database for the island included a total of 369 specimens over 12 decades (1900-2013), including a total of 51 taxa, most of which were known previously [2], and in Zizka's [13, 14] papers.

Most of the species of grasses of the island are introduced, some of them cited very early in the botanical literature, such as *Cynodon dactylon, Cenchrus echinatus* and *Sorghum halepense*, as well as some cultivated species, such as *Zea mays, Triticum aestivum* and *Arundo donax* [10]. Some native species were collected very early by R. and G. Forster in 1774 [15], for example, *Paspalum forsterianum*, dedicated to them by Flüggé. In Rapa Nui, Forster also collected *Sporobolus indicus, Dichelachne micrantha, Bromus catharticus* and the type specimen of *Agrostis avenacea* (=*Lachnagrostis filiformis*) [14]. *Stipa horridula* was collected by Skottsberg in Mount Katiki in 1917; this specimen (Skottsberg 660) became the type (lectotype) of *Stipa horridula* published by Pilger in 1922, and considered endemic to the island for a long time. In 1990, Everett and Jacob [30] reduced it to the synonymy of *Stipa scabra* Lindl., later transferred to *Austrostipa* (*A. scabra*). Skottsberg also collected, in 1917, the type specimen (Skottsber 658) of *Danthonia paschalis* Pilg. [=*Rytidosperma paschale* (Pilg.) C.M. Baeza] [17]. This species is one of the two recognized endemic Poaceae from Rapa Nui.

In 1911, Francisco Fuentes collected, among other plants, a specimen published as the holotype of *Paspalum paschale*. This name was soon transferred to genus *Axonopus* (*A. paschalis*), a species considered for a long time endemic to the island. In addition, another 18 species of Poaceae were collected by Fuentes. In 1936, Guillaumin collected eight grass species, most of them

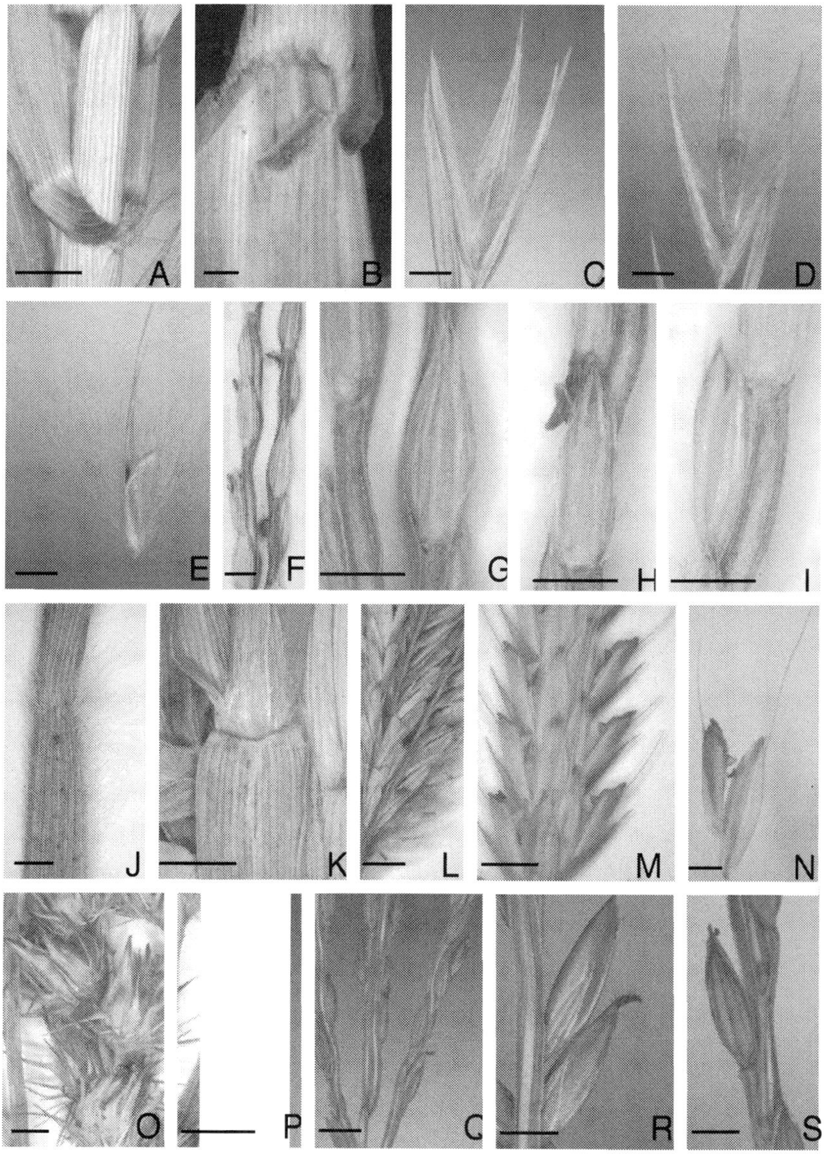

Figure 2. A-E.*Arundo donax* (Alves 34). F-J. *Axonopus compressus* (Zizka 357). K. *Cenchrus clandestinus* (Alves 57). L-N. *Chloris gayana* (Zizka 562). O-P. *Cenchrus echinatus* (Rodríguez 2202); Q-S. *Digitaria ciliaris* (Alves 13[a]). Scale bars: A-B=5 mm; C-J, N, R-S=1 mm; K, O-P=2 mm; L-M, Q=3 mm

Figure 3. A-C. *Dichelachne micrantha* (Lücke 15). D-F. *Eleusine indica* (Zizka 330). G-J. *Eragrostis tenuifolia* (Alves 30). K-M. *Lachnagrostis filiformis*. N-P. *Melinis repens* (Zizka 586). Q. *Bothriochloa ischaemum* (Zizka 330). Scale bar: Q=2 mm. Scale bars: B-F, I-L=1 mm; G-H, M, P=2 mm; A, N-O=3 mm

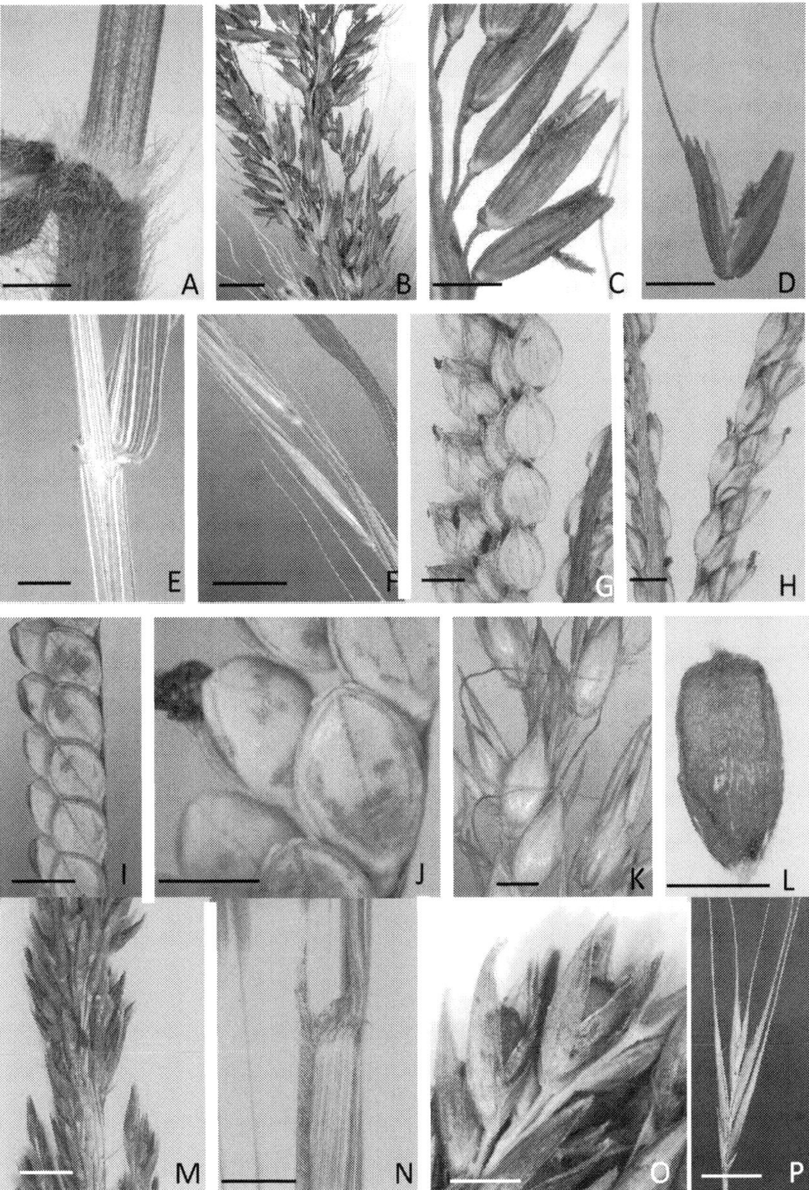

Figure 4. A-D. Melinis minutiflora (Alves 60). E-F. Ehrharta stipoides (Alves 32). G. Paspalum dilatatum (Stuessy 11008). H. Paspalum forsterianum (Vidal s.n.). I-J. Paspalum scrobiculatum (Alves 98). K. Sorghum halepense. L-O. Sporobolus indicus. Scale bars: A, C-E, J, L=1 mm; G-I, K, M-O=2 mm; B=3 mm

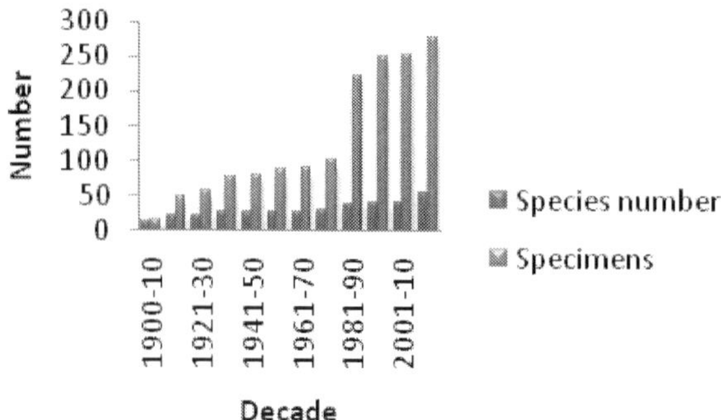

Figure 5. Number of specimens and species collected in Rapa Nui in 12 decades between 1904 and 2013

previously known from Fuentes' collection. Guillaumin collected, probably for the first time, *Briza minor*, nowadays naturalized all over the world [16]. Zizka [13, 14] collected and described 46 species representing the most important contribution to the knowledge of the grass family in Rapa Nui.

5. Taxonomic distribution of Poaceae in Rapa Nui

The taxonomic distribution of the species of Poaceae in Rapa Nui is shown in Table 1. Seven subfamilies are represented. In continental Chile, eight subfamilies are present [30], of which only species of the subfamily Aristidoideae are absent in Rapa Nui.

The subfamily Arundinoideae comprises only one species: *Arundo donax*, a perennial C3 reed-like species introduced from southern Europe, probably at the beginning of the 20th century or before. It was mentioned as a component of the Rapa Nui flora by Francisco Fuentes in 1913 [10]. According to our records and the literature [4], it is restricted to the crater of the Rano Kau volcano as a remnant of cultivation [14]. This species is recognized as invasive in Southern Africa, Australia, North America and the Pacific Islands [3].

The subfamily Bambusoideae is represented only by cultivated bamboos of the genus *Bambusa*.

The subfamily Chloridoideae comprises six genera and six species in Rapa Nui (Table 1), all belonging to the subtribe Cynodonteae, most introduced from Tropical Africa, such as *Chloris gayana*, *Cynodon dactylon*, *Eragrostis atrovirens* and *Sporobolus indicus*. Accordig to Zizka [13], the identity of the species of *Eragrostis* in Rapa Nui is not clear. This author mentions *E. spartinoides* (=*E. brownii*) and *E. leptostachya*. In this paper, we follow the revision of the genus *Eragrostis* by Escobar *et al.* [34]. According to these authors, the species of *Eragrostis* inhabiting Rapa Nui correspond to *E. atrovirens* and *E. tenuifolia*. *Eleusine indica* is a cosmopolitan Chloridoideae reported in Rapa Nui early in the 20th century.

The subfamily Danthonioideae comprises only two species in Rapa Nui. One of the species was mentioned only once in the literature dealing with Rapa Nui flora [11], under the name

Gynerium argenteum a synonym for *Cortaderia selloana*. This species is native to South America and is probably alien in Rapa Nui. It seems that this plant was introduced for erosion control [3]. The second species is *Ritydosperma paschale*, endemic to the island. As was established by Zizka [13], and according to our database, this species is restricted to the slopes of the Rano Kau volcano. It was also collected previously in Pua Katiki [4], but it seems to now be restricted to Rano Kau [14].

The only species of the subfamily Ehrhartoideae known in Rapa Nui is *Ehrharta stipoides* (=*Microlaena stipoides*), a species growing in Africa, Tropical Asia, Australasia and the Pacific [32]. This species is important as forage, but it was reported as invasive in Hawaii and Réunion Island. It was reported in Rapa Nui as a species widely distributed in the island, advantaged by overgrazing [4].

The subfamily Panicoideae comprises 12 genera and 23 species, representing about 44 % of the grass flora of Rapa Nui (Table 1). This clearly contrasts with the total grass flora of Chile, where Panicoideae represents only *ca.* 10 % [30]. *Paspalum*, *Setaria* and *Digitaria* are the most speciose genera.

Panicoideae includes the second endemism of this family from the island, *Paspalum forsterianum*, a species whose conservation status is "Vulnerable" [3]. To this subfamily also belongs *Axonopus paschalis*, long considered endemic to the island. This species was recently included as a synonym of *A. compressus*. According to Morrone [unpublished data, Flora de Chile], *A. paschalis* is morphologically similar to *A. compressus* from which it differs by the size of the leaf blades, the longer, narrower and more rigid leaves, by the hairiness of the spikelets and by the brown superior floret. *Bothriochloa ischaemum* was collected for the first time shortly after it had been introduced to the island; this species behaves invasively [4]. Several other Panicoideae are found, some of them widely recognized as invasive (v. gr. *Setaria parviflora*, *Cenchrus echinatus*, *C. clandestinus*, *Paspalum scrobiculatum*, etc.), others cultivated, such as *Saccharum officinarum*.

Subfamily	Tribes	Genera	Species number	Species (%)
Arundinoideae	1	1	1	1.96
Bambusoideae	1	1	1	1.96
Chloridoideae	1	6	7	13.73
Danthonioideae	1	2	2	3.92
Ehrhartoideae	1	1	1	1.96
Panicoideae	2	12	23	45.09
Pooideae	4	15	16	31.37
Total	11	37	51	100.00

Table 1. Number of tribes, genera and species of Poaceae in Easter Island

The subfamily Pooideae comprises 16 species from 15 genera, being the second most diverse subfamily. Whereas Pooideae include most of the grasses of the Chilean flora (74.65 %) [30], in Rapa Nui it represents only *ca.* 31 % (Table 1). A list of the species, homeland, photosynthetic pathway, life cycle and classification (subfamily, tribe) is given in Table 2.

6. Phytogeographical origin of Poaceae in Rapa Nui

As in other oceanic islands, grasses are the most common alien plants occurring in Rapa Nui [3, 33, 35]. The number of endemic, native and alien species of the grass flora in the island is shown in Fig. 6, where we can see that alien plants represent the vast majority of the grass flora (Meyer's secondarization index=0.24). Introduced species are mainly of African, European and Asian origin (Fig. 7, Table 2).

The proportion of alien species in six Poaceae subfamilies (Bambusoideae was not included as it contains only one cultivated species) is shown in Figure 8. All Chlorodoideae seem to be alien, most of African origin. However, the identity of some species, mainly those of *Eragrostis*, is difficult to elucidate. *Eragrostis atrovirens*, *E. spartinoides* and *E. tenuifolia* have been mentioned, however, only *E. atrovirens* and *E. tenuifolia* were recently recognized in Rapa Nui [34]. *Sporobolus indicus* (Chloridoideae) was collected in cultivated field as agrestal weed (vineyards, pineapple).

Most of the species of subfamily Panicoideae and subfamily Pooideae, which constitute the bulk of the grass flora of the island, are alien. The only species of subfamily Ehrhartoideae, *Ehrharta stipoides*, seems to be invasive; this species is widely distributed in the island and expands in cases of overgrazing [4].

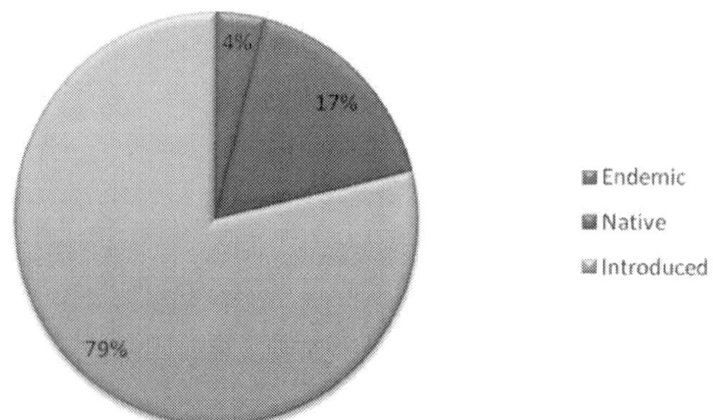

Figure 6. Percentage of native, endemic and introduced grass species in Rapa Nui

Rapa Nui belongs to the Polynesian Biogeographic Province [36] or Polynesian Floristic Region [37], included among the 25 biodiversity hotspots of the world [38, 39]. Specifically, it belongs to the Eastern Polynesia Subregion of Polynesia and represents the driest island of the subregion, with 1,325 mm of annual precipitation. In effect, there is no native vegetation on the island today and introduced plants outnumber native species due to the deforestation that occurred soon after the arrival of the first Polynesian inhabitants (38).

Introduced species that became naturalized or invasive represent one of the major threats to native species. It has been proposed that aboriginal people significantly modified the vegetation of the island [6] and the original vegetation communities were replaced by grasslands. In these plant communities, introduced species became increasingly abundant [40]. According to Aldén [2], the composition of the flora underwent a rapid change from the 18th century, when European people begin to visit the island.

On the other hand, around 70,000 tourists visit the island each year, causing environmental deterioration [41]. As shown in Figure 9, a sharp increase in the amount of alien species occurs in the 1990s; by this time, there are three times the number of alien species present compared to when Francisco Fuentes visited the island in 1911; nevertheless, in the 1910s, aliens already exceed the number of native species. Zizka [14] collected six species of grasses for the first time, all of them introduced. In some cases, nevertheless, we cannot be absolutely sure if certain species are indigenous or were accidentally introduced by man.

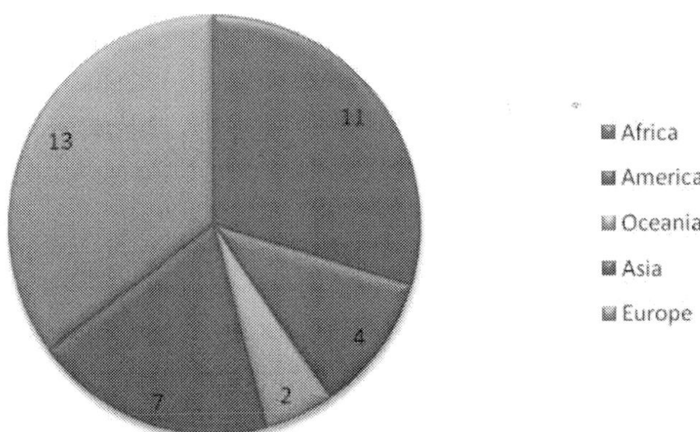

Figure 7. Geographic origin of the species introduced to Rapa Nui

From a physiological point of view, introduced species mostly show C4 photosynthesis (Fig. 10), and most of the species, both alien and native, are perennial (Fig. 11). It has been demonstrated that alien species distantly related to the native flora are more likely to become harmful weeds for regional ecosystems, supporting Darwin's naturalization hypothesis; thus, special attention should be paid to newly introduced species for which there are no close relatives in the regional flora [42]. In Rapa Nui, our data show that species from 30 genera and two subfamilies that do not include native species have been introduced.

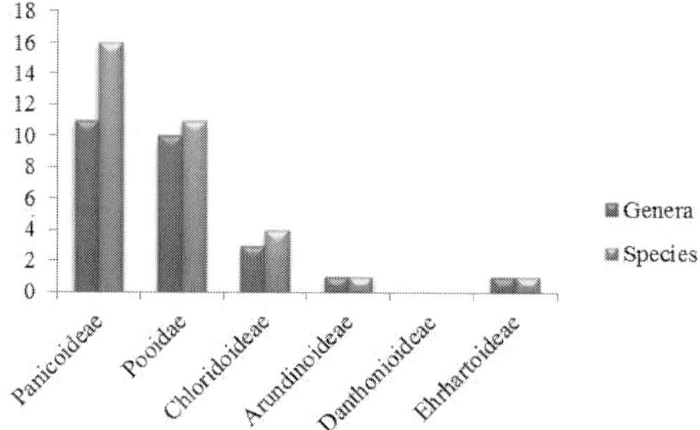

Figure 8. Taxonomic distribution of the alien species recorded for Rapa Nui

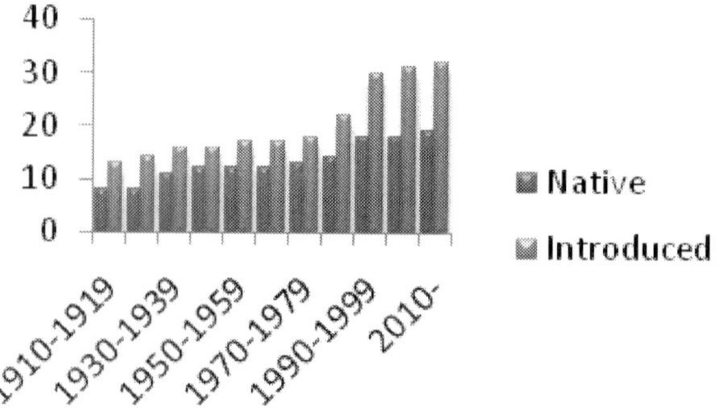

Figure 9. Number of native and introduced grass species in Rapa Nui in 12 decades of botanical collections

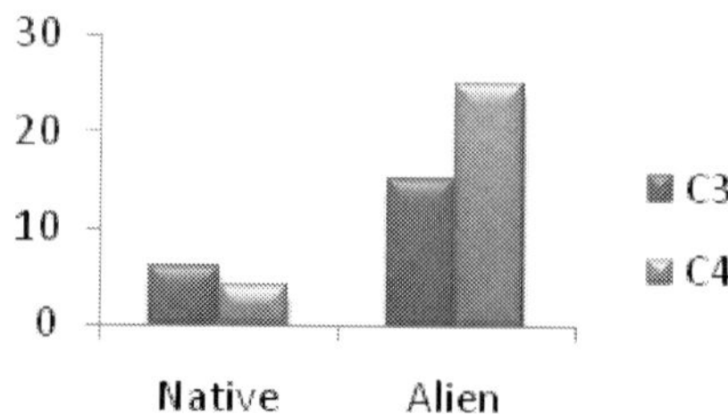

Figure 10. Number of C3 and C4 grass species in Rapa Nui

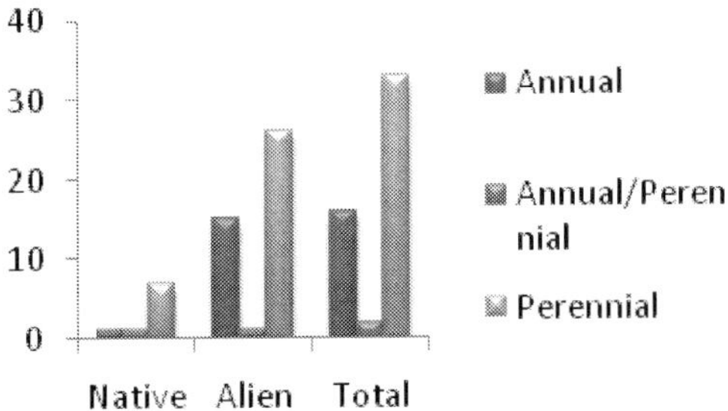

Figure 11. Life cycle of native and alien species of Poaceae in Rapa Nui

7. Comparing the diversity of Poaceae in Rapa Nui with other Pacific Islands

The diversity of grasses in Rapa Nui, calculated as the number of species per area expressed in square kilometres (regional diversity index) is slightly lower than that of the Juan Fernández Archipelago (Fig. 12). According to the literature, Poaceae in Juan Fernández Archipelago comprises 32 genera and 53 species [33, 43]. This number is relatively small compared with Hawaii (216 species) and Galápagos (94 species). From Desventuradas islands (San Félix, San Ambrosio), only two species have been recorded (*Eragrostis kuschelii* and *E. peruviana*), the first one endemic to Chile (Desventuradas Islands). If the identity of the species is considered, Rapa Nui is still more similar to Juan Fernández than to other Pacific islands (Fig. 13). However, taxonomic distribution of the flora of Poaceae is different in these two Chilean islands. C3 Pooideae dominated Poaceae in Juan Fernández, whereas in the more tropical Rapa Nui, C4 Panicoideae are more abundant. In both islands, perennial grasses dominate over annuals.

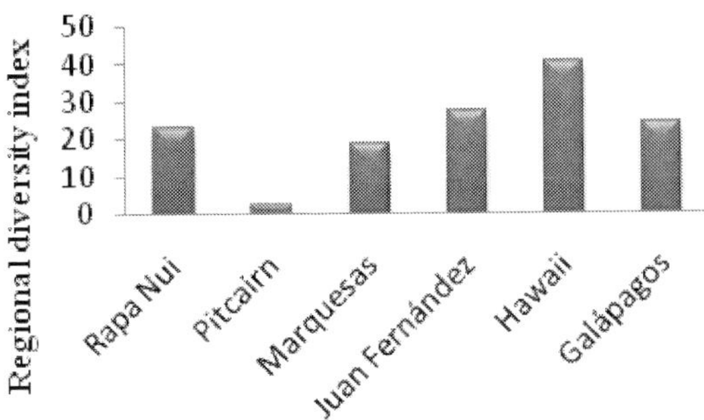

Figure 12. Diversity of Poaceae of Rapa Nui (Easter Island) compared to other oceanic islands

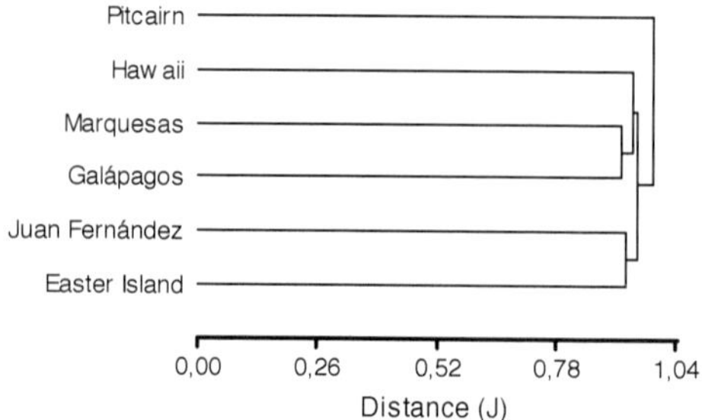

Figure 13. Poaceae floristic similarity between Rapa Nui (Easter Island) and other Pacific islands.

Species	Origin	Homeland	C3/C4	Life cycle	Subfamily	Tribe
1. *Agrostis stolonifera* L.	I	Europe	C3	P	Pooideae	Poeae
2. *Arundo donax* L.	I	Europe	C3	P	Arundinoideae	Arundineae
3. *Austrostipa scabra* (Lindl.) S.W.L. Jacobs & J. Everett	I	Australia	C3	P	Pooideae	Stipeae
4. *Avena fatua* L.	I	Europe	C3	A	Pooideae	Poeae
5. *Axonopus compressus* (Sw.) P. Beauv.	N	South America	C4	P	Panicoideae	Paniceae
6. *Bambusa* sp.	C	Asia	C3	P	Bambusoideae	Bambuseae
7. *Bothriochloa ischaemum* (L.) Keng	I	Asia	C4	P	Panicoideae	Andropogoneae
8. *Briza minor* L.	I	Europa	C3	A	Pooideae	Poeae
9. *Bromus catharticus* Vahl	N	America	C3	P	Pooideae	Bromeae
10. *Cenchrus clandestinus* (Hochst. Ex Chiov.) Morrone	I	Africa	C4	P	Panicoideae	Paniceae
11. *Cenchrus echinatus* L.	I	Cosmopolitan	C4	A	Panicoideae	Paniceae
12. *Chloris gayana* Kunth	I	Africa	C4	P	Chloridoideae	Cynodonteae
13. *Coix lacryma-jobi* L.	I	Asia	C4	P	Panicoideae	Andropogoneae
14. *Cortaderia selloana*	I	South America	C3	P	Danthonioideae	Danthonieae
15. *Cynodon dactylon* (L.) Pers.	I	Tropical Africa	C4	P	Chloridoideae	Cynodonteae
16. *Dichelachne crinita* (L. f.) Hook. f.	N	Australia, Asia & Pacific	C3	P	Pooideae	Poeae
17. *Dichelachne micrantha* (Cav.) Domin	N	Australia, Asia & Pacific	C3	P	Pooideae	Poeae
18. *Digitaria ciliaris* (Retz.) Koeler	N?	South America	C4	A	Panicoideae	Paniceae
19. *Digitaria setigera* Roth ex Roem. & Schult.	I	Australia, Asia & Pacific	C4	A	Panicoideae	Paniceae

Species	Origin	Homeland	C3/C4	Life cycle	Subfamily	Tribe
20. *Digitaria violascens* Link	I	South America	C4	A	Panicoideae	Paniceae
21. *Eleusine indica* (L.) Gaertn.	I	Africa	C4	A	Chloridoideae	Cynodonteae
22. *Eragrostis atrovirens*	I	Australia	C4	P	Chloridoideae	Cynodonteae
23. *Eragrostis tenuifolia* (A. Rich.) Steud.	I	Africa	C4	P	Chloridoideae	Cynodonteae
24. *Gastridium ventricosum* (Gouan) Schinz & Thell.	I	Europa	C3	A	Pooideae	Poeae
25. *Hordeum murinum* L.	I	Europa	C3	A	Pooideae	Triticeae
26. *Lachnagrostis filiformis* (G. Forst.) Trin.	N	New Zealand, Australia, New Guinea, Rapa Nui	C3	A/P	Pooideae	Poeae
27. *Lepturus repens* (G. Forst.) R. Br.	I	Africa	C4	P	Chloridoideae	Cynodonteae
28. *Lolium perenne* L.	I	Europe	C3	P	Pooideae	Poeae
29. *Megathyrsus maximus* (Jacq.) B. K. Simon & S. W. L. Jacobs var. *pubiglumis* (K. Schum.) B. K. Simon & S. W. L. Jacobs	I	Africa	C4	P	Panicoideae	Paniceae
30. *Melinis minutiflora* P. Beauv.	I	Africa	C4	P	Panicoideae	Paniceae
31. *Melinis repens* (Willd.) Zizka	I	Africa	C4	A/P	Panicoideae	Paniceae
32. *Ehrharta stipoides* Labill.	I	Australia	C3	P	Ehrhatoideae	Ehrharteae
33. *Paspalum conjugatum* P. J. Berg.	I	America	C4	P	Panicoideae	Paniceae
34. *Paspalum dilatatum* Poir.	I	America	C4	P	Panicoideae	Paniceae
35. *Paspalum forsterianum* Flüggé	E	Rapa Nui	C4	P	Panicoideae	Paniceae
36. *Paspalum notatum* Flüggé var *saurae* Parodi	N?	South America	C4	P	Panicoideae	Paniceae
37. *Paspalum orbiculare* G. Forst.	I	Africa	C4	P	Panicoideae	Paniceae
38. *Poa annua* L.	I	Europe	C3	A	Pooideae	Poeae
39. *Rostraria cristata* (L.) Tzvel.	I	Europe	C3	A	Pooideae	Poeae
40. *Rytidosperma paschale* (Pilg.) C. M. Baeza	E	EI	C3	P	Danthonioideae	Danthonieae
41. *Saccharum oficinarum* L.	I	Asia	C4	P	Panicoideae	Andropogoneae
42. *Setaria* cf. *palmifolia* (Koenig) Stapf	I	Asia	C4	P	Panicoideae	Paniceae
43. *Setaria parviflora* (Poir.) Kerguélen var. *parvilfora*	I	America	C4	P	Panicoideae	Paniceae
44. *Setaria sphacelata* (Schumach.) Stapf & C. E. Hubb. ex M. B. Moss	I	Africa	C4	P	Panicoideae	Paniceae
45. *Sorghum bicolor* (L.) Moench.	I	Asia	C4	A	Panicoideae	Andropogoneae
46. *Sorghum halepense* (L.) Pers.	I	Europe	C4	P	Panicoideae	Andropogoneae

Species	Origin	Homeland	C3/C4	Life cycle	Subfamily	Tribe
47. *Sporobolus indicus* (L.) R. Br.	I	Africa	C4	P	Chloridoideae	Cynodonteae
48. *Triticum aestivum* L.	I	Asia	C3	A	Pooideae	Triticeae
49. *Vulpi amyuros* (L.) C. C. Gmel var. *myuros*	I	Europe, Asia, Africa	C3	A	Pooideae	Poeae
50. *Vulpia myuros* var. *megalura* (Nutt.) Auquier	I	Europe, Asia, Africa	C3	A	Pooideae	Poeae
51. *Zea mays* L.	I	America	C4	A	Panicoideae	Andropogoneae

Table 2. List of the species of Poaceae in Rapa Nui. Life cycle: A=annual; P=perennial; Origin: EI=Rapa Nui; e=endemic; i=introduced; n=native

8. Taxonomic sampling effort

The collection effort (sampling) is an important part of the taxonomic work on which the knowledge of species richness and, ultimately, the knowledge of biodiversity are based. Although collectors try to find all the species in a region, this goal is almost impossible, or at least very difficult to achieve; thus, the real number of species can only be estimated from the number of observed (collected) species (48). As shown in Table 3 and Figure 14, for a total of 50 observed species, the species richness estimated by different estimators ranges from 58.3 (Bootstrap) to 86.8 (Chao). An increased collection effort for Poaceae in Rapa Nui could yield between eight and 36 additional hitherto unsampled species. As shown in Fig. 15, collections are concentrated only in a few localities, chiefly in Rano Kau, near Hanga Roa, Anakena, and Rano Raraku.

Diversity Estimator	Species Richness
Sobs (Mao Tau)	50
ACE	70.6
ICE	73.7
Chao 1	86.8
Chao 2	86.8
Jack 1	69.3
Jack 2	82.2
Bootstrap	58.3
Michaelis-Menten	66.06

Table 3. Estimated species richness of Poaceae in Rapa Nui, using eight different estimators.

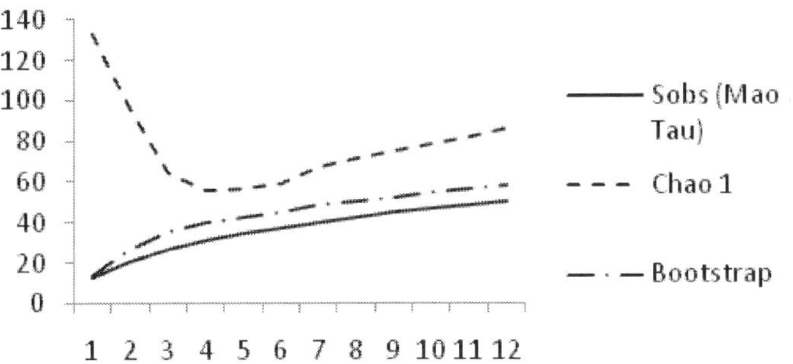

Figure 14. Species accumulation curve (Sobs) and estimated species curves for 12 decades of sampling, based on Chao 1 and Bootstrap estimators.

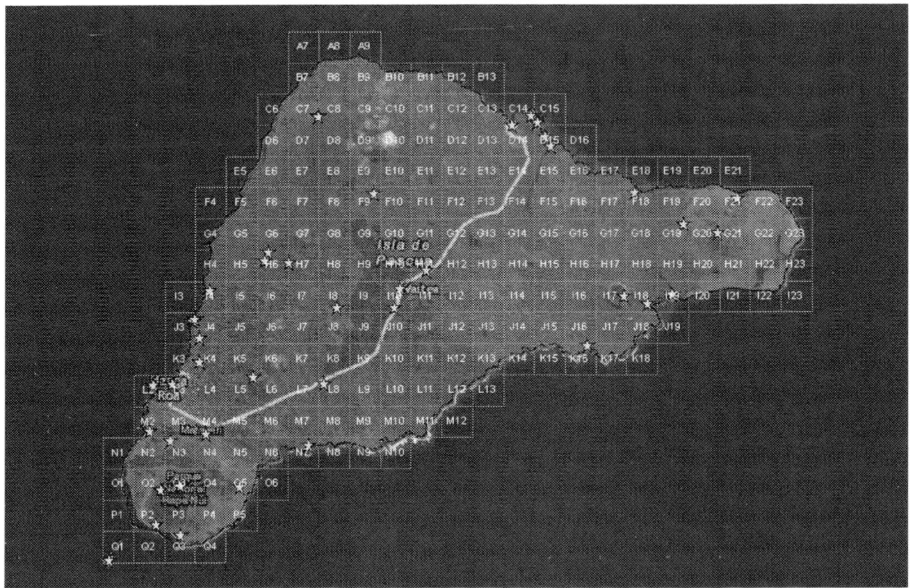

Figure 15. Map of the localities of Poaceae collected in Rapa Nui. Each point represents at least one collected specimen.

9. Concluding remarks

The floras of the oceanic islands are especially prone to serious threats from alien invaders [44], because they have a propensity to include highly adapted specialist rather than generalist species [45]. This is of particular interest, as these ecosystems comprise high numbers of endemic plants, in contrast with continental regions of similar size [6]. For these reasons, it is necessary to have on hand complete lists of the flora indicating alien plants that could be agricultural weeds and invasive species that can put pressure on native ecosystems.

Poaceae contains an important number of species that behave as weeds in natural environments (invasive), as well as in ruderal and agricultural habitats all over the world. In Rapa Nui, grasses are the most diverse family of vascular plants; nearly 90 % of the island is covered by grasslands and alien grasses represent nearly 80 % of the Poaceae of the island. A similar situation occurs in other oceanic islands [33]. Several alien species introduced to Rapa Nui are noxious weeds (*Agrostis stolonifera*, *Hordeum murinum*, *Sorghum halepense*, *Lolium perenne*, *Setaria parviflora*), probably introduced accidentally, mostly from Europe. As was previously established [38], Rapa Nui shows the greatest (<1) secondarization index (number of native species/number of naturalized species=0.68) when compared with many other Easter Polynesian islands. For Poaceae, this index is greater still (0.24).

On the other hand, only two Poaceae endemic to Rapa Nui are recognized. Another two Poaceae considered endemic in previous studies have been the object of taxonomic studies that demonstrate their non-endemic status: *Axonopus paschalis* (=*Axonopus compressus*) and *Stipa scabra* (*Austrostipa scabra*).

Herbarium specimens provide valuable information to appreciate regional plant diversity, as well as to understand plant invasions and geography [46, 47]. A wide array of data (phenology, flowering periodicity, distribution, altitude, morphometry, minimum residence time, new distributional records, etc.), that are important to biodiversity monitoring can be obtained from herbarium specimens [49]. However, herbarium collections in Rapa Nui are relatively scarce.

Author details

Víctor L. Finot[1], Clodomiro Marticorena[2], Alicia Marticorena[2], Gloria Rojas[3] and Juan A. Barrera[4]

*Address all correspondence to: victorfinotsaldias@gmail.com

1 Department of Animal Production, Faculty of Agronomy, University of Concepción, Chillán, Chile

2 Department of Botany, Faculty of Natural and Oceanographic Resources, University of Concepción, Concepción, Chile

3 National Museum of Natural History, Santiago, Chile

4 Department of Soil Science and Natural Resources, Faculty of Agronomy, University of Concepción, Chillán, Chile

References

[1] UNESCO (United Nations Educational Scientific and Cultural Organization). Rapa Nui National Park. http://whc.unesco.org/en/list/715 (accessed 02 June 2014).

[2] Aldén B. Wild and introduced plants on Easter Island: A report on some species noted in February 1988. In Esen-Baur, H. M. (ed.), State and perspectives of scientific research in Easter Island culture. Courier Forschungsinstitut Senckenberg, Frankfurt, Germany, 1990, pp. 209-216.

[3] Dubois A., Lenne P., Nahoe E., Rauch M. Plantas de Rapa Nui. Guía ilustrada de la flora de interés ecológico y patrimonial. Umangamo te Natura, CONAF, ONF Int., Santiago, 2013, pp. 1-132.

[4] Etienne M., Faúndez L. Gramíneas de Isla de Pascua. Facultad de Ciencias Agrarias, Veterinarias y Forestales, Universidad de Chile, Boletín Técnico 1983; 12: 1-55.

[5] Rull V., Cañellas-Bolta N., Sáez A., Giralt S., Pla S., Margalef O. Paleoecology of Easter Island: Evidence and uncertainties. Earth Sciences Reviews 2010; 99: 50-60.

[6] Castro S., Muñoz M., Jaksic F. Transit towards floristic homogenization on oceanic islands in the south-eastern Pacific: Comparing pre-European and current floras. Journal of Biogeography 2007; 34: 213-222.

[7] National Geographic Society, Ocena Chile and Armada de Chile. Expedición a Isla de Pascua y Salas y Gómez. Informe científico, p. 53. http://oceana.org/sites/default/files (accessed 04 Jul 2014).

[8] Mann D., Edwards J., Chase, W. Beck, R. Reanier, M. Mass, B. Finney & J. Loret. Drought, vegetation change, and human history on Rapa Nui (Isla de Pascua, Easter Island). Quaternary Research 2008; 69: 16-28.

[9] Skottsberg C. La población botánica y zoological de las Islas Chilenas del Pacífico. Revista Chilena de Historia y Geografía 1936; 78: 139-153.

[10] Fuentes, F. Reseña botánica sobre la Isla de Pascua. Botanische Skizze der Osterinsel. Inst. Cetr. Meteorol. Chile, Publ. 1913; 4: 140-149.

[11] Fuentes F. Reseña botánica sobre la Isla de Pascua. Boletín del Museo Nacional 1913 [1914]; 5(2): 320-337.

[12] Zizka, G. Naturgeschichte der Osterinsel. In: Bothmer-Plates, A. von, Esen-Baur, H.M., Sauer, D.F. (eds.), 1500 JahreKultur der Osterinsel. Schätzeausdem Land des Hotu Matua: 20-38, Mainz.

[13] Zizka G. Changes in the Easter Island flora; comments on selected families. Courier Forsch. Inst. Senckenberg, 1990; 125: 189-207.

[14]   Zizka G. Flowering plants of Easter Island. Palmarum Hortus Francofurt, 1991;3: 1-108.

[15]   Forster G. Florulae insularum australium prodomus. J. C. Dieterich, Göttingen, 1786.

[16]   Guillaumin A., Camus A., Tardieu-Blot, M. L. Plantes vasculaires récoltées à l'Ile de Pâquespar la misión franco-belge. Bull. Mus. Natl. Hist. Nat. Sér., 1936; 2, 8: 552-556.

[17]   Skottsberg, C. The natural history of Juan Fernandez and Easter Island, edited by Carl Skottsberg. Uppsala. Vol. II, Botany, 1921; pp.1920-1956.

[18]   Kotzebue O. von, Chamisso, A. von, Engelhardt, M. van, Eschscholtz, Horner J. C. A voyage of discovery into the South Sea and Beering's [sic] Straits, for the purpose of exploring a north-east passage undertaken in the years 1815-1818, at the expense of His Highness the chancellor of the empire, Count Romanzoff, in the ship Rurick, under the command of the lieutenant in the Russian Imperial Navy, Otto von Kotzebue. Longman, Hurst, Rees, Orme and Brown, 1821.

[19]   Marticorena C. Historia de la exploración botánica de Chile. In: C. Marticorena & R. Rodríguez (eds.), Flora de Chile. 1-62. Editorial Universidad de Concepción. Chile, 1995.

[20]   Endlicher S. Bermerkungenüberdie Flora der Südseeinseln, 1836.

[21]   Hemsley, W. B. Report on present state of knowledge of various insular floras, being an introduction to the botany of the Challenger expedition. Rep. Sci. Results Voyage H.M.S. Challenger. Botany, 1885; 1: 1-75.

[22]   Klopper R. R., Gautier L., Chatelain C., Smith G. F., Spichiger R. Floristics of the angiosperm flora of Sub-Saharan Africa: An analysis of the African plant checklist and database. Taxon 2007; 56(1): 201-208.

[23]   Di Rienzo J. A., Casanoves F., Balzarini M. G., Gonzalez L., Tablada M., Robledo C. W. InfoStat versión Grupo InfoStat, FCA, Universidad Nacional de Córdoba, Argentina, 2013. http://www.infostat.com.ar (accessed 08 Jul 2014).

[24]   Kingston N., Waldren S., Bradley U. The phytogeographical affinities of the Pitcairn Islands – a model for south-eastern Polynesia? Journal of Biogeography 2003; 30: 1311-1328.

[25]   Waldren, S., M. I. Weisler, K. C. Hather & D. Morrow. The non-native vascular plants of Henderson Island, South Central Pacific Ocean. Atoll Research Bulletin n°.463, National Museum of Natural History, Smithsonian Institution, Washington DC, USA, 1999.

[26]   Wagner W. L., Herbst D. R., Khan N., Flynn T. Hawaiian vascular plant updates: A supplement to the Manual of the flowering plants of Hawai'i's ferns and fern allies, 2012.   http://botany.si.edu/pacificislandbiodiversity/hawaiianflora/supplement.htm (accessed 03 Jul 2014).

[27] Robinson B .L. Flora of the Galapagos Islands. Proceedings of the American Academy of Arts and Sciences 1902; 38: 77-270.

[28] Jaramillo-Díaz O., Guézou A., Mauchamp A., Tye A. CDF Checklist of Galapagos flowering plants. In: Bungartz, F., H. Herrera, P. Jaramillo, N. Tirado, G. Jiménez-Uzcátegui, D. Ruiz, A. Guézou & F. Ziemmeck (eds.), Charles Darwin Foundation Galapagos species list. http://www.darwinfoundation.org/datazone/checklist/ vascularplants/magnoliophyta/ last updated 16 Apr 2014 [02 Jul 201].

[29] Finot, V. L., Barrera, J. A., Marticorena, C. & Rojas, G. Systematic diversity of the family Poaceae (Gramineae) in Chile, pp. 71-108. In: Grillo, O. & G. Venora (eds.), The Dynamical Processes of Biodiversity - Case Studies of Evolution and Spatial Distribution, In-Tech, 2011.

[30] Everett, J., Jacobs, S. W. L. Notes on Stipa (Poaceae) in Australia and Easter Island. Telopea 1990; 4(1): 7-11.

[31] Ausgrass2. Australian Grasses. http://ausgrass2.myspecies.info/category/australian-grasses/australian-grasses (accessed 03 Jul 2014).

[32] Baeza C. M., Stuessy T., Marticorena C. Notes on the Poaceae of the Robinson Crusoe (Juan Fernández) Islands, Chile. Brittonia 2002; 54: 154-163.

[33] Escobar I., Ruiz E., Finot V. L., Negritto M. A., Baeza C. M. Revisión taxonómica del género Eragrostis Wolf en Chile, basada en análisis estadísticos multivariados. Gayana Botánica 2011; 68: 49-85.

[34] Rodgers J. C., Parker C. Distribution of alien plants in relation to human disturbance on the Georgia Sea Islands. Diversity and Distributions 2003; 9: 385-398.

[35] Thorne R. F. 1963. Biotic distribution patterns in the tropical Pacific. In: Gressitt J. L. (ed.), Pacific basin biogeography, a symposium. Proceedings, 10th Pacific Science Congress, 1961, Honolulu, Hawai'i. Bishop Museum Press, Honolulu, pp. 311–354

[36] Takhtajan A. Floristic regions of the world (translated by T. J. Crovello and A. Cronquist). University of California Press, Berkeley, 1986.

[37] Meyer, J. Y. Threat of invasive alien plants to native flora and forest vegetation of Eastern Polynesia. Pacific Science 2004; 58: 357-375.

[38] Myers N., Mittermeier R. A., Mittermeier C. G., Da Fonseca G. A. B., Kent J. Biodiversity hotspots for conservation priorities. Nature (Lond.) 2000; 403: 853–858.

[39] Hoffmann A., Marticorena C. La Vegetación de las islas oceánicas chilenas. In: Castilla J. C. (ed.), Islas oceánicas chilenas: Conocimiento científico y necesidades de investigaciones. Ediciones Universidad Católica de Chile, Santiago, Chile, pp. 127-165. 1987.

[40] ez M., Rodríguez C. Impactos ambientales generados por el desarrollo turístico en la Isla de Pascua. Revista Interamericana de Ambiente y Turismo 2011; 7: 42-48.

[41] Strauss S. Y., Webb C.O., Salamin N. Exotic taxaless related to native species are more invasive. PNAS 2006; 103: 5841-5845.

[42] Marticorena C. Contribución a la estadística de la flora vascular de Chile. Gayana Botánica 1990; 47: 85-113.

[43] Whittaker R. J. Island biogeography. Oxford Univ. Press, Oxford, 1998.

[44] Milne R. I., Abbott R. J. Geographical origin and taxonomic status of the invasive Privet, *Ligustrumrobustum* (Oleaceae) in the Mascarene Islands, determined by chloroplast DNA and RAPDs. Heredity 2004; 92: 78-87.

[45] Wu S. H., Rejmánek M., Grotkopp E., Di Tomaso J. M. Herbarium records, actual distribution, and critical attributes of invasive plants: genus Crotalaria in Taiwan. Taxon 2005; 54: 133-138.

[46] Park I. W., Schwartz M. D. Long-term herbarium records reveal temperature-dependent changes in flowering phenology in the southeastern U.S.A. International Journal of Biometeorology 2014; http://download.springer.com/static/pdf/.

[47] Colwell R. K., Coddington J. A. Estimating terrestrial biodiversity through extrapolation. Philosophical Transactions: Biological Sciences 1994; 101-118.

[48] Miller-Rushing A. J., Primack R. B., Primack D., Mukunda S. Photographs and herbarium specimens as tools to document phenological changes in response to global warming. American Journal of Botany 2006; 93(11):1667–1674.

Chapter 16

Plant Structure in the Brazilian Neotropical Savannah Species

Suzane Margaret Fank-de-Carvalho,
Nádia Sílvia Somavilla,
Maria Salete Marchioretto and Sônia Nair Báo

Additional information is available at the end of the chapter

http://dx.doi.org/10.5772/59066

1. Introduction

This chapter presents a review of some important literature linking plant structure with function and/or as response to the environment in Brazilian neotropical savannah species, exemplifying mostly with Amaranthaceae and Melastomataceae and emphasizing the environment potential role in the development of such a structure.

Brazil is recognized as the 17th country in megadiversity of plants, with 17,630 endemic species among a total of 31,162 Angiosperms [1]. The focus in the Brazilian Cerrado Biome (Brazilian Neotropical Savannah) species is justified because this Biome is recognized as a World Priority Hotspot for Conservation, with more than 7,000 plant species and around 4,400 endemic plants [2-3].

The Brazilian Cerrado Biome is a tropical savannah-like ecosystem that occupies about 2 millions of km² (from 3-24° Latitude S and from 41-43° Longitude W), with a hot, semi-humid seasonal climate formed by a dry winter (from May to September) and a rainy summer (from October to April) [4-8]. Cerrado has a large variety of landscapes, from tall savannah woodland to low open grassland with no woody plants and wetlands, as palm swamps, supporting the richest flora among the world's savannahs-more than 7,000 native species of vascular plants-with high degree of endemism [3, 6]. The "cerrado" word is used to the typical vegetation, with grasses, herbs and 30-40% of woody plants [9-10] where trees and bushes display contorted trunk and branches with thick and fire-resistant bark, shiny coriaceous leaves and are usually recovered with dense indumentum [10]. According to [8], natural fires and anthropogenic fires coexisted for thousands of years and, together with the seasonality of

rainfall and the poverty of nutrients in the soil are the responsible for the phytophysiognomy of Cerrado.

One of the first systematized studies of Cerrado Biome was the one done in Lagoa Santa, Minas Gerais State, around the year of 1892, by Warming [11], who described the place in aspects of soil, temperature, water precipitation and vegetation. When he [11] described the vegetation of flat grassland, he emphasized the thickness and toughness of Poaceae and Cyperaceae leaves and the abundance of perennial herbs or subshrubs with large lignified underground organs, multiple shoots growing from an underground stem and xeromorphic characteristics, as dense pilosity, coriaceous leaves positioned in acute angle and with reduced size. His [11] conclusion was that the dryness of the air, the harsh and dry clay soil and eventually, the fire occurrence, were responsible for these xeromorphic features of the plants.

Since then, a lot of work has been done to explain some contradictions such as the abundant flowering and budding and no signs of turgor loss during the dry season [10]. In [12] linked the plants physiognomy with the occurrence of fire and proposed an ecological classification of the Cerrado plants: plants which survive only during the rainy season, without any bud or leaf during the dry season (winter); grasses with superficial roots, like *Echinolaena inflexa* (Poir.) Chase and *Tristachya chrysothrix* Nees, which wither when the water is gone in the superficial soil; bushes and small trees with deep roots (up to 11 meters), usually green during all the dry season, which represent the typical vegetation. Leaves of the specimens observed [12] never closed completely their stomata: *Kielmeyera coriacea* Mart., *Annona coriacea* Mart., *Annona furfuracea* A.St.-Hil., *Palicourea rigida* Kunth, *Stryphnodendron obovatum* Benth.(syn=*S. barbati-mao*), *Didymopanax vinosum* (Cham. & Schltdl.) Marchal, *Byrsonima coccolobifolia* Kunth, *Cocos leiospatha* Barb. Rodr., *Echinolaena inflexa* (Poir.) Chase , *Andira laurifolia* Benth., *Anacardium pumilum* Walp., *Neea theifera* Oerst. and two species of *Erythroxylon* genus. In [13], perennial species with deep roots were associated to the ability of regenerating the aerial parts after a long dry season or after fire; these type of plants were designated as periodics, because they reduce or eliminate their leaves and branches during the winter, when the available water is rare at the soil surface.

The work [14] indicated that Cerrado soils are deep, with pH between 4,0 and 5,5 (acid) and connected the xeromorphic features in trees to nitrogen deficiency, because the studies done in Cerrado showed that water was not a limiting factor to these plants. In [15] it was added another important aspect to explain xeromorphic features in Cerrado plants: the high levels of aluminum would be a principal cause of mineral deficiency which would affect all Cerrado vegetation. Soils under Cerrado are usually poor, acid, well drained, deep, and show high levels of exchangeable aluminum [16-17]. The soil of the low grassland in the area of the old Experimental Station of hunting and fishing Emas (Pirassununga, São Paulo State) can be as deep as 20 meters and the groundwater is at 17-18 meters below the surface; only the first one to 1.5 meters dries during the winter and roots of at least one tree (*Andira* sp.) can reach the deepest groundwater; a shrub species, *Anacardium humile* A. St.-Hil., with aerial parts reaching 0.5 meters high, can have roots with over three times its shoot length [12]. The underground systems of roots and stems are so big in some species, such as in *Andira laurifolia* Benth., that in [11] it was called an "underground tree". Low concentration of nutrients in the leaves of native species is related with the low concentration of nutrients of the dystrophic soils [18] and the floristic composition and dominance of species is a reflection of it. Cerrado plants absorb significant amount of aluminum and when the leaf concentration is over 1,000 mg Kg^{-1} the species is referred as Al-accumulative [19]. It is still unknown if this amount of aluminum have any physiological or structural significance in the metabolism of the native plants [20], but the translocation of this element is showed by the presence of aluminum in the phloem and other metabolically active tissues of leaves and seeds and there are at least two plants which cannot survive in medium without aluminum: *Miconia albicans* (Sw.) Steud. (Melastomataceae) and *Vochysia thyrsoidea* Pohl (Vochysiaceae), woody species from Cerrado [18, 20]. Another curious aspect about Cerrado plants is that there are species that only live in calcareous or acid soils and there are also those which are indifferent to soil fertility [21].

The occurrence of wildfire is a common and important factor to be considered in the studies of this Biome vegetation, because it selects structural and physiological features of the plants and act as a renewal element [22]. In [13] were described some strategies which could help the perennial smaller plants to survive fire; during the dry season, some of them reduce or eliminate leaves and shoots and rely on their extensive underground system to re-sprout the aerial system after the dry season or after fire. As examples of fire resistance, in [13] were quoted the plants studied by [11], *Andropogon villosus f. apogynus* (Hack.) Henrard (Poaceae), *Scirpus warmingii* Boeckeler, *Scirpus paradoxus* (Spreng.) Boeckeler and *Rhynchospora warmingii* Boecheler (Cyperaceae), as well as *Aristida pallens* Cav., explaining that these plants have buds in the base of the aerial system well protected by some layers of sheath blades; the old ones are more external and will burn first, always protecting the newest ones and the internal buds.

An extensive review of the morphological and ecological studies is given in [10], whose author considered the Cerrado a great environment for scientific discussion and discuss the vegetation in a broader perspective, and in [23-24], whose author is more centered in anatomical aspects of Cerrado species.

Although the Cerrado Biome is a hotspot for the conservation of global biodiversity which shelters species fully adapted to survive under harsh conditions of soil and climate of this savannah-like environment [2], only 30% of its biodiversity is reasonably known [25]. Considering that the open environments in this Biome are subject to high luminosity and seasonal variation in the rain, how do plants react to adapt themselves? Considering that fire is also a natural event during the dry season, is there any morphological and/or anatomical variations developed to survive? Considering that the groundwater level of some areas can vary in a high degree among the two seasons, how do plants manage to survive? Some of these questions will be addressed and data about it will be shown.

2. Methodology

In order to perform studies about morphology, anatomy or cell biology, as well as when the flora is been studied, it is usual to collect control or testimony material to guarantee species identification and further studies [26]. Vegetative and flowering plant branches are collected, pressed, dried and deposited as control material at some Brazilian Herbaria, following usual techniques [27]. The previous identification of the species is done with the aid of a stereomicroscope, identification keys and specialized literature [28-38]. After previous identification, plant vouchers stay preserved to the study of a taxonomist specialized on the family and for future references, including of the place of occurrence, under a specific number of the principal collector, normally not only in one Herbarium (duplicates are usually distributed).

When studying the leaf anatomy, **histological** samples are obtained from visually healthy green and completely developed leaves, usually from 3rd to 5th node, of pre-identified species; samples can be or not submitted to different fixatives and paraffin embedding medium [39] before slicing it to be studied under an optical microscope (light microscopy). **Cell samples in tissue** for ultrastructural studies can be smaller pieces of the aimed organ submitted to a fixative and post-fixed in heavy metals in the dark, followed by in-block staining [40]. Later, plant pieces are dehydrated and slowly embedded in a harder medium (epoxy or epon resin), to be sliced for observation. Semi-thin sections can be obtained with an ultramicrotome using glass knives, stained and analyzed under the optical microscope in order to localize the cells in the tissues; ultra-thin sections of the same material are obtained with a diamond knife, collected in copper grids and analyzed under a transmission electron microscope (TEM), with or without any additional staining. To be studied under a scanning electron microscope (SEM), sections of the plant are also fixed and post-fixed as indicated for TEM analysis [40], with some modifications because of plant characteristics, but the use of control pieces [41] is necessary to avoid interpretation errors. After that, fixed pieces are dehydrated in ethanol or acetone solution and critical point dried in the proper device, attached to a stub and gold sputtered to be observed under SEM [40].

3. Morphology, histology and cell biology studies

Morphological studies are used to identify and characterize the plant species in taxonomy, but it is also important to understand the behaviour of plants in nature. The first hint on the function is based on the external morphology of the organs. Different plant species can be very alike in habit and vegetative morphology, especially in some plant families as Amaranthaceae, which rely upon some flower details to be truthfully identified, demanding a highly specialized work [33-38, 42-44].

Anatomy and cell biology studies aim to describe and understand the species organs and cells and help taxonomy to define affinities and parental relationships among plant groups. When combined with histochemical analysis they can lead to a better understanding of the cell, tissue or organ function and the interaction between the plant and its environment.

It is usual, in plant structural studies, to bear a description of the aimed plant or organs of interest, assuming that the function is already explained enough by the function of the organ in the plant or by previous researches. As results are subject to interpretation and there are some variables to be considered, it is not usual to connect the structure to the function.

However, in the Cerrado species case, since the first studies there is an attempt to explain the structure and relate it to the unique environmental conditions, which was detailed in the introduction of this work, helping to give a broader significance to the structure.

Studying the so called xeromorphic features of three leaves, [14] concluded that they could be explained by the soil oligotrophic conditions, given raise to the theory of oligotrophic sclero-morphism: the mineral elements deficiency would be the main responsible for the plant characteristics, by limiting its grow; the carbohydrate accumulation is then converted in deposits of thick cuticle, thicker cell walls, wax deposits over the epidermis and other sclero-morphic features. High level of aluminum in Cerrado soils, another cause for the oligotrophic scleromorphism [15] is considered the main responsible for the acidity of Cerrado soils [45]. Through the study of eleven Cerrado plants, [46] it was indicated the constant presence of fungi over its leaves, mostly on species without epicuticular wax, and connected this outermost layer over the cuticle layer to environmental adaptation, as protection against any fungus hypha.

Sclerenchymatous elements, fibers and sclereids are distinctive structural features in vegeta-tive organs of Cerrado plants, and the presence of gelatinous fibers is frequent, associated or not with the tension wood; besides, it is also constant the impregnation of silica and siliceous bodies, not only in Poaceae and Cyperaceae species, but also in leaf and stem epidermis, roots and xylopodium of *Brasilia sickii* G. M. Barroso (Asteraceae) [23-24]. In epidermis, [24] silica is connected to the protection against excessive transpiration and as a defense mechanism against fungi. In [24], author also explained why the aperture or closure of stomata can be slower in Cerrado plants: it would be due to the thickening of the guard cells walls of a stoma, which can be impregnated with lignin, as in *Ouratea spectabilis* (Mart. ex Engl.) Engl. or have silica incorporated, like in *Esterhazya splendida* J. C. Mikan, *B. sickii* and *Casearia grandiflora* Cambess.

Amaranthaceae family is considered a good representative of the herbs and subshrubs of Cerrado due to its morphology and adaptations that promote survival in adverse conditions (drought and fire), such as tuberous or woody roots, xylopodium, herbaceous or subshrub habit, dense pubescence in aerial portions, senescence of shoots and leaves during the driest season, dependence on rain or fire to re-sprout and/or flowering, fruiting followed by wind dispersion, thick cuticle on epidermis and C_4 photosynthetic metabolism [37, 47-48]. The knowledge of the reproductive structures in Amaranthaceae Brazilian species is mostly restricted to the obtained during floristic survey and with taxonomic purpose, with few additional studies of reproductive structures, such as [49], who studied the flower vascular pattern in *Pfaffia jubata* Mart., *Gomphrena macrocephala* A. St.-Hil. and *Froelichia interrupta* (L.) Moq. and [50], with the study of pollen from Cerrado species, helping to understand the phenology of them through the analysis of herbarium species.

In this section, some morphological, histological or cellular aspects of reproductive and vegetative (aerial and underground) organs will be exemplified, discussing the aspects related to the environment where these plants grow and survive and to the function in the plant species, whenever possible.

3.1. Reproductive organs — Flower, fruits and seeds

Amaranthaceae flowers are generally small and densely clustered in terminal or axillary inflorescences (figure 1), pollinated by the wind or by insects, with self-pollinating or out-crossing [51]. Due to the hairy perianth, small and dry fruits or seeds, the dispersion is usually done by wind or water [51]. In some genus, small seeds fall from the parent plant and germinate only when the site is again disturbed; seeds can be, also, eaten and dispersed by browsing animals [51].

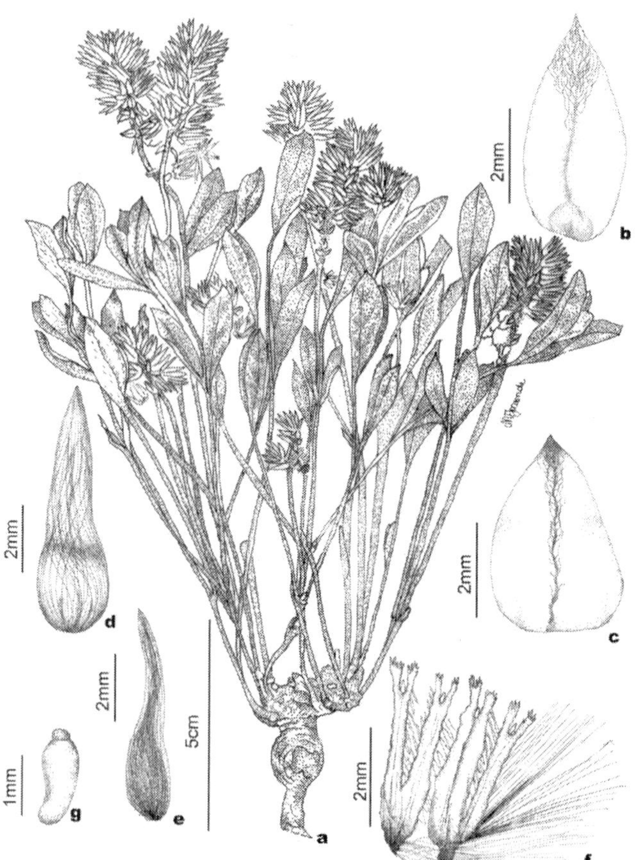

Figure 1. *Pfaffia jubata* Mart. **a**: habit; **b**: median bract; **c**: lateral bract; **d-e**: sepals; **f**: staminal tube; **g**: ovary (*Hatschbach et al. 53625*, MBM). Reproduced from [37] with permission of Hoehnea publisher. The species was considered a good representative of Amaranthaceae family: habit from herb to subshrub, 0.10-0.20 m high, stems erect, densely villous and woody root. Leaves have the upper side densely villous and lower side tomentose. Inflorescence is a spike, isolated, simple, terminal, with ferruginous trichomes.

Brazilian species of *Celosia, Chamissoa* and *Amaranthus* have circumscissile capsule fruits, with only one seed in the two last genera and more seeds in the first one; capsules are surrounded by the perianth and parts of the androecium and gynoecium, except in *Amaranthus* where the filaments are free and there is no staminate tube [52]. Dehiscence can be median (*Amaranthus*) or semi basal (*Celosia*) and dispersal is probably done by autochory [52], although [53] considers the accidental endozoochory more probable in *Amaranthus* case, because ruminants usually eat *Amaranthus* inflorescences.

Fruits of *Alternanthera, Pfaffia, Gomphrena, Blutaparon, Achyranthes, Cyathula* and *Pseudoplantago* are one-seeded capsules with two valves included in the perianth sepals which are glabrous to hairy or spine-like; these fruits can carry part of the androecium and gynoecium and sometimes the perianth displays external bristle tufts or modified bracts [34-35, 52]. Wind seems to be responsible for the dispersal of *Alternanthera, Pfaffia, Gomphrena* and *Blutaparon* fruits because of the hairiness of the perianth, which presents long trichomes; species of *Froelichia, Froelichiella, Pfaffia* and *Gomphrena* genera from open grassland have their fruit easily dispersed by the wind [34-35, 52].

In [54] wind dispersal of fruits in species which occur in Cerrado regions that were affected by fires, mostly *Gomphrena macrocephala* associating the natural fire with the easier dispersal of its fruits and seed germination. The passage of fire burns grasses and help dispersal of the *G. macrocephala, G. pohlii* Moq. and *G. virgata* Mart. fruits; fire also promotes the dehiscence of the fruits, leaving the seeds nearer to the soil [42, 55]. The maturation and dispersal of fruits is exemplified in *G. arborescens* L.f. (figure 2), a native Cerrado plant which behaves the same way of *G. macrochepala* [54]: the inflorescence opens for pollination and closes after that due to a growth of its bracts for the maturation of the fruits; after that, the shoot inclines towards the soil and the inflorescence structure reopens to release the dispersal units in the soil, formed by the seed and parts of the perianth, until the wind carries it. The same phenomenon was observed in *Pfaffia argyrea* Pedersen, *P. cipoana* Marchior. et al., *P. denudata* (Moq.) Kuntze, *P. elata* R. E. Fr., *P. hirtula* Mart., *P. jubata* Mart. *P. minarum* Pedersen, *P. rupestris* Marchior. et al., *P. sarcophylla* Pedersen, *P. siqueiriana* Marchior. et al, *P. townsendii* Pedersen, *P. tuberculosa* Pedersen, *P. velutina* Mart. and *Froelichiella grisea* (Lopr.) R.E.Fr. all Cerrado species which occur in areas subject to burning [37, 55]. *Froelichia* has one-seeded nutlet involved by parts of the gynoecium and androecium, partially because of the connated perianth and presents a wing structure densely hairy which favours the wind dispersal [34, 52].

Although *Alternanthera* and *Blutaparon* genera fruits are usually wind dispersed, *Alternanthera pungens* Kunth have its perianth highly modified, presenting uneven sepals, with the outer two sharply pointed, which can help its adhesiveness in animal skin or fur to be dispersed [52]. The same occurs with *Achyranthes* and *Cyathula* genera, which fruits are dispersed by animals through adhesiveness structures [53]. *Achyranthes* two perianth lateral bracts are thorn-like, the same as in *Pseudoplantago friesii* Suess [52]. *Cyathula* have a set of uncinate bristles which can be considered sterile flowers and play a special function in helping the fruit dispersal attached to skin or fur of animals, a phenomenon called epizoochory [52].

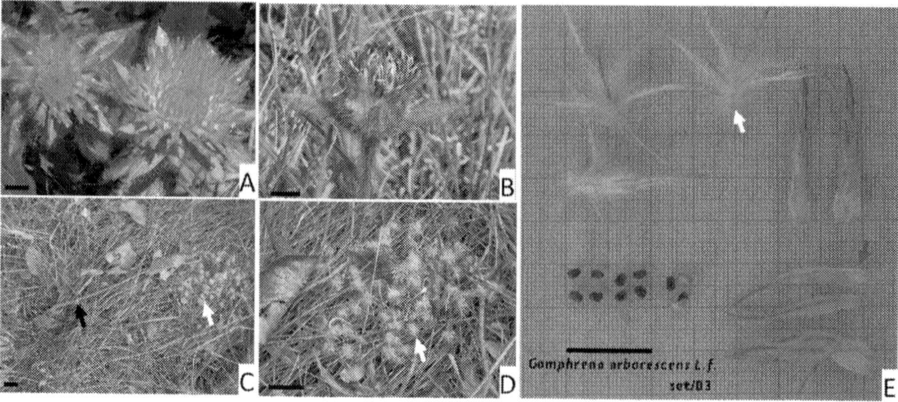

Figure 2. Fruits of *Gomphrena arborescens* L.f. are dispersed by the wind. The inflorescence opens for pollination (**A**) and closes after due to a growth of its bracts (**B**) for the maturation of the fruits; after that, the shoot (black arrow) inclines towards the soil (**C**) and the infructescence structure reopens to release the diaspores (white arrow in **C**, **D**, **E**) in the soil, formed by one seed (red arrow in **E**) and parts of the perianth (blue arrow in **E**), until the wind carries it. **Scale bars:** A-E: 2 cm.

Amaranthaceae species are well adapted to Cerrado environment and some species display different strategies to survive during the markedly seasonal climate of the Cerrado Biome [47]. Using only data obtained about perennial species, an interesting case is the aerial life cycle of *Froelichiella grisea*, which is endemic of the rocky fields of Chapada dos Veadeiros, Goiás State [34]; this species was registered during the onset of the flowering stage in the field, almost after two years of searching for it, only 20 days after a fire that burned out the vegetation (in August, during the dry season); *Gomphrena lanigera* Pohl ex Moq. also was found only at the same day, at the fruiting stage, revealing an even faster life cycle of the aerial parts [48, 57]. On the other hand, the species *Pfaffia townsendii* and *G. hermogenesii* J.C. Siqueira were found in the same region all year round; whilst the first one was always bearing flowers (it is a well-branched shrub that stands out in the middle of the rocks), the second one was usually found in the vegetative stage, more or less hidden among the surrounding Gramineae (=Poaceae) and Cyperaceae; only after the fire grazed all the grasses of the area, *G. hermogenesii* re-sprouted its aerial organs and was found in the onset of the flowering stage [48, 57]. Another species, *G. arborescens* is usually found at bloom time at RECOR/IBGE and at the Olympic Center of Universidade de Brasília, in Brasília, Federal District, during the rainy station (from November to April) and at vegetative stage from August to October (during winter); in this 0.5 meter high species, leaves were always attached to the shoot at the same time as the flowers and fruits, although in the end of the fruiting stage, some or all the shoots bearing fruits can dry out, in order to release the fruits [58]. Another species, *G. virgata*, found in the same locations than *G. arborescens*, reaches 2.0 m high and is found at vegetative stage from March to July (middle of the dry season) and fruiting goes until September (end of the winter); in the beginning of the bloom period, leaves enter in a senescence process and are detached from the shoot; after dispersing seeds, its aerial parts also dry out, re-sprouting around March (near the end of rainy season) [58]. Found only at the Olympic Center, *G. pohlii* develops its aerial organs up to 1.8 m high from August to November, starting the blooming period in December and finishing fruit dispersion around April, drying out its aerial portions during the winter [58].

As an agent of perturbation in the vegetation of Cerrado, fire can produce variable effects in the flowering and fruiting patterns: whilst flowering is more intense in the herbaceous layer after a fire breaks out, the same phenophases are not affected in trees and shrubs [59]. A thicker pericarp in dry fruits may provide greater protection for seeds, acting as a barrier against high external temperatures, such as in *Kielmeyera coriacea* winged seeds inside dry fruits [60]. Fire increases the dehiscence in anemochoric species [54, 60].

Would be necessary more research in order to understand not only the structures of repro-ductive organs, but also describe the relation among flowers and its pollinators in Cerrado species and to understand the phenology of the species, which have different strategies to survive natural and eventual events, since dropping leaves during the flowering/fruiting phase, recycling all aerial parts after completing the fruiting phase, among others.

3.2. Vegetative structures of Amaranthaceae and Melastomataceae species

During the study of Brazilian Amaranthaceae species some morphological characteristics stood out in Cerrado species: well developed subterranean systems with xylopodium, high level of endemism and hairy stem, leaves, flowers and fruits [34, 37, 47, 56], which indicates adaptations of these plants to the environment. Xeromorphy and scleromorphy are common features in leaves of Cerrado species [61]. Although the two terms describe similar morphology results, a xeromorphic plant is adapted to withstand drought and a scleromorphic plant is the result of other limiting factors to its growth instead of water, for instance a restricted nutrient intake [14] or aluminum toxicity [15]. Some aspects of the plants can be genetically determined, developed as a selective advantage, such as the development of xylopodia in *Clitoria guianen-sis* Benth. and *Calliandra dysantha* Benth. [62], both Cerrado plants.

Scleromorphism is precocious in all organs, especially the vegetative ones [24], which is why the most aimed organs to study structure are leaves, stem and roots, although the identification of a plant is usually obtained by the study of its reproductive organs. Field observations, morphological, anatomical and cellular data on aerial structures of Amaranthaceae and Melastomataceae species will be emphasized in order to improve the understanding of the surviving strategies used by some species of these families [35, 47, 61].

3.2.1. Leaf structure

Leaf anatomical traits are useful to infer adaptations to a specific environment [63-64] and are good predictors of performance [65] because of their common and strong relation-ships with functional parameters such as photosynthesis, leaf nutrient content and radial growth [66-69]. Studies with *Macairea radula* (Bonpl.) DC. and *Trembleya parviflora* (D. Don) Cogn. (Melastomataceae) showed quantitative anatomical plasticity in different environ-ments [61]: plants on open flooded area of palm swamp had significantly smaller values of specific leaf area and significantly higher values of leaf mesophyll thickness when compared with leaves partially shaded on non-flooded soil of the Cerrado *sensu stricto*, indicating that the former are smaller and thicker than the latter. Quantitative plasticity may also appear in the leaf flush as an influence of seasonality. Leaves which flushed during dry or wet season, in *Gochnatia polymorpha* (Less.) Cabrera (Asteraceae), showed statistically significant differences in the cuticle, mesophyll and abaxial epidermis thickness, stomatal size, stomata and trichomes density, indicating a probable water-status control and an adaptation to seasonality of rainfall in the Cerrado [70]. Furthermore, in [70] the authors emphasized a positive correlation between the increase in the number of trichomes and stomata density in the dry season, and suggested that this relationship would promote a highest control over stomatal conductance and transpiration.

Plants of different physiognomy of Cerrado showed that leaves are hypostomatic in most of the species, whilst there are also amphistomatic species [61, 71-73]. Stomata only on the abaxial surface are an advantageous trait for plants on low relative humidity and high temperature environment because it could reduce the loss of water vapour as the temperature on the abaxial side of the leaf is lower [74-75]. On the other hand, stomata on both surfaces makes it easier the intercellular diffusion of CO_2 in mesophyll of thicker leaves [76] and amphistomatic leaves are characteristics of plants living in high-light environments and with high photosynthetic capacity [74]. The same species can display stomata on both or only on one leaf surface in response to the light intensity under which they are grown, which can be related with leaf thickness, photosynthetic capacity and maximum stomatal conductance [75]. Leaves of the same species which were grown on high-light environment can be amphistomatous, thicker and with higher rates of photosynthesis, stomatal conductance for CO_2 uptake and loss of water vapour, whilst leaves of plants grown under low-light intensity are hypostomatous and show lower values for the same variables [74-76], demonstrating the plasticity of this feature and its influence on the hydric balance and gas exchange of the plants.

Ericaceae species *Gaylussacia brasiliensis* (Spreng.) Meisn. (figure 3 C and 3E), and species of Myrtaceae [77] and Melastomataceae [61] of cerrado *sensu stricto* and palm swamps have epidermis cells that present simultaneously evaginations of protoplasm and invaginations of the external periclinal cell wall (figure 3A-D). In frontal view the cell walls are sinuous and evagination points are usually shinier (figure 3E-F). Although the function of this feature is unproven, similar structures were found in *Drosera* [78] and named miniature papillose processes, which would be functioning as sensors for mechanical or chemical stimuli.

Emergences are structures of mixed protoderm and ground meristem origin, and are generally found in Melastomataceae leaves (figure 4). These structures are related with the vascular system and ultrastructural and histochemical analyses of the cell walls revealed micro channels permeable to water and nutrients, indicating that these structures are related with the transport of substances and may absorb or exude solutions [61, 79-80].

Phenolic compounds are regarded as protective against the incidence of UV-B radiation and could act as filters or antioxidants [81-84]. These secondary metabolites are also considered inhibitors of herbivory which, along with radiation, function as a stimulator in the biosynthesis of phenolic compounds [85]. Phenolic compounds are very common in leaves of Cerrado plants of diverse physiognomy such as palm swamp, a flooded and open soil, in a dry forest

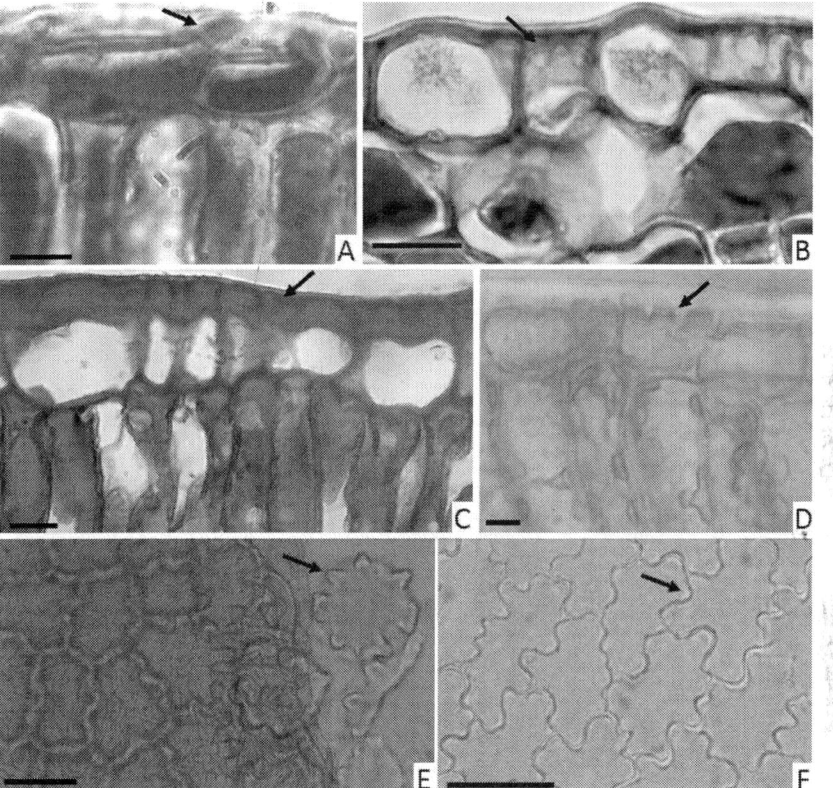

Figure 3. Leaf epidermis with evaginations and invaginations of external periclinal cell wall. **A-D:** cross section; **E-F:** frontal view. **A:** *Myrcia cordifolia* DC. (Myrtaceae). **B:** *Gomidesia pubescens (DC.) D. Legrand* (Myrtaceae). **C, E:** *Gaylussacia brasiliensis* (Spreng.) Meisn. (Ericaceae). **D, F:** *Lavoisiera bergii* Cogn. (Melastomataceae). Arrow show evagination points in dissociated of epidermal cells (E) and shiny dots in frontal view (F). **Scale bars:** A-D: 20 μm, E-F: 50 μm.

on limestone outcrops or in open areas of Cerrado [26, 61, 73], where the leaves are exposed to high irradiance and high herbivory or fungi infection.

Although the photosynthesis is highly dependent on structural and ultrastructural coordination in leaves, the environment is responsible for the relative abundance of a determined subtype of the C_4 pathway [86]. The C_4 photosynthesis pathway evolved in a great diversity of Kranz anatomy forms, biochemical routes and dimorphism of chloroplast ultrastructure [87-89] and is broadly dispersed among Angiospermae plants, including in the Amaranthaceae family [90-91].

The parenchyma bundle sheath and the mesophyll cell arrangement are the most usual anatomic pattern to determine the C_4 photosynthesis pathway [92]. However, the high degree

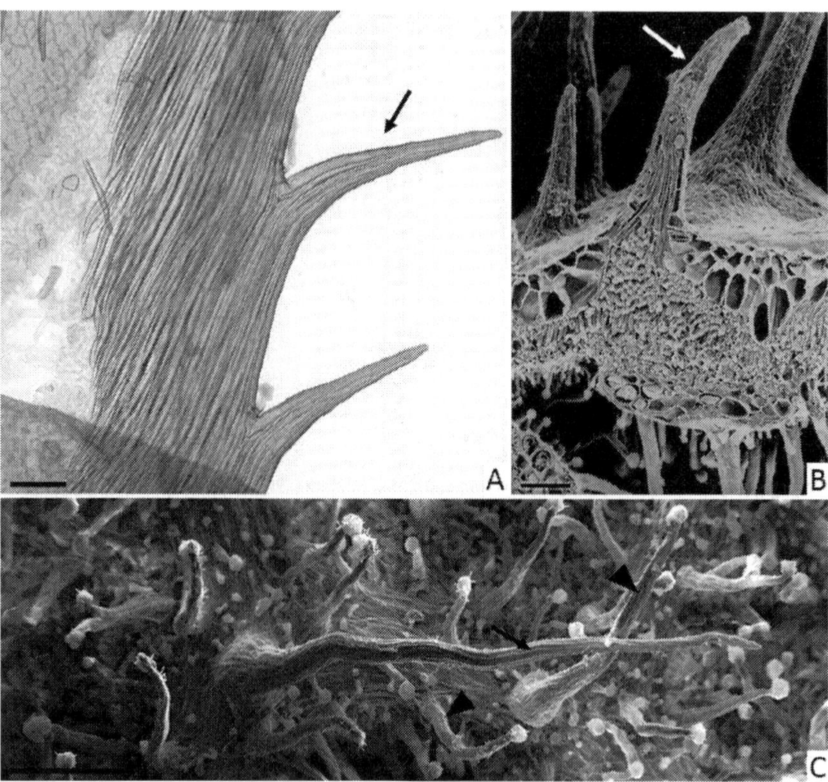

Figure 4. Leaf emergences in Melastomataceae. **A**: Non-glandular emergence (arrow) in margin of *Lavoisiera bergii* Cogn. **B**: Non-glandular emergence (arrow) in adaxial surface of *Macairea radula* (Bonpl.) DC. **C**: Non-glandular (arrow) and glandular (arrowhead) emergences in abaxial surface of *Macairea radula* **Scale bars**: 100 μm

of evolutionary convergence does not guarantee a unique pattern at biochemical or cellular and subcellular levels [86]. The structural type of leaves (Kranz or non-Kranz), the chloroplasts position, the absence or presence of stacked disks (grana) in the thylakoid membranes of chloroplasts and the number of mitochondria are important characteristics to know the photosynthetic metabolism of a plant species [89, 93]. At ultrastructural level, number and concentration of chloroplasts, mitochondria and peroxisomes in the bundle sheath cells are the most reliable criteria to determine the photosynthetic capacity of a plant [94]. However, recent studies showed that the C_4 photosynthesis can operate by dimorphic chloroplasts located in different regions of the same cell, as demonstrated in *Orcuttia* sp. and in *Borszczowia aralocaspica* Bunge and *Bienertia cycloptera* Bunge [87, 95-97]. This way, the ultrastructural study of leaves can be a key element to understand the plant metabolism.

Since the first works, leaf anatomical studies done in Brazilian Amaranthaceae species showed a well-developed vascular bundle in *Gomphrena* and *Froelichia* genera, but not in *Alternan-*

thera and *Pfaffia* genera [26, 98-101]. Intracellular studies are rarer [56, 102-103]. In [103] it is showed a gradual change in the stacked thylakoids of dimorphic chloroplasts in *Gomphrena macrocephala, G. prostrata* Desf. and *G. decipiens* Seub. connecting it with the NADP-ME subtype of C_4 photosynthesis pathway, which was the same preliminarily observation done in *G. arborescens* [56]. *Pfaffia jubata* displays the same type of chloroplasts in different leaf tissues, indicating the operation of the C_3 pathway of photosynthesis [102]. In [103] are described dimorphic chloroplasts in *G. scapigera* Mart. leaves which ultrastructure was considered compatible with NADP-ME subtype of C_4 photosynthesis. These results are coherent with the ones showed in Australian *Gomphrena* species: *G. celosioides* Mart., *G. globosa* L., *G. conica* (R. Br.) Spreng., *G. brachystylis* F. Muell., *G. brownie* Moq., *G. flaccida* R. Br., *G. canescens* R. Br. [93].

Leaves of 13 Amaranthaceae species-*Alternanthera brasiliana* (L.) Kuntze, *A. paronychioides* St.-Hil, *Froelichiella grisea, Gomphrena arborescens, G. hermogenesii, G. lanigera, G. pohlii, G. prostrata, G. virgata, Hebanthe eriantha* (Poir.) Pedersen, *Pfaffia glomerata* (Spreng.) Pedersen, *P. gnaphaloides* (L.f.) Mart. and *P. townsendii* were studied in order to understand their metabolic pathway of photosynthesis [47, 56]. All the species are native to Brazil and occur in the Cerrado Biome; six of them are endemic to Brazil and one is endemic to the Brazilian Cerrado Biome [38, 48]. All leaves are hairy (trichomes are rarer in *Alternanthera* species), amphistomatic (except *Pfaffia townsendii*), have one cell thick epidermis with stomata more or less leveled to the surrounding epidermal cells and dorsiventral mesophyll (except *F. grisea*, which has isobilateral mesophyll) with collateral bundles. All six *Gomphrena* species have thick-walled parenchymatous bundle sheath and organelles ultrastructure compatible with the operation of NADP-ME subtype of C_4 pathway of photosynthesis. Whilst *G. pohlii* and *G. virgata* have a more classical type of Kranz anatomy (figure 5) [47], *G. arborescens, G. hermogenesii, G. lanigera* and *G. prostrata* have the same type of Kranz anatomy found in most *Gomphrena* species [93, 98, 102-105], classified as "*Gomphrena* type" [89] but all share dimorphic chloroplasts (figure 6) [47]. *Alternanthera* species presented variable anatomy, with organelles positioned towards the vascular bundle in *A. paronychioides* and in the peripheral position in *A. brasiliana* bundle sheath cells. The first *Alternanthera* species was characterized as a C_3-C_4 intermediate species [106] based on its leaf anatomy, CO_2 compensation point and activity of key photosynthetic enzymes, but the authors did not mention stomata on both epidermis leaf surfaces, which is considered another fundamental feature to lower the CO_2 compensation point [74]. Thus, the position of organelles in bundle sheath cells can be a key element in determining the intermediary metabolic type in *Alternanthera* species (figure 7) [47]. *Froelichiella, Hebanthe* and *Pfaffia* species have leaf anatomy and ultrastructure compatible with C_3 metabolism [47]. Chloroplasts of the Kranz cells of C_4 plants usually have no grana, present little PSII activity and a larger amount of starch [107] whilst, in C_3 plants, the palisade cells show a larger amount of starch than the ones of the spongy parenchyma [108].

If the evolution of C4 metabolism is associated to the weather and ecological disturbance, is it possible to link some structural changes in leaves of Cerrado plant species to these evolution factors? The evolution of C_4 metabolism in these Amaranthaceae species can be related to the development of amphistomatic leaves, associated with increased leaf thickness, thicker bundle sheath cell walls, fast lifespan of the aerial system and well-developed gemmiferous underground system as adaptation to an open shinny environment with seasonal rain and oligotrophic acid soil, at least partially. Species' survival in adverse environments can be achieved by the operation of C_4 photosynthesis and carbon accumulation [109], which is associated with high photosynthesis rate [74]; the accumulated carbon is stored and protected in the underground organs, explaining the energy source to re-sprout of Cerrado species. Although in [92] trichomes are believed to affect the gas exchange and leaf temperature, reducing the light incidence, *Gomphrena arborescens* developed a translucid tissue surrounding the large trichome bases, which allows the light to reach internal photosynthetic tissue [26], indicating that its trichomes can be more a restriction to herbivory than to reduce leaf temperature or light incidence, the same way as in *G. pohlii* and *G. virgata* [58]; another aspect shared by all Amaranthaceae Cerrado´ species is the constant presence of calcium oxalate druses inside leaves and shoots [26, 47, 57-58], which is considered a highly specialized way of sequester and immobilise calcium [110].

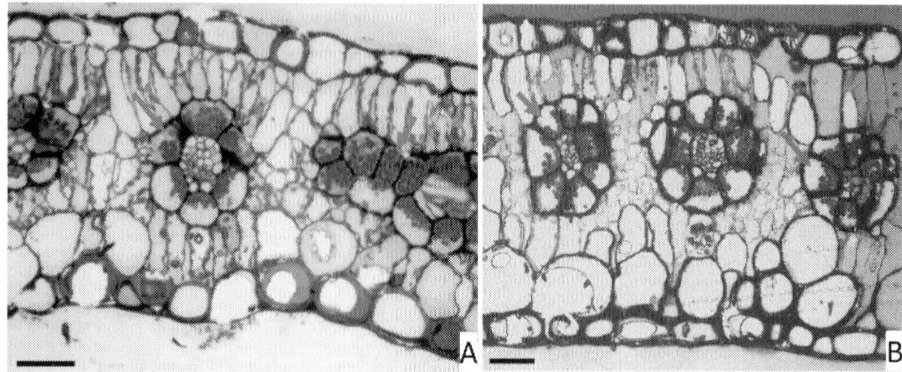

Figure 5. *Gomphrena pohlii* Moq. (A) and *G. virgata* Mart. (**B**) (Amaranthaceae) leaves in cross sections. Red arrow indicates bundle sheath cells with organelles positioned near the inner cell walls, towards the vascular bundle. **Scale bars:** 50 μm.

Large plastoglobuli were found in chloroplasts *F. grisea, G. arborescens, G. hermogenesii, G. pohlii* and *G. virgata*. According to [111], plastoglobule consists of an outer lipid monolayer containing neutral lipids and proteins/enzymes related to lipid metabolism; its dimensions vary from 30 nm to several micrometers. Plastoglobuli shape and size change during development and plastid differentiation, and under stress conditions, clustering of large groups of connected plastoglobuli were observed [111-112]. Lipid and protein storage inside the chloroplasts could support plants' fatty acid regulation, unsaturation and mobilization in response to the stress caused by biotic interactions, especially due to the presence of plastoglobulin among its proteins [111, 113]. Although higher plastoglobuli content in chloroplasts could be linked to plant senescence [114], this may not be the case of these Amaranthaceae plants because all leaf samples were visually healthy and green when collected. In *G. hermogenesii* leaves which were infected for some sort of septate endophytic organism (figure 8),

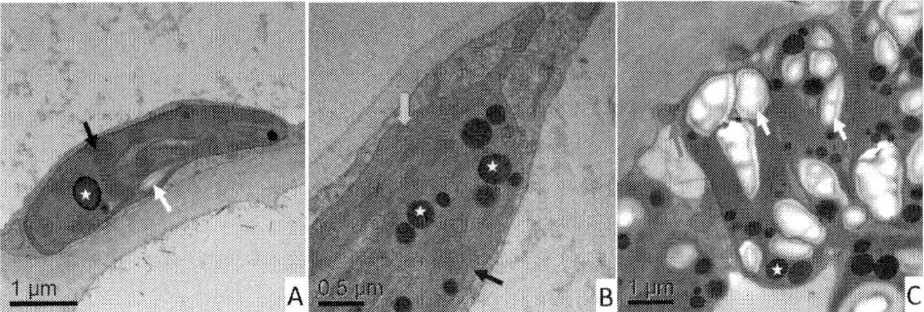

Figure 6. Dimorphic chloroplasts in *Gomphrena* species (Amaranthaceae) observed through an electron transmission microscope. **A:** Granal (black arrow) chloroplast of *Gomphrena arborescens* L.f. spongy parenchyma, with few starch granules (white arrow) and a large plastoglobule (white star). **B:** Granal chloroplast in *G. pohlii* Moq. palisade parenchyma, with well-developed peripheral reticulum (orange arrow) and large plastoglobuli (white stars). **C:** Organelles in *G. arborescens* bundle sheath cell, which are positioned towards the vascular bundle, with chloroplasts with no stacked disks (grana) of thylakoid membranes, but large amount of starch granules (white arrow), large plastoglobuli (white star), and mitochondria (red arrow). Scale bars: A, C: 1 μm; B: 0.5 μm.

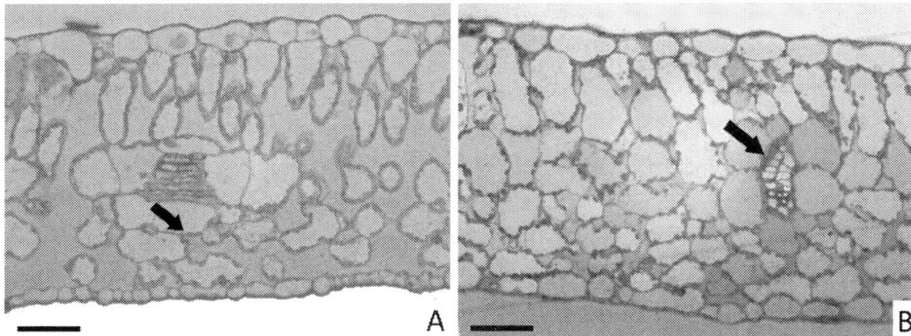

Figure 7. Leaf anatomy of *Alternanthera* species (Amaranthaceae) observed in cross sections. **A:** *Alternanthera brasiliana* (L.) Kuntze (syn=*A. dentata*). **B:** *A. paronychioides* A. St.-Hil. (syn=*A. ficoidea*). Black arrow indicates the bundle sheath cell with organelles near the outer cell wall in **A** (C_3 species) and towards the vascular bundle in **B** (C_3-C_4 intermediary species). **Scale bars:** 50 μm.

when compared with non-infected leaves (figure 8), plastoglobuli were reduced in size, which suggests a mobilization of its content due to the interaction with this microorganism.

Leaf surfaces of some Cerrado species of the genus *Gomphrena* presented epicuticular wax crystals in platelet form, oriented in parallel [58], an aspect previously described only in Chenopodiaceae species. If these platelets are present only in Cerrado species, it could be explained as a way of immobilize the excess of carbohydrates produced by a species with high photosynthesis rates and limited growth – as preconized by the theory of oligotrophic scleromorphism [14]. Crystalloid wax projections were found on both leaf surfaces of *Gomphrena arborescens*, *G. pohlii* and *G. virgata*, as platelets distributed in different densities and

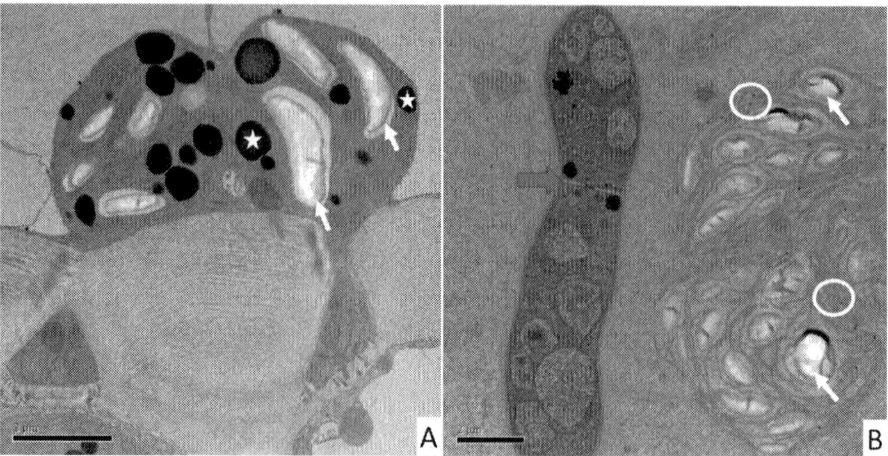

Figure 8. Chloroplasts in *Gomphrena hermogenesii* J.C. Siqueira bundle sheath cell. **A**: Non-infected sample show the chloroplast with starch (white arrow) and large plastoglobuli (white star); mitochondria (red arrow). **B**: Sample with a septate endophytic microorganism (blue arrow), shows that chloroplast's large plastoglobuli were mobilized (white circle in B) and the membrane system was disrupted, while starch (white arrow) was preserved. Scale bars: 2 μm.

patterns; epicuticular wax did not prevent the development of fungi hyphae on leaves of none of these three Cerrado native species; however, the development of such a structure should be one adaptation of the Cerrado species, because they present more platelets and ridges of epicuticular waxes in relation to *G. globosa*, an introduced species also studied [58].

There is much work to be done in order to understand all aspects connected to leaf function in Cerrado plants, because more than been the primary photosynthetic organ which produce carbohydrate for the plant, leaves also give place to biotic interactions (it is common to find fungi or insects larvae in leaves) which can affect the plant life, including the onset of the production of complexes chemical compounds of interest because of its biologic activity (alkaloids, tannins or other phenolic compounds, sterols, saponin).

3.2.2. Root and stem structure

Melastomataceae species are found in several physiognomies in the Cerrado, from well drained to periodically or permanently flooded soils [115], displaying anatomical features which give them the ability of survive in different environments. Species of the palm swamps with periodically flooded soils produce an aerenchymatous tissue in roots and stem during the primary and secondary growth [116]. During the primary growth of root and stem the tissue is a schyizolisigenous aerenchyma and schyzogen aerenchyma, respectively. During the secondary growth, root and stem develop phellogen from division of pericicle cells, deriving two cells types, one with square or rectangular shape (compact cells positive for suberin in histochemical test under light microscopy) and another cell with "T" shape (negative for suberin), which are disposed with intercellular spaces, naming the tissue aerenchymatous polyderm. The polyderm of the same organ and species found in well drained soils or emerged in the flooded soils does not have this aerenchymatous aspect [116]. According to [117], these intercellular spaces are filled of gases and longitudinally interconnected emerged parts with immersed parts, providing a low resistance way which facilitates internal diffusion of these gases at long distances throughout the organs of the plant. Although the epidermic cell walls and cuticle of primary roots generally are thin [63, 118], any species show thickness of external periclinal and anticlinal walls of epidermic cells in plants submitted to flooding in palm swamps [116]. This thickness could provide protection against adverse conditions near to root surface in the flooded soils [119].

Gelatinous fibers are different of other sclerenchyma fibers because they have a cellulosic thickening in the inner cell walls (figure 9) which, due to artifact of manufacturing of the blade, disconnects from lignified cell walls and stands out [120]. The most accurate way to observe the gelatinous fibers is the color technique using dyes that differs lignin of cellulose. For example, acid floroglucine will stain only the external layer of wall where there is lignin and safranine-fast green, a double staining, that will stain red the lignified wall and green the gelatinous layer, indicating presence of cellulose [121]. This layer is also called mucilaginous layer or "G" layer and generally occurs in tension wood and underground organs [23, 122-123]. Gelatinous fibers are very common in the Cerrado plants and they usually appear associated with the secondary xylem of stem, mainly in the initial layers of the growth ring [120, 124] but also may appear on other organs such as petiole and raquis of leaves [121]. Generally, the mucilaginous aspect of "G" layer is linked to the ability of aggregate water because the structure is highly hygroscopic [23].

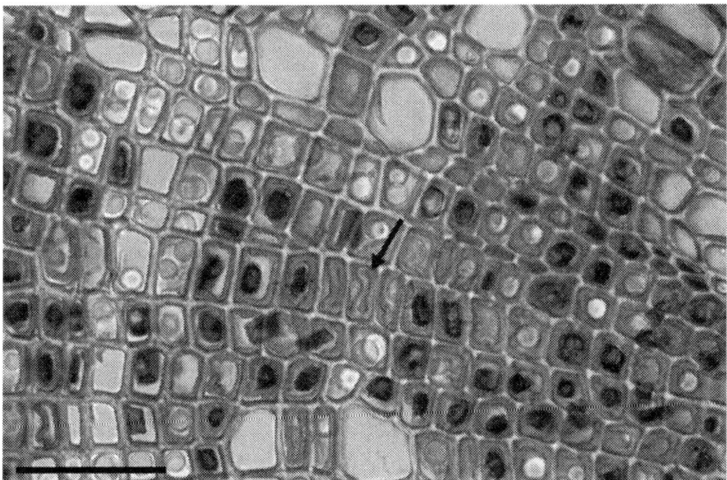

Figure 9. Gelatinous fibers in the secondary xylem of *Macairea radula* (Bonpl.) DC. (Melastomataceae) stem. Arrow show the "G" layer with cellulosic thickness. **Scale bar:** 50 μm.

In Amaranthaceae species it is common to find a secondary thickening formed by a series of vascular cambia arising successively farther outward from the center of the stem, each producing xylem toward the inside and phloem toward the outside – an anomalous secondary thickening [125-126]. *Amaranthus* spp. can have two or more rings of primary vascular bundles and a complex organization of leaf traces associated with leaf gaps [127]. Some members of the family have vascular bundles included in the medullary tissue of the shoot, with unknown function, which can contribute to leaf vascularization, as in Melastomataceae and Piperaceae species [128]. The anomalous secondary thickening was found in all investigated members of Amaranthaceae family, which is considered an important group to understand the origin of successive cambia and its products and the variation in the wood anatomy and stem in dicots [126, 129].

Preliminary study of *Gomphrena arborescens* [56] did not show successive vascular cambia in the secondary thickening of the shoot, but the bidirectional activity of a singular vascular cambia adding more cells in the secondary xylem than in the secondary phloem. There were found nucleated fibers and perimedullary amphicribral vascular bundles near the node regions (figure 10) [56], which were connected to the vascularization of a deriving leaf. Further studies are necessary to determine the function of these elements and the reason to this kind of secondary thickening, although it seems to indicate that this species is in transition from herbal to subshrub habit.

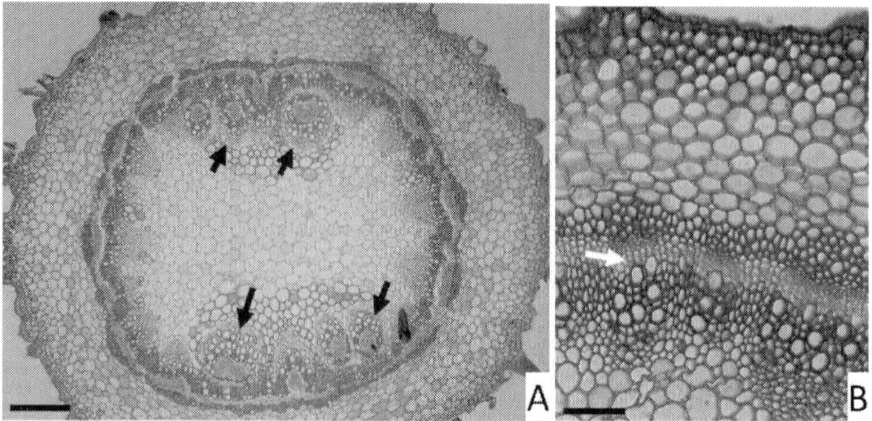

Figure 10. Transverse sections of *Gomphrena arborescens* L.f. stem. **A**: Perimedullary amphicribral vascular bundles (black arrow) near the node regions, which were connected to the vascularization of a deriving leaf. **B**: Bidirectional activity of a singular vascular cambia (white arrow). **Scale bars**: **A**: 500 μm. **B**: 100 μm.

Although the Raunkiaer system [130] is widely used to classify the life form of plants based on the level of protection of budding structures, new ecological classifications were proposed in Brazil due to the diversity of the subterranean systems found in our flora [131]. The first researcher to use the term "xylopodium" [132], around the year of 1900, described a lignified structure responsible for the regeneration of the aerial parts of a plant, during his studies of the ecology on open fields of Rio Grande do Sul State, in Brazil. Around the year of 1908, [11] another researcher noticed the same structure in Lagoa Santa plants (in Minas Gerais State), but did not attempt to define it. Since then, all the studies trying to understand this organ were concentrated in Cerrado plants [131].

Studying plants with xylopodium in São Paulo State, [133] author concluded that the superficial portion of the subterranean system, which originates the first aerial sprouts after burning, is a small subterranean stem – and sometimes is difficult to understand where the stem finishes and where the root starts. Naturally, the subsequent studies were focused in understanding the ontogenesis and the environment were these plants grow. Another study [134] determined that the xylopodium can be formed by the tuberous growth of the primary root near the soil or by the tuberous growth of the hypocotyl; the first one is due to a disturbed environment which prevents the plant to grow naturally (like in *Mimosa multipinna* Benth, *Stryphnodendron adstringens* (Mart.) Coville, *Palicourea rigida* H.B.K and *Kielmeyera coriacea* (Mart.) var. *glabripes* N. Saddi) and the second one is genetics and independent of the environmental conditions (like in *Clitoria guianensis* Benth. and *Calliandra dysantha* Benth.). Through the study of the xylopodium of *Brasilia sickii* (Asteraceae) [23, 135], it was understood that this organ could be considered a morphological unit, but not an anatomical one; it was stated the need of ontogenetic studies to really understand the origin of any plant xylopodium, since it can be originated from the root in a young xylopodium and from the stem in an older one, always with the predominance of xylem tissue, including gelatinous fibers with the ability of storing water. In [62] authors theorized that the cutting-off of the shoots, at the end of every dry season, would favour the development of xylopodium as an adaption to the conditions prevailing in the grassland or "*campos*" – where these plants are well established. This cutting-off of the shoots could, also, be provided by fire and the hard lignified xylopodium would survive as underground persistent organs of plants that dwell in savannah-like regions with a dry season lasting from 4-6 months [134]. In [136], *Mandevilla illustris* (Vell.) R.E. Woodson and *M. velutina* K. Schum. (Apocynaceae) from Cerrado were studied and authors concluded that a same plant can have an underground organ formed by a xylopodium (the hard superior portion) and by a fleshy tuberous root; the xylopodium is formed by a cambium tissue, in the junction of the hypocotyl and the primary root, without the participation of the shoot and retaining the capacity of re-sprout the shoot.

In [23], soboles are indicated as common feature in Cerrado plants, an underground horizontal stem which grow out as an erect plant [137]. Sobole of *Annona pygmaea* Warm. can be originated from the hypocotyl, in the begging of the development, or from the top of the root when the primary stem is destroyed; the organ can have aerial portions of leaves and is usually a storage organ with well-developed starchy parenchyma; this kind of plant can perform vegetative reproduction, the same way as the other type of root suckers, the gemmiferous root [62].

In this study [131] are described *Bauhinia forficata* Link, *Centrolobium tomentosum* Guill. ex Benth, *Inga laurina* Willd. and other Cerrado tree species with long roots running in parallel to the surface and showing budding shots. The point of origin to the aerial parts was a typical root without medullary tissue and with primary xylem with centripetal maturation, usually storing starch [131]. In these cases and in herbal or subshrub species from Cerrado, the underground organ is called gemmiferous roots, the second of the root sucker types [62].

The underground system is very important in Cerrado plants, being linked not only to the anchorage, support and water absorption, but also to carbohydrates and water storage and to vegetative reproduction [23, 138]. This Biome is subject to fires since remote ages, which could be caused by electrical discharge or by the primitive men as a strategy to hunt and, most recently, in order to open areas to grow crops [22]; in experimental burned Cerrado areas, the surface temperature is about 74 ºC but, under the soil, heating is drastic not so high, varying from 55 ºC one centimeter below and reaching only few degrees at 5 cm under the surface. Certain savannah species are ephemeral but most of Brazilian Cerrado's species are perennial [22]. Probably, the temperature difference during the fire can allow underground organs to survive, although the aerial parts are burned out; adding to this the budding characteristic of the xylopodium, soboles and gemmiferous roots, the well-developed underground organs of Cerrado's species can explain the prevailing perennial habit.

The underground organ of *Gomphrena macrocephala* (Amaranthaceae), a Cerrado species, revealed fructan as the main storage carbohydrate; it was the first reference to this polysaccharide in a plant of the superorder Caryophyllidae [139-140]. Fructans are fructose based polysaccharides, usually found in Asteraceae and Gramineae, two of the most evolved families, which indicates it to be a selective advantage [108]. In *G. macrocephala*, the capacity to store fructans instead of starch was considered an advantage developed in response to the environmental stress of the dry season and eventual fires [141]. The fluctuation in fructan content of *G. macrocephala* is connected to photoperiodism: shorter days, typical from the dry winter, induce senescence of the aerial organs and increase in the fructan content whilst long days, typical of the rainy summer, stimulated the development of the aerial organs and resulted in shortage of fructan content [142]. A well-marked seasonality of fructan accumulation was found in tuberous roots of *Gomphrena marginata* Seub. [143] and authors correlated it with seasonal changes in the availability of water in the soil; the content of fructans decreased during the rainy season and increased during the dry season, keeping almost steady the relative water content of the underground organs. Preliminary data show that tuberous roots of *G. arborescens* only stores fructan (figure 11), whilst in the shoot there was found starch [56]. As *G. arborescens* roots are used in folk medicine to reduce fever, against asthma and bronchitis or as tonic [144-147], along with other Amaranthaceae species used for the same purposes, the isolation and study of their fructans could help to understand the origin of its folk medicinal properties. In roots of *Arctium lappa* L., var. *herkules* there were fructans from inuline series which were proved to be biologically active as cough-suppressing agent in a cat model *in vivo*; the activity of these fructans in suppressing cough was compared to the parameters stablished for antitussive efficiency of drugs commonly used in clinical practice [148]. So, more than a challenge to understand the morphology and anatomy of the underground system in Amaranthaceae species, it is also necessary a more comprehensive study of the carbohydrates and phenology of another species of this family, even to determine if the presence of fructan is widespread or restricted to some members or genera of the family.

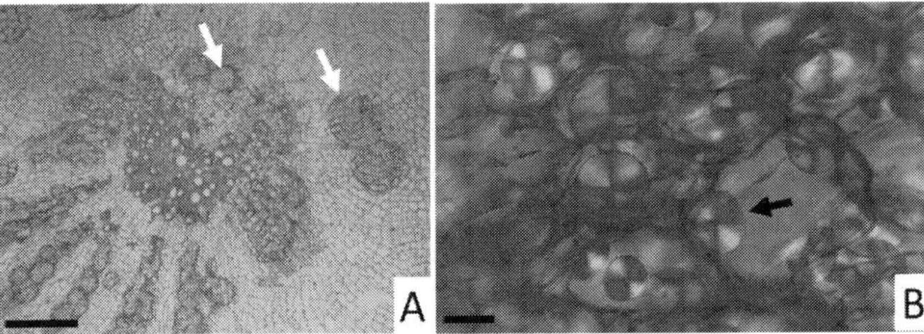

Figure 11. Transverse sections of *Gomphrena arborescens*L.f. root. A: After treatment in pure ethanol for at least 7 days, clusters of spherical fructan crystals (white arrow) appear in the cortex region. B: Spherical fructan crystals (black arrow) under polarized light, after dying the section with safranine ethanolic solution. **Scale bars: A:** 500 μm. **B:** 100 μm.

4. Retrospective and perspectives about Cerrado knowledge and structural studies

Much of the actual knowledge about the Brazilian savannah or Cerrado Biome and its vegetation is derived from a group of researchers who shared their point of views in a series of Symposia, organized initially by them in order to gather efforts and develop multidiscipli‐ nary research in networks, mostly by initiative of teachers of USP – the São Paulo University [10], who also republished in Portuguese some very important works done by the first foreign researchers [10-11, 132]. The first Symposium was realized in 1962, in São Paulo city: the *"Simpósio Nacional do Cerrado"* was intended to improve the knowledge to grow crops and raise cattle in the region [149]; the second one was realized in Rio de Janeiro, in 1965, where people concluded that most of the knowledge about Cerrado was related to the plant biology and that would be necessary to realize a multidisciplinary approach to create a national policy and also to undertake basic and applied research. The third event was again in São Paulo, in 1971, and gathered so many researchers that it was necessary to realize simultaneous meetings and extend the event in order to allow everybody to present their works [10, 149].

From the year 1975 on, the federal government created a set of programs to speed up the development of federal States in the center of the Cerrado Biome (Goiás, Minas Gerais, Mato Grosso and Federal District) through financial aid for the construction of roads, schools and warehouses, funding agricultural research, providing technical assistance to incorporation of new areas into the production process and encouraging the use of limestone and phosphate to correct the soil pH, among others [150]. More than that, Brazilian Enterprise for Agricultural Research – EMBRAPA – a state-owned company affiliated with the Brazilian Ministry of Agriculture, created its unit Embrapa Cerrados (CPAC) with the aim of developing agricul‐ tural systems viable to the Cerrado Biome and to give technical support to farmers. So, from the fourth Symposium on, realized in 1976, these events were done in Brasília, with the incentive of Embrapa Cerrados; collecting important data for agricultural development, all the work done in this event was of great value to improve the newly approved policy of the Program for Development of Cerrados [149].

According to [149], all the information gathered wasn't still enough to support the region development, mostly because they were generalized. Among other problems [149], there were: irregular distribution of rain (a challenge to grow crops), the soils low level of fertility, inadequate methods used to cultivate soils leading to soil exhaustion, incidence of illness in monotypic crops and few knowledge about environmental, economic and social peculiarities of the core region of the Cerrado Biome. Embrapa Cerrado leaded the development of networks with institutes, universities, other Embrapa units and state companies to obtain systematic data on every field of interest to understand and complete the knowledge gaps in order fulfill its own mission. The knowledge gained through political, technical and economic focus, transferred as technical support to the farmers, created a scale gain which benefited all the participants; the technology incorporated by the farmers implicated in rapid increase of cultivated area [149], at a speed that is not currently possible to maintain without a huge loss of biodiversity.

In 1979 the Symposium theme was "Cerrado: use and management" and in 1982 it was the begging of the international comprisement of the event, which was themed "Savannah: food and energy" and shared the concerns in the use of this kind of environment around the world [149]. In 1989, the seventh Symposium was done in order to gather data on the increasing efficiency to produce crops and, in 1996, the VIII National Symposium and the I International Symposium were done in a year where the cultivated area in Cerrado Biome was four times more efficient, during the onset of environmental damages such as soil degradation, weed spreading and pests; from then to now, the rational usage of savannah areas is the main concern [149-150].

In 2006 [149], the Cerrado region contributed to 33% of the Brazilian Gross Domestic Product, employing around 40% of the labor force. So, in 2008 the theme "Challenges and strategies for the equilibrium between society, agribusiness and natural resources" [149] was chosen to delineate the main discussions of the IX National Symposium of Cerrado and the II International Symposium of Tropical Savannas; in the third chapter [149] there is a review, in English, about the importance of savannah environments to the global climate change, emphasizing the distribution of this kind of environment in the world, not only in tropical regions of Africa, South America, Asia and Pacific, but also in temperate climate regions of North America (prairies) and derived savannahs in Europe.

According to [151] tropical savannahs are characterized by physiognomies with trees and shrubs and abundance of herbs from Poaceae and Cyperaceae families over dystrophic and sandy soils under a climate with seasonal rainfall. The predominance of bushes and trees over grasses depends on the soil fertility and fire as a natural or anthropomorphic phenomenon, among other environmental characteristics [151]. However, savannah flora presents differences: whilst Australian and African savannahs have more deciduous species among bushes and trees, evergreen species are the main representatives of these groups in Brazilian savannah [151-152]. African species can close leaf stomata very rapidly, but this is not the rule for Brazilian species, although there are some exceptions; this characteristic and the deciduous-ness can be linked to the shallow root system of African species [151]. Similar to the subterranean organs of Brazilian species earlier cited in this chapter, Australian savannah species develop lignotubers [153-154] with regenerative and storage functions which allows the resprout after fire. Tropical savannahs are considered more suitable to intensive grain cropping and livestock production, but it is necessary to ponder the need of food production and give value to the ecosystem services of this environment, as the maintenance of fresh water resources and moderation of the Carbon cycle, in order to create another income source for farmers [150] over preserved land.

Although there is a lot of data obtained already for Cerrado species, due to the research network which leaded to the creation of Embrapa Cerrados, and later by the increase of the number and quality of the networks created by Embrapa itself and by other Federal Government Agencies policies (as the Milenium Institutes Program, the National Institutes for Science and Technology-INCT Program and the Long-Term Ecological Research Program – PELD, from CNPq), some basic structural studies are still needed to improve the knowledge about the huge diversity of Cerrado's species (plants, fungi and fauna), preferably multidisciplinary ones, ranging from ecology, morphology and taxonomy to the anatomy and cell biology of species. The structural knowledge is the basis to further development of applied studies (preservation, investigation of pharmacological properties and others). For example, histochemical investigation in plants can help taxonomy [155-156] and the establishment of patterns for quality control of drugs or micro-scale identification of the potential origin of pharmacological properties in Folk medicinal plants, but it is necessary special preparation and fresh material at your disposal [26]. Mostly because of the time consuming and the high cost of pharmacological and pharmaceutical studies, in Brazil there are a great amount of plants used by the population as medicinal [144-147] without almost any scientific study to confirm it.

Since the high humidity and intense heat of the Cerrado's rainy season favours the development of fungal hyphae on leaf surfaces, including on the Amaranthaceae species *Gomphrena arborescens*, *G. pohlii* and *G. virgata* [46, 58], the study of the plant-microorganism interaction also can lead to a wide range of applications; for example, health risks to human or stock farming animals' can be avoided simply by preventing the consumption of contaminated food or the medicinal plant. Plants offer a wide range of habitats for microorganisms, including its aerial parts, rhizosphere and internal transport system [157]. This kind of interaction contributes to the environmental equilibrium and can play essential roles in agricultural and food safety [157-158]. The plant metabolites against endophytic invaders could be isolated and used for the genetic improvement of crop biochemical defenses; selected microorganism metabolites can be isolated to act as a biological control of crop diseases and herbivores [157].

These and other fields of study demands the basic studies of taxonomy, morphology and anatomy in order to be properly interpreted and, later, lead to application not only on the increase of crop production, but also in the conservation of the few areas of the Cerrado Biome which are still preserved, mostly due to some Conservation Units created to integrate the Conservation Unit System of Brazil. Goiás State is in the center region of the Cerrado Biome and only 15% of the natural savannah was protected in 2002 [159]; originally, savannah vegetation represented 50% of the State territory, and the author claims that the remaining species biodiversity will only be found in Conservation Units about a hundred years from now.

According to [160] the apparent dichotomy between food production and preservation of the natural vegetation is not impossible, because Brazil has already cleared enough area to support all food, fiber and bioenergy production that is necessary to meet not only the own country needs but also the global market. So, maybe it is time to set a new policy not only for agricultural and livestock development, but also to improve infrastructure and the efficiency of these activities and for encouraging and expanding the Conservation Unit System in order to better preserve the huge biodiversity of flora and fauna and its direct and indirect benefices to Brazilian people, now and through the significant amount of research that is still to be done.

Acknowledgements

CAPES, CNPq, FAPDF, FINEP

Author details

Suzane Margaret Fank-de-Carvalho[1*], Nádia Sílvia Somavilla[2], Maria Salete Marchioretto[3] and Sônia Nair Báo[4]

*Address all correspondence to: suzifankcarvalho@gmail.com

1 Bioscience Coordination, Conselho Nacional de Desenvolvimento Científico e Tecnológico – CNPq, Brasilia, Brazil

2 Botany Department, Institute of Biological Sciences, Universidade Federal de Juiz de Fora – UFJF, Juiz de Fora,, Brazil

3 PACA Herbarium, Instituto Anchietano de Pesquisas, São Leopoldo, Brazil

4 Cell Biology Department, Institute of Biological Sciences, Universidade de Brasília – UnB, Brasília, Brazil

References

[1]   Forzza RC., Leitman PM., Costa AF., Carvalho Jr. AA., : Peixoto AL., Walter BMT., Bicudo C., Zappi D., Costa DP., Lleras E., Martinelli G., Lima HC., Prado J., Stehmann JR., Baumgratz JFA., Pirani JR., Sylvestre L., Maia LC., Lohmann LG., Queiroz LP., Silveira M., Coelho MN, Mamede MC, Bastos MNC, Morim MP, Barbosa MR,

Menezes M, Hopkins M, Secco R., Cavalcanti TB., Souza VC. Catálogo de Plantas e Fungos do Brasil-Vol. 1. Rio de Janeiro: Andrea Jakobsson Estúdio/Instituto de Pesquisa Jardim Botânico do Rio de Janeiro; 2010.

[2] Myers N., Mittermeier RA., Mittermeier CG., Fonseca GAB., Kent J. Biodiversity hotspots for conservation priorities. Nature 2000;403 853-858.

[3] Klink CA., Machado RB. A conservação do Cerrado brasileiro. Megadiversidade 2005;1(1) 147-155.

[4] Dias BFS. Alternativas de desenvolvimento dos cerrados: manejo e conservação dos recursos naturais renováveis. Brasília: FUNATURA/IBAMA; 1992.

[5] Miranda H., Miranda AC.. Queimadas e estoques de carbono no Cerrado. In: Moreira AG., Schwartzman S. (eds). As Mudanças Climáticas e os Ecossistemas Brasileiros, Brasília: Ed. Foco; 2000. p75-81.

[6] Simon MF., Proença C. Phytogeographic patterns of Mimosa (Mimosoideae, Leguminosae) in the Cerrado biome of Brazil: an indicator genus of high-altitude centers of endemism? Biological Conservation 2000;96 279-296.

[7] Eiten G. Vegetação natural do Distrito Federal. Brasília: SEBRAE/UNB; 2001.

[8] Miranda HS., Sato MN., Nascimento Neto W., Aires FS. Fires in the cerrado, the Brazilian savanna. In: Cockrane MA. (ed.) Tropical Fire Ecology-climate change, land use, and ecosystem dynamics. Berlin: Springer; 2009. p427-450.

[9] Eiten G. Cerrado vegetation of Brazil. The Botanical Review 1972;38(2) 201-341.

[10] Ferri MG. A vegetação de Cerrados Brasileiros. In: Warming E., Ferri MG. Lagoa Santa e a vegetação dos Cerrados Brasileiros. Belo Horizonte: Itatiaia; São Paulo: EDUSP; 1973. p285-386.

[11] Warming E. Lagoa Santa. In: Warming E., Ferri MG. Lagoa Santa e a vegetação dos Cerrados Brasileiros. Belo Horizonte: Itatiaia; São Paulo: EDUSP; 1973. p1-284.

[12] Rawitscher F., Ferri MG., Rachid M. Profundidade dos solos e vegetação em campos cerrados do Brasil Meridional. Anais da Academia Brasileira de Ciência 1943;15(4) 267-294.

[13] Rachid-Edwards M. Alguns dispositivos para proteção de plantas contra a seca e o fogo. Boletim da Faculdade de Filosofia Ciências e Letras da Universidade de São Paulo-Botânica 1956;58(5) 39-68.

[14] Arens K. O cerrado como vegetação oligotrófica. Boletim da Faculdade de Filosofia Ciências e Letras da Universidade de São Paulo-Botânica 1958;15 59-77.

[15] Goodland R. Oligotrofismo e alumínio no Cerrado. In: Ferri MG. (coord). III Simpósio sobre o Cerrado. São Paulo: Ed. USP; São Paulo: Ed. Edgard Blücher; 1971. p44-60.

[16] Queiroz-Neto JP. Solos da região dos cerrados e suas interpretações (revisão de literatura). Revista Brasileira de Ciência do Solo 1982;6 1-12.

[17] Reatto A., Correia JR., Spera ST. Solos do bioma Cerrado: aspectos pedológicos. In: Sano.M., Almeida SP.(eds.) Cerrado: Ambiente e Flora. Planaltina: Embrapa; 1998. p47-86.

[18] Haridasan M. Aluminum accumulation by some cerrado native species of central Brazil. Plant and Soil 1982;65 265-273.

[19] Chenery EH. Aluminum in the plant world. Part I. General survey in the dicotyledons. Kew Bulletin 1948;3 173-183.

[20] Haridasan M. Alumínio é um elemento tóxico para as plantas nativas do cerrado? In: Prado CHBA, Casali CA. (eds.) Fisiologia Vegetal: práticas em relações hídricas, fotossíntese e nutrição mineral. Barueri: Manole; 2006.

[21] Ratter JA., Richards PW., Argent G., Gifford DR. Observations on the forests of some mesotrophic soils in central Brazil. Revista Brasileira de Botânica 1978;1 47-58.

[22] Coutinho LM. As queimadas e seu papel ecológico. Brasil Florestal 1980;10(44) 7-23.

[23] Paviani TI. Anatomia vegetal e cerrado. Ciência e Cultura 1978;30(9) 1076-1082.

[24] Paviani TI. Situação da anatomia ecológica no Brasil. Ciência e Cultura 1984;36(6) 927-932.

[25] Paiva PHV. A Reserva da Biosfera do Cerrado: fase II. In: Cavalcanti TB., Walter BMT. (eds.) Tópicos atuais em botânica-palestras convidadas do 51° Congresso Nacional de Botânica. Brasília: Sociedade Brasileira de Botânica/Embrapa-Cenargen; 2000. p332-334.

[26] Fank-de-Carvalho SM. Graciano-Ribeiro D. Arquitetura, anatomia e histoquímica das folhas de Gomphrena arborescens L. f. (Amaranthaceae). Acta Botanica Brasilica 2005;19(2) 377-390.

[27] Bridson D., Forman L. The herbarium handbook. Richmond: Royal Botanic Gardens Kew; 1992.

[28] Joly AB. Botânica: Chaves de identificação das plantas vasculares que ocorrem no Brasil, baseadas em chaves de Franz Thomer. 3. ed. São Paulo: Cia Ed. Nacional; 1977.

[29] Barroso GM., Guimarães EF., Ichaso CLF., Costa CG., Peixoto AL. Sistemática de Angiospermas do Brasil. v1. Rio de Janeiro: Livros Técnicos e Científicos Editora SA; 1978.

[30] Barroso GM., Peixoto AL., Ichaso CLF., Costa CG., Guimarães EF., Lima HC. Sistemática de Angiospermas do Brasil. Vol. 2. Viçosa: Imprensa Universitária, Universidade Federal de Viçosa; 1984

[31] Barroso GM., Peixoto AL., Ichaso CLF., Costa CG., Guimarães EF., Lima HC. Sistemática de Angiospermas do Brasil. Vol. 3. Viçosa: Imprensa Universitária, Universidade Federal de Viçosa; 1984

[32] Agarez FV., Pereira C., Rizzini CM. Botânica Angiospermae: Taxonomia, morfologia, reprodução: chave para determinação das famílias. 2ed. Rio de Janeiro: Âmbito Cultural; 1984.

[33] Siqueira JC. Amaranthaceae. In: Wanderley MGL., Shepherd G., Giulietti AM. (eds.) Flora Fanerogâmica do Estado de São Paulo. São Paulo: FAPESP-HUCITEC; 2002. p11-30.

[34] Marchioretto MS., Windisch PG., Siqueira JC. Os gêneros Froelichia Moench e Froelichiella R.E. Fries (Amaranthaceae) no Brasil. Pesquisas-Botânica 2002;52 7-46.

[35] Marchioretto MS. Os gêneros Hebanthe Mart. e Pfaffia Mart. (Amaranthaceae) no Brasil.PhD thesis. Universidade Federal do Rio Grande do Sul; 2008.

[36] Marchioretto MS., Miotto STS., Siqueira JC. O gênero Hebanthe (Amaranthaceae) no Brasil. Rodriguésia 2009;60(4) 783-798.

[37] Marchioretto MS., Miotto STS., Siqueira JC. O gênero Pfaffia Mart. (Amaranthaceae) no Brasil. Hoehnea 2010;37(3): 461-511.

[38] Jardim Botânico do Rio de Janeiro. Amaranthaceae. In: Marchioretto MS., Senna L., Siqueira JC. Lista de Espécies da Flora do Brasil. http://floradobrasil.jbrj.gov.br/jabot/ floradobrasil/FB42 (accessed 20 May 2014)

[39] Kraus JE., Arduin M. Manual básico de métodos em morfologia vegetal. Rio de Janeiro: EDUR; 1997.

[40] Souza W. (ed.) Técnicas básicas de microscopia eletrônica aplicadas às ciências biológicas. 3ed. Rio de Janeiro: Sociedade Brasileira de Microscopia; 2010.

[41] Paiva EAS. Effect of sample preparations for SEM studies of epicuticular wax in Tradescantia pallida (Commelinaceae) leaves. Brazilian Journal of Morphological Sciences supplement 2005; 258.

[42] Siqueira JC. O gênero Gomphrena L. (Amaranthaceae) no Brasil. Pesquisas-Botânica 1992; 43: 5-197.

[43] Siqueira JC. Fitogeografia das Amaranthaceae Brasileiras. Pesquisas-Botânica 1995; 45: 5-21.

[44] Siqueira JC. Amaranthaceae: padrões de distribuição geográfica e aspectos comparativos dos gêneros Africanos e Sulamericanos. Pesquisas-Botânica 2004; 55: 177-185,

[45] Moraes Neto SP. Acidez, alcalinidade e efeitos da calagem no solo. Planaltina, DF: Embrapa Cerrados. 2009. http://www.cpac.embrapa.br/noticias/artigosmidia/publicados/112/ (accessed 21 July 2014).

[46] Salatino A., Montenegro G.; Salatino MLF. Microscopia eletrônica de varredura de superfícies foliares de espécies lenhosas do cerrado. Revista. Brasileira de Botânica 1986; 9: 117-124.

[47] Fank-de-Carvalho SM. Contribuição ao conhecimento da anatomia, micromorfologia e ultraestrutura foliar de Amaranthaceae do Cerrado. Thesis. Instituto de Biologia, Doutorado em Biologia Celular e Estrutural, Universidade Estadual de Campinas, Campinas, SP, Brazil; 2011.

[48] Fank-de-Carvalho SM., Báo SN., Marchioretto MS. Amaranthaceae as a bioindicator of neotropical savannah diversity. In: Lameed GA. (ed.). Biodiversity enrichment in a diverse world. Rijeka: InTech; 2012. p235-262.

[49] Monteiro-Scanavacca WR. Vascularização floral em Amaranthaceae. Ciência e Cultura 1971; 23(3) 339-349.

[50] Laboriau MLS. Pollen grain of plants of the "Cerrado"-I. Anais da Academia Brasileira de Ciências 1961; 33(1) 119-130.

[51] Judd WS., Campbell CS., Kellogg EA., Stevens PF., Donoghue MJ. Plant systematics. A Phylogenetic approach. 2ed. Sunderland, Sinauer Associates; 2002.

[52] Siqueira JC. Frutos e unidades de dispersão em Amaranthaceae. Eugeniana 1984; 7 3-11.

[53] Pijl, L. van der. (1982). Principles of dispersal in higher plants. Berlim Springer-Verlang, Heidelberg; 1982.

[54] Coutinho ML. Aspectos ecológicos no Cerrado II. As queimadas e a dispersão de sementes em algumas espécies anemocórias do estrato herbáceo-subarbustivo. Boletim Botânico da Universidade de São Paulo 1977; 5 57-64.

[55] Marchioretto MS., Windisch PG., Siqueira JC. Problemas de conservação das espécies dos gêneros Froelichia Moench e Froelichiella R. E. Fries (Amaranthaceae) no Brasil. Acta Botânica Brasílica 2005; 19(2) 215-219.

[56] Fank-de-Carvalho SM. (2004). Contribuição ao conhecimento botânico de Gomphrena arborescens L.f. (Amaranthaceae)-estudos anatômicos e bioquímicos. (Dissertation). Instituto de Ciências Biológicas, Mestrado em Botânica, Universidade de Brasília, Brasília, DF, Brazil; 2005.

[57] Fank-de-Carvalho SM., Marchioretto MS., Báo SN. Anatomia foliar, morfologia e aspectos ecológicos das espécies da família Amaranthaceae da Reserva Particular do Patrimônio Natural Cara Preta, em Alto Paraíso, Goiás, Brasil. Biota Neotropica 2010; 10 77-86.

[58] Fank-de-Carvalho SM., Gomes MRA., Silva PIT., Báo SN. Leaf surfaces of Gomphrena spp. (Amaranthaceae) from Cerrado biome. Biocell 2010; 34(1) 23-35.

[59] Miranda HS., Sato MN. Efeitos do fogo na vegetação lenhosa do Cerrado. In: Scariot A., Sousa-Silva JC., Felfilli JM. (eds.). Cerrado: ecologia, biodiversidade e conservação. Brasília: Ministério do Meio Ambiente; 2005. p 95-105.

[60] Cirne P. Efeitos do fogo na regeneração da lenhosa Kielmeyera coriacea em áreas de cerrado sensu stricto: mecanismos de sobrevivência e época de queima. Thesis. Universidade de Brasília, Brasília, Brasil; 2000.

[61] Somavilla NS., Graciano-Ribeiro G. Análise comparativa da anatomia foliar de Melastomataceae em ambiente de vereda e cerrado sensu stricto. Acta Botanica Brasílica 2011; 25 764-775.

[62] Rizzini CT., Heringer EP. Underground organs of plants from some southern Brazilian savannas, with special reference to the xylopodium. Phyton 1961; 17 105-124.

[63] Fahn A. Structural and functional properties of trichomes of xeromorphic leaves. Annals of Botany 1986; 57 631-637.

[64] Dickinson WG. 2000. Integrative plant anatomy. San Diego: Academic Press; 2000.

[65] Poorter L., Bongers F. Leaf traits are good predictors of plant performance across 53 rain forest species. Ecology 2006; 87 1733-1743.

[66] Kitajima K., Mulkey SS., Wright SJ. Decline of photosynthetic capacity with leaf age in relation to leaf longevities for five tropical canopy tree species. American Journal of Botany 1997; 84 702-708.

[67] Reich PB., Wright IJ., Cavender-Bares J., Craine JM., Oleksyu J., Westtoby M., Walter MB. The evolution of plant functional variations: traits, spectra, and strategies. International Journal of Plant Sciences 2003; 164 (S3) S143-S164.

[68] Wright IJ., Reich PB., Westoby M., Ackerly DD., Baruch Z., Bongers F., Cavender-Bares J., Chapin FS., Cornelissen JHC., Diemer M., Flexas J., Garnier E., Groom PK., Gulias J., Hikosaka K., Lamont BB., Lee T., Lee W., Lusk C., Midgley JJ., Navas M-L.,Niinemets U., Oleksyn J., Osada N., Poorter H., Poot P., Prior L., Pyankov VI., Roumet C., Thomas SC., Tjoelker MG., Veneklaas EJ., Villar R. The worldwide leaf economics spectrum. Nature 2004; 428 821-827.

[69] Rossatto DR., Hoffmam WA., Franco AC. Differences in growth patterns between co-ocurring forest and savanna trees affect the forest-savanna bondary. Functional Ecology 2009; 23 689-698.

[70] Rossatto DR., Kolb RM. An evergreen Neotropical savanna tree (Gochnatia polymorpha, Asteraceae) produces different dry-and wet-season leaf types. Australian Journal of Botany 2009; 57 439-443.

[71] Reis C., Bieras AC., Sajo MG. Anatomia foliar de Melastomataceae do Cerrado do Estado de São Paulo. Revista Brasileira de Botânica 2005; 28 451-466.

[72] Bieras AC., Sajo MG. Leaf structure of the cerrado (Brazilian savanna) woody plants. Trees 2009; 23 451-471.

[73] Somavilla NS., Kolb RM., Rossatto DR. Leaf anatomical traits corroborate the leaf economic spectrum: a case study with deciduous forest tree species. Brazilian Journal of Botany 2014; 37 69-82.

[74] Mott KA., Gibson AC., O'Leary JW. The adaptive significance of amphistomatic leaves. Plant, Cell and Environment 1982; 5: 455-460.

[75] Mott KA., Michaelson O. Amphistomy as an adaptation to high light intensity in Ambrosia cordifolia (Compositae). American Journal of Botany 1991; 78(1) 76-79.

[76] Parkhurst DF. The adaptative significance of stomatal occurrence on one or both surfaces of leaves. Journal of Ecology 1978; 66(2) 367-383.

[77] Gomes SM., Somavilla NSD., Gomes-Bezerra KM., Miranda SC., De-Carvalho PS., Graciano-Ribeiro D. 2009. Anatomia foliar de espécies de Myrtaceae: contribuições à taxonomia e filogenia. Acta Botanica Brasilica 2009; 23 (1) 223-238.

[78] Haberland, G. Physiological plant anatomy. London, MacMilan Company Ltd.; 1928.

[79] Sousa HC. Estudo comparativo de adaptações anatômicas em órgãos vegetativos de espécies de Lavoisiera DC. (Melastomataceae) da Serra do Cipó, MG. Thesis. Universidade de São Paulo, São Paulo; 1997.

[80] Milanez CRD., Machado SR. Leaf emergences in Microlepis oleaefolia (DC.) Triana (Melastomataceae) and their probable function: an anatomical and ultrastructural study. Micron 2007; 39 (7) 884-890.

[81] Landry LG., Chapple CCS., Last RL. Arabidopsis mutants lacking phenolic sunscreens exhibit enhanced Ultraviolet-B injury and oxidative damage. Plant Physiology 1995;109 1159-1166.

[82] Booij-James IS., Dube SK., Jansen MAK., Edelman M., Mattoo AK. Ultraviolet-B radiation impacts light-mediated turnover of the photosystem II reaction center heterodimer in Arabidopsis mutants altered in phenolic metabolism. Plant Physiology 2000; 124 1275-1284.

[83] Bieza K., Lois R. An Arabidopsis mutant tolerant to lethal ultraviolet-B levels shows constitutively elevated accumulation of flavonoids and other phenolics. Plant Physiology 2001; 126 1105-1115.

[84] Figueroa FL., Korbee N., Carrillo P., Medina-Sánchez JM., Mata M., Bonomi J., Sánchez-Castillo PM. The effects of UV radiation on photosynthesis estimated as chlorophyll fluorescence in Zygnemopsis decussata (Chlorophyta) growing in a high mountain lake (Sierra Nevada, Southern Spain). Journal of Limnology 2009; 68 206-216.

[85] Izaguirre MM., Mazza CA., Svatos A., Baldwin IT., Ballar CL. Solar ultraviolet-B radiation and insect herbivory trigger partially overlapping phenolic responses in Nicotiana attenuata and Nicotiana longiflora. Annals of Botany 2007; 99 103–109.

[86] Press MC. The functional significance of leaf structure: a search for generalizations-research reviews. New Phytologist 1999; 143 213-219.

[87] Edwards GE., Franceschi VR., Voznesenskaya EV. Single-cell C4 photosynthesis versus dual-cell (Kranz) paradigm. Annual Review of. Plant Biology. 2004; 55 173-96.

[88] Sage RF. The Evolution of C4 photosynthesis. New Phytologist. 2004; 161 (2) 341-370.

[89] Kadereit G., Borsh T., Weising K., Freitag H. Phylogeny of Amaranthaceae and Chenopodiaceae and the evolution of C4 photosynthesis. International Journal of Plant Sciences 2003; 164 (6) 959-986.

[90] Laetsch WM. The C4 syndrome: a structural analysis. Annual Review of Plant Physiology 1974; 25 27-52.

[91] Sage RF., Monson RK. (eds). C4 Plant Biology. United States of America: Academic Press; 1999.

[92] Gutschick VP. Biotic and abiotic consequences of differences in leaf structure-research reviews. New Phytologist 1999; 143 3-18.

[93] Carolin RC., Jacobs, SWL., Vesk M. Kranz cells and mesophyll in the Chenopodiales. Australian Journal of Botany 1978; 26 683-698.

[94] Black CC. Jr., Mollenhauer HH. Structure and distribution of chloroplasts and other organelles in leaves with various rates of photosynthesis. Plant Physiologist 1971; 47 15-23.

[95] Edwards GE., Franceschi VR., Ku MSB., Voznesenskaya HV., Pyankov VI., Andreo CS. Compartmentation of photosynthesis in cells and tissues of C4 plants. Journal of Experimental Botany 2001; 52(536) 577-590.

[96] Sage RF. C4 photosynthesis in terrestrial plants does not require Kranz anatomy. Trends in Plant Science 2002; 7 283-285.

[97] Voznesenskaya EV., Edwards GE., Kiirats O., Artyusheva EG., Franceschi VR. Development of biochemical specialization and organelle partitioning in the single-cell C4 system in leaves of Borszczowia aralocaspica (Chenopodiaceae). Annals of Botany 2003; 90(12) 1669-1680.

[98] Handro W. Contribuição ao estudo da venação e anatomia foliar das amarantáceas dos cerrados. Anais da Academia Brasileira de Ciências 1964; 36(4) 479-499.

[99] Handro W. Contribuição ao estudo da venação e anatomia foliar das amarantáceas dos cerrados. II-Gênero Pfaffia. Anais da Academia Brasileira de Ciências 1967; 39(3/4) 495-506.

[100] Gavilanes ML. Estudo anatômico do eixo vegetativo de plantas daninhas que ocorrem em Minas Gerais. 1. Anatomia foliar de Gomphrena celosioides Mart. (Amaranthaceae). Ciência e Agrotecnologia 1999; 23(4) 881-898.

[101] Duarte MR., Debur MC. Characters of the leaf and stem morpho-anatomy of Alternanthera brasiliana (L.) O. Kuntze, Amaranthaceae. Brazilian Journal of Pharmaceutical Sciences 2004; 40(1) 85-92.

[102] Estelita-Teixeira ME., Handro W. Leaf ultrastructure in species of Gomphrena and Pfaffia (Amaranthaceae). Canadian Journal of Botany 1984;.62(4) 812-817.

[103] Antonucci NP. Estudos anatômicos, ultra-estruturais e bioquímicos da síndrome Kranz em folhas de duas espécies de Gomphrena L. (Amaranthaceae). Dissertation. Universidade Estadual de São Paulo; 2010.

[104] Ueno O. Immunogold localization of photosynthetic enzymes in leaves of various C4 plants, with particular reference to pyruvate orthophsphate dikinase. Journal of Experimental Botany 1998; 49(327) 1637-1646

[105] Muhaidat R., Sage RF., Dengler NG. Diversity of Kranz anatomy and biochemistry in C4 eudicots. American Journal of Botany 2007; 94(3) 362-381.

[106] Rajendrudu G., Prasad JSR., Rama Das VS. C3-C4 species in Alternanthera (Amaranthaceae). Plant Physiology 1986; 80 409-414.

[107] Buchanan BB., Gruissem W., Jones RL. Biochemistry & Molecular Biology of Plants. U.S.A: American Society of Plant Physiologist; 2000.

[108] Lewis DH. Occurrence and distribution of storage carbohydrates in vascular plants. Pp. 1-52. In.: Lewis DH. (ed) Storage carbohydrates in vascular plants. Cambridge University Press: Society for Exp. Biology, Seminar; 1984. p1-52. Series 19.

[109] Borsch T. Clemants S., Mosyakin S. Symposium: Biology of the Amaranthaceae-Chenopodiaceae alliance. Journal of the Torrey Botanical Society 2001; 128(3) 234-235.

[110] Volk GM., Goss LJ., Franceschi VR. Calcium Channels are Involved in Calcium Oxalate Crystal Formation in Specialized Cells of Pistia stratiotes L. Annals of Botany 2004; 93 741-753.

[111] Bréhelin C., Kessler F. The plastoglubule: a bag full of lipid biochemistry tricks. Photochemistry and Photobiology 2008; 84(6) 1388-1394.

[112] Grennan AK. Plastoglobule proteome. Plant Physiology 2008; 147 443-445.

[113] Upchurch R.G. Fatty acid insaturation, mobilization and regulation in the response of plants to stress. Biotechnology Letters 2008; 30 967-977.

[114] Kutík J. The development of chloroplast structure during leaf ontogeny. Photosynthetica 1998; 35 (4) 481-505.

[115] Mendonça RC., Felfili JM., Walter BMT., Silva-Junior MC., Rezende AV., Filgueiras TS., Nogueira PE., Fagg CW. Flora vascular do Bioma Cerrado: checklist com 12.356

espécies. In: Sano SM., Almeida SP., Ribeiro JF. (Eds.) Cerrado: ecologia e flora. Vol.. 2. Brasília: Embrapa Cerrados; 2008. p 421-1181.

[116] Somavilla NS., Graciano-Ribeiro G. Ontogeny and characterization of aerenchymatous tissues of Melastomataceae in the flooded and well-drained soils of a Neotropical savanna. Flora 2012; 207 2012-222.

[117] Armstrong W., Brele R., Jackson MB. Mechanisms of flood tolerance in plants. Acta Botanica Neerlandica 1994; 43 307-358.

[118] Esau K. Anatomia das plantas com sementes. São Paulo: Edgard Blücher; 1976.

[119] Ponnamperuma, F.N. Effects of flooding soils. In: Kozlowski TT. (ed.). Flooding and Plant Growth. Califórnia: Academic Press; 1984. p9-45.

[120] Marcati CR., Angyalossy-Alfonso V., Benetati L. Anatomia comparada do lenho de Copaifera langsdorfii Desf. (Leguminosae-Caesalpinioideae) de floresta e cerradão. Revista Brasileira de Botânica 2001; 24(3) 311-320.

[121] Mendes ICA., Paviani TI. Morfo-anatomia comparada das folhas do par vicariante Plathymenia foliolosa Benth. e Plathymenia reticulata Benth. (Leguminosae-Mimosoideae). Revista Brasileira de Botânica 1997; 20 185-195.

[122] Evert RF. Esau's Plant Anatomy: meristems, cells, and tissues of the plant body: their structure, functiion and development. 3ed. New Jersey: John Wiley & Sons; 2006.

[123] Scatena VL., Dias ES. Parênquima, colênquima e esclerênquima. In: Appezzato-da-Glória B., Guerreiro SMC. (eds.). Anatomia Vegetal. Viçosa: UFV. 2ed; 2006. p109-141.

[124] Marcati CR., Oliveira IS., Machado SR. Growth rings in cerrado woody species: occurrence and anatomical markers. Biota Neotropica; 2006 6(3) http://www.biotaneotropica.org.br/v6n3/PT/abstract?article+bn00206032006. (Accessed 24 July 2014)

[125] Esau K. Anatomy of Seed Plants. U.S.A.: John Wiley and Sons Inc. 2ed; 1977.

[126] Rajput KS., Rao KS. Secondary growth in the stem of some species of Alternanthera and Achyrantes aspera (Amaranthaceae). IAWA Journal 2000; 21(4) 417-429.

[127] Costea M., DeMason DA. Stem morphology and anatomy in Amaranthus L. (Amaranthaceae)-taxonomic significance. Journal of the Torrey Botanical Society 2001; 128 254-281.

[128] Mauseth JD. Plant Anatomy. California: The Benjamin/Cummings Publishing Company, Inc; 1988.

[129] Carlquist S. Wood and stem anatomy of woody Amaranthaceae s.s.: ecology, systematics and the problems of defining rays in dicotiledons. Botanical Journal of the Linnean Society 2003; 14: 1-19.

[130] Raunkiaer C. The life forms of plants and statistical plant geography. Oxford: Clarendon Press; 1934.

[131] Appezzato-da-Glória B. Morfologia de sistemas subterrâneos-histórico e evolução do conhecimento no Brasil. Ribeirão Preto, Brazil: A.S. Pinto Editor; 2003.

[132] Lindman CAM., Ferri MG. A vegetação do Rio Grande do Sul. Belo Horizonte: Ed. Itatiaia; São Paulo: EDUSP; 1974.

[133] Rachid M. Transpiração e sistemas subterrâneos da vegetação de verão dos campos Cerrados de Emas. Boletim da Faculdade de Filosofia Ciências e Letras da Universidade de São Paulo, 80 (Botânica) 1947; 5 5-140.

[134] Rizzini CT., Heringer EP. Studies on the underground organs of trees and shrubs from some southern Brazilian savannas. Anais da Academia Brasileira de Ciências 1962; 34 235-247.

[135] Paviani TI. Estudos morfológico e anatômico de Brasilia sickii G. M. Barroso. II: anatomia da raiz, do xilopódio e do caule. Revista Brasileira de Biologia 1977; 37(2) 307-324.

[136] Appezzato-da-Glória B., Estelita MEM. The developmental anatomy of the subterranean system in Mandevilla illustris (Vell.) Woodson and M. velutina (Mart. ex Stadelm.) Woodson (Apocynaceae). Revista Brasileira de Botânica 2000; 23 7-35.

[137] Bell AD., Brian A. Plant form-an illustrated guide to flowering plant morphology. Portland, USA; London, England: Timber Press, Inc; 2008.

[138] Hoffmann WA. The relative importance of sexual and vegetative reproduction in Cerrado woody plants. In: Cavalcanti TB., Walter BMT. (eds.), Tópicos atuais em botânica-palestras convidadas do 51° Congresso Nacional de Botânica. Brasília: Sociedade Brasileira de Botânica/Embrapa-Cenargen; 2000. p231-234.

[139] Figueiredo-Ribeiro RCL., Dietrich SMC., Carvalho MAM., Vieira CCJ., Isejima EM., Dias-Tagliacozzo GM., Tertuliano MF. As múltiplas utilidades dos frutanos-reserva de carboidratos em plantas nativas do cerrado. Ciência Hoje 1982; 14(84) 16-18.

[140] Figueiredo-Ribeiro RCL. Distribuição, aspectos estruturais e funcionais dos frutanos, com ênfase em plantas herbáceas do cerrado. Revista Brasileira de Fisiologia Vegetal 1993; 5(2) 203-208.

[141] Vieira CCJ., Figueiredo-Ribeiro RCL. Fructose-containing carbohydrates in the tuberous root of Gomphrena macrocephala St.-Hil. (Amaranthaceae) at different phenological phases. Plant, Cell and Environmentm 1993; 16 919-928.

[142] Moreira MF., Vieira CCJ., Zaidan LBP. Efeito do fotoperíodo no crescimento e no padrão de acúmulo de frutanos em plantas aclimatizadas de Gomphrena macrocephala St.-Hil. (Amaranthaceae). Revista. Brasileira de Botânica 1999; 22(3)3 397-403.

[143] Silva FG., Cangussu LMB., Paula SLA., Melo GA., Silva EA.. Seasonal changes in fructan accumulation in the underground organs of Gomphrena marginata Seub.

(Amaranthaceae) under rock-field conditions. Theoretical and Experimental Plant Physiology 2013; 25(1) 46-55.

[144] Barros MGAE. Plantas medicinais-usos e tradições em Brasília-DF. Oréades 1982; 8 (14/15) 140-149.

[145] Pio Corrêa M. Dicionário de plantas úteis do Brasil e das exóticas cultivadas. Brasília: Ministério da Agricultura/IBDF; 1984.

[146] Siqueira JC. Importância alimentícia e medicinal das amarantáceas do Brasil. Acta Biologica Leopoldinensia 1987; 9 (1) 99-110.

[147] Lorenzi H., Matos FJA. Plantas medicinais no Brasil-nativas e exóticas. Nova Odessa, Brazil: Editora Plantarum, 2ed.; 2008.

[148] Kardošová A., Ebringerová A., Alföldi J., Nosál'ová G., Fraňová S., Hřibalová V. A biologically active fructan from the roots of Arctium lappa L., var. Herkules. International Journal of Biological Macromolecules 2003; 33 (Nov 2003) 135–140.

[149] Faleiros FG., Farias Neto, AL. Savanas – desafios e estratégias para o equilíbrio entre a sociedade, agronegócio e recursos naturais. Planaltina-DF, Brazil: Embrapa Cerrados, 2008.

[150] Embrapa Cerrados, Unidade, História. 2012. Brasília: Embrapa. http://www.cpac.embrapa.br/unidade/historia/ (acessed 25 August 2014).

[151] Pinheiro, MHO. Formações savânicas mundiais: uma breve descrição fitogeográfica. Brazilian Geographical Journal: Geosciences and Humanities research medium 2010; 1 (2) 306-313.

[152] Williams RJ., Myers, BA., Muller WJ., Duff GA., Eamus D. Leaf phenology of woody species in a north Australian tropical savanna. Ecology 1997; 78(8) 2542-2558.

[153] Kerr LR. The lignotubers of Eucalyptus seedlings. Proceedings of the Royal Society of Victoria 1925; 37 79-97.

[154] Mibus R., Sedgley M. Early lignotuber formation in Banksia-Investigations into the anatomy of the cotyledonary node of two Banksia (Proteaceae) species. Annals of Botany 2000; 86 575-587.

[155] Johansen DA. Plant Microtechnique. London: McGraw-Hill; 1940.

[156] Harborne JB. The evolution of flavonoid pigments in plants. In: T. Swain (ed.). Comparative phytochemistry. London: Academic Press; 1966. p271-295.

[157] Montesinos E. Plant-associated microorganisms: a view from the scope of microbiology. International Microbiology 2003; 6 221-223.

[158] Rocha R., Luz DE., Engels C., Pileggi SAV., Jaccoud Filho DS., Matiello RR., Pileggi M. Selection of endophytic fungi from comfrey (Symphytum officinale L.) for in vitro

biological control of the phytopatogen Sclerotinia sclerotiorum (LIB.). Brazilan Journal of Microbiology 2009; 40 73-78.

[159] Siqueira JC. O Bioma Cerrado e a preservação de grupos taxonômicos: um olhar sobre as Amaranthaceae. Pesquisas, Botânica 2007; 58 389-394.

[160] Martinelli LA., Joly CA., Nobre CA., Sparovek G. A falsa dicotomia entre a preservação da vegetação natural e a produção agropecuária. Biota Neotropica 2010; 10(4) http://www.biotaneotropica.org.br/v10n4/pt/abstract?point-of-iew+bn00110042010 (accessed 27 July 2014).

Chapter 17

Social Perception and Supply of Ecosystem Services — A Watershed Approach for Carbon Related Ecosystem Services

Antonio J. Castro, Caryn C. Vaughn, Jason P. Julian,
Marina García Llorente and Kelsey N. Bowman

Additional information is available at the end of the chapter

http://dx.doi.org/10.5772/59280

1. Introduction

Over the past two decades research on ecosystem services, i.e. the benefits that humans derive from natural systems [1], has gained importance among scientists, managers, and policy-makers worldwide as a way to communicate societal dependence on ecological life support systems integrating both natural and social science perspectives [2]. Ecosystem services can be direct benefits, such as food or freshwater for drinking, or indirect benefits through provisioning of services such as carbon sequestration [1]. Ecosystem services include 1) provisioning services obtained directly from the ecosystem such as food provision, 2) regulating services such as water regulation, habitat, air quality, and water quality, and 3) cultural services, which are the benefits that people obtain through tourism, aesthetic values, spiritual enrichment, and sense of place [3, 4].

The ecosystem services approach is useful for decision-making in conservation and natural resource management [5] because it assigns value to nature by translating ecosystem proper-ties into human needs [6]. Ecosystem services can be valued using different approaches ranging from biophysical quantifications to sociocultural surveys to economic assessment. Biophysical quantification of services such as carbon storage and sequestration have recently been used extensively in conservation applications. However, to conserve biodiversity, we need to move beyond narrow studies of species or habitat status and increase social awareness of the broader importance of conservation [2]. A key challenge in implementing this approach is identifying an ecosystem's capacity to provide services (supply side) and the social demand for those services (demand side). Addressing both the supply and demand for ecosystem

services underscores the fact that the importance of an ecosystem service to people is influenced not only by the ecosystem's properties but also by societies need for that service and how that need is perceived.

The Kiamichi River watershed, in southeastern Oklahoma (USA), provides many direct and indirect ecosystem services to stakeholders that live in or visit the area. This watershed and the area surrounding it is a national biodiversity hotspot, meaning it is biologically rich, yet threatened. This area is also at the center of a highly politicized debate between different stakeholders' plans for the use of the watershed's ecosystem services and activities that may affect those services [7]. The Kiamichi watershed not only provides many important freshwater services (e.g., drinking water, water filtration or recreation), but it provides numerous terrestrial ecosystem services as well such as habitat for species and food production. The land is relatively undeveloped with few urban areas and extensive tracts of second growth, forested landscapes [8] that provide carbon storage and sequestration. Carbon sequestration is considered an optimal descriptor of ecosystem functioning [9, 10, 11]. It is a current focus in climate change studies and is classified as an intermediate service [12] or as supporting the delivery of other regulating services [13]. Most people are unaware that carbon sequestration provides direct benefits such as erosion control and soil fertility and indirect benefits such as air quality and habitat for species.

Here, we used the Kiamichi River watershed as a case study to examine the social perception and biophysical supply of carbon related services. We first assessed the social perception of the general public regarding a variety of ecosystem services provided by the Kiamichi watershed in southeastern Oklahoma, including direct and indirect benefits related to the carbon cycle. We used a carbon sequestration model to quantify the spatial distribution of carbon storage and sequestration across the watershed. We used these results along with the social perception of services and the watershed capacity for carbon sequestration to analyze the supply-demand framework of ecosystem services [14]. Finally, we discuss the implications for linking the structure and functioning of biodiversity within the watershed.

2. Problem statement

Changes in land use-land cover are recognized as one of the most important direct drivers in ecosystem services delivery [15]. Landscapes across the U.S. are changing with human population growth and increased development. These land use changes alter the natural sinks and pools of carbon in the environment, but are often not included in land management or planning. Different land use types and dominant vegetation differ in their storage capacity and sequestration rate [15]. To better understand the impacts of land cover-use changes in relatively undeveloped areas such as the Kiamichi watershed, research is needed on different land uses and land changes and their impacts on carbon storage and sequestration.

Carbon sequestration can be viewed as an optimal descriptor of ecosystem functioning [10,11], and human-derived carbon fluctuations [16] in the atmosphere affect many other services such as air quality and biomass production. Changes in air quality are one of the carbon related ecosystem services that is most easily recognized by the public. Thus, understanding how the public perceives the status and importance of air quality can help inform resource management. Our study compares perceptions of Kiamichi watershed stakeholders with the actual state of carbon sequestration services and land use practices in the watershed.

3. Application area

The Kiamichi River watershed in southeastern Oklahoma, with a drainage area of 4,650 km^2, is a major tributary of the Red River, which flows into the Mississippi River and Gulf of Mexico (Figure 1). The watershed is 64% forest, 18% pasture, 11% grassland/shrubland, 3% urban, 3% open water, and 1% wetlands according to the 2006 National Land Cover Dataset. While most of the watershed is temperate deciduous forest (primarily oak-hickory), there are several conifer plantation forests across the watershed. Its steep and rugged terrain has limited major row crop agriculture, there are no nearby major cities or interstates, and human population density is low [5.6 people / km^2] [17]. This lack of development in the watershed has left the Kiamichi River with relatively pristine water and high aquatic biodiversity, containing 86 fish species and 31 mussel species, three of which are federally endangered [18,19,17, 20].

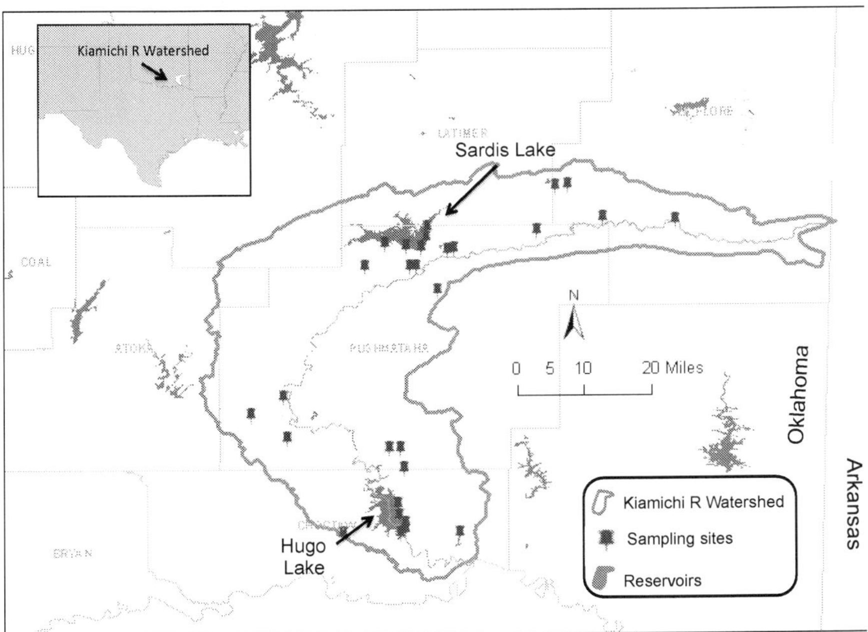

Figure 1. Kiamichi River watershed study area and sampling sites.

4. Method used

4.1. Social sampling and analysis of perceptions of ecosystem services

We conducted social sampling regarding public perceptions of a suite of ecosystem services provided by the Kiamichi watershed. In summer 2013, we conducted 304 random, individual, face-to-face surveys across the watershed. Interviewees included stakeholders residing in the watershed, tourists, and people working within the watershed. Sampling was conducted at 30 sites in the watershed (Figure 1). Social preferences regarding the variety of ecosystem services provided by the Kiamichi River were explored through ranking [21]. Our study included eight categories of carbon-and water-related ecosystem services in three classes: provisioning (freshwater provision), regulating (water regulation, water quality, air quality, and habitat for species), and cultural services (recreation, cultural heritage, and local identity). We asked interviewees if they felt that the Kiamichi River provided benefits that contribute to human well being (very much, much, not very much, and nothing), and asked them to provide examples of potential benefits. All respondents were asked to indicate the relative importance and perceived trend of each service over the last 10 years. To do this, they were asked to select the four services most important to them and to rank them from 1 to 4 (important to essential services). From this information, we created an ordinal measure of the importance of each service to each respondent [22].

4.2. Mapping the distribution of carbon storage and sequestration

To model carbon storage we used InVEST (Integrated Valuation of Environmental Services and Tradeoffs). InVEST is a family of GIS tools designed by the Natural Capital Project to inform decisions about natural resource management and provides an effective tool for evaluating trade-offs among ecosystem services under different scenarios [23]. InVEST models are spatially explicit and return results in either biophysical (e.g., tons of carbon stored) or economic terms (e.g., net present value of that sequestered carbon). We used the InVEST carbon sequestration model to quantify and map the current (i.e., 2006) spatial distribution of carbon sequestration across the Kiamichi watershed. Here, the carbon model estimates for each pixel (30-meter resolution) a value that represents the change in storage between two time periods. Negative values represent a loss in carbon sequestering capacity, and positive values represent areas that have gained more capacity to sequester carbon.

We used InVEST Terrestrial Toolboxes (version 2.5.6) in ArcMap (10.2) to generate a map of the balance of carbon sequestration in the Kiamichi watershed. The model needs several inputs to successfully estimate carbon sequestration including land use-land cover (LULC) maps for the two years of comparison and data on each LULC's capacity to stock carbon in four fundamental carbon pools: above ground biomass, below ground biomass, soil, and dead matter. These data can be collected from real time monitoring of carbon levels or from the literature. We obtained carbon pool values from the 2006 IPPC Guidelines for National Greenhouse Gas Inventories report by the Intergovernmental Panel on Climate Change

[24]. According to this source, southeastern Oklahoma is considered a subtropical steppe climate. Estimated carbon values for LULC types for a subtropical steppe climate were derived from various IPCC tables in Volume 4 of the report. For the five LULC types selected we calculated the mean value when multiple values were available. Not all four of the required carbon pools were listed for each LULC category in the IPCC report; so additional literature searches were conducted [25,26]. Finally, all carbon pool values were converted into metric tons (or Mega grams) per hectare (Mg ha^{-1}) and formatted in a table, as per InVEST model requirements.

4.3. Land use-Land Cover (LULC) maps

We compared changes in LULC between 1898 and 2006. The LULC map for 1898 is the earliest complete data set for the Kiamichi watershed and served as the reference year for the carbon model. LULC in 1898 largely represents the potential natural vegetation and pre-European landscape of southeastern Oklahoma [27]. We created the 1898 LULC map using data from [28], which was derived from Public Land Survey System records made available by the Bureau of Land Management's General Land Office. Our 1898 map included four LULC categories: cropland, forest, grassland, and wetland. The 2006 LULC map was derived from the National Land Cover Database [29], which contained over twenty LULC categories. To make the two datasets compatible with each other and InVEST, we grouped LULC as follows: Urban-Barren, Water-Wetland, Forest, Shrub-Grassland-Pasture, and Cropland. The 1898 dataset was converted to a 30-meter raster to match the 2006 NLCD.

5. Status and results

5.1. Social perception of watershed services

Of the 304 respondents, 300 (99%) answered that the Kiamichi River is "providing benefits that are contributing to human wellbeing," with 86% answering that it provides substantial benefits (i.e., very much, Figure 2a). Only one respondent said that no benefits were provided by the Kiamichi, and three respondents did not answer the question. When asked to give an example of a benefit provided by the Kiamichi, virtually all of those who responded gave an example related to water resources (i.e., drinking water, fishing, recreation). Air quality was not mentioned by any of the respondents as a watershed benefit.

The ecosystem service with the highest average importance among all respondents was habitat for species, followed by recreation and water quality (Figure 2b). Ecosystem services considered less important were local identity, followed by cultural heritage and air quality. Most respondents thought that many of the services they considered most important to human wellbeing (habitat for species and water quality) had declined, while those services that were not considered as important (cultural heritage and local identity) had remained stable or increased (Figure 3). Air quality was considered to be the most stable ecosystem service.

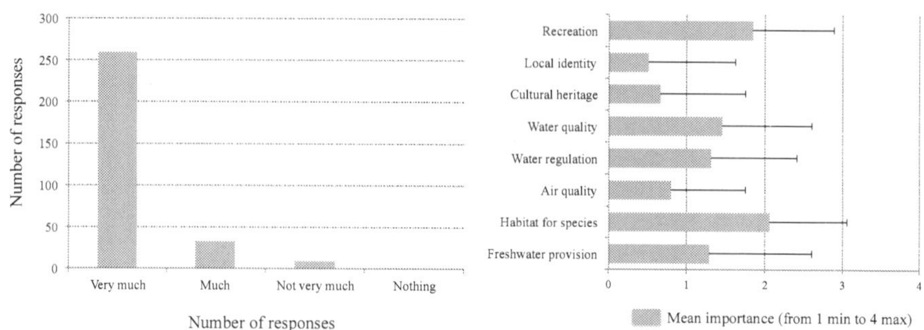

Figure 2. Perception of Kiamichi watershed benefits and social importance of supplied ecosystem services

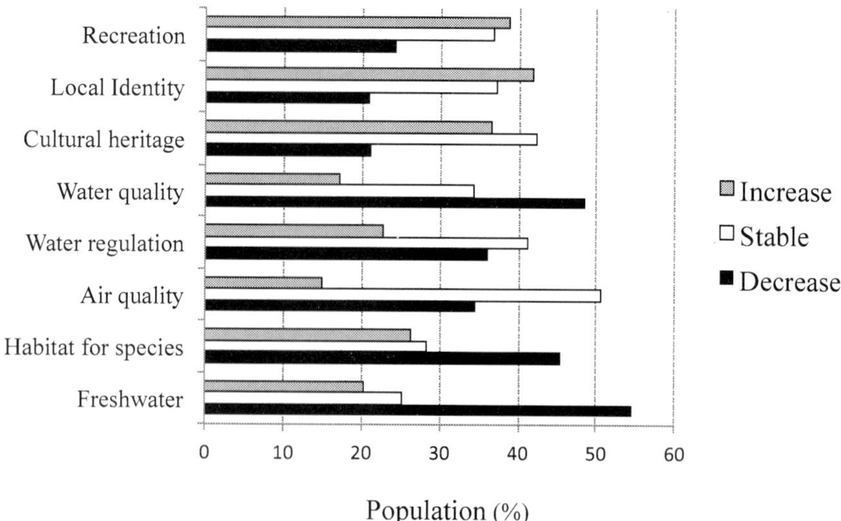

Figure 3. Ecosystem services trends in the Kiamichi Watershed.

5.2. Land use-Land Cover (LULC) change between 1898 and 2006

Changes in LULC between 1898 and 2006 are important to understanding the carbon seques‐
tration balance in the Kiamichi watershed. To run the sequestration model, LULC datasets for
1898 and 2006 were reclassified into five categories: urban-barren, cropland, forest, shrub-
grassland-pasture, and water-wetland (Figure 4). In 1898, 93.9% of the Kiamichi watershed was
covered in forest, and 5.4% was covered in shrubland, grassland, and pasture. Only a fraction
of a percent of the land was covered by cropland (0.6%) or water-wetland (0.1%). In 2006, the
Kiamichi watershed represented a rural landscape, with many of the forests replaced with

pastures. The 30.3% decline in forest was largely accounted for by the 23.6% increase in shrub-grassland-pasture. Urban development (via 4 small towns) and water reservoir creation (via two large dams) accounted for the rest of the lost forests. Between 1898 and 2006, cropland decreased from 0.6% to 0.1%.

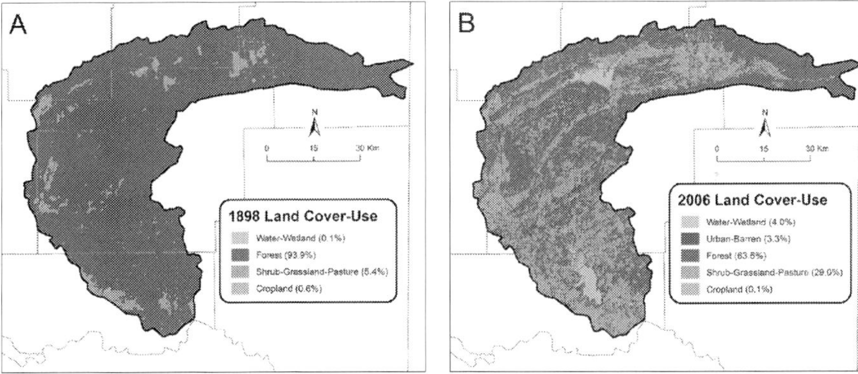

Figure 4. Land cover-use maps for the Kiamichi River watershed in southeastern Oklahoma for 1898 (A) and 2006 (B). Oklahoma county boundaries are included for reference.

5.3. Carbon storage and sequestration in the Kiamichi watershed

Based on carbon stocks data collected for each LULC type (Table 1), we observed a significant decrease between 1898 and 2006 in the capacity of the watershed to store carbon (Figure 5a-b). Because of this, there are areas, mostly in the upper and lower basin, where the conversion of forests to shrub-grassland-pastures produced a decrease of total carbon storage (Figure 5B). This involved a reduction in the aboveground and soil stocks, producing a difference of over 30 Mg ha^{-1} of carbon per hectare in both the aboveground and soil carbon stocks.

Land use-land cover	Above ground biomass	Below ground biomass	Soil	Dead matter
Cropland	1.67	4.52	17.80	0.00
Forest	37.60	7.52	48.50	3.45
Shrub-Grassland-Pasture	1.27	5.08	24.05	0.13
Water-Wetland	0.00	0.00	68.25	0.15
Urban-Barren	0.00	0.00	0.00	0.00

Table 1. Carbon stock related to the four carbon pools required for the InVEST carbon model. Data are converted into metric tons of carbon per hectare (Mg ha^{-1})

The results from the carbon model output clearly show an overall trend of decreasing and null carbon sequestration in most of the Kiamichi watershed (Figure 5c). Considering the positive and negative carbon sequestration estimations, the carbon model obtains a watershed total of 9.197.087 metric tons of carbon. This result shows that the watershed stored 9.1 million less metric tons of carbon in 2006 than it did in 1898. There are small patches of positive carbon sequestration (green area in Figure 5c) due to recent reforestations around Sardis and Hugo reservoirs. The areas experiencing the most negative carbon sequestration (red area in Figure 5c) are those areas converted from forest and grasslands into urban-barren land. The lower watershed area has experienced the most loss of carbon sequestration capacity. One explanation for this pattern of agricultural land conversion is that the lower watershed is flatter and more suitable for pasture while the steeper slopes of the upper watershed limit pasture development. However, there is still a loss in sequestration as the forested mountain slopes are being thinned for timber production.

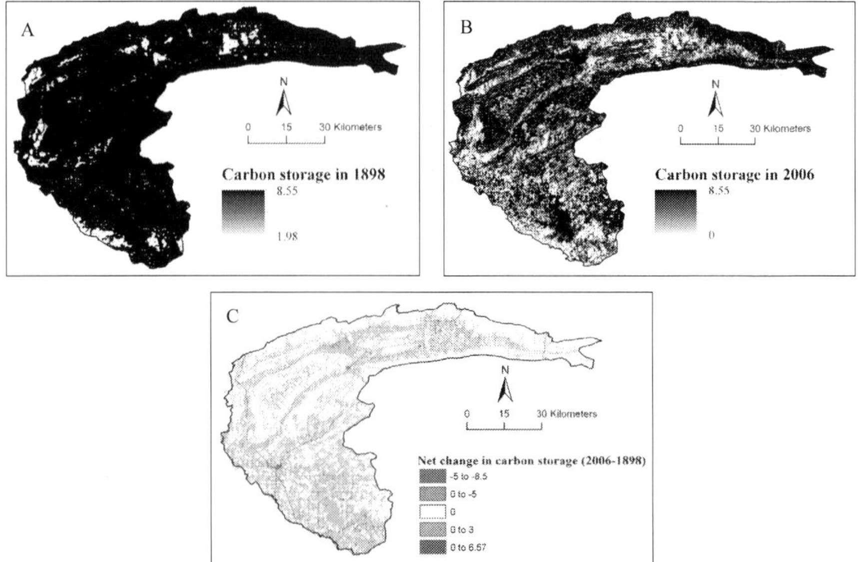

Figure 5. Maps of carbon storage in 1898 (A) and 2006 (B), and net change in carbon storage during this period (C) for the Kiamichi River watershed. Positive values indicate a net-gain in carbon sequestration (e.g., cropland to forest), whereas negative values indicate lost carbon sequestration (e.g., forest to pasture). The values are in Mg/km^2.

6. Conclusions

Conserving ecological processes is necessary to maintain human wellbeing. The ecosystems services approach allows for quantification of the importance of ecological processes to humans. Such quantification should include multiple dimensions including biophysical, socio-cultural and economic valuations. Our study provides a multidimensional valuation of carbon- and water-related ecosystem services in a large rural watershed. Carbon sequestration is an optimal ecosystem service because it ensures the supply of other ecosystem services such as food production, green areas for recreation and better air quality [13,11]. Our results show that people living in the watershed think the area provides ecosystem services, but that air quality is not as important as services such as habitat for species, water quality, and recreation. Ecosystem services associated with water resources are highly visible (i.e. water availability, recreation on lakes) and these are the most highly valued by our survey respondents. Unlike water related services, air quality is less tangible and difficult for people to visualize in areas without heavy air pollution. However, changes in carbon storage (the watershed lost the capacity to store and sequester 9.1 million metric tons of carbon since 1898) reflect conversion of natural forests into agricultural lands or timber production stands. Stakeholders in the watershed need to understand that in the long term, continuing this land conversion trend will decrease carbon sequestration and potentially air quality. We think our novel, multidimensional approach combining both biophysical supply and social perception of carbon related ecosystem services will help stakeholders and managers make more informed land use decisions in the future.

For future research, as climate change and human development continue to interact and affect the delivery of ecosystem services, other valuation practices including mapping the biophysical supply of other ecosystem services such as biodiversity conservation or water regulation will provide practical results for landscape management and planning. Currently, other mapping tools such as the Artificial Intelligence for Ecosystem Services (ARIES) [30] or POLYSCAPE [31] are applied to landscapes of all sizes and are expected to work well with each unique scenario [2]. Many researchers in the field of biology, ecology, and environmental studies are calling for a focus on multidimensional approaches that include both a natural valuation component along with a social one [28].

7. Study limitations, assumptions, and future work

There were limitations to this study and assumptions made for the InVEST model. Regrouping and simplifying LULC classes obviously generalized carbon storage losses/gains. When reclassifying the 2006 LULC map, some reclassifications were obvious by the descriptions, but some others required assumptions. For example, those LULC types classified as Central Oak-Hardwood and Pine Forest by the National Vegetation Classification were reclassified into simply Forest. Reclassification of other National Vegetation Classification LULCs, such as Recently Disturbed or Modified were assumed, and requires further investigations.

Carbon pool data collection also presented some challenges. Because the available IPCC carbon data values were based on broadly generalized values for each climate division, many assumptions were made as to vegetation types in the area. In this sense, further research for carbon pools for each dominant vegetation species per LULC is needed to obtain a value that is more indicative of the watershed itself, not just the climate region.

Further, this study only looked at two years to derive carbon storage estimates. Southeastern Oklahoma is a dynamic landscape that can change at monthly and annual timescales due to timber harvesting, fire, drought, and insect infestations [32]. Some studies have characterized this region as having one of the highest annual rates of land cover change in the U.S. [33] and as being one of the most sensitive to climate change [34]. If we want to capture these land cover changes at higher spatio-temporal resolutions, new techniques will be needed [e.g., 32, 35]. These frequent and intense changes to forest cover have many implications for carbon storage dynamics, which was also beyond the scope of our study.

Acknowledgements

We thank all of the people in the Kiamichi watershed that kindly responded to the questionnaire. We thank Melanie Lawson and Joseph Sardasti for assisting in fieldwork, and Bruce Hoagland and Todd Fagin for assisting in land use-land cover maps. This research is based on work partially funded by The University of Oklahoma through the Oklahoma Biological Survey and the South Central Climate Science Center.

Author details

Antonio J. Castro[1*], Caryn C. Vaughn[1], Jason P. Julian[2], Marina García Llorente[3] and Kelsey N. Bowman[4]

*Address all correspondence to: acastro@ou.edu

1 Oklahoma Biological Survey, Department of Biology and Ecology and Evolutionary Biology Graduate Program, University of Oklahoma, Norman, OK, USA

2 Department of Geography, Texas State University, San Marcos, TX, USA

3 Sociology of Climate Change and Sustainable Development research group, Department of Social Analysis, University Carlos III, Madrid, Spain

4 Department of Geography and Environmental Sustainability, University of Oklahoma, USA

References

[1] Daily G, Alexander S, Ehrlich, P, Goulder L, Lubchenco J, Matson P, Woodwell G. Ecosystem Services: Benefits Supplied to Human Societies by Natural Ecosystems. Issues in Ecology; 1997.

[2] Castro A, Garcia-Llorente M, Martin-Lopez B, Palomo I, Iniesta-Arandia I. Multidimensional approaches in ecosystem services assessment. In: Earth Observation of Ecosystem Services, 2013a. 441-468.

[3] Perrings C, Naeem S, Ahrestani, FS, Bunker E, Burkill P, Canziani G, Elmqvist T, Fuhrman JA, Jaksic FM, Kawabata Z, Kinzig A, Mace GM, Mooney HM, Prieur-Richard AH, Tschirhart J, Weisser A.. Ecosystem services, targets, and indicators for the conservation and sustainable use of biodiversity. Frontiers in Ecology and the Environment 2011;9:512-520

[4] Wainger L, Mazzotta M. Realizing the potential of ecosystem services: A framework for relating ecological changes to economic benefits Environmental Management 2011;48:710-733.

[5] Harrison PA. Ecosystem services and biodiversity conservation: an introduction to the RUBICODE project. Biodiversity Conservation 2010;19:2767-2772

[6] Castro AJ, Martín-López B, García-Llorente M, Aguilera PA, López E, Cabello J. Social preferences regarding the delivery of ecosystem services in a semiarid Mediterranean región. Journal of Arid Environment 2011;75:1201–1208

[7] Oklahoma Water Resources Board vs. Choctaw and Chickasaw Nations of Oklahoma. Oklahoma Supreme Court, case 110375. 2012.

[8] Mast M, Turk J. Environmental characteristics and water quality of Hydrologic Benchmark Network stations in the West-Central United States, 1963-95. U.S. Geological Survey Circular. 1999.

[9] Castro AJ, Paruelo JM, Alcaraz-Segura D, Cabello J, Oyarzabal M, López-Carrique E. Missing gaps in the estimation of the carbon gains service from Light Use Efficiency models In: Alcaraz-Segura, Di Bella, C., D. (Eds). Earth Observation of Ecosystem Services, 2013b.105-124

[10] Paruelo JM, Piñeiro G, Baldi G, Baeza S, Lezama F, Altesor AI, Oesterheld M. Carbon Stocks and Fluxes in Rangelands of the Río de la Plata Basin. Rangeland Ecosystem Management 2009;63.94-108.

[11] Cabello J, Fernández N, Alcaraz-Segura D, Oyonarte C, Piñeiro G, Altesor A, Delibes M, Paruelo JM. The Ecosystem Functioning Dimension in Conservation: insights from remote sensing. Biodiversity and Conservation 2012;21:3287-3305.

[12] Fisher B, Turner RK, Morling P. Defining and classifying ecosystem services for decision making. Ecological Economics, 2008;643-653.

[13] (MA) Millennium Ecosystem Assessment. Washington, DC: Island Press 2005.

[14] Martín-López BE, Gómez-Baggethun M, García-Llorente M, Montes C. Trade-offs across value-domains in ecosystem services assessment. Ecological Indicators 2014;37:220–228

[15] Vitousek PM, Mooney HA, Lubchenco J, Melillo JM. Human domination of Earth's ecosystems. Science 1997;277(5325), 494

[16] Alcaraz-Segura D, Paruelo JM, Epstein HE, Cabello J. Environmental and Human Controls of Ecosystem Functional Diversity in Temperate South America. Remote Sensing 2013;5(1):127-154

[17] Matthews WJ, Vaughn CC, Gido, KB, Marsh-Matthews E. Southern plains rivers, In: Rivers of North America, eds A. Benke, C.E. Cushing. Academic Press 2005.

[18] Vaughn CC, and M. Pyron Population ecology of the endangered Ouachita Rock Pocketbook mussel, *Arkansia wheeleri* (Bivalvia: Unionidae), in the Kiamichi River, Oklahoma. American Malacological Bulletin 1995;11:145-151.

[19] Vaughn CC, 2000. Changes in the mussel fauna of the Red River drainage: 1910 – present. In: Proceedings of the First Freshwater Mussel Symposium, edited by R. A. Tankersley, D. I. Warmolts, G. T. Watters, B. J. Armitage, P. D. Johnson, and R. S. Butler, Ohio Biological Survey, pp. 225-232 Columbus, Ohio.

[20] Galbraith HS, Spooner DE, Vaughn CC. Status of rare and endangered freshwater mussels in southeastern Oklahoma rivers. Southwestern Naturalist 2008;53: 45-50.

[21] Agbenyega O, Burgess PJ, Cook M, Morris J. Application of an ecosystem function framework to perceptions of community woodlands. Land Use Policy 2009;26:551-557

[22] Winkler R. Valuation of ecosystem goods and services, An integrated dynamic approach. Ecological Economics 2006;59:82-93

[23] Nelson E, Sander H, Hawthorne P, Conte M, Ennaanay D. Projecting Global Land-Use Change and Its Effect on Ecosystem Service Provision and Biodiversity with Simple Models. PLoS ONE 2010;5(12): e14327. doi:10.1371/journal.pone.0014327

[24] Eggleston S, Buendia L, Miwa, K, Ngara T, Tanabe K. IPCC Guidelines for National Greenhouse Gas Inventories, Volumen 4. Japan: Institute for Global Environmental Strategies. 2006.

[25] Potter KN, Derner JD. Soil carbon pools in central Texas: Prairies, restored grasslands, and croplands. Soil Water Conservation 2006;124-128.

[26] Smith J, Heath L, Skog K, Birdsey R. Methods for calculating forest ecosystem and harvested carbon with standard estimates for forest types of the United States. Newtown Square, PA: US Department of Agriculture 2006.

[27] Goins CR, and Goble D. Historical Atlas of Oklahoma, fourth ed., 286 pp., University of Oklahoma Press, Norman. 2006

[28] Watkins BW. Reconstructing the Choctaw Nation of Oklahoma, 1894-1898: Landscape and Settlement on the Eve of Allotment, 202 pp, Oklahoma State University, Stillwater, 2007

[29] Fry JG, Xian S, Jin J, Dewitz C, Homer L, Yang C, Barnes N, and J. Wickham. Completion of the 2006 National Land Cover Database for the Conterminous United States. PE&RS 2011;858-864.

[30] Villa F, Bagstad KJ, Voigt B, Johnson GW, Portela R, Honzak M, Batker D. A methodology for adaptable and robust ecosystem services assessment. PLoS ONE 2014;9(3):e91001.

[31] Jackson B, Timothy P, Sinclair F. Polyscape: A GIS mapping framework providing eficient and spatially explicit landscape-scale valuation of multiple ecosystem services. Landscape and Urban Planning 2013;112:74–88

[32] Tran TV. Measuring land cover change at high spatio-temporal scales, 95 pp, University of Oklahoma; 2013.

[33] Sleeter BM, Sohl TL, Loveland TR, Auch RF, Acevedo W, Drummond MA, Sayler KL, and Stehman SV. Land-cover change in the conterminous United States from 1973 to 2000, Global Environmental Change, 2013;23(4), 733-748

[34] Reker RR, Sayler, KL, Friesz AM, Sohl TL, Bouchard MA, Sleeter BM, Sleeter RR, Wilson TS, Griffith GE, and Knuppe ML. Mapping and modeling of land use and land cover in the eastern United States from 1992 through 2050, in Baseline and projected future carbon storage and greenhouse gas fluxes in ecosystems of the eastern United States: U.S. Geological Survey Professional Paper 1804, edited by Z. Zhu and B. C. Reed, 2014;pp. 27-54.

[35] Jawarneh RN, and Julian JP. Development of an accurate fine-resolution land cover timeline: Little Rock, Arkansas, USA (1857–2006). Applied Geography 2012;35(1–2)104-113.

Chapter 18

The Importance of Large Trees in Shrine Forests for the Conservation of Epiphytic Bryophytes in Urban Areas

Yoshitaka Oishi and Keizo Tabata

Additional information is available at the end of the chapter

http://dx.doi.org/10.5772/59074

1. Introduction

1.1. Shrine forests

Shrines in Japan are often surrounded by forests known as *chinju-no-mori* (shrine forests; Figure 1). Previous studies have shown that shrine forests contribute to the conservation of biodiversity, particularly in urban areas where green area is severely limited, although these forests are often fragmented by the city matrix. For example, shrine forests promote the diversity of birds [1, 2], trees [3], grasses [4], and ferns [5]. These forests can thus be regarded as important for biodiversity in urban areas.

1.2. Bryophytes in shrine forests

Shrine forests are also important habitats for bryophytes in urban areas [6-10]. Oishi [8] found approximately 30–60 epiphytic bryophyte species, including endangered species, on tree bark in several shrine forests.

Bryophytes are unique among plants in that they lack vascular bundles and cuticle layers on their leaves (Figure 2); they absorb water and nutrients through their leaf surfaces instead of through roots [11]. This character allows bryophytes to grow on tree bark where soils are scarce; some bryophytes strongly prefer to grow on tree bark [10]. Thus, tree bark is an important habitat for bryophytes.

Figure 1. Shrine forests: Left: Nakaragi shrine and its shrine forest, Kyoto prefecture; Right: Enlarged view of the shrine forest

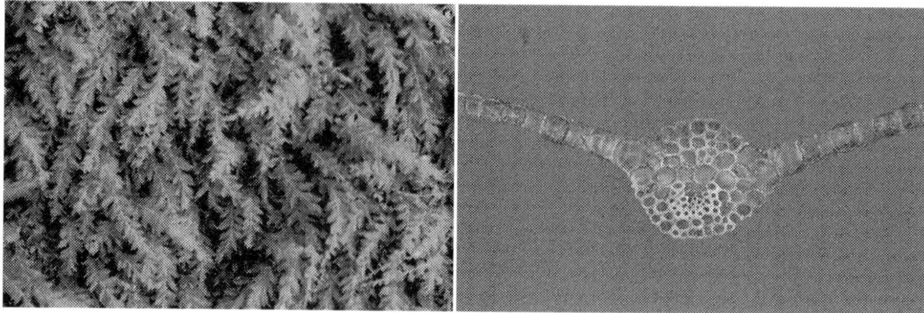

Figure 2. *Plagiomnium actum* (left) and a leaf section (right). As the leaf section shows, the body structure of bryophytes is very simple and lacks vascular bundles.

1.3. Environmental factors for bryophytes in shrine forests

Several studies have examined epiphytic bryophytes in fragmented forests, including shrine forests. Oishi [8] showed that species richness in fragmented forests is strongly affected by patch size and maintenance. In fragmented forests, the forest edge dries more quickly because of its greater exposure to strong wind and light intensity (edge effects); therefore, patch size is closely connected with drought stress [12-13], which impacts bryophyte diversity [8]. This drought stress causes severe damage to species vulnerable to desiccation, such as bryophytes on tree bases [8-9]. In another study, Hylander et al. [14] found that bryophytes on convex forms (e.g., logs, tree bases, and mesic ground) are more vulnerable to desiccation than those on concave forms.

Conversely, some bryophytes prefer to grow at sunny sites. For these species, maintenance such as tree cutting or trimming is necessary to increase light intensity in the forest interior [8]. Previous studies [6-10] have partly revealed the effects of environmental conditions on bryophytes; however, the effects of forest structure on these species have not been sufficiently addressed.

To understand the relationship between forest structure and bryophyte diversity, we focused on the diameter at breast height (DBH) of trees in shrine forests for the following reasons. First, large trees in shrine forests are often regarded as sacred and are preferentially preserved. Second, DBH is deeply related to bryophyte diversity because the nature of tree bark changes with DBH, thereby impacting bryophyte diversity [15-18]. Therefore, revealing the relationship between DBH and epiphytic bryophyte diversity is useful for understanding the effects of shrine forests on these species.

2. Objective

In this study, we examined the role of shrine forests in the conservation of epiphytic bryophytes. Based on our results, we discuss the effective conservation methods for epiphytic bryophytes in fragmented forests.

3. Methods

3.1. Study site

The study was conducted at a shrine forest of the Shimogamo Shrine, Kyoto, Japan (Figure 3). This shrine may have been founded before 8th century [19] and is designated a World Heritage Site. The shrine forest is known as the "Tadasu-no-mori" and covers approximately 12.4 ha. One of the dominant trees is *Aphananthe aspera* (Thunb.) Planch, of which the forest contains more than 300 individuals. However, the numbers of *Celtis sinensis* Pers. and *Cinnamomum camphora* (L.) J. Presl have recently increased [20]. The dynamics of the trees in this forest have been reported by Tabada et al. [20-23].

3.2. Bryophyte survey

In 2006, we surveyed the epiphytic bryophyte flora at the study site. We recorded the occurrence of species and their cover on each *A. aspera* individual. The DBH of *A. aspera* was measured in 2002 by one of the authors. The average DBH was 161.5 ± 82.4 cm (mean ± standard deviation), the maximum was 420.0 cm, and the smallest was 27.0 cm. To understand the relationship between bryophyte diversity and DBH, we analyzed the changes in bryophyte life forms and reproductive strategies in addition to those of species richness and cover.

3.3. Bryophyte life form

Bryophytes change their forms according to light intensity and humidity [24]. For example, in sunny and dry environments, bryophytes maintain water content in their bodies by forming contact mats similar to cushions [24]. Conversely, in dark and humid environments, bryo-

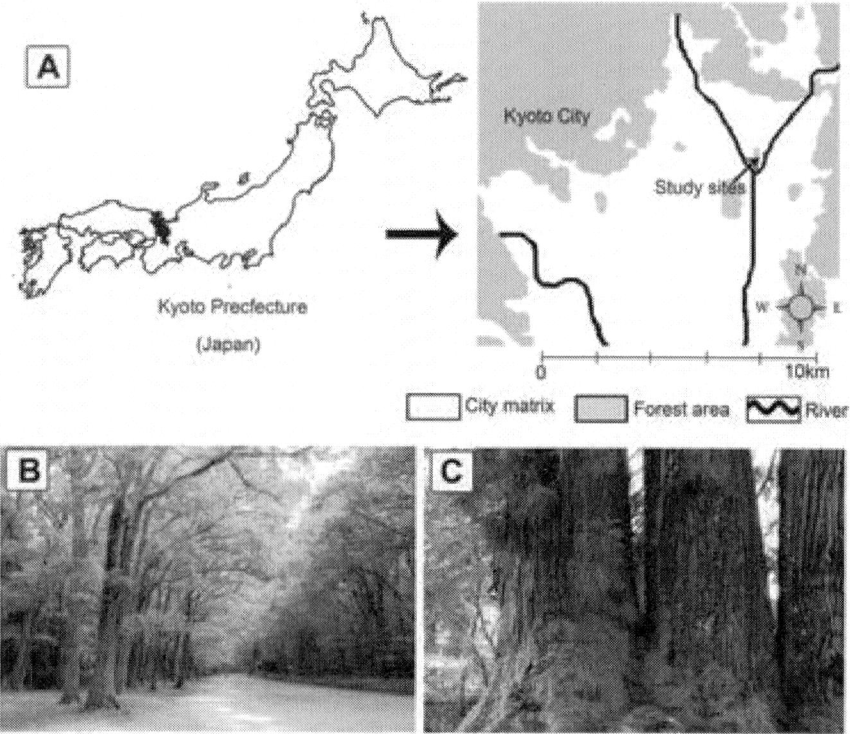

Figure 3. Study site (Tadasu-no-mori, Shimogamo Shrine). A. Location of the study site , revised from Fig. 1 in Oishi [9]; B. Tadasu-no-mori forest; C. Bryophytes on tree trunks

phytes increase photosynthetic efficiency by forming flat mats similar to fans [24]. Therefore, bryophyte life forms are useful for evaluating habitat environmental conditions, and several studies have used them for this purpose [8-10, 25-26].

3.4. Reproductive strategy

Bryophytes have two main reproductive strategies: sexual reproduction by spores and asexual reproduction by gemmae, fragile body parts, etc [27]. Sexual reproduction may be further classified into monoicous and dioicous types. Monoicous bryophytes have both antheridia and archegonia on the same shoot, while dioicous bryophytes have these organs on different shoots. Therefore, monoicous bryophytes have more opportunities for fertilization than do dioicous ones. Bryophytes with asexual reproduction can also reproduce more frequently than dioicous species. We hypothesized that this difference in reproductive frequency would affect the habitat preferences of bryophytes.

4. Analysis

4.1. DBH and bryophyte diversity

First, we compared the DBH values of trees with and without epiphytic bryophytes using the
t-test to reveal the characteristics of trees with bryophytes. We then examined the effects of
DBH on the diversity at both the community and species levels. The flow chart of this study
is presented as Figure 4.

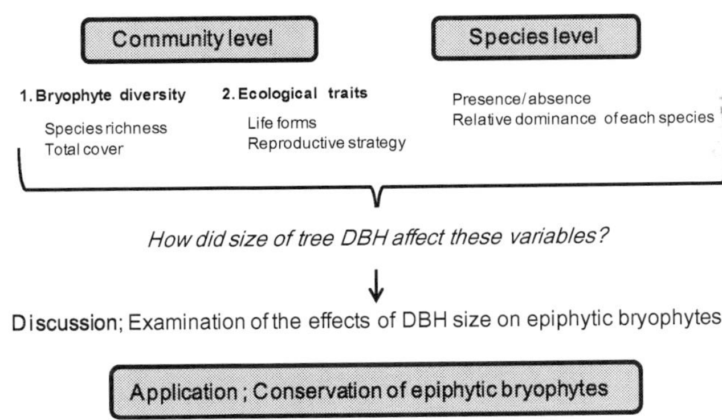

Figure 4. Flow chart of this study

4.2. Community level

The relationships between bryophyte diversity (total species richness and cover) and the DBH
of *A. aspera* were examined using Pearson's product-moment correlation coefficients. Additionally, we examined the correlations of DBH with both the richness of the life forms and the
species richness of each life form. The life forms recorded at the study site were short turfs (t),
small cushions (cu), dendroids (D), rough mats (Rm), smooth mats (Sm), thalloid mats (Th),
thread-like forms (Tl), wefts (W), and fans (F). These forms were classified according to the
system of Bates [24]. Finally, we examined the correlation of DBH with the ratio of the species
richness of dioicous species (RDi). This ratio was calculated as follows:

RDi=species richness of dioicous bryophytes/total species richness

The value of RDi therefore increases with the dominance of dioicous species.

4.3. Species level

To understand the preferences of each species for DBH, we examined the changes in the
relative dominance of each species (RDo) as DBH increased. To clarify the relationship between
RDo and DBH, the *A. aspera* trees with epiphytic bryophytes were evenly divided into three
categories (small, medium, and large). The relative dominance was calculated as follows:

RDo=cover of each species on trees of one category (small, medium, or large)/total cover of the species

This analysis was conducted for species that occurred more than 10 times at the study site.

5. Results

5.1. Presence/absence of bryophytes

Bryophytes were found on 181 of the 313 *A. aspera* trees at the study site. We compared the DBH of trees with and without bryophytes using the *t*-test. The DBH values of trees with bryophytes were significantly higher than those without bryophytes (t=-5.4, d.f.=311, $p < 0.01$; Figure 5). In the following analysis, we examined the relationships between tree DBH and bryophyte diversity in trees with bryophytes.

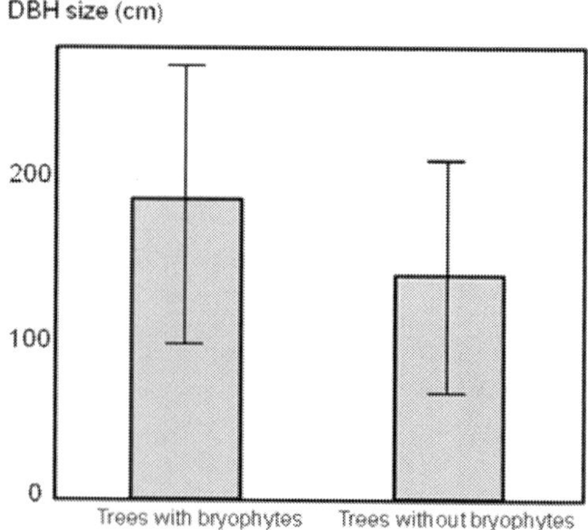

Figure 5. Comparison of DBH between trees with/without epiphytic bryophytes

5.2. Bryophyte flora

We found 42 bryophyte species (28 mosses and 14 liverworts) on the *A. aspera* trees, including two endangered species [*Leskeella pusilla* (Mitt.) Nog. and *Hypnodontopsis apiculata* Z. Iwats. & Nog.]. Figure 6 displays several species found at the study site. The most frequently observed species was *Trocholejeunea sandvicensis* (Gottsche) Mizut. (73 times), followed by *Metzgeria lindbergii* Schiffn. (64 times), *Rhynchostegium pallidifolium* (Mitt.) A. Jaeger (62 times), *Macvicaria ulophylla* (Steph.) S. Hatt. (58 times), and *Frullania parvistipula* Steph. (58 times). The species with the largest total cover was *T. sandvicens*, followed by *R. pallidifolium*, *M. lindbergii*,

Rhynchostegium inclinatum (Mitt.) A. Jaeger, and *M. ulophylla*. The complete species list is presented in the Appendix.

Figure 6. Several bryophyte species found at the study site

5.3. Bryophyte diversity

The relationships between the species richness/cover and DBH were examined using Pearson's product-moment correlation coefficients. Both species richness and bryophyte cover were significantly and positively correlated with DBH (Table 1).

Variables	Pearson's product-moment correlation coefficients
Species richness	0.22**
Cover	0.28**
Life forms	
Life form richness	0.19*
Short turfs	-0.05
Small cushions	0.07
Dendroids	0.20**
Rough mats	0.17*
Smooth mats	0.14
Thalloid mats	0.18*
Thread–like forms	0.14
Wefts	0.14
Fans	-0.05
Ratio of the species richness of Dioicous species	0.25**

**; $p < 0.01$, *; $p < 0.05$

Table 1. Relationships between bryophyte diversity and DBH

6. Life forms and reproductive strategy

The Pearson's product-moment correlation coefficients between DBH and life form diversity are shown in Table 1. The richness of life forms also increased with increasing DBH. The species richness of dendroids, rough mats, and thalloid mats were significantly and positively correlated with DBH. RDi was also significantly and positively correlated with DBH.

7. Preference of each species for large trees

Bryophytes were observed on 181 *A. aspera* trees, which were divided into three categories based on DBH (small, medium, and large) containing 60, 60, and 61 trees, respectively. The changes in the relative dominance of each species are shown in Figure 7. As seen in the figure, the bryophyte species could be classified into four types based on dominance pattern. Type 1 (three species) preferred to grow on trees with small DBH, type 2 (three species) on trees with middle DBH, and type 3 (nine species) on trees with large DBH. Type 4 (five species) grew almost exclusively on trees with large DBH.

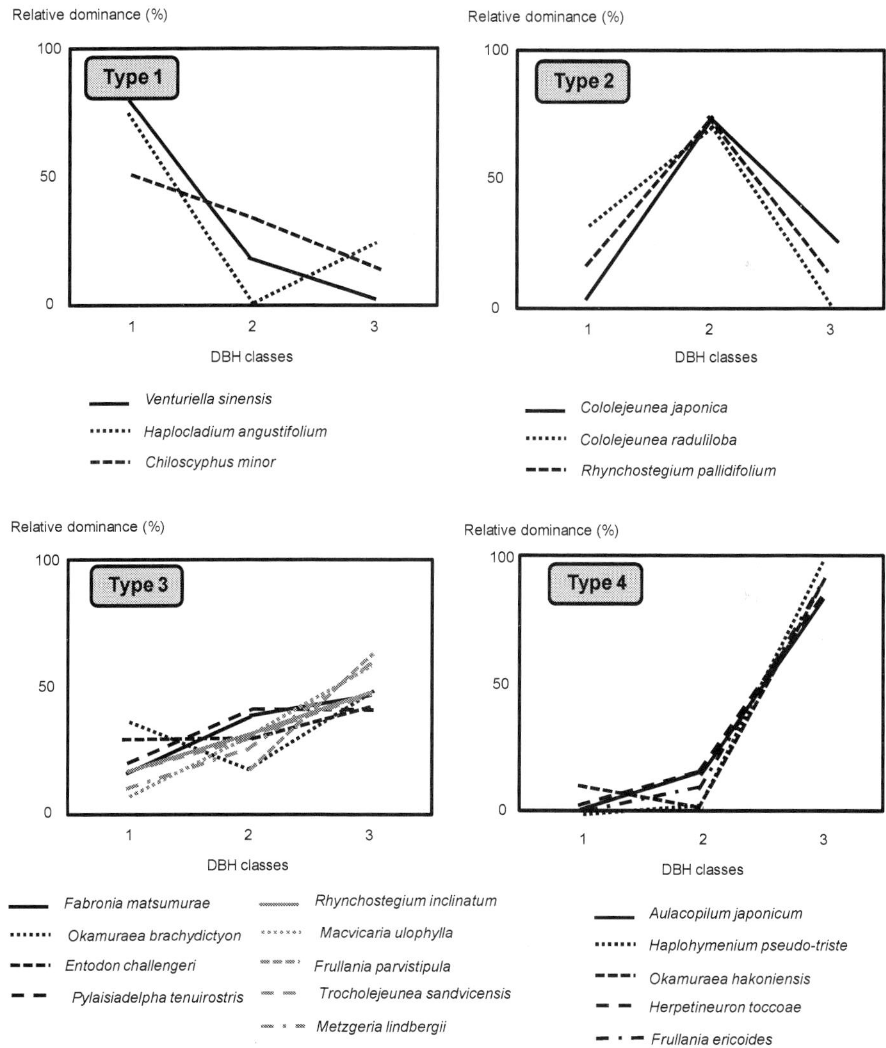

Figure 7. Relationships between the dominance of each species and DBH. DBH class; 1=small, 2=medium, 3=large. The classification of species into types 1–4 is as follows: Type 1: prefer to grow on trees with small DBH, Type 2: prefer to grow on trees with medium DBH, Type 3: prefer to grow on trees with large DBH, Type 4: almost exclusively grow on trees with large DBH

8. Discussion

8.1. Bryophyte diversity

We found 42 species on the bark of *A. aspera* alone (313 trees), while Oishi [8-10] found 57 species on the bark of all trees at the site. Approximately, two-thirds of the epiphytic bryophytes at the study site (containing more than 3000 total trees) were found on *A. aspera*, which indicates the high diversity of bryophytes on this species.

Notably, two endangered species (*L. pusilla* and *H. apiculata*) were found at the site. *L. pusilla*, which is classified as an endangered species on the red list of Kyoto prefecture [28], grows in large forests where desiccation stress is low [6]. Therefore, the large patch sizes of the study site support the occurrence of this species. *H. apiculata* is endemic to Japan and has severely limited habitat; therefore, this species is designated as "critically threatened" on the red list of Japan [29]. Why does this rare species grow at the study site? This species may be threatened by habitat losses caused by development [29]. As mentioned in the introduction, the study was conducted in an area that has long been preserved as a shrine forest. The preservation history of the study site likely contributed to the survival of this species.

8.2. Bryophyte diversity and DBH

Our results indicate that the diversity of both epiphytic bryophytes and life forms increased with DBH. Additionally, the relative dominance of 14 species (Types 3 & 4) increased with DBH; notably, five species (Type 4) occurred almost exclusively on large trees. These results indicate that the presence of large trees can increase the diversity of epiphytic bryophytes and are necessary for the conservation of these species. These relationships may be explained by (1) changes in tree bark and (2) the longer lives of large trees.

8.2.1. Changes in tree bark

The features of tree bark change with tree size: the bark surface of large trees has a higher moisture content and is rougher than that of small trees [30]. This higher moisture content can be important for bryophytes that grow in forms vulnerable to desiccation, such as fans, dendroids, and wefts [24], as reflected in the positive significant correlations of DBH with both the richness values of these life forms and total life form richness.

Furthermore, the rough bark surface of large trees may be more suitable for capturing bryophyte spores/gemmae than is the smooth surface of small trees. At our study site, the dominance of bryophytes with low reproductive frequency (dioicous species) increased with DBH. This result indicates that large trees are especially important for the establishment of bryophytes with low reproductive frequency due to their higher capture ability.

McGee & Kimmerer [31] showed that the occurrence and abundance of epiphytic bryophytes on large maple trees are likely regulated to a greater extent by factors such as dispersal or protonemal establishment than by the habitat requirements of mature gametophytes. Although we cannot directly apply the results of McGee & Kimmerer [31] to our study because of differences in species and environmental conditions, their results suggest that changes in bark features more strongly affect dispersal or protonemal establishment than mature gametophytes.

8.2.2. Longer lives

Generally, larger trees live for longer periods than do small trees in similar environments, which provide comparatively more opportunities for bryophyte spore/gemma establishment. This effect may be partly reflected in the positive correlations between DBH and both the species richness and dominance of bryophytes with low reproductive frequency.

8.3. The significance of shrine forests

McGee & Kimmerer [17] described the importance of large trees for the conservation of epiphytic bryophytes in hardwood forests. This study shows that large trees can also contribute to the conservation of epiphytic bryophytes in shrine forests through their preferential bark features and longer lives. The effects of large trees are reflected in the changes of bryophyte life forms diversity and reproductive strategy according to DBH. In shrine forests, large trees are preferentially conserved because they are regarded as sacred. And, the management of shrine forests is effective for the conservation of epiphytic bryophytes.

Contrary to these results, several authors have reported that tree DBH does not strongly impact epiphytic bryophytes [32-33]. The possible explanations for the differences between this study and previous studies are as follow.

1. This study analyzed the epiphytic bryophytes on *A. aspera* alone. Therefore, other tree factors (e.g., bark pH) were relatively uniform, isolating the effects of DBH on epiphytic bryophytes.

2. The large differences of DBH in this study (minimum=27.0 cm, maximum=420.0 cm) clarified the effects of tree DBH on epiphytic bryophytes

8.4. History of shrine forests

This study also indicates that the long history of shrine forests contributes to the conservation of epiphyte diversity. Although this study did not gather sufficient data to examine the relationship between forest history and epiphytic bryophyte diversity, previous work has shown the importance of history for these species [18]. This conservation effect has also been reported in a fragmented forest in Kyoto city [34].

8.5. Epiphytic bryophytes and ecosystem

Epiphytic bryophytes play important roles in water storage [35, 36], nutrient cycling [37], and the retention of inorganic nitrogen [38] in forest ecosystems. These functions of epiphytic bryophytes have been examined not in urban forests but in tropical montane or old growth Douglas fir and western hemlock forests, in which epiphyte biomass is relatively high. The biomass of epiphytic bryophytes in urban forests is comparatively small; however, these organisms may also be important in urban ecosystems. In particular, the role of bryophytes in water storage may contribute to the conservation of biodiversity, as the drought stress caused by edge effects is severe in fragmented urban forests [13].

9. Conclusion

The results of this study indicate that large trees in shrine forests can provide suitable habitats for epiphytic bryophytes and enhance their diversity in urban environments where green area is limited. These trees are especially effective for the conservation of species that are vulnerable to desiccation and/or have low reproductive frequency.

Epiphytic bryophytes are affected by environmental factors such as tree density [15, 33], past landscape structure [18, 34], bark type [39], silvicultural disturbance [40], air pollution [41], etc. By examining the influence of these factors on bryophytes in future studies, we can propose more effective methods for the conservation of these species. Furthermore, we should also examine the ecological roles of epiphytic bryophytes (e.g., water storage) in fragmented forests to understand the importance of their conservation.

Appendix

Appendix: Species list

The bryophyte nomenclature follows that reported by Iwatsuki [27].

Moss	Frequency	Cover (cm^2)
Anomodon giraldii Müll. Hal.	1	<100
Aulacopilum japonicum Broth.ex Card.	21	25000
Brachymenium nepalense Hook.	1	<100
Brachythecium buchananii (Hook.) A.Jaeger	1	<100
Brachythecium populeum (Hedw.) Schimp.	1	800
Bryum capillare Hedw.	4	<100
Entodon challengeri (Paris) Card.	30	30000
Entodon sullivantii (Müll. Hal.) Lindb.	1	400
Fabronia matsumurae Besch.	35	2600
Haplocladium angustifolium (Hampe & Müll.Hal.) Broth.	20	600

Moss	Frequency	Cover (cm²)
Haplocladium microphyllum (Hedw.) Broth.	2	<100
Haplohymenium pseudo-triste (Müll. Hal.) Broth.	12	1700
Haplohymenium triste (Ces.) Kindb.	1	<100
Herpetineuron toccoae (Sull. & Lesq.) Card.	34	20700
Hypnodontopsis apiculata Z.Iwats.& Nog.	1	<100
Hypnum plumaeforme Wilson	1	<100
Leskeella pusilla (Mitt.) Nog.	2	<100
Neckera humilis Mitt.	1	<100
Okamuraea brachydictyon (Card.) Nog.	15	<100
Okamuraea hakoniensis (Mitt.) Broth.	12	800
Orthotricum consobrinum Card.	3	<100
Pylaisiadelpha tenuirostris (Bruch & Schimp.) W.R.Buck	34	5700
Rhynchostegium inclinatum (Mitt.) A.Jaeger	35	41400
Rhynchostegium pallidifolium (Mitt.) A.Jaeger	62	53600
Schwetschkea matsumurae Besch.	2	400
Sematophyllum subhumile (Müll. Hal.) M.Fleisch.	1	2000
Trachycystis microphylla (Dozy & Molk.) Lindb.	1	<100
Venturiella sinensis (Vent.) Müll. Hal.	14	200
Liverwort		
Acrolejeunea pusilla (Steph.) Grolle & Gradst.	6	600
Chiloscyphus minor (Nees) J.J.Engel & R.M.Schust	37	4800
Cololejeunea japonica (Schiffn.) S.Hatt. ex Mizut.	18	800
Cololejeunea raduliloba Steph.	15	2600
Frullania diversitexta Steph.	1	<100
Frullania ericoides (Nees) Mont.	12	11500
Frullania muscicola Steph.	8	3210
Frullania parvistipula Steph.	58	35550
Lejeunea japonica Mitt.	2	2400

Moss	Frequency	Cover (cm^2)
Lejeunea ulicina (Tayl.) Gottsche, Lindenb.& Nees	1	<100
Macvicaria ulophylla (Steph.) S.Hatt.	58	38700
Metzgeria lindbergii Schiffn.	64	41500
Radula constricta Steph.	5	900
Trocholejeunea sandvicensis (Gottsche) Mizut.	73	59800

Acknowledgements

The author thanks Professor Yukihiro Morimoto for providing critical comments and suggesting improvements to this chapter, as well as Kaori Kuriyama for assistance with the bryophyte survey. This research was supported by a Grant-in-Aid for Scientific Research (A) (No. 18201008) from the Japan Society for the Promotion of Science.

Author details

Yoshitaka Oishi[1*] and Keizo Tabata[2]

*Address all correspondence to: oishiy@shinshu-u.ac.jp

1 Department of Forest Science, Faculty of Agriculture, Shinshu University, Japan

2 Faculty of Agriculture, Kinki University, Japan

References

[1] Hashimoto H, Murakami K, Morimoto Y. Relative species-area relationship and nestedness pattern of woodland birds in urban area of Kyoto City. Landscape Ecology and Management 2005; 10(1) 25-35.

[2] Hashimoto H, Sawa K, Tabata K, Morimoto Y, Nishino S. Characteristics of Legacy Trees with Hollows in the Urban Area of Kyoto City, Japan. Journal of the Japanese Institute of Landscape Architecture 2006; 69(5) 529-32.

[3] Murakami K, Morimoto Y. Landscape ecological study on the woody plant species richness and its conservation in fragmented forest patches in the Kyoto city area. Journal of the Japanese Society of Revegetation Technology 2000; 25(4) 345-50.

[4] Imanishi A, Imanishi J, Murakami K, Morimoto Y, Satomura A. Herbaceous plant species richness and species distribution pattern at the precincts of shrines as non-forest greenery in Kyoto city. Journal of the Japanese Society of Revegetation Technology 2005; 31(2) 278-83.

[5] Murakami K, Matsui R, Maenaka H, Morimoto Y. Relationship between species composition of pteridophytes and micro-landform types in fragmented forest patches in Kyoto city area. Journal of the Japanese Institute of Landscape Architecture 2003; 66(5); 513-6.

[6] Oishi Y, Murakami K, Tabata K, Hashimoto H, Morimoto Y. Environmental factors for endangered epiphytic bryophyte distribution in fragmented forests in Kyoto city. Journal of the Japanese Institute of Landscape Architecture 2007; 70(5) 483-6.

[7] Oishi Y, Murakami K, Morimoto Y. A distribution pattern of an exotic moss *Tortula pagorum* (Milde) De Not. in Kyoto City. Journal of the Japanese Society of Revegetation Technology 2008; 34(1) 81-4.

[8] Oishi Y, Morimoto Y. Fragmented forests effective for epiphytic bryophyte conservation. Journal of the Japanese Institute of Landscape Architecture 2009; 72(5) 503-6.

[9] Oishi Y. A survey method for evaluating drought-sensitive bryophytes in fragmented forests: a bryophyte life-form based approach. Biological Conservation 2009; 142(12) 2854-61.

[10] Oishi Y. Identifying indicator species for bryophyte conservation in fragmented forests Landscape and Ecological Engineering 2014; in press.

[11] Proctor MCF. The physiological basis of bryophyte production. Botanical Journal of the Linnean Society 1990; 104 (1-3) 61-77.

[12] Gignac LD, Dale MRT. Effects of fragment size and habitat heterogeneity on cryptogam diversity in the low –boreal forest of western Canada. The Bryologist 2005; 108(1) 50-66.

[13] Murcia C. Edge effects in fragmented forests: implications for conservation. Trends in Ecology and Evolution 1995; 10(2) 58-62.

[14] Hylander K, Nilsson C, Jonsson BG, Göthner T. Substrate form determines the fate of bryophytes in riparian buffer strips. Ecological Applications 2005; 15(2) 674-88.

[15] Hazell P, Kellner O, Rydin H, Gustafsson L. Presence and abundance of four epiphytic bryophytes in relation to density of aspen (*Populus tremula*) and other stand characteristics. Forest Ecology and Management 1998; 107 (1-3) 147-158.

[16] Holz I, Gradstein R. Cryptogamic epiphytes in primary and recovering upper montane oak forests of Costa Rica-species richness, community composition and ecology. Plant Ecology 2005; 178(1) 89-109.

[17] McGee GG, Kimmerer RW. Forest age and management effects on epiphytic bryophyte communities in Adirondack northern hardwood forests, New York, USA. Canadian Journal of Forest Research 2002; 32(9) 1562-76.

[18] Snäll T, Hagström A, Rudolphi J, Rydin H. Distribution pattern of the epiphyte *Neckera pennata* on three spatial scales-importance of past landscape structure, connectivity and local conditions. Ecography 2004; 27(6) 757-66.

[19] Kawamoto H. A study of two KAMO-JINJA positions: Empirical research that relates to relation of position of Shinto shrine part 13. Proceedings of AIJ Shikoku Chapter architectural research meeting 2009; 9 89-90.

[20] Tabata K, Hashimoto H, Morimoto Y. Growth and dynamics of dominant tree species in Tadasu-No-Mori forest. Journal of the Japanese Institute of Landscape Architecture 2004; 67(5) 499-502.

[21] Tabata K, Hashimoto H, Morimoto Y, Maenaka H. The effect of neighboring tree size and distance on mortality of trees in Tadasu-No-Mori Forest. Journal of the Japanese Society of Revegetation Technology 2004; 30(1) 27-32.

[22] Tabata K, Morimoto Y, Maenaka H. Simulation model of size frequencies of saplings in restored natural habitat in urban area by using exponential distribution. Journal of the Japanese Institute of Landscape Architecture 2005; 68(5) 679-82.

[23] Tabata K, Hashimoto H, Morimoto Y, Maenaka H. Dead forms of canopy trees and regeneration process in Tadasu-No-Mori Forest. Journal of the Japanese Society of Revegetation Technology 2007; 33(1) 53-8.

[24] Bates JW. Is "life-form" a useful concept in bryophyte ecology? Oikos 1998; 82(2) 223-37

[25] Pardow A, Gehrig-Downie C, Gradstein R, Lakatos M. Functional diversity of epiphytes in two tropical lowland rainforests, French Guiana: using bryophyte life-forms to detect areas of high biodiversity. Biodiversity and Conservation 2012; 21(14) 3637-55.

[26] Strazdiņa L, Brūmelis G, Rēriha I. Life-form adaptations and substrate availability explain a 100-year post-grazing succession of bryophyte species in the Moricsala Strict Nature Reserve, Latvia. Journal of Bryology 2013; 35(1) 33-46.

[27] Iwatsuki Z., editor. Mosses and liverworts of Japan. Tokyo: Heibonsha; 2001.

[28] Kyoto prefecture, editor. Red data book of Kyoto prefecture Vol.1. Kyoto prefecture: Kyoto; 2002.

[29] Environment Agency of Japan. Threatened wildlife of Japan – Red Data Book 2nd edition. Vol. 9, Bryophytes, algae, lichens, fungi. Tokyo: Japan Wildlife Research Center; 2000.

[30] Studlar SM. Host specificity of epiphytic bryophytes near Mountain Lake, Virginia. The Bryologist 1982; 85(1) 37-50.

[31] McGee GG, Kimmerer RW. Size of *Acer saccharum* Hosts Does Not Influence Growth of Mature Bryophyte Gametophytes in Adirondack Northern Hardwood Forests. The bryologist 2004;107(3) 302-11.

[32] Mezaka A, Zontina V. Epiphytic bryophytes in old growth forests of slopes, screes and ravines in north-west Latvia. Acta Universitatis Latviensis 2006; 710 103-16.

[33] Ojala E, Mönkkönen M, Inkeröinen J. Epiphytic bryophytes on European aspen *Populus tremula* in old-growth forests in northeastern Finland and in adjacent sites in Russia. Canadian Journal of Botany 2000; 78(4) 529-36.

[34] Kuriyama K. Effect of forest types on the diversity of epiphytic bryophytes in Kyoto prefectural botanical garden. MS thesis. Kyoto Prefectural University Kyoto; 2007.

[35] Pypker TG, Unsworth MH, Bond BJ. The role of epiphytes in rainfall interception by forests in the Pacific Northwest. I. Laboratory measurements of water storage. Canadian Journal of Forest Research 2006; 36(4) 809–18.

[36] Pypker TG, Unsworth MH, Bond BJ. The role of epiphytes in rainfall interception by forests in the Pacific Northwest II. Field measurements at the branch and canopy scale. Canadian Journal of Forest Research 2006; 36(4) 819–32.

[37] Nadkarni NM. Epiphyte biomass and nutrient capital of a neotropical elfin forest. Biotropica 1984; 16(4) 249–256

[38] Clark KL, Nadkarni NM, Gholz HL. Retention of inorganic nitrogen by epiphytic bryophytes in a tropical montane forest. Biotropica 2005; 37(3) 328-36.

[39] Kenkel NC, Bradfield GE. Ordination of epiphytic bryophyte communities in a wet-temperate coniferous forest, South-Coastal British Columbia. Vegetatio 1981; 45(3) 147-54.

[40] Newmaster SG, Bell FW. The effects of silvicultural disturbances on cryptogam diversity in the boreal mixedwood forest. Canadian Journal of Forest Research 2002; 32(1) 38-51.

[41] Giordano S, Sorbo S, Adamo P, Basile A, Spagnuolo V, Cobianchi RC. Biodiversity and trace element content of epiphytic bryophytes in urban and extra-urban sites of southern Italy. Plant Ecology 2004; 170(1) 1-14.

Chapter 19

Miombo Woodlands Research Towards the Sustainable Use of Ecosystem Services in Southern Africa

Natasha Sofia Ribeiro, Stephen Syampungani,
Nalukui M. Matakala, David Nangoma and
Ana Isabel Ribeiro-Barros

Additional information is available at the end of the chapter

http://dx.doi.org/10.5772/59288

1. Introduction

The Miombo woodlands are the most extensive warm dry forest type in southern Africa [1], covering ca. 2.7 million km^2 across seven countries: Tanzania and the Democratic Republic of Congo (DRC) in the north, Angola and Zambia in the east, and Malawi, Zimbabwe and Mozambique in the south [2-4] (Figure 1). It is one of the most important ecosystems in the world, playing an important role at the social, economic and environmental levels. Being an important center of plant biodiversity Miombo is a key provider of goods and services, supporting the livelihoods of more than 65 million of people in the region [4]. The woodlands are also very important to the national economies as they provide timber for exportation. From the environmental point of view Miombo is determinant to energy, carbon and water balance [3,5].

The ecological dynamics of Miombo is strongly influenced by their woody component, particularly by large trees, which play a key role in ecosystem function, primarily in nutrient cycling, accounting for a great deal of the carbon pool. This component is in turn influenced by a combination of climate, disturbances [e.g. drought, fire, grazing and herbivory primarily by elephants (*Loxodonta africana* Blumenbach)] and human activities [6-7]. Growing population in the region over the last 20-25 years has resulted in increased woodland degradation and deforestation. Slash and burn agriculture and charcoal production are the major causes of forest loss and degradation in the Miombo Ecoregion [8-10]. Additionally, the region is experiencing several major investments in mining, commercial agriculture and infrastructures, which have further increased the pressure on the woodlands. In Zambia for example, where

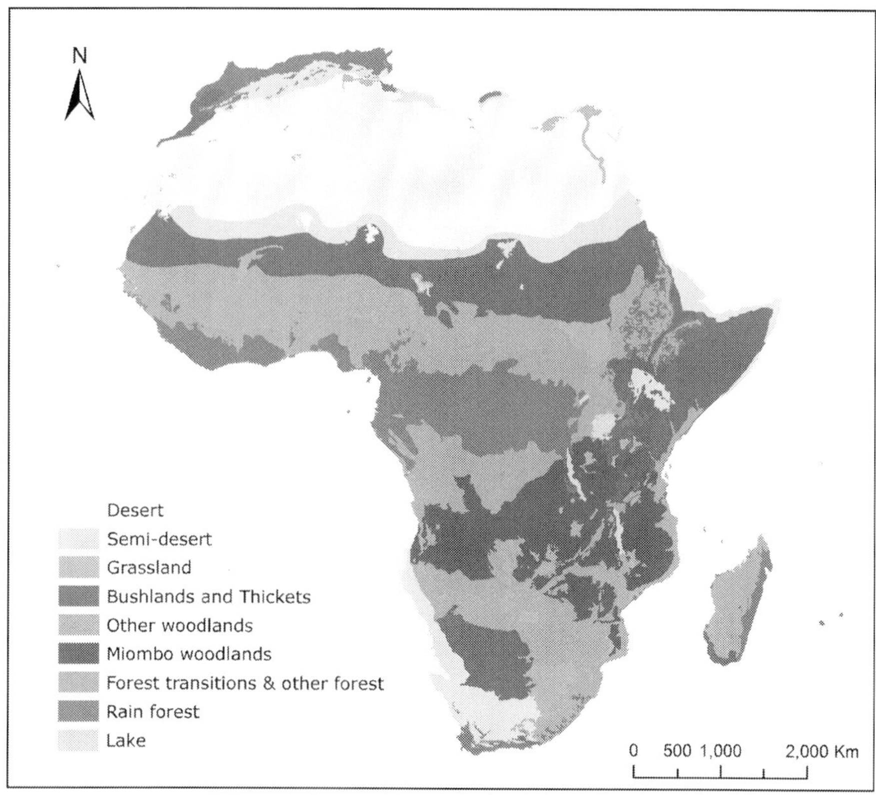

Figure 1. Map of African vegetation, showing the Miombo woodlands in dark green (Source: White, 1983).

there is large-scale mining for copper, the mining sector has greatly contributed to forest cover and biomass losses. Often, huge tracts of land are cleared to provide space for mining infrastructures; at Kalumbila Concession, Solwezi alone, infrastructure development resulted in the loss of more than 7,000 hectares of land [11]. This is often followed by an increased demand for construction timber, creating further pressure on forests.

Changes in the global climatic pattern, *e.g.* 5-15% predicted reduction in precipitation for southern Africa [12], constitute another major threat across the various global ecosystems. In the Miombo Woodlands, they are mainly associated with more extreme wet and dry seasons, *i.e.* drier interior regions, wetter coastal regions, as well as extreme temperatures, which may change disturbances regimes (fire, shifting cultivation, amongst others) and thus the prevailing biodiversity status. According to [13], the combined effect of climate change and disturbances may cause the loss of ca. 40% of the woodlands by the middle of the century. There is an increased concern that the loss of mature trees in landscapes may result in the transformation of the woodlands into scrub or grasslands. This may impose changes in biodiversity and biomass with associated modifications on the pattern of goods and services offered by this ecosystem.

It is widely recognized that Miombo Woodlands have great potential to provide financial resources through Carbon-based Payment for Ecosystem Services (PES) [14], but their function as dynamic C-pools in biogeochemical cycles is largely unknown [15]. In this context, understanding biodiversity and carbon variations under different land use scenarios as well as the rates and the extent to which Miombo recover from disturbances has important implications in the emerging C-based PES schemes [16], which are taking center-stage in the United Nations Framework Convention on Climate Change (UNFCCC) through mechanisms such as Reducing Emission from Deforestation and Forest Degradation (REDD+). On the other hand, such assessments will be crucial for future land use decisions to ensure optimal land use benefits, hence ensuring forest conservation and sustainable management [15].

Under the scenario described above, current research efforts in the region aim at understanding the ecology of Miombo including its biodiversity, biomass production and carbon sequestration, as well as the role of disturbances and its socio-economic relevance. In this chapter we summarize the existing information on the dynamics of biodiversity and biomass /carbon), in order to identify research gaps and needs. It is our intention to contribute towards a research agenda for the Miombo Woodlands, which is being developed under the context of the Miombo Network of Southern Africa, an alliance of scientists for informed research to decision making in the region.

2. Biodiversity dynamics

Miombo has an estimated diversity of 8,500 plant species, of which ca. 54% are endemic. Together with Mopane, it is amongst the five high biodiversity wilderness areas in the world whose conservation should be prioritized because of their irreplaceability in terms of species endemism [17-18]. The woodlands are characterized by the overwhelming dominance of *Brachystegia* (*Miombo* in Swahili), *Julbernardia* and *Isoberlinia* tree species belonging to the Fabaceae (legume) family [19-20], associated with a variety of other woody plants, such as *Pseudolachnostylis maprouneifolia* Pax., *Burkea africana* Hook and *Diplorhynchus condylocarpon* (Müll.Arg.) Pichon. In mature Miombo these species comprise an upper canopy layer made of 10-20 m high trees and a scattered layer of sub-canopy trees. The understorey is discontinuous and composed of broadleaved shrubs of the genera *Eriosema, Sphenostylis, Kotschya, Dolichos* and *Indigofera*, among others, and suppressed saplings of canopy trees. A sparse but continuous herbaceous layer of grasses, forbs and sedges composed of *Hyparrhenia, Andropogon, Loudetia, Digitaria* and *Eragrostis* dominate the ground-layer [2,21]. Miombo is usually referred as a homogenous ecosystem, but differences in species composition, diversity and structure occur at a local scale [22]. The origin of these differences is unclear but, geomorphic evolution of the landscape [23], soil moisture and nutrients [24], land uses changes and other anthropogenic disturbances [25] have all been indicated. Figure 2 illustrates Miombo structure and composition and fire occurrence.

Figure 2. Illustrative exemple of the Miombo Woodlands. (A) Diversity of tree species; (B) Close up pf *Brachystegia bohemii*; (C) Regrowing area; (D) Recently burned area. Credits to Isabel Moura, Tropical Research Institute, Portugal (A, B) and Ivete Maquia, Biotechnology Center, Eduardo Mondlane University (C, D)

Only a few Miombo biodiversity studies have been recently published and most of them were focused either on a limited number of tree species and/or on specific geographical locations. Hence, the information on the conservation status of Miombo plant species is scarce. For example, based on the existing national surveys, the number of threatened plant species is difficult, if not impossible, to estimate, but the Sudan-Zambezian zone, to which the Woodlands belong, is reported to have the highest values of threatened species [18]. Since the conservation status of a particular species is a good indicator of the impact of threats and its capacity to provide goods and services [26], site-specific studies confined to one or to a small group of species are of utmost importance to upgrade the existing information and thus help future planning and management programs.

The establishment and management of fully protected areas such as National Parks are often assumed to be the best strategy for conserving species diversity and maintain forest composition and structure. To evaluate this assertion, Banda and co-authors [27] conducted a study in western Tanzania in areas with four different levels of protection: a National Park (high protection level), a Game Controlled Area (with tourist hunting of big game animals), a Forest

Reserve (with selective harvest of trees), and an Open Area (unrestricted access to forest resources). The authors observed that the forest structure was quite similar in the four sites and that species richness was significantly higher in the Game Controlled Area and Forest Reserve than in the other areas. More recently, Giliba and collaborators [22] assessed species richness, diversity, dominance and exploitation in Bereku Forest Reserve, northern Tanzania, concluding that the use of Miombo products and services by the surrounding communities does not compromise the stability of the woodlands, which are fairly stocked with high tree and shrub species diversity. These studies suggest that National Parks do not always host the greatest diversity of trees or unique species. This may imply that a suite of different types of protection strategies may be the key for conservation in African dry tropical forests [27].

The effect of environmental factors, particularly soil and disturbance history, on tree diversity and size structure was analysed by [19] in and around the Ihombwe village, Kilosa District,Tanzania where shifting cultivation is practiced. The authors observed that there was a considerably high capacity for tree species regeneration, partly due to the relatively isolated position of the village and also due to the fact that local communities recognize the importance of the sustainable ecosystem use. However, fires were pointed as the main driver of species composition change as they tend to support the proliferation of fire tolerant species, such as *P. maprouneifolia, Pterocarpus angolensis* DC and *D. condylocarpon* at the expense of dominant Miombo species. Similarly, [27] observed the dominance of *Terminalia sericea* Burch. ex DC., *Combretum adenogonium* Steud. ex A.Rich., and *C. colinum* Fresen. in dry Miombo of the Katavi-Rukwa ecosystem of western Tanzania due to frequent burning. Also Williams and co-authors [28], working on the effects of slash-and-burn agriculture in the Nhambita community, Sofala Province, Mozambique, observed that in abandoned (regrowing) sites, defining Miombo species were replaced by secondary dominant trees. However, the biodiversity of woody species (i.e. Shannon index and species richness) in older abandonments (>10 years old) and intact woodlands were similar. In general tree biodiversity has not been degraded by the slash-and-burn disturbance, but the time-scale of recovery of defining Miombo species was unclear. In Zambia [29-31] have also suggested that though Miombo systems recover relatively fast in terms of species diversity, species composition takes longer to recuperate. In another study in Mozambique, [32] observed that fire and herbivory by elephants are the main drivers of ecosystem structure and composition. For example, places with high fire frequency and elephant density were dominated by fire-resistant species such as *T. stenostachya, Combretum* spp. and *D. condylocarpon*. Similar results were obtained by [33] in the Miombo Woodlands of northwestern Zimbabwe, where elephants and fires reduced the proportions of large trees, tree heights, stem basal area and densities of all trees. Besides that, tree species frequencies dropped 28–89.6% and the most visible floristic alteration was the replacement of the typical *Brachystegia boehmii* Taub. by *P. maprouneifolia* and Combretaceae species. The results are in line with those from [19,27-28] and are corroborated by the prediction model developed by [34] regarding the impact of elephants and fires on the structure of semi-arid Miombo Woodlands of north-western Zimbabwe. The author hypothesized that elephants alone at a density of 0.27 km^{-2} would convert the woodland into coppice in 120 years due to massive declines of large trees; the same result would be achieved in 10 years if elephant density increased to 2 km^{-2}; the pattern would remain similar if simultaneous fire occurred once every 4.7 years with elephants at 0.27 km^{-2}. Thus, it was predicted that elephants alone can degrade and maintain semi-arid Miombo Woodlands into coppice, largely due to their damaging impacts on mature canopy trees. Fire may also speed up the process by suppression of an already low recruitment. However, this driver alone had less influence on the woodland structure than elephants because of low fuel loads due to heavy grazing and low grass production as a result of low rainfall and inherently poor soils in the area.

Another important aspect in understanding the biodiversity dynamics in the Miombo Woodlands and in assisting conservation programs, is the application of molecular markers (MM). MM are essential tools to analyse population structure and genetic diversity as well as to identify particular traits (including genotypes and genes) associated with outstanding performances and resilience to extreme environments (*e.g.* fire, drought, high temperatures) [35-37]. The use of MM to understand the dynamics and potentialities of Miombo species is still incipient and only two studies have been published. The first reported on the use of Amplified Fragment Length Polymorphisms (AFLP) to assess the genetic diversity of natural populations of *Uapaca kirkiana* Muel. Arg. from three geographical regions of Malawi, in relation to deforestation, fragmentation and wildfires [38]. AFLP markers revealed moderate differentiation among the studied populations, but very high variation among individuals within populations. The second study was based on the use Inter Simple Sequence Repeat (ISSR) markers to assess genetic diversity in *B. boehmii* and *B. africana* across a fire gradient in the Niassa National Reserve (NNR) [39]. Although fire differentially affected the biodiversity in each species, in general, the overall genetic diversity was high and their survival did not seem to be compromised by the frequency of fires, agreeing with the fact that NNR is one of the least disturbed areas of deciduous Miombo. The results point also to a link between fire-tolerance and genetic diversity, as judged by the higher diversity levels observed in *B. africana* (fire-tolerant) in comparison to *B. boehmii* (fire-sensitive). Furthermore, *B. boehmii* presented an evolutionary response to fire, i.e. the levels of diversity were lower in frequent fire prone areas than in areas of low fire frequency, a phenomena attributed to the pyrodiversity-like effect [40]. In both papers, the authors emphasize the need for more intensive genetic studies spanning other populations of these and other important tree species to produce a wider picture of the levels of distribution of genetic diversity across the Miombo Ecoregion and its relation to major threats.

In conclusion, the available literature generally suggests that biodiversity in the miombo woodlands is shaped by disturbances, including anthropogenic actions, and to some extent may be compromised by the ongoing pressures. Despite the risks to which the woodlands are exposed, the species diversity and the levels of genetic diversity are considerably high. This seems to be particularly associated with the apparent resiliency of Miombo to various disturbances. However, there are evidences that typical species not always recover and in some cases may be replaced by secondary species. As a consequence, the range and type of goods and services provided by the woodlands may be altered. This calls for the implementation of management strategies that are appropriate for conserving biodiversity of Miombo.

2.1. Biomass and carbon dynamics

Estimations of biomass and carbon stocks are an essential step in accounting for ecosystem goods and services particularly when considering land use options and strategies to promote carbon sequestration. This is relevant for implementing carbon credit market mechanisms such as REDD+, which seeks to mitigate climate change through enhanced CO_2 storage in terrestrial ecosystems.

Biomass and carbon stocks have a pronounced variation across the Miombo Ecoregion. This has been mainly associated to: i) soil fertility and plant nutrition; ii) fires and herbivory; and iii) age and status of the woodland. Woody biomass was observed to range from 1.5 Mg ha^{-1} (3-6 years old coppice) to 144 Mg ha^{-1} (mature wet Miombo) [9,41-45]. Dry Miombo ranges between 53-55 Mg ha^{-1} [45-48]. It is confirmed that wood and soil compartments are the most important of these stocks [48-49], but grass, litter and root may contribute significantly to carbon sequestration. Table 1 presents comparative results of carbon stock density in different compartments across different sites.

Ecosystem compartment	Carbon Stock Density (Mg C ha^{-1})	Localization	Reference
Soil	57.90	Gorongosa, Mozambique	[28]
	34.72 ± 17.93	Niassa Reserve, Mozambique	[48]
	31.04	Dombe, Manica, Mozambique	[47]
Trees	19.00 ± 8.00	Gorongosa, Mozambique	[28]
	13.17 - 32.10	Beira Corridor, Mozambique	[50]
	20.88	Niassa Reserve, Mozambique	Sitoe, Unpublished data
	26.48	Dombe, Manica, Mozambique	[47]
	29.88 ± 13.07	Niassa Reserve, Mozambique	[48]
Grasses	1.2	Niassa Reserve, Mozambique	Sitoe, Unpublished data
	0.65	Dombe, Manica, Mozambique	[47]
	030 ± 0.89	Niassa Reserve, Mozambique	[48]
Litter	0.80	Niassa Reserve, Mozambique	Sitoe, Unpublished data
	3.00	Dombe, Manica, Mozambique	[47]
	0.06 ± 0.03	Niassa Reserve, Mozambique	[48]
Dead Trees	0.06 ± 0.19	Niassa Reserve, Mozambique	[48]
Herbaceous	0.02 ± 0.01	Niassa Reserve, Mozambique	[48]
	0.55 ± 0.02	Eastern Arc Mountains, Tanzania	[49]
Total carbon	10.13-79.69	Niassa Reserve, Mozambique	[48]
	13 - 30	Eastern Arc Mountains, Tanzania	[48]

Table 1. Comparative results of Carbon Stock Density across the Miombo Ecoregion. Source: Adapted from [48].

The dynamics of Miombo is in general influenced by its tree component given its dominance. Wood vegetation is in turn affected by environmental and disturbance factors [1]. Fire is particularly an important factor in Miombo as its behavior, timing, intensity and frequency vary greatly across the ecosystem, thus affecting vegetation structure and biomass differently. Frequent late dry season fires can transform woodland into open tall grass savanna with isolated fire-tolerant canopy trees and scattered understorey trees and shrubs [52] thereby reducing woody biomass. The impact of fires on biomass and carbon stocks has been addressed in a few countries. [53] in Zimbabwe, [54] and [31] in Zambia, and [46] and [55] in Mozambique have observed that fire protected sites had more woody biomass than frequently burned sites. [55] also noted that annual fire suppressed woody biomass development (up to 38 Mgha^{-1} in the studied area of central Mozambique) while low intensity fires at lower frequencies promoted biomass accumulation. Many studies have reported that once trees reach a certain height, they are less susceptible to fire [54 and references therein]. However, in his 22-year period study in Zambia, [31] found that large and tall trees were just as susceptible to fire as small trees, but their death was gradual and occurred over longer periods of time. In this area, fire alone was responsible for more than 25% of the observed biomass losses. The author concluded that avoided forest degradation at the study sites would have increased standing woody biomass up to 4.0 t ha^{-1} year^{-1} over the 22-year period. Recently, [45] found that carbon storage in the tree-dominated ecosystems of the Tanzanian Eastern Arc Mountains has decreased at a mean rate of 1.47 Mg C ha^{-1} yr^{-1} (ca. 2% of the stocks of carbon per year) due to 74% forest area loss driven by 5-fold increase in cropland area.

The interactive effect of fire and herbivory by elephants is quite interesting in Miombo. In general, elephants uproot, de-branch and/or debark large trees, increasing fuel-load in the forest ground due to intensified light intensity. Higher fuel loads result in frequent and fierce fires that influence the woodland. [42] and [56] have studied the effect of elephants in Sengwa National Park, Zimbabwe. The study compared areas inside the National Park (high elephant density and fire occurrence) and outside the National Park (low elephant density and fire occurrence) and revealed a reduction in biomass up to 31.8 t ha^{-1} for the area inside the national park due to elephant grazing. Fires inside the park leveraged elephant's effect by killing young sapling and debarked susceptible trees. [32] and [46] analised the combined effect of fires and elephants in NNR, northern Mozambique, revealing denser woodlands and higher wood biomass in places with low fire frequency and low animal densities. Recently, [48] studied the dynamics of the biomass in the Miombo woodlands in NNR and observed that woody biomass had a net increase of 3 Mg ha^{-1} in a 5-year period of study. However, when looking at the species level, *Diplorhynchus condylocarpon* presented the highest growth (a net increase of 0.54 Mg ha^{-1}). This species has been reported elsewhere in the region has fire indicator due to its capacity to thrive in high fire frequency environments and to the fact that it is less preferred by elephants. *Julbernardia globiflora* (Benth.), on the other hand, experienced a net decrease in biomass of 0.09 Mg ha^{-1}. The reason might be associated with fire susceptibility as well as high preference by local population and elephants and. As referred above, not many studies have addressed the specific responses to disturbances, though it should be considered fundamental to understand the ecosystem trends. In fact, species dynamics may disclose particular behaviors that are not seen at the ecosystem level, but are important in defining conservation and management strategies, which are not just ecosystem but also species oriented.

Charcoal production is one of the main drivers of Miombo degradation but has been poorly accounted for in biomass and carbon studies. Only one study [57] was found in the literature. This study was conducted in Zambia, by comparing a protected area with a highly disturbed site. The results revealed considerably reduced biomass after logging for charcoal production -150 t ha^{-1} within *versus* 24 t ha^{-1} outside the protected area. The authors discuss that better inventory data is urgently required to improve knowledge about the current state of the woodland usage and recovery after logging. They further argue that net greenhouse gas emissions could be reduced substantially by improving the post-harvest management, charcoal production technology and/or providing alternative energy supply.

Although soil is one of the main carbon pools in Miombo, studies that deal with this component are limited [28,48,58-59]. [28] observed that woodland soils were capable of storing >100 t C ha^{-1}, whereas in re-growing areas soil carbon stocks did not exceed 74 t C ha^{-1}. The study concluded that there was a potential for C sequestration in soils on abandoned farmlands. However, there was no discernible increase in soil C stocks within the period of re-growth, suggesting that the rate of accumulation of organic matter in these soils was very slow. On the other hand, [58] observed that agricultural soils in Malawi had 40% less carbon than mature Miombo Woodlands. The authors stated that as the area of land converted to agriculture increases in the region, land in this re-growth state will most likely become the dominant form of Miombo. Therefore studies of the nutrient dynamics in this type of land cover will be essential.

Understanding biomass and carbon recovery (along with biodiversity) rates is essential to predict future scenarios of ecosystem stock densities and thus, its capacity to provide goods and services. Short to medium term (16-50 years) studies in the region reveal a capacity for stock regeneration between 1.0 and 1.8 M g ha^{-1} yr^{-1} [1,28,43,60]. In Zambia, [31] reported net changes in aboveground biomass over a 22-year period of -113.4 Mg (-5.16 Mg ha^{-1} year^{-1}) and 25.7 Mg (1.17 Mg ha^{-1} year^{-1}) associated with old-growth and re-growth sites, respectively. Biomass loss in old-growth sites was driven by agriculture and fire. The conclusion drawn from these studies indicated that Miombo has capacity to recover after disturbances but at slow rates. The latter can be exacerbated or reverted by recurrent disturbances, compromising ecosystem resiliency. However, given the limited number of studies and the associated short to medium time spans, there are still knowledge gaps such as: i) which species recover and at which rate; ii) what are the thresholds of changes relation to disturbances; iii) what are the rates of soil carbon recovery. Improving the knowledge on recovery rates and patterns is important given the complexity of the ecosystem associated with the varied environmental gradients across the region.

3. Research gaps and management needs

It is evident that there have been a considerable amount of studies undertaken in the Miombo Woodlands. In July 2013 the Miombo Network of Southern Africa met in Maputo, Mozambique to discuss the existing knowledge and gaps. In general, there is a consensus that much is still to be investigated.

Miombo displays complex vegetation patterns in which dense vegetation alternates with sparsely populated or bare soil in response to environmental and disturbance (deforestation/ degradation, fires and herbivory) factors. Low vegetation cover, in some places, and small-scale variations in others, can produce unpredictable errors in the quantification of ecosystem dynamics. Ignoring this spatial variation can produce inaccurate results, even in fairly homogeneous environments [61-62].

Miombo complexity has introduced limitations in the past in terms of accurate estimations/ mapping of Land Cover and Land Cover Change (LCLCC), biomass/carbon and biodiversity. In fact, there have been several attempts to estimate LCLCC and biomass at the local and national scale, but at the regional level there is still a need to improve and update the existing products. Land cover mapping is important to delineate LC types associated with degradation levels and the role of the associated drivers. The latter is highly relevant in determining the role of ecosystem in the carbon cycle as well as in defining appropriate rehabilitation and conservation strategies. These are particularly important in the context of REDD+ as it would be important to demarcate areas of interest to develop REDD+ projects.

Ecosystem rehabilitation requires a good understanding of its past and present status including the specific and interactive role of the drivers (fire, herbivory, slash and burn agriculture and climate change) as well as of its recovery patterns across environmental gradients. It also requires a better understating of its biodiversity beyond floristic surveys. In this context, the following questions need to be answered:

- What are the impacts of the different ecosystem drivers on biodiversity?

- What is the capacity of biodiversity to supply and underpin goods and services (current and future)?

- What are the patterns of genetic diversity of important species across environmental gradients?

- How different land cover types affect the existing patterns of biodiversity?

- How these changes in biodiversity affect the availability and accessibility of resources to rural and urban dwellers?

It is important to recognize that biomass and carbon estimations are very scattered in terms of methods and sampling efforts recalling a need to perform harmonized estimations to better position the region in the international context. Hence, finding benchmark sites is vital as it allows determination of deviations under different land uses. This is particularly important given the fact that the diversity of soils, climate, hydrology and disturbances return highly variable biomass and carbon densities making a comparison among sites not always possible [28,49]. Biomass estimations are also relevant to understand the contribution of different pools (soils, grasses, litter, etc) as well as the role of drivers in the ecosystem biomass/carbon sequestration. Particularly in the case of soil carbon, efforts should focus on identifying and protecting C-rich soils. It is also important to investigate whether fire control on recovering woodlands can stimulate the accumulation of soil C and tree biomass, and hence restore defining Miombo species.

Finally, the use of modern (*e.g.* remote sensing and molecular markers) and harmonized sampling data collection and analysis techniques across the region would contribute to the robustness of data and support improved ecosystem management and conservation strategies.

4. The role of the Miombo Network in promoting the Miombo Woodlands sustainability

Founded in 1995 by a group of regional and international scientists, the Miombo Network is under the auspices of the IGBP/IHDP Land Use and Cover Change (LUCC) Project and the IHDP/IGBP/WCRP Global Changes System for Analysis, Research and Training (START). The Network's goal was to support the development of sustainable Miombo Woodlands management policies and practices through the collaborative data acquisition, from land-based research, monitoring, remote sensing and other geospatial information technologies. The membership of the network is drawn from government, university and research institutions of the Miombo Ecoregion countries namely: Malawi, Mozambique, Tanzania, Zambia and Zimbabwe. However, there are also member institutions outside Africa due to their passion for Miombo management.

Being a collaborative alliance, the Miombo Network aspires to conduct joint research that contributes to forest policy definition and decision-making. The entry point for this is a strong link with the SADC forestry program, which intends to develop harmonized policies for the region. The Miombo Network has also potential to contribute for the establishment of the REDD+ programme - a programme that has great potential to turn around the economic and environmental value of the ecosystem across the region.

5. Final considerations

Despite being considered the most important ecosystem of southern Africa, the Miombo Woodlands face some risks. Although policies may be supportive as far as Miombo management is concerned, the woodlands continue to be degraded and deforested. Partly, this is due to the fact that institutions that are responsible for managing the forests have limited human and financial capacity. Additionally, Community-Government partnerships for woodland management need to be enhanced in the region. It would therefore be important that national, regional and international institutions put more effort to establish effective collaborations in order to understand the interplay of issues that affect the management of Miombo Woodlands.

Acknowledgements

The authors thank Isabel Moura (Tropical Research Institute, Portugal) and Ivete Maquia (Biotechnology Center, Eduardo Mondlane University, Mozambique) for providing the photographs for Figure 2.

Author details

Natasha Sofia Ribeiro[1], Stephen Syampungani[2], Nalukui M. Matakala[2], David Nangoma[3] and Ana Isabel Ribeiro-Barros[4*]

*Address all correspondence to: aribeiro@itqb.unl.pt

1 Faculty of Agronomy and Forest Engineering, Eduardo Mondlane University, Campus Universitário Principal, Edifício 1, Maputo, Mozambique

2 Copperbelt University, School of Natural Resources, Kitwe, Zambia

3 Mulanje Mountain Conservation Trust, Mulanje, Malawi

4 PlantStress&Biodiversity Group, Biotrop Center, Tropical Research Institute, Quinta do Marquês, Oeiras, Portugal

References

[1] Frost P. The Ecology of Miombo Woodlands. In: Campbell B. (ed) The Miombo in Transition: Woodlands and Welfare in Africa. Bogor: CIFOR; 1996. p11-55.

[2] Desanker P., Frost PGH., Frost CO., Justice CO., Scholes RJ. The Miombo Network: Framework for a Terrestrial Transect Study of Land-Use and Land-Cover Change in the Miombo Ecosystems of Central Africa. Stockholm: The International Geosphere-Biosphere Programme (IGBP); 1997.

[3] Ribeiro N., Cumbana M., Mamugy F., Chaúque A. Remote Sensing of Biomass in the Miombo Woodlands of Southern Africa: Opportunities and Limitations for Research. In: Fatoyinbo L. (ed.) Remote Sensing of Biomass – Principles and Applications. Rijeka: InTech; 2012. p77-98.

[4] WWF. Miombo Eco-Region "Home of the Zambezi" Conservation Strategy 2011-2020. Harare: WWF-World Wide Fund for Nature; 2012.

[5] Chidumayo E., Marunda C. Dry Forests and Woodlands in Sub-Saharan Africa: Context and Challenges. In: Chidumayo EN., Gumbo D. (eds.) The Dry Forests and Woodlands of Africa: Managing for Products and Services. London: Earthscan; 2010. p1-10.

[6] Ribeiro NS. Interaction between Fires and Elephants in Relation to Vegetation Structure and Composition of Miombo Woodlands in Northern Mozambique. PhD Thesis, University of Virginia; 2007.

[7] Gumbo D., Chidumayo E. Managing Dry Forests and Woodlands for Products and Services: a Prognostic Synthesis. In: Chidumayo EN., Gumbo D. (eds.) The Dry Forests and Woodlands of Africa: Managing for Products and Services. London: Earthscan; 2010. p261-279.

[8] Stromgaard P. Early Secondary Succession on Abandoned Shifting Cultivator's Plots in the Miombo of South Central Africa. Biotropica 1987;18(2), 97–10.

[9] Chidumayo EN. Woody Biomass Structure and Utilization for Charcoal Production in a Zambian Miombo Woodland. Bioresources Technology 1991;37(1) 43–52.

[10] Malambo FM., Syampungani S. Opportunities and Challenges for Sustainable Management of Miombo Woodlands: the Zambian Perspective. In: Varmola M., Valkonen S., Tapaninen (eds.) Research and Development for Sustainable Management of Semi-Arid Miombo Woodlands in East Africa. Vantaa: Finnish Forest Research Institute; 2008. p125-130.

[11] Vinya R., Syampungani S., Kasumu EC., Monde C., Kasubika R. Preliminary Study on the Drivers of Deforestation and Potential for REDD+in Zambia. A Consultancy Report Prepared for Forestry Department and FAO under the National UN-REDD +Programme. Lusaka: Ministry of Lands and Natural Resources; 2011.

[12] Chidumayo EN. Effects of Climate on the Growth of Exotic and Indigenous Trees in Central Zambia. Journal of Biogeography 2005;32(1) 111-120.

[13] Green JMH., Larrosa C., Burgess ND., Balmford A., Johnston A., Mbilinyi BP., Platts PJ., Coad L. Deforestation in an African Biodiversity Hotspot: Extent, Variation and the Effectiveness of Protected Areas. Biological Conservation 2013;164 62–72.

[14] Baker TR., Jones JPG., Thompson ORR., Cuesta RMR., del Castillo D., Aguilar IC., Torres J., Healey JR. How Can Ecologists Help Realise the Potential of Payments for Carbon in Tropical Forest Countries? Journal of Applied Ecology 2010;47(6) 1159–1165.

[15] Schöngart J., Arieira J., Felfili Fortes C., Arruda Cunha EC, Nunes da Cunha CN. Carbon Dynamics in Aboveground Coarse Wood Biomass of Wetland Forests in the Northern Pantanal, Brazil. Biogeosciences 2008;5(3) 2103–2130.

[16] Mwampamba TH., Schwartz MW. The Effects of Cultivation History on Forest Recovery in Fallows in the Eastern Arc Mountain, Tanzania. Forest Ecology and Management 2011;261 1042–1052.

[17] Mittermeier R., Mittermeier C., Brooks TM., Pilgrim JD., Konstant WR., da Fonseca GAB., Kormos C. Wilderness and Biodiversity Conservation. Proceedings of National Academy of Sciences 2003;100(18) 10309-10313.

[18] Shumba E., Chidumayo E., Gumbo D., Kambole C., Chishaleshale M. Biodiversity of Plants. In: Chidumayo EN., Gumbo D. (eds.) The Dry Forests and Woodlands of Africa: Managing for Products and Services. London: Earthscan; 2010. p 43-61.

[19] Backéus I., Petterson B., Strömquist L., Ruffo C. Tree Communities and Structural Dynamics in Miombo (*Brachystegia-Julbernardia*) Woodland, Tanzania. Forest Ecology and Management 2006;230 171-178.

[20] Dewees P., Campbell B., Katerere Y., Sitoe A., Cunningham AB., Angelsen A., Wunder S. Managing the Miombo Woodlands of Southern Africa: Policies, Incentives, and Options for the Rural Poor. Washington: Program on Forests (PROFOR); 2011.

[21] Campbell B., Frost P., Byron N. Miombo Woodlands and their Use: Overview and Key Issues. In: Campbell B. (ed) The Miombo in Transition: Woodlands and Welfare in Africa. Bogor: CIFOR; 1996. p1-5.

[22] Giliba RA., Boon EK., Kayombo CJ., Musamba EB., Kashindye AM., Shayo PF. Species Composition, Richness and Diversity in Miombo Woodland of Bereku Forest Reserve, Tanzania. Journal of Biodiversity 2011;2(1) 1-7.

[23] Cole M. The Savannas: Biogeography and Geobotany. London: Academic Press; 1986.

[24] Campbell BM., Swift MJ., Hatton J., Frost PGH. Small-scale Vegetation Pattern and Nutrient Cycling in Miombo Woodland. In: Verhoeven JTA., Heil GW., Werger MJA. (eds.) Vegetation Structure in Relation to Carbon and Nutrient Economy. The Hague: SPB Academic Publishing; 1988. p69-85.

[25] Chidumayo EN. Early Post-Felling Response of *Marquesia* Woodland to Burning in the Zambian Copperbelt. Journal of Ecology 1989;77(2) 430-438.

[26] Hamilton A., Hamilton P. Plant Conservation: an Ecosystem Approach. London: Earthscan; 2006.

[27] Banda T., Schwartz MW., Caro T. Woody Vegetation Structure and Composition along a Protection Gradient in a Miombo Ecosystem of Western Tanzania. Forest Ecology and Management 2008;230 179-185.

[28] Williams M., Ryan C., Rees RM., Sambane E., Fernando J., Grace J. Carbon Sequestration and Biodiversity of Re-Growing Miombo Woodlands in Mozambique. Forest Ecology and Management 2008;254 145-155.

[29] Chidumayo EN. Development of *Brachystegia-Julbernardia* Woodland after Clear-Felling in Central Zambia: Evidence for High Resilience. Applied Vegetation Science 2004;7(2) 237-242.

[30] Kalaba FK., Quinn H., Dougill AJ., Vinya R. Floristic Composition, Species Diversity and Carbon Storage in Charcoal and Agriculture Fallows and Management Implications in Miombo Woodlands of Zambia. Forest Ecology and Management 2013, 304 99-109.

[31] Chidumayo E. Forest Degradation and Recovery in a Miombo Woodland Landscape in Zambia: 22 Years of Observations on Permanent Sample Plots. Forest Ecology and Management 2013; 291 154–161.

[32] Ribeiro NS., Shugart HH., Washington-Allen R. The Effects of Fire and Elephants on Species Composition and Structure of the Niassa Reserve, Northern Mozambique. Forest Ecology and Management 2008;255 1626-1636.

[33] Mapaure I., Moe SR. Changes in the Structure and Composition of Miombo Woodlands Mediated by Elephants (*Loxodonta africana*) and Fire Over a 26-year Period in North-Western Zimbabwe. African Journal of Ecology 2009;47(2) 175–183.

[34] Mapaure I. A Preliminary Simulation Model of Individual and Sinergistic Impacts of Elephants and Fire on the Structure of Semi-Arid Miombo Woodlands in Northwestern Zimbabwe. Journal of Ecology and the Natural Environment 2013;5(10) 285-302.

[35] De Vicente MC., Guzmán FA., Engels J., Rao VA. Genetic Characterization and its Use in Decision-Making for the Conservation of Crop Germplasm. In: Ruane J., Sonnino A. (eds.) The Role of Biotechnology in Exploring and Protecting Agricultural Genetic Resources. Rome: FAO of the United Nations; 2006. p129-138.

[36] Fraleigh B. Global Overview of Crop Genetic Resources. In: Ruane J., Sonnino A. (eds.) The Role of Biotechnology in Exploring and Protecting Agricultural Genetic Resources. Rome: FAO of the United Nations; 2006. p 21-32.

[37] Gepts P. Plant genetic resources conservation and utilization: the accomplishments and future of a societal insurance policy. Crop Science 2006;46(5) 2278–2292.

[38] Mwase WF., Bjørnstad A., Stedje B., Bokosi JM., Kwapata MB. Genetic Diversity of *Uapaca kirkiana* Muel. Arg. Populations as Revealed by Amplified Fragment Length Polymorphisms (AFLPs). African Journal of Biotechnology 2006;5(13) 1205-1213.

[39] Maquia I., Ribeiro NS., Silva V., Bessa F., Goulao LF., Ribeiro A. Genetic Diversity of *Brachystegia boehmii* Taub. and *Burkea africana* Hook. f. Across a Fire gradient in Niassa National Reserve, Northern Mozambique. Biochemical Systematics and Ecology 2013;48 238–247.

[40] Parr CL., Andersen AN. Patch Mosaic Burning for Biodiversity Conservation: a Critique of the Pyrodiversity Paradigm. Conservation Biology 2006;20(6) 610–619.

[41] Malaisse FP., Strand MA. A Preliminary Miombo Forest Seasonal Model. In: Kern L. (ed.) Modeling Forest Ecosystems, I.B.P. Woodland Workshop: Oak Ridge; 1973. p291-295.

[42] Guy PR. Changes in the Biomass and Productivity of Woodlands in the Sengwa Wildlife Research Area, Zimbabwe. Journal of Applied Ecology 1981;18(2) 507-519.

[43] Chidumayo EN. Miombo Ecology and Management: an Introduction. Stockholm: Stockholm Research Institute; 1997.

[44] Luoga EJ, Witkowski ETF., Balkwill K. Harvested and Standing Wood Stocks in Protected and Communal Miombo Woodlands of Eastern Tanzania. Forest Ecology and Management 2002;164 15-30.

[45] Willcock S., Phillips OL., Platts PJ., Balmford A., Burgess ND., Lovett JC., Ahrends A., Bayliss J., Doggart N., Doody K., Fanning E., Green JMH., Hall J., Howell KL., Marchan R., Marshall AR., Mbilinyi B., Munis PKT., Owen N., Swetnam RD., Topp-Jorgensen EJ., Lewis SL. Quantifying and Understanding Carbon Storage and Sequestration within the Eastern Arc Mountains of Tanzania, a Tropical Biodiversity Hotspot. Carbon Balance and Management 2014;9: 2.

[46] Ribeiro NS., Saatchi SS., Shugart HH., Washington-Allen RA. Aboveground Biomass and Leaf Area Index (LAI) Mapping for Niassa Reserve, Northern Mozambique. Journal of Geophysical Research 2008;113 G02S02.

[47] Sitoe A. Baseline Carbon Estimation in Dombe, Manica Biofuel Production Area. Maputo: Faculty of Agronomy and Forestry Engineering, Eduardo Mondlane University; 2009.

[48] Ribeiro NS., Matos CN., Moura IR., Washington-Allen RA., Ribeiro AI. Monitoring Vegetation Dynamics and Carbon Stock Density in Miombo Woodlands. Carbon Balance and Management 2013;8:11.

[49] Shirima DD., Munishi PKT., Lewis SL., Burgess ND., Marshall AR., Balmford A., Swetnam RD., Zahabu EM. Carbon Storage, Structure and Composition of Miombo Woodlands in Tanzania's Eastern Arc Mountains. African Journal of Ecology 2008;40(3) 332-342.

[50] Tchaúque FDLJ. Avaliação da Biomassa Lenhosa Aérea no Corredor da Beira. BSc Thesis. Eduardo Mondlane University; 2004.

[51] Timberlake J., Chidumayo E., Sawadogo L. Distribution and Characteristics

[52] of African Dry Forests and Woodlands. In: Chidumayo EN., Gumbo D. (eds.) The Dry Forests and Woodlands of Africa: Managing for Products and Services. London: Earthscan; 2010. p11-41

[53] Bond WJ., Keeley JE. Fire as a Global 'Herbivore': the Ecology and Evolution of Flammable Ecosystems. Trends in Ecology and Evolution 2005;20 (7) 387-394.

[54] Furley PA., Rees RM., Ryan CM., Saiz G. Savanna Burning and the Assessment of Long-Term Fire Experiments with Particular Reference to Zimbabwe. Progress in Physical Geography 2008;32(6) 611–634.

[55] Ryan C., Williams M., Grace J. Above-and Belowground Carbon Stocks in a Miombo Woodland Landscape of Mozambique. Biotropica 2011;43(4) 423-432.

[56] Guy PR. The Influence of Elephants and Fire on a *Brachystegia-Julbernardia* woodland in Zimbabwe. Journal of Tropical Ecology 1989;5(2) 215-226.

[57] Kutsch WL., Merbold L., Ziegler W., Mukelabai MM., Muchinda M., Kolle O., Scholes RJ. The Charcoal Trap: Miombo Forests and the Energy Needs of People. Carbon Balance and Management 2011;6:5.

[58] Walker SM., Desanker PV. The Impact of Land Use on Soil Carbon in Miombo Woodlands of Malawi. Forest Ecology and Management 2004;203 345-360.

[59] Shelukindo HB., Msanya B., Semu E., Mwango S., Singhand BR., Munishi M. Characterization of Some Typical Soils of the Miombo Woodland Ecosystem of Kitonga Forest Reserve, Iringa, Tanzania: Physico-Chemical Properties and Classification. Journal of Agricultural Science and Technology 2014;A4 224-234.

[60] Stromgaard P. Biomass Estimation Equations for Miombo Woodland. Agroforestry Systems1985;3(1) 3-13.

[61] Aubry P., Debouzie D. Estimation of the Mean from a Two-Dimensional Sample: the Geostatistical Model-Based Approach. Ecology 2001;82(5) 1484–1494.

[62] Haining R. Spatial Data Analysis in the Social and Environmental Sciences. Cambridge: Cambridge University Press; 1990.

Chapter 20

Ecological Restoration in Conservation Units

Suzane Bevilacqua Marcuzzo and Márcio Viera

Additional information is available at the end of the chapter

http://dx.doi.org/10.5772/59090

1. Introduction

Forest ecosystems have rapidly been transformed into areas for the occupation of the human population and for economic purposes [1]. Rainforests distributed around the planet have been cleared because of a complex set of factors, which vary according to the characteristics of each region [2]. Among the factors are deforestation for alternative use of the soil (pasture, commercial or subsistence farming, and biofuel generation), tree cutting for the timber industry, wood extraction for energy biomass and poaching [3]. The reduction of vegetation cover leads to a decrease of biodiversity and to an increase of carbon emissions, changing the global climate [1].

The creation and establishment of conservation units is one of the main strategies to ensure biodiversity [4], allowing governments to tackle climate changes and, in the process, ensure biodiversity [5]. Conservation units are protected areas, established to maintain biological diversity and genetic resources, to protect endangered species, to conserve and restore diversity in natural ecosystems [4]. About 10% of the land surface of the planet is under some form of protected area [6]. However, the challenges are huge, because many protected areas are not yet fully implemented or adequately managed [6]. In Brazil, protected areas account for 17% of the Earth's surface [7].

The process of biodiversity preservation in protected areas is not always efficient, leading to lack of connectivity between the different forest remnants. This lack of connectivity affects the movement of organisms between the different environments, influencing the stability of populations, communities and ecological processes. In addition, areas defined as priorities for conservation may show significant environmental changes due to changes in land use [8]. Still, due to the great threat to biodiversity caused by the conversion of natural areas into production systems, conservation units provide adequate guarantees to ensure protection to the environment.

Thus, areas altered because of changes in land use can be restored to recover the ecological interactions necessary for biodiversity maintenance. The use of restoration techniques to recover altered ecosystems is considered a fundamental strategy for biodiversity conservation. Ecological restoration has been widely used in Brazil as a measure to reverse the degradation process and enhance biodiversity conservation, ensuring ecosystem services.

In this context, arboreal forest species play a fundamental role in the reconstruction of the three-dimensional and functional structure of the forest (canopy, understory, strata, biomass, carbon, etc.). These species define local patterns of both organic matter accumulation in the soil [9] and nutrient cycling [10]. They help soils against the effects of erosion processes, favoring water infiltration (less runoff) and the definition of the microclimate standards of the habitat (shading, air and soil temperature, etc.) [11]. The tree species increase the abundance and diversity of shelters and foods for the fauna, enhancing the capacity to attract seed dispersers [12].

In this chapter, we will present the bases and methodological strategies of forest restoration in altered areas of the Atlantic Forest, Brazil. We will also address the potential of biodiversity conservation in protected areas. We will show results of restoration actions in altered areas in a conservation unit, focusing on ecosystem biodiversity and its applicability in other biomes.

2. Fundaments and strategies of forest restoration in Brazil

Actions for forest restoration in Brazil were first introduced in areas affected by works of public interest, especially in places altered by the construction of large hydroelectric plants. Restoration actions were based on the planting of forest species without considering the ecological criteria, including the use of exotic species. These procedures led to the formation of mixed forests with low diversity and potential for land occupation [12]. Later, the original structure of a preserved forest fragment (diversity and successional groups) started to be considered [13]. This analysis served as a basis for the adoption of the recovery method to be applied in the altered environment. Currently, it is proposed that the succession process in restoration areas may occur following multiple trajectories and not a predefined template. These trajectories exhibit a dynamic equilibrium, where each final community will have its phyto-sociological and structural peculiarities according to the environment and the history of the ecosystem usage [14].

Defining a strategy for environmental restoration in altered ecosystems requires accurate indicators of the ecology in the biosystem. These indicators allow the use of specific methodologies for each type of tree formation, ensuring a more effective restoration process, regardless of the recovery speed of the ecosystem, which vary enormously among forest ecosystems [15]. One of the main attributes of ecosystems is their ability to change in time. All ecosystems, terrestrial or aquatic, are subject to natural or human disturbances that inflict changes to a greater or lesser degree in time [16].

An ecosystem can be considered stable when it reacts to a disturbance, keeping a state of dynamic equilibrium [17]. However, a degraded ecosystem has undergone disturbances leading to the decrease of its resilience, with consequent loss of important species and interactions [17]. Resilience is the ability of a system to restore its balance after being disrupted by a condition, that is, its ability to recover [18]. When the ecosystem undergoes severe damage, such as extinction of key-species and intensification of degradation (diseases, erosion, leaching and inbreeding), human intervention is required [19]. According to the authors, this intervention must reverse the degradation processes leveraging local characteristics of auto-restoration.

For the success of a restoration process, some factors must be taken into account, namely the historical use of the area, the degradation intensity, the degree of forest fragmentation and the preservation degree of the surrounding vegetation. Thus, altered environments that have little or no vegetation preserved in their surroundings have less capacity to recover, once, mainly in tropical areas, seed dispersal occurs predominantly by animals. The animals hardly ever leave forested areas toward open agricultural areas [20]. Therefore, the presence of forest remnants facilitates the movement of seed dispersers [21].

Another important factor is the length of time and the intensity of the area use. Areas with consolidated agricultural and livestock activities recover more slowly than areas used for itinerant agriculture for short periods of time. In general, more intensely degraded areas have a seed bank with low diversity, limiting self-healing. Additionally, compacted soils and soils with low natural fertility limit the emergence and growth of seedlings.

Thus, for the recovery of natural ecosystems, success lies in the restoration of ecological processes responsible for the reconstruction and maintenance of a functional community [22]. The authors highlight that the effectiveness of this process depends on the use of high biodiversity involving species of trees, shrubs, vines as well as lianas, in addition to the fauna and the interactions between living beings that inhabit that environment. This diversity can be obtained through direct restoration actions of altered environments and guaranteed over time by the natural dynamics of the community restored [22].

The recovery of a degraded environment can be understood as reconstructions of its function and its structure [23]. According to the author, several optional objectives guide the recovery of a degraded ecosystem, namely: the reproduction of the exact original condition of the site (structure and function); the reproduction of conditions similar the conditions before degradation, enabling the balance of environmental processes; the development of an alternative activity suitable to human use and not simply the reconstitution of the original vegetation, provided this process is carried out to prevent negative environmental impacts; and abandonment, which can lead to a normal succession process or to future degradation if the ecosystem is subject to erosion or other debilitating agent.

3. Biodiversity conservation in conservation units

The Atlantic Forest and the Amazon Rainforest, historically, have had periods of connection, interspersed by periods of isolation. This alternation of isolation and connection with other biotas and the combination with striking geographic factors resulted in high biological diversity and endemic occurrences in these biomes [24]. Although these biomes show immense biological wealth, a significant part of biodiversity is still unknown. Between 1990 and 2006, [25] indicated the discovery of over 1,190 plant species by the scientific community for the Atlantic Forest.

The Atlantic Forest, first region colonized in Brazil, has undergone continuous deforestation and, currently, it is estimated that only 11-16% of the original forest cover remains [26]. Deforestation has resulted in severe changes in ecosystems, especially high fragmentation and degradation of native vegetation and loss of regional species of flora and fauna [27]. In the last decade, the deforestation rate has reduced because of numerous ordinances created to protect the biome, at different levels of government. Among the ordinances, we highlight the "Lei da Mata Atlântica" Law number 11,428 of December 22, 2006, which addresses the use and protection of native vegetation of the Atlantic Forest biome.

The definition of new protected areas represents an important strategy for biodiversity preservation. At the end of the 1990's, Brazil held more than 1,000 public and private conservation units, totaling approximately 76 million hectares [28]. In the Atlantic Forest biome, protected areas also increased during that period. The main category created was of Environmental Protection Areas (EPA). This category represents 91% of the total area of sustainable use in conservation units in the biome [29]. However, whole protection areas are not as effective, as they accept human occupation. In the Atlantic Forest biome, it is estimated that 40% of the area of sustainable use in conservation units is already occupied by human population and has no forest cover. Nevertheless, the whole protection areas account for 88% of its total area covered by preserved natural vegetation.

According to recent data from the Ministry of the Environment [7], Brazil has more than 2,100 conservation units, totaling circa 150 million hectares, which account for 17% of the Brazilian territory. The Amazon biome has approximately 73% of the total cover area in conservation units in Brazil, while the Atlantic Forest contributes to 7% [7]. The Atlantic Forest has approximately 9.7% (107,242 km2) of its territory in protected areas, being only 2% in whole protection conservation units [7].

A large number of rare and/or endangered species, reported on the so-called "red lists" [30, 31], are restricted only to protected areas. Thus, their existence is greatly linked to the future of these conservation units. The Official List of Threatened Species of the Brazilian Flora [31] contains 472 species, four-folds of the previous list of 1992. Of these, 276 species (more than 50%) belong to the Atlantic Forest. The list includes species that have been the most economically exploited over time, such as pau-brasil (*Caesalpinia echinata*), palmito juçara (*Euterpe edulis*), araucaria (*Araucaria angustifolia*), jequitibá (*Cariniana ianeirensis*), jaborandi (*Pilocarpus jaborandi*), xaxim (*Dicksonia sellowiana*), jacarandá-da-bahia (*Dalbergia nigra*), canela-sassafrás (*Ocotea odorifera*) and various orchids and bromeliads.

The Official List of Threatened Species of the Brazilian Fauna [30] contains 633 species, including fish and aquatic invertebrates. Seven species have already been listed as extinct in the wild. Of these endangered species of the Atlantic Forest, 185 are vertebrate species (69.8%

of all threatened species in Brazil), represented by 118 species of birds, 16 amphibians, 38 mammals and 13 reptiles. In addition, there are 59 fish species facing extinction [30]. A significant part of these endangered species is endemic, such as muriqui-do-sul (*Brachyteles arachnoides*), muriqui-do-norte (*Brachyteles hypoxanthus*) and papagaio-da-cara-roxa (*Amazona brasiliensis*).

4. Restoration of the Atlantic Forest

The Atlantic Forest is one of the so-called global hotspots of biodiversity. The biome comprises 34 regions with high richness of endemic species; however, it is seriously threatened by significant loss of forest cover [32]. Originally distributed in more than 1.3 million km^2, in the eastern side of Brazil, the Atlantic Forest is home to at least 60% of the Brazilian population and approximately 70% of the national GDP is concentrated in this region of the country. Currently, there are degraded and isolated forest fragments with predominantly less than 50 ha [26]. Restoration of altered and fragmented areas is essential to ensure biodiversity maintenance, because protected areas must be connected to suit the functionality of the landscape ecosystem.

In this sense, we conducted a study on the Parque Estadual Quarta Colônia (Figure 1), a state protected area that was created in 2005. Its creation is attributed to a compensatory measure for the construction of a hydroelectric power plant. It is the largest established conservation unit of the "Deciduous Seasonal Forest" in the central region of Rio Grande do Sul State (1,847 ha), part of the Atlantic Forest Biome. A park is a category that aims the conservation of the ecosystem characteristic of a region and the practice of environmental education and recreation [4]. The Parque Estadual Quarta Colônia houses a floral species, *Dyckia agudensis* Irgang & Sobral (Figure 2), seriously threatened of extinction [33]. This species is lithophyte growing on basaltic formations among xerophytic vegetation. *D. agudensis* is at risk of extinction due to habitat fragmentation caused by agricultural activities in the surroundings.

Rugged-to-flat relief comprises the topography of the conservation unit. In some areas of the unit, there used to be small rural properties that were expropriated during the construction of the dam. There are several altered areas, decommissioned parts of the construction site and functional facilities of the plant power. Even before the creation of the conservation unit, some degraded areas of the park had been recovered with the planting of seedlings of native species in the year 2001. Other altered areas that were once abandoned are currently in early regeneration stages, influenced by the natural forest matrix in the surroundings.

The most preserved areas of the park feature a succession mosaic due to anthropogenic interference in the area. The vegetation is classified as medium-to-advanced stage of secondary succession. The early secondary species contribute to greater diversity and the late secondary species appear less pronounced. Understory species possess the greatest number of individuals and have a characteristic occupation of greater range of luminosity. Thus, they suffer greater influence of soil variables in the definition of ecological niches of plant species [34].

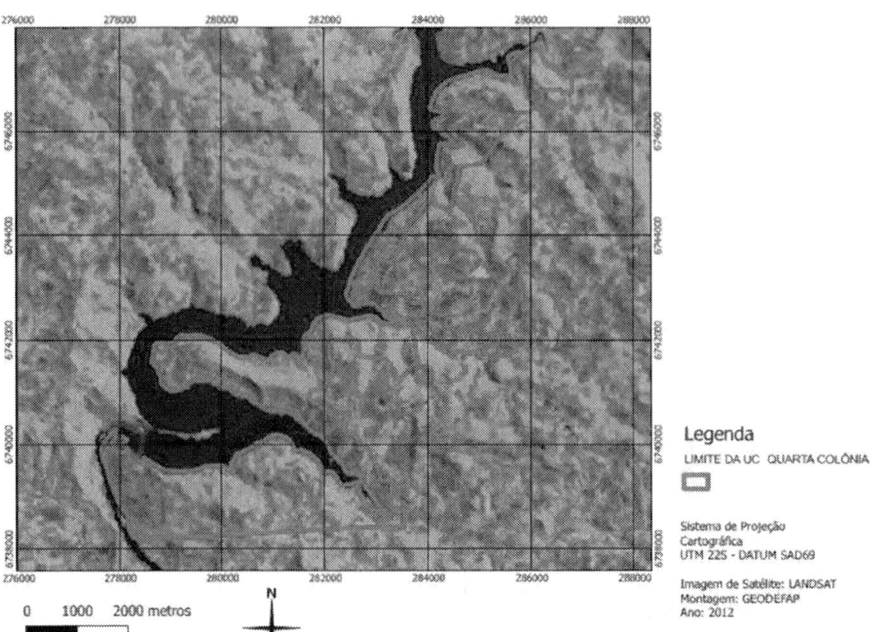

Figure 1. Limits of the conservation united (Parque Estadual Quarta Colônia), southern Brazil.

Figure 2. Species critically endangered of extinction (*Dyckia agudensis* Irgang & Sobral) found in the area of the Parque Estadual Quarta Colônia. Photo 2: Büncher (2011) - Digital Flora of Rio Grande do Sul.

However, for an ecosystem to be considered restored, it is necessary to analyze its biodiversity and compare it with preserved environments. Based on the principles of the Society for Ecological Restoration [35], a restored ecosystem should present diversity and structure similar to a reference ecosystem. Diversity is commonly measured by determining the richness and abundance of organisms. Similar to the specific composition of species, the vegetation structure is usually analyzed by its density, biomass, and canopy coverage or by structural aspects of the vegetation, and these measurements are useful to predict the direction of plant succession [36]. Additionally, ecological processes, such as nutrient cycling and soil enzymatic activity [37], are related to stabilization and soil fertility [38]. In the same region of the conservation unit, [10] found that with leaf deposition of the leguminous tree *Parapiptadenia rigida*, soil nutrients returned to 32.4 kg ha^{-1} yr^{-1} of Ca, followed by N (26.1), K (3.2), Mg (2.1), S (1.3) and P (1.0 kg ha^{-1} yr^{-1}).

From 2010 onwards, some attributes were evaluated to verify the recovery of degraded environments. We analyzed the vegetation structure, the diversity and ecological processes, which served as a parameter to evaluate different areas under restoration in the conservation unit. We initially characterized the vegetation in relation to environmental variables of the reference ecosystem to evaluate and monitor the areas under the restoration process. The reference area has been free of anthropic interventions for more than 20 years, and before that, there used to be small farms in the less steep slopes. Currently, it forms a mosaic of different successional stages [39].

Two groups of species composition characterize the forest. One group consists of understory species, which, due to the smaller size, establish on sloping and stable terrain. In this group, *Trichilia clausseni* is the indicator species that exerts a strong influence on forest succession and on the community due to its high density and frequency in forest regeneration. The other group of species is formed by *Nectandra lanceolata* and *Nectandra megapotamica* as dominant in the forest structure. In addition, early secondary species such as *Cupania vernalis*, *Ocotea puberula* and *Casearia sylvestris* indicate the dynamics of clearings in the area.

The monitoring of restoration was carried out in different altered areas of the Parque Estadual Quarta Colônia. The areas monitored (A1 and A2), both with seven years of planting, feature the following characteristics:

A1 – covers an area of 2.21 ha. It is a reminiscent of ancient successive crops of tobacco (*Nicotiana tabacum* L.) with about 560 m of the reference area. In the planting, 12 species were used with five pioneers (*Schinnus terebinhtifolius* Raddi, *Inga vera* Willd., *Parapiptadenia rigida* (Benth.) Brenan, *Ateleia glazioviana* Baill., *Psidium cattleyanum* L.) and seven early secondary (*Prunus myrtifolia* (L.) Urb., *Vitex megapotamica* (Spreng.) Moldenke, *Cedrela fissilis* Vell., *Ficus luschnathiana* (Miq.) Miq., *Luehea divaricata* Mart., *Peltophorum dubium* Sprengel. e *Ocotea puberula* (Rich.)). The soil was prepared by means of subsoiling at an average depth of 35 cm. Afterwards, trenches were opened along the lines of grooves, and seedlings were planted at spacing of 2.5 m x 2.5 m. Cultural practices were performed for a period of 24 months.

A2 – covers an area of 2.27 ha. It is about 615 m of the reference area and 78 m far from a slope area with secondary forest in the middle stage of succession. The soil was compacted with presence of construction waste (75% of particle size > 200 mm) [40], originating from the demolition of old facilities. In this region, 24 species were planted, being five pioneers (*Parapiptadenia rigida*, *Psidium cattleyanum*, *Schinus terebinthifolius*, *Enterolobium contortisiliquum* (Vell.) Morong, *Calliandra brevipes* (Spreng.) J. F. Macbr), 15 early secondary species (*Allophylus edulis* (A.St.-Hil., Cambess. & A. Juss.) Radlk., *Strychnos brasiliensis* (Spreng.) Mart.,

Cordia americana (L.) Gottshling & J.E.Mill., *Luehea divaricata, Peltophorum dubium, Cedrela fissilis* Vell., *Schizolobium parahyba, Cabralea canjerana* (Vell.) Mart, *Handroanthus heptaphyllus, Handroanthus chrysotrichus, Jacaranda micrantha, Eugenia uniflora, Campomanesia xanthocarpa* O. Berg, *Vitex megapotamica, Cordia trichotoma* (Vell.) Arráb. ex Stend.) and four late secondary species (*Ficus luschnathiana, Eugenia involucrata* DC, *Annona rugulosa* (Schltdl.) H. Rainer, *Myrcianthes pungens* (O. Berg.) D. Legrand). The spacing used was 4 m x 4 m with planting in trenches without subsoiling and cultural practices.

The structure of the vegetation diversity and enzymatic activity of the soil are significantly different between the areas under the restoration process and the reference area. It is observed, for example, when comparing the high proportion of pioneering species and reduced basal area growth in areas A1 and A2 (Table 1).

	Arboreal Component			Natural Regeneration		
	A1	A2	RA	A1	A2	RA
Age (years)	7	7	± 20	-	-	-
Planting space (m)	2 x 2	4 x 4	-	-	-	-
Average height (m)	3.15	4.30	9.30	0.44	1.00	2.80
Basal area (m²/ha⁻¹)	4.13	4.27	27.13	-	-	-
Density (plants.ha⁻¹)	1,741	297	3,408	23,333	11,388	15,909
Canopy cover (%)	109.3	35.7	-	-	-	-
Richness	19	29	49	21	16	42
Diversity (H')	2.31	2.86	3.00	2.23	2.29	2.60
Equability (J')	0.78	0.85	0.78	0.73	0.82	0.69
Zoochoric plants (%)	58.0	70.0	75.2	62.0	56.2	79.5
Anemochoric plants (%)	42.0	30.0	24.8	38.0	43.7	20.5
Sucessional group (% P:NP)	47:53	48:52	18:82	52:48	44:56	12:88
Exotic species (%)	26.3	17.2	0	28.5	31.2	0

Where: (% P:NP)=percentage of pioneer species (P) in relation to non-pioneer species (NP).

Table 1. Structure and diversity in Subtropical Seasonal forest natural area (RA) and areas under restoration process (A1 and A2) in Parque Estadual Quarta Colônia.

Area A1 is in intermediate stage in relation to the other two areas. In area A2, the low density of plants is related to three factors: larger planting spacing; presence of restrictive layers to root growth; and lack of management after planting. The absence of weed control favored the permanent presence of invasive grasses, competing with arboreal individuals and preventing the development of some native species. The lower initial spacing in area A1 provided higher density of plants in the area, which resulted in the rapid canopy coverage in relation to area A2.

Area A1 has a higher possibility of achieving the objectives of restoration due to increased canopy coverage. The greater shading of the canopy enabled grass reduction and, consequently, the establishment of a greater number of regenerating individuals. The vegetation cover controls the quantity, quality and distribution of light, influencing the growth and survival of seedlings and determining vegetable composition [41]. The importance of richness of tree species and regeneration in the areas undergoing restoration was lower than that observed in the reference area (RA). Because the RA represents secondary forest, attract avifauna, which favors forest development. However, there is a need to manage the areas through the eradication of exotic species. The exotic species with most occurrence medium-to-advanced stage of succession, may display predominance of some species, resembling the diversity index of a deployed area.

In natural regeneration, the floristic richness of the RA was enough to enable the development of various species, allowing a higher diversity index in relation to areas under restoration. This was attributed to the increased shading and flow of diaspores of species in reproductive stage. In the regeneration process of areas A1 and A2, zoochoric pioneering species predominated, with great capacity to was *Psidium guajava*, pioneer species with zoochoric dispersal (Table 2).

Local	Specie	AD	AF	SG
	Allophylus edulis (A.St.-Hil., et al.) Hieron. Ex Niederl.	958.3	58.3	ES
	Cupania vernalis Cambess.	2750.0	83.3	ES
RA	*Gymnanthes concolor* (Spreng.) Müll. Arg.	6000.0	83.3	ES
	Trichilia clausseni C.DC.	1166.7	58.3	ES
	Trichilia elegans A. Juss.	1083.3	75.0	LS
	Hovenia dulcis Thunb.*	2916.7	16.7	P
	Inga vera Willd.	1805.6	27.8	P
A1	*Ocotea puberula* (A.Rich.) Ness	1527.8	33.3	P
	Psidium guajava L.*	1944.4	38.9	P
	Schinnus terebinthifolius Raddi	8750.0	61.1	P
	Caliandra brevipes (Spreng.) J. F. Macbr.	1805.6	33.3	P
	Cordia trichotoma (Vell.) Arráb. ex Stend.	833.3	5.6	ES
A2	*Pinus elliottii* Engelm.*	694.4	11.1	P
	Psidium guajava L.*	3333.3	44.4	P
	Syzygium cumini (L.) Skeels*	1388.9	11.1	P

AD: Absolute Density; AF: Absolute Frequency; SG: Succession Group; P: Pioneer; ES: Early Secondary; LS: Late Secondary. *Exotic Species.

Table 2. Five species better ranked in the regeneration process in the natural reference area (RA) and in areas under restoration process (A1 and A2).

Soil enzymes (amidase, urease, acid phosphatase, and arysulfatase) in the RA, at 0-5 cm of depth, presented higher values than those in restoration areas. Ground cover possibly influenced the enzymatic activity, since the restoration areas feature the presence of invasive grasses. However, in the RA, we can observe a dense layer of litterfall, which can reach 10.9 Mg ha^{-1} [42] in this type of forest formation. Still, the enzymatic activity is observed in all areas, although with significant differences between the restoration areas and the RA.

In area A2, a significant regeneration was verified under the canopy of *Inga vera*. This indicates that the *Inga* is a key species or a facilitator in the process of ecological restoration. In the restoration process of degraded areas, facilitators are species that, at an early stage of succession, alter the conditions of the community, allowing better establishment of other species [43]. A species capable of forming aggregates of other species is considered a facilitator. The colonization processes that occur in the surroundings of this species are called nucleation [44], which occurs mainly by zoochoric dispersal [23].

Among several facilitators, *Inga vera* stands out by offering features that promote an improvement in environmental conditions, namely large tree crowns, rapid growth [23] and biological nitrogen fixation [45]. Its fruit is a hairy yellowish pod, measuring from 4-12 cm long with white pulp, sweet and edible, which makes it attractive to frugivorous animals, allowing zoochory [46]. The *Inga* is classified as a pioneer species in the ecological succession group. It has wide geographical distribution and is found mainly in the Atlantic Forest biome in Brazil.

Therefore, we evaluated regeneration under the canopy of *Inga vera* (50 plants) 10 years after the planting in area A2. We identified the presence of 756 individuals, belonging to 47 species and distributed among 25 families. The families Fabaceae (five species; 183 individuals), Myrtaceae (four species; 157 individuals) and Solanaceae (four species; five individuals) were the most representative in natural regeneration. Table 3 shows the main species found in natural regeneration.

Regarding the ecological groups, 26 species are pioneers (55.3%), 16 early secondary (34%) and two late secondary (4.3%), two unidentified (4.3%) and a mix of ES/LS (2.1%). In terms of seed dispersion, 29 species are zoochoric (61.7%), 13 are anemochoric (27.7%), two barochoric (4.3%), two unidentified (4.3%) and one authochoric (2.1%).

The Shannon diversity index (H′) found was medium (2.68). For high diversity, the index must be greater than 3.0; medium, between 3.0 and 2.0; low, between 2.0 and 1.0 and very low, smaller than 1.0 [47]. The Pielou evenness index (J′) was 0.7. This value indicates that some species have high densities, and others have few individuals [48]. The species *Ligustrum lucidum* (21.7%), *Inga vera* (17.9%), *Syzygium cumini* (12.8%), *Baccharis semiserrata* (7.7%), *Psidium guajava* (6.35%) and *Allophylus edulis* (5.3%) altogether represented 71.7% of the density of natural regeneration (542 individuals), a fact that explains the low evenness. The analysis of the index of importance value (IIV) shows that the species *Ligustrum lucidum* (30.2%), *Inga vera* (27.4%), *Syzygium cumini* (19.2%), *Baccharis semiserrata* (18%), *Psidium guajava* (14.9%) and *Allophylus edulis* (12.7%) have the highest values.

The species *Inga vera* and *Allophylus edulis* had fruits most attractive to frugivorous animals [46, 49]. The presence of *Inga* in the degraded area, for its characteristics, has the ability to form

Scientific name	Ni	Np	D (cm)	H (m)	RD	RF	IIV	EG	Disp.
Ligustrum lucidum W. T. Aiton*	164	24	1,92	1,89	21,69	8,51	30,20	P	Zoo
Inga vera Willd.	135	27	3,00	2,19	17,86	9,57	27,43	P	Zoo
Syzygium cumini (L.) Skeels.*	97	18	1,40	1,08	12,83	6,38	19,21	P	Zoo
Baccharis semiserrata DC.	58	29	2,22	2,16	7,67	10,28	17,96	P	Anemo
Psidium guajava L.*	48	24	2,84	2,52	6,35	8,51	14,86	P	Zoo
Allophylus edulis (A.St.-Hil.) Radlk.	40	21	1,53	1,74	5,29	7,45	12,74	SI	Zoo
Calliandra brevipes Benth.	39	9	2,00	1,76	5,16	3,19	8,35	P	Anemo
Cupania vernalis Cambess.	24	15	1,31	1,10	3,17	5,32	8,49	SI	Zoo
Nectandra megapotamica Mez	13	6	1,74	2,00	1,72	2,13	3,85	SI	Zoo
Schinus terebinthifolius Raddi	13	12	2,20	2,04	1,72	4,26	5,97	P	Zoo

Ni=number of individuals; Np=number of plots; D=diameter at 5 cm above the soil (cm); H=height (m); RD=relative density; RF=relative frequency; IIV=index of importance value; EG=ecology group; Disp=dispersion; Zoo=zoochoric; Anemo=anemochoric; Baro=barochoric; Auto=authochoric; *exotic species.

Table 3. Main species found in natural regeneration under the canopy of *Inga vera*, in the restoration area.

nuclei of native and exotic species. This formation is mostly attributed to its great attraction to frugivorous vertebrates, primarily birds and bats. Its main attractive features for the fauna are the fleshy and sweet fruits. In addition, the *Inga* species is capable of forming a large crown, serving as a natural perch for birds that end up defecating or regurgitating in the site.

The negative aspect observed in the study was the presence of exotic species, which account for 19.1% of the total number of species in the regeneration areas, however with 43.6% of the number of individuals. It is highlighted the presence of *Ligustrum lucidum, Syzygium cumini* and *Psidium guajava*, invasive exotic species with high zoochoric seed dispersion and a high number of individuals (309) and density (40.9%). Conservation units with total protection should be representative of native species and ecosystems, therefore, the existence of invasive exotic species is not desirable nor permitted [50]. The main management strategies involve the eradication and/or control to contain the spread of exotic species, reducing their abundance and their density and/or mitigating their impacts [51].

Additionally, it is possible to affirm that the two areas under restoration (A1 and A2) are returning to natural succession, given that the diversity, structure and ecological processes show a growing trend in relation to the RA. The enzymatic activity can be considered a good indicator of ecological restoration, evidencing that the two areas under restoration resumed the succession process. However, to allow a greater complexity of the ecosystem, the areas should be managed to remove the exotic species.

5. Conclusion

The effectiveness of ecological restoration is largely attributed to the resilience capacity of an ecosystem, to the restoration actions and to the monitoring of recovery indicators. In this sense, the focus on ecological restoration should take into account that the areas are part of an integrated system, requiring the knowledge of its structure and functions for its sustainability, as well as the individual role of each species, especially those that play a fundamental role in strong interactions and in the resumption of ecological succession.

Temporal analysis of ecosystem attributes comprises the basis for the evaluation of the restoration process, also for the comparison of the speed and direction of its performance in different environments and geographical regions. The use of smaller spacing enables faster recovery of altered areas, because the plants shade the soil more quickly, reducing competition for invasive exotic grasses.

It is essential to take into consideration the performance of key species and the arrangements of functional species, because they keep the ecosystem balanced on several levels, both biotic and abiotic. This fact prevents exotic species from becoming invasive by occupying an ecological emptiness (absence of natural predators and competitors) and from settling in areas under the restoration process.

Author details

Suzane Bevilacqua Marcuzzo and Márcio Viera*

*Address all correspondence to: marcio.viera@ufsm.br

Universidade Federal de Santa Maria, Silveira Martins, Brasil

References

[1] Allen C.D., Macalady A.K., Chenchouni H., Bachelet D., McDowell N., Vennetier M., Kizberger T., Rigling A., Breshears D.D., Hogg E.H., Gonzalez P., Fensham R., Zhang Z., Castro J., Demidova N., Lim J.H., Allard G., Running S.W., Semerci A., Cobb N. A global overview of drought and heat-induced tree mortality reveals emerging climate change risks for forests. Forest Ecology and Management 2010; 259(4): 660-684. http://www.sciencedirect.com/science/article/pii/S037811270900615X (accessed 15 April 2014).

[2] Asner G.P., Rudel T.K., Aide T.M., Defries R., Emerson R. A contemporary assessment of change in humid tropical forests. Conservation Biology 2009; 23(6):

1386-1395.   http://onlinelibrary.wiley.com/doi/10.1111/j.1523-1739.2009.01333.x/full (accessed 15 July 2014).

[3] Holl K.D., Aide T.M. When and where to actively restore ecosystems? Forest Ecology and Management 2011; 261(10): 1558-1563. http://www.sciencedirect.com/science/article/pii/S0378112710003750 (accessed 2 June 2014).

[4] IUCN, International Union for Conservation of Nature. Global Protected Areas Programme Species. Protected Areas Category II. 2014 http://www.iucn.org/about/work/programmes/gpap_home/gpap_quality/gpap_pacategories/gpap_pacategory2   (accessed 2 July 2014).

[5] Hoeflich E.E. An easy win. World Conservation 2011; 41(1): 12-13. http://cmsdata.iucn.org/downloads/wcm_magazine_eng.pdf (accessed 7 July 2014).

[6] Dudley, (https://portals.iucn.org/library/efiles/documents/PAG-021.pdf) Dudley, N. Ed. Guidelines for Applying Protected Area Management Categories. Gland: IUCN; 2008.

[7] MMA, Ministério do Meio Ambiente. Espécies ameaçadas de extinção. http://www.mma.gov.br/biodiversidade. (accessed 4 July 2014)

[8] Marcuzzo, SB. Métodos e espécies potenciais à restauração de áreas degradadas no Parque Estadual Quarta Colônia, RS. Thesis. Universidade Federal de Santa Maria; 2012.

[9] Schumacher M.V., Trüby P., Marafiga J.M., Viera, M., Szymczak D.A. Espécies predominantes na deposição de serapilheira em fragmento de Floresta Estacional Decidual no Rio Grande do Sul. Ciência Florestal 2011; 21(3): 479-486. http://cascavel.ufsm.br/revistas/ojs-2.2.2/index.php/cienciaflorestal/article/view/3805   (accessed 21 July 2014).

[10] Marafiga J.S., Viera M., Szymczak D.A., Schumacher M.V., Trüby P. Deposição de nutrientes pela serapilheira em um fragmento de Floresta Estacional Decidual no Rio Grande do Sul 2012; 59(6): 765-771. http://www.scielo.br/pdf/rceres/v59n6/05.pdf (accessed 24 July 2014).

[11] Brancalion H.P., Rodrigues R.R., Gandolfi S., Kageyama P.Y., Nave A.G., Gandara F.B., Barbosa L.M., Tabarelli M. Instrumentos legais podem contribuir para a restauração de Florestas Tropicais biodiversas. Revista Árvore 2010; 34(3): 455-470. http://www.scielo.br/pdf/rarv/v34n3/a10v34n3.pdf (accessed 11 July 2014).

[12] Rodrigues RR., Brancalion PHS., Isernhagen I. Pacto pela restauração da mata atlântica: referencial dos conceitos e ações de restauração florestal. São Paulo: LERF/ESALQ; 2009.

[13] Barbosa K.C., Pizo M.A. Seed Rain and Seed Limitation in a Planted Gallery Forest in Brazil. Restoration Ecology 2006; 14(4): 504-515. http://onlinelibrary.wiley.com/doi/10.1111/j.1526-100X.2006.00162.x/pdf (accessed 2 July 2014).

[14] van Andel A. Challanges for ecological theory. In: van Andel A., Aronson J. (ed.) Restoration Ecology: the new frontier. Blackwell Publishing, Oxford, 2005. p223-233.

[15] Holl K.D. Oldfiel vegetation succession in the Neotropics. In: Hobbs R.J., Cramer V.A. (ed.) Old fields. Washingon: Island Press; 2007. p93-117.

[16] Uhl C., Nepstad D., Buschbacher R., Clark K., Kauffman B., Subler S. Studies of ecosystem response to natural and anthropogenic disturbances provide guidelines for designing sustainable land-use systems in Amazonia. In: Anderson A.B. (ed.). Alternatives to deforestation: steps toward sustainable use of the Amazon rain forest. New York: Columbia University Press; 1990. p24-42.

[17] Gandolfi S., Joly C.A., Rodrigues R.R. Permeability-Impermeability: canopy trees as biodiversity filters. Scientia Agrícola 2007; 64(4): 433-438. http://www.scielo.br/pdf/sa/v64n4/14.pdf (accessed 11 July 2014).

[18] Gunderson L.H. Ecological resilience-in the theory and application. Annual Review of Ecology and Systematics 2000; 31(1): 425-439. http://www.annualreviews.org/doi/pdf/10.1146/annurev.ecolsys.31.1.425 (accessed 11 July 2014).

[19] Rudolf W.H.V., Lafferty D.K. Stage structure alters how complexity affects stability of ecological networks. Ecological Letters 2011; 14(1): 75-79. http://onlinelibrary.wiley.com/doi/10.1111/j.1461-0248.2010.01558.x/full (accessed 12 July 2014).

[20] Jordano P., Galetti M., Pizo MA., Silva W.R. Ligando frugivoria e dispersão de sementes à biologia da conservação. In: Rocha C.F., Bergallo H.G., Sluys M.V., Alves M.A.S. (ed.) Essências em Biologia da Conservação. Rio de Janeiro: RIMA Editora; 2006. p411-436.

[21] Harvey C.A., Komar O., Chazdon R., Ferguson B.G., Finegan B., Griffith D.M., Martínes-Ramos M., Morales H., Nigh R., Soto-Pinto L., Breugel M.V., Wishnie M. Integrating agricultural landscapes with biodiversity conservation in the Mesoamerican hotspot. Conservation Biology 2008; 22(1): 8-15. http://onlinelibrary.wiley.com/enhanced/doi/10.1111/j.1523-1739.2007.00863.x/ (accessed 11 July 2014).

[22] Gandolfi S., Rodrigues R.R., Martins S.V. Theoretical bases of the Forest ecological restoration. In: Rodrigues R.R., Martins S.V. High diversity Forest restoration in degraded areas. New York: Nova Science Publishers; 2007. p280.

[23] Carpanezzi A.A. Fundamentos para a reabilitação de ecossistemas florestais. In: Galvão A.P.M., Silva V.P. Restauração florestal: fundamentos e estudos de caso. Colombo: Embrapa; 2005. p27-46.

[24] Fernandes A.M. Fine-scale endemism of Amazonian birds in a threatened landscape. Biodiversity and Conservation 2013; 22(11): p2683-2694. http://link.springer.com/article/10.1007/s10531-013-0546-9 (accessed 01 August 2014).

[25] Sobral M., Stehmann J.R. Na analysis of new angioperm species discoveries in Brazil (1990-2006). Taxon 2009; 58(1): 227-231. http://www.iapt-taxon.org/index_layer.php?page=s_taxon (accessed 12 July 2014).

[26] Ribeiro M.C., Metzger J.P., Martensen A.C., Ponzoni F.J., Hirota M.M. The Brazilian Atlantic Forest: How much is left, and how is the remaining forest distributed? Implications for conservation. Biological Conservation 2009; 142(6): 1141–1153. http://www.sciencedirect.com/science/article/pii/S0006320709000974 (accessed 01 August 2014).

[27] Canale G.R., Peres C.A., Guidorizzi C.E., Gatto C.A.F., Kierulff M.C.M. Pervasive defaunation of forest remnants in a tropical biodiversity hotspot. Plos One 2012; 7(8) e41671 http://www.plosone.org/article/info%3Adoi%2F10.1371%2Fjournal.pone.0041671 (accessed 01 August 2014).

[28] Fonseca M., Lamas I., Kasecker T. O papel das unidades de conservação. Scientific American; 2009. p18-23.

[29] Cunha A.A. Expansão da rede de unidades de conservação da Mata Atlântica e sua eficácia para a proteção das fitofisionomias e espécies de primatas: análises em sistemas de informação geográfica. Thesis. Universidade Federal de Minas Gerais; 2010.

[30] MMA, Ministério do Meio Ambiente. Instrução Normativa MMA nº 03, de 27 de maio de 2003-Lista Oficial das Espécies da Fauna Brasileira Ameaçadas de Extinção. 2003. http://www.mma.gov.br/estruturas/179/_arquivos/179_05122008034002.pdf (accessed 12 July 2014).

[31] MMA, Ministério do Meio Ambiente. Instrução Normativa MMA nº 06, de 23 de setembro de 2008-Reconhece como espécies da flora brasileira ameaçada de extinção. 2008. http://www.mma.gov.br/estruturas/179/_arquivos/179_05122008033615.pdf (accessed 12 July 2014).

[32] Mittermeier R.A., Gil P.R., Hoffman M., Pilgrim J., Brooks T., Mittermeier R.A., Lamoreux J., Fonseca G.A.B. Hotspots revisited: earth's biologically richest and most endangered terrestrial ecoregions. Cidade do México: Cemex & Agrupacion Sierra Madre; 2004.

[33] Martinelli G., Moraes A.M. Livro vermelho da Flora do Brasil. (1° ed.) Rio de Janeiro: CNCFLORA; 2013.

[34] Lieberman M., Lieberman D., Peralta R., Hartshorn G.S. Canopy closure and the distribution of tropical forest tree species at La Selva, Costa Rica. Journal of Tropical Ecology 1995, 11(2): 161-178.http://journals.cambridge.org/action/displayAbstract?fromPage=online&aid=5252900&full textType=RA&fileId=S0266467400008609 (accessed 3 July 2014).

[35] SER, Society For Ecological Restoration International and Policy Working Group. The SER International Primer on Ecological Restoration International. 2004. http://www.ser.org/pdf/primer3.pdf (accessed 4 June 2014)

[36] Engel V.L., Parrotta J.A. Definindo a restauração ecológica: tendências e perspectivas mundiais. In: Kageyama P.Y., Oliveira R.E., Morais L.F.D., Engel V.L., Gandara F.B. (ed.). Restauração ecológica de ecossistemas naturais. Botucatu: FEPAF; 2008. p1-26.

[37] Han X., Tsunekawa A., Tsubo M., Shao H. Responses of plant-soil properties to increasing N deposition and implication for large-scale eco-restoration in the semiarid grassland of the northern Loess Plateau, China. Ecological Engineering 2013; 60(1): 1-9. http://www.sciencedirect.com/science/article/pii/S0925857413003042 (accessed 01 August 2014).

[38] Rodríguez-Loinaz G., Onaindia M., Amezaga I., Mijangos I., Garbisu C. Relationship between vegetation diversity and soil functional diversity in native mixed-oak forests. Soil Biology & Biochemistry 2008; 40(1): 49-60. http://www.sciencedirect.com/science/article/pii/S0038071707001745 (accessed 6 June 2014)

[39] Marcuzzo S.B., Araújo M.M., Longui S.J. Estrutura e relações ambientais de grupos florísticos em fragmento de Floresta Estacional Subtropical. Revista Árvore 2013; 37(2): 275-287. http://www.scielo.br/pdf/rarv/v37n2/09.pdf (accessed 4 June 2014)

[40] Pedron F.A., Fink J.R., Dalmolin R.S.D., Azevedo A.C. Morfologia dos contatos entre solo-saprolito-rocha em Neossolos derivados de arenitos da Formação Caturrita no Rio Grande do Sul. Revista Brasileira de Ciência do Solo 2010; 34(6): 1941-1950. http://www.scielo.br/pdf/rbcs/v34n6/19.pdf (accessed 3 June 2014)

[41] Melo A.C.G. Guia para monitoramento de reflorestamentos para restauração. Circular Técnica, n.1, São Paulo, 2010. (Projeto Mata Ciliar).

[42] Viera M, Caldato SL, Rosa SF, Kanieski MR, Araldi DB, Santos R, et al. Nutrientes na serapilheira em um fragmento de floresta estacional decidual, Itaara, RS. Ciência Florestal 2010; 20(4): 611-619. http://cascavel.ufsm.br/revistas/ojs-2.2.2/index.php/cienciaflorestal/article/view/2419 (accessed 4 July 2014).

[43] Ricklefs R.E.A Economia da Natureza: um livro-texto em ecologia básica. Rio de Janeiro: Guanabara/ Koogan, 2003. 3. ed, p357-358.

[44] Baylão Júnior H.F., Valcarcel R., Roppa C. Influência da regeneração espontânea no processo de restauração ecológica sob copa de Guarea guidonia (L.) Sleumer em ecossistemas perturbados na Serra do Mar, Pirai-RJ. In: Simpósio Nacional de Recuperação de Áreas Degradadas, 2012, Rio de Janeiro, 2012.

[45] Faria S.M.D., Franco A.A. Identificação de bactérias eficientes na fixação biológica de nitrogênio para espécies leguminosas arbóreas. Seropédica: Embrapa Agrobiologia, 2002. 16p. (Embrapa Agrobiologia. Documentos, 158). http://ainfo.cnptia.embrapa.br/digital/bitstream/CNPAB-2010/27435/1/doc158.pdf (accessed 2 July 2014).

[46] Carvalho P.E.R. Espécies arbóreas brasileiras. Colombo: Embrapa Floresta; 2008.

[47] Cavalcanti E.A.H., Larrazábal M.E.L.D. Macrozooplâncton da zona econômica exclusiva do Nordeste do Brasil (segunda expedição oceanográfica – REVIZEE/NE II) com ênfase em Copepoda (Crustaceae). Revista Brasileira de Zoologia 2004; 21(3): 467-475. http://www.scielo.br/pdf/rbzool/v21n3/21895.pdf (accessed 11 July 2014).

[48] Budke J.C., Giehl E.L.H., Athayde E.A., Eisinger S.M., Záchia R.A. Florística e fitossociologia do componente arbóreo de uma floresta ribeirinha, arroio Passo das Tropas, Santa Maria, RS, Brasil. Acta Botanica Brasilica 2004; 18(3): 581-589. http://www.scielo.br/pdf/abb/v18n3/v18n3a16.pdf (accessed 10 July 2014).

[49] Carvalho P.E.R. Espécies arbóreas brasileiras. Colombo: Embrapa Floresta; 2006.

[50] BRASIL. Lei nº 9985 de 18 de julho de 2000. Institui o Sistema Nacional de Unidades de Conservação da Natureza e dá outras providências. Diário Oficial da União, Brasília, DF, 19 jul. 2000. http://www.planalto.gov.br/ccivil_03/leis/l9985.htm (accessed 11June 2014).

[51] Ziller S.R. Espécies exóticas da flora invasora em Unidades de Conservação. In: Campos J.B., Tossulino M.D.G.P., Müller C.R.C. (ed.). Unidades de Conservação: ações para valorização da biodiversidade. Curitiba: Instituto Ambiental do Paraná, 2005. p34-52.

Chapter 21

Actions for the Restoration of the Biodiversity of Forest Ecosystems in Cuba

Eduardo González Izquierdo, Juan A. Blanco, Gretel Geada López, Rogelio Sotolongo Sospedra, Martín González González, Barbarita Mitjans Moreno, Alfredo Jimenez González and José Sánchez Fonseca

Additional information is available at the end of the chapter

http://dx.doi.org/10.5772/59333

1. Introduction

Human will and interests have used landscapes without limits for different purposes, usually for the economic benefit of a minority [1]. Nowadays, the Earth is threatened daily by the degradation of its ecosystems due to fragmentation. One of the main consequences is biodiversity loss. Despite the economic progress and conservation actions carried out in many countries, the planet is losing genuine tropical forest, which is distributed mainly in the "low and middle income countries". The reasons are diverse: inappropriate use of extractive practices in forestry related to wood and non-wood products, land use change when clearing the forest for agriculture and cattle ranching, tourism development, and others. These reasons have facilitated the introduction of new species that then behave as invasive species [2], which usually produce strong competition with local species, reducing biomass and the forest's productivity.

Ecological restoration of disturbed areas is one of the most important and complex issues that forestry faces, due to the lack of knowledge on the ecological functioning of populations, communities, ecosystems, and natural landscapes. In addition, we must consider other components such as social, political, and economic interests of the local communities [3]. In ecological restoration, we need to keep in mind four elements as priorities: to develop the conservation of biodiversity; to maintain human use; to empower the local communities in the management of the area; and, at the same time, to improve the productivity of an ecosystem. Thus, ecological restoration can be considered the main component for conservation and for sustainable management programmes, particularly in tropical areas [4].

Ecosystems can restore themselves if there are no barriers (biotic or abiotic) that hamper the natural process of passive restoration (natural succession). When ecosystems are too degraded they will not overcome such a state over time, and consequently it will be necessary to implement actions to move the ecosystem towards a succession pathway. This is denominated active or assisted restoration [5]. To improve the design of such active restoration programmes, the study of the vegetable communities contributes important data about the phenological and demographical patterns of species, and species suitable for replacement. Such knowledge will allow estimations of how easy it might be to recover and develop the affected ecosystems [6], and how effective the assisted restoration actions might be.

In this chapter we present the results obtained during the restoration of three tropical forests in Cuba: 1) the mesophyll semi-deciduous forest in the western sector of the Biosphere Reserve"Sierra del Rosario" (BRSR); 2) the riverside forest of the Cuyaguateje River in western Cuba; and 3) the exploited native rainforests of the sector Quibiján-Naranjal of the River Toa in eastern Cuba. The BRSR presents a high variety of ecosystems. Several vegetable formations can be distinguished in the reserve [7–8]: evergreen forest, semi-deciduous forest, and pine forests. The largest formation is semi-deciduous forest, with 40–65% of deciduous trees, shrubs, herbs, scarce epiphytes, and an abundance of climbers. The predominant variant is the mesophyll semi-deciduous forest [9]. The 22 ha of the riverside forest of the Cuyaguateje River belonging to the cooperative "Menelao Mora" is placed in the Pinar del Río province. The forest is classified as a typical riverside forest with an arboreal stratum of 15–20 m of shrubs, herbs, scarce epiphytes, and climbers [10]. The native rainforests of the sector Quibiján-Naranjal of Toa are considered to be true rainforests in Cuba [11], where there are not deciduous elements with an abundance of epiphytes with two main tree strata from 20–25 m and 8–15 m.

1.1. Problem statement

The BRSR is classified according to the International Union for the Conservation of the Nature (UICN) as a Protected Area of Managed Resources: which means that this area type is accorded a larger flexibility for management, conservation, and also some productive activities and services, if done in a sustainable way [6]. This forest has been subjected to extensive exploitation since the sixteenth century, contributing to its degradation and the lack of valuable timber trees and other important species [12]. The riverside forest of the Cuyaguateje was selected because the landownership system (Cooperative farm) supports active agriculture activities. It has the highest degradation grade among the Cuban riverside forests. It is also disturbed by recurrent inundation episodes [13]. The native rainforest was chosen due to its previous intensive exploitation, the current anthropogenic pressure imposed by the adjacent communities [14], and its condition as a rainforest from the Sagua-Baracoa region [11].

At all these sites, disturbances can combine and produce many factors modifying the structure, composition, and functioning of populations, communities, and ecosystems [15], thus changing the availability of resources and habitats. The identification, characterization, and understanding of the communities or types of forests are fundamental in order to manage and conserve forest biodiversity [16]. However, although there are some studies to help us identify and define types of forests in the Neotropic, the information is still limited and more research is needed in this area [17], especially in the Caribbean tropical forests. Therefore, conservation and restoration programmes need to include diagnosis in their analysis, for the evaluation of the structure of populations, communities, and ecosystems, and their correlation with present and past disturbances. In this research, we introduce such studies for three different Caribbean forest types.

2. Material and methods

2.1. Application area

The Biosphere Reserve "Sierra del Rosario" (BRSR) occupies an area of 25,000 ha at 600 metres above sea level (m.a.s.l) in the Artemisa province (western Cuba) from 22°45′–23°00′N to 82°50′–83°10′W. It is part of the National System of Protected Area in Cuba [8]. In the BRSR, the semi-deciduous forest has special importance due to its large area. It constitutes the natural vegetation of Cuba, as high as approximately 600 m.a.s.l. Trees reach a height of 20 to 30 m, the canopy is constituted by two arboreal layers and a shrubby understory, with leaves of approximately 13 to 26 cm of longitude, mostly compound. The herbaceous layer is usually missing. The highest trees usually lose their leaves during the driest period, while those of the second arboreal layer usually conserve their leaves for the entire year [10].

The riverside forest of the Cuyaguateje is located in the Guane municipality of Pinar del Rio Province, from 22°11′–22°13′N to 84°03′–84°05′W, at 10 to 30 m.a.s.l. The native rainforests of the sector Quibiján-Naranjal belong to the mountain formation of Nipe-Sagua-Baracoa, in Toa′s river basin in the Guantánamo province (Figure 1). The research site is in the riverside forests of the Cuyaguateje, in the river′s middle reaches. This forest′s limits to the west are marked by the urban perimeter of Guane, to the north by plantations of the Forest Enterprise Macurije, to the south by the Cooperative of Credits and Strengthened Services "Secundino Serrano", and to the east by the end of the Sierra Cerro of Guane.

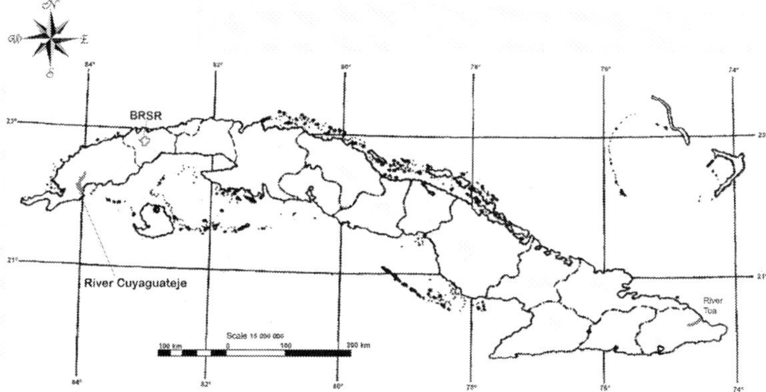

Figure 1. Geographic location of the study cases. BRSR: Biosphere Reserve "Sierra del Rosario".

2.2. Field sampling

First, the study areas were selected as representative units of the described forest type. Later on, sampling plots were chosen where floristic inventories were carried out. Environmental and dasometric variables were also measured. A group of variables related to human interference and its impact on the structure and quality of the forests was assessed, along with the environmental variables (soil, elevation, and distances from different human activities). Finally, a proposal of an action plan for the conservation and restoration of these forests was designed.

A random stratified design was used for field sampling, setting down 0.1 hectares plots (50 m x 20 m) following the "Methodology of Quick Inventory" [18–19]. In each plot the diameter and height of all the examples of arboreal species (Height > 2 m and over 5 cm of $D_{1.3}$, diameter at breast height) were identified and measured. Environmental variables considered were: nutrient content of soil (ppm) of Na, Mg, K, and Ca; pH; content of organic matter in soil (MO); and distance from the centre of the plot to the areas of human activity (cultivated lands, housings, and tourist facilities). Sampling data were validated using the curved area species method.

Diversity indexes were calculated using the floristic data from the inventory. Beta diversity (β) was estimated using hierarchical cluster analysis, using Sorensen distance (Bray-Curtis). This distance was estimated as the floristic similarity among the identified groups with the previous analysis calculated with Jaccard's index for qualitative data and the Morisita-Horn index for quantitative data. To identify the indicator species of each one of the identified groups through the cluster analysis, the Dufrene and Legendre method was used [20]. Diversity alpha (α) was calculated with the reciprocal of Simpson index (C inv.) [21], and an unbiased estimator of diversity was calculated using the jack-knife technique.

The horizontal structure was described with the relative values of abundance, dominance, and frequency of each species. In addition, the tree diametric class distributions were described for each plot. The ecological importance index value (IVIE), was calculated for each species as the sum of the parameters in the horizontal structure [22]. To describe the relationships among the variables, a principal components analysis (PCA) was performed, while to determine the association among environmental variables with the distribution and abundance of species for plots a canonical correspondence analysis (CCA) was done.

Key or vulnerable species for high-priority consideration in restoration programmes were identified based on their abundance, dominance, commercial wood potential, and dasometric variables. The design of the restoration proposal was based on the approaches of [5] who suggest 13 steps for restoration strategies.

3. Results

3.1. Study of case No. 1: Mesophyll semi-deciduous forest of the Biosphere Reserve "Sierra del Rosario" (BRSR)

We identified 36 families, 75 genera, and 91 species, with a total of 7,799 individuals registered from 30 sampled plots. The endemism rate was 11.24%, a similar value reported for the complete BRSR (from 11 to 34% [7]). The cluster analysis showed the presence of three groups among the plots, according to the flora composition. This result was tested by the MRPP test, which revealed differences among groups (p< 0.001) and supported the classification into three clusters. The analysis of species indicator [18] in each group is shown in Table 1.

Indicator species	Groups	IVI	p	Indicator species	Groups	IVI	p
Bursera simaruba		72.3	0.01	*Oxandra lanceolata*		90,9	0.01
Guettarda sp.		68.4	0.01	*Laurocerasus occidentalis*		74,2	0.01
Matayba apetala	1	68.0	0.01	*Zanthoxylum martinicense*		65,7	0.01
Andira inermis		61.7	0.02	*Pithecellobium arboreum*		60,5	0.02
Nectandra coriacea		59.6	0.02	*Caesalpinia bahamensis*		60,0	0.01
Trophis racemosa		72.4	0.01	*Swietenia mahagoni*		60,0	0.01
Casearia mollis		67.9	0.01	*Tabebuia shaferi*	3	57,0	0.01
Cupania americana		62.9	0.01	*Cedrela odorata*		56,9	0.04
Guazuma ulmifolia	2	52.8	0.02	*Didymopanax morototoni*		50,8	0.02
Roystonea regia		52.6	0.04	*Rauwolfia nitida*		49,7	0.01
Trichilia hirta		50.1	0.04	*Erythrina poeppigiana*		48,7	0.03
Spondias mombin		47.1	0.02	*Poeppigia procera*		45,0	0.02
				Eugenia maleolens		40,0	0.01

Table 1. Indicator species for the three groups, ordered by their IVI (p <0.05), obtained in the floristic inventory carried out in the western sector of BRSR.

According to these results, species related to the secondary forest that could be associated with a disturbance like timber extraction predominate in groups 1 and 2. Group 3 contained species from preserved sites, which correspond with plots in the Natural Reserve El Mulo. The results of the CCA analysis were globally significant. The first three axes offered a good solution to the ordination of the sampling units and of the species, because due to the present total variability in the data of abundance of the species (inertia = 2.55) it was possible to explain 23.7% by means of the group of this axes. The analyses reveal that the effect of the soil is not significant in the distribution and presence of species and therefore in the classification of the samples. The variable distance to human establishments has a bigger effect, mainly related with the plots of group 1 that are the furthest away and therefore less affected by the anthropic action. Group 3 is separated by the composition of species in the parcels or plots of the El Mulo located in more conserved area corresponding to the area nucleus of the reserve (Figure 2).

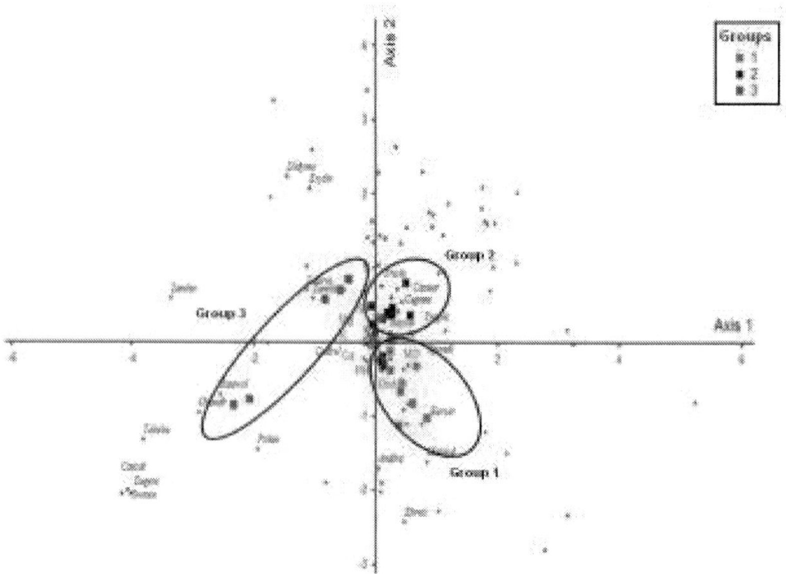

Figure 2. Projection of environmental variables, sampling units, and species in the plane defined by the axes CCA1 and CCA2. The continuous explanatory variables are shown as lines, the categorical explanatory variable is indicated by the colour of sampling unit, and the species with the codes given to their names.

The diversity of species and the equitability (alpha diversity) did not show significant differences (p>0.05) among localities. Therefore, in general, both variables can be considered to have high values. Numerically the areas of El Mulo and Mogote have a higher value that corroborates the characteristic vegetation for the complex of vegetation of Mogote [9]. In the case of El Mulo, the category of "Natural Reserve" favours its conservation and therefore displayed a higher diversity value (Table 2).

Sites	Simpson (1/C) VPi	Equitativity (E) VP
San Ramón	17.51 ± 3.09	.8510 ± 0.05
Mogote	18.52 ± 8.54	.8521 ± 0.03
Brazo Fuerte	16.73 ± 2.03	.8581 ± 0.03
Los Hondones	16.33 ± 1.05	.8518 ± 0.04
El Mulo	20.31 ± 2.13	.8231 ± 0.06
Average	17.88 ± 3.50	.8500 ± 0.04

Table 2. Average of dear diversity by means of the method of "Calculation Jump" (jack-knifing) for mesophyll semi-deciduous forest in the western sector of the BRSR.

No	Species	AR	FR	DmR	IVIE
1	*Roystonea regia*	2,9	93,3	6,23	102,48
2	*Trophis racemosa*	7,1	90,0	0,75	97,87
3	*Matayba apetala*	9,5	83,3	1,93	94,75
4	*Pseudolmedia spuria*	5,4	86,7	0,50	92,53
5	*Bursera simaruba*	4,1	83,3	2,53	89,94
6	*Guarea guidonia*	8,3	80,0	1,57	89,88
7	*Ficus aurea*	0,7	66,7	21,22	88,55
8	*Calophyllum antillanun*	4,8	80,0	1,03	85,83
9	*Andira inermis*	2,2	80,0	1,92	84,11
10	*Cecropia schreberiana*	1,5	80,0	2,18	83,70
11	*Syzygium jambos*	6,7	70,0	1,08	77,79
12	*Cupania americana*	2,4	70,0	0,29	72,70
13	*Matayba domingensis*	4,6	63,3	1,72	69,67
14	*Cinnamomun elongatum*	2,0	66,7	0,37	69,08
15	*Guazuma ulmifolia*	2,1	63,3	1,88	67,29

Table 3. First 15 arboreal species located by their Value of Ecological Importance in mesophyll semi-deciduous forest in the western sector of the BRSR.

3.1.1. Horizontal structure

The tree species with a higher ecological importance were those with a higher frequency (over 60%), so abundance and dominance are more important to the IVIE. *Roystonea regia* appears in the first position as the typical species in this forest. Other species, such as *Trophis racemosa* and *Matayba apetala,* are important because of their abundances, and *Ficus aurea* for its dominance (Table 3).

In the El Mogote, the species *Guarea guidonia, Roystonea regia, Bursera simaruba,* and *Matayba apetala* were present in the lowest and the middle parts, showing how their distributions are central to the altitudinal distribution of this formation [10]. The species *Cecropia schreberiana* was located among the most important by its relative dominance. It has very few individuals with small diameters that inhabit the most exposed places to light. This distribution favours the establishment of early secondary communities. Such communities evolve to establish a homeostasis in an approximately ten-year period. Then they stabilize the canopy in places that have suffered natural or anthropogenic disturbances. This behaviour is typical of pioneer species that will be later substituted in the successional process. *Syzygium jambos* is among the most abundant species, and it shows a high migration capacity, confirming their invasive condition. The abundance of these species demonstrates an increase in the populations of this group of plants and they indicate an altered ecological integrity.

Within the species with intermediate abundance, *Calycophyllum candidissimun*, *Dendropanax arboreus*, and *Samanea saman* hold special interest for future conservation strategies. The species *Chione cubensis*, *Lagetta wrightiana*, and *Terminalia chicharronia* (classified as endemic in "Sierra del Rosario") have a low abundance. However, those species were reported as being very abundant in this formation [10]. In our inventories, their presence only in the areas of San Ramón and Brazo Fuerte gave them the classification of being rare species. Among the most dominant species, the high presence of *Ficus aurea* is the outcome of selective felling being carried out in these forests, due to the scarce commercial value of *Erythrina poeppigiana*. *Roystonea regia*, *Mangifera indica*, *Laurocerasus occidentalis*, *Swietenia mahagoni*, and *Zanthoxylum martinicense* all reached 59% dominance.

The biggest values of basal area (m²/ha) for the mesophyll semi-deciduous forest were found for the western sector of the reserve, in the areas of El Mulo and Brazo Fuerte. The species *Ficus aurea* with 23.7 m²/ha and *Erythrina poeppigiana* with 18.9 m²/ha showed the biggest values in this parameter.

3.1.2. Vertical structure

These forests presented a high height, with two strata whose emergent trees are *Calophyllum antillanun*, *Andira inermis*, *Roystonea regia*, *Pseudolmedia spuria*, and *Matayba apetala*. All these species can reach up to 35 m in the hollows of San Ramón de Aguas Claras, Los Hondones, Brazo Fuerte, and in the low altitudes of the western part of "El Mogote de Soroa". The canopy was composed of individuals with heights from 20 to 30 m, with slight differences between the areas of hillsides and the summits. In the nature reserve El Mulo, the forest has two arboreal floors. The top stratum (more than 25 m in height) contained emergent *Ficus aurea*, *Eritrina poeppigiana*, *Cecropia schreberiana*, *Didymopanax morototoni*, *Trichospermun mexicanum*, and *Roystonea regia*, among other species. The trees in this layer reach up to 30–35 m in height. The intermediate stratum of mesophyll semi-deciduous forest in the studied areas is occupied by trees between 15 to 10 m tall. The lower stratum was integrated by evergreen species that reached heights of 6 to 12 m. The lower stratum is composed of juvenile individuals of the most abundant and frequent species, such as: *Trophis racemosa*, *Guarea guidonia*, *Bursera simaruba*, *Pseudolmedia spuria*, *Syzygium jambos*, *Calophyllum antillanun*, and *Dendropanax cuneifolius*. Trees in the lower stratum are usually younger individuals. Regeneration is fundamentally of species typical of secondary forests, except in El Mulo where *Matayba apetala* had regenerated. The similar abundance of sharing species among strata of the forest (Table 4) was determined by the Morisita-Horn index giving analogous values (≥ 80 %).

	$D_{1,3}$ <5 cm height <1,5 m	$D_{1,3}$ <5 cm height ≥1,5 m	$D_{1,3}$ ≥5 ≤ 10 cm height ≥1,5 m	$D_{1,3}$ ≥10 cm height "/ >1,5 m
$D_{1,3}$ <5 cm height <1,5 m		0,92	0,87	0,80
$D_{1,3}$ <5 cm height ≥1,5 m			0,90	0,85
$D_{1,3}$ ≥5 ≤ 10 cm height ≥1,5 m				0,80
$D_{1,3}$ ≥10 cm height "/>1,5 m				

Table 4. Morisita-Horn index of the components of the vertical structure of mesophyll semi-deciduous forest

3.1.3. Disturbances and relationship with the forest status

Ecological disturbances were classified according to their intensity in a scale from 1 to 4, where 1 = without disturbance; 2 = light; 3 = moderate; and 4 = high disturbance. Selective felling, alterations for the construction of roads, clearings, and the felling of trees due to winds, firewood extraction, and for other non-timber forest products were evaluated according to this scale. It was proven that the anthropogenic alterations prevailed: the most intense were related with selective felling, one of the factors that alters the dynamics of the regeneration more than others, changing the structure and composition of the forest.

Results of the principal components analysis were displayed on a correlation matrix between variables describing disturbance and variables describing species structure (Table 5). This table reveals that the first three orthogonal components explain 67% of the present variability. The variables that contributed more to the segregation of the components were: maximum number of individuals, selective felling, total number of individuals, and dominance. The first component reveals the inverse relationship between species number and basal area versus environmental variables such as the distance of the sampling places from the populated areas and forest clearings. The second and third components confirmed the direct relationship between the quantity of individuals and variables such as the intensity of selective felling, extraction of non-timber forest products, and road construction. The high density of individuals in the perturbed places is related to canopy opening and high regeneration of heliophill (sun tolerant) species, which usually have small diameters and therefore low values of basal area.

Variable	Communality	Component		
		1	2	3
Specie richness	,725	**-,773**	,306	-,183
Clearings and fall of trees	,617	**,766**		-,157
Diversity (1/D)	,776	**,624**		,620
Distance	,609	**,607**		-,489
Basal area	,414	**-,601**	-,231	
Selective felling	,860		**,919**	
Firewood extraction and non-timber forest product (NTFP)	,638	,236	**,753**	,121
Total of individual numbers	,777	-,295	**,635**	,535
Individual numbers of the most abundant species	,877		,338	**,869**
Total Roads impact	,408	-,158	-,355	**,507**
Auto value		2,486	2,2111	2,003
% of variance		24,862	22,105	20,031
Σσ²		24,862	46,967	**66,998**

Table 5. Main components analysis based on the correlation matrix between the disturbance variables and structural variables.

The relationship among sites, species, and some environmental variables indicates that *Swietenia mahagoni, Oxandra lanceolata, Tabebuia shaferi*, and *Caesalpinia bahamensis*, among others, reached the biggest values of abundance in environments with low disturbance. This fact indicates that these taxa are characteristic of places only slightly altered, like in the parcels of El Mulo.

Figure 3 shows the results of classifying the sampling units (plots) according to their disturbance grade.

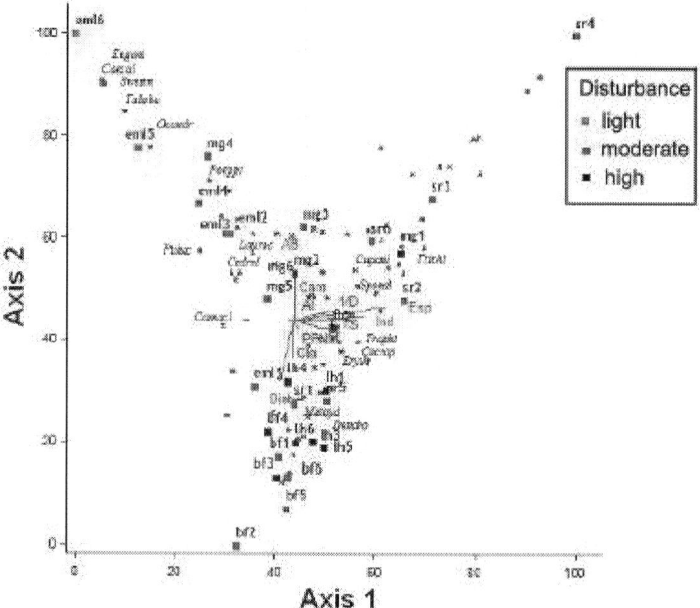

Figure 3. Projection of sampling units (squares) and species (asterisks) in the plane defined by the first two axes of the ACP. The continuous explanatory variables are shown as lines and the categorical explanatory variable is shown according to the colour of the symbol of the sampling unit. Sampling places: (sr1-sr6) — plots of San Ramón; mg1-mg6 — plots of El Mogote; bf1-bf6 — plots of Brazo Fuerte; lh1 a lh6 — plots of Los Hondones; eml1-eml6 — plots of El Mulo. Variables: Dist — distances to populational establishments and other human activities; AB — Basal area; Cam — roads; AL — height of sea level; 1/D — diversity; TS — selective felling; Esp — specie richness; Ind — individuals' maximum number; PFNM — Non-timber forest products; Cla — Forest spacements. Species codes: (*Eugeni* — *Eugenia maleolens*); (*Caesal* — *Caesalpinia bahamensis*); (*Tabebu* — *Tabebuia shaferi*); (*Pithec* — *Pithecellobium arboreum*); (*Oxandr* — *Oxandra lanceolata*); (*Poepi* — *Poeppigia procera*); (*Cedrel* — *Cedrela odorata*); (*Comocl* — *Comocladia dentata*); (*Lauroc* — *Laurocerasus occidentalis*); (*Sweten* — *Swietenia mahagoni*); (*Matayd* — *Matayba domingensis*); (*Cupani* — *Cupania americana*); (*Trophi* — *Trophis racemosa*); (*Cecrop* — *Cecropia shreberiana*); (*Erythr* — *Erythroxyllum havanense*); (*Spondi* — *Spondias mombin*); (*Trichi* — *Trichilia havanensis*); (*Dendro* — *Dendropanax arboreus*).

Secondary species characteristic of mesophyll semi-deciduous forest that generally have little commercial value, such as *Matayba apetala* and *Trophis racemosa*, were more abundant in places with moderate to high disturbance levels. These sites coincide with the plots in Brazo Fuerte,

Los Hondones, and San Ramón, characterized by the presence of the biggest diversity alpha, and the largest presence of species and abundance. According to these results the hypothesis of intermediate disturbance is corroborated [23]. Such a hypothesis states that the opening of forest gaps favours a much higher level of diversity (at local and regional scales) that would be presented if they lacked those disturbances.

On the other hand, the species *Swietenia mahagoni* and *Caesalpinia bahamensis* have a perfect correlation with sites with the lowest disturbance level, according to the test of Dufrene and Legendre [20]. Such correlation was demonstrated in the field, as these species were located very close to the plots of El Mulo. The opposite result was found for *Cupania americana* and *Cecropia schreberiana*, which were mostly present in sites with a high degree of disturbance. They were observed very near the most perturbed places of Brazo Fuerte, Los Hondones, El Mogote, and San Ramón.

In accordance with the results, indicator species were not present under conditions of moderate disturbance. Such species are fundamentally pioneer species of very wide distribution that surpass different ecological conditions, related with the alteration of environment. Indicator species with a significance level (p <0.05) are shown in Table 6. *Swietenia mahagoni* and *Caesalpinia bahamensis* could be considered as perfect indicators of the lowest disturbance level. On the other hand, *Cupania americana* and *Cecropia schreberiana* were the best indicators of a high degree of disturbance.

Indicator species	Disturbance	IVI	p*	Indicator species	Disturbance	IVI	p*
Caesalpinia bahamensis		100,0	0,001	*Cecropia schreberiana*		71,0	0,001
Swietenia mahagoni		100,0	0,001	*Cupania americana*		69,9	0,001
Tabebuia shaferi		97,5	0,002	*Trichilia havanensis*		68,4	0,021
Oxandra lanceolata		91,2	0,002	*Trophis racemosa*		63,6	0,003
Poeppigia procera	Light	81,3	0,002	*Dendropanax cuneifolius*	High	62,4	0,019
Pithecellobium arboreum		80,9	0,003	*Matayba apetala*		53,4	0,032
Comocladia dentata		68,6	0,013	*Roystonea regia*		53,3	0,047
Eugenia maleolens		66,7	0,013	*Erythroxyllum havanense*		50,0	0,042
Laurocerasus occidentalis		62,6	0,019	*Spondias mombin*		47,1	0,032
Cedrela odorata		61,3	0.020				

Factor level: 2 — Light; 4 — High.

Table 6. Lists of the main indicator species ordered for IVI (p <0.05) according to the disturbance level.

3.1.4. Diameter structure of forest species in the reserve

The number of woody species of local and commercial interest was reduced, and their diameter distributions presented few individuals in superior categories (Table 7). This can be the result of selective commercial logging, described as one of the main disturbances in the region. These results corroborate those outlined in [24], for "Sierra del Rosario". These authors concluded that in that region, only a few individuals end up having diameters bigger than 20 cm, due to topography and shallow soils. Although the diversity of trees is high, the abundance of forestry species' regeneration with commercial value is low in the mesophyll semi-deciduous forest of the reserve.

Species	Number of trees for diameter class $d_{(1,3)}$ (cm)											
	Total	2,5–9	10–19	20–29	30–39	40–49	50–59	60–69	70–79	80–89	90–99	≥100
Matayba apetala	502	333	128	40	1							
Guarea guidonia	471	346	67	35	10	6	2	1	2	1	1	
Bursera simaruba	420	188	172	44	9	3		2		2		
Calophyllum antillanum	347	254	48	27	10	2	3	1	1		1	
Trichospermun mexicanum	206	57	89	42	14	3	1					
Andira inermis	189	111	51	22	5							
Oxandra lanceolata	149	99	30	12	6	2						
Laurocerasus occidentalis	141	91	32	12	4	1	1					
Mangifera indica	130	56	3		7	4	7	31	8	2	7	5
Zanthoxylum martinicense	122	52	20	21	21	3	2	1			2	
Talipariti elatum	116	62	19	12	7	11	2	3				

Table 7. Structure of the diameter class of arboreal species with more than 100 individuals in the sampling places in the western sector of the BRSR.

It seems that in the reserve there is a considerable quantity of species tolerant to disturbances, such as *Talipariti elatum, Bursera simaruba, Calophyllum antillanun, Laurocerasus occidentalis,* and *Guarea guidonia.* They regulate the diametric structure that favours their presence in the forest. This phenomenon could indicate their possible use in reforestation programmes. On the other hand, species of high commercial value like *Swietenia mahagoni* and *Cedrela odorata* are only represented by 28 and 45 individuals, respectively. The stand diametric structure reveals very few individuals in the regeneration category, and also a very minor presence in other diametric classes. This situation puts the future existence of these species at risk.

3.1.5. Indicator species

The regular form in the diametric distribution observed in *Matayba apetala, Guarea guidonia, Bursera simaruba,* and *Calophyllum antillanun,* with an abundance of juvenile individuals in the inferior categories, suggests tolerance to the competition caused by disturbances in the forest. Therefore, some of these species could be incorporated into a monitoring programme for these ecosystems. Due to their quick growth, they could also be considered for inclusion in a restoration programme in the region. In this context, among the arboreal species suggested as key species are: *Guarea guidonia, Matayba apetala, Pseudolmedia spuria, Calophyllum antillanun, Laurocerasus occidentalis, Mangifera indica, Cecropia schreberiana,* and *Bursera simaruba.* These species possess additional attributes that allow them to provide stability to the ecosystem before the disturbances [21]. The approaches that endorse this selection are related in the following.

Large trees of *Mangifera indica* found in San Ramón de Aguas Claras, El Mogote, Brazo Fuerte, and Los Hondones support a high epiphytic diversity, with flowers like orchids, bromeliads, and others. In addition, such epiphytes are also an important element for wildlife food sources (mainly birds and small mammals), whose populations could be at risk in this reserve. Many fruits are also collected by the residents of the local communities for their own consumption. In "Sierra del Rosario", the presence of *Cecropia schreberiana* allows the establishment of early secondary communities. Such communities develop a homeostasis in around a 10-year cycle, stabilizing the canopy in places that have suffered natural or anthropogenic disturbances. Other species like *Trophis racemosa* and *Syzygium jambos* could be suggested as indicator species of highly disturbed forests in this region. *Trophis racemosa* is located mainly in the medium to high altitudes, while *Syzygium jambos* is distributed in the medium to low areas near rivers or streams.

3.1.6. Vulnerable species

In an alarming way, individuals of some commercially important species such as *Lagetta wrightiana* and *Terminalia chicharronia,* which were previously plentiful in these forests (as local residents remember), were present only in the inventories of San Ramón de Aguas Claras and Brazo Fuerte, respectively. For the last species, individuals were not observed in natural regeneration inside the sampled plots. This result supports its consideration of species under threat, according to the species Red List of Cuba [9].

3.1.7. Restoration proposal

The purpose of the proposal is to suggest a group of actions to guide restoration and conservation of forests in the western sector of the BRSR and adjacent areas. In the elaboration of this proposal, the approaches by [5] have been considered. These authors suggest 13 steps or main elements to consider in the elaboration of the plan. These are:

1. To define the ecosystem or reference community;

2. To evaluate the current state of the ecosystem or community;

3. To define the scales or levels of organization of the ecosystem;

4. To establish the scales and disturbance hierarchies;

5. To achieve the local human community's participation;

6. To evaluate the potential of regeneration of the ecosystem;

7. To establish the restoration barriers at different scales;

8. To select the appropriate species for restoration;

9. To spread and to manage the selected species;

10. To select the intervention sites;

11. To design strategies to overcome the barriers to the restoration;

12. Monitoring the restoration processes;

13. To consolidate the restoration process.

According to the previous steps, the actions for the restoration have been based on the following aspects: 1) it has been kept in mind the previous main scientific research studies carried out in the area that today occupies the Biosphere Reserve of "Sierra del Rosario". Among these, the book *Ecology of Rainforests of Sierra del Rosario* [7] constitutes a summary of the main research carried out in this ecosystem. Other important studies considered aspects of the geology in relation with the presence of hydrocarbons in the eastern part of the reserve [26], agrobiodiversity that the farmers manage in homemade orchards and properties in protected areas [27], and restoration of arboreal diversity of rainforests in the reserve [6]. Also, analysis of research needs in the protected area [30], and the floristic characterization of mesophyll semi-deciduous forest in the western sector of the BRSR, as well as a group of indicator environmental variables of disturbances.

When proposing the restoration plan, we also kept in mind the results of floristic inventories carried out in the Biosphere Reserve of "Sierra del Rosario", as well as the impacts of the traditional extraction of products of forest and other land uses. We also tried to incorporate the perceptions and identification of the environmental services of forest by the local residents of the reserve, according to the approaches of [28]. Opinions of the technical staff of the Ecological Station (the state entity responsible for the protection of the BRSR), and recommendations of experts on the basic approaches for the conservation and the sustainable forest administration [27, 6] were also taken into account. In addition, guidelines by International Organization of Tropical Wood for restoration, ordination, and rehabilitation of secondary and degraded tropical forests [29], as well as recommendations on the Methodology for the Elaboration of Management Plans for Protected Areas of Cuba [30], were incorporated. Finally, the Management Plan of the Biosphere Reserve of "Sierra del Rosario" for the period 2011–2015 and the operative Plan of the Reserve were consulted [8].

After this review, it was determined that the main barriers to the restoration are:

• Natural barriers: dominance of species with little commercial value, invasion of exotic species, irregularities in the diameter classes in distribution of species of high commercial value, and gaps in the forest as a result of the wind and the falling of trees.

• Social barriers: selective logging, firewood extraction, non-timber forest products harvesting, and road construction.

Recommended actions for forest restoration:

• Adoption of modern silviculture techniques, such as favouring passive reforestation (natural regeneration), enrichment of the natural forest, reforestation with native species, and implementation of agroforestry or silvopastoral systems.

- To reach higher percentages of natural regeneration in places affected by windthrow, selective logging, road construction, or agricultural cultivation, the following species are recommended for restoration: *Matayba apetala, Guarea guidonia, Bursera simaruba,* and *Calophyllum antillanun.* The species with commercial use are: *Andira inermis, Swietenia mahagoni, Laurocerasus occidentalis, Terminalia chicharronia,* and *Gerascanthus gerascanthoides.*

- A high-priority is given to the recovery of native forest species whose populations are in critical state, such as *Lagetta wrightiana* and *Terminalia chicharronia.* Similarly, reintroduction of trees with beautiful wood that have been extinguished locally, such as *Cedrela odorata* and *Swietenia mahagoni,* is also a high-priority.

3.2. Study of case No. 2: Riparian forest of the River Cuyaguateje

3.2.1. Biodiversity

For alpha diversity, floristic inventories in both borders of the studied area were done every 323 m and natural regeneration was estimated. The specie richness was 38 species, including 21 families and 36 genera. Table 8 shows the 860 individuals of the species (> 2 m of height). Similar results were reported in previous studies [31–32] and also referred to by local farmers since the 1970s. The most representative species found were *Bambusa vulgaris, Guazuma ulmifolia, Samanea saman, Sapindus saponaria, Lonchocarpus domingensis, Spondias mombin, Roystonea regia, Trichilia hirta,* and *Bursera simaruba.* Natural regeneration in passive restoration allows abundant recruitment of autochthonous species such as *G. ulmifolia, S. saponaria, L. domingensis, T. citrifolia, S. mahagoni, T. hirta, T. elatum,* and *S. mombin.* These species can create more suitable habitats for other typical species, reducing the impacts of human activities [33] and increasing ecological resilience.

Species	Individuals	Dr	Species	Individuals	Dr
Bambusa vulgaris Schrader ex Wendland	191	22,2	*Psidium guajava* L.	15	1,74
Guazuma ulmifolia Lam.	87	10,1	*Cephalanthus occidentalis* L.	14	1,63
Samanea saman (Jacq.)	79	9,19	*Roystonea regia* HBK O. F. Cook.	13	1,51
Sapindus saponaria L.	46	5,35	*Mangifera indica* L.	12	1,40
Lonchocarpus domingensis (Pers). DC.	42	4,88	*Swietenia macrophyla* King.	10	1,16
Tabernaemontana citrifolia L.	42	4,88	*Gliricidia sepium* Jacq. Urb.	7	0,81
Swietenia mahagoni L	37	4,30	*Tabebuia angustata* Britt.	7	0,81
Trichilia hirta L.	35	4,07	*Bursera simaruba* (L). Sargent.	6	0,70

Species	Individuals	Dr	Species	Individuals	Dr
Gmelina arborea Roxb	34	3,95	*Terminalia catappa* L	6	0,70
Talipariti elatum Frixell (Sw.)	32	3,72	*Andira inermis* (W. Wright) Kunth ex DC.	4	0,47
Spondias mombin L	26	3,02	*Casearia hirsuta* Sw	3	0,35
Guarea guidonia L. Sleumer	21	2,44	*Melicoccus bijugatus* Jacq	3	0,35
Cordia collococca L	17	1,98	*Annona reticulata* L	2	0,23
Cedrela odorata L	16	1,86	*Comocladia dentata* Jacq	2	0,23
Gerascanthus gerascanthoides L.	16	1,86	*Cupania americana* L.	2	0,23
Dichrostachys cinerea (L) Wigth et Arm	16	1,86	*Cocos nucifera* L	1	0,12
Acacia mangium Willd.	15	1,74	*Khaya senegalensis* Desr (A. Juss.)	1	0,12
			Total	860	100

Table 8. Arboreal and shrub species identified in the area in the riversides of the River Cuyaguateje (> 2 m height).

During the first year (2000) the area could be recognized by the presence of scarce trees of eight species along the riverside. Figure 4 illustrates how specie richness increased during the rehabilitation process, simultaneously with the reduction of the human cultivation practices, extensive shepherding, and sand extraction from the riverbed.

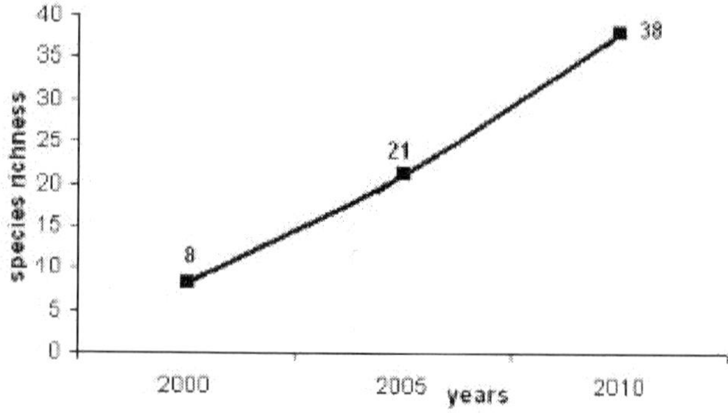

Figure 4. Specie richness trends during the period (2000–2010) according to [13].

Table 9 represents biodiversity expressed by Simpson's index and its reciprocal in each research plot. The highest values of diversity corresponded to plots 3, 5, and 8 in the border 1, and plots 18, 24, 25, and 30 in the border 2. These plots correspond to locations where local farmers had implemented forest rehabilitation measures by themselves. Generally speaking, biodiversity values per plot varied from 1.29 to 12.61 in border 1, and from 0.62 to 14.25 in border 2. Such a trend clearly shows the differences among sites with and without human activity in both borders, since the topography and soils are similar in the whole studied area.

Border 1			Border 2		
Plots	Dominance (D)	Diversity (I/D)	Plots	Dominance (D)	Diversity (I/D)
1	0.22	4.50	18	0.08*	12.20*
2	0.21	4.75	19	0.14	7.35
3	0.12*	8.61*	20	0.09	10.60
4	0.22	4.61	21	0.27	3.72
5	0.11*	9.23*	22	0.22	4.62
6	0.40	2.80	23	0.24	4.11
7	0.20	6.51	24	0.07*	14.25*
8	0.08*	12.61*	25	0.08*	12.05*
9	0.50	2.20	26	0.61	0.62
10	0.27	3.67	27	0.19	5.60
11	0.39	2.60	28	0.12	8.56
12	0.57	1.74	29	0.16	6.18
13	0.51	1.97	30	0.19*	11.27*
14	0.78	1.29	31	0.15	6.30
15	0.40	2.50	32	0.09	10.50
16	0.39	3.05	33	0.16	8.11
17	0.35	2.30	34	0.15	6.66

Table 9. Dominance index and their reciprocal one, of the individuals with d1.3≥ 2.5 cm for plots in each border

The plots with the highest diversity values (8, 18, 24, 25, and 30) could be used as examples of the expected restoration targets for riparian forests in terms of floristic composition. However, they show some signs of past anthropogenic intervention. Ecologically speaking, these plots form a complex mosaic as a result of the processes that favour ecological resilience, thus increasing biodiversity values.

As for beta-diversity and Jaccard's index of similarity, they are very important for assessing the degree of similarity among communities and the degree of exclusion between two species within the same community, especially when the species become dominant [34–35]. Jaccard's index indicates similarity between rehabilitated borders (0.78) that have 29 species in common (Table 10). Indeed, the high value in flora similarity between borders (0.78) is probably due to the response to similar climatic factors (temperature, humidity, and precipitation), soil type, and latitudinal position, while the differences could be related to structural elements of common species.

Number	Species	Number	Species
1	*Acacia mangium* Willd	16	*Melicoccus bijugatus* Jacq.
2	*Andira inermis* (W. Wright) Kunth ex DC.	17	*Psidium guajava* L.
3	*Anonna reticulata* L	18	*Roystonea regia* HBK O. F. Cook.
4	*Bambusa vulgaris* Schrader ex Wendland	19	*Samanea saman* Jacq.
5	*Bursera simaruba* (L) Sargent.	20	*Sapindus saponaria* L.
6	*Casearia hirsuta* Sw	21	*Simaruba glauca* D.C.
7	*Cordia collococca* L	22	*Spondias mombin* L
8	*Cupania americana* L.	23	*Swietenia mahagoni* L
9	*Dichrostachys cinerea* (L) Wigth et Arm.	24	*Swietenia macrophyla* King.
10	*Gerascanthus gerascanthoides* L.	25	*Tabernaemontana citrifolia* L.
11	*Gmelina arborea* Roxb.	26	*Talipariti elatum* Frixell (Sw)
12	*Guarea guidonia* L. Sleumer	27	*Terminalia catappa* L
13	*Guazuma ulmifolia* Lam.	28	*Trichilia havanensis* Jacq.
14	*Lonchocarpus domingensis* (Pers). DC.	29	*Trichilia hirta* L.
15	*Mangifera indica* L.		

Table 10. Species in common in both borders that been rehabilitated.

In addition, another element that should be considered in support of this analogy is the occurrence of floods that favour seed dispersion and natural regeneration of pioneer species, while reducing anthropogenic pressure and the presence of typical vegetation such as *G. ulmifolia*, *L. domingensis*, *T. citrifolia*, *T. hirta*, *S. mombin*, and *S. saponaria*. Among both communities (rehabilitated and control sites) an index of similarity of 0.28 with 14 common species was reported (Table 11). The small similarity between the control and rehabilitated sites could be caused by the short time of the rehabilitation process, the differences include the environmental conditions, and the human cultivation activities that enhance the presence of species, many of which exhibit invasive behaviour.

If agriculture, extensive shepherding, and sand extraction activities were controlled, while local actors managed the natural regeneration by increasing species numbers (even with exotic species), better ecological conditions will develop in the long term for the transition towards a rehabilitated site, improving the ecosystem's functionality. In addition, such favourable conditions (fertility, humidity, and deposition of seeds) for natural regeneration and the plantation of species like *T. elatum* and *S. mahagoni* would improve the ecosystem's productivity. In this sense, in some cases ecosystems degraded anthropogenically can be restored when the external stresses are reduced, with the reintroduction of native species, the removal of the exotic species, and the beginning of the passive restoration processes [36].

Species (natural area)	Species (rehabilitated area)	Common in both areas
Alibertia edulis (L. C. Richt.)	*Acacia mangium* Willd	*Bursera simaruba* L. Sargent.
Calycophyllum candidissimun (Vahl) DC.	*Andira inermis* (W. Wright) Kunth	*Casearia hirsuta* Sw.
Cinnamomun elongatum (Nees) Kostermans	*Anonna reticulata* L.	*Comocladia dentata* Jacq
Cochlospermun vitifolium (Willd.)	*Bambusa vulgaris* Schrader ex Wendland	*Cupania americana* L.
Chrysophyllum cainito L.	*Calophyllum antillanum* (Britt.) Standl.	*Guarea guidonia* L. Sleumer
Erythroxylum havanense Jacq.	*Cedrela odorata* L.	*Guazuma ulmifolia* Lam.
Gossypiospermum praecox (Gris.) P. Wilson	*Cephalanthus occidentalis* L.	*Lonchocarpus domingensis* (Pers). DC.
Luehea speciosa Willd.	*Cocos nucifera* L.	*Roystonea regia* HBK O. F. Cook.
Matayba apetala Macf. RDKL.	*Cordia collococca* L.	*Sapindus saponaria* L.
Oxandra lanceolata Sw. Baill.	*Dichrostachys cinerea* (L.) Wigth et Arm.	*Spondias mombin* L.
Protium cubense (Rose)	*Gerascanthus gerascanthoides* L.	*Syzygium jambos* L. Alston in Trimen
Zanthoxylum elephantiasis Macfd.	*Gliricidia sepium* Jacq. Urb.	*Tabernaemontana citrifolia* L.
	Gmelina arborea Roxb.	*Trichilia hirta* L.
	Khaya senegalensis Juss	*Trichilia havanensis* Jacq.
	Mangifera indica L.	
	Melicoccus bijugatus Jacq.	
	Psidium guajava L.	
	Samanea saman (Jacq.)	
	Simaruba glauca D.C.	
	Swietenia mahagoni L.	
	Swietenia macrophyla King.	
	Tabebuia angustata Britt.	
	Talipariti elatum Frixell (Sw.)	
	Terminalia catappa L.	

Table 11. Species identified in the study area (rehabilitated and control)

Figures 5 and 6 show a diagram of range/abundance of the species in the borders. At both sites, the highest abundance value corresponds to *S. jambos*, which in Cuba is an invasive species linked to human activities. This species is a fruit-bearing shrub introduced in Cuba. It later colonized some mountainous areas along riversides and constitutes a serious obstacle for reforestation.

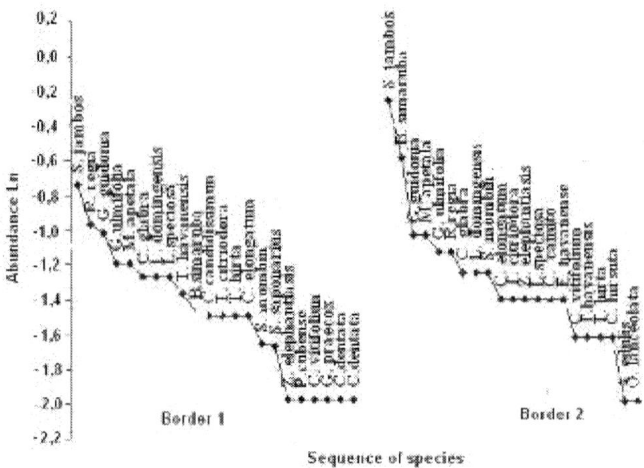

Figure 5. Diagram of the range/abundance of identified species in riversides (1 and 2) of the Cuyaguateje's riversides according to [13].

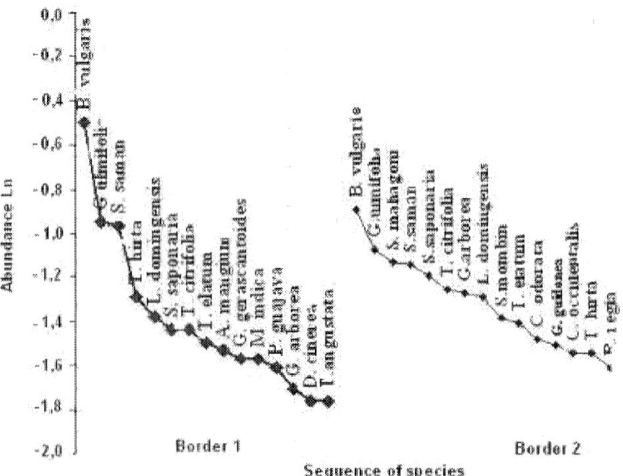

Figure 6. Diagram of the range/abundance for the 15 more representative species of the area that is rehabilitated in borders (1 and 2) of the Cuyaguateje's riversides according to [13]

3.2.2. Forest structure

Table 12 shows the results of horizontal structural parameters of rehabilitated areas under local communities' participation (dominance (D), abundance (A), and frequency (F)), and the ecological importance value index (IVIE) of the species in each border. The species with low IVIE are those typical of the riparian forest, such as: *B. simaruba, A. inermis, C. dentata, C. americana, S. mahagoni,* and *R. regia.* According to their relative abundance, the species most sensitive to environmental or anthropogenic disturbances were identified: *A. inermis, C. dentata, S. macrophyla, S. mahagoni, A. reticulata,* and *R. regia.* This parameter can be used as an indicator of the degrading process [34, 37]. Similarly, in border 2 the species with low IVIE had more commercial importance, such as *G. arborea, S. saman, G. ulmifolia, L. domingensis, S. mahagoni, T. elatum, S. saponaria,* and *R. regia* (Table 12). In the case of *G. arborea,* the value of the IVIE was due to its dominance. This species is not recommended for planting along riversides because it is considered as potentially invasive [38]. It is also not recommended for planting in sites prone to floods due its low survival rate under such conditions.

Border 1					Border 2				
Species	Dr	Fr	Ar	IVIE	Species	Dr	Fr	Ar	IVIE
Bambusa vulgaris	6,93	9,01	31,70	47,63	*Bambusa vulgaris*	11,53	5,52	13,69	30,73
Samanea saman	20,78	11,71	10,81	43,30	*Gmelina arborea*	19,63	4,83	5,74	30,20
Guazuma ulmifolia	17,47	9,01	11,30	37,78	*Samanea saman*	13,39	8,28	7,73	29,39
Lonchocarpus domingensis	3,99	7,21	4,18	15,37	*Guazuma ulmifolia*	9,99	6,21	9,05	25,25
Sapindus saponaria	2,87	7,21	3,69	13,76	*Lonchocarpus domingensis*	5,00	7,59	5,52	18,10
Acacia mangium	8,75	1,80	2,95	13,50	*Swietenia mahagoni*	4,14	5,52	7,95	17,61
Trichilia hirta	3,42	4,50	5,16	13,08	*Talipariti elatum*	8,25	4,83	4,19	17,27
Spondias mombin	7,89	3,60	1,47	12,97	*Sapindus saponaria*	3,49	6,21	6,84	16,54
Gmelina arborea	5,65	3,60	1,97	11,22	*Roystonea regia*	8,82	4,14	2,65	15,61
Mangifera indica	2,70	5,41	2,70	10,81	*Spondias Bombin*	3,63	6,21	4,42	14,25
Tabernaemontana citrifolia	1,18	5,41	3,69	10,27	*Tabernaemontana citrifolia*	0,52	6,21	5,96	12,69
Talipariti elatum Frixell	3,97	2,70	3,19	9,87	*Trichilia hirta*	0,29	4,83	3,09	8,21
Gerascanthus gerascanthoides	2,64	4,50	2,70	9,85	*Cedrela odorata*	1,83	2,07	3,53	7,44
Psidium guajava	1,47	3,60	2,46	7,53	*Cordia collococca*	0,81	3,45	2,43	6,69
Cordia collococca	3,15	2,70	1,47	7,33	*Swietenia macrophyla*	1,81	2,76	1,99	6,56
Tabebuia angustata	1,36	2,70	1,72	5,78	*Guarea guidonia*	0,30	2,76	3,31	6,37
Guarea guidonia	0,97	2,70	1,47	5,15	*Cephalanthus occidentalis*	0,18	2,07	3,09	5,33
Bursera simaruba	1,21	1,80	1,23	4,24	*Dichrostachys cinerea*	0,90	2,07	1,99	4,95
Terminalia catappa	0,89	1,80	1,23	3,92	*Gliricidia sepium*	1,25	1,38	1,55	4,17

Border 1					Border 2				
Species	Dr	Fr	Ar	IVIE	Species	Dr	Fr	Ar	IVIE
Andira inermis	0,58	1,80	0,49	2,87	*Acacia mangium*	0,95	2,07	0,66	3,68
Dichrostachys cinerea	0,08	0,90	1,72	2,70	*Psidium guajava*	0,33	2,07	1,10	3,50
Comocladia dentata	1,15	0,90	0,49	2,54	*Andira inermis*	1,43	1,38	0,44	3,25
Casearia hirsuta	0,03	0,90	0,74	1,67	*Gerascanthus gerascanthoides*	0,52	1,38	1,10	3.00
Cupania americana	0,15	0,90	0,49	1,54	*Melicoccus bijugatus*	0,03	2,07	0,66	2,76
Swietenia macrophyla	0,31	0,90	0,25	1,46	*Cocos nucifera*	0,37	0,69	0,22	1,28
Swietenia mahagoni	0,22	0,90	0,25	1,36	*Khaya senegalensis*	0,37	0,69	0,22	1,28
Annona reticulata	0,14	0,90	0,25	1,29	*Bursera simaruba*	0,21	0,69	0,22	1,12
Roystonea regia.	0,05	0,90	0,25	1,20	*Annona reticulata*	0,03	0,69	0,22	0,94
Totals	100	100	100	300	*Mangifera indica*	0,01	0,69	0,22	0,92
					Terminalia catappa	0,01	0,69	0,22	0,92
					Totals	100	100	100	300

Table 12. Phytosociological Parameters of the riparian forest of the River Cuyaguateje, border (1 and 2). Dr = Relative Dominance; Fr = Relative Frequency; Ar = Relative Abundance; IVIE = Ecological Importance Value Index

Fruit tree species were present with very low values in frequency, dominance, and abundance. Only *P. guajava*, *M. indica*, *A. reticulata*, and *M. bijugatus* were observed. Local farmers did not accept *M. bijugatus* because this species is a potentially invasive species [38]. The scarcity in fruits could also be related to the low germination and development rates of *Pouteria campechiana* (canistel) and *Calocarpum sapota* (Red Mammee) and the low presence of this species in natural areas. Incorporating fruit trees to reforestation plans could be used as an incentive for local farmers to protect riversides.

Natural regeneration reflects ecosystem fitness. Our results show a good condition with 3,952 in border 1 and 4,564 seedlings in border 2. Nonetheless, it is important to compare the species in the upper canopy stratum with those that are regenerating because their relative frequencies are directly associated to successional processes [36]. Our results indicate that favourable conditions exist for the regeneration and the recruitment of arboreal and shrub species that can increase or diminish in the measure that restrictive factors changes. Logically, it is also related to the approval of local farmers of rehabilitation processes [13].

At both borders, regenerated individuals in the herbaceous stratum were found for *G. ulmifolia*, *S. saman*, *L. domingensis*, *S. saponaria*, *G. guidonia*, *R. regia*, *T. citrifolia*, *S. mombin*, *S. saponaria*, *T. elatum*, *C. collococca*, *L. domingensis*, and *B. simaruba*, which are typical species of this forest type. The relative natural regeneration (RNRi) index, which combines abundance and frequency, showed the higher values for *T. citrifolia*, *S. saman*, *G. ulmifolia*, *S. mombin*, *L. domingensis*, and *S. saponaria*. Of the nine species that were used at the beginning of the reforestation [39] only four were presented in this stratum: *G. arborea*, *T. elatum*, *M. indica*, and *P. guajava*. However, very good germination rates of *G. arborea* were observed, a typical feature of invasive species. Its successful establishment will depend on flooding events and silvicultural management.

Table 13 shows the values of the Modified Ecological Importance Value Index (IVIEA) for both borders. The IVIEA is the most important indicator in evaluating forest dynamics [40]. It integrates horizontal and vertical structures in the mature mass, as in the natural regeneration. *S. saman, B. vulgaris, G. ulmifolia, G. arborea,* and *S. saponaria* were the most important species. Likewise, there are correspondences among plants that are identified in the arboreal stratum and natural regeneration processes in both borders. The appearance of new species in the inferior stratum, not identified in the mature state, indicates the existence of seeds and environmental factors that favour the ultimate rehabilitation of the forest. Source [29] considers that the readiness of different regeneration mechanisms is a crucial factor in the speed and course of secondary succession. Nevertheless, reproduction by seeds is the main mechanism of regeneration of the widely dispersed pioneer species, especially after repeated cultivation-fallow cycles over long periods.

	Border 1						Border 2					
Species	Ar	Fr	Dr	IVIr	RNRi	IVIEA	Species	Ar	Fr	Dr	IVIr	RNRi	IVIEA
Samanea saman	20,8	11,7	10,8	43,3	8,7	52,0	*Gmelina arborea*	19,6	4,8	5,7	30,2	13,1	43,3
Bambusa vulgaris	6,9	9,0	31,7	47,6	2,3	49,9	*Samanea saman*	13,4	8,3	7,7	29,4	5,5	34,9
Guazuma ulmifolia	17,5	9,0	11,3	37,8	6,1	43,9	*Bambusa vulgaris*	11,5	5,5	13,7	30,7		30,7
Tabernaemontana citrifolia	1,2	5,4	3,7	10,3	11,4	21,7	*Guazuma ulmifolia*	10,0	6,2	9,1	25,2	3,6	28,9
Sapindus saponaria	2,9	7,2	3,7	13,8	6,5	20,2	*Sapindus saponaria*	3,5	6,2	6,8	16,5	9,5	26,0
Lonchocarpus domingensis	4,0	7,2	4,2	15,4	4,2	19,6	*Talipariti elatum*	8,2	4,8	4,2	17,3	5,1	22,4
Trichilia hirta	3,4	4,5	5,2	13,1	6,5	19,5	*Lonchocarpus domingensis*	5,0	7,6	5,5	18,1	3,3	21,4
Spondias mombin	7,9	3,6	1,5	13,0	6,1	19,0	*Roystonea regia*	8,8	4,1	2,6	15,6	4,7	20,4
Gmelina arborea	5,6	3,6	2,0	11,2	5,7	16,9	*Tabernaemontana citrifolia*	0,5	6,2	6,0	12,7	7,3	20,0
Acacia mangium	8,7	1,8	2,9	13,5		13,5	*Swietenia mahagoni*	4,1	5,5	7,9	17,6		17,6
Talipariti elatum	4,0	2,7	3,2	9,9	3,4	13,3	*Spondias Bombin*	3,6	6,2	4,4	14,3	2,9	17,2
Gerascanthus gerascanthoides	2,6	4,5	2,7	9,9	3,0	12,9	*Trichilia hirta*	0,3	4,8	3,1	8,2	6,9	15,1
Mangifera indica	2,7	5,4	2,7	10,8	0,8	11,6	*Guarea guidonia*	0,3	2,8	3,3	6,4	7,3	13,7

Border 1

Species	Ar	Fr	Dr	IVIr	RNRi	IVIEA
Psidium guajava	1,5	3,6	2,5	7,5	3,0	10,6
Cordia collococca	3,2	2,7	1,5	7,3	2,7	10,0
Cupania americana	0,1	0,9	0,5	1,5	8,4	9,9
Tabebuia angustata	1,4	2,7	1,7	5,8	3,0	8,8
Guarea guidonia	1,0	2,7	1,5	5,2	2,3	7,4
Terminalia catappa	0,9	1,8	1,2	3,9	2,7	6,6
Bursera simaruba	1,2	1,8	1,2	4,2	1,5	5,8
Comocladia dentata	1,1	0,9	0,5	2,5	3,0	5,6
Andira inermis	0,6	1,8	0,5	2,9	1,1	4,0
Roystonea regia	0,1	0,9	0,2	1,2	2,3	3,5
Dichrostachys cinerea	0,1	0,9	1,7	2,7	0,8	3,5
Trichilia havanensis					2,3	2,3
Melicoccus bijugatus					2,3	2,3
Casearia hirsuta	0,0	0,9	0,7	1,7		1,7
Swietenia macrophyla	0,3	0,9	0,2	1,5		1,5
Swietenia mahagoni	0,2	0,9	0,2	1,4		1,4
Anonna reticulata	0,1	0,9	0,2	1,3		1,3
Total	100	100	100	300	100	400

Border 2

Species	Ar	Fr	Dr	IVIr	RNRi	IVIEA
Cordia collococca	0,8	3,4	2,4	6,7	1,8	8,5
Cephalanthus occidentalis	0,2	2,1	3,1	5,3	2,9	8,3
Cedrela odorata	1,8	2,1	3,5	7,4		7,4
Swietenia macrophyla	1,8	2,8	2,0	6,6		6,6
Dichrostachys cinerea	0,9	2,1	2,0	5,0	1,1	6,0
Cupania americana					5,5	5,5
Terminalia catappa	0,0	0,7	0,2	0,9	4,0	4,9
Gliricidia sepium	1,2	1,4	1,5	4,2	0,7	4,9
Psidium guajava	0,3	2,1	1,1	3,5	1,1	4,6
Gerascanthus gerascanthoides	0,5	1,4	1,1	3,0	1,5	4,5
Andira inermis	1,4	1,4	0,4	3,3	1,1	4,3
Melicoccus bijugatus	0,0	2,1	0,7	2,8	1,1	3,9
Acacia mangium	0,9	2,1	0,7	3,7		3,7
Bursera simaruba	0,2	0,7	0,2	1,1	1,8	2,9
Mangifera indica	0,0	0,7	0,2	0,9	1,8	2,7
Casearia hirsuta					2,2	2,2
Trichilia havanensis					1,5	1,5
Cocos nucifera	0,4	0,7	0,2	1,3		1,3
Khaya senegalensis	0,4	0,7	0,2	1,3		1,3
Simaruba glauca					1,1	1,1
Anonna reticulata	0,0	0,7	0,2	0,9		0,9
Syzygium jambos					0,7	0,7
Calophyllum antillanun					0,7	0,7
Total	100	100	100	300	100	400

Table 13. Phytosociological parameters of riparian forest of the River Cuyaguateje area that becomes rehabilitated border (1 and 2), including (IVIEA). Ar = Relative Abundance; Fr = Relative Frequency; Dr = Relative Dominance IVIE = Relative Importance Value Index; RNRi = Relative Natural Regeneration, IVIEA = Enlarged Ecological Importance Value Index

Figure 7 represents the horizontal structure of this forest, expressed by their distribution in three diameter classes. It shows irregularity in their distribution, characteristic of forests that recover from disturbances mostly by natural regeneration. The biggest frequency values were registered in the diameter classes from 2 to 10 cm and the smallest frequencies were found for diameter classes of 20 cm and bigger. This is the typical histogram in an inverted "J". Such a result agrees with [41], who states that forests age irregularly, and that species have the biggest frequency of individuals in small diameter classes. In general, it seems that the ecosystem is formed by heterogeneous populations with irregular diameter classes. This has been previously reported for most forests with a complex structure [42–43]. This result also indicates that previous reforestation efforts by state institutions using forest plantations were not suitable, as they did not mimic natural diameter distributions but provided the opportunity for natural regeneration to begin.

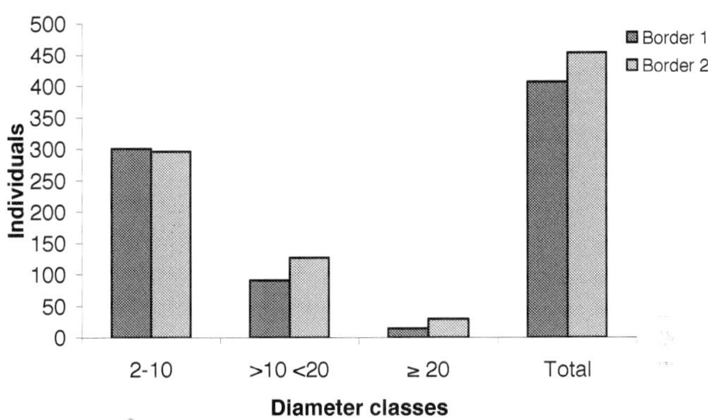

Figure 7. Distribution of all individuals in three diameter classes (Borders 1 and 2 of the area that is rehabilitated) in the Cuyaguateje's riversides.

As for the vertical structure, we found different strata (Figure 8). Although tree ages did not exceed 10 years, disparity in height was observed. Lower in the canopy were those species that present less resilience under disturbances (natural and artificial), because environmental conditions favour pioneer species. Tree height in border 2 is higher, mainly due to the presence of *G. arborea* and *B. vulgaris*. In general, frequency distribution in the form of an inverted "J" is certain, fundamentally for low seedling survival rates in plantations during the first years. The largest quantity of individuals have settled down through the natural regeneration in the last few years, although there is the presence of some individuals in the superior strata (up to 20 m in height) as a result of the plantations carried out during the early years.

At both borders, the biggest abundance of individuals was found for the smallest height classes (2 to 5 m), at 56%. As tree height increased, the number of individuals proportionally decreased, reaching less than 8% in the class for 10 m or more of height. At the two borders, 69%

of the individuals were present in the smallest heights. Only 5% were found reaching heights above 20 m. In this case, they were individuals of *R. regia*, *S. saman*, or *S. mombin*. These species coincided with those that farmers reported as existing in the riverside before 1970. This fact confirms the date when the water dam "The Cuyaguateje" was built and intensive sand extraction with heavy equipment began along the whole riverside [44].

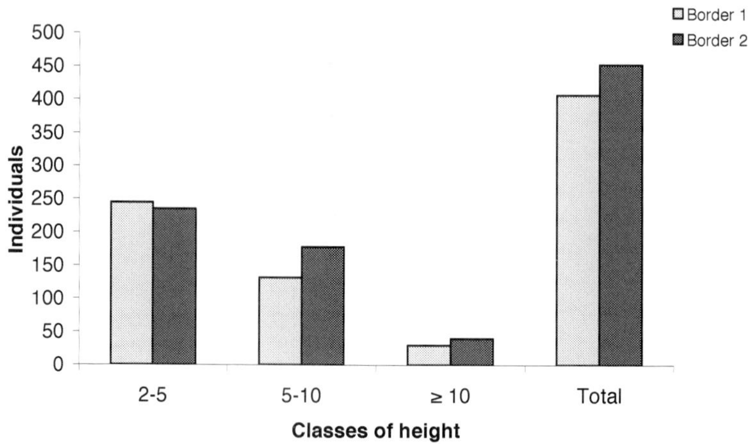

Figure 8. Vertical structure distributed in three classes of height (Borders 1 and 2 of the area that is rehabilitated) in the riverside of the Cuyaguateje.

Therefore, the biggest heights and diameters were represented by the smallest quantity in individuals. Such results coincide with reports by [32], who found low percentages of achievements and survivals in the plantations carried out by States Enterprise, as a consequence of the social insubordination of the residents, the inadequate selection of species and plantation methods during the reforestation, together with only partial participation of the local farmers in the reforestation plans. The biggest values of abundance found in the inferior stratum (a result of natural regeneration), indicate that the area has the potential to recover naturally. As [45] states, a high number of juvenile and young adults can be indicative of a stable or even an expansive population.

3.2.3. Conservation degree of the rehabilitated area

The conservation grade that is presented to an ecosystem is conditioned by the different indicators evaluated. Consideration of the behaviour of these indicators allowed for the inclusion of various elements in planning the complete forest rehabilitation. Identifying the causes of deterioration constitutes a fundamental link in their later management. Starting from the results obtained in the indicators evaluated according to [13], it was proven that the ecosystem is at the beginning of the rehabilitation stage: forest cover reached 72% of the surface, the modification degree was in the stocking category, the forest has recovered 26% of the original species, impacting on the recovery of the secondary forest. The summary of the evaluated aspects is shown in Table 14. If the different conservation categories suggested by [46] are used, and the sum of the values is assigned to each one of the evaluated parameters, the conservation degree of the vegetation cover was in the range of 10, indicating that it was fairly conserved (Table 15).

Indicators	Riparian forest of "River Cuyaguateje"	
	Index	Evaluation
Original species grade	High — Media	2
Stratification grade	Low Media	1
Cover grade	Very high	3
Modification grade	Media	2
Synanthropic index	0,19	1
Invader species grade	Media — low	1
Total of accumulated points		10

Table 14. Summary of indicators evaluated to determine the conservation degree of the riparian forest alongside in the River Cuyaguateje according to [13 and 46]

Sum of total values	Rehabilitation grade	Category of conservation
10	Beginning of rehabilitation	Fairly conserved

Table 15. Rehabilitation grade of riparian forest in its middle reaches.

The arguments mentioned previously present the result of the rehabilitation of the riversides of the Cuyaguateje, achieved through the participation of the executer actors (farmers and families). Study [5] shares similar approaches, indicating that rehabilitation does not imply that the site achieves an original, pre-disturbance ecological state. As the author explains, it is possible that a forest can recover its ecosystem function without recovering its structure completely. In many cases, the plantation of native trees or dominant pioneer species of ecological importance can help to start the rehabilitation process. Source [47] considers that ecological rehabilitation is an intermediate level between a degraded system and a restored ecosystem, with a composition and structure that can be similar or dissimilar to that of pre-disturbance. The restored system can be self-sustaining and used to provide ecological services, as in wooden production, medicinal products, and food, among others.

3.2.4. Participative strategy for the rehabilitation of the Cuyaguateje's riversides

Local actors' perception about the riparian forest of the River Cuyaguateje is very important for successful biodiversity restoration. As outlined by [47]: "In Cuba the care and the conservation of the biodiversity has advanced, and the existence of legal, logistical and infrastructural means that guarantee the operation of a good part of the environmental system, but this is still not enough". Participative methods applied in the present research contributed towards positive positions about the economic, environmental, and social perception of the local actors that live and work adjacently to the riparian forest. This study promoted environmental education as a result of the training that was implemented in this area. Local residents also have a particular perception of the legislation for the protection of the riverside forest. Figure 9 shows that the biggest incidence of infractions made by the local actors in the riversides of the River Cuyaguateje was during the period 2001–2004. A gradual decrease was observed, starting from 2004, when the participative work took place. A change in local residents' perspective was probably a consequence of acquired knowledge and environmental consciousness.

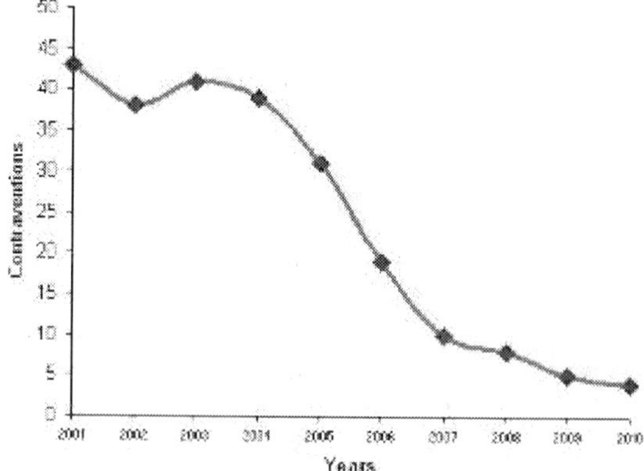

Figure 9. Trends of the infractions made by the local actors in the Cuyaguateje's riversides in the period 2001–2010.

Frequency analysis on the perception of local actors about regulations showed that 83% (24/29) of the interviewed actors could be classified as having the positions of advanced perception (Figure 10). It is very promising to realize that 78% (18/23) of the local residents are in advanced positions, or in other words, they know the regulations applicable in the riverside (minimum forest width, compulsory reforestation, and silvicultural management to be carried out, etc.). Local actors that only recognize the authority of the Forest State Service (FSS) and the fact that it has more than enough forest patrimony 22% (5/23) are classified as having intermediate positions. The lack of local residents classified in last group indicates a change in the perception and knowledge of legislation from the date of the initial surveys. This demonstrates the influence of the training actions and the high motivation showed by locals at the participative workshops and visits to the field.

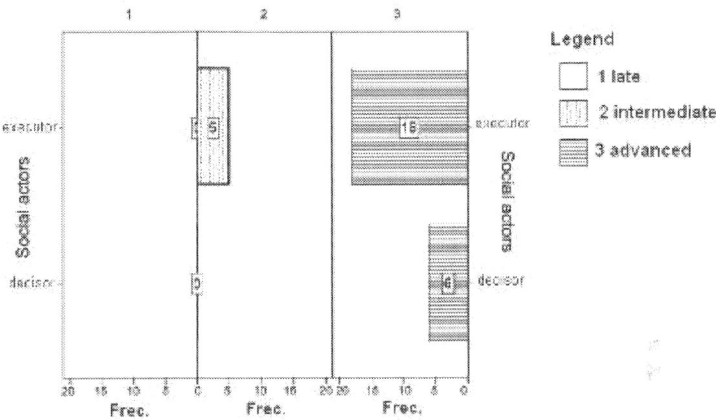

Figure 10. Distribution of frequencies obtained according to the perception of local actors on legislation governing the Cuyaguateje's riversides.

Our results contrasted with other studies without previous participative work from the local communities, such as the one carried out for [48] on the riversides of the River Caona. In that case, frequencies of high awareness positions were considerably low, about 30% (4/13). Therefore, it seems likely that the inclusion of a Participative Action Investigation (PAI) allowed the local actors to acquire knowledge on regulations for the protection of the riverside forests. Our results were also validated by the positions of the farmers and officials of the Forest State Service (FSS) during interviews and fieldwork. A good example is the change perceived in the farmer Noel Pérez. He became the informal leader of the community. A qualitative jump was perceived in his knowledge of the Forest Law, because he made several infractions until the year 2002, due to ignorance of the established regulations. After the training carried out in the participative workshops, the sense of ownership and commitment toward the riverside forest increased. This was evidenced in the results obtained in reforestation and establishment of the plantations, with the area belonging to Mr Pérez becoming the forest laboratory of the University Campus of Guane municipality and the Pinar del Río University.

Another example that confirms the necessity of introducing participative methods in the recovery of riparian forests is the study carried out by [49], in the basin of the River Sesesmiles (Honduras). These authors declared that 90% (18/20) of the local farmers did not know the legislation about the width of riverside fringes, although 75% (15/20) of the residents recognized the necessity of receiving training on the topic, and that is the state that watches over the execution of the established protection laws.

The approach in our research is that tree populations are the main agent to be managed when transforming the ecosystem. With an active participation of local residents it was possible to use their experience and to discuss with the scientists which tree species were more important for ecosystem rehabilitation. Most of the local residents showed a high degree of awareness

(Figure 11). About 78% (18/23) of the interviewed farmers showed knowledge on the species that should be planted: they recognized the species typical of the ecosystem, and they identified the characteristics of such species with regard to their function in the riverside forests. Participative workshops together with the application of PAI created learning environments that allowed forest scientists to explain the reality better, and at the same time to find solutions to the problems identified by the local actors. In other words, PAI was used as a mean of social mobilization. The importance of traditional, experience-based knowledge by farmers was demonstrated by the selection of species and management regimes. Preference was observed towards tree species and autochthonous shrubs that can establish and grow by passive reforestation. The knowledge of the present species in the ecosystem is necessary. According to [33], the presence of multiple species provides the security that ecosystem health will be maintained during disturbances or other environmental changes.

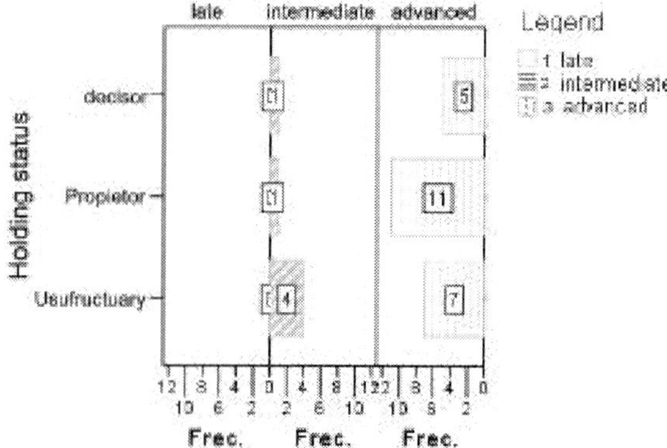

Figure 11. Distribution of frequencies of awareness of local residents on the appropriateness of forest species for riversides' restoration.

Local residents also showed knowledge on plantation times needed to obtain the best survival rates. Such times were related to previous results in riparian forests [10, 43–44, 50]. It is also important to recognize ownership status. Up to 92% (11/12) of the landowners behaved in terms of positions of advanced knowledge. They outlined as the main difficulties of achieving reforestation: the inadequate selection of species, the time for plantation, and especially the extensive impact of equine livestock. This livestock belongs to people that come from urban areas but who are not landowners.

Native farmers' knowledge of ecological communities and populations was of key importance. Their experiences contributed valid approaches about the species, reforestation methods, and the management of the restoration process. For example, the local Antonio Santoyo served as a guide to identify the species during the surveys of the riversides. He also indicated in workshops and interviews the most suitable species and the best methods for planting them.

Local empirical knowledge should be combined with the techniques of modern science to search for management regimes most suitable for the local ecological conditions. Farmers have accumulated knowledge that can be analysed by forest scientists, who in turn have scientific theories that can be put into practice. This is why it is important to implement participative methodologies. Social-participative ecosystem rehabilitation, with appropriate foundations, can be realizable and constitute an answer to some of the issues Cuba is facing at the moment. The decentralization of the environmental administration reinforces the grade of responsibility and the local residents' rights to forests. The knowledge that people have about their region, traditional uses of natural resources, location of the species, and in some cases the form of propagation of the plants, are important questions to consider in forest management plans. Indeed, to guarantee the success of reforestation it is important to combine academics' and farmers' knowledge on forest restoration practices. Unfortunately, this view has been usually dismissed in current research process not only for restoration but also in general silviculture.

Local residents were also aware of the socio-economic functions that riparian forests offer. Figure 12 shows that only 20% (6/29) of the local residents viewed the forest as useful only for timber. However, 72% (21/29) of local residents attributed to forests other functions of economic and social importance. This denotes that a transformation has been achieved as to the economic perception of the forest by locals. This result contrasted with those obtained by [48], where 77% of the interviewees assumed the forest to be exclusively a source of timber. In the same way, [49] diagnosed that only 45% of the farmers of the riverside of the River Sesesmiles (Honduras) perceived the riparian forests as having both economic and social value, because they offered products like fruits, firewood, and wood for the consumption of the family. According to [37], one of the goals of forest restoration could be the sustainable supply of goods and specific natural services for the social benefit of local residents.

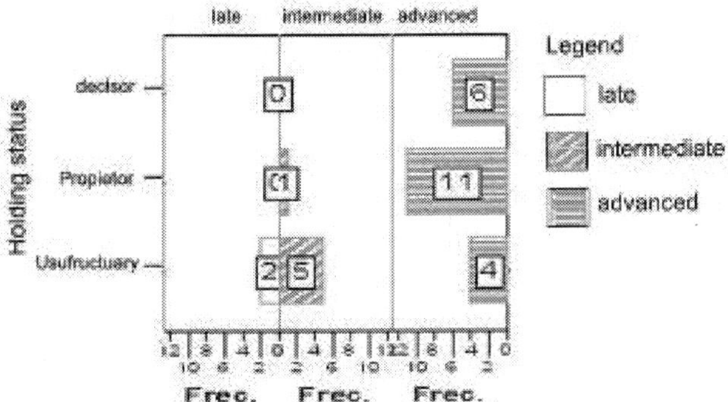

Figure 12. Distribution of frequencies obtained according to the perception of local actors about the socio-economic importance of riparian forest

As for the perception of the environmental function of the riparian forest, 52% (12/23) of local residents interviewed had an environmental knowledge of the protection of the soil (an intermediate position). This understanding seems to be transmitted from ancestors, but it could be changed by the participatory research methods. Nowadays, 48% (11/23) of the farmers (owners and tenants) that are adjacent to the riversides of the River Cuyaguateje are in positions of advanced awareness, identifying the forest function on controlling soil erosion. They relate this ecosystem service to water quality and biological diversity. Local residents also recognize the impact that forests have beyond their boundaries (Figure 13).

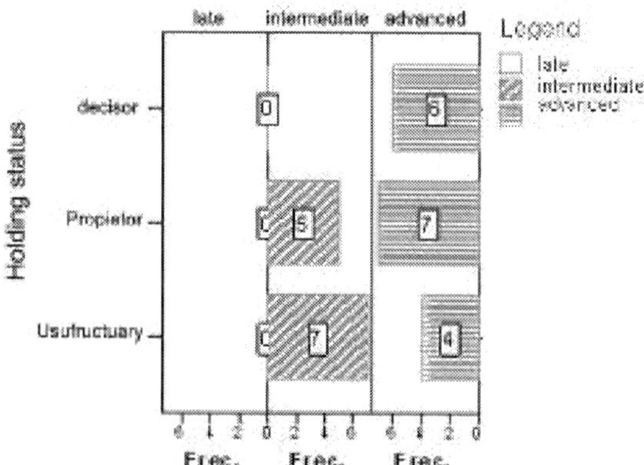

Figure 13. Distribution of frequencies obtained according to the perception of local actors about the environmental function of riparian forest.

In general, all the local actors recognized the forest's environmental functions. A group of them perceived the forest as an important element in the control of the current soil erosion in their farms. This aspect can be linked to the economic importance of forests, because farmers do not only see the forest as a source of timber, but rather they appreciate it as a tool for soil recovery that favours their crops and, therefore, their economy. No interviewed farmers were in late positions of awareness. The biggest proportion of advanced awareness was found for farmers that have lived all their lives in areas adjacent to the riverside, showing a sense of ownership toward this ecosystem. Therefore, the ownership status also influences the change of the perception in this variable. It highlights the polarization of the answers of the studied variable. Such a phenomenon especially affected the positions of advanced awareness, in which the interviewees identified the function of the forest in controlling the erosion of soil, and they related the forest with water quality and fauna migration. They also recognized the impact of the forest on the recovery of soil nutrients and food. As a consequence, there is an increment cultivation practices in bordering areas the forest. The testimonies aired in the workshops showed that farmers have appropriated the knowledge because they already attributed importance to the necessity of riverside protection. Therefore, they have not removed the soils of riverbanks; nor has the riverbed been altered by the formation of gullies in the areas lacking vegetation.

Local farmers were also asked "in their opinion, what are the causes of the present deterioration in the Cuyaguateje's riversides: direct or indirect human activity for long periods?" It was observed that all the local actors knew the causes of deterioration; these were identified as external and internal causes. In Figure 14, the positions of advanced knowledge are 69% (20/29), with an important representation of those who have always lived in this community. The four tenants represented in the figure, are also from this location: they gave up their lands in a moment of their life and after Cuban´s "special period" they got them back as tenants.

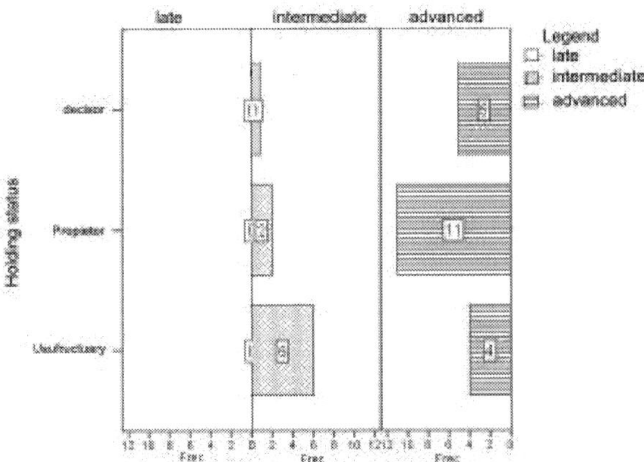

Figure 14. Distribution of frequencies obtained according to the perception of local actors on the deterioration causes of the Cuyaguateje's riversides.

As for the native farmers, cited deterioration causes were: dike construction, tree logging, sand extraction, and agricultural activities for tobacco production and other crops. Interestingly, they identified these causes as external, even the use of the soils for agricultural cultivation is for them a problem caused by their predecessors. Such a position shows low self-awareness of their impact on the environment, because they only identified responsibility in other actors different from themselves. Similar results were obtained by [48], who reported that 46% of the interviewees considered that the causes of forest deterioration are multiple, but never included the rural owners of these lands. Most of the farmers perceive the borders of the river as being their own property, or at least they think that they are entitled to use them and to manage them, in spite of the fact that the land is state property. Such aspect coincided with the results of [48–49].

Using available historical data and taking into account the riversides' situation in the year 2000 (the presence of scarce isolated trees), it can be assumed that the riversides were subjected to anthropogenic pressures, larger even than the natural causes mentioned above. Besides deforestation of the riversides, the riverbed has also been distorted, conditioned fundamentally by the expansion of agriculture, the lack of appropriate conservation techniques, and dike construction. In a general way, it is evident that the deterioration of native forests conditioned irreversible changes in the riverbed, both in the width of the river and in the formation of bends in the waterways [31–32].

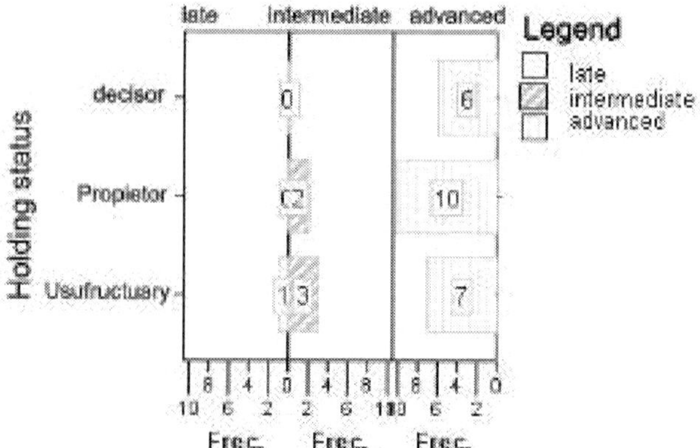

Figure 15. Distribution of frequencies obtained according to the perception of local actors on the most interested in reforestation in the Cuyaguateje's riversides.

Farmers that inhabit and work in the Cuyaguateje's riversides have always witnessed a deforested landscape. According to their recollections, the first reforestation actions began in the year 2000, with an audit procedure in which the Forest State Service, and businesses (Forest Enterprise Macurije, Tobacco Enterprise, and Enterprise of Several Cultivations) participated. It was coordinated by the Municipal Government's Direction. In that year, 22 hectares were planted for the first time. The biggest incidences of social indiscipline also happened. Starting from the year 2006, a gradual recovery of the riverside forest was observed, with the establishment of some trees that were planted and others that have colonized the site in a natural way as a consequence of the decrease of anthropogenic pressure (resulting from the knowledge acquired in the training workshops that began in the year 2004).

When local farmers were asked who is most interested in reforesting the riversides of the River Cuyaguateje, 96% of the interviewees (28/29) were considered in one way or another to have an interest in the reforestation, locating them in positions of intermediate and advanced awareness (Figure 15).

The behaviour of the local residents during the different participatory activities indicates a high disposition for auto-transformation. They proposed collective actions for their own community. They recognized the importance of their approaches in the rehabilitation process and that they can be part of decision-making about the tasks that impact the solutions of the outlined problems. In the same way, they recognized the necessity of collective action and participation: the systemic integration of everybody is necessary to implement actions. The previous position indicates the need to focus on the relationships among the local actors from a sociological perspective, where the cooperation of everybody prevails. Previous results on top-down approaches were not positive; it is necessary to integrate local residents' opinions and decide with them what is conceived for their environment.

It is important in the rehabilitation process to focus on science, technology, and society. This requires that rehabilitation plans should have a range of political and strategies designed to incorporate feedback from local actors' experiences. To achieve the effective contribution of the executive managers, they have to be motivated and prepared by education. Therefore, one of the first actions to implement should be training stakeholders and decision-makers on the principles of sustainability. The decision-making actors should not adopt a technocratic and authoritarian position; they should frame their administration in the socio-cultural and natural context in which the problem is found. Coinciding with [48], we think this is one of the reasons why the traditional approaches to stop the problems associated with riverside management have, in most cases, failed. Therefore, to get sustainable forest administration of the riverside forests, the institutions, especially the Forest State Service, have to get adopt the paradigm 'Science, Technology, and Society'.

As for the local residents' perceptions about the actions of changes toward the reforestation, it was observed that 78% of the executive actors (18/23) have carried out positive actions to the benefit of the riversides, including the decrease of cultivation and cattle ranching activities (Figure 16). A participative forest administration has become a primordial element in the strategies of forest administration: a structured collaboration between the government and the local actors (farmers) to manage forest resources in order to obtain common and sustainable objectives. Previous works have been executed, although not all of them have been favourable in outcomes. The failures have been mainly caused by the lack of participation of the local residents, such as the case of disproportionate plantation of exotic species (*G. arborea* and *A. mangium*) which have disordered the natural landscape. Such inadequate species selection produced low survival rates of *G. arborea* in the areas where floods last the longest. It is therefore necessary to increase the local knowledge and perception of riverside forests, since the points of view of the community actors are fundamental for achieving sustainable management. In this aspect, it is important to highlight the case of the rural community leader Noel Pérez for the good results obtained in that group's property. The Forest State Service has already certified four hectares of forest established on their property.

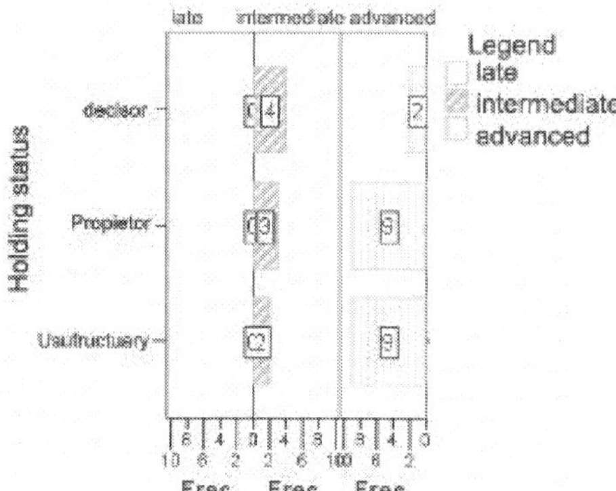

Figure 16. Distribution of frequencies obtained according to the perception of local actors on their tendency to the change in the Cuyaguateje's riversides.

The final variable tested was the local perception of the most suitable method for reforestation. Figure 17 shows the results of perception analysis, with most of the actors leaning towards natural regeneration (74%, 17/23). The main condition of achieving successful implementation is to eliminate the barriers that prevent natural regeneration. Regarding this issue, [51] outlined that when the stressing factors are eliminated in a degraded ecosystem, there is a trend that restoration follows. These authors asserted that success of the restoration does not only depend on the costs, funding sources, or the political will of the interested institutions. The main issue is the participation of the local communities with the power of decision over restoration plans.

As for the selected species, all the ones established during the regeneration or the plantations are valid. The improvement of a degraded system can also begin by means of the plantation of native trees, of dominant pioneer species and those of more ecological weight: all those that protect the riversides of the river. In this approach, local residents outlined the necessity of reforestation with native species, in addition to just leaving the land without agricultural management. Among the species that were identified as suitable for natural regeneration are: *L. domingensis, G. ulmifolia, S. saman, T. catappa, T. hirta, S. saponaria, T. angustata, B. simaruba, C. collococca, T. citrifolia, R. regia, C. dentata, A. reticulata, C. hirsuta, A. inermis*, and *S. mombin*. It was also identified *G. sepium*, to be regenerated through direct seeding or by planting stakes.

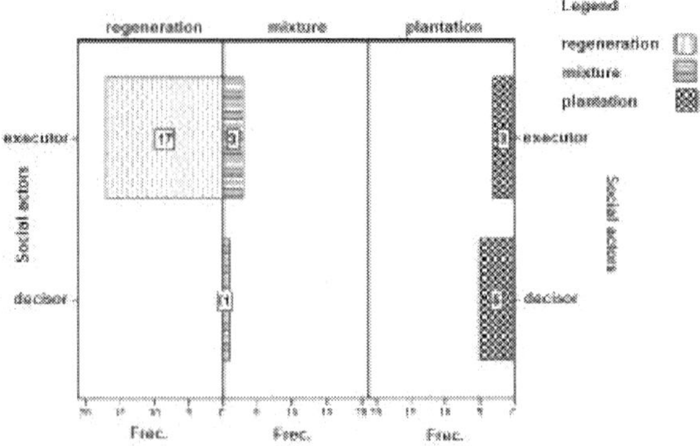

Figure 17. Distribution of frequencies obtained according to the perception of local actors on the most suitable reforestation method in the Cuyaguateje's riversides.

3.2.5. Correlation between studied variables

Table 16 presents the correlation analysis of the variables studied, applying the correlation coefficient Spearman's Rho. Significant relationships among several variables were found: local actors, perception of the environmental function, and the tendency to change. The variable "interest for the reforestation" was highly correlated with the perception of the environment, and less significantly with the perception of the suitable species. Correlation was also observed among the following variables: local actors, reforestation methods, ownership status, and origins of those interviewed.

	AL	ET	PE	PSLF	PSFA	PSEI	PSCD	PSIE	IR	TC	MR
AL	1,000	-,754(**)	-,637(**)	-,233	-,429(*)	-,051	,127	-,312	-,260	,393(*)	-,658(**)
ET	-,754(**)	1,000	,302	,393(*)	,454(*)	,240	,132	,601(**)	,346	-,340	,464(*)
PE	-,637(**)	,302	1,000	,077	,362	,125	-,313	-,100	,314	-,253	,408(*)
PSLF	-,233	,393(*)	,077	1,000	,173	-,008	,019	,286	,201	-,306	,167
PSFA	-,429(*)	,454(*)	,362	,173	1,000	,262	,209	,150	,606(**)	,193	,428(*)
PSEI	-,051	,240	,125	-,008	,262	1,000	,302	,267	,390(*)	,393(*)	-,087
PSCD	,127	,132	-,313	,019	,209	,302	1,000	,152	,145	,244	-,145
PSIE	-,312	,601(**)	-,100	,286	,150	,267	,152	1,000	,343	-,057	,295
IR	-,260	,346	,314	,201	,606(**)	,390(*)	,145	,343	1,000	,051	,298
TC	,393(*)	-,340	-,253	-,306	,193	,393(*)	,244	-,057	,051	1,000	-,167
MR	-,658(**)	,464(*)	,408(*)	,167	,428(*)	-,087	-,145	,295	,298	-,167	1,000

Legend

	P S L. Perception about legislation	PSIE. Perception about economic importance
AL. Local actors	PSFA. Perception of environmental function	IR. Interest in the reforestation
ET. Ownership status	PSEI. Perception of suitable species	TC. Tendency to change
PE. Origin of those interviewed	PSCD. Perception of deterioration causes	MR. Reforestation method

* The correlation is significant at the level 0.05 (bilateral).

** The correlation is significant at the level 0.01 (bilateral).

Table 16. Correlation analysis among the studied variables. Correlation coefficient Spearman's Rho.

In other words, the local actors that are interested in reforestation also attribute importance to protection of the environment and at the same time they are those that have the best knowledge on the most suitable species for reforestation. Obviously, there was a change of attitude in the local actors related with the assimilation of knowledge. This behaviour is achieved when there is an empowerment of local actors regarding the problem to be solved, and they are motivated to promote the commitment and the responsibility of each one of them towards the local forest. In such a sense, [47] stated: "among social actors, psychological and social elements mediate. This does not mean that a concern for the environmental questions necessarily implies pro-environment behaviour. To achieve this point, a committed participation of the involved actors is needed". Those individuals that manifest sensitivity towards the environment are the most interested in reforesting the area and at the same time they plead for the insertion of the participative silviculture for the rehabilitation. This indicates a change of attitude and sense of ownership toward the riverside.

3.2.6. Participative strategy and work lines for the rehabilitation of the riparian forest of the River Cuyaguateje

The information needed to elaborate the participative strategy was:

- Characterization of the structure and composition of the rehabilitated forest with the participation of local actors.

- The results of encounters, interviews, workshops, visits, participative observations, and consultation of the different local actors, decision-makers, and executives.

- Meetings with technicians and specialists at municipal, provincial, and national levels.

- The results of the work carried out during a ten-year period in a case study of rehabilitation and "Participative Action Investigation", which served as a basis for the feedback of the strategy.

The objective of the strategy was to rehabilitate the riparian forest of the River Cuyaguateje in their middle reaches through participative silviculture and the method Participative Action Investigation, strengthening the capacities of the local actors, guiding them towards sustainable forest management, and to be based on the perceptions of the local actors, institutions, and technologies. The specific objectives were:

- To achieve the success of the strategies, methodologies, and programmes proposed by the state for the recovery of the riparian forest of the River Cuyaguateje in its middle reaches.

- To improve the ecological state of the riversides and the River Cuyaguateje, reforesting it with the participation of the local actors, with preferably autochthonous fruit-bearing and wood species, although exotic ones can be used if their adaptation has been proved, whenever the established technical and juridical norms are followed.

- To encourage the integration of local actors in the rehabilitation of this fluvial ecosystem, and their use of politics and management with sustainability approaches.

- To contribute with information and experiences to improve the performances that are carried out in the rehabilitation of the riparian forest of the River Cuyaguateje and other rivers in the country.

To achieve this participative rehabilitation, several strategic lines were proposed:

1. *Analysis and real organization of the context.* This line is focused on two basic objectives: one to locate the area and the other to organize the context. Having broad information on the topic is needed, which should include theoretical foundations and existing regulations,

as well as the practical experience on previous programmes and strategies related to the topic.

2. *Training and education.* Both are the roads to obtaining knowledge, abilities, and skills. Also, to propitiate the change of attitudes towards favourable values for the rehabilitation, sustainable management, administrative capacity, and local organization on the riverside forest. Several processes were carried out at the same time: encounters with farmers, families, decision-makers, and researchers. The experiences of the local actors were exposed in the gully rectification, selection of species, and plantation methods. In the same way, decision-makers presented to the farmers scientific results on species and methods used for the rehabilitation of the riverside forests.

3. *The committed participation of local actors.* To achieve this three key elements are needed: local actors admit to being part of the problem, are allowed to participate which helps the acquisition of knowledge, and to play a part in deciding what bears on values formation. It is therefore necessary to link the ecological rehabilitation with the biocultural rehabilitation, such that the historical, cultural, social, aesthetic, and moral dimensions of the involved actors are included.

4. *Silvicultural management under the modality of community participation.* Social or participative silviculture is achieved with the inclusion of executor actors in the rehabilitation process, from the farmer and the community managing the forest to actions by the state and the community. The individual forest management for the farmer's family benefit should be established.

5. *Self-management by the farmers of the riparian forest.* This leads to better planning of the natural resources, transforming each farmer into the administrator of his/her riverside tract. They also become the main decision-maker in the selection of species and suitable methods for the reforestation, as well as the different silvicultural practices after forest establishment.

6. *Protection activities and surveillance.* Regulations concerning the protection of the riparian forests are indispensable tasks to obtain better results. Good technical results cannot be applied to contexts if agreements and regulation mechanisms are not in place.

7. *Participative evaluation.* Evaluation constitutes a basic requirement to measure the advances or limitations of the execution of the carried-out activities. It is a necessary feedback mechanism to identify obstacles and to look for alternative solutions in a timely fashion. It also allows reaching the goals agreed by all the parts (executers and decision-makers). The main purpose is to stimulate those who intervene in the rehabilitation to stop and meditate on what has happened in the past, with the purpose of making better decisions in the future. When carrying out the evaluation, the actors find out what has worked well or not, and why. By means of this process, there is the greater chance people will take better corrective measures, given that the actors are those that discover and understand these measures.

Conglomerates	Plots	Total
Conglomerate 1	1; 8;11; 2; 36; 3; 28; 15; 34; 13; 14; 16; 21; 35; 5; 19; 31; 29; 32	19
Conglomerate 2	6; 7; 12; 10; 30	5
Conglomerate 3	4; 18; 17; 27; 33; 9; 24;26	8
Conglomerate 4	20; 22; 23; 25	4

Table 17. Resulting conglomerates by means of Ward's linking method.

3.3. Study of case No. 3: The native rainforests of the Toa´s sector Quibiján-Naranjal

3.3.1. Diversity of species in native forests exploited in the basin of the River Toa (sector Quibiján-Naranjal)

A total of 36 plots were sampled that represented the overall ecosystem. The cluster analysis based on Sorensen's similarity showed four clusters (Table 17).

The first conglomerate accounted for plots containing species of high economic and ecological value, such as *B. buceras*, *C. antillanum*, *H. elatus*, and *Purdiaea velutina*. These plots, although distant from each other, presented very similar flora. The second conglomerate was also notorious for the presence of species with high commercial value and ecological importance, such as: *Castilla elastica*, *G. guara*, *Spondias mombin*, *Cedrela odorata*, and *Carapa guianensis*. An important proportion of the present species in the study, are heliophytes species (which grow best under direct sunlight), many of them commercially important and reported in Neotropical forests such as *C. antillanum*, *C. utile*, and *B. capitata*. Conglomerates III and IV presented a mix of species (*C. antillanum*, *C. peltata*, *H. elatus*, *C. odorata*, *C. guianensis*, *A. inermis*, and *J. vulgaris*) that varied in abundance. The presence of pioneer species clearly indicates anthropic activities. These species had regenerated easily because the seeds are big and heavy, which favours germination. In general, the four groups share almost all the species, the rare species being *Guazuma tomentosa*, *L. domingensis*, and *G. sepium*.

3.3.2. Structure of the native forests in the sector Quibiján-Naranjal of Toa

Regarding the horizontal structure of the forest, Table 18 shows the 10 species with the highest abundance, frequency, dominance, and IVIE. The species with more IVIE were *H. elatus*, *C. antillanum*, *S. laurifolium*, *G. guara*, and *T. catappa*. This ecological importance value index represents the intricate relationships that the species maintain with other species of plants and organisms that help to maintain the dynamic and functional balance of the ecosystems [36].

No.	Species	Abundance (%)	Frequency (%)	Dominance (%)	IVIE (%)
1	*Hibiscus elatus* Sw.	9,54	63,9	11,0	84,38
2	*Calophyllum antillanum* Britton	8,11	61,10	2,81	72,04
3	*Sapium laurifolium* Griseb.	5,19	61,10	1,26	64,78

No.	Species	Abundance (%)	Frequency (%)	Dominance (%)	IVIE (%)
4	*Guarea guara* (Jacq.) P.	7,01	50,00	4,70	58,93
5	*Terminalia catappa* L.	5,00	47,20	3,26	55,48
6	*Syzygium jambos* L.	5,65	44,40	1,29	51,38
7	*Cecropia peltata* L.	2,66	47,20	2,87	49,98
8	*Spondias mombin* L.	3,11	4,44	3,29	48,07
9	*Carapa guianensis* Aubl.	2,79	33,30	8,92	45,05
10	*Zanthoxylum martinicense* Lam.	2,99	38,90	4,04	43,14

Table 18. Abundance, Frequency, Dominance, and IVIE of the native forests of the sector Quibiján-Naranjal of Toa.

Table 19 shows how *Jambosa vulgaris* is one of the most abundant species, and it is also recognized in Cuba as invasive taxa [38]. This species is altering the structure and function of the forest. That fact is corroborated by [41–42], who stated that for this plant community valuable species are *C. antillanum, A. inermis*, and *H. elatum*, but generally in a smaller proportion. Owing to the anthropic pressure of the local communities, the species that are more plentiful are of scarce woody value; prevailing in higher proportions are *Syzygium jambos* and other species like *C. peltata, G. guidonia, L. domingensis*, and *Bursera simaruba*. Source [55] confirmed that these species present good adaptation to the soil and climatic conditions of the strip forest of the River Toa.

Species	Inferior Stratum		Medium Stratum		Superior Stratum	
	Number of trees	PS	Number of trees	PS	Number of trees	PS
Syzygium jambos L.	60	13,69	24	2,98	0	0
Guarea guara (Jacq.) P.	34	7,76	58	7,21	3	3,44
Terminalia catappa L.	32	7,30	41	5,09	0	0
Calophyllum antillanum Britton.	28	6,39	91	11,31	1	1,14
Hibiscus elatus Sw.	26	5,93	49	6,09	25	28,70
Dendropanax arboreus L.	20	5,56	16	1,99	0	0
Sapium laurifolium Griseb	24	5,47	47	5,84	8	9,19
Casasia calophylla A. Rich.	5	1,14	5	0,62	0	0
Mangifera indica L.	6	1,36	13	1,61	0	0
Acalypha diversifolia Jacq.	7	1,59	18	2,23	1	1,14
Andira inermis (W. Wright) DC	9	2,05	28	3,48	1	1,14
Zanthoxylum martinicense L.	10	2,28	30	3,73	5	5,74
Ocotea leucoxylon Sw.	12	2,73	3	0,37	0	0
Roystonea regia HBK	17	3,88	32	3,98	0	0

Table 19. Main species best represented in terms of sociological position in the native forests of the sector Quibiján-Naranjal of Toa.

Regarding the vertical structure of the forest, Table 19 shows that the main species with better sociological positions are *Syzygium jambos* in the lower canopy layer and *C. antillanum* in the intermediate canopy layer. In the superior canopy stratum, *C. guianensis* stands out together with *H. elatus*. These results indicate that in the forest there are not stable relationships in all the strata, nor is there evidence of the application of silvicultural systems. It seems that all canopy layers are affected by traditional management with a high intensity of indiscriminate tree felling, either for housing, firewood, or gap opening for agricultural cultivation. The superior stratum presents few species of economic importance [43]. Many of the existent species are not represented in all the strata. On the other hand, [53] outlines that a certain species takes an important role in the structure and composition of the forest when it is represented in all its strata. The author also said that the more regular the distribution of the individuals of a species in the vertical structure of a forest (gradual decrease of the number of trees as you ascend from the inferior stratum to the superior), so higher value in the phytoso-ciological position.

3.3.3. Influences of environmental variables in the structure of the native forest.

The results of canonical correspondence analysis (CCA) were globally significant (trace = 1.876, F = 2.27, and P = 0.002). The first four axes of the CCA offered a solution to the ordination of units of samplings and of species. Total variability in the data of species abundance (inertia = 3.609) explained 49.2% of the relationship between environmental variables and species distribution, and 26% of the variance of species distributions in each group. These results indicate a strong gradient (Table 20), because for ecological data the value of inertia is typically low (smaller than 10%), especially when they present strong gradients [50]. The significance test of the first canonical axis demonstrated that this was also statistically significant, with auto-value = 0.316, F = 2.15, and P = 0.0020 (Figures 18 and 19).

Number of canonical axis: 4	Axis 1	Axis 2	Axis 3	Axis 4	Total Inertia
Auto-values:	0,316	0,250	0,198	0,173	3,609
Correlation species — environmental values:	0,956	0,931	0,884	0,938	
Accumulated percentage of the variance					
of data of species :	8,8	15,7	21,2	26,0	
of relation of species-environmental values:	16,6	29,8	40,1	49,2	
Sum of auto-values					3,609
Sum of canonical auto-values					1,902

Table 20. Result of the canonical correspondence analysis (CCA) of species abundance, transformed logarithmically in each one of the 36 sampling units in the function of their environmental variables.

The negative end of the axis 1 (CCA 1) shows a relationship with "distance to the highway (DCAR)" and with "distance to housing (DVIV)". Although in a smaller proportion, the negative end of the axis 2 (CCA 2) is also related to increases of the altitude (Alt), and slope (PEN). The positive end of axis 3 (CCA 3) is associated to match increase (P), distance to the highway (DCAR), and organic matter (MO), and in its negative end it is related to distance to the highway (Figure 19). Plots of group I correspond to slope gradient and altitude (Figure 18 and 19), while plots of group III are ordered following a gradient of distance to the highway and housing. The results of the ordination using the CCA 1 and CCA 3 show an evident distinction for the separation among plots of groups I, II, and IV for the variables' explanatory measures and for the separation of group III. While *C. elastica, G. guara, C. odorata, T. catappa,* and *S. mombin* species are present in plots of group III, they are associated with distance to the highway and distance to roads, coinciding with observations in the research area that research plots of group III have bigger anthropogenic activity that facilitates the development of secondary species.

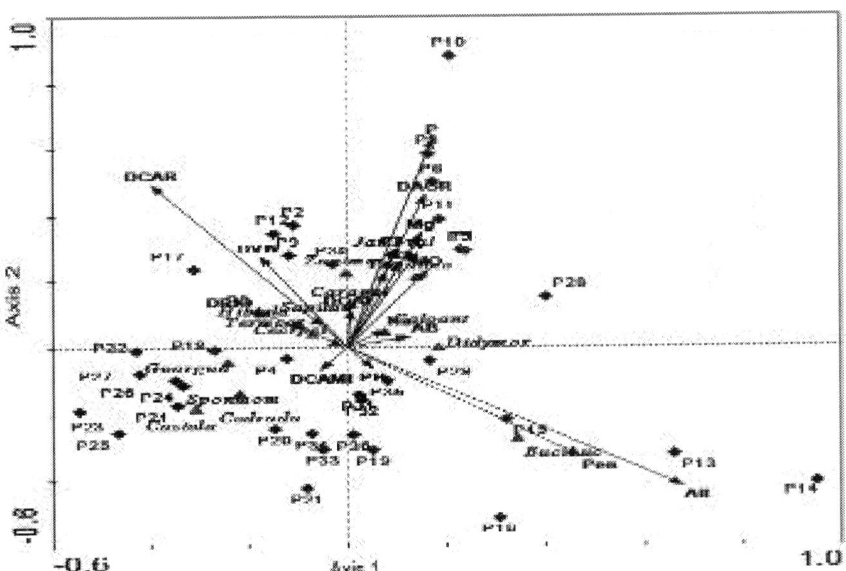

Figure 18. Projection of environmental variables, in census units and 15 species of bigger IVIE of the analysis of canonical correspondence in relation to the axes ACC1 and ACC2. The plots are the rhombuses, the species are the triangles, and the ecological variables are the arrows. Codes: Hibiela = *Hibiscus elatus,* Caloant = *Calophyllum antillanum,* Sapilau = *Sapium laurifolium,* Guargua = *Guarea guara,* Cecrpel = *Cecropia peltata,* Termcat = *Terminalia catappa,* Jambvul = *Jambosa vulgaris,* Sponmom = *Spondias mombin,* Andiine = *Andira inermis,* Zantmar = *Zanthoxylum martenicense,* Caragui = *Carapa guianensis,* Didymor = *Didymopanax morototonii,* Cedrodo = *Cedrela odorata,* Castela = *Castilla elastica,* Buchecap = *Buchenavia capitata.*

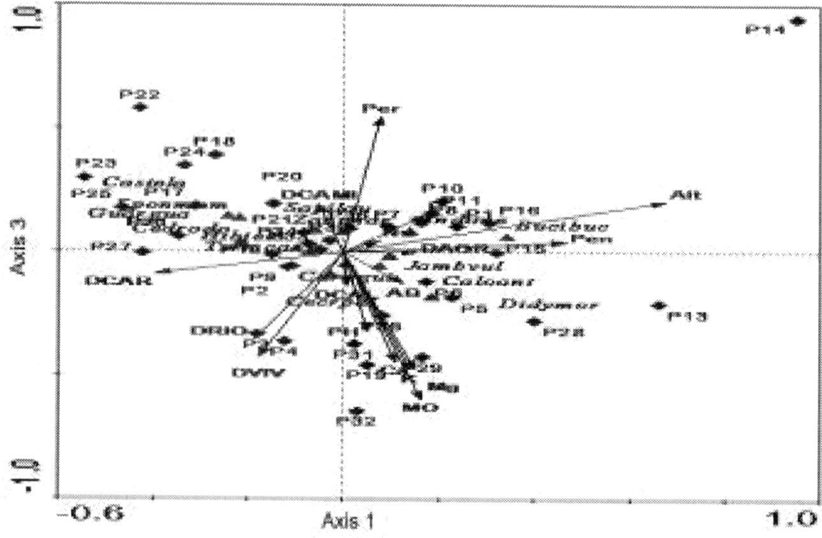

Figure 19. Projection of environmental variables, plots, and species of canonical correspondence analysis in relation to the axes ACC1 and ACC3. Codes: Hibiela = *Hibiscus elatus*, Caloant = *Calophyllum antillanum*, Sapilau = *Sapium laurifolium*, Guargua = *Guarea guara*, Cecrpel = *Cecropia peltata*, Termcat = *Terminalia catappa*, Jambvul = *Jambosa vulgaris*, Sponmom = *Spondias mombin*, Andiine = *Andira inermis*, Zantmar = *Zanthoxylum martinicense*, Caragui = *Carapa guianensis*, Didymor = *Didymopanax morototonii*, Cedrodo = *Cedrela odorata*, Castela = *Castilla elastica*, Buchecap = *Buchenavia capitata*.

3.3.4. Proposal for rehabilitation of native forests in the sector Quibiján-Naranjal of Toa

To design the restoration proposal, we have considered the approach by [5], who suggested 13 steps for the restoration plan.

- *Step 1: Definition of the reference ecosystem.* Knowledge of the region and its land-use history has been kept in mind when planning research to be carried out in the area. These are reflected in [11, 50, and 55].

- *Step 2: Evaluate the current state of the ecosystem or community.* Evaluate the structure, composition of species, and their ecological function. The study identified 24 families, 49 genera, and 52 species of angiosperm plants, for a total of 1507 individuals. The families with higher abundance increased the biodiversity, but they did not contain most of the individuals. This fact evidenced changes in the structure and composition of the species in the study area, as a consequence of selective logging, timber and firewood extraction, gap opening for subsistence agriculture, and road opening. These were the main sources of disturbances in the secondary forest to the riverside of the River Toa.

- *Step 3: Definition of the scales at organization levels.* Scales: Regional — Basin Toa, Local — Quibiján-Naranjal of Toa, Levels of organization community (biological) — secondary forest.

- *Step 4: Establishing the scales and hierarchies of disturbances.* Scales: 1: selective logging; 2: extraction of firewood: 3: gap opening for subsistence agriculture, and 4: total roads impact. Consequences of such disturbances: 1) irregularities in the diameter structure of species of high commercial value, 2) abundance of species of low commercial value, and 3) dominance of exotic species.

- *Step 5: Achieving local community participation.* Ecological restoration is an activity with different spatial and temporal scales, in which the anthropogenic disturbances play an important part in any scale that is chosen [5]. The interest and the consciousness of the different actors that are involved in the management of natural resources in Basin Toa form a very important aspect in the coordination and execution of the projects and programmes, according to the strengths and possibilities that they present. Training to guide the individuals of these communities towards a more scientific-based knowledge, development, and strengthening their abilities should be one of the objectives proposed in the action plan.

- *Step 6: Evaluating regeneration potential.* As for the evaluation of the regeneration potential, one should keep in mind the location, readiness of the species in the forest, abundance, and successional stage, with the objective of selecting a group of pioneer, dominant, and co-dominant species with high economic values. Tree species that present better sociological positions are in Table 19.

- *Step 7: Identifying barriers to restoration.* Natural barriers: 1) dominance of species with little commercial value, 2) abundance of exotic species, and 3) irregularities of the diameter classes of the species of high commercial value. Social barriers: 1) selective logging, 2) firewood extraction, 3) gap opening for subsistence agriculture, and 4) total roads impact.

- *Step 8: Selecting the appropriate species for forest restoration.* This is a very important step for the success of the restoration plan. It is the fundamental axis in any reforestation project that seeks to be carried out in a certain area with the objective of re-establishing ecological values and of conservation of the ecosystem in general. The application of modern silvicultural techniques that seek to go beyond the traditional ones, such as: the spaced group technique or planted selection forest, reforestation with native species, passive reforestation (using natural regeneration), and using agroforestry systems (the method of Taungya). This will frame a new vision in the silvicultural development of the local Enterprise for the sake of increasing the technical personnel's skill and training level. Combined with the nucleation strategy described by [55–57] that also contributes to the recruitment of native species, it will increase the effectiveness of the restoration of tropical forests. With the objective of achieving the quick recovery of the forest, species recommended [5] are: *H. elatus, C. antillanum,* and *B. capitata.* These species present high percentages of natural regeneration in places that have been affected by the action of the winds, selective logging, and the effect of borders of roads or agricultural cultivations. It is recommended to use species of high economic value and those that present high values of relative abundance, relative frequency, relative dominance, and importance value index, such as *H. elatus, C. antillanum, C. guianensis, D. cubensis,* and *T. dubia.*

- *Step 9: Propagating and handling of the species.* The species selected present their own qualities for forestry. Therefore, it is important to have knowledge of their characteristics, how to spread them, and how they are to be managed. The success of the programmes and plantation projects depends on this knowledge.

- *Step 10: Selecting sites.* Selected sites were chosen as product of field research that could give us precise information regarding scale levels and their regime of disturbances, as well as the level of anthropization in the ecosystem.

- *Step 11: Defining the strategy to overcome the barriers to the restoration.* In this step it should be determined how to use the so-called modern silviculture techniques that go beyond traditional forestry, such as: the technique of spaced dense groups for plantation, reforestation with native species, passive reforestation (using natural regeneration), and the agroforestry systems (the method of Taungya).

- *Step 12: Monitoring of the restoration process.* Monitoring is fundamental to understanding the behaviour of the ecosystem over time, to predict and/or to prevent unwanted changes, and for evaluating if objectives are met or whether pertinent modifications should be made.

- *Step 13: Consolidating the restoration process.* The consolidation of a restoration project implies that most of the barriers to restoration have been identified and overcome, and that the ecosystem evolves according to the outlined objectives. The maintenance works and monitoring programmes should indicate that the process goes on in a satisfactory way and the ecosystem begins to show self-sustaining properties, such as the enrichment of species, wildlife recovery, and re-establishment of environmental services related with the quality of the water and the soil [55–57].

4. Conclusions

The three study cases introduced in this chapter shows how restoration of degraded tropical forests is possible, if management plans are implemented in a thorough way and they involve local residents and administrations at different levels. The assessment of forest ecosystems condition is complex, and changing it will have a multifactorial impact that depends not only on the forest's structure but also on ecosystem resilience and on-going human pressure. The natural and induced changes in the flora have an influence on the sequence of ecological succession. Therefore, to design a potential restoration strategy, planning must begin from the holistic assessment of the ecosystem functioning (composition and structure) and the participative action of local communities.

Based on our results we can conclude that: 1) the illegal selective logging and the exploitation of forest for wood and non-woody forest products are the main stresses placed on the Cuban forest ecosystems in the last hundred years. 2) The implementation of modern silviculture techniques that use key species identified during intensive forest assessment should be the starting point for restoration of tropical forests. 3) The participative techniques during the rehabilitation and restoration process should play a crucial role in the cases where the local communities govern the area to be restored. We think that such conclusions are applicable to most of the tropical forests around the world, without forgetting their own local particularities.

Author details

Eduardo González Izquierdo[1*], Juan A. Blanco[2], Gretel Geada López[3],
Rogelio Sotolongo Sospedra[4], Martín González González[5], Barbarita Mitjans Moreno[6],
Alfredo Jimenez González[7] and José Sánchez Fonseca [8]

*Address all correspondence to: eduardo@upr.edu.cu

1 Forest Research Centre, Pinar del Río University, Pinar del Río, Cuba

2 Universidad Pública de Navarra, Pamplona, Spain

3 Biology Department, Pinar del Río University, Pinar del Río, Cuba

4 Forestry Department, Pinar del Río University, Pinar del Río, Cuba

5 Sociocultural Studies Department, Pinar del Río University, Pinar del Río, Cuba

6 Municipal University Centre of Guane, Pinar del Río University, Guane, Cuba

7 Agricultural Department, Artemisa University, Artemisa, Cuba

8 Agroforestry Department, Guantánamo University, Guantánamo, Cuba

References

[1] Blanco J. A. Aplicaciones de modelos ecológicos a la gestión de recursos naturales. OmniaScience. Monographs 2013. http://dx.doi.org/10.3926/oms.60

[2] Food and Agricultural Organization (FAO) Situación de los bosques del mundo 2007.

[3] Barrera I. and Valdés C. Herramientas para abordar la Restauración Ecológica de áreas disturbadas en Colombia. Revista de la Facultad de Ciencias Edición especial II, 2007; 12:11–24.

[4] Gann G. D. and Lamb D. La restauración ecológica — un medio para conservar la biodiversidad y mantener los medios de vida. http//:www.ser.org (Accessed June 2012) 2009.

[5] Vargas O. Los pasos fundamentales en la restauración ecológica. Universidad Nacional de Colombia. Guía Metodológica para la Restauración Ecológica del Bosque Altoandino. Grupo de Restauración Ecológica, Universidad Nacional de Colombia, 2008.

[6] Hernández F. Restauración de la diversidad vegetal arbórea de los bosques siempreverdes de la Reserva de la Biosfera Sierra del Rosario (RBSR), Pinar del Río, Cuba. PhD thesis. Universidad de Pinar del Río, Cuba and Universidad de Alicante España. 2010.

[7] Herrera R., Menéndez L., Rodríguez M. E., and García E. E. (Eds.) Ecología de los Bosques Siempreverdes de la Sierra del Rosario, Cuba Proyecto MAB No. 1, 1974–1987. Instituto de Ecología y Sistemática. Academia de Ciencias de Cuba. 1988.

[8] ECOVIDA Plan de Manejo Reserva de la Biosfera Sierra del Rosario. Centro de Investigaciones y Servicios Ambientales ECOVIDA. Ministerio de Ciencias, Tecnología y Medio Ambiente, Pinar del Río Período 2011–2015. 2011.

[9] Berazaín R.; Areces F.; Lazcano J. C. and González L. R. Lista roja de la Flora Vascular Cubana. Documentos del Jardín Botánico Atlántico (Gijón) 2005; 4:1–86.

[10] Capote R. and Berazaín R. Clasificación de las formaciones vegetales de Cuba. Revista Jard. Bot. Nac. 1984; 5 (2): 26–37.

[11] Reyes O. J. and Acosta F. C. Vegetación en Cuba: Parque Nacional: Alejandro de Humboldt: Rapid Biological Inventories. 2005; (14): 54–69.

[12] Jimenez A. Contribución a la ecología del bosque semideciduo mesófilo en el sector oeste de la Reserva de la Biosfera "Sierra del Rosario", orientada a su conservación. PhD thesis. Universidad de Pinar del Río, Cuba; 2012.

[13] Mitjans, B. Rehabilitación del bosque de ribera del río Cuyaguateje, en su curso medio. Estrategia participativa para su implementación. PhD thesis. Universidad de Pinar del Río, Cuba; 2012.

[14] Sánchez J., González E., and Ferro J. Estructura de los bosques nativos explotados en la cuenca hidrográfica del Toa y su relación con variables ambientales. In: Proceedings of the VIII International Symposium on Sustainable Forest Management, Pinar del Río, Cuba. 2014.

[15] Pickett A. and White S. The Ecology of Natural Disturbance and Patch Dynamics. Academic Press, EE.UU. 1985.

[16] Finegan B and Bouroncle B. Patrones de fragmentación de los bosques de tierras bajas, su impacto en las comunidades y especies vegetales y propuestas para su mitigación. In: Harvey, C. (Eds.). Evaluación y conservación de biodiversidad en paisajes fragmentados de Mesoamérica. Santo Domingo de Heredia, Costa Rica, 2008.

[17] Berry P. Diversidad y endemismo en los bosques neotropicales de Bajura, In: Guariguata, M.; Kattan, G. (Eds.). Ecología y Conservación de bosques neotropicales. 2002.

[18] Gentry H. Changes in Plant Community Diversity and Floristic composition on Environmental and Geographical Gradients. Ann. Missouri Bot. Gard. 1988; 75(1): 2–34.

[19] Keels S.; Gentry A. and Spinzi L. Using vegetation analysis to facilitate the selection of conservation sites in eastern Paraguay. (Biodiversity Measuring and Monitoring Certification Training, Volume 2). Washington: SI/MAB. 1997.

[20] Dufrene M. and Legendre P. Species assemblages and indicator species: the need for a flexible asymmetrical approach. Ecological Monographs 1997; 67:345–366.

[21] Feinsinger P. El diseño de estudios de campo para la conservación de la biodiversidad. Editorial FAN. Santa Cruz de la Sierra, Bolivia. 2003.

[22] Curtis J. T. and Mcintosh R. P. The interrelation of certain analytic and synthetic phytosociological characters. Ecology 1950; 31: 434–445.

[23] Connell H. Diversity in tropical rain forest and coral reefs. Biotropica 1978; 12: 47–55.

[24] Menéndez L., Capote P., and González V. La Reserva de la Biosfera. Áreas de Estudio. Capítulo 2. Ecología de los Bosques Siempreverdes de la Sierra del Rosario. Instituto de Ecología y Sistemática, Academia de Ciencias de Cuba. Proyecto MAB No. 1, 1988.

[25] Davic D. Linking keystone species and functional groups: a new operational definition of the keystone species concept. Cons. Ecol. 2003; 7:11.

[26] Cofiño C. E. Características microestucturales de las secuencias del Jurásico Superior-Cretácico y su relación con la potencialidad de Hidrocarburos en la parte Oriental de Sierra del Rosario. PhD thesis. Universidad de Pinar del Río, Cuba. 2002.

[27] García M. Conservación y Manejo *in situ* de la Biodiversidad en Huertos Caseros y Fincas de Cuba. PhD thesis. Universidad de Pinar del Río, Cuba and Universidad de Alicante España. 2006.

[28] Garibaldi C. Efectos de la extracción y uso tradicional de la tierra sobre la estructura y dinámica de bosques fragmentados en la Península de Azuero, Panamá. PhD thesis. Universidad de Pinar del Río, Cuba; 2008.

[29] OIMT (Organización Internacional de Maderas Tropicales). Directrices de la OIMT para la restauración, ordenación y rehabilitación de bosques tropicales secundarios y degradados, Serie de políticas forestales No. 13. 2002.

[30] Gerhartz J. L., Estrada R., Hernández E., Hernández A., and González A. Metodología para la elaboración de los planes de manejo de las áreas protegidas de Cuba. Centro Nacional de Áreas Protegidas de Cuba (CNAP), Editorial Feijóo, 2007.

[31] Mitjans B., Lago A., Beato S., González E., and Sánchez J. Estudio del comportamiento de la *Acacia mangium* en la faja hidrorreguladora del río Cuyaguateje. In: Proceed-

ings of the V International Symposium on Sustainable Forest Management, Pinar del Río, Cuba. 2008.

[32] Mitjans B., Bonilla M., Suárez A. G., González E., and González M. Estado de conservación del bosque de ribera del río Cuyaguateje (Municipio Guane). In: Proceedings of III International Symposium of Ecological Restoration, Santa Clara, Cuba. 2010.

[33] Barrera J.; Contreras N.; Rodríguez V.; Moreno A. and Montoya S. Manual para la Restauración Ecológica de los Ecosistemas Disturbados del Distrito Capital. Secretaría Distrital de Ambiente (SDA), Pontificia Universidad Javeriana (PUJ). Bogotá, 2010.

[34] Magurran A. E. Ecological diversity and its measurement. Princeton University Press, New Jersey, 1988.

[35] Moreno C. E. Métodos para medir la biodiversidad. M&T — Manuales Tesis SEA, Vol.1. Zaragoza. 2001.

[36] Vales M., Vilamajó D., and Herrera P. Especies forestales e integridad ecológica en fragmentos de bosques semideciduos de la provincia de La Habana, Cuba. In: Proceedings of the V Forestry Congress of Cuba. 2011.

[37] SER Principios de SER Internacional sobre la restauración ecológica. Society for Ecological Restoration (SER) Internacional. Grupo de trabajo sobre ciencia y políticas. www.ser.org y Tucson: Society for Ecological Restoration International. 2004.

[38] González L. R., Rankin R., and Palmarola A. (Eds.) Plantas invasoras en Cuba. Bissea 2012; 6(1):22–96.

[39] Servicio Estatal Forestal (SEF) Informe Balance anual de la dinámica forestal y la reforestación de las riberas del río Cuyaguateje. 2005.

[40] Jaramillo H. Guía para estudiar estructura de Bosque Natural UTLVT/FACAAM/ Escuela de Ciencias Forestales y Ambientales. http//:www.scrib.com/doc/27722564. (Accessed September 2010) 2009.

[41] Lamprecht H. Silvicultura en los trópicos. Los ecosistemas forestales en los bosques tropicales y sus especies arbóreas. Posibilidades y métodos para un aprovechamiento sostenido. Deutsche Gesellschaft für Technische Zusammenarbeit (GTZ) Gmobh. Alemania. 1990.

[42] Samek V. Regiones Fitogeográficas de Cuba. Dpto. de Ecología Forestal. Academia de Ciencias de Cuba. Serie Forestal 1973, (15).

[43] Álvarez P. A. and Varona J. C. Silvicultura, 2da reimpresión. Editorial Félix Varela, La Habana, 2007.

[44] Mitjans B., Bonilla M., Suárez A. G., González E., and González M. Propuesta participativa para la rehabilitación del bosque de ribera del Río Cuyaguateje. In: Proceedings of the V Forestry Congress of Cuba. 2011.

[45] Primack R., Roíz R., Feinsinger P., Dirzo R., and Massardo F. Fundamentos de Conservación Biológica. Perspectivas latinoamericanas. Fondo de Cultura Económica. México, 2001.

[46] Matos J. and Ballate D. ABC de la restauración ecológica. Editorial Feijoó. Cuba. 2006.

[47] Núñez L. Las percepciones ambientales de actores locales en áreas protegidas cubanas. Ventajas y desventajas para asumir la sostenibilidad. http://dlc.dlib.indiana.edu/archive/00001456/00/NúñezMorenoPercepciones_040512_Paper 583.pdf, (Accessed December 2010). 2004.

[48] Eupierre H. Adecuación de la metodología FFH al contexto sociocultural del sector campesino. El caso del río Caonao. MSc. thesis. Universidad de Pinar del Río. 2008.

[49] Arcos I., Jiménez F., and León A. Percepción local acerca del papel de los bosques ribereños en la conservación de los recursos naturales en la microcuenca del río Sesesmiles, Copán, Honduras. In: Recursos Naturales y Ambiente, CATIE. Costa Rica. 2006; (48): 118–122.

[50] Sánchez J. Criterios e indicadores en la faja hidrorreguladora del río Cauto. In: Proceedings of the VI Forestry Congress of Cuba. 2011.

[51] Vargas O. and Mora F. La restauración ecológica, su contexto, definiciones y dimensiones. In: O. Vargas (Ed). Estrategias para la restauración ecológica del bosque Alto andino. Universidad Nacional de Colombia. 2007.

[52] Corbin J. D. and Holl K. D. Applied nucleation as a forest restoration strategy. Forest Ecology and Management 2012; (265): 37–46. Doi: 10.1016/f.foreco 2011.10.013.

[53] Pouyú E. Synanthropic Liliatea and some other minor groups. Fontqueria 1995; 42:367–429.

[54] Ter Braak C. J. and Verdonshot F.M. Canonical correspondence analysis and related multivariate methods. Aquatic Ecology. Ecology Sciences 1995; 57: 255–286.

[55] Rodríguez Y. and Sánchez J. Diseño sostenible para la recuperación y conservación de las Fajas Forestales Hidrorreguladoras del río Toa. Guantánamo. In: Proceedings of DEFOR, La Habana, Cuba, 2005.

[56] Holl K. D., Zahawi R. A., Cole R. J., Ostertag R., and Cordell S. Planting seedlings in tree islands versus plantations as a large-scale tropical forest restoration strategy. Restoration Ecology 2011; (19): 470–479.

[57] Zahawi R. A., Holl K. D., Cole R. J., and Reid L. Testing applied nucleation as a strategy to facilitate tropical forest recovery. Journal of Applied Ecology 2013; (50): 88–96, doi 10.1111/1365-2664.12014.

Chapter 22

Biodiversity in Metal-Contaminated Sites – Problem and Perspective – A Case Study

E. Roccotiello, P. Marescotti, S. Di Piazza, G. Cecchi,
M.G. Mariotti and M. Zotti

Additional information is available at the end of the chapter

http://dx.doi.org/10.5772/59357

1. Introduction

Metal contamination is one of the major environmental problems in the world, posing significant risks to ecosystems and human population in abiotic environmental compartments (soil, water, air) and in the related biota (e.g., uptake by fungi and plants).

The primary causes of soil contamination are intensive industrial activities and inadequate waste disposal and treatment (although these categories vary widely across Europe) [1]. Good knowledge of the content and variability of metals in soils linked to both the contribution of parent rock (lithogenic sources) and human activities (anthropogenic sources) is also necessary for evaluating metal pollution. These tasks are particularly difficult to achieve in ancient populated areas, such as the European Mediterranean region, where unpolluted soils are almost impossible to find [2, 3]. Among anthropogenic sources, mining activities are the fourth largest source of land pollution (e.g., 7% of the National Priority (Superfund) Sites in the USA; [4]).

About 2.5 million sites in Europe produce potentially polluting activities [5]. With the help of improved data collection, the number of recorded polluted sites is expected to increase, as are research studies on the topic. Considering present tendencies and laws, contaminated sites will presumably rise by up to 50% by 2025 [5]. Although still at a low rate, there has nonetheless been progress in the remediation of contaminated sites.

Around 59,000 sites have already been remediated. Nonetheless, it should be noted that around 340,000 sites may need urgent remediation [5]. However, conventional technologies for metal-contaminated soil remediation have often been expensive and disruptive [6].

A good tool for evaluating environmental damage might be the Ecological Risk Assessment (ERA), which is already used in the US [7, 8] and in some countries in the European Union (e.g., Netherlands, [9] and the UK [10]), while in Italy, it still remains largely unknown. Consequently, threshold contaminant concentration, according to Italian law [11], is a human-based risk that lacks consideration in terms of biodiversity and ecosystem services.

Before starting to plan restoration of polluted industrial environments, the 'target condition' should first be assessed. However, the remediation of a contaminated site with all its particular biotic and abiotic characteristics seems unrealistic. Several authors have indicated the lack of data providing detailed characterization of the biotic and abiotic components of environments in a pre-industrial condition and have highlighted that communities did not return to their previous states [12-15].

The best approach for preserving or restoring the biodiversity of metal-contaminated environments should be site-specific characterization. For this purpose, the evaluation of the surrounding area's biodiversity, a choice of best reasonable target condition for biotic components and the selection of tolerant organisms occurring on the polluted sites are essential. Among the biotic components of soil, fungi and plants are the pioneer organisms that play a key role in the colonization of contaminated sites.

1.1. Biodiversity in metal-contaminated sites

The physicochemical properties of metal-contaminated environments tend to inhibit soil-forming processes and affect the area's biodiversity by exerting a strong selective pressure on fungi and plants [16-19]. Specifically, the bulk metal content of soils and its metal releasability are among the most important edaphic factors determining vegetation composition. Other than metal toxicity, vegetation successions are also retarded by low nutrient status, poorly developed soil structure and water-restricted conditions [20].

In addition, microorganism communities play a significant part in the detoxification of noxious chemicals and in the control of plant growth [21], and also provide pivotal information about soil metal bioavailability [22]. Metalliferous biota is increasingly exploited for the stabilization or active remediation of the metal-contaminated ecosystems and represents an important research topic in the contemporary field of green technology [23, 24].

It is well-known that a number of plants and fungi are able to survive and actively grow in metal-contaminated soils. For instance, recent studies [25] have shown that some arbuscular mycorrhizal fungi from Cu-contaminated soils [*Claroideoglomus claroideum* (N.C. Schenck and G.S. Sm.) C. Walker and A. Schüßler in association with *Imperata condensata* Steud. and *Rhizophagus irregularis* (Błaszk., Wubet, Renker and Buscot) C. Walker and A. Schüßler in association with carrot roots] are able to compartmentalize Cu in spores as a survival strategy in polluted environments. Additionally, microfungi are essential in colonizing and detoxifying metal-contaminated soil ecosystems and consequently have environmental and economic significance [16, 26-28].

Mine dumps cause high selective pressure, enabling bacteria and microfungi to be the first organisms able to re-colonize mine soils [29]. Under this pressure, microfungal communities change their composition and several resistant strains are selected [28]. In addition, plant communities have established a primary succession on mine wastes [30] and can be exploited for biogeochemical prospecting and mine stabilization (e.g., abandoned mines contaminated with arsenic, antimony and tungsten [17]).

1.2. Selecting native fungi and plants for bioremediation

Critical environmental conditions related to high metal concentration are present either in natural soils (e.g., serpentinitic and ultramafic soils; [31, 32 and references therein]) and in anthropogenic contaminated sites (such as industrial, agricultural and mining sites [33, 1 and references therein]).

Fungi and plants from metal-rich soils develop specific strategies to cope with metals via avoidance, accumulation or hyperaccumulation [34]. Care should be taken in choosing the right species for the application of bioremediation techniques, because the introduction of alien fungi and plants may alter and disrupt indigenous ecosystems [35], or may be unsuitable for local climate conditions [36]. Therefore, a more appropriate option is to find native hyperaccumulator fungi and plants that have adapted to growing on metalliferous sites, and use them for soil bioremediation in the same region [37, 38].

Bioremediation process consists of two main approaches: 1) myco- and phytostabilization; 2) myco- or phyto-extraction. Myco- and phyto-stabilization are mechanisms that immobilize pollutants – mainly metals – within the root zone, by adsorption, chelation and metal ion precipitation, thus preventing migration of contaminants by erosion, leaching and runoff [39,40].

Myco- and phyto-extraction are typically used to uptake metals, metalloids and radionuclides. The metals accumulate in the fruit-bodies or in the plant's aboveground biomass and can be removed or recovered by harvesting and burning the biomass.

Several organisms, including microbes, micro- and macro-fungi, agricultural and vegetable crops, ornamentals and wild metal hyperaccumulating plants have been tested both in laboratory and field conditions for selecting and providing organisms able to clean-up metalliferous substrates [24].

In fact, recent studies have shown how macrofungi such as *Trametes versicolor* (L.) Lloyd [41] and microfungi such as *Aspergillus niger* Tiegh. [42], *A. terreus* Thom [43], *A. versicolor* (Vuill.) Tirab. [44], *Penicillium notatum* Westling [45], *Rhizopus arrhizus* A. Fisch. [46], *Trichoderma atroviride* P. Karst. [47], *T. viride* Pers. [48] are able to absorb Cu from contaminated liquid and solid matrices.

Similarly, plant taxa naturally occurring in metalliferous soils have been selected, tested and confirmed as hyperaccumulators under experimental conditions for different metals like Ni (e.g., *Alyssum bertolonii* Desv., *A. murale* Waldst. & Kit., *A. lesbiacum* (Candargy) Rech.f., *A.*

corsicum Duby) and Cd (*Thlaspi caerulescens* J.Presl & C.Presl and *Arabidopsis halleri* (L.) O'Kane & Al-Shehbaz) [49-54].

Little information is available about the processes occurring at the soil-rhizosphere level. Though roots are the only organ directly interacting with soil trace elements, most of the studies on hyperaccumulation by plants have focused on above-ground organs. Less than 10% of the known hyperaccumulator species have been investigated at the rhizosphere level [55].

Bacteria and fungi in the hyperaccumulator rhizosphere may exhibit increased metal tolerance by i) acting as a plant growth promoting microorganisms; ii) modifying metal speciation and solubility; iii) influencing plant trace element concentrations [55, 56-58].

Fungal species not identified as mycorrhizae have also been found in the hyperaccumulator rhizosphere [59, 60]. The role of these organisms still needs to be established, but some have been identified as able to concentrate and volatilize pollutants [61]. The knowledge about hyperaccumulators and associated microorganisms continuously increases, thus suggesting the significant roles of bacteria and fungi in hyperaccumulation.

2. Case study – Multidisciplinary investigations on biodiversity into a sulphide-rich waste-rock dump

The present review illustrates the results of a six-year multidisciplinary study aimed at understanding the relationships among the mineralogy and chemistry of the Libiola mine (eastern Liguria, Italy, Fig. 1) and the metal uptake by fungi and plants spontaneously growing in the mine-waste dump. The mine is located in a moderately steep mountainous terrain at an altitude between 40 and 400 m asl, close to the Liguria sea coast.

This mine had already been known during the Bronze Age [63] and was industrially exploited from 1864 until 1962. During this period, it produced over 1 Mt of Fe-Cu sulphides with an average grade ranging from 7 to 14 Cu wt%, thus representing one of the most important Fe-Cu sulphide mines [64, 65] in Italy. Sulphide mineralization occurs within the Jurassic ophiolites of the Northern Apennines (Vara supergroup; [66]) and is geologically characterized primarily by pillow basalts with minor serpentinites, gabbros and ophiolitic breccias. During exploitation, five major waste-rock dumps were built up through the progressive accumulation of heterogeneous sterile rocks (derived from galleries and open-pit excavations) and non-valuable ore-fragments, with metal concentrations below the economic cut-off produced during beneficiation processes [67]. The soils of the dumps are characterized by severe edaphic conditions due to their peculiar physical (steep slopes, low moisture retainability, impermeabilization due to cementification and hardpan formation; [68]) and chemical (high Cr-, Cu-, Co-, Ni- and Zn-concentrations, low pH values and the low availability of essential macronutrients) properties. This site presents several environmental problems as a result of active acid mine drainage (AMD) processes, which determines the water acidification and metal pollution of soils and waters.

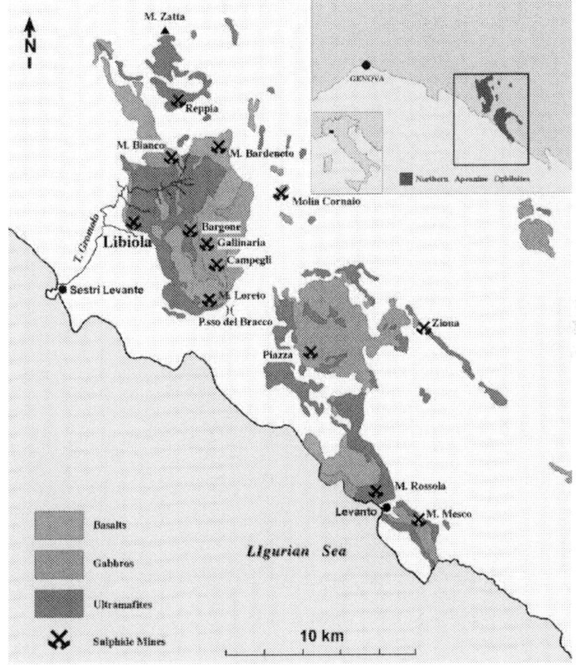

Figure 1. Location of Libiola mine site (from [62], modified).

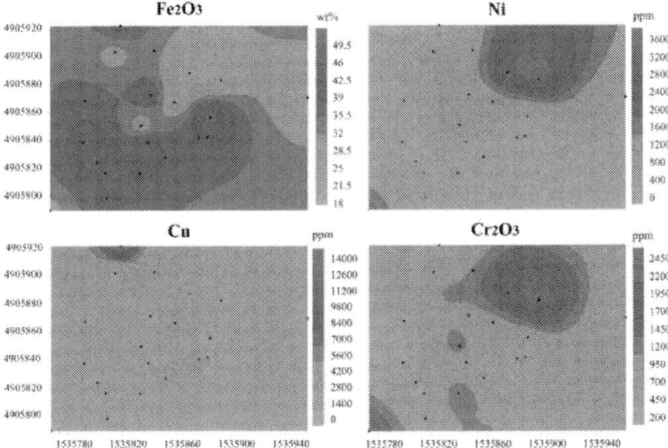

Figure 2. Contour maps of selected elements (Fe2O3, Cr2O3, Cu and Ni) in the main waste-rock dump of the Libiola mine (from [67], modified).

We evaluated the plant and fungal diversity in these contaminated soils in order to 1) identify factors that influenced the pioneer fungi and plants colonizing this stressed environment; 2) identify and select tolerant and hyperaccumulating plants and fungal strains suitable for mine remediation.

The waste-rock materials deposited in the mining area are mainly gravely-sandy sediments with a relatively uniform particle size distribution in the range 2-64 mm; the silt and clay fractions are subordinate components and vary from 5% to 26%. Most of the dumps evidence strong superficial cementation induced by Fe-oxides precipitating from acid sulphate water seepage, which determines the formation of centimetre-thick impermeable hardpans on several parts of the exposed surface of the dumps [67, 68].

Due to active and widespread AMD processes, the pH of waste-dump soils is generally acidic (3.5-4.3) and significantly low, compared to the surrounding serpentinitic and basaltic soils (6.2-6.8) [30, 69, 68].

The mineralogical and lithological composition of the waste-rock materials is quite homogeneous throughout the mine area, though the relative proportions of the detected lithotypes and mineral species significantly vary from site to site. The studied samples are mainly composed of polycrystalline rock fragments, which can be grouped into the following lithological classes [65, 67, 68]: 1) serpentinites 20%-50%; 2) basalts 5%-10%; 3) sulphide mineralizations (2%-10%); 4) massive Fe-oxyhydroxides and -oxides clasts (35%-65%). Other subordinate components (3%-6%) are represented by garnet- and epidote-rich rodingites, polygenic ophiolitic breccias and brecciated basalts.

The mineral species occurring within the waste-rock dump [65, 67, 68] can be divided into three major groups on the basis of their origin and/or origin: 1) host rocks and gangue minerals; 2) ore minerals with different degrees of alteration; 3) authigenic secondary minerals. Serpentine minerals (60%-70% of the recognized mineral species) and Fe-oxyhydroxides (mainly goethite) are by far the most abundant species, respectively representing the main rock-forming minerals of the lithotypes of the surrounding area and the main authigenic minerals forming as a result of the ongoing AMD processes. Sulphides (mainly pyrite and chalcopyrite) are subordinate but important components from an environmental point of view, either for their role in triggering the AMD processes and/or for the release of metals (particularly Fe, Cu and Zn). Magnetite, Cr-bearing magnetite and chromian spinels are the only Cr-bearing minerals, whereas serpentines are the main Ni-bearing minerals. Nevertheless, they are stable mineral species, even in this highly reactive environment, that likely do not contribute significantly to the bioavailable fraction of toxic metals.

According to the mineralogical and lithological composition, the waste rock dumps of the Libiola mine are invariably characterized by very high concentrations of several potentially toxic metals (such as Ti, Mn, Co, Ni, V, Cr, Cu, Zn and Cd). Although a notable variability is always present and several hot spots have been found (see, for example, Fig. 2), most of the detected metals (in particular Cr, Cu, Co, Ni and Zn) strongly exceed the Italian limits for residential and industrial sites [11].

Xero-acidophilous plant communities characterize the areas surrounding the mine site. These are different aspects of sclerophyllous evergreen maquis and mixed sclerophyllous evergreen and deciduous shrub thickets (pseudomaquis) formed by *B. sempervirens* L. and/or *Genista desoleana* Valsecchi, *Erica arborea* L., *Calicotome spinosa* (L.) Link., *Juniperus oxycedrus* L. subsp. *oxycedrus* and *Arbutus unedo* L. Chamaephytic and sub-shrubby layers are well represented by *Euphorbia spinosa* L. subsp. *ligustica* (Fiori) Pignatti, *Helichrysum italicum* (Roth) G. Don, *Minuartia laricifolia* (L.) Schinz. and Thell subsp. *ophiolitica* Pignatti, *Thymus* sp.pl., and *Satureja montana* L.; Maritime pine (*Pinus pinaster* Aiton) old reforestations are also present. Less frequent are relics of holm oak (*Quercus ilex* L.), pubescent oak (*Quercus pubescens* Willd.) copses and thermophile mixed woods [3, 70, 71].

The study area is characterized by bare soils or by different successions of plant communities ranging from herbaceous to arboreal layers (Fig. 3). In all the plots, species richness and vegetation cover were extremely low and the flora showed acidophilous traits [30]. The bare soil is a substrate with an almost complete absence of vegetation (Fig. 3B). The herbaceous layer has pioneer vegetation dominated by discontinuous communities of low-sized grasses and herbs such as *Deschampsia flexuosa* (L.) Trin., *M. laricifolia* subsp. *ophiolitica*, *Sesamoides interrupta* (Boreau) G.López, *Festuca robustifolia* Markgr.-Dann. and, in sites with more developed soil, also by *Cerastium ligusticum* Viv. and *Asplenium adiantum-nigrum* L.. The subsequent layers are colonized by *Thymus vulgaris* L., *S. montana*, *E. spinosa* subsp. *ligustica*.

The shrub layer is mainly located on serpentine debris on the edge of the landfill areas and colonized by semi-natural plant communities dominated by *E. spinosa* subsp. *ligustica*, *Alyssoides utriculata* (L.) Medik., *T. caerulescens*, *Silene paradoxa* L., *F. robustifolia* and *H. italicum* (Fig. 3C).

Despite harsh environmental conditions, the waste-rock dump has been progressively colonized since 2008 by several plants of *Pinus pinaster* Aiton, already found naturally on metal-rich sites, thereby establishing ectomycorrhizal symbiosis (ECM) [72]. The maritime pine populations constituting the tree layer derive from seeds dispersed by the surrounding plants and employed for the revegetation of fired areas (Fig. 3D). A few herbaceous species such qw *D. flexuosa*, *M. laricifolia* subsp. *ophiolitica* and *F. robustifolia* are associated with *P. pinaster*. The absence of the shrub layer in the mine dump and the presence of the tree layer strictly composed of pine is particularly uncommon, and the success of *P. pinaster* colonization is mainly due to the presence of *Scleroderma polyrhizum* (J.F. Gmel.) Pers (Fig. 4A) and *Telephora terrestris* Ehrh. (Fig. 4B) ectomycorrhizic with pine (see Fig. 4C-D) [69]. Maritime pine is known to be able to cope with some limiting factors such as a low level of macronutrients, a lack of organic matter and water stress, typical of dismissed mining areas [73] such as the one in our study. In addition, we found that *P. pinaster* is able to completely exclude toxic metals from its tissue (Fig. 5), thereby acting as a phytostabilizer, as demonstrated by bioaccumulation (i.e., BF = shoot:soil metal concentration) and translocation (i.e., TF = shoot:root metal concentration) factors (BFs>1 and TFs<1, respectively [68]). Where metal concentrations decrease [30, 68], plants constitute semi-natural Mediterranean serpentine vegetation, typical to NW Italy [70, 71]

Figure 3. Libiola mine site **A)** View of the mine waste rock dump; **B - D)** sampling sites; **B)** bare soil with no vegetation; **C)** shrub layer; **D)** tree layer with macrofungi.

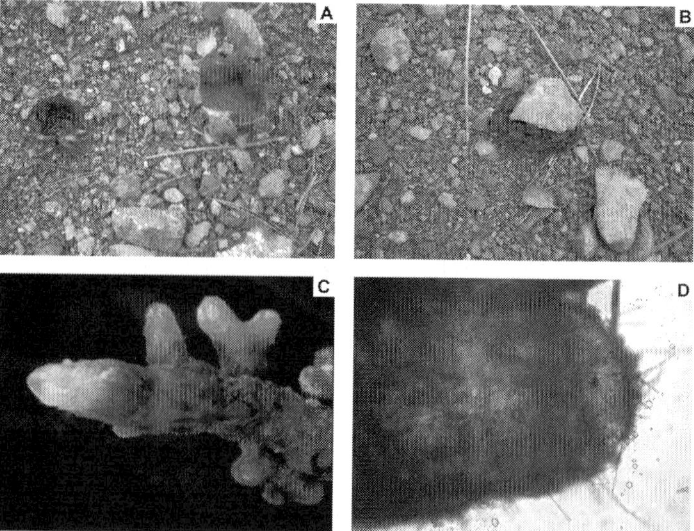

Figure 4. Macrofungi at the Libiola mine and their ECM symbiosis. **A)** *S. polyrhizum*; **B)** *T. terrestris*; **C)** *P. pinaster* roots with ECM fungi; **D)** details of root apex with ECM fungi; blue cotton staining.

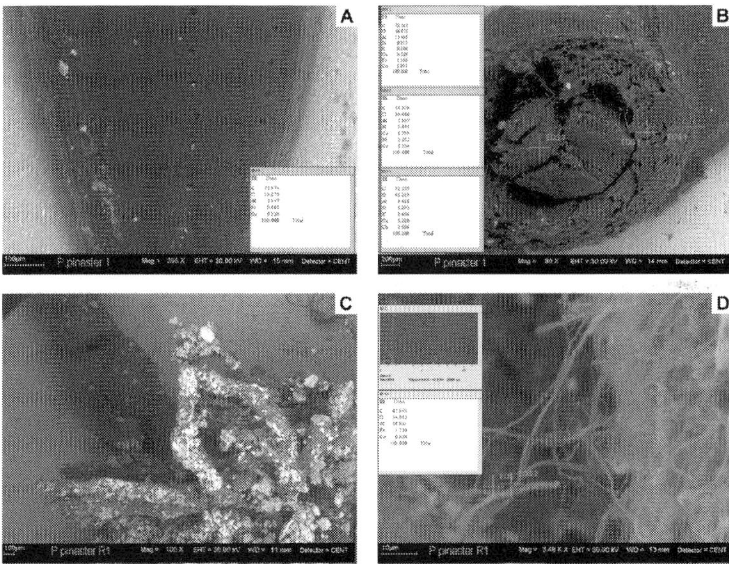

Figure 5. ESEM micrographs of *P. pinaster* with EDS spectra. A) *P. pinaster* leaves showed soil particles on stomata, but no metals were detected inside tissues; Bar = 20 μm. B) *P. pinaster* stems; Bar = 200 μm C). *P. pinaster* roots with ECM fungi; Bar = 100 μm. D) details of root hairs; Bar = 10 μm. EHT: 30 KV, WD: 14 mm, detector: Centaurus.

Among the mine plants screened for Ni accumulation in plant tissue, only the well-known metal hyperaccumulator *Thlaspi caerulescens* J. & C. Presl and *A. utriculata* yielded a positive response (Fig. 6). The latter was recently confirmed as a new Ni facultative hyperaccumulator, able to concentrate more than 1000 mg kg⁻¹ Ni DW in leaves [74]. Plant efficiency tests were carried out on native soils to evaluate the growing ability and the ecophysiology of this promising species, and recent experiments have confirmed its suitability for phytoextraction (data not shown).

Soil samples collected from the *A. utriculata* rhizosphere and barren mine soils were examined to determine microfungal flora. On the whole, the majority of isolated colonies belonged to the genus *Aspergillus, Botrytis, Clonostachys, Penicillium* and *Trichoderma* (Fig. 7). Regarding *A. utriculata,* the rhizosphere were isolated species such as *Eurotium amstelodami* L. Mangin, *Aspergillus carbonarius* (Bainier) Thom, *A. tubingensis* Mosseray, *Penicillium waksmanii* K.M. Zalessky and *Rhodotorula* spp. not found in other mine soil samples. We can hypothesize that these microfungi, growing in the hyperaccumulator rhizosphere, may function as plant growth promoting factors altering element solubility and increasing plant metal uptake.

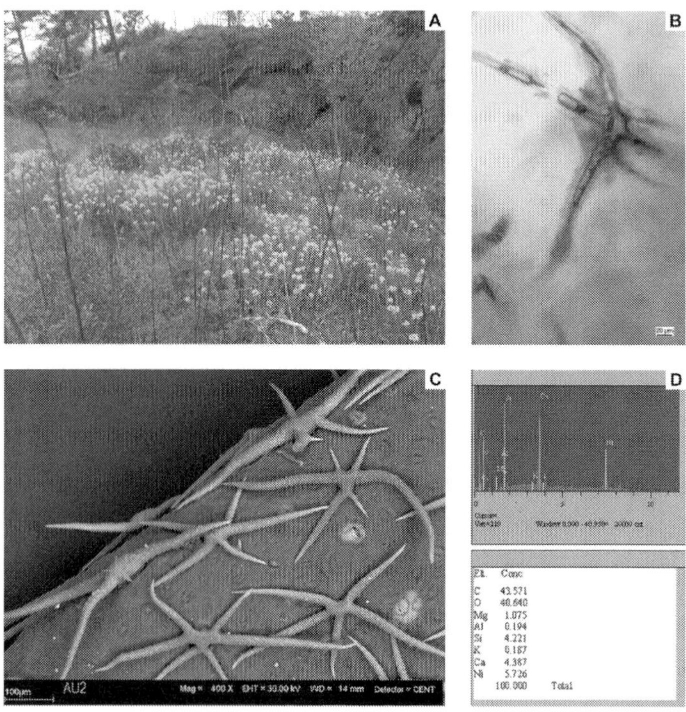

Figure 6. *Alyssoides utriculata* **A)** in the mine dump; **B)** positive leaf trichome DMG test, light microscopy micrographs; **C, D)** ESEM micrographs and EDS spectrum of leaf trichomes storing Ni, up to 8%; Bar = 100 µm. EHT: 30 KV, WD: 14 mm, detector: Centaurus.

The ectomychorrizal macrofungi collected reveal a highly significant metal accumulation, in particular Cu>1000 mg kg^{-1} in *Telephora terrestris* and Ag>50000 µg kg^{-1} in *Scleroderma polyrhizum*. When present in mine sites, these fungi are able to actively absorb most of the potential toxic elements in the sites' basidiomata. The absorption sequence Cu>Zn>Cr>Ni>Co obtained for these macrofungi overlaps well with the sequences obtained using EDTA extractions and water leaching tests [69, 68]. Both species also established ECM symbiosis with pine and we could not exclude that they played a role in the phytostabilization process at the root level.

Finally, we studied soil microfungi to test the growth responses of culturable isolated microfungal strains in copper enriched media and to evaluate their potential use in mycoremediation. The species most recurrent were filamentous microfungi: *Trichoderma harzianum*, *Clonostachys rosea* and *Aspergillus alliaceus*. We hypothesized that these fungi were particularly tolerant/resistant to copper. The Cu tolerance level of *T. harzianum* and *C. rosea* were tested *in vitro* at increasing Cu(II) concentrations. The tests showed a Cu(II)-tolerance capability ranging from 100 to 400 mg L^{-1} [75]. These preliminary analyses proved that several fungal species were able to grow in Cu-contaminated media, thereby underlying the importance of selecting new

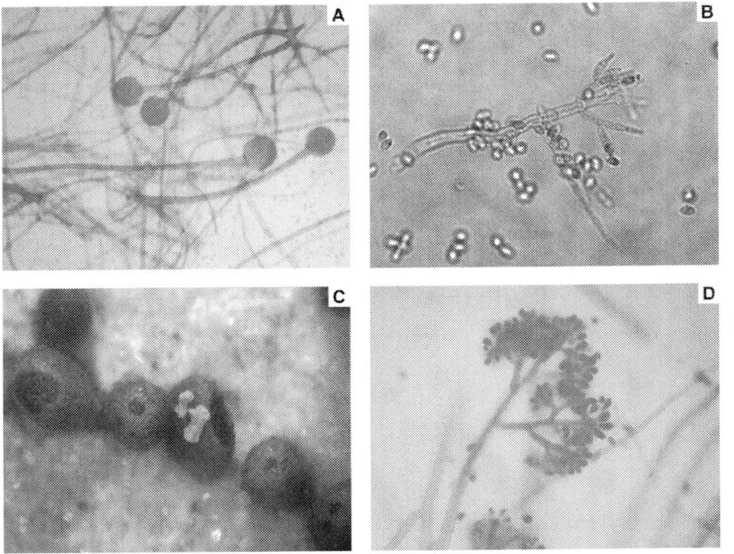

Figure 7. Microfungi isolated from mine site: **A)** *Rhizopus oryzae* Went & Prins. Geerl., acid fuchsine staining, light microscopy; **B)** *Clonostachys rosea* (Link Schroers, Samuels, Seifert & W. Gams, acid fuchsine staining, light microscopy; **C)** sclerotia of *Aspergillus alliaceus* Thom & Church, stereomicroscopy; **D)** *Botrytis cinerea* Pers. ex Nocca & Balb., acid fuchsine staining, light microscopy.

tolerant strains and testing their potential metal uptake capabilities for application to mycoremediation protocols.

Silver, a noble metal of historical and economic importance, also indicated a high concentration in the Libiola mine soil. Even if silver is usually not considered an environmental contaminant, Ag^+ represents one of the most toxic metals to bacteria, algae and fish [76], and can damage cellular components and reduce enzymatic activities. Some studies have shown that bacteria, yeasts and macrofungi can accumulate silver; however, the chemical form of Ag in the macrofungal fruit-bodies was not investigated in detail [77, 78, 76]. In this context, we have tested the potential Ag^+ accumulation by microfungi isolated from the Libiola mine soils. First, we tested the *in vitro* growth capabilities on Ag^+-enriched media by *Aspergillus alliaceus*, *Trichoderma* sp., *C. rosea* in order to select the most tolerant microfungal strain. *Trichoderma* sp. showed the best and speediest capabilities for growing *in vitro* on media spiked with 400 mg kg^{-1} of Ag^+, uptaking 150 mg kg^{-1} dry weight (Fig. 8), as confirmed by ICP-MS analysis.

The considered contaminated environment chiefly affected the biodiversity of the area and exerted a strong selective pressure on the local flora and mycoflora. These results suggest the use of *P. pinaster*, *A. utriculata*, *T. terrestris*, *S. polyrhizum*, *T. harzianum* and *C. rosea* for developing experimental protocols of bioremediation and habitat restoration for avoiding ecosystem disruption.

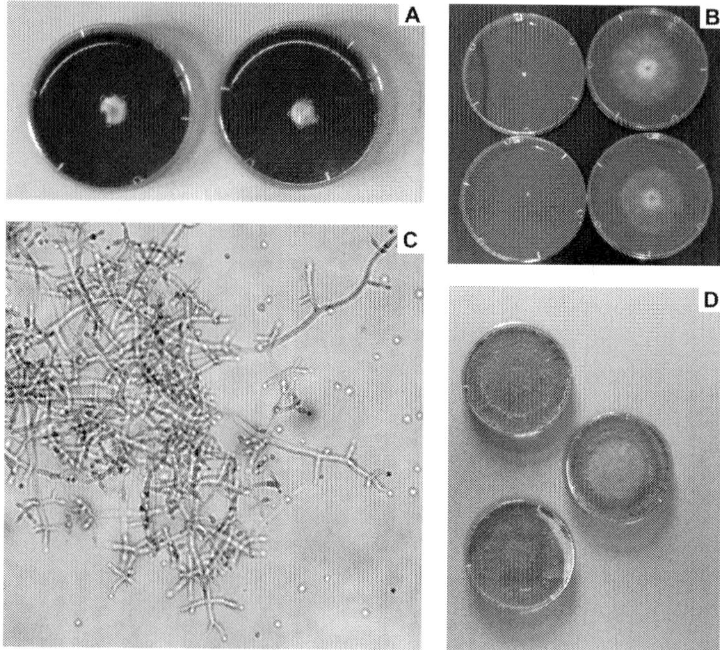

Figure 8. Ag accumulation tests. A-B) Screening test of microfungal growth capability on silver enriched media; C) microscopic detail of the selected *Trichoderma* sp. strain D) *Trichoderma* sp. growth.

The study highlighted differences between mineralogy, geochemistry, flora and mycology among strongly polluted selected sites of the Libiola mine, which are key points for future reclamation of the area. In particular, our results evidenced the significant control of soil mineralogy and chemistry on the biodiversity of the mining area, as well as on the capacity of mycoflora and flora to accumulate specific metals. Knowing what the factors are influencing the first colonization by plants and the interaction among plants, fungi and soils, allow us to develop a method for the land restoration of metal polluted sites in a manner that minimizes interventions and costs.

3. Future perspectives – How to apply what we have learned

Due to the complexity of soil and *in-situ* conditions, each contaminated site requires its own strategy and site-specific designs for decontamination, especially in Mediterranean areas. Multi-element contaminated soils contain several pollutants; consequently, it is necessary to screen out fungi and plants that can survive on different pollutants simultaneously and to accumulate or stabilize some of them.

The use of metal tolerant species adapted to native conditions can assist in balancing the ecological pressure generated by soil pollution. Consequently, it is necessary to evaluate the potential bioremediation of native fungi and plants from contaminated sites before choosing other species suitable for bioremediation.

Phytostabilization may be employed as a temporary solution until new techniques are available. However, for large contaminated sites (e.g., mining or industrial sites), phytostabilization likely represents the best option for ecosystem restoration [79]. Moreover, below-ground restoration success involves the employment of native microfungi for developing and improving the soil microbial biomass [80].

In the joint use of plants and soil microorganisms, plants provide a C source for microorganisms, which absorb, degrade or release elements for plant absorption [81]. The plant allocates most of the metals to their roots so that plant shoots can be more efficient under metal stress [82]. Rhizospheric fungi are able to alleviate the stress of metals on plant growth through soil bioremediation (bioaugmentation) and can sometimes alleviate the unfavourable effects of metal on plant growth by the process of phytostabilization [83]. With the employment of the right soil, microfungi and ECM fungi, we can avoid amendments to soil, thereby improving organic matter and soil-forming processes that are essential for the colonization of pioneer plant species. These species will guarantee the durable, sustainable and ecological restoration of polluted mine sites, thereby increasing soil fertility.

The enormous potential of native fungi and plants that are able to colonize metal-contaminated soils need to be studied in-depth in order to preserve the natural genetic resources of metal-liferous habitats and to increase our basic knowledge about the natural adaptation mechanisms of hyperaccumulators in order to employ them in phytoremediation purposes.

Acknowledgements

The authors wish to thank Carmela Sgrò for technical support provided during the laboratory work and Cristina Brusco for reviewing the paper for the English language.

Author details

E. Roccotiello[1*], P. Marescotti[2], S. Di Piazza[1], G. Cecchi[1], M.G. Mariotti[1] and M. Zotti[1]

*Address all correspondence to: enrica.roccotiello@unige.it

1 Polo Botanico Hanbury, DISTAV-Department of Earth, Environment and Life Sciences, University of Genoa, Corso Dogali, Genoa, Italy

2 DISTAV-Department of Earth, Environment and Life Sciences, University of Genoa, Corso Europa, Genoa, Italy

References

[1] Alloway BJ. Heavy Metals in Soils: Trace Metals and Metalloids in Soils and their Bioavailability. Dordrecht: Springer; 2013.

[2] Facchinelli A, Sacchi E, Mallen L. Multivariate statistical and GIS-based approach to identify heavy metal sources in soils. Environmental Pollution 2001;114: 313-324.

[3] Marsili S, Roccotiello E, Carbone C, Marescotti P, Cornara L, Mariotti MG. Plant colonization on a contaminated serpentine site. Northeastern Naturalist. Special issue: Soil Biota Serpentine: A World View 2009;16: 297-308.

[4] Allen HE, Huang CP, Bailey GW, Bowers AR. Metal Speciation and Contamination of Soil. Boca Raton: Lewis Publishers; 1995.

[5] European Environment Agency, (EEA) Progress in management of contaminated sites (LSI003) Assessment May 2014. http://www.eea.europa.eu/data-and-maps/indicators/progress-in-management-of-contaminated-sites-3/assessment

[6] European Environment Agency state of the environment report (SOER). The European environment - state and outlook – SOIL. 2010. http://www.eea.europa.eu/soer/europe/soil/download

[7] U.S. Environmental Protection Agency Guidelines for Ecological Risk Assessment, EPA/630/R-95/002F 1998. http://www.epa.gov/raf/publications/guidelines-ecological-risk-assessment.htm

[8] ASTM Standard Guide for Risk-Based Corrective Action for Protection of Ecological Resource, ASTM E 2205/E2205M – 02 (Reapproved 2009); 2009. http://www.astm.org/Standards/E2205.htm

[9] Honders T. New risk-based soil quality standards in the Netherlands: conference proceedings, seventh International conference on the Environmental and Technical Implications of Construction with Alternative Materials (WASCON, 2009). Lyon, France; 2009.

[10] Environment Agency An ecological risk assessment framework for contaminants in soil Science report SC070009/SR1 2008. https://www.gov.uk/government/publications/contaminants-in-soil-assessing-the-ecological-risk (accessed 1 October 2008).

[11] Legislative Decree n. 152 del 3 Aprile (2006) Norme in materia ambientale. Gazzetta Ufficiale della Repubblica Italiana, n. 88 del 14/4/2006, Supplemento ordinario, n. 96. http://www.camera.it/parlam/leggi/deleghe/06152dl.htm

[12] Batty LC, Hallberg KB. Consequences of living in an industrial world. In: Batty LC, Hallberg KB. (eds.) Ecology of Industrial Pollution. Cambridge: Cambridge University Press; 2010. p1-6.

[13] Langford TEL, Shaw PJ, Howard SR, Ferguson AJD, Ottewell D, Eley R. Ecological recovery in a river polluted to its sources: the River Tame in the English Midlands.

In: Batty LC, Hallberg KB. (eds.) Ecology of Industrial Pollution. Cambridge: Cambridge University Press; 2010. p255-275.

[14] Tibbett M. Large-scale mine site restoration of Australian eucalypt forests after bauxite mining: soil management and ecosystem development. In: Batty LC, Hallberg KB. (eds.) Ecology of Industrial Pollution. Cambridge: Cambridge University Press; 2010. p309-326.

[15] Williams AE, Waterfall RJ, White KN, Hendry K. Manchester Ship Canal and Salford Quays: industrial legacy and ecological restoration. In: Batty LC, Hallberg KB. (eds.) Ecology of Industrial Pollution. Cambridge: Cambridge University Press; 2010. p276-308.

[16] Gadd GM. Interactions of fungi with toxic metals. New Phytologist 1993;124: 25-60.

[17] Pratas J, Prasad MNV, Freitas H, Conde L. Plants growing in abandoned mines of Portugal are useful for biogeochemical exploration of arsenic, antimony, tungsten and mine reclamation. Journal of Geochemical Exploration 2005;85: 99-107.

[18] Crane S, Dighton J, Barkay T. Growth responses to and accumulation of mercury by ectomycorrhizal fungi. Fungal Biology 2010;114: 873-880.

[19] Ceci A, Maggi O, Pinzari F, Persiani AM. Growth responses to and accumulation of vanadium in agricultural soil fungi. Applied Soil Ecology 2012;58 :1-11.

[20] Baker AJM, Ernst WHO, van der Ent A, Malaisse F, Ginocchio R. Metallophytes: the unique biological resource, its ecology and conservational status in Europe, central Africa and Latin America In: Batty LC, Hallberg KB. (eds.) Ecology of Industrial Pollution. Cambridge: Cambridge University Press; 2010. p7-40.

[21] Filip Z. International approach to assessing soil quality by ecologically related biological parameters. Agriculture Ecosystems & Environment 2002;88(2): 169-174.

[22] Saeki K, Kunito T, Oyazu H, Matsumoto S. Relationships between Bacterial tolerance Levels and Forms of Copper and Zinc in Soils. Journal of Environmental Quality 2002;32: 1570-1575.

[23] Adriano DC. Trace Elements in Terrestrial Environments: Biogeochemistry, Bioavailability and Risks of Metals. New York: Springer; 2001.

[24] Prasad MNV, Freitas H. Metal hyperaccumulation in plants - Biodiversity prospecting for phytoremediation technology. Electronic Journal of Biotechnology 2003;6: 275-321. http://www.ejbiotechnology.info/index.php/ejbiotechnology/article/view/v6n3-6/617 (accessed 15 December 2003)

[25] Cornejo P, Pérez-Tienda J, Meier S, Valderas A, Borie F, Azcón-Aguilar C, Ferrol N. Copper compartmentalization in spores as a survival strategy of arbuscular mycorrhizal fungi in Cu-polluted environments. Soil Biology and Biochemistry 2013;57: 925-928.

[26] Fomina MA, Alexander IJ, Colpaert JV, Gadd G.M. Solubilization of toxic metal minerals and metal tolerance of mycorrhizal fungi. Soil Biology & Biochemistry 2005;37: 851-866.

[27] Fomina MA, Gadd GM, Burford EP. Toxic Metals and Fungal Communities. In: Dighton J, White JF, Oudemans P. (eds.) The fungal community: its organization and role in the ecosystem, 3rd edition. Boca Raton: CRC Press; 2005. p733-758.

[28] Gadd GM. Geomycology: biogeochemical transformations of rocks, minerals, metals and radionuclides by fungi, bioweathering and bioremediation. Mycological Research 2007;111: 3-49.

[29] Novàkovà A. Soil microfungi in two post-mining chronosequences with different vegetation types. Restoration Ecology 2001;9: 351-358.

[30] Micò C, Recatalà M, Peris M, Sánchez J. Assessing heavy metal sources in agricultural soils of an European Mediterranean area by multivariate analysis. Chemosphere 2006;65: 863-872.

[31] Oze C, Fendorf S, Bird DK, Coleman RG. Chromium geochemistry of serpentine soils. International Geology Review 2004;46: 97-126.

[32] Alexander EB, Coleman RG, Keeler-Wolf T, Harrison SP. Serpentine geoecology of western North America: Geology, Soils, and Vegetation. New York: Oxford University Press; 2007.

[33] Jambor JL, Blowes DW. Short course handbook on environmental geochemistry of sulfide mine-waste. Mineralogical Association of Canada, Waterloo, Ontario; 1994.

[34] Van der Ent A, Baker AJM, Reeves RD, Pollard AJ, Schat H. Hyperaccumulators of metal and metalloid trace elements: facts and fiction. Plant and Soil 2013;362: 319-334.

[35] Angle J, Chaney R, Li YM, Baker A. The risk associated with the introduction of native and hyperaccumulators plants. Abstract. USDA, Agricultural Research Service, USA. 2001.

[36] Vangronsveld J, Herzig R, Weyens N, Boulet J, Adriaensen K, Ruttens A, Thewys T, Vassilev A, Meers E, Nevajova E, van der Lelie D, Mench M. Phytoremediation of contaminated soils and groundwater: lessons from the field. Environmental Science and Pollution Research 2009;67: 765-794.

[37] Pilon-Smits EAH, Freeman JL. Environmental cleanup using plants: biotechnological advances and ecological considerations. Frontiers in Ecology and the Environment 2006;44: 203-210.

[38] Carrillo Gonzalez R, Gonzalez-Chavez MCA. Metal accumulation in wild plants surrounding mining wastes. Environmental Pollution 2005;144: 84-92.

[39] Pivetz BE. Phytoremediation of Contaminated Soil and Ground Water at Hazardous Waste Sites. Ground Water Issue, U.S. Environmental Protection Agency, Office of

Research and Development and Office of Solid Waste and Emergency Response. 2001.

[40] Bystrzejewska-Piotrowska G, Pianka D, Bazała MA, Stęborowski R, Manjon JL, Urban PL. Pilot study of bioaccumulation and distribution of cesium, potassium, sodium and calcium in king oyster mushroom (*Pleurotus eryngii*) grown under controlled conditions. International Journal of Phytoremediation 2008;10(6):503-514.

[41] Şahan T, Ceylan H, Şahiner N, Aktaş N. Optimization of removal conditions of copper ions from aqueous solutions by *Trametes versicolor*. Bioresource Technology 2010;101: 4520-4526.

[42] Dursun AY, Uslua G, Cucia Y, Aksub Z. Bioaccumulation of copper(II), lead(II) and chromium(VI) by growing *Aspergillus niger*. Process Biochemistry 2003;38: 1647-1651.

[43] Cerino-Córdova FJ, García-León AM, Soto-Regalado E, Sánchez-González MN, Lozano-Ramírez T, García-Avalos BC, Loredo-Medrano JA. Experimental design for the optimization of copper biosorption from aqueous solution by *Aspergillus terreus*. Journal of Environmental Management 2012;95 Suppl.: S77-82.

[44] Taştan BE, Ertuğrul S, Dönmez G. Effective bioremoval of reactive dye and heavy metals by *Aspergillus versicolor*. Bioresource Technology 2010;101: 870-876.

[45] Anjum F, Bhatti HN, Ghauri MA, Bhatti IA, Asgher M, Asi MR. Bioleaching of copper, cobalt and zinc from black shale by *Penicillium notatum*. African Journal of Biotechnology 2009;8: 5038-5045.

[46] Sağ Y, Akçael B, Kutsal T. Ternary biosorption equilibria of chromium(VI), copper(II), and cadmium(II) on *Rhizopus arrhizus*. Separation Science and Technology 2002;37: 279-309.

[47] Yap CK, Yazdani M, Abdullah F, Tan SG. Is the High Cu Tolerance of *Trichoderma atroviride* Isolated from the Cu-Polluted Sediment Due to Adaptation? An *In Vitro* Toxicological Study. Sains Malaysiana 2011;40(2): 119-124.

[48] Wang B, Wang K. Removal of copper from acid wastewater of bioleaching by adsorption onto ramie residue and uptake by *Trichoderma viride*. Bioresource Technology 2013;136: 244-250.

[49] Robinson BH, Chiarucci A, Brooks RR, Petit D, Kirkman JH, Gregg PEH, De Dominicis V. The nickel hyperaccumulator plant *Alyssum bertolonii* as a potential agent for phytoremediation and phytomining of nickel. Journal of Geochemical Exploration 1997;59: 75-86.

[50] Robinson BH, Leblanc M, Petit D, Brooks RR, Kirkman JH, Gregg PEH. The potential of *Thlaspi caerulescens* for phytoremediation of contaminated soils. Plant and Soil 1998;203: 47-56.

[51] Reeves RD, Schwartz C, Morel JL, Edmondson J. Distribution and metal-accumulating behavior of *Thlaspi caerulescens* and associated metallophytes in France. International Journal of Phytoremediation 2001;3: 145-172.

[52] Bert V, Bonnin I, Saumitou-Laprade P, De Laguérie P, Petit D. Do *Arabidopsis halleri* from nonmetallicolous populations accumulate zinc and cadmium more effectively than those from metallicolous populations? New Phytologist 2002;155: 47-57

[53] Singer AC, Bell T, Heywood CA, Smith JAC, Thompson IP. Phytoremediation of mixed-contaminated soil using the hyperaccumulator plant *Alyssum lesbiacum*: evidence of histidine as a measure of phytoextractable nickel. Environmental Pollution 2007;147: 74-82.

[54] Broadhurst CL, Tappero RV, Maugel TK, Erbe EF, Sparks DL, Chaney RL. Nickel and manganese accumulation, interaction and localization in leaves of the Ni hyperaccumulators *Alyssum murale* and *Alyssum corsicum*. Plant and Soil 2008;314: 35-48.

[55] Alford ER, Pilon-Smits EAH, Paschke MW. Metallophytes – a view from the rhizosphere. Plant and Soil 2010;337: 33-50.

[56] Turnau K, Mesjasz-Przybylowicz J. Arbuscular mycorrhiza of *Berkheya coddii* and other Ni-hyperaccumulating members of Asteraceae from ultramafic soils in South Africa. Mycorrhiza 2003;13: 185-190.

[57] Abou-Shanab RAI, Angle JS, Chaney RL. Bacterial inoculants affecting nickel uptake by *Alyssum murale* from low, moderate and high Ni soils. Soil Biology & Biochemistry 2006;38: 2882-2889.

[58] Farinati S, Dal Corso G, Bona E, Corbella M, Lampis S, Cecconi D, Polati R, Berta G, Vallini G, Furini A. Proteomic analysis of *Arabidopsis halleri* shoots in response to the heavy metals cadmium and zinc and rhizosphere microorganisms. Proteomics 2009;9: 4837-4850.

[59] Jankong P, Visoottiviseth P, Khokiattiwong S. Enhanced phytoremediation of arsenic contaminated land. Chemosphere 2007;68: 1906-1912.

[60] Wangeline AL, Reeves FB. Two new *Alternaria* species from selenium-rich habitats in the Rocky Mountain Front Range. Mycotaxon 2007;99: 83-89.

[61] Wangeline AL. Fungi from seleniferous habitats and the relationship of selenium to fungal oxidative stress. PhD thesis. Colorado State University, Fort Collins; 2007.

[62] Brigo L, Ferraio A. Le mineralizzazioni della Liguria Orientale. Rendiconti S.I.M.P. 1974;30: 305-316.

[63] Campana N, Maggi R, Pearce M. Ricerche archeologiche nelle miniere di Libiola e di M. Loreto (Genova). Paleoexpress 1999;3: 9-10.

[64] Carbone C, Di Benedetto F, Marescotti P, Martinelli A, Sangregorio C, Cipriani C, Lucchetti G, Romanelli M. Genetic evolution of nanocrystalline Fe oxide and oxy-

hydroxide assemblages from the Libiola mine (Eastern Liguria, Italy): structural and microstructural investigations. European Journal of Mineralogy 2005;17:785-795.

[65] Marescotti P, Carbone C, De Capitani L, Grieco G, Lucchetti G, Servida D. Mineralogical and geochemical characterisation of open-air tailing and waste-rock dumps from the Libiola Fe-Cu sulphide mine (Eastern Liguria, Italy). Environmental Geology 2008;53: 1613-1626.

[66] Abbate E, Bortolotti V, Principi G. Appennine Ophiolites: a peculiar oceanic crust. Ofioliti Special Issue 'Thethian Ophiolites: 1, western area' 1980;1: 59-96.

[67] Marescotti P, Azzali E, Servida D, Carbone C, Grieco G, De Capitani L, Lucchetti G. Mineralogical and geochemical spatial analyses of a waste-rock dump at the Libiola Fe-Cu sulphide mine (Eastern Liguria, Italy). Environmental Earth Sciences 2010;61: 187-199.

[68] Marescotti P, Roccotiello E, Zotti M, De Capitani L, Carbone C, Azzali E, Mariotti MG, Lucchetti G. Influence of soil mineralogy and chemistry on fungi and plants in a waste-rock dump from the Libiola mine (eastern Liguria, Italy). Periodico di Mineralogia 2013;82: 141-162.

[69] Roccotiello E, Zotti M, Mesiti S, Marescotti P, Carbone C, Cornara L, Mariotti MG. Biodiversity in metal-polluted soils. Fresenius Environmental Bulletin 2010;19(10b): 2420-2425.

[70] Mariotti MG. Osservazioni sulle formazioni a *Buxus sempervirens* e a *Genista salzmannii* della Liguria orientale. Memorie della Accademia Lunigianese di Scienze 1994;59: 77-125.

[71] Vagge I. Le garighe a *Genista desoleana* Valsecchi ed *Euphorbia spinosa* L. subsp. *ligustica* (Fiori) Pign. della Liguria orientale (Italia NW). Fitosociologia 1997;32: 239-243.

[72] Zotti M, Di Piazza S, Ambrosio E, Mariotti MG, Roccotiello E, Vizzini A. Macrofungal diversity in *Pinus nigra* plantations in Northwest Italy (Liguria) Sydowia 2013;65(2): 223-243.

[73] Bradshaw AD, Chadwick MJ. The restoration of land. Berkeley: University of California Press; 1980.

[74] Roccotiello E, Serrano HC, Mariotti MG, Branquinho C. Nickel phytoremediation potential of the Mediterranean *Alyssoides utriculata* (L.) Medik. Chemosphere 2015;119: 1372–1378.

[75] Zotti M, Di Piazza S, Roccotiello E, Lucchetti G, Mariotti MG, Marescotti P. Microfungi in highly copper-contaminated soils from an abandoned Fe-Cu sulphide mine: growth responses, tolerance and bioaccumulation. Chemosphere 2014;117: 471-476.

[76] Borovička J, Kotrba P, Gryndler M, Mihaljevič M, Řanda Z, Rohovec J, Cajthaml T, Stijve T, Dunn CE. Bioaccumulation of silver in ectomycorrhizal and saprobic macro-

fungi from pristine and polluted areas. Science of the Total Environment 2010;408: 2733-2744.

[77] Gomes NCM, Rosa CA, Pimentel PF, Mendonça-Hagler LCS. Uptake of free and complexed silver ions by different strains of *Rhodotorula mucilaginosa*. Brazilian Journal of Microbiology. 2002;33: 62-66.

[78] Borovička J, Randa Z, Jelinek E, Kotrba P, Dunn CE. Hyperaccumulation of silver by *Amanita strobiliformis* and related species of the section *Lepidella*. Mycological research 2007;III: 1339-1344.

[79] Schwitzguébel JP, Kumpiene J, Comino E, Vanek T. From green to clean: a promising and sustainable approach towards environmental remediation and human health for the 21st century. Agrochimica 2009;53: 209-237.

[80] Harris J. Microbiological tools for monitoring and managing restoration progress. In: Moore HM, Fox HR, Elliot S. (eds.) Land Reclamation. Lisse: A.A. Balkema Publishers; 2003. p201-206.

[81] Miransari M. Hyperaccumulators, arbuscular mycorrhizal fungi and stress of heavy metals. Biotechnology Advances 2011;29: 645-653.

[82] Audet P, Charest C. Allocation plasticity and plant-metal partitioning: Meta-analytical perspectives in phytoremediation. Environmental Pollution 2008;156: 290-296.

[83] Janouskova M, Pavlıkova D, Vosatka M. Potential contribution of arbuscular mycorrhiza to cadmium immobilisation in soil. Chemosphere 2006;65: 1959-1965.

Chapter 23

Underutilized Crops and Intercrops in Crop Rotation as Factors for Increasing Biodiversity on Fields

Franc Bavec and Martina Bavec

Additional information is available at the end of the chapter

http://dx.doi.org/10.5772/59131

1. Introduction

Protection of biodiversity in intensive agricultural areas was one of the priorities of Common Agricultural Policy (CAP) and Agri-environmental payment (AEP 2007-2013) [1]. In general, AEP measures include crop rotation, the greening of fields, organic agriculture, soil coverage in water protected areas, etc., which depend on detailed national measures. In the new EU perspective of the Greening CAP reform 2014-2020 [2], three basic measures are foreseen in the first pillar: (i) maintaining permanent grassland; and (ii) crop diversification (a farmer must cultivate at least 2 crops when his arable land exceeds 10 hectares and at least 3 crops when his arable land exceeds 30 hectares. The main crop may cover at most 75% of arable land, and the two main crops at most 95% of the arable area); (iii) maintaining an "ecological focus area" of at least 5% of the arable area of the holding for farms with an area larger than 15 hectares (excluding permanent grassland) – i.e. field margins, hedges, trees, fallow land, landscape features, biotopes, buffer strips, afforested area. This figure will rise to 7% in 2017 according to the Commission report and the legislative proposal (CAP reform [2], MEMO 13/621, 26.6.2013). In the 2nd pillar documents (Regulation EU 1305/2013, 17.12.2013),'biodiversity' is mentioned 12 times and is mainly focused on conversion into organic farming, the protection of biodiversity in Nature 2000 areas and forest biodiversity conservation statuses of species and habitats, as well as enhancing the public amenity value of Nature 2000 areas or other high value nature areas. In these documents, we were mainly faced with general terms of biodiversity, however the detailed measures in the second pillar should be defined on a national level. There is also the EU Biodiversity Strategy to 2020, which is trying to halt the loss of biodiversity and the degradation of ecosystem services in the EU by 2020. It is also trying to feasibly restore biodiversity, while stepping up the EU contribution to averting global biodiversity loss. In case of pollinators and their vs. pollinators' plants, they are mentioned

only in introduction of document. In the agricultural part the focus is only on maximising agricultural areas across grasslands, arable land and permanent crops that are covered by biodiversity-related measures under the CAP so as to ensure the conservation of biodiversity and to bring about a measurable improvement in the conservation status of species and habitats affected by agriculture and in the provision of ecosystem services as compared to the EU 2010 Baseline, thus contributing to enhance sustainable management; without mentioned intercrops and diverse – alternative crops.

2. Aim and methodology

The Biodiversity Strategy to 2020, compared to previous documents, places more emphasis on field crop areas, but the measurements for 55.9 mio ha of field crop areas in Europe (FAO stat, 2012) are still not defined clearly enough, which is why special measurements for increasing the biodiversity of crops are needed on the intensive fields. As we can see in the Greening CAP reform 2014-2020 [2], these actions manage only a small portion of the 2nd and/or 3rd crop, and all of those can be produced in monoculture. Further on, the 1st Pillar supports more green washing than actual change of crop rotation, crop structure and diversity of field crops on the fields. Pillar 2, which depends on national regulations and is more open to potentially important measurements (like intercropping and introduction of alternative - rare - underutilized crops into field crop production for increasing biodiversity) is rarely mentioned in regulations and strategies. Sometimes its importance is unknown to administration workers, which is probably the reason why officers are not interested in their implementation (a good example is the Slovenian Ministry for Agriculture and Environment). A review of relevant literature, research and practical experiences of the authors will be present in this chapter.

3. Crop rotation

It is a fact that intensive agriculture and dominant monoculture based on high chemical inputs and even genetically modified organisms in some countries are destroying the fields biodiversity. The decreasing soil biodiversity does not support the natural cycles for sustaining good soil characteristics and natural plant nutrition pathways. For those in the system, like the US Corn Belt where these issues are key environmental problems include water contamination by nutrients and herbicides emitted from cropland, a lack of non-agricultural habitat to support diverse communities of native plants and animals and a high level of dependence on petrochemical energy in the dominant cropping systems. In addition, projected changes in this region's climate, which include increases in the proportion of precipitation coming from extreme events, could make soil and water conservation in existing cropping systems more difficult [3]. In spite of the professionals' scepticisms about intensive systems, the authors Liebman et al. [3] concluded that increasing biodiversity through the strategic integration of perennial plant species can be a viable strategy for reducing the reliance on purchased inputs and for increasing agro ecosystem health and resilience in the US Corn Belt. It is our concern that the measures in the EU Greening CAP reform 2014-2020 in the first pillar will not solve these kinds of problems in European countries either.

Biodiversity and plant protection against plant diseases and pests research is becoming an important issue in the world. For example, crop diseases are the most important natural disasters for food production and food safety, they are also one of the main reasons confining sustainable development of agricultural production. The discovery of the biodiversity mechanism (instead of chemicals and genetic modified organisms) to control crop diseases can reasonably guide the rational deployment and rotation of different crops and establish an optimization of different crop combinations [4].

The transformation of agriculture in the past half-century has triggered a decline in bees and other insect pollinators. In this case it is concluded [5] that areas of cultivated land farming, field margins, field edges and paths, headlands, fence-lines, rights of way, and nearby uncultivated patches of land are important refuges for many pollinators. According to the authors of this chapter, the yield of buckwheat increased twice in cases of pollination by honey bees compared to isolated plants by covering plots without honey bees (not published data). Likewise, crops including at least 800 cultivated plants depend on animal pollination in a functional biodiversity, which can be increased by intercrops and more diverse crop species in crop rotation. These solutions were not clearly defined and supported by professionals and policy makers, not even in the EU CAP reform and at the country levels as separate measures in Pillar 2 (like in Slovenia).

4. Biodiversity and the multifunctional importance of alternative field crops [1]

Intensive agriculture based on monoculture evidently reduces fields' biodiversity (Fig. 1) including the number of utilized crops and in consequence reduces natural health and nutritional compounds in the food. As reported by Jacobsen et al. [6], we have to safeguard both biodiversity and arable land for future agricultural food production, and we need to protect genetic diversity to safeguard ecosystem resilience. They conclude that majority of the research funding currently available for the development of genetically modified crops would be much better spent in other research areas of plant science, e.g., nutrition, policy research, governance, and solutions close to local market conditions if the goal is to provide sufficient food for the world's growing population in a sustainable way.

Introduction of alternative (rare, underutilized, disregarded, neglected, new and alternative GMO free) crops into the structure of field crop rotation can increase plant biodiversity and the nutritional and health value of food. Alternative crops are rich natural resources of essential amino acids, antioxidants, minerals, stimulators and other usable compounds, which are often limited to products from just a few main crops produced over the world. Production of

---

1 The section based on research of Bavec F. presented in the Agrosym 2011 paper, which content is republished with permission.

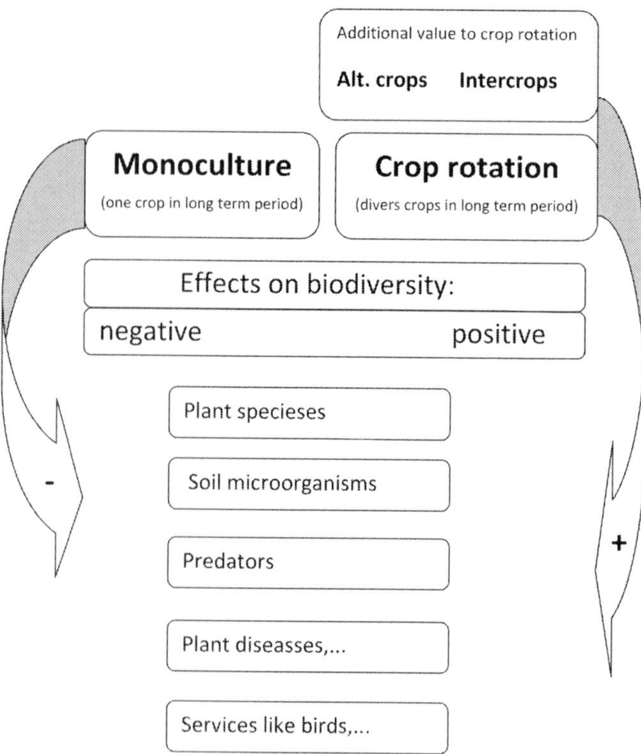

Figure 1. Comparison of negative effects of monoculture and positive effects of crop rotation, alternative crops and intercrops on biodiversity parameters

underutilized crops shall help to increase resistance to plant diseases, predators, and helping us to produce food without synthetic pesticides [7]. The described organic production of underutilized field crops represents a very important option for an environmentally acceptable crop production and a niche for 'special organic' products. The selling of this kind of products is of special interest for small scale farms, because it is a better solution compared to producing and selling cheaper products on global markets. But in this case the consumers, advisers and farmers need professional knowledge about the preferences of underutilized crops, production characteristics, clear guidelines for organic production, post-harvest technology, food processing including product certification and clear marketing strategies. The support given by educational, research and governmental institutions should also be in accordance with these needs. Each specific possibility and activity of each country can influence the consumer-producer relationships and the effective marketing by specific or/and niche products based on underutilized crops [8, 9].

A rich biodiversity of produced edible underutilized field crops encompasses cereals and pseudo cereals including millets, pulse crops, root and tuber crops, oil seed crops and dyes, some of which (including fibre crops) are usable for creating new market niches based on small scale production and processing. Furthermore, some of them are also suitable for industrial processing. Depending on the country, some of these plants are indigenous, and based on spread secondary diversity, some are completely new, sometimes even exotic. The understanding and use of underutilized crops is based mainly on tradition and the specifics of their growth circumstances. Most of them are unknown to a great percentage of agronomists. But the interest for underutilized crops is also increasing with publishers, because during the last decades a few publications spread the knowledge about underutilized field crops [10, 11, 12] with a special attention to temperate climate [7].

For example, in Slovenia temperate climate circumstances are predominant and just a small part is Mediterranean. For those, the tropic crops should be introduced into the temperate climate with special attention to a growth period of less than 160 frost free days, such as those for genotypes of sweet corn (*Zea mays* L. var. *saccharata*), batata (*Ipomea batata* L.) and other tropical tuber crops, specific genotypes of grain amaranths (*Amaranthus* sp.), quinoa (*Chenopodium quinoa* L.), groundnut (*Arachis hypogea* L.), vignas (*Vigna* ssp.), etc. The next factor is the system of reproduction based on plant parts growth in greenhouse conditions during winter time like in the case of batata. The spelt which is well adapted to temperate climates (*Triticum aestivum* L. ssp. *spelta* MacKey) has been forgotten, however todays introduced into crop rotations on many organic farms with field crop production during the last decade. Other farro group cereals such as einkorn (*Tritium monocccum* L.), emmer (*Triticumdico*ccum L.), etc. are introduced into organic farming just like sample crops on a few farms. Buckwheat (*Fagopyrum esculentum* L.), proso millet (*Panicum milliaceum* L.) and oil seed pumpkins (*Cucurbita pepo* L. group Pepo) were traditional, but neglected until the last decade when their production started to increase. The group of alternative oil crops such as false flax (*Camelina sativa* L.), saflower (*Carthamus tinctorius* L.), garden poppy (*Papaver somniferum* L. ssp. *somniferum* Kadereit) are being researched and considered for eventual introduction into crop rotations. We are also looking at some legumes and the group of millets from Africa-potential crops for dry conditions described in the book Organic production and use of alternative crops by Bavec and Bavec [7]. In changing climate conditions some crops like millets from Africa and quinoa from the Andes might play an important role for creating new stress-tolerant species and genotypes for future agriculture. However, quinoa is described as a crop with high biodiversity value, which maintains productivity even on rather poor soils and high salinity [13].

Traditional cropping systems of undeveloped countries contain numerous genotypes of domesticated crop species, as well as their wild relatives. The richness of plant biodiversity of traditional agro-ecosystems is comparable with natural systems. It is one of the reasons why underutilized crops have to play a greater role, especially in organic farming. Underutilized crops bring diversity into crop rotations and provide new possibilities for soil cultivation. Organic farming, which is based on traditional farming systems, offers a way of promoting the diverse food and food risks, it increases pollinator insects and reduces plant insects and disease incidence, it is efficient in labour use and it also brings an intensification of production with acceptable resources, a maximization of returns and stability under responsible technologies. Underutilized crops help local communities to be more independent while using the local resources for production and transport expense reduction. A similar option might be used for organically produced underutilized crops [14]. The use of underutilized field crops has resulted in an in ceased product competitiveness, a rich nutritional and health value of food, in tradition, locality, special quality according to organic production guidelines and even in market attractions. The health and nutritional rich products, especially if they are produced according to organic farming guidelines, represent a special niche in the market of the developed world.

Knowledge about food health and nutritional attributes based on underutilized crops is very useful for promotion, decision support for producers and for the buying motivation of consumers. Special attention needs to be given to the coexistence of pollinator insects and buckwheat, to antioxidants in food (tocopherols in oil crops, squalen in grain amaranths, anthocyanins in sweet potato, etc.), to rich amino acid compositions (grain amaranths, quinoa, partly buckwheat, partly legumes, etc.), to gluten free foods for people with celiac disease (buckwheat, grain amaranth, quinoa, millets), to good quality fibre food (whole grained spelt and other cereals), to food rich in minerals, vitamins and their good balance, etc. Many of them are used in pharmacology and alternative medicine, like oil seed pumpkins [15], buckwheat [16, 17, 18], amaranths [19, 20], etc.

The above information show a very interdisciplinary approach to alternative field crops, which can help to change the structure of crop diversity with clear steps towards a better social and economic behaviour. At the field level the rich crop diversity supports multifunctional processes like more sustainable nutrient cycles, caused by different root systems and different nutrient uptakes. A bigger variety of crops in crop rotation leads to more permanent soil covers of plants during growth periods, especially if they are grown in stubble crops. If some of them are grown as green manure they can have an important influence on soil fertility, because of more rich micro-organisms activity. In cases such as *Brasicaceaes*, for example white mustard (*Sinapis alba*), the crop also has a phytosanitary effect. Diverse crops encourage more sustainable production systems because of the positive effects on relationships between predators and pests (Fig. 1). They also decrease plant diseases in crop rotation, because of their life cycle breaks.

5. Intercropping — Unexploited beneficial measure [2]

Intercropping (sowing two or more crops together) represents a high valued strategy for long term sustainable plant production management, due to its many beneficial effects like effects on increasing diversity of cultivated crops, nitrogen fixations by legumes [21, 22, 23, 24] instead of synthetic N fertilisers, weed control, yield stability, inter-specific complementarity, a more efficient use of environmental sources, soil coverage in under-sown crops, a higher protein

---

2 The section based on research of Bavec F. presented in the Agrosym 2011 paper, which content is republished with permission.

content in seeds used for grain feed or silage mixture (especially important after BSE crises, etc.). Because of the complexity of these systems, intercropping has been neglected in practice and just partly researched as a plant production system under climates and cultivation circumstances worldwide. The research is deficient, especially in the case of intercropping impacts on biodiversity [25].

Farmers and researchers carry out different cropping systems to increase productivity and sustainability by using crop rotations, relay cropping, and intercropping of different annual crops. An associated culture often involves cereals and legumes due to its advantages for soil conservation, weed control, lodging resistance, yield increment, hay curing, forage preservation over pure legumes, high crude protein content and protein yield. Different seeding ratios or planting patterns have been practised for cereal-legume intercropping. Bean yields in an intercrop culture are usually less sown than those obtained from sole bean stands. It is possible to increase yields with suitable management practices such as the use of optimum plant population and improved bean cultivars. However, bean yields in intercrops represent a surplus to the main maize crop yield. In EU countries, cowpea is rarely used in intercropping with cereals on small-scale farms. A number of indices such as land equivalent ratio, relative value total and monetary advantage have been proposed to describe the competition within and economic advantages of intercropping systems (from unpublished review, Bavec et al., Univ. Maribor). However, such indices have not been used for climbing bean (*Phaseolus vulgaris* L.) and maize intercropping to evaluate the competition among species and to evaluate the economic advantages of each intercropping system.

Intercropping of climbing bean and maize is a common production system on small scale farms and of interest to researchers in Latin America, as well as in South Africa, Ethiopia and other African countries. In a temperate climate, this type of intercropping has traditionally been practised 30 years ago, also on small-scale farms in Slovenia, Romania and in other Middle and Eastern European countries. Despite the fact that intercropping systems should involve integrating crops, using space and labour more efficiently, recommendations supporting good sole cropping systems, in which net incomes are also higher.

In Slovenia, two cases of conversion from a manual to mechanized production system of intercropped bean and maize production was established on an approximate 4 to 6 ha planting area per year, where the bean seeds are used for silage fed to ruminants (farm Jankovič, Vihre/ Krško), and for human consumption (farm Jakob, Lipovci). In the other European countries we also practically lost this traditional production system, although it still exists in some poor and self-sufficient small-scale farms, with some attention on the dry climate due to recent climate changes [26].

In general, important benefits of intercropping cereals with legumes are the following: more available nitrogen due to nitrogen fixation with legumes-with up to 84 % of nitrogen may be derived from fixation by the climbing bean, maize and bean intercropping may help converse a deficiency of bean production in European countries, in that it involves integrating crops using space and labour more efficiently, increased efficient competitions of cereals with weeds, improved soil structure, reduced loss of plant nutrients, less damage of plants due to pathogens and insects, especially in organic farming systems. Based on the available literature, most researches have focused on intercropping bush beans in non-European growing conditions. Because of the different canopy characteristics of bush beans, data for bush beans are not comparable with climbing bean maize intercropping – this has only been reported by Gebeyehu et al. [27]. However, there is a lack of scientifically relevant information about promising plant ratios of maize-climbing bean intercropping systems, especially for the ones produced in European temperate zones under integrated or organic production systems.

In this paper we want to focus our attention on two additional reasons for the proposed intercropping climbing bean/maize in temperate climate vs. marginal regions for maize (to FAO group 400) production [28], which are also suitable for growing climbing beans, due to temperatures and humidity, enabling a simple and environmentally friendly production. In this case the soil preparation is conventional (ploughing in autumn, pre-sowing soil preparation in spring), the same machine is used for sowing maize and bean seeds. The bean seeds need to be sown close to the maize strips at the stage when the maize has few true leaves, after the 1st or 2nd inter-row mechanical hoeing, close to the strips of the maize plants. For harvesting the maize bean whole plant mixture a silage combine is used, but for the seed harvesting a cereal combine and eventual separation of the bean and maize seeds could take place. The second benefit is the production of protein rich silage caused by bean grains, which contain approximately 20% of crude proteins. These kinds of proteins are good and might be a relatively cheap replacement for animal source proteins, which are not allowed for ruminants feed after the appearance of BSE 'mad cow disease'.

In cereal-legume intercropping, cereal crops establish uniform canopy structures and then legume crops and the roots of cereal crops grow to a greater depth than those of legume crops with less lodging consequences [29]; however the agronomic traits of genotypes need to be well adapted for intercropping. Climbing bean cultivars need specific adaptations to intercrops using predominantly morphological maize types grown in the area [27]. Somewhat earlier maize cultivars can give an improved net income when intercropped with climbing beans, because they are resistant to stem lodging. This indicates that the component crops probably have differing spatial and temporal use of environmental resources such as radiation, water and nutrients. Therefore, this cropping system may help improve productivity of low external input farming, which depends largely on natural resources such as rainfall and soil fertility. The intercropping productivity is largely dependent on planting date and plant population of its components. Small-scale farmers have practised traditional cropping techniques, such as intercropping, in which they unknowingly manipulate the plant population [28] because that way, interspecies competition is stronger.

The effects of intercrops on weed communities were characterised in terms of growth, species diversity (richness and evenness), and floristic and functional composition. Intercrops and barley monocultures generally produced similar effects on the companion weed communities, whereas pea effects were less suppressive and more variable. Spring-emerging species generally increased its relative importance in the intercrop weed communities; whereas winter-emerging species were usually less abundant in intercrops. Divergence in the abundance of winter and summer emerging weeds could be attributed to the different canopy dynamics of intercrops and monocultures [30].

Plant diversity based on intercropping systems includes potentially important mechanism for chemically mobilizing nutrients in otherwise-unavailable forms of one or more limiting soil nutrients, such as phosphorus (P) and micronutrients (iron (Fe), zinc (Zn) and manganese (Mn)). In case of phosphorus-mobilizing, crop species improve P nutrition for themselves and neighbouring non-P-mobilizing species by releasing acid phosphatases, protons and/or carboxylates into the rhizosphere which increases the concentration of soluble inorganic P in soil. Similarly, on calcareous soils with a very low availability of Fe and Zn, Fe-and Zn-mobilizing species, such as graminaceous monocotyledonous and cluster-rooted species, benefit themselves, and also reduce Fe or Zn deficiency in neighbouring species, by releasing chelating substances [31].

Only in one EU research programme a survey was carried out within five European countries with regard to the practice of cereal grain legume intercropping. The most commonly used combination was spring barley-spring pea with 27 other combinations between pulses and cereals. 72 % of all examples consisted of spring varieties and the rest of winter varieties, mainly a special case of the French South West with a mild winter climate. Intercrops were mainly used for feeding purposes. Yield stability, effective weed suppression, and good quality of feed were reported as the best outcomes. The negative outcomes were complicated mechanical weed regulation, unequal maturation and additional costs for separation. The interviewed farmers showed predominantly positive prospects for the development of intercropping on their farms, problems with sowing techniques being the only importance [32].

Based on the statements in this chapter, we can once again underline the importance of the use of well-known [33] and new findings of beneficial intercropping effects on productivity and biodiversity in different farming systems [34] which is especially very important in changing climates [35]. Because the fact that intercropping is a more expensive and complicated cultivation than sole crop production, intercrops need wider support (like new research and simulation models) [36] that will be included in farming systems as a basic environmental measure at the field production level.

6. Conclusions

Based on the lack of environmental indicators which influence the functional biodiversity in the field, a precisely described importance of crop rotation, introduction of underutilized crops and intercrops, is given. Due to their many beneficial effects (crop rotation, nitrogen fixations by legumes instead of synthetic N fertilisers, weed suppression, yield stability, inter-specific complementarily, more efficient use of environmental sources, soil cover at under-sown crops, higher protein content in the seeds for grain feed or silage mixture, especially important after BSE crises), inter-cropping and underutilized crops represents a high valued strategy for long term biodiversity and sustainable plant production management. Because of the farmland intensification and complexity of biodiversity, intercropping and underutilized crops have been neglected in practice and only partly researched as a plant production system under different cultivation circumstances (sometimes site specific). More diverse crops and intercropping support a more stable ecosystem productivity, especially in the case of intercrops with legumes; here the inputs of artificial nitrogen can be significantly reduced. However, ecological intensification of agriculture depends on simple and clear ecologically oriented agro-environmental policies all over the world, which will not support »green washing« of conventional agriculture or »conventionalisation« of organic farming. Because of climate changes, the alternative field crops and intercrops need more political support and should be taken into account in EU CAP and OECD policies, which do not include suggested measures on national levels as part of environmental payments, not even outside Europe. Because of their importance, intercrops and alternative crops need to be a part of biodiversity indicators at field and landscape levels.

Acknowledgements

The idea for this paper came as part of the project 'CRP V4-1137 Alternative field crops in different production systems and crop rotations as a base for adaptation to climate changes and food supply with food and feed' supported by the Ministry of Education, Science and Sport and the Ministry of Agriculture and the Environment, for which we are thankful.

Author details

Franc Bavec* and Martina Bavec

*Address all correspondence to: franci.bavec@um.si

University of Maribor, Faculty of Agriculture and Life Sciences, Hoce/Maribor, Slovenia

References

[1] AEP 2007-2013 – Agri-environmental payments, Ministry of Agriculture, Forestry and Food, Slovenia, Ljubljana, RDPRS, 2007: Rural Development Programme of the Republic Slovenia 2007-2013. Ministry of Agriculture, Forestry and Food, Ljubljana, 20.07.2007, 232 p.

[2] Greening CAP reform 2014-2020, and NAT/471, 2010-European Economic and Social Committee. 9 p. Available at (http://www.eesc.europa.eu/ressources/docs/ces1178-2010_ac_en.doc) Accessed15.09. 2010.

[3] Liebman M., Helmers M. J., Schulte L. A., Chase C. A. Using biodiversity to link agricultural productivity with environmental quality: Results from three field experiments in Iowa. Renew. Agric. Food Syst. 2013;28115-128.

[4] Yang J., Shi Z-F., Gao D., Liu L., Li C. Y. Mechanism on biodiversity managing crop diseases. Yichuan. 2012;34 1390-1398.

[5] Nicholls C. I., Altieri M. A. Plant biodiversity enhances bees and other insect pollinators in agro ecosystems. A review. Agron. Sust. development. 2013;33257-274.

[6] Jacobsen S. E., Sorensen M., Pedersen S. M., Weiner J. Feeding the world: genetically modified crops versus agricultural biodiversity. Agron. Sust. development. 2013;33 651-662.

[7] Bavec F., Bavec M. Organic production and use of alternative crops. Boca Raton, New York, London : Taylor & Francis: CRC Press. 2006

[8] Williams J.T., Hag N. Global research on underutilized crops. An assesment of current activities and proposals for enhanced cooperation. ICUC.Sauthhampton, UK, 46 p. www.biodiversityinternational.org/Publications/, 2002; accessed 18.05.2014.

[9] Bavec F. Grobelnik Mlakar, S., Rozman, Č. Turinek, M., Bavec, M. How to create an efficient organic production and market for underutilized field crops : a review on the Slovenian case. In: JAENICKE, Hannah (ed.). Proceedings of the International Symposium on Underutilized Plants for Food Security, Nutrition, Income and Sustainable Development, Arusha, Tanzania, March 3-6, *2008*, (Acta horticulturae 2009;806 443-449.

[10] Duke J., duCellier J. CRC handbook of alternative cash crops. Boca Raton: CRC Press. 1993

[11] Williams J.T. (ed.). Underutilised Crops: Cereals and Pseudocereals, London: Chapman & Hall. 1995

[12] Belton P., Taylor J. (eds.). Pseudocereals and less common cereals. Springer-Verlag, Berlin, Germany. 2002.

[13] Ruiz K. B., Biondi S., Oses R., Acuna-Rodriguez I.S., Antognoni F., Martinez-Mosqueira E. A., Coulibaly A., Canahua-Murillo A., Pinto M., Zurita-Silva A., Zurita-Silva A., Bazile D. Jacobsen E.-S., Molina-Montenegro M. Quinoa biodiversity and sustainability for food security under climate change. A review. Agron. Sust. Development. 2014;34 349-359.

[14] Dixon J., Hellin J. Erestein O., Kosina P.U-imapct pathway for diagnosis and impact assessment of crop improvement. J. Agric. Sci. 2007;145195-206.

[15] Zuhair H. A. Pumpkin-seed oil modulates the effect of felodipine and captopril in spontaneously hypertensive rats. Pharmacol. Res. 2000;41 555-561.

[16] De Francischi M. L., Salgado J.M., Da Costa C.P. Immunological analysis of serum for buckwheat fed celiac patients. Plants. Foods Hum. Nutr. 1994;46 207-211.

[17] Li S. Q., Zhang Q.H. Advances in the development of functional foods from buckwheat. Crit. Rev. Food Sci. Nutr. 2001;41 451-459.

[18] Krkoškova B., Mrazova Z. Prophylactic components of buckwheat. Food Res. Int. 2005;38 561-568.

[19] Rangarajan A., Chenoweth E. A., Kelly J. F., Age K. M. Iron bioavailability from amaranthus species: 2 Evaluation using haemoglobin repletition in anaemic rats. J. Sci. Food Agric. 1998;78 274-280.

[20] Prokopowicz D. Health promoting attributes of amaranths (*Amaranthus cruentus*). Met. Vet. 2001;57 559-561.

[21] Hauggaard-Nielsen H., Jørnsgaard H. Kinane J., Steen Jensen E. Grain legume-cereal intercropping: The practical application of diversity, competition and facilitation in arable and organic cropping system. Renewable Agriculture and Food Systems 2007;23 3-12.

[22] Bergkvist G. Stenberg M., Wetterlind J., Bath B., Elfstrand S. Clover cover crops under-sown in winter wheat increase yield of subsequent spring barley-Effect of N dose and companion grass. Field Crops Research.2011;120 292-298.

[23] Lithourgidis A. S., Vlachostergios D. N., Dordas C. A., Damalas C. A. Dry matter yield, nitrogen content, and competition in pea-cereal intercropping systems. Eu. Jour. Agron. 2011;34 278-294.

[24] Sun Y. M., Zhang N. N., Wang E. T., Yuan H. L., Yang J. S. and Chen W. X. Influence of intercropping and intercropping plus rhizobial inoculation on microbial activity and community composition in rhizosphere of alfalfa (*Medicago sativa* L.) and Siberian wild rye (*Elymus sibiricus* L.). FEMS microbiology ecology. 2009;7062-70.

[25] Shili-Touzi I., De Tourdonnet, S., Launay, M., Dore, T. (2011): Does intercropping winter wheat (*Triticum aestivum*) with red fescue (*Festuca rubra*) as a cover crop improve agronomic and environmental performance? A modeling approach Original Research Article. Field Crops. 2011;116218-229.

[26] Badgley C., Moghtader J., Quintero E., Zakem E., Chappell M.J., Aviles Vazquez K., Samulon A., Perfecto I. Organic Agriculture and the Global Food Supply. Renewable Agriculture and Food Systems 2007;22 86-108.

[27] Gebeyehu S., Simane B., Kirkby R. Genotype x cropping system interaction in climbing bean (*Phaseolus vulgaris* L.) grown as sole crop and in association with maize (*Zea mays* L.). Eur. J. Agron. 2006;24 396-403.

[28] Bavec F., Bavec M. Effect of plant population on leaf area index, cob characteristics and grain yield of early maturing maize cultivars (FAO 100-400). Eur. J. Agr. 2002;16 151-159.

[29] Kontturi M., Laine A., Niskanen M., Hurme T., Hyovela M., Peltonen-Sainio P. Pea-oat intercrops to sustain lodging resistance and yield formation in northern European conditions Acta Agric. Scandinavica 2011;61612-621

[30] Poggio S. L. Structure of weed communities occurring in monoculture and intercropping of field pea and barley. Agric. Ecosyst. & Env. 2005;109 48-58.

[31] Li L., Tilman D., Lambers H., Zhang F. S. Plant diversity and over yielding: insights from belowground facilitation of intercropping in agriculture. New phytologist. 2014; 203 63-69.

[32] vonFragstein und Niemsdorff P., Knudsen D., Marie T., Gooding Mike J., Dibet A., Monti, M. Experiences with intercropping design – a survey about pulse cereal-combinations in Europe. Paper presented Jun 18-20 at ISOFAR 2008 in Modena, Italy.

[33] Willey R.W. Intercropping: its importance and research needs. Part 1. Competition and yield advantages. Field Crop Abstracts, Amsterdam, 1979;32 1-10.

[34] Bavec M., Grobelnik Mlakar S., Rozman Č., Pažek K., Bavec F. Sustainable agriculture based on integrated and organic guidelines: understanding terms : The case of Slovenian development and strategy. Outlook Agric.2009;38 89-95.

[35] Tsubo M., Walker S., Ogindo H. O. A simulation model of cereal-legume intercropping systems for semi-arid regions, II. Model application. Field crop research 2005;93 23-33.

[36] Baldy C., Stigter C.J. Agrometeorology of multiple cropping in Warming climates, INRA. 1997.

Chapter 24

Protected Areas and Ecosystem Services — Integrating Grassland Conservation with Human Well-Being in South Africa

André Pelser, Nola Redelinghuys and
Anna-Lee Kernan

Additional information is available at the end of the chapter

http://dx.doi.org/10.5772/59015

1. Introduction

In recent years development agencies and conservation organizations such as the World Conservation Union, World Bank, Birdlife International, the United Nations, the World Wide Fund for Nature and Fauna as well as Flora International, have served to reinforce a number of conservation practices and policies in which the link between natural conservation and improving the lives of rural communities has been piquantly accentuated. The central emphasis that has emerged from these accents is that protected areas – and national parks in particular-cannot be viewed as isolated from the economic and social context within which they are located. Worldwide – and particularly in the developing world – protected areas are progressively expected to navigate past the conventional primary focus on biodiversity protection to also, through the process of conserving biodiversity, contribute to improving the well-being of those communities adjacent to conservation areas through the delivery of social and economic benefits [1]. To be more precise, it has become essential that the goals of protected-areas management and biodiversity conservation become acquiescent with the socio-economic expectations and needs of local communities [2,3,4]. The very survival of such areas and the people surrounding it depends on a mutually beneficial interaction. In fact, protected areas have a powerful potential to markedly influence human well-being through the generation of social, environmental and economic initiatives that may benefit both protected areas as well as the local communities [5].

One example in South Africa where protected areas have been influential in attempting to improve the well-being of neighbouring communities is the People and Parks Programme of

South African National Parks (SANParks), which was implemented as an intermediary that endeavours to address the various socio-economic tribulations that were often ignored or sidelined in favour of conservation during the Apartheid rule. The post-apartheid policy of SANParks is entrenched in the conviction that biodiversity conservation should be directly linked with the needs of neighbouring communities, thus opening up possibilities for augmenting the well-being of communities neighbouring national parks in the country [6]. Some of the initiatives aimed at improving the well-being of neighbouring communities include health programmes, the development of cultural resources, heritage management, environmental education, the interpretation of medicinal plant use, the unlocking of economic opportunities in the form of job creation, and the carrying out of an assortment of arts and crafts projects [3,6].

Emanating from the above, this chapter reflects on a study conducted in the Golden Gate Highlands National Park (Golden Gate) in the Eastern Free State of South Africa, and the role of the park as a vehicle for improving the well-being of those living within the surrounding communities by means of the latter's participation in a grass harvesting programme in the park. Essentially, the broad aim of this research venture was to assess to what extent the thatch harvesting programme at Golden Gate had impacted on human well-being within the park's neighbouring communities. More specifically, this study set out to explore and answer the following interrelated research questions: To what extent has the thatch harvesting programme at Golden Gate benefited the communities bordering the park, and particularly the most vulnerable and poorest section of the community? What evidence is there to indicate that the thatch harvesting programme has improved the community's well-being? What interventions are needed to strengthen and maximise the impact of the said programme in order for it to effectively enhance the well-being of those within the target community? To what extent, if any, has this programme impacted the park's conservation mission?

2. About the project

This section firstly provides a broad overview of the general state of the grassland biome in South Africa, followed by a more detailed discussion of the grass-harvesting programme at Golden Gate.

2.1. Setting the scene: The grassland biome in South Africa

Globally the grassland biome covers about 40% of the earth's surface, is home to more than one billion people in the world and provides many essential ecosystem services required to support these people and many others who are not living inside this biome [7]. Grasslands are the largest of South Africa's nine biomes and cover roughly one third of the country [8]. South African grasslands constitute a complex ecosystem that includes amongst others 42 river systems, five Ramsar wetlands and three World Heritage Sites. There are more than 3,000 plant species found in these grasslands, and only one in six of them are grasses. Grasslands are the habitat for a wide variety of wild life, and provide many crucial ecosystem services that are essential for human development and well-being. Apart from providing grazing for millions of cattle and sheep, the grasslands biome also offers all-important services in water production, wetland functioning, flood attenuation, recreational amenities and support for livelihoods such as thatch for housing, grass for weaving and medicinal plants [8]. South African grasslands play a critical role in the hydrological cycle by reducing erosion and runoff, and by storing runoff as either groundwater or in wetlands, thereby contributing to water supply and freshwater ecosystem services [7].

The grasslands biome is one of the most threatened biomes in South Africa as a result of population increase, rapid urbanisation, expanding mining operations, increased forestry and commercial agriculture. Approximately 35% of this biome has been irreversibly transformed and less than 2% is officially conserved [7,9]. The current state of South African grasslands, as well as expected future developments, means that the important biodiversity and ecosystem services in the grasslands are being degraded to such an extent that human well-being is threatened. As a result, the importance of protecting the grassland biome for both biodiversity and economic development reasons has been recognized by the National Biodiversity Strategy and Action Plan that has identified this biome as a spatial priority for conservation action in South Africa [9].

2.2. Grassland conservation and grass harvesting at the Golden Gate Highlands National Park

Golden Gate) is situated in the foothills of the Maloti Mountains in the north-eastern part of the Free State Province (Figure 1), and plays a critical role in the country's grassland conservation strategy. Established in 1963, Golden Gate comprises more than 30 000 hectares of highland habitat, is home to a large variety of mammals, antelope and bird species, and is renowned for its sandstone formations and important paleontological discoveries [3]. The park is home to more than 60 species of grasses, and is currently the only national park in South Africa that protects the Afromontane grassland biome. The grass species include the red *Themeda triandra*, which is a highly nutritious grass for grazing antelope and widely regarded as an indicator of a healthy ecosystem [10]. Much of the grasslands outside the park have been permanently lost as a result of overgrazing and soil erosion. The larger Golden Gate region is also one of the most important water-catchment areas in South Africa, with more than half of the country's freshwater supply coming from this area [3].

Since the proclamation of the first national park in South Africa in 1926, no form of resource utilization was allowed in any of the 22 national parks, including grass harvesting at Golden Gate. This conventional policy of SANParks changed in 2003 when national legislation was amended to provide for communities to access resources from protected areas. The changed legal provision subsequently called for a revision of SANParks' own policy on resource use, and introduced a new resource use policy that regulates standard operating procedures for resource use in all South African national parks. In a broader context, the new policy on resource utilisation in national parks serves to confirm many initiatives since the mid 1990s that have served to underline the importance of the role of national parks with regard to

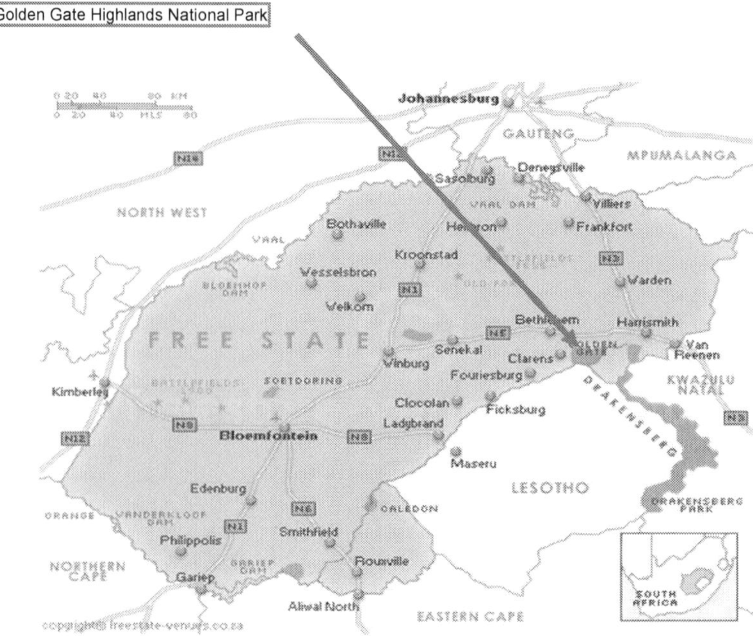

Figure 1. Locality of the Golden Gate Highlands National Park [11]

sustainable economic development and their augmentation of the well-being of their neighbouring communities.

The thatch harvesting programme at Golden Gate has been one of several projects for resource use within SANParks aimed at transferring social and economic benefits accruing from biodiversity protection to the impoverished surrounding communities through prospective employment opportunities by means of commercial access permits and park assisted entrepreneurial endeavours [12]. For many generations QwaQwa National Park, which amalgamated with Golden Gate in 2009, offered a rich source of accessible and harvestable grasses for communities residing in the area. These grasses were used to produce a wide variety of items such as brooms, hats, baskets, roof thatching, decorations and floor mats [12]. However, in accordance with the National Environmental Management Protected Areas Act (Act 57 of 2003), Golden Gate was obliged to restrict harvesting activities within its borders, which as a result cut off natural resources otherwise used by local community members. Recognising the financial consequences of such constraints, and in conjunction with SANParks Resource Use Policy which was signed into effect in March 2010, Golden Gate began exploring the possibility for regulated and controlled access and use of harvestable grass within the park. In June 2011, the necessary documents pertaining to the application for access, the access permits, the conditions for entry and harvesting within the park as well as the monitoring document for harvesting, were conceptualised and submitted for evaluation. In September 2011, a draft needs analysis report was also submitted for review [12]. Upon consideration and acceptance of these supporting documents, a pilot project for the proposed thatch harvesting programme was subsequently launched in 2012.

3. Conceptual framework

3.1. Ecosystem services and human well-being

In recent years, the need for more efficient management of ecosystem services, coincided with the needs and values of neighbouring communities, has become increasingly acknowledged by numerous governments as a means for improving the quality of life and well-being of their respective populations [13]. It is widely agreed that poverty and well-being are commonly experienced and expressed as counter extremes of one another, with the 2000/01 World Development Report further strengthening this concept by defining poverty as "the pronounced deprivation of well-being" [13]. Adding to this, the experience of well-being or ill-being is strongly dependent on the situation and context in which local personal and social factors such as ecology, gender, age, geography and culture play a large and very important role [13].

Both the ecosystem and human well-being are directly interdependent in that ecosystem services provide humans with the necessary resource opportunities they require to survive and improve their quality of life, and the availability of these resources can profoundly affect aspects such as health, the rate of economic growth, the frequency and persistence of poverty, livelihood security and so forth. The ecosystem also offers human beings nonmaterial benefits such as education, recreational and spiritual services. On the other hand, ecosystems are impinged upon by human activity through the need of ecosystem services such as fuel wood, food, fresh water, fibre and grass. [13]. It clearly follows from this interaction that nature is often valued for its usefulness: it satisfies a predilection, provides a function, and meets human needs [14]. These values are assigned to something because of the satisfaction and enjoyment that can be obtained through the use of biological resources. When an object is utilized as a method to satisfy a need or as a means to achieve an end, either the relation or entity can then be classified as an instrumental value. Thus through the *economic/utilitarian* perception of the value of nature, the efficacy of the environment is articulated through individual preferences or an accumulation of preferences [14,15]. In addition to this, the consumption of environmental resources refers to *consumptive use values* which are the values placed on those resources which are consumed directly without having passed through a market. Consumptive use values are especially significant to the rural populace in developing countries where these biological resources are used and collected as a source of subsistence. Pressures to conserve biodiversity have consequently resulted in reduced access to these resources and for the poor and politically weak, this has typically impacted them severely [15]. Put differently, the erosion of natural capital has serious consequences for human quality of life, and particularly that of poor, rural communities.

Natural capital can be defined as those goods and services supplied by ecosystems that are both renewable and non-renewable, including the ecological practices regulating their use and existence that may serve to meet various human needs [16, 17]. Natural capital plays a fundamental role in determining the well-being of both individuals as well as groups, in that it provides a number of essential elements such as air quality, the reduction of greenhouse gases, water quantity, quality of soil and landscape, but to name a few [13, 18]. In addition to this, ecological services play a fundamental role in providing the necessary resources required to live a life of normal length through medicines for diseases, freshwater, foods, and the regulation of threatening human diseases [19]. Thus, natural capital impacts all communities, most especially those communities surrounding protected areas wherein healthy, sustainable ecosystems with numerous community benefits are essential to their well-being and quality of life [13,20].

3.2. Measuring quality of life linked to ecosystem services

The search for a conceptual clarification of "quality of life" has seen the development of two essential methodologies of measurement, namely subjective well-being and "objective" or social indicators of well-being [17,21]. Objective well-being is quantifiably assessed by making use of both economic, social and health indicators, as well as observable variables such as life expectancy, literacy levels, and economic production that reflect the degree to which human needs have been met and which are deemed essential for a good life. However, whilst these measurements may provide researchers with an indication of the extent to which the social and physical needs are met, they are limited, and do not encompass other elements essential to quality of life such as psychological security and life satisfaction [17]. Thus, by analysing the quality of life of a society solely in terms of economic, social and health indicators, it clearly depreciates fundamental elements such as self-development, love, and acquiring meaning in life [21].

Consequently, to successfully measure quality of life it is necessary to also consider individual perceptions of well-being, which leads us to the second measurement, namely subjective well-being. The latter pertinently focuses on individually reported levels of contentment, happiness, fulfilment, pleasure and other such forms of human experience and cognitive satisfaction [17,21]. This indicator is grounded on the supposition that in order for researchers to understand the individual's or group's empirical quality of life, it is necessary to diametrically investigate how they feel about life within the perspective of their own standards and values [21]. The overall quality of life is thus determined by both the degree to which groups or individuals are content in their life experiences as well as the level to which their needs are met. By incorporating both "objective" and "subjective" variables, it becomes possible to gain a clearer picture of the true meaning of quality of life on both temporal and multiple spatial scales [17]. It is thus argued that constituents such as subjective well-being, objective well-being, human needs, values and the supply of ecosystem services are needed to form an integrated approach in order to understand human quality of life and how it might be obtained at the interface of people and protected areas.

4. Methods

4.1. The study site and target population

Golden Gate falls within the boundaries of the Thabo Mofutsanyana District Municipality (TMDM) in the QwaQwa region of the Free State. TMDM has the second largest population (736 238 in 2011) of the five districts in the Free State, with an average household size of 3.3, which is more or less equal to the national average of 3.4 [22]. Almost one third (31.9%) of the population of the TMDM is younger than 15 years. When it comes to socio-economic development and human well-being, the district is characterised by a high unemployment rate of 44.3% (2013) that translates into a staggering poverty rate of 69.1% (2011) – the highest of all districts in the province. The high poverty and unemployment rates have propelled an out-migration of male labour that in turn has resulted in a skew gender distribution of 87.3 males per 100 females in the district [22]. Overall, the district is thus hamstrung by low levels of human development and a low quality of life, low literacy and/or education levels and a high unemployment rate. Under these conditions, and more so in this area, grass has been known to have important livelihood functions, as traditionally it has been used for grazing, thatching, weaving and the manufacturing of household items such as brooms and mats [23].

4.2. Research design

As an analytical framework for the evaluation of the thatch harvesting programme, an outcome analysis was used in order to ascertain to what extent the objectives of the programme have been achieved. Elements highlighted in the outcome analysis included assessing how successful the programme has been, what obstacles this programme has faced, the levels of satisfaction among the direct beneficiaries of the programme, to what extent this programme has effectively reached its target population, and finally, to ascertain how this programme might be improved for future use. Both desk-top and empirical components have been incorporated within a mixed method design of quantitative and qualitative approaches. During the desk-top phase of the study, a theoretical basis was established that ascertained the relative interface between communities and the protected ecosystem which they neighbour. During the empirical phase various data gathering methods such as individual interviews, a focus group session and in-depth interviews with key informants were employed.

Analytically, the concept of well-being and the perceptions attached to this concept played a significant role in the development of the research design and methodology for this study. The methodology was developed in analogy of the five dimensions of well-being as proposed by the Millennium Ecosystem Assessment [13], which includes both the quantitative and qualitative components of well-being alluded to in section 3 of this paper. The first component is that of *material* well-being wherein an individual experiences a good and secure life through prospects such as income, assets, livelihoods, shelter, clothing and access to goods. Secondly, the *health* component pertains to living in a healthy physical environment, feeling well and being strong. The third component is that of good *social relations* which includes mutual respect, good family and gender relations, social cohesion and the ability to provide, when needed, for friends and children. The fourth component of well-being portends to that of *security* in which secure access to natural or other resources, living in a controllable environment and having security from natural and human-made disasters are vital. The final key dimension of human well-being is *freedom and choice* in which the individuals must have control over their lives and their values or being. Accordingly, these five dimensions may serve to either positively or negatively reinforce one another, thus changes in one may bring about changes in others. Concurrently, these essential elements of well-being were pertinently and comparatively utilized and assessed throughout this study in order to gauge the degree of well-being for those stakeholders directly benefitting from the thatch harvesting programme established at Golden Gate, all of which were used to suitably address the complexities of human endeavor, human capability, and human life [13, 24].

Methodologically, the five dimensions of human well-being were operationalised in two separate, yet concurrently running, stages for the purposes of programme evaluation: a primary and secondary stage. The primary evaluation focused on those directly benefiting from the programme as well as the potential benefits for the park itself. (The concept of direct beneficiaries did not only allow for the inclusion of the individual harvesters, but also for their households). The secondary stage of the impact evaluation explored the impact of the programme on the broader community, as well as the business sector.

4.3. Sampling and sample sizes

In order to understand the machinations of the thatch harvesting programme, and subsequently it's potential strengths, weaknesses and opportunities, it was necessary to not only interview those directly benefitting from the programme, but also those directly involved in the development and running of the programme. Additionally, in order to ascertain possible secondary or multiplier impacts, those commercial companies involved in purchasing the thatch after harvesting of the grass were also interviewed. Consequently, three samples were drawn: one from the harvesters (direct beneficiaries), a second sample from park officials and a third from those commercial companies who purchase the thatch immediately after harvesting.

A total of 34 harvesters – i.e. everybody who were involved in the 2012 pilot programme-were selected and interviewed through the use of a purposive sampling method. The park officials in Golden Gate directly involved in the running and support of the thatch harvesting programme were sampled by means of a non-probability purposive sampling method. These key informants included the People and Parks Manager and the Community Facilitator based at the park. However, due to unforeseen circumstances, the People and Parks Manager was unable to attend the focus group session, but the Park Manager of Golden Gate was able to participate in her stead. During the secondary stage of impact evaluation, two commercial companies were identified and contacted, which served to ascertain possible potential multiplier effects of the programme within the neighbouring social and economic environment. The first company interviewed was *Biggarsberg Thatchers,* and the second company *Thatch Craft.* Both companies are located in the neigbouring KwaZulu Natal province (Figure 1). Official representatives of both these companies were interviewed telephonically due to a limited project budget. Interviews with the harvesters and park officials were conducted

http://dx.doi.org/10.5772/59015

between October and December 2013, while the two companies were contacted and interviewed during May 2014.

4.4. Data collection mechanisms and measuring instruments

Data for the 34 harvesters was collected by means of both structured and semi-structured individual interviews, while a focus group session was conducted with the two park officials. Instruments that were utilized during data collection included a structured questionnaire set for the harvesters and semi-structured questionnaires for both the park officials and the representatives of the commercial companies that purchased the thatch. The structured questionnaire developed for the harvesters served to assess to what extent and in what way the programme had positively contributed towards the well-being of not only the direct beneficiaries, but their household members as well. In addition to this, the questionnaire also served to ascertain the harvesters' perceptions regarding both the programme as well as Golden Gate itself, the application process, in what ways they benefitted from being a part of the programme, the challenges they faced in the past, and their perceptions regarding possible solutions to these challenges. Furthermore, the questionnaire also served to identify potential social networks and established social ties between the community and the protected area. Due to the anticipated low levels of literacy amongst the harvesters, a Sesotho-speaking facilitator was used to translate the English constructed questionnaire items during the interviews with the harvesters, in order that the validity and reliability of the measuring instruments could be enhanced. All interviews were recorded and later re-evaluated by another Sesotho-speaking facilitator.

Following the interviews conducted with the harvesters, a focus group session was conducted with the two park officials at Golden Gate mentioned earlier, who not only provided insight into the machinations of the programme, but also served to confirm and clarify main issues raised by the harvesters. Areas outlined during the focus group session included the logistics pertaining to those responsible for the running of the programme, in-depth information regarding the selection and sustainable use of harvestable grass found in Golden Gate, the application process for direct beneficiaries, the exploration of established/potential networks, the exploration of facilities offered to direct beneficiaries, the challenges Golden Gate has faced since the conception of the programme, and possible recommendations regarding issues revealed during the interviews with the direct beneficiaries. The interviews with the park officials as well as those with the respective companies were conducted in English, and thus no translation of the measuring items was necessary. Lastly, electronic correspondence was conducted with the specialist scientist: vegetation ecology in SANParks' Division of Scientific Services to determine how the grassland ecosystem in the park has been affected (if any at all) by the harvesting programme.

Analysis of the data sets was conducted thematically and descriptively to create an incorporated and holistic view of the progress of the thatch harvesting programme, as well as the potential opportunities it has to offer for future beneficiaries. Specific data-sets relative to the quantitative principles within this study were analysed through the use of predictive analytics software, namely the Statistical Package for the Social Sciences (SPSS), version 21.

5. Findings and discussion

The findings of the study commence with an overview of the socio-economic status of the households to which the respondents belonged. This socio-economic profile provides insight into the dire socio-economic circumstances of the communities that these respondents reside in. An overview of the socio-economic context enables the assessment of the contribution of the thatch harvesting programme to the overall well-being of the respondents and their households. The assessment of the programme's contribution to the well-being of respondents and their households follows the dimensions of the Millennium Ecosystem Assessment [13], as previously outlined in the methodology section of this chapter. More specifically, the findings assess the extent to which the thatch harvesting programme has benefited the most vulnerable and poorest section of the community and explores whether the programme has, as perceived by the respondents, served to improve individual and household well-being. Lastly, challenges experienced by beneficiaries in this programme are discussed and interventions proposed by them to strengthen and maximise the impact of the programme are outlined.

5.1. Socio-economic status of households

Households represented by the respondents are fairly large, with more than half of the households (55.9%) having between five and eight household members, and a further 8.7% of households comprising of between nine and thirteen members (Table 1). Household members were defined as those who sleep at the dwelling for at least four nights a week, share physical resources (i.e. food and income) and eat together with the rest of the household.

Members per household	Number of households	
	N	%
1-4	12	35.3
5-8	19	55.9
9-13	3	8.7
Total	34	100

Table 1. Household size of respondents

The average household size for this sample of respondents is 5.3. This is much higher than the average household size for the larger Qwa Qwa area, which is 3.3 as mentioned earlier. The households represented by the programme beneficiaries are among the poorest households in the community. Poorer households are generally characterised by larger household numbers due to factors such as higher fertility rates and poverty, compelling people to pool resources. When analysing the household age structure, it transpires that 76.5% of households had children under 15 years of age, while almost one third of the households interviewed (32.3%) had at least one household member older than 65 years. Almost half of the households interviewed (47.1%) had two children under 15 years, while 23.5% of the households had between four and five children under 15 years of age. In total, the 34 households represented in the sample had 72 children under the age of 15, and 14 adults over the age of 65 (Figure 2).

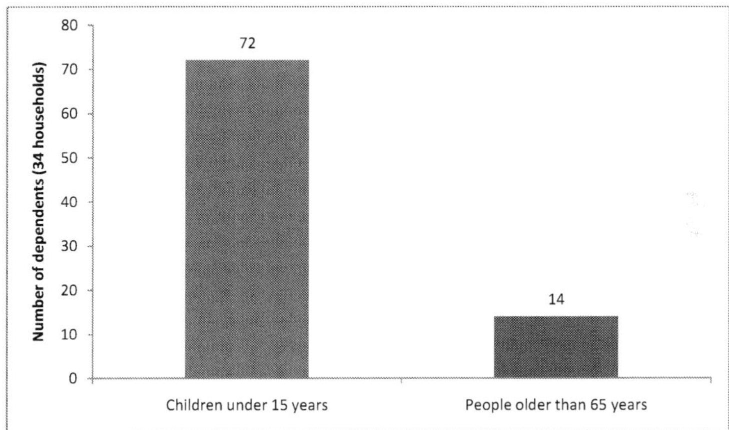

Figure 2. Total number of dependents per age category

The age structure of the households points towards a high dependency ratio and provides further insight into the overall profile of the households that are targeted by the thatch harvesting programme. The household size and the number of dependents per household present a population profile peculiar to poverty-stricken households in rural areas in South Africa and other developing countries, namely larger households with a large number of dependents. This profile is further strengthened by data on the total monthly income for the households in the sample (Table 2).

Monthly household income	N	%
Less than R1000 (US$95)	7	21%
R1001-R2000 (US$96-189)	14	41%
R2001-R3000 (US$190-284)	5	15%
R3001-R4000 (US$285-380)	3	9%
R4001-R5000 (US$381-475)	3	9%
R5001 and more (US$476 and more)	2	6%
Total	34	100%

Table 2. Total monthly household income (excluding contribution of thatch harvesting programme)

From table 2 it is evident that 62% of households as represented by the harvesters interviewed earned less than ZAR 2000.00 per month. This translates to approximately US$189.00 per month or US$6.3 per day per household. Three respondents (8.8%) reported household incomes lower than ZAR450.00 per month per household (or US$1.19 per household per day), placing these households below the upper bound poverty line of ZAR620.00 per capita per month [25]. The sources of household income in the sample comprised a combination of welfare grants, sporadic employment, self-employment and in one case formal, permanent employment.

Child care grants to the amount of ZAR300.00 per child were reported as sources of income by 26 of the households and 11 households reported that they benefited from a monthly old age pension of ZAR 1200.00 received by one or more of their family members. Occasional employment offers a limited contribution to the economic well-being of households. In some cases, occasional employment contributes to as little as ZAR100.00 per month, with the maximum amount earned through this form of employment being ZAR1500.00 per month. In six (17.7%) of the households, respondents indicated that self-employed individuals contributed to the household income, but the contribution was highly variable and ranged between ZAR 300.00 and ZAR 5000.00 per month. In one household, apart from the respondent, there was another member of the household who was part of a wetland rehabilitation and poverty alleviation programme run by Golden Gate, from which she received approximately ZAR 3500.00 per month. Notwithstanding these other sources of income, for 52.9% of households represented in this study, the only income that they received came from the involvement of one of their household members in the thatch harvesting programme.

Household expenditure is another indicator of the socio-economic well-being of households. Poverty-stricken households' consumption patterns are focused on day-to-day survival. A large proportion of household expenditure satisfies subsistence needs such as food and energy, with the consumption of higher-end consumer products such as electronic equipment and household appliances not forming part of the day-to-day household expenditure. In poverty-stricken households, even consumption of electricity is often regarded as a luxury, with energy needs being satisfied by relying more on freely available, or cheaper natural resources such as wood, animal dung, coal or paraffin. The data confirms that most, if not all, of the household income reported by the respondents in the sample was absorbed by day-to-day living expenses such as food and energy, with a small proportion of the household income going towards other needs such as transport and schooling. No household represented in the sample was required to pay rent for their dwellings, therefore no household expenditure went towards securing shelter. Electricity was purchased by 55.9% of households, but judging from the amount of purchased electricity (ZAR 100.00 per month), this was not the primary source of energy used by households. A fairly large number of households (41.2%), indicated that they did not spend any of their income on transport costs. This may again point to the fact that these households were characterised by low levels of economic well-being. Low transport costs may be indicative of an inability to afford transport, but may also reveal high unemployment, as households do not need to make use of transport to travel to work. Those households that did report transport costs as part of their expenditure spent relatively little (less than ZAR 600.00 per month) on transport. The linkage between transport expenses and poverty is further substantiated by data on how one of the beneficiaries transported thatch harvested for personal use. This respondent indicated that she carried the bundles that she harvested home on foot, opting to not make use of other forms of transportation in order to save costs.

Households do not generally spend money on luxury items such as furniture, with furniture purchases rather being reserved for when extra cash was available. The four households that do spend money on furniture on a monthly basis all indicated that they are paying off store accounts for furniture purchases. Even expenditure on cell phones is not a regular household expense with only 5.9% of households purchasing air time on a monthly basis. Household expenditure on cell phone air time is very little, ranging from between ZAR 12.00 to ZAR 75.00 per month. Two households indicated that they paid clothing accounts on a monthly basis and only six (17.7%) respondents contributed to a funeral scheme on a monthly basis. Thus, it seems that households live from hand-to-mouth, with very few of the households being able to purchase consumer items such as furniture and clothing on credit, or, more importantly being able to make a monthly commitment towards their future financial security. None of the respondents indicated spending household income on any form of leisure or recreational activities such as family vacations. This does not, however, suggest that households do not fulfil the need for *play and leisure*, which according to Nussbaum (2007: 21) is regarded as a basic human right. Households partake in leisure activities such as community gatherings or cultural events that are not dependent on an economic contribution.

Another indication of the low level of socio-economic well-being experienced by these households is seen in the level of educational attainment of the respondents. For South Africa as a whole, there is a close correlation between the educational level of the household head and poverty, with 65% of households where the head had no formal education, compared with 2,8% of households where the head had a post secondary school qualification [25]. Only 9% of the respondents in the sample completed their secondary schooling, with 41% having partly completed their secondary schooling (Table 3). Low educational attainment is linked to lower economic prospects and reduces the ability of respondents to contribute to the material well-being of their households. Low educational attainment also has an impact on the future educational prospects of children growing up in these households, which then impacts on their future employment prospects. Thus, low educational attainment contributes to perpetuating the cycle of poverty and low levels of well-being that these households are subjected to.

Educational attainment	N	%
None	6	18
Completed primary school	11	32
Partly completed secondary school	14	41
Completed secondary school	3	9
Total	34	100

Table 3. Respondents' level of educational attainment

Low educational attainment does not only impact on current and future material well-being, but also constrains the day-to-day functioning of people. This is evident in the data on literacy-related questions asked to respondents. With regard to the literacy levels of those interviewed, the majority of the respondents (85.3%) reported having no difficulty in writing their own names. However, the ability to read, write and consequently, the ability to fill out forms, ranged from no difficulty to being unable to do this at all (Table 4).

Literacy ability	No difficulty	Some difficulty	A lot of difficulty	Unable to	Total
Reading	8 23.5%	12 35.3%	8 23.5%	6 17.6%	34
Writing	7 20.6%	14 41.2%	7 20.6%	6 17.6%	34
Filling out forms	7 20.6%	8 23.5%	7 20.6%	12 35.3%	34

Table 4. Respondents' ability to read, write and fill out forms

The majority of respondents experienced at least some difficulty in performing the skills of reading and writing, which in turn translated into a lower ability to fill out forms. Only between 20% and 23% of respondents indicated that they didn't have any difficulty with these three skills. While six (17.6%) of the respondents were unable to read and write at all, and consequently were unable to fill out forms, a further 17.6% of respondents also indicated an inability to fill out forms, despite their ability to at least read and write to some extent. This is an indication of low educational attainment as well as low skill levels that in turn impacts the respondents' ability to find stable and secure employment. Consequently, it can be assumed that due to these low levels of education and literacy, coupled with unemployment and underemployment, respondents and their household members are seriously constrained by their socio-economic circumstances to achieve higher levels of well-being.

The following sections serve to ascertain to what extent the thatch harvesting programme has positively contributed towards raising the level of well-being of its beneficiaries, and subsequently the households of which they form a part of.

5.2. The health and well-being of beneficiaries to the Thatch Harvesting Programme

The results presented with regards to well-being pertain to the 2012 harvesting season. For the 2013 harvesting season, half the respondents who harvested during the 2012 season re-applied and were granted permits to harvest again in 2013. The other half did not apply for this particular year and gave two reasons for this. These respondents stated that they either did not apply on time, or they did not profit sufficiently from harvesting in the previous year and therefore ventured into other areas of employment. However, during the 2013 season thatch harvesting was stalled due to two massive fires that destroyed the areas allocated for harvesting. This resulted in beneficiaries not generating any income for that year.

With regards to the 2012 season, all of the respondents indicated that the programme has benefited them in some way, even though they only participated in one season of harvesting (during 2012). Most respondents remarked that their lives before participating in the programme were difficult and that their lives improved as a result of their involvement in the programme. Only one respondent expressed the opinion that her quality of life had not changed much since participating in the programme. Additionally, when asked whether the programme had in general affected them negatively in any way, 79.4% respondents indicated that it had not. The benefits of the programme for the participants, and consequently for their households, become more nuanced when gauged according to the dimensions of well-being of the Millennium Ecosystem Assessment.

5.2.1. Material well-being

Material well-being, according to the definition of this dimension [13], is the individual's experience of a good and secure life through prospects such as income, assets, livelihoods, shelter, clothing and access to goods.

For the 2012 harvesting season, most respondents did not harvest large volumes of thatch. Almost half of the respondents (45.5%) harvested an average of 5 to10 bundles per day, whilst a further 30.3% of respondents averaged 11 to 15 bundles per day. This amount was harvested over a 30 day period allotted by the park's management. However, even though a 30 day period was allotted for harvesting, this included weekends when transport was difficult to obtain, and subsequently respondents were actually only able to harvest for 20 days during this allotted period. Only 6% of respondents managed to harvest more than 25 bundles per day (Figure 3).

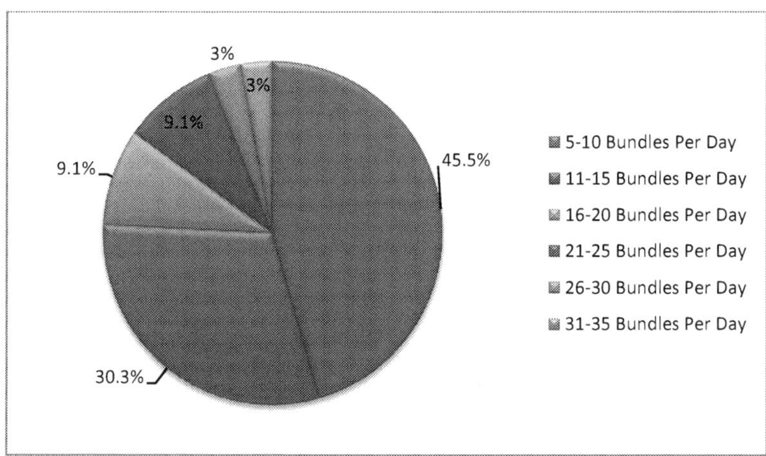

Figure 3. Average Number of Bundles Harvested by Respondents per Day (N=33)

Many beneficiaries were unable to indicate the actual amount that they had earned during the harvesting season as they were paid either daily or weekly for the number of bundles they harvested. This may enforce the earlier analysis that highlighted the hand-to-mouth existence of beneficiaries to this programme. The total income was calculated according to the average number of bundles that each respondent was able to harvest within a day. Each bundle was sold at approximately ZAR12.00. The equation used to calculate the total thatch harvest of respondents is as follows: (Number of Bundles per day X 20 days) X ZAR 12.00=Total individual income. Based on this calculation, the total income generated from the thatch harvesting programme approximated to ZAR104,580 for the 2012 season. This amounts to an average of ZAR3,076 for each of the 34 respondents in the sample, although eventually the per capita income depended on the actual number of bundles harvested per person per day.

Thirty three (33) of the 34 respondents actively harvested thatch, while one respondent was contracted as a driver by a harvesting coordinator to collect and transport the thatch harvested. Most of the respondents (91.2%) sold their harvest to the harvesting coordinator. These beneficiaries indicated that they were recruited by the harvesting coordinator to take part in the programme. The harvesting coordinator bought the thatch bundles from the beneficiaries and in turn sold this harvest to commercial thatching companies. One respondent indicated that the thatch harvested was used to repair the roof of their dwelling, while another respondent harvested thatch to make brooms and small carpets to sell to tourists and community members. Thus, only two of the respondents did not form part of the economic supply chain involving the harvesters, the harvesting coordinator and the thatching companies. The respondents therefore seem to prefer the security offered by having an immediate buyer for their thatch, rather than using the income obtained for the funding of entrepreneurial enterprises, which may prove to be more uncertain in terms of securing material well-being-especially in the short term.

One respondent, as indicated above, used the thatch as input material for a small entrepreneurial enterprise. Three other respondents indicated that the money received from selling thatch contributed to start-up capital for new businesses. One respondent used her money to fund the start-up of a small sewing enterprise. Another respondent purchased fresh produce to sell at the local markets, enabling the start-up of a sustainable small business supplying local markets with fresh produce. One other respondent was able to purchase enough stock to start a tuck shop close to one of the local schools in Qwa Qwa. Although at a very small scale, these cases are indicative of the potential of the programme to stimulate entrepreneurship and as such to contribute to a more sustainable economic well-being of beneficiaries. The number of respondents who saw the thatch harvesting programme as an opportunity for starting a new business is low, although this is on par with the general trend in entrepreneurship in South Africa. In a recent study on entrepreneurship in South Africa [26], it was found that only 37.8% of South Africans were of the opinion that there will be good opportunities to start businesses in the area in which they live within the next six months. This is much lower than the average of 74.5% for Sub-Saharan Africa as a whole. The same study [26] also revealed that only 42.7% of the South African adult population believe that they have the knowledge, skills and experience to start a new business.

The ability of respondents to purchase assets with the incomes they obtained from selling their harvested grass is indicative of an improved ability to gain *materialistic control over their environment* [17,24]. When analysing what respondents spent the money on which they received from selling the thatch that they had harvested, their improved material well-being is evident. Only four respondents reported that the incomes generated from the thatch harvesting programme were used towards purchasing basic necessities such as food and toiletries, while 38% of the items purchased were consumable items such as blankets, clothes and shoes. Respondents indicated, among others, that they purchased electronic equipment, furniture, household appliances and livestock. Over half of the expenditure (52%) mentioned by the respondents could be characterised as spending on household assets, while 6% of the items mentioned could be classified as spending towards improving existing assets, i.e. purchasing of building materials or vehicle parts. Interestingly, most respondents did not mention that the money received was used for subsistence needs such as food and transport, but rather emphasised their improved ability to purchase items that would not have been possible if they did not have the added income received from thatch harvesting. Thus the programme seems to have contributed to improving the material well-being of those house-holds benefiting from the programme.

However, respondents did not include expenditure for items that would improve their quality of life in the long term, such as education. It appears that the satisfaction of short-term material needs was more of a consideration for respondents than working towards obtaining long-term and sustainable material well-being that would be achieved by contributing to savings plans, or pursuing further education. Only one respondent used his income from harvesting to improve his prospects for finding permanent employment as a truck driver in the foreseeable future by utilising some of the money from harvesting to go for driving lessons. While the programme has therefore managed to improve the short-term material position of the beneficiaries, the long-term material well-being of these people did not seem to improve markedly. At least 65% of the respondents indicated that they struggled financially and could not find employment. Some respondents (17.6%) indicated that they were offered sporadic employment by the park, i.e. working in the stable yards, repairing perimeter fencing, or as part of other poverty alleviation programmes run by the park. It can therefore be concluded that the programme has not benefited the long-term employment prospects of the beneficiaries significantly.

5.2.2. Health dimension

The health dimension of the Millennium Ecosystem Assessment [13] pertains to living in a healthy physical environment and to feeling well and being strong. For the purposes of this study, the analysis of the contribution of the programme is assessed in terms of physical as well as psychological well-being.

With regards to physical well-being, 82.3% of the respondents indicated that the programme had positively contributed towards their physical well-being. Of this group, 64.3% experienced being physically fitter and healthier, while 35.7% indicated that they felt physically stronger after participating in the programme. Some respondents, however, indicated that the programme impacted negatively on their physical health. More specifically, they pointed at health issues such as allergic reactions to the grass (5.9%) as well as severe cuts and wounds on their legs that took long to heal (11.8%). The harvesters were not provided with protective clothing such as safety boots and gloves that would prevent such injuries from occurring. One respondent indicated that she had problems with her blood pressure and that the hard labour of harvesting worsened her condition. She resignedly stated: *"But what choice do I have? I must work"*. These negative impacts on health were, however, not experienced by the majority of the respondents. The latter did not mention any negative health impacts as a result of their involvement in the programme.

The grass harvesting programme does seem to have significant benefits for the psychological well-being of participants. Fifty nine percent (59%) of respondents indicated that the programme had positively contributed towards their psychological well-being. Half of the respondents who indicated a psychological benefit specifically pointed out that the involvement in the programme made them feel more positive about their future, while the other 50% mostly experienced emotional relief over their ability to cope with their financial pressures. Additionally, the consensus among respondents (67.6%) was that they were very happy to be able to work in the thatch harvesting programme and that the programme contributed to their sense of pride, dignity and independence (32.4%). These positive perceptions of subjective well-being since joining the programme indicate the fulfilment of the need for identity with regards to feelings of differentiation and recognition. Two of the respondents specifically pointed out that the programme boosted their confidence and self-worth, while one respondent stated that by being a part of the programme, he was able to improve his communication skills and this consequently boosted his confidence as well.

5.2.3. The dimension of good social relations

The dimension of good social relations includes aspects such as mutual respect, good family and gender relations, social cohesion and the ability to provide, when needed, for friends and children [13].

An important component of social cohesion is affiliation. Affiliation can be conceptualised as the capability of humans to be able to envision the circumstances of another entity, and to acknowledge and display concern for this entity as well [17,24]. Without a sense of affiliation, group cohesion is not attainable. Respect, dignity, equality and receptiveness are key factors in this need. The grass harvesting programme contributed towards satisfying beneficiaries' need for affiliation on two levels: Firstly, in relation to the communities of which the beneficiaries form part, and secondly, in relation to Golden Gate itself.

Overwhelmingly positive sentiments were expressed when respondents were asked about how their community perceived their involvement in the thatch harvesting programme. Most of the respondents (73.5%) stated that the community was very proud of them for working in the thatch harvesting programme. Almost one in every four respondents (23.5%) nevertheless reported that many community members were jealous because they (community members) had not been able to obtain permits to harvest as the beneficiaries had. The predominantly positive perception about the beneficiaries' involvement in the programme may serve to bolster feelings of affiliation with the community and generate better group cohesion.

It also transpired that Golden Gate serves as a vital cohesive element in the lives of the communities surrounding the park. A large number of respondents (76.5%) often travelled through Golden Gate to reach the nearby towns of Clarence and Bethlehem, which means that the park serves to connect people from different surrounding communities to one another. The park is also utilised by community members for cultural and spiritual activities as well as for recreation and leisure purposes. One fifth of the respondents (20.5%) had used the park for cultural and spiritual activities such as initiation ceremonies and meditation, while 8.8% of the respondents had used Golden Gate for leisure and recreational purposes. Although the latter proportion might appear to be very small, it should be interpreted in the context of the high levels of poverty and unemployment that prevail in the region.

Figure 4 illustrates the respondents' perceptions regarding the importance of Golden Gate as a conservation area. Respondents were allowed to offer more than one response in this section.

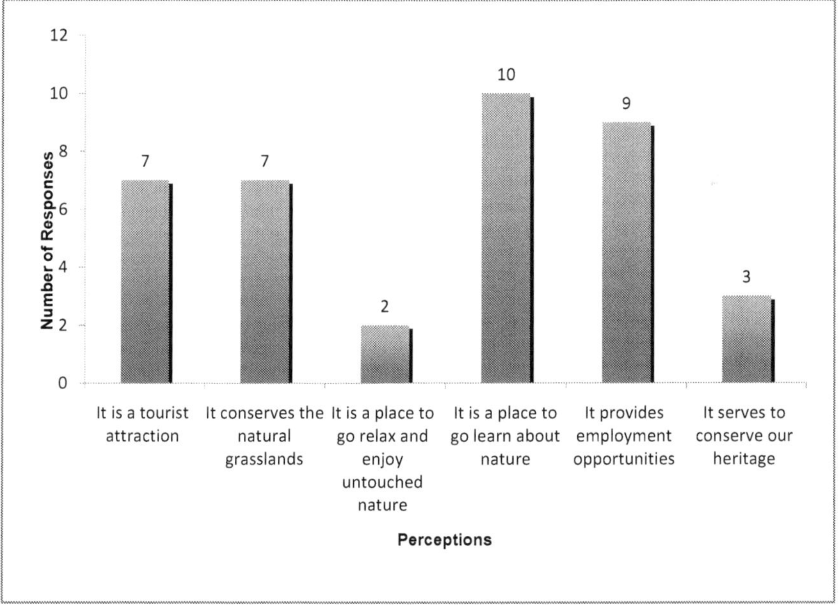

Figure 4. Respondents' perception about Golden Gate

All the respondents believed that the park is an important entity. The two most frequent responses given to substantiate this sentiment were that the park provided a place to go and learn about nature and that it provided employment opportunities. This was followed by responses such as "It is a tourist attraction" and "It conserves the natural grasslands". Notably, three of these four categories mentioned are either directly or indirectly related to the park's contribution to employment and economic opportunities. Tourism was regarded as important by respondents because it provided them with an opportunity to sell their wares in the form of pots, brooms, baskets, mats and jewellery. The protection of grasslands was regarded as important by the respondents, since it is a direct source of income for them. Subsequently, several respondents stated that if everyone was allowed to graze their cattle in the park, live there and/or harvest the grass whenever they wanted to, then the opportunity to harvest good quality grass would be reduced. These respondents also indicated that it is important to set rules and boundaries in the park's conservation policy in order to ensure the future sustainable utilisation of resources and protection of the ecosystem services. Three respondents indicated that Golden Gate also served to conserve and protect their heritage-a heritage which they felt was an essential part of their culture and which they hoped their children and future generations might enjoy as well. Finally, two respondents felt that the park was an important place because it is where one can go to relax and enjoy the beauty of untouched nature.

Most of the respondents (94.1%) felt that the land should remain a protected area, despite the fact that this means that access to the park's resources are restricted. Only two respondents (5.9%) felt that the land should be utilized for economic practices rather than for conservation. These respondents felt that there was not enough grazing for cattle and that the land should be put to use for that purpose. The majority of the respondents therefore experienced a sense of affiliation towards Golden Gate. They were aware of the need for the land to be protected, the reasons thereof, and the benefits they gained from having a protected area so close to their local community.

The thatch harvesting programme also contributed to respondents being relieved at their ability to provide financially for their families. The majority of the respondents (85.0%) reported experiencing a sense of relief knowing that they were able to provide for their families. Poverty and a lack of employment are significant sources of family conflict. Thus, increased material well-being may serve to improve family relations. Interestingly, four respondents (11.8%) believed that some of their family members were jealous of the work they had found. This jealousy could again increase tension and impact negatively on family relations in these families. However, 30 (88.8%) of the respondents expressed that their family members were very proud of them because of the income they were able to generate from the project. Thus, overall, the conclusion can be drawn that the programme has contributed towards improving family relations and social cohesion in the neighbouring community.

5.2.4. The security dimension

This dimension refers to the ability to secure access to natural or other resources, living in a controllable environment and having security from natural and human-made disasters. The programme has to some extent contributed to improving the ability of respondents to secure access to natural resources by allowing them to harvest a natural resource for household use, as well as to improve their material well-being. Through their involvement in the programme, the respondents' knowledge of the natural environment, as well as the importance of conservation was somewhat improved. While only four of the respondents reported having received some form of environmental education from the park, another seven indicated that they had received information from the harvesting coordinator in this regard. The information provided to the respondents included common rules applicable within many protected areas such as 'do not kill the animals', 'do not litter rubbish in the park', 'you may not start fires in the park', and lastly, 'do not destroy other plant life within the park'. Information such as this is vital in assisting beneficiaries to secure access to natural resources, in this case thatch, and also empowered respondents to secure themselves from the possibility of natural and human-made disasters such as veld fires – a very real hazard in a grassland environment. However, 67.6% of the respondents indicated that they did not receive any form of environmental education while being involved in the programme.

With regards to the correct techniques and procedures to harvest thatch, the overwhelming majority of respondents (87.9%) had prior knowledge of this activity. This knowledge is vital in enabling respondents to effectively access the thatch resources. Of these respondents with prior knowledge, 22 were taught by family members how to harvest while growing up, while seven respondents indicated that the harvesting coordinator taught them how to harvest the grass, how to cut, tie, and/or store the grass after harvesting and the appropriate length and thickness of the grass that should be cut. Some of the respondents expressed their gratitude towards the harvesting coordinator who imparted this knowledge to them, since they would have harvested the wrong types of grasses, or the wrong length and thickness without his assistance.

Thus, it appears that involvement in the programme has, at least to some extent, enabled beneficiaries to gain access to natural resources. With regards to having security from natural and human-made disasters, the programme did, in the context of the activity of harvesting itself, provide beneficiaries with knowledge to secure them from veld fires which are among the most commonly expected natural disasters in a grassland environment. Security from disasters, however, extends further than the day-to-day harvesting. As was previously discussed under material well-being, one respondent indicated that she used the thatch harvested for repairs on her roof, while three others used the money received from the thatch that they sold to buy building materials with which to repair and improve their dwellings. Through these activities, households are provided with the opportunity to enhance their security from some environmental hazards that plague households that are not able to afford proper dwellings.

5.2.5. The dimension of freedom and choice

The dimension of freedom and choice refers to individuals having control over their lives and their values. From the data it transpires that 32% of the respondents reported that, before working on the programme, they felt helpless because they stayed at home doing nothing while their families had to struggle to find money to sustain the basic needs of those living within their household. Through the income provided by the programme, beneficiaries could expand the choices that they made about their immediate consumption patterns as well as their future well-being. This is evident in the different ways in which beneficiaries opted to spend the income they received, i.e. improving their dwellings, buying appliances and electronic equipment, enrolling for driving lessons, using the money as start-up capital for small businesses and buying equipment such as sewing machines which would enable them to expand their future choices.

5.2.6. Responses from commercial companies

One of the key issues identified during the interviews with the two commercial companies that purchase the grass harvested at Golden Gate, was the lack of knowledge, skills and training of harvesters with regards to correct methods of harvesting thatch. This has resulted in both these companies receiving, at some point in time, bundles of thatch not suitable for use. Challenges included the following: the grass still being green when harvested; it was the wrong species of grass; the thatch was not straight; it was too thick, and/or it had not been cleaned properly. These challenges pose as major concerns regarding the sustainability and potential opportunities of this programme in the future. For instance, grass that is still green when cut means that the seeds have not yet had time to dry and drop from the stalk. Consequently, the premature harvesting of grass which may result in the absence of future re-growth could severely jeopardise the availability and sustainability of harvestable grass at Golden Gate in the future.

In addition to this, both companies strictly conform to guidelines set by the South African Bureau of Standards wherein the thickness, length, species and quality of the thatched bundles are core principles and must be stringently adhered to. Subsequently, these companies are forced to return grass that is unsuitable for use without payment or transport subsidy. Not only is this a waste of natural resources, but it also threatens the livelihood of these companies in that they rely heavily on the supply of thatch from harvesting coordinators. Augmenting this is also the negative impact this will have on those harvesting coordinators who had provided the thatch. The cost of transporting the grass from Golden Gate to the aforementioned companies is only viable if the grass can be sold upon arrival, and the return of unsuitable grass can result in harvesting coordinators such as the one previously mentioned, facing disgruntled labourers coupled with payment disputes. These issues can serve to heavily undermine the development of budding entrepreneurs such as this, and may result in the harvesting coordinator being forced to cease his/her operations. Even more worrying in a situation like this, is the fact that those labourers who had vested their time and physical energy to harvest the grass, must return to their homes empty-handed. Subsequently, lack of knowledge, skills and training has the potential to create this trickle-down effect and poses as a major challenge to the sustainability of this programme.

In order to prevent a situation such as this, it became clear that an intervention of sorts would be necessary. Upon enquiry, one of the commercial thatching companies indicated they would be willing to provide training sessions to those beneficiaries who have been granted permits to harvest in the park, wherein the beneficiaries will be provided information regarding matters such as the environmental impact of harvesting, how to identify the correct species of grass, the correct way to cut the grass, the required length and thickness of the grass, and how to properly clean the bundles for sale. Not only will this improve the knowledge base and skills of the beneficiaries, but it will also serve to enhance the sustainability and viability of this project in the future.

Taking the above findings into consideration, the following section will serve to highlight the challenges faced by the thatch harvesting programme and the beneficiaries' responses to possible ways in which the programme can be improved.

5.3. Challenges faced by beneficiaries to the thatch harvesting programme

While the programme seems to have contributed to improving the overall well-being of respondents and their families, respondents also experienced some challenges while being involved in the programme and offered some suggestions for improving the programme for future beneficiaries (Table 5).

Challenges	N*	Suggestions for improvement	N*
Insufficient time to harvest grass	18	More time should be given to harvest	16
Rangers treat us badly when we are there to harvest	5	The park should provide tools/equipment for harvesting of thatch	14
Fires destroy our income we rely on being able to cut grass	5	The park should provide toilet facilities	12
The park does not advertise the programme early enough	2	The park should burn fire breaks earlier to protect the grass	6
It is difficult to find buyers	1	The park should provide training to improve harvesting skills	4
They (the park) do not provide tools/ equipment	1	The park should help us find people to buy our bundles of grass	3
		The park should advertise the programme earlier	2

* The n-values in table 5 indicate the number of respondents who identified each issue. Respondents could indicate more than one challenge or suggestion, or nothing at all.

Table 5. Challenges experienced and suggestions for improvement

From the data above, the biggest issue faced by respondents relates to insufficient time for harvesting. Eighteen of the respondents highlighted that the time allocated for harvesting was too short. This was followed by the issues of rangers treating them badly while harvesting and the issue of fires that diminish their potential to harvest. The respondents pointed out that fires destroyed the viable grass allocated for each season, forcing them to harvest in areas that were not designated by the park for harvesting. Park officials have indicated that they were aware of this challenge and, with the assistance of the harvesting coordinator, would choose harvesting areas more carefully for the coming seasons, and would also demarcate the allotted areas better to prevent people from harvesting in undesignated areas.

Some respondents felt that the park does not do enough to advertise the programme in a timeous manner. This leaves people little time to apply for the programme. When asked if they had experienced any problems with the application process, seven (20.6%) respondents indicated that they had not experienced any problems, whilst 19 (55.9%) felt that the process took too long. Other respondents added to this by stating that, by the time the permits were granted, the period for harvesting had already begun, and that this increased the risk of fires destroying the grass before they could harvest. The remaining eight (23.5%) respondents expressed having felt frustrated during the application process because they did not know when to pick up their permits. The park officials reported that during 2012, they noted a number of individuals that had come to harvest before and during the time allotted for harvesting who did not have permits. This made it difficult to ascertain and monitor who had permits to harvest and who did not. It must also be noted that during the interviews with the beneficiaries it transpired that a few of those who had harvested in 2012 were individuals who did not reside in the local community as defined by the park. It was reported that these individuals borrowed identity documents from members of the local community to pass off as their own in order that they might harvest. This challenge is an important one, as the purpose of the programme is to benefit members of the local communities only. Subsequently, illegal harvesting has posed as a major challenge for the park and for local communities who should benefit from access to the natural resources in the park.

Furthermore, there appeared to be miscommunication between the park management and the local community with regards to the nature of the programme. This came in the form of local community members perceiving the thatch harvesting programme to be a source of employ-ment, whereas this programme is only offered as an opportunity to utilise the park's natural resources for their own benefit. Lastly, the respondents raised the issue of the park not providing them with tools or equipment with which to harvest, and a large number of respondents (n=14) suggested that the park should equip them with the necessary harvesting tools. Also, during the focus group session with the park officials, it was indicated that Golden Gate had established networks that formed part of a park forum wherein there are various traditional leaders that act as representatives within their local communities and serve to communicate issues of mutual concern. However, when asked; none of the respondents were aware of any community representatives, nor of any community meetings held with regards to projects made available by the park. In a similar vein, none of the beneficiaries interviewed reported having heard of any community members being involved in decisions regarding the thatch harvesting programme.

6. Conclusion

Due to the poor socio-economic conditions surrounding the park, most respondents and their households depend heavily on the income earned from their involvement in the thatch harvesting programme. In fact, more than half of the households represented in the sample have no other source of income except for the employment of one of the household members on the programme. Thus, although the immediate benefits of the programme are limited to only a tiny proportion of the community, these benefits still make a significant and tangible difference to the well-being of those households living on the edge of subsistence. As has been confirmed previously by other outreach programmes in protected areas [3], this 'limitation' should nevertheless not be seen as a defect or an impediment of the thatch harvesting programme, but should serve as a constant reminder of what is realistically achievable with programmes of this kind offered by national parks and other protected areas in developing countries. Arguably, the main strength and impact of the programme – and other programmes of this kind – is not so much to significantly reduce poverty among a large proportion of households, but rather its ability to cultivate positive perceptions regarding conservation, sustainable utilisation of ecosystem services and the specific protected area in particular, among the local population.

The thatch harvesting programme, at this stage, seems to be constrained by logistical and administrative challenges such as permits not being granted in time for harvesting, an unclear selection process and poor supervision of park officials to ensure that harvesting does not impede on the conservation function of the park. Anecdotal evidence from the interviews suggest that in some cases grass is harvested illegally, thus limiting the benefits that should trickle to local communities. This has also been found in a previous study conducted in the same park [27]. Although the current park management plan (compiled in 2011) provides the legal framework for the managing of natural resources at Golden Gate, the plan fails to quantify and account for the resources that are being harvested by adjacent communities. More specifically, the park's management plan does not adequately demonstrate *what* is being harvested, or the *extent* and *impact* of grass harvesting in the park. If managed properly, grass and grass harvesting can provide a long-term sustainable benefit to neigbouring communities and economic institutions, but the guidelines for such harvesting need to be set clearly in the park's management plan. Consequently, as previously pointed out [27], there is a clear need to monitor, evaluate and set the boundaries for grass harvesting in the park, and to clearly stipulate these limitations in the management plan. This problem, however, is not unique to Golden Gate, as there is a general lack of published research on resource extraction from national parks in South Africa, as well as from protected areas in general.

Based on the findings of the study, a small proportion of the community does seem to benefit from their involvement in the thatch harvesting programme. The data offers evidence of improved material well-being, better physical and psychological health, enhanced group cohesion, environmental security and more freedom of choice for beneficiaries. The impacts of the programme are however, for most respondents, short term. Only a limited number of respondents have used the money obtained from harvesting to enable the fulfilment of sustainable long term economic pursuits as is evidenced by the four respondents who managed to start small businesses and the one respondent who used the money to obtain a drivers licence.

In conditions of severe poverty and high levels of unemployment such as those that prevail in the area surrounding Golden Gate, natural resources play a crucial role in sustaining people's livelihoods. Under these conditions, the harvesting of grass for a commercial market presents an opportunity for the local community to increase their income base and improve their well-being. However, as previously concluded [23], more grass would have to be harvested to meet the demands of a commercial market than would be required for household use or producing items for a local market. In other words, although an increase in grass harvesting holds potential benefits for increased human well-being in the local community, an increase in the commercialization of harvesting at the same time requires strict monitoring and evaluation mechanisms to ensure a sustainable supply of raw materials and mitigation regarding the impact on the protected area. Since none of the businesses interviewed are involved in grass management and protection, they are potential victims of overharvesting and resource depletion as much as the members of the local community. Resource harvesting in a protected area that supplies the demands of a commercial market thus clearly requires different rules and monitoring mechanisms, than rules aimed at the regulation of such activities at a local level and only for the strict benefit of the local community.

With reference to the impact of the thatch harvesting programme on the ecosystem of the targeted areas allocated, the results remain indefinite. The reason for this being that the programme only became active in 2012, and in 2013 a massive fire swept through the parks grasslands, subsequently also destroying the areas allocated for harvesting. As a result of this, coupled with the fact that this programme is relatively new, a detailed analysis of these areas regarding the grass species composition, vegetation structure and biomass measure following the harvesting in 2012 has not yet been finalised. SANParks (Division of Scientific Services) has initiated a vegetation monitoring project in two of the areas that form part of the harvesting programme. It is, however, a long term monitoring process and no informed conclusions could be made in the relatively short period that the monitoring project has been running in the park. Early indications are nevertheless that the grassland ecosystem in the park, as well as the patterns and processes that are associated with it, have not been negatively affected by the harvesting programme. In areas where the grasses have been harvested the height of the grassland is lower than the conventional 1.8 meters (Species *H. dregeana*), but apart from this visual impact it appears that the species composition of the grassland has not changed and the same grass species still dominates these areas. Currently harvesting is taking place on old agricultural lands that were previously ploughed and grazed in the time of commercial farming activities in the area. The main two grasses that are being collected are *Hyparrhenia* cf. *hirta* (common thatching grass) and *Hyparrhenia* cf. *dregeana* (thatching grass) which are often found in disturbed and degraded areas such as these. The sustainable manner in which these grasses are harvested also contributes to the stability of the degraded land that it occupies. In fact, the harvesting of these grasses improves the palatability for other grazers of the wildlife group within the park, and assists in supporting a natural succession process in these degraded areas. The harvesting (clearing of grasslands) also allows for other plant species to thrive within an area usually dominated by one or two plant species.

However, there were some concerns regarding the use of some of these areas by grass owls (*Tyto capensis*) for nesting. Consequently, in order to determine the impact of the harvesting on this species, a habitat assessment of possible areas has been proposed. Practices in other protected areas have nevertheless shown that, despite all efforts of national parks to conserve biodiversity and ecosystem integrity, unsustainable resource use remains a threat because ecological functions and processes often occur over larger spatial scales [28]. To ensure that an ecosystem such as the grassland biome retains the ability to renew itself, additional land is needed for the expansion of national parks. In South Africa, national population policy drivers such as social redress and poverty alleviation, strongly influence resource use in national parks. This means that localized management solutions for ecosystem integrity and resource use should be embedded in a broader systems approach that recognizes the interface between protected areas and their surrounding communities, while also acknowledging the complex, multiple and reciprocal relationships of sustainability between ecological and socio-economic components in the environment.

Acknowledgements

The authors wish to thank South African National Parks, and particularly the management of the Golden Gate Highlands National Park, for their support during the planning and field work stages of this study. We also gratefully acknowledge the contribution made by the Vegetation Specialist of SANParks with regards to the current status of the grasslands as a result of the harvesting programme. Opinions expressed in this paper and conclusions that drawn, however, are those of the authors and do not necessarily represent the official views of either South African National Parks or that of the management of Golden Gate Highlands National Park.

Author details

André Pelser*, Nola Redelinghuys and Anna-Lee Kernan

*Address all correspondence to: pelseraj@ufs.ac.za

Department of Sociology, University of the Free State, Bloemfontein, South Africa

References

[1] Dudley N, Mansourian S, Stolton S, Suksuwan S. Protected Areas and Poverty Reduction. Gland: World Wide Fund for Nature (WWF); 2008.

[2] Ghimire KB. Parks and people: livelihood issues in national parks management in Thailand and Madagascar. In: Ghai D. (ed). Development and Environment: Sustaining People and Nature. Oxford: Blackwell; 1995, p. 195-227.

[3] Pelser AJ, Redelinghuys N, Velelo N. Protected areas as vehicles in population development: lessons from rural South Africa. Environment, Development and Sustaina-

bility 2013;15(5):1205-1226. http://link.springer.com/article/ 10.1007%2Fs10668-013-9434-4 (accessed 12 June 2014)

[4] Roe D., Walpole MJ. Whose value counts? Trade-offs between biodiversity conservation and poverty reduction. In: Williams N., Adams WM, Smith RJ (eds.) Trade-offs in Conservation: Deciding What to Save. Oxford: Blackwell; 2008. p157-174.

[5] International Union for Conservation of Nature and Natural Resources, 2005, Benefits Beyond Boundaries: Proceedings of the Vth IUCN World Parks Congress, Cambridge: IUCN.

[6] South African National Parks (SANParks). People and Conservation: A brief history. http://www.sanparks.org/people/about/history.php. (accessed 6 July 2014).

[7] Egoh BN, Reyers B, Rouget M, Richardson DM. Identifying priority areas for ecosystem service management in South African grasslands. Journal of Environmental Management 2011; 92: 1642-1650.

[8] Grasslands Programme. What is the grasslands biome? http://www.grasslands.org.za/the-biodome (accessed 14 June 2014).

[9] Grasslands Programme. Grasslands under pressure. http://www.grasslands.org.za/the-biodome (accessed 16 June 2014).

[10] Ramsay S. Golden Gate Highlands National Park. Getaway 2013; 77-83.

[11] COPA-Academy. COPA-Academy-Areas: Free State map. http://copaacademy.files.wordpress.com/2013/07/free-state-map.gif (accessed 20 July 2014)

[12] South African National Parks. Golden Gate Highlands National Park Grass Harvesting Project: Progress Report. Pretoria: South African National Parks; 2012

[13] Millennium Ecosystem Assessment. *Ecosystems and Human Well-being: A Framework for Assessment.* Washington: Island Press; 2003.

[14] Bengston DN. Environmental Values Related to Fish and Wildlife Lands. In: Department of Forest Resources (ed.) Human Dimensions of Natural Resource Management: Emerging Issues and Practical Applications. St Paul: University of Minnesota; 2000. p. 126 – 132.

[15] Blaikie P, Jeanrenaud S. Biodiversity and Human Welfare. In Ghimire KB, Pimbert M (eds.) Social Change and Conservation. London: Earthscan; 2000. p 46-69. Available from http://books.google.co.za/books?hl=en&lr=&id=UvcymNufO8AC&oi=fnd&pg=PA46&dq=±instrumental±and±noninstrumental±human±values&ots=NqxYkD1Pdp&sig=ZrT0lV4PE6gl3i7UAiHgUQbMbsM#v=onepage&q=instrumental%20and%20non-instru mental%20human%20values&f=false. (Accessed 12 June 2013).

[16] Throsby D. Cultural Capital. Journal of Cultural Economics 1999; 23:3-12.

[17] Constanza R, Fisher B, Ali S, Beer C, Bond L, Boumans R, Danigelis NL, Dickinson J, Elliot C, Farley J, Gayer DE, Glenn LM, Hudspeth T, Mahoney D, McCahill L, McIntosh B, Reed B, Rizvi SAT, Rizzo DM, Simpatico T, & Snapp R. Quality of Life: An Approach Integrating Opportunities, Human Needs, and Subjective Well-being. Vermont: Elsevier; 2006.

[18] Haider S, Jax K. The application of environmental ethics in biological conservation: A case study from the southernmost tip of the Americas. Biodiversity Conservation 2007; 16: 2559 – 2573.

[19] Holland B. Justice and the Environment in Nussbaum's "Capabilities Approach": Why Sustainable Ecological Capacity Is a Meta-Capability. Political Research Quarterly 2008;61(2) 319 – 332.

[20] Flora CB. 2000. Measuring the Social Dimensions of Managing Natural Resources. In: Department of Forest Resources (ed.) Human Dimensions of Natural Resource Management: Emerging Issues and Practical Applications. St Paul: University of Minnesota; 2000. p. 83-99.

[21] Diener ED, Suh E. Measuring Quality of Life: Economic, Social, and Subjective Indicators. Social Indicators Research 1997; 40: 189 – 216.

[22] South African Institute of Race Relations. South Africa Survey 2013. Johannesburg: SAIRR; 2013.

[23] Mwalukomo H. The Role of Traditional Leaders in Environmental Governance in the Context of Decentralization: A Case Study of Grass Utilization in QwaQwa, Eastern Free State. MSc dissertation. University of the Witwatersrand; 2008.

[24] Nussbaum MC. *Creating Capabilities: The Human Development Approach*. Cambridge: Harvard University Press; 2011. http://books.google.co.za/books?id=Gg7Q2V8fi8gC&printsec=frontcov er&dq=capabilities±approach±±Nussbaum&hl=en&sa=X&ei=goPlULjCGseShge0y4DoBQ&ved=0CDcQ6AEwAA (accessed 27 June 2014)

[25] Statistics South Africa. Poverty trends in South Africa: An examination of poverty between 2006 and 2011. Pretoria: StatsSA; 2014.

[26] Herrington M, Kew J. GEM 2013 South African Report: 20 Years of Democracy. Cape Town: IDRC/ CRDI; 2013.

[27] Taru P, Chingombe W, Mukwada G. South Africa's Golden Gate Highlands National Park management plan: Critical reflections. South African Journal of Science 2013; 109(11/12): 1-3.

[28] Scheepers K., Swemmer L. Vermeulen WJ. Applying adaptive management in resource use in South African National Parks: A case study approach. Koedoe 2011: 53(2). doi:10.4102/koedoe.v53i2.999

Chapter 25

Shortage of Biodiversity in Grassland

Ricardo Loiola Edvan, Leilson Rocha Bezerra and
Carlo Aldrovandi Torreão Marques

Additional information is available at the end of the chapter

http://dx.doi.org/10.5772/59755

1. Introduction

In this chapter, we report existing distortions between the cultivation of native and exotic forage plants for breeding at pasture, a practice that reduces the biodiversity in grassland environments.

Raising animals on pasture with native forage species is cheaper than the cost of confinement and the use of exotic species. The introduction of exotic plants in pastures may increase production costs for farmers when introducing forage species that are not adapted to the system, in addition to causing environmental damage due to the reduction of biodiversity and the failure to consider the soil and climate of the region.

An approach will be made in relation to the climate-soil-plant-animal complex, emphasising the economic feasibility of using forage species with high genetic potential for native and exotic pastures and aiming to assist in the selection of the most appropriate forage species for grazing production systems and to avoid problems with biodiversity in grasslands.

2. Grass production

Ruminant grazing on native forage is the most economical way of producing meat for human consumption. Livestock grazing is generally a more profitable approach than raising confined livestock [1]. The use of concentrate feeds in the diet increases the cost of production, as does the use of exotic forage that is not adapted to the environment. In Ethiopia, pastures are not only used as an economical source of food for herbivores; most are also used for recreation, because in this country, there are many natural parks, reserves and sanctuaries [2]. The management of rangelands with native species goes beyond animal production, because there

are many factors in this diverse ecosystem with different purposes. Reducing the biodiversity of natural pastures to cultivate only a single species can cause serious environmental problems.

Native forage species reduce problems with invasive species, improve the wildlife habitat, contribute to carbon sequestration and increase the stability of the ecosystem [3], because they maintain the biodiversity of the environment. According to [4], the effect of cattle on native pasture is important for management because it prevents excess dead biomass, which would increase the risk of fires.

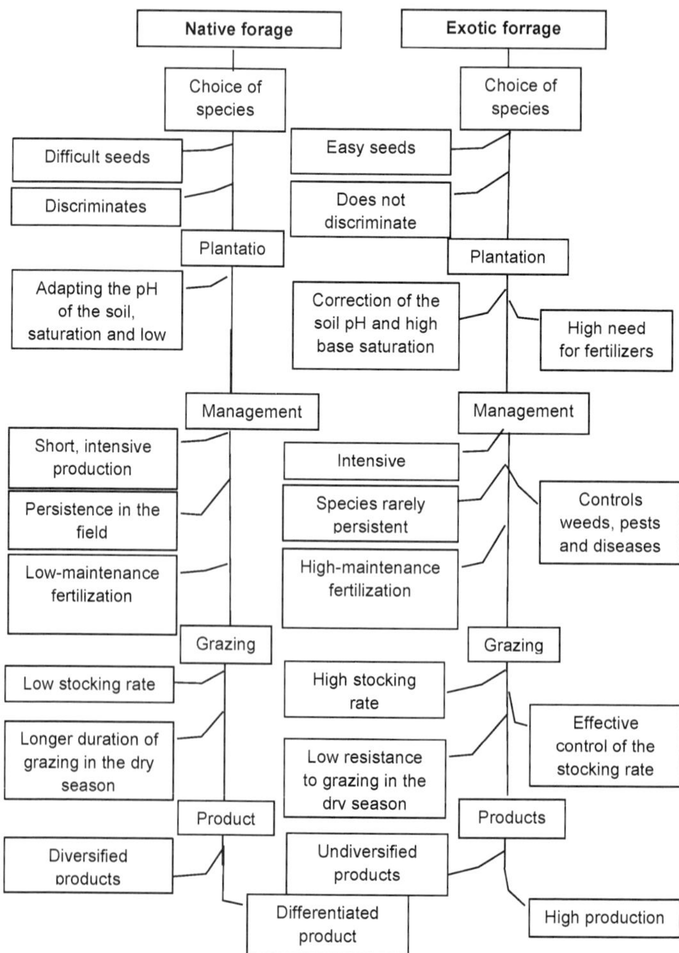

Figure 1. Livestock system with native and exotic (introduced) fodder plants

Note that when working with agriculture, all of the investment is returned quickly because the product is generated by the plant, while livestock products come from the animal that consumes the plant forage in pastures, which represents an additional cost to the production system compared to agriculture. Therefore, systems that integrate agriculture with livestock represent a viable option for the present and future of both agriculture and livestock, mainly due to increased environmental biodiversity and to the reduced production costs associated with this type of system.

The farmer must carefully analyse every investment that will be made in the cultivation of forage species, because the introduction of species into unnatural environments can require expensive treatments to render the soil suitable for their cultivation; livestock production also renders this task more expensive because of the cost of the animals (Figure 1).

In India, only 4% of the land requires pasture to increase the area under cultivation with forage species; however, there are fewer seeds of forage quality that are available in the local market. The authors note that the percentage of germination of available tropical grass seeds averages 20-30% [5]. The acquisition of forage seeds for planting pasture is expensive, and farmers in this country prefer to work with agricultural crops, which is concerning because the shortage of quality in the seed market is the decisive point in the selection of forage species.

The native species have little or no available seed in trade, as multinational companies acting in tropical climate countries restrict the distribution and sale of seeds of forage species to a few cultivars, which are often of the same species. In the selection of forage species, several criteria are considered rather than only assessing certain aspects, such as productivity and nutritional value. Table 1 presents the main criteria for the selection of forage species, comparing between the native and exotic species.

	Native forage	Exotic forage
Adaptation to soil	+	-
Seed quality	-	+
Seek marketing	-	+
Fertiliser requirement	-	+
Growth	+/-	+
Productivity	+/-	+
Persistence in pasture	+	-
Technicalisation level	-	+
Resistance to pest and diseases	+	-
Acceptance of producer	-	+
Nutritional value	+	+/-
Production cost	-	+

Table 1. Comparison between native and exotic (introduced) species of grasses

3. Forage species

Forage species are of fundamental importance for livestock everywhere on the planet because these species are responsible for feeding the herds of herbivores that in turn feed a large part of the world's population and some carnivorous animals. The use of native species is important because these plants are already adapted to inhabit regions that, due to a long process of natural selection carried out over time by nature, maintain the natural biodiversity of the grassland ecosystem. The variation in the wealth of native forage species is a valuable source for selecting the best kind [6].

The adaptation of forage plants to their natural environment occurs in the same way as for any other plant species in which the most suitable species for situations of soil and climate prevail in the region, along with different animals, such as insects, birds and herbivores. Long-term research must be conducted in native environments because there are several factors that affect the biodiversity and productivity of native pastures, and potential carbon sequestration is more complex and unpredictable than previously believed [7].

Native grasses have the potential to revegetate degraded land; however, due to little knowledge about their biology, preference is given to the use of exotic species that can be invasive, thus affecting local biodiversity [8]. The establishment and spread of plant species from other regions in natural ecosystems can reduce, disrupt or terminate the original flora populations and thus alter the balanced ecosystem, which today is one of the most significant environmental problems [9]. The planting and management of native grasses are difficult because information about these plants is scarce [10]. According to these authors, the native grasses of Brazil *Andropogon bicornis*, *Andropogon leucostachyus*, *Echinolaena inflexa* and *Setaria parviflora*, among others, show morphological and physiological characteristics that allow them to survive in environments with different stages of degradation, making these species good options for pasture recovery.

The genus *Paspalum* is promising for this country, but there are few studies of this genus. According to Barreto [11], 75% of the described *Paspalum* species occur in Brazil under a wide range of ecological conditions and as part of several plant communities. The importance of the genus *Paspalum* for Brazil is not only due to the production potential and quality of the species but also because it has a high potential for use in the recovery and conservation of degraded soils [12]. *Paspalum nicorae* is usually found in sandy soils, indicating a potential to tolerate drought and low soil fertility, and has a high response to fertilisation [13]. *Paspalum nicorae* has a wealth of natural morphological variation, which is a valuable source for the selection of new varieties for the native forage for Brazil [6].

In Congo, it is necessary to conduct studies with native and adapted species; the use of species that are not adapted to their conditions promotes food shortages at certain times of the year due to a lack of food during the dry season, which was a major cause of the reduced herd of the Sud-Kivu region in this country [14]. The identification of native forage species is essential for the establishment of small traditional farmers as agro-pastoralists [14]. In Nepal, species that are regarded as weeds in agricultural crops are native forage species; according to the authors, these species demonstrate an adaptability that could be used with greater frequency in animal feed [15].

The climate-soil-animal-plant complex must be considered when seeking to utilize grassland; it is not a simple task to introduce a system of livestock production using species from other regions for the purpose of feed because the native or exotic livestock will therefore have the same soil and climatic conditions of the place of origin in the new habitat. The use of exotic forage species is disturbing when introduced irrationally without considering the large climate-soil-plant-animal complex, and the introduction of these species can influence the biodiversity of the environment.

The control of native species to introduce exotic species is generally done with fire or herbicide use, causing environmental damage [16]. Santos et al. [17] argue that the reasons for using fire in rangelands are diverse. Fire is one of the practices that causes environmental damage to the soil and is rarely carried out in pastures with exotic cultures, showing that not even the crop treatment applies. Another important aspect is the production cost of these exotic forage species when accounting for a system of livestock production. The production and introduction of new forage species are frequent; every year private sector and public companies launch new varieties of these species which, in most cases, are exotic species that originated in another region. The improvement being realised is that the Kikuyu (*Pennisetum clandestinum*) originates from material coming from local farms, as a previous bid improved the tropical grass *Digitaria milanjiana* by giving it more leaves and a higher grass digestibility, but this species did not persist in the same pasture under proper management.

The *Brachiaria* genus originated in Africa and has resistance to acid soils and low fertility [19], allowing it to easily spread by seed and to be highly competitive with weeds [20], which explains the expansion of this type of grass to tropical and subtropical regions of the world [21]. However, cultivated species currently belonging to these genera demand large amounts of fertiliser to persist in the field and to compete with the native species of each region. Exotic species represent the high cost of deployment and maintenance. In Brazil, cultivated forage species originate from Africa (Table 2). Research of these species at the intuited has improved the forage potential of these species. The most cultivated species and that which is of greatest importance to Brazilian livestock are, respectively, *Brachiaria decumbens* and *Brachiaria brizantha* [22]. In Brazil, there is a lack of diversity in the cultivation of fodder plants in 45% of the areas, and 60% of the produced seeds are of cv. Marandu (*Brachiaria brizantha*) [23].

Species	Origin
Brachiaria decumbens	Africa
Brachiaria humidicula	Africa
Brachiaria brizantha	Africa
Panicum maximum	Africa
Penissetum purpureum	Africa
Cynodon sp.	Africa
Andropogon gayanus	Africa

Table 2. Origin of the forage species that are grown in Brazil

A lack of diversity can lead to future problems with pests and diseases, as these species are not natural in that particular region. Due to the large amount of area that is currently planted with these species in Brazil, future losses may occur in the livestock sector due to the lack of diversity of forage species. Brazil has 190 million hectares of grasslands that sustain 209 million cattle, the largest export of meat and the largest commercial cattle herd in the world [24]. According to these authors, the *Brachiaria brizantha, B. decumbens, B. humidicola* and *Panicum maximum* are the primarily used pastures (Figure 2); therefore, of the few cultivars that occupy a large amount of grazing area, only *Brachiaria brizantha* cv. Marandu occupies 50 million hectares of area in this country. Marandu is apomictic and resistant to leafhoppers-of-pastures [25], which has made this species the most produced in this country. The same mistakes were made as before, with areas being planted with cv. Marandu and with *Brachiaria decumbens,* which is susceptible to this pest. With the leafhopper attack that reduced pasture productivity in Brazil came the need to replace this forage species; the same may happen again with the Marandu. In this country only 0.8% of the grassland areas are planted with native species that are resistant to these natural pest attacks (Figure 2).

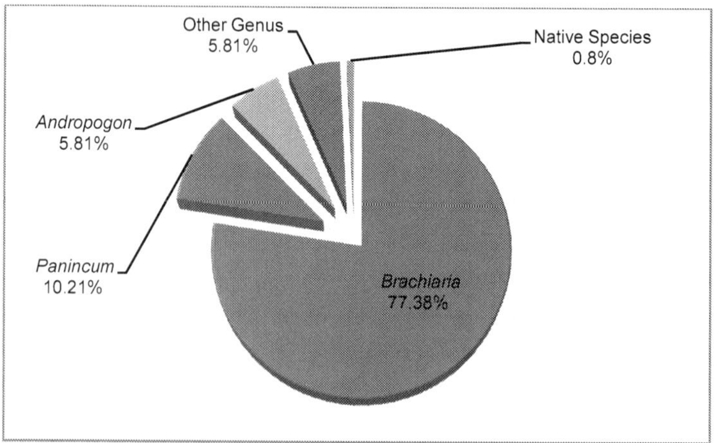

Adapted from www.rallydapecuária.com.br

Figure 2. Genus of the forage crops that were grown on farms in Brazil in 2012

In Brazil, there are five species of *Brachiaria* genus that are native and, according to these authors, do not possess forage potential [26]. Conclusions such as this undermine the achievement new research and the use of native species, as studies presenting such species do not evaluate all of the criteria to determine the actual potential forage and thus cannot conclude that these species do not have forage potential.

Studies with native forage species must be conducted to improve the production and nutritional characteristics of these plants so as not to harm their hardiness in relation to soil fertility.

This fact is identified mainly in tropical and subtropical regions of the world, in which several authors have reported the need for studies with native species in each region. In this context, Hacker et al. [27] report that Australia is required to conduct studies that are related to the improvement of tropical and subtropical grasslands, mainly related to forage species.

There are efforts by multinational companies and researchers to assess fodder plant species in different regions of the world. Care must be taken in this type of approach, as it creates a need the evaluation of a region, which is more profitable for multinational companies that hold the production of these seeds in its area of operation and sale of seeds. In their study, Hare [28] report that the same efforts for distributing forage seeds are being addressed in more than 20 countries in tropical regions of Asia, Africa, the Pacific and Central and South America with the species *Brachiaria ruziziensis*, *B. Decumbens*, *B. Brizantha*, *Panicum maximum*, *Stylosanthes guianensis* and *Paspalum atratum*. The lack of diversity of cultivated forage species in extensive grazing area represents a risk to livestock production in any country that adopts this practice [24].

There is not a single grass species or group of fodder species that is cultivated in any region as a standard species in grazing production systems. Although these species are tolerant of different climate regions, the soil is the determining factor in modifying their fertility to adapt, which is costly. Temperate countries do not seek alternatives elsewhere, but rather study the development of technology with their own native species. In Canada, since 2001, efforts for the reestablishment of native forage species have successfully developed sustainably systems that use native pasture [29]. This example should be followed by countries in tropical and subtropical climates.

In Australia, efforts are being made to encourage the planting of the native species *Themeda australis*, which is considered valuable forage that country, as it is food source for wild and domestic animals, conserves soil and water, decreases the use of exotic species and contributes to the rehabilitation of degraded and polluted [habitats 30]. Studies are underway in Africa and India to spread *Themeda triandra* species, which has physiological and morphological characteristics that are similar to those of Australia.

4. Importance of soil in the pasture production system

The soils in tropical regions are easily eroded and degraded by inappropriate land use, especially when trying to deploy a production system with fodder plants that are not suitable for the region without correcting the soil acidity and without adequate levels of fertiliser use, resulting in the depletion of pasture over time and thereby causing degradation in these areas. The loss of pasture productivity is mainly due to inadequate livestock management and the lack of nutrient replenishment [31]. Pastures in tropical regions are characterised by extensive grazing systems with the application of low levels of nitrogen fertilisers mainly due to unfavourable economic returns and the limited availability of fertilisers [32]. For planting forage species, farmers usually choose soils with severe limitations with regard to features such as natural chemical fertility, acidity and topography [33], because the best soils in terms of fertility and topography are intended for agricultural crops.

The costs of the cultivation of natural species of the region are always lower, although these plants often fail to achieve the production of commercialised exotic species that have high genetic potential. Planted under the same soil and fertility conditions, native and exotic forage plants have differential potentials. Fertile soils with high levels of chemical fertiliser for exotic forage with high genetic potential present greater production compared to that for native forage but with high costs to acquire seeds to achieve the high genetic potential of these plants. Comparisons between native and exotic species are often performed incorrectly; for example, genetically crafted plant species will produce more forage mass per plant than is produced by native plants, and many studies do not assess the cost of production, neglecting the costs of fertiliser. Despite the African continent having a similar climate as that of Brazil, the soil characteristics are different. The nutritional requirements in terms of exotic forage crops that are currently grown are high, excessively increasing the cost of production. Many farmers do not use the recommended fertiliser for forage plants, damaging the permanence of these species in the pasture. The cost fertilising is high because the grassland species that are selected for implantation are discerning. Studies that are performed with these species use high fertilisation, and the species that are evaluated have excellent results in this type of evaluation; however, economic feasibility studies are not performed considering that grazing areas are generally of large tracts.

The degree of adaptation to soil fertility of *Brachiaria brizantha* cv. Marandu leads to the recommendation of a medium nitrogen level of 200-250 kg ha^{-1} (10,000.00 m^2) per year for the Brazilian Cerrado soils [34]. An extensive grazing system with a carrying capacity from 0.5 to 0.8 AU (animal unit=450 kg animal live weight) ha^{-1} according to these authors for the same species would require only 50 kg of nitrogen per hectare (ha) and may reach 350 to 400 kg N ha^{-1} per year for irrigated grazing systems with a carrying capacity of 6-7 AU ha^{-1}.

The cost of fertiliser must be evaluated because every region of the world will have different recommendations for the correction of soil acidity using nitrogen, phosphorus, potassium and micronutrients. It can be seen in Table 3 that the cost of nitrogen fertiliser is different in Brazil and Thailand using the same species of forage crop. In 1980, Seifert reported that in Brazil, the species that were at their peak at the time were *Brachiaria decumbens* and *Brachiaria humidicula* and that these species reached high production on fertile soils. Costa et al. [34] reported that these species are considered to have low soil fertility requirements, with a recommendation of 100 kg ha^{-1} of nitrogen for both of these species. The recommendation of fertilisers has increased over time for the same species; this fact is related mainly to the impoverishment of the soil due to consecutive cultivation using species with high fertility requirements and to the non-replenishment of soil fertility.

In Brazil, due to the use of forages that are exacting with regard to soil fertility and to the inadequate handling of these plants, the degraded areas are increasing, and every year, approximately 8 million hectares are renewed or recovered in this country [24]. Low nitrogen availability in the soil is among the major limitations to the production of forage in tropical and subtropical areas and is one the greatest causes of pasture degradation [32]. Nitrogen is the nutrient that limits productivity, being rapidly depleted by the cultivation mainly of forage species with a C4 metabolic cycle that are characterised by rapid growth and high yields.

Species	Kind	Nitrogen (kg ha^{-1} year^{-1})	U$ ha^{-1} year^{-1}	Author
		BRAZIL		
Panicum maximum	Mombaça	307	367.70	Mello et al., 2008
Panicum maximum	Tanzânia	360	431.18	Corrêa et al., 2003
		THAILAND		
Panicum maximum	Mombaça	30	35.93	Hare et al., 2013
Panicum maximum	Tanzânia	30	35.93	Hare et al., 2013

ha: (10,000.00 m^2)

Value of U$: day 30/07/2014 in Brazil

Table 3. Recommendation and cost of fertilisation with nitrogen for *Panicum maximum* in Brazil and Thailand

Soil is an element that must be considered with great caution, because its management is expensive when seeking to make changes in fertility traits. The deployment of species elsewhere may represent higher production costs, as soil suitable must become suitable for the development of exotic species. The production of natural species for grazing is inexpensive because these species are already adapted to the soil and climate of the region, making it unnecessary to correct the soil acidity frequently to raise the base saturation of the soil to high levels, which is a common practice when planting exotic species and represents high costs to the livestock production system. This fact does not indicate that native species do not respond to the use of fertilisers.

In Brazil, many find that exotic forage plants have better forage potential than native or natural species. It is noteworthy that the forage potential encompasses several aspects and not only productivity and chemical composition. In reality, plants with forage potential should submit economic returns to farmers without compromising the integrity of the soil or the environment. How much more productive the most demanding fertility forage plant will be generally occurs with exotic forage species, in addition to the fact that these species are not adapted to the local soil, making it necessary to correct the soil pH to achieve the full potential of these species. Pastures with exotic species that are fertilised with nitrogen have a higher stocking rate than do native pasture species without nitrogen fertiliser [35]. This type of comparison is incipient because the native pasture was not fertilised and cannot be compared to a species under the influence of fertilisation. The native forage species tolerate low soil fertility and have a high response to fertilisation [36].

Intensive soil tillage may represent high costs for a production system that still has expenses for animal production. Rating adaptations to acidity and soil fertility were performed with the plant species that have been classified according to the degree of need for adaptation and fertiliser. This criterion is valid for identifying classes of soils in relation to base saturation and levels of phosphorus and potassium in the soil for many forage species [37]. Exotic species with high genetic potential are generally classified as the most demanding.

5. Nutritional differences

The interrelationship among the elements existing in an ecosystem for a forage species in a given region is more complex than for human consumption species because the animals of each region depend on this interrelationship to survive and influence the natural selection of species through grazing. The nutritional value of exotic species that are grown is undoubtedly superior to native species because these species have been enhanced to provide greater production and nutritional value for animals, while the native have been neglected by research. The nutritional value of exotic forage species does not justify the replacement of native species, although this value is used by many researchers and technicians to justify the introduction of exotic species in systems of grazing production [38]. More than one variable should be considered in the choice of forage; even if that species has a superior nutritive value than that of the native species, the nutritional parameter of an individual plant cannot define which exotic species is a better choice than the native species.

The existence of anti-nutritional factors in native pastures has been questioned by many researchers who report that these factors may be obstacles to animal nutrition at pasture [39], but research should be performed to improve these species to decrease the levels of these compounds, as is done with the exotic species with a high genetic potential. In a study evaluating the quality of native forages in Canada, native species during the dry season of the year were enough to maintain livestock [40]. Native species provide a differential characteristic in the products that are generated by grazing. Native species in a region provides character-istics and peculiar flavours to the products that are generated by the animals, valuing the product in the marketing. An appreciation of animal products exists in which animals graze native forages and are internationally known and marketed for a higher price. Thus, the low productivity of cattle is the main problem when trying to produce animals on pasture with native species, but this can be overcome by asking the highest price when marketing the generated products.

6. Conclusion

The cost of production is the decisive factor in choosing a forage species, especially in relation to the requirement of the plant in relation to soil fertility. The choice of forage species must be made with caution, especially when using exotic species with a high genetic potential. Native forage species are viable options for use in a production system with grazing animals. These species are already adapted to the characteristics of the soil and do not require changes to soil fertility, unlike exotic species. It is necessary to conduct further studies with native forage species in each region, especially in regions with a tropical climate. The use of native species in grasslands maintains an ecologically balanced environment because it preserves local biodiversity.

Author details

Ricardo Loiola Edvan*, Leilson Rocha Bezerra and Carlo Aldrovandi Torreão Marques

*Address all correspondence to: edvan@ufpi.edu.br

Department of Zootecnia, University Federal of Piauí, Bom Jesus, Piauí, Brasil

References

[1] Ziliotto MR, Silveira C, Camargo ME, Motta MEV, Priesnitz Filho W. Comparação do Custo de Produção de Bovinocultura de Corte: Pasto versus Confinamento. VII SE-GeT – Simpósio de Excelência em Gestão e Tecnologia, 12.p,2010.

[2] Abate T, Ebro A, Nigatu L. Pastoralists perceptions and rangeland evaluation for livestock production in South Eastern Ethiopia. Livestock Research for Rural Development, 2009;21;7.

[3] Jefferson PG, Iwaasa AD, Mcleod JG. Re-evaluation of native plant species for seeding and grazing by livestock on the semiarid prairie of western Canada. In: Managing: changing prairie landscapes. Radenbaugh, T.A.; Sutter, G.C. 2005, Canadian plains Research Center, university of Regina Copyright Notice. 2005. Disponível em: <http://www.albertapcf.org/rsu_docs/re-evaluation-of-native-plant-species-for-seeding-and-grazin.pdf>. Acesso em: abril em 25 de 2014.

[4] Pozer CG, Nogueira F. Flooded native pastures of the northern Region of the pantanal of mato grosso: biomass and primary productivity variations. Braz. J. Biol., 2004;64;4;859-866.

[5] Malaviya DR, Vijay D, Gupta CK, Roy AK, Kaushal P. Quality seed production of range grasses – A major constraint in revitalizing tropical pastures. Tropical Grasslands – Forrajes Tropicales, 2013;1;97–98.

[6] Reis CAO, Dall'agnol M, Nabinger C, Schifino-Wittmann MT. Morphological variation in *Paspalum nicorae* Parodi accessions, a promising forage. Sci. Agric., 2010;67;2;143-150.

[7] Iwaasa AD, Schellenberg MP, Mcleod JG. Re-establishment of Native Mixed Grassland Species into Annual Cropping Land. Prairie Soils & Crops Journal, 2012;5;85-96.

Disponível em: <http://www.prairiesoilsandcrops.ca/articles/volume-5-9-screen.pdf>. Acesso em: 25 de abril de 2014.

[8] Figueiredo MA, Baêta HE, Kozovits AR. Germination of native grasses with potential application in the recovery of degraded areas in Quadrilátero Ferrífero, Brazil. Biota Neotrop. 2012;12;3;118-123,. Disponível em: http:// www.biotaneotropica.org.br/v12n3/en/abstract?article±bn02912032012. Acesso em: 27 de abril de 2014.

[9] Freitas GK, Pivello VR. A ameaça das gramíneas exóticas a biodiversidade. In. O cerrado Pé-de-Gigante: ecologia e conservação – Parque Estadual de Vassununga.(Pivello, V.R.; Varanda, E.M., ed.) SMA, São Paulo, p. 233-348, 2005.

[10] Filgueiras TS, Fagg CW. Gramíneas nativas para a recuperação de áreas degradadas no cerrado. In: Bases para a recuperação de áreas degradadas na Bacia do São Francisco (Felfili, J.M., Sampaio, J.C. & Correia, C.R.M. de A., ed.). Centro de Referência em Conservação da Natureza e Recuperação de Áreas Degradadas (CRAD), Brasília, p.89-108. 2008.

[11] Barreto IL. O gênero *Paspalum* (Gramineae) no Rio Grande do Sul. UFRGS, Porto Alegre, RS, Brazil. 1974.

[12] Dall'Agnol M, Steiner MG, Baréa K, Scheffer-Basso SM. Perspectivas de lançamento de cultivares de espécies forrageiras nativas: o gênero *Paspalum*. p.149-162. In: Dall'Agnol M, Nabinger C, Rosa LM, Silva JLS, Santos DT, Santos RJ. eds. 2006. Anais do I Simpósio de Forrageiras e Produção Animal. Faculdade de Agronomia, UFRGS, Porto Alegre, RS, Brasil. 2006.

[13] Nabinger C, Dall'agnol M. Principais gramíneas nativas do RS: características gerais, distribuição e potencial forrageiro.p. 7-54. In: Dall'Agnol M, Nabinger C, Santos RJ. eds. 2008. Anais... do III Simpósio de Forrageiras e Produção Animal. Faculdade de Agronomia, UFRGS, Porto Alegre, RS, Brasil, 2008.

[14] Bacigale SB, Paul BK, Muhimuzi FL, Mapenzi N, Peters M, Maass BL. Characterizing feeds and feed availability in Sud-Kivu province, DR Congo. Tropical Grasslands – Forrajes Tropicales, 2014;2;9–11.

[15] Bhatta KP, Chaudhary RP. Species diversity and distribution pattern of grassland and cultivated land species in upper manang, Nepal Trans-Himalayas. Scientific World, 2009;7;7;76-79.

[16] Gomar EP, Reichert JM, Reinert DJ, Prechac FG, Marchesi EBC. Semeadura direta de forrageiras de estação fria em campo natural com aplicação de herbicidas: I. Produção de forragem e contribuição relativa das espécies. Ciência Rural, 2004;34;3;761-767.

[17] Santos AB, Quadros FLF, Rossi GE, Pereira LP, Kuinchtner BC, Carvalho RMR. Valor nutritivo de gramíneas nativas do Rio Grande do Sul/Brasil, classificadas segundo uma tipologia funcional, sob queima e pastejo. Ciência Rural, 2013;43;2;342-347.

[18] Lowe KF, Bowdler TM, Sinclair K, Holton TA, Skabo SJ. Phenotypic and genotypic variation within populations of kikuyu (*Pennisetum clandestinum*) in Australia. Tropical Grasslands, 2010;44;84–94.

[19] Macedo WR, Fernandes GM, Possenti RA, Lambais GR, Camargo PRC. Responses in root growth, nitrogen metabolism and nutritional quality in Brachiaria with the use of thiamethoxam. Acta Physiol Plant 2013;35;205–211.

[20] Valle CB, Macedo MCM, Euclides VPB, Jank L, Resende RMS Gênero Brachiaria. In: Fonseca DM, Martuscello JA (eds) Plantas Forrageiras, 1st edn. UFV, Viçosa, pp.30–77. 2010.

[21] Guenni O, Marı́n D, Baruch Z. Responses to drought of five Brachiaria species. I: biomass production, leaf growth, root distribution, water use and forage quality. Plant Soil, 2002; 243:229–241. doi:10.1023/A:1019956719475

[22] Alves GF, Figueiredo UJ, Pandolfi Filho AD, Barrios SCL, Do Valle CB. Breeding strategies for *Brachiaria* spp. to improve productivity – an ongoing project. Tropical Grasslands – Forrajes Tropicales, 2014;1;1-3.

[23] Euclides VPB, Valle CB, Macedo MCM, Almeida RG, Montagner DB, Barbosa RA. Brazilian scientific progress in pasture research during the first decade of XXI century. R. Bras. Zootec., 2010;39;151-168.

[24] Jank L, Barrios SC, Valle CB, Simeão RM, Alves GF. The value of improved pastures to Brazilian beef production. Crop and Pasture Science, 2014;10;14-15.

[25] Basso KC, Resende RMS, Valle CB, Gonçalves MC, Lempp B. Avaliação de acessos de *Brachiaria brizantha* Stapf e estimativas de parâmetros genéticos para caracteres agronômicos. Acta Scientiarum. Agronomy, 2009;31;1;17-22.

[26] Pizarro EA, Valle do CB, Keller-Grein G, Schultze-Kraft R, Zimmer AH. Regional Expertice with Brachiaria: Tropical America-Savannas. Brachiaria: Biology, Agronomy, and Improvement. (CIAT & Embrapa). P.225-246. 1996.

[27] Hacker A, et al. Potential of Australian bermudagrasses (*Cynodon* spp.) for pasture in subtropical Australia. Tropical Grasslands – Forrajes Tropicales, 2013;1;81–83.

[28] Hare MD. Village-based tropical pasture seed production in Thailand and Laos – a success story. Tropical Grasslands – Forrajes Tropicales, 2014;2;165–174.

[29] Iwaasa AD, Schellenberg MP, Mcconkey B. Re-establishment of Native Mixed Grassland Species into Annual Cropping Land. Prairie Soils & Crops Journal, 2012;5;85-95.

[30] Dell'Acqua M, Gomarasca S, Porro A, Bocchi S. A tropical grass resource for pasture improvement and landscape management: *Themeda triandra* Forssk. Grass and Forage Science, 2012;68;205–215.

[31] Lima SS, Alves BJR, Aquino AM, Mercante FM, Pinheiro EFM, Sant'anna SAC, Urquiaga S, Boddey RM. Relação entre a presença de cupinzeiros e a degradação de pastagens. Pesquisa Agropecuária Brasileira, 2011;46;12;1699-1706.

[32] Vendramini JMB, Dubeux Jr JCB, Silveira ML. Nutrient cycling in tropical pasture ecosystems. Revista Brasileira de Ciências Agrária, 2014;9;2;308-315.

[33] Martha Junior GB, Vilela L. Pastagens no Cerrado: baixa produtividade pelo uso limitado de fertilizantes em pastagens. Planaltina: Embrapa Cerrados, 2002. 32 p. (Embrapa Cerrados. Documentos, 50).

[34] Costa KAP, Oliveira IP, Faquin V. Adubação nitrogenada para pastagens do gênero Brachiaria em solos do Cerrado. – Santo Antônio de Goiás: Embrapa Arroz e Feijão, 2006. 60p.: il. – (Documentos/Embrapa Arroz e Feijão, ISSN 1678-9644; 192).

[35] Brambilla DM, Nabinger C, Kunrath TR, Carvalho PCF, Carassai IJ, Cadenazzi M. Impact of nitrogen fertilization on the forage characteristics and beef calf performance on native pasture overseeded with ryegrass. R. Bras. Zootec. 2012;41;3;528-536.

[36] Nabinger C, Dall'Agnol M. Principais gramíneas nativas do RS: características gerais, distribuição e potencial forrageiro.p. 7-54. In: Dall'Agnol M, Nabinger C, Santos RJ. eds. 2008. Anais do III Simpósio de Forrageiras e Produção Animal. Faculdade de Agronomia, UFRGS, Porto Alegre, RS, Brasil. 2008.

[37] Euclides V PB, Do Valle CB, Macedo MCM, Almeida RG, Montagner DB, Barbosa RA. Brazilian scientific progress in pasture research during the first decade of XXI century. R. Bras. Zootec., 2010;39;151-168.

[38] Deschamps FC, Tcacenco FA. Parâmetro nutricionais de forragerias nativas e exóticas no Vale do Itajaí, Santa Catarina. Pesq. Agropec. Bras. 2000;35;2;457-465.

[39] Guimarães-Beelen PM, Berchielli TT, Beelen R, Filho JA, Oliveira SG. Characterization of condensed tannins from native legumes of the brazilian Northeastern semi-arid. Sci. Agric. 2006;63;6;522-528.

[40] Jefferson PG, McCaughey WP, May K, Woosaree J, Mc Farlane L. Forage quality of seeded native grasses in the fall season on the Canadian Prairie Provinces. Can. J. Plant Sci. 2004;84;503–509.